PROGRESS IN BRAIN RESEARCH

VOLUME 153

HYPOTHALAMIC INTEGRATION OF ENERGY METABOLISM

Other volumes in PROGRESS IN BRAIN RESEARCH

PROGRESS IN BRAIN RESEARCH

VOLUME 153

HYPOTHALAMIC INTEGRATION OF ENERGY METABOLISM

Proceedings of the 24th International Summer School of Brain Research, held at the Royal Netherlands Academy of Arts and Sciences, Amsterdam, The Netherlands 29 August–1 September 2005

EDITED BY

A. KALSBEEK

Netherlands Institute for Brain Research, Meibergdreef 33, 1105 AZ Amsterdam, The Netherlands

E. FLIERS

Academic Medical Center, University of Amsterdam, Department of Endocrinology and Metabolism, Meibergdreef 9, 1105 AZ Amsterdam, The Netherlands

M.A. HOFMAN
D.F. SWAAB
E.J.W. VAN SOMEREN
R. M. BUIJS

Netherlands Institute for Brain Research, Meibergdreef 33, 1105 AZ Amsterdam, The Netherlands

ELSEVIER

AMSTERDAM – BOSTON – DUSSELDORF – LONDON – NEW YORK – OXFORD
PARIS – SAN DIEGO – SAN FRANCISCO – SINGAPORE – SYDNEY – TOKYO

Elsevier
Radarweg 29, PO Box 211, 1000 AE Amsterdam, The Netherlands
The Boulevard, Langford Lane, Kidlington, Oxford OX5 1GB, UK

First edition 2006

Library of Congress Cataloging-in-Publication Data
A catalog record for this book is available from the Library of Congress

British Library Cataloguing in Publication Data
A catalogue record for this book is available from the British Library

ISBN-13: 978-0-444-52261-0 (this volume)
ISBN-10: 0-444-52261-1 (this volume)
ISSN: 0079-6123 (series)

International Summer School of Brain Research (24th : 2005 : Amsterdam, Netherlands)
  Hypothalamic integration of energy metabolism : proceedings of the 24th International
  Summer School of Brain Research, held at the Royal Netherlands Academy of Arts and
  Sciences, Amsterdam, the Netherlands, 29 August-1 September 2005.
  (Progress in brain research; v. 153)
  1. Hypothalamus - Congresses
  I. Title II. Kalsbeek, A.
  612.8'262

  ISBN-13: 9780444522610
  ISBN-10: 0444522611

For information on all Elsevier publications
visit our website at books.elsevier.com

Printed and bound in The Netherlands

06 07 08 09 10   10 9 8 7 6 5 4 3 2 1

List of Contributors

R.S. Ahima, Department of Medicine, Division of Endocrinology, Diabetes and Metabolism, University of Pennsylvania School of Medicine, Philadelphia, PA 19104, USA

S.F. Akana, Department of Physiology, Box 0444, University of California San Francisco, 513 Parnassus, HSW-747, San Francisco, CA 94143-0444, USA

A. Alkemade, Department of Endocrinology and Metabolism, Academic Medical Center, University of Amsterdam and Netherlands Institute for Brain Research, Meibergdreef 33, 1105 AZ Amsterdam, The Netherlands

P. Barrett, Rowett Research Institute, Greenburn Road, Bucksburn, Aberdeen AB21 9SB, UK

M.E. Bell, Department of Physiology, Box 0444, University of California San Francisco, 513 Parnassus, HSW-747, San Francisco, CA 94143-0444, USA

S. Bhatnagar, Department of Physiology, Box 0444, University of California San Francisco, 513 Parnassus, HSW-747, San Francisco, CA 94143-0444, USA

N. Bos, Netherlands Institute for Neurosciences, Meibergdreef 47, 1105 BA Amsterdam, The Netherlands

R.M. Buijs, Netherlands Institute for Brain Research, Meibergdreef 33, 1105 AZ Amsterdam, The Netherlands

C. Cailotto, Netherlands Institute for Brain Research, Meibergdreef 33, 1105 AZ Amsterdam, The Netherlands

P.E. Cryer, Division of Endocrinology, Metabolism and Lipid Research, and the General Clinical Research Center and the Diabetes Research and Training Center, Washington University School of Medicine, Campus Box 8127, 660 South Euclid Avenue, St. Louis, MO 63110, USA

M.F. Dallman, Department of Physiology, Box 0444, University of California San Francisco, 513 Parnassus, HSW-747, San Francisco, CA 94143-0444, USA

I.S. Farooqi, Departments of Medicine and Clinical Biochemistry, University of Cambridge, Addenbrooke's Hospital, Cambridge, UK

H.L. Fehm, Medizinische Klinik I, Universität Lübeck, Ratzeburger Allee 160, D-23538 Lubeck, Germany

C. Fekete, Department of Endocrine Neurobiology, Institute of Experimental Medicine, Hungarian Academy of Sciences, Budapest 1083, Hungary and Tupper Research Institute and Department of Medicine, Division of Endocrinology, Diabetes and Metabolism, Box 268, Tufts New England Medical Center, Boston, MA 02111, USA.

E. Fliers, Department of Endocrinology and Metabolism, Academic Medical Center, University of Amsterdam, 1105 AZ Amsterdam, The Netherlands

T.-M. Fong, Department of Metabolic Research, Merck Research Laboratories, 126 Lincoln Avenue, Rahway, NJ 07065

L. Fu, Department of Molecular and Human Genetics, Bone Disease Program of Texas, Baylor College of Medicine, One Baylor Plaza, Houston, TX 77030, USA

A.B. Ginsberg, Department of Physiology, Box 0444, University of California San Francisco, 513 Parnassus, HSW-747, San Francisco, CA 94143-0444, USA

A.P. Goldstone, Imaging Sciences Department, MRC Clinical Sciences Centre, Faculty of Medicine Imperial College, Hammersmith Hospital Campus, Du Cane Road, London, W12 ONN, UK

V.D. Goncharuk, Cardiovascular Research Center, Moscow, Russia

X.-M. Guan, Department of Metabolic Research, Merck Research Laboratories, 126 Lincoln Avenue, Rahway, NJ 07065, USA

A. Guijarro, Surgical Metabolism and Nutrition Laboratory, Neuroscience Program, University Hospital, SUNY Upstate Medical University, 750 Adams Street, Syracuse, NY 13210, USA

M.H. Hastings, MRC Laboratory of Molecular Biology, Hills Road, Cambridge CB2 2QH, UK

M.A. Hofman, Netherlands Institute for Neuroscience, Meibergdreef 47, 1105 BA Amsterdam, The Netherlands

T.L. Horvath, Section of Comparative Medicine, and Departments of Ob./Gyn. & Reproductive Sciences and Neurobiology, Yale University School of Medicine, 375 Congress Ave LSOG 339, New Haven, CT 06520, USA

H. Houshyar, Department of Physiology, Box 0444, University of California San Francisco, 513 Parnassus, HSW-747, San Francisco, CA 94143-0444, USA

A. Kalsbeek, Netherlands Institute for Brain Research, Meibergdreef 33, 1105 AZ Amsterdam, The Netherlands

A. Kanatani, Tsukuba Research Institute, Banyu, Japan

G. Karsenty, Department of Molecular and Human Genetics, Baylor College of Medicine, One Baylor Plaza, Houston, TX 77030, USA

W. Kern, Medizinische Klinik I, Universität Lübeck, Ratzeburger Allee 160, D-23538 Lübeck, Germany

F. Kreier, Netherlands Institute for Brain Research, Meibergdreef 33, 1105 AZ Amsterdam, The Netherlands

S.E. La Fleur, Netherlands Institute for Brain Research, Meibergdreef 33, 1105 AZ Amsterdam, The Netherlands

K.C. Laugero, Department of Physiology, Box 0444, University of California San Francisco, 513 Parnassus, HSW-747, San Francisco, CA 94143-0444, USA

A. Laviano, Department of Clinical Medicine, University of Romea 'La Sapienza', Viale dell'Universita 37, 00185 Rome, Italy

R.M. Lechan, Tupper Research Institute and Department of Medicine, Division of Endocrinology, Diabetes and Metabolism, Box 268, Tufts New England Medical Center and Department of Neuroscience, Tufts University School of Medicine, 750 Washington Street, Boston, MA 2111, USA

D. MacNeil, Department of Metabolic Research, Merck Research Laboratories, 126 Lincoln Avenue, Rahway, NJ 07065, USA

E.S. Maywood, MRC Laboratory of Molecular Biology, Hills Road, Cambridge CB2 2QH, UK

M.M. Meguid, Surgical Metabolism and Nutrition Laboratory, Neuroscience Program, University Hospital, SUNY Upstate Medical University, 750 Adams Street, Syracuse, NY 13210, USA

J.G. Mercer, Rowett Research Institute, Greenburn Road, Bucksburn, Aberdeen AB21 9SB, UK

P.J. Morgan, Rowett Research Institute, Greenburn Road, Bucksburn, Aberdeen AB21 9SB, UK

R. Nargund, Department of Medicinal Chemistry, Merck Research Laboratories, 126 Lincoln Avenue, Rahway, NJ 07065, USA

J. O'Neill, MRC Laboratory of Molecular Biology, Hills Road, Cambridge CB2 2QH, UK

M.S. Patel, Department of Molecular and Human Genetics, Bone Disease Program of Texas, Baylor College of Medicine, One Baylor Plaza, Houston, TX 77030, USA

N.C. Pecoraro, Department of Physiology, Box 0444, University of California San Francisco, 513 Parnassus, HSW-747, San Francisco, CA 94143-0444, USA

A. Peters, Medizinische Klinik I, Universität Lübeck, Ratzeburger Allee 160, D-23538 Lübeck, Germany

Y. Qi, Department of Medicine, Division of Endocrinology, Diabetes and Metabolism, University of Pennsylvania School of Medicine, Philadelphia, PA 19104, USA

A.B. Reddy, MRC Laboratory of Molecular Biology, Hills Road, Cambridge CB2 2QH, UK

A.W. Ross, Rowett Research Institute, Greenburn Road, Bucksburn, Aberdeen AB21 9SB, UK

M. Ruiter, Netherlands Institute for Brain Research, Meibergdreef 33, 1105 AZ Amsterdam, The Netherlands

C.B. Saper, Department of Neurology, and Program in Neuroscience, Beth Israel Deaconess Medical Center, Harvard Medical School, 330 Brookline Avenue, Boston, MA 02215, USA

F.A. Scheer, Medical Chronobiology Program, Harvard Medical School, Boston, USA

U. Schibler, Department of Molecular Biology, Sciences III, University of Geneva, 30, Quai Ernest Ansermet, CH-1211, Geneva 4, Switzerland

N.S. Singhal, Neuroscience Graduate Group, University of Pennsylvania School of Medicine, Philadelphia, PA 19104, USA

D. Spanswick, Division of Clinical Sciences, Warwick Medical School, The University of Warwick, Coventry CV4 7AL, UK

A. Strack, Department of Pharmacology, Merck Research Laboratories, 126 Lincoln Avenue, Rahway, NJ 07065, USA

A.M. Strack, Department of Physiology, Box 0444, University of California San Francisco, 513 Parnassus, HSW-747, San Francisco, CA 94143-0444, USA

D.F. Swaab, Netherlands Institute for Neuroscience, Meibergdreef 47, 1105 BA Amsterdam, The Netherlands

M. van den Top, Division of Clinical Sciences, Warwick Medical School, The University of Warwick, Coventry CV4 7AL, UK

L.H.T. Van Der Ploeg, Merck Research Laboratories, 33 Avenue Louis Pasteur, MRL B3-406, Boston, MA 02115, USA

E.J.W. Van Someren, Netherlands Institute for Neuroscience, Meibergdreef 47, 1105 BA Amsterdam, The Netherlands

J.P. Warne, Department of Physiology, Box 0444, University of California San Francisco, 513 Parnassus, HSW-747, San Francisco, CA 94143-0444, USA

W.M. Wiersinga, Department of Endocrinology and Metabolism, Academic Medical Center, University of Amsterdam, 1105 AZ Amsterdam, The Netherlands

G.K.Y. Wong, MRC Laboratory of Molecular Biology, Hills Road, Cambridge CB2 2QH, UK

C. Yi, Netherlands Institute for Neurosciences, Meibergdreef 47, 1105 BA Amsterdam, The Netherlands

Preface

The 24th International Summer School of Brain Research was held in Amsterdam from August 29 to September 1, 2005 at the Auditorium of the Royal Netherlands Academy of Arts and Sciences (KNAW). The Summer School was organized by the Netherlands Institute for Brain Research (NIBR), one of the institutes of the KNAW, according to a long-standing biennial tradition. The history of the NIBR dates back to the beginning of the last century. At the meeting of the International Association of Academies held in Paris in 1901, the anatomist Wilhelm His proposed that research on the nervous system should be placed on an international footing. In 1904 this resulted in the formation of the International Academic Committee for Brain Research, which set itself the task of "… organizing a network of institutions throughout the civilized world, dedicated to the study of the structure and functions of the central organ …". Several governments responded to this ambition by founding brain research institutes, among which was the Netherlands (Central) Institute for Brain Research, which opened its doors on 8 June 1909. Professor C.U. Ariëns Kappers (1877–1946) became the first director of the institute, a position he held until his death. In honor of its first director the institute decided to create the C.U. Ariëns Kappers Award, which during this Summer School was presented to Dr. Clifford B. Saper (Boston, USA) for his outstanding achievements in deciphering the neuroanatomy of the mammalian hypothalamus and its intricate pathways involved in the control of behavior and physiology of the organism.

The focus of this 24th Summer School was on the mammalian hypothalamus, and especially its involvement in the physiology and pathology of the control of energy metabolism. The awareness of the important role of the hypothalamus in food intake and energy metabolism dates back to 1840, when Mohr described a case of hypothalamic obesity associated with a rapid gain of body weight in a patient with a pituitary tumor. It was only after 100 years, however, that the animal experiments of Hetherington and Ranson in 1940 showed that obesity resulted from lesions restricted to the hypothalamus, independent of pituitary damage. Our understanding of the hypothalamic control of energy metabolism was given a second boost another 50 years later, when Friedman et al. in1994 discovered the leptin gene, i.e. the long sought for hormonal factor from the adipose tissue that informs the brain, and especially the hypothalamus, about peripheral fat stores. More recently it has become clear that, in addition to leptin, many more peripheral signals feed back to the hypothalamus in order to enable it to monitor the status of peripheral energy stores and fuel availability. Thus, although the key function of the hypothalamus in energy metabolism was already clear a long time ago, a more detailed understanding of the hypothalamic pathways involved and their neurochemical make-up have only evolved in the past few years. Main question is still, however, how the hypothalamus is able to integrate all the "sensory" information from the internal and external environment and to fine-tune the integrated information with the intricate balance between the needs and demands of different physiological systems. For this reason the focus of interest in the present volume of *Progress in Brain Research* is on the following topics: (1) the emerging role of the hypothalamus in the control of energy metabolism, (2) the hypothalamic sensing of metabolic factors, (3) the integrative role of the hypothalamus in thyroid and bone metabolism, (4) the interaction between circadian information and energy metabolism, and (5) the important interplay between the immune system, energy metabolism and the autonomic nervous system.

The enthusiasm with which a great number of scientists have agreed to come to Amsterdam and to contribute to this volume is very gratifying. We would like to acknowledge the generosity of both the Royal Netherlands Academy of Arts and Sciences and the Graduate School Neurosciences Amsterdam, under whose joint auspices this Summer School was being held as well as many other generous financial supporters. Finally, we would like to express our special gratitude to Tini Eikelboom, Henk Stoffels, Wilma Verweij and Wilma Top for their invaluable organizational and editorial assistance.

Andries Kalsbeek, Eric Fliers,
Michel A. Hofman, Dick F. Swaab,
Eus J.W. Van Someren and Ruud M. Buijs

Acknowledgments

The 24th International Summer School of Brain Research has been made possible by financial support from:

- Elsevier
- Ferring BV
- Graduate School Neurosciences Amsterdam (ONWA)
- Hersenstichting Nederland
- Merck Research Laboratories, Boston
- Nederlandse Vereniging voor Endocrinologie
- Nuclilab
- Numico Research BV
- NV Organon
- Royal Netherlands Academy of Arts and Sciences (KNAW)
- Solvay Pharmaceuticals BV
- Stichting Diabetes Fonds Nederland
- Stichting C.H. Van den Houtenfonds
- ZonMw

Contents

Section I.   Hypothalamic Integration of Energy Metabolism

Section II.   Hypothalamic Integration of Blood-borne Signals

xiv

SECTION I

Hypothalamic Integration of Energy Metabolism

Kalsbeek, Fliers, Hofman, Swaab, Van Someren & Buijs
Progress in Brain Research, Vol. 153
ISSN 0079-6123

CHAPTER 1

The human hypothalamus in metabolic and episodic disorders

D.F. Swaab*

Netherlands Institute for Neuroscience, Meibergdreef 47, 1105BA Amsterdam, The Netherlands

The hypothalamus in disorders of eating and metabolism

Obesity is one of the most pressing health problems in the Western world. It is, among other things, responsible for 65–75% of essential hypertension, diabetes mellitus and cardiovascular problems (Hall et al., 2001). This epidemic is in need of therapeutics, but only limited progress has been made as far as the pharmacotherapy of this condition is concerned (Van der Ploeg, 2000; Clapham et al., 2001). The etiologies of eating disorders, such as anorexia nervosa, bulimia nervosa and obesity, are poorly understood. Inherited vulnerabilities, cultural pressures and adverse individual and family experiences are presumed to have a part in the pathogenetic mechanisms (Walsh and Devlin, 1998; Polivy and Herman, 2002), while biological factors have only recently become the subject of study.

Some clinical observations illustrate the importance of hypothalamic mechanisms for governing satiety and hunger (Lustig et al., 1999). In Diencephalic syndrome or hypothalamo-optic-pathway glioma, emaciation of the entire body is found in infancy and childhood. Lesions in the ventromedial hypothalamic area cause increased appetite and obesity (Fig. 1), whereas tumors in the lateral hypothalamic area (LHA) may cause anorexia. In addition, hypercortisolism in Cushing's syndrome or due to corticosteroid therapy may cause obesity accompanied by depression, hypertension and circadian disturbances (Fig. 1). Uncommon, intractable hypothalamic obesity syndrome occurs after cranial insults. It is often coupled with other hypothalamo–pituitary disturbances that may exacerbate obesity, such as growth hormone deficiency or hypothyroidism, but the obesity remains after hormone replacement. Rare is neurocystiscerosis in the anterior hypothalamus, an infection caused by the presence of Taenia larvae, that can be accompanied by obesitas and hyperphagia (Lino et al., 2000).

Human functions and behavior in health and symptoms in disease show circadian and circanual fluctuations, while also disorders of these episodic alterations are encountered. The physiological basis for these rhythmic phenomena is situated in the biological clock of the brain, the suprachiasmatic nucleus (SCN). In postmortem tissue of a group of young subjects (6–47 years of age), we observed significant fluctuations in the number of vasopressin- and vasoactive-intestinal polypeptide (VIP) expressing SCN neurons over the 24-h period (Hofman and Swaab, 1993; Hofman, 2003). During the daytime, for example, the SCN contained twice as many vasopressin-expressing neurons as during the night, with peak values in vasopressin cell number occurring in the early morning.

In contrast to the general belief that human beings have few, if any, seasonal rhythms (Lewy and Sack, 1996), we observed strong seasonal

---

*Corresponding author.; E-mail: d.f.swaab@nih.knaw.nl

DOI: 10.1016/S0079-6123(06)53001-8

4

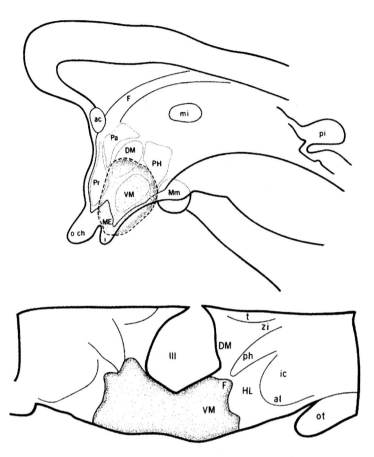

Fig. 1. In the course of several years, a young woman developed marked obesity and hyperphagia, associated with aggressive behavior. At autopsy, she was found to have a hamartoma that destroyed the ventromedial nucleus. Diagrammatic representation of the tumor projected on midsagittal plane. ac, anterior commissure; al, ansa lenticularis; DM, dorsomedial nuclear region; F, fornix; HL, lateral hypothalamus; I, infundibular stalk; ic, internal capsule; mi, massa intermedia; Mm, mamillary body; ME, median eminence; o ch, optic chiasm; ot, optic tract; Pa, paraventricular nucleus; ph, pallidohypothalamic tract; PH, posterior hypothalamus; pi, pineal body; Pr, preoptic region; t, thalamus; VM, ventromedial nuclear region; zi, zona incerta; and III, third ventricle (From Reeves and Plum, 1969; with permission).

fluctuations in the SCN. The number of vasopressin- and VIP-containing neurons in the SCN was found to alter in the course of a year, with August–September values being two times higher than April–May values (Hofman and Swaab, 1992b; Hofman, 2001; Fig. 2). Photoperiod seems to be the major Zeitgeber (pacemaker) for the observed annual variations in the SCN (Hofman et al., 1993). The hypothalamic levels of serotonin and dopamine, neurotransmitters known to innervate the SCN, show diurnal rhythms and seasonal rhythms as well (Carlsson et al., 1980; Figs. 3 and 4). In addition, binding to the serotonin receptor is

higher in summer than it is in winter in the hypothalamus of healthy subjects (Neumeister et al., 2000). How these seasonal fluctuations causally relate to the SCN circannual rhythms has not been determined. However, the fact that both aminergic rhythms are observed in the hypothalamus indicates that at least in this respect the SCN drives the monoaminergic systems instead of the other way around. In addition, we observed a notable seasonal variation in the volume of the paraventricular nucleus (PVN) in our material indicating activity changes. This volume reached its peak during the spring (Hofman and Swaab, 1992a).

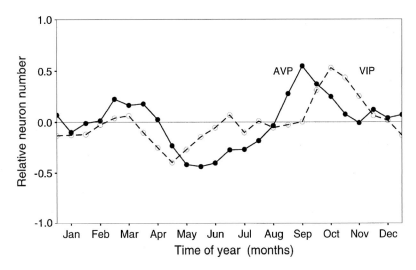

Fig. 2. Normalized values of the annual cycles of the AVP- and VIP-expressing neurons in the human suprachiasmatic nucleus (SCN). Cross-correlation analysis revealed that the two series are positively correlated at lag zero, indicating that the cycles reach their peaks and troughs in the same periods of the year (From Hofman, 2001; with permission).

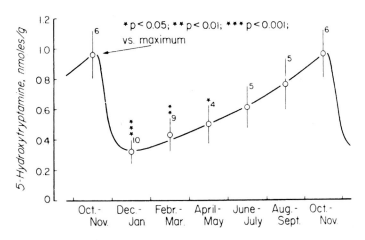

Fig. 3. Seasonal fluctuations in the level of 5-hydroxytryptamine (5-HT) in human hypothalamus determined postmortem, in relation to the month of death. Shown are the means $\pm$ S.E.M. (n) of pooled values of two consecutive months. Statistics: Student's t-test (From Carlsson et al., 1980; with permission).

One can assume that these hypothalamic seasonal fluctuations are the basis for the fall/winter increases in eating behavior and body weight, that are considered to reflect the human expression of a basic evolutionary process, present across multiple species, ensuring maximum conservation of energy when food supplies are becoming scarce (Rosenthal et al., 1987). The marked seasonal rhythm of nutrient intake in human consists of an increased total caloric intake, especially of carbohydrate, in

the fall, associated with an increased meal size and a greater rate of eating (De Castro, 1991). Although adaptive in evolution, the same process has become maladaptive in human when highly palatable, high-caloric foods are readily available, as nowadays in the Western world and may lead, e.g., to the seasonal weight gain in seasonal affective disorder and obesity (Levitan et al., 2004). Reduced dopaminergic tone in the hypothalamic nuclei may contribute to the "thrifty" genotype

6

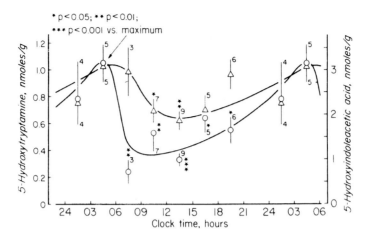

Fig. 4. Diurnal fluctuations in the levels of 5-hydroxytryptamine (5-HT) (O) and 5-hydroxindole-acetic acid (Δ) in human hypothalamus determined postmortem, in relation to clock time of death. Shown are the means ± S.E.M. (n) of pooled values of seven 3-h intervals. Statistics: Student's *t*-test (From Carlsson et al., 1980; with permission).

that has the ability to grow obese and insulin resistant in times of food abundance, which has a survival advantage in times of famine (Pijl, 2003).

Ventromedial hypothalamus syndrome

One of the disorders illustrating the involvement of the human hypothalamus in eating and metabolism is the ventromedial hypothalamus syndrome. Following invasion of a tumor into the area of the ventromedial hypothalamic nuclei (VMN), a tetrad of symptoms has been described, i.e. (i) episodic rage, (ii) emotional lability, (iii) hyperphagia with obesity and (iv) intellectual deterioration. Memory loss is the most prominent feature of intellectual decline. Lesion of the descending columns of the fornix and mamillary bodies may be important in this respect, but a primary role for the VMN in memory has also been postulated (Reeves and Plum, 1969; Climo, 1982; Flynn et al., 1988; Fig. 1). Indeed, experimental lesions of the VMN area in animals produce rage and excessive eating, and, in a child with massive leukemia infiltration at the level of the VMN of the hypothalamus, violent hunger and obesity were reported. In addition to tumors, encephalitis, tuberous sclerosis and vascular lesions have been found to cause hypothalamic obesity (Bastrup-Madsen and Greisen, 1963; Coffey, 1989). A 3-year-old boy who developed obesity on the

basis of encephalitis had a severe bilateral outfall of the neurons of the VMN (Wang and Huang, 1991), which argues for some role of the VMN in eating behavior. It should be noted, though, that experimental lesions that are restricted to the VMN do not produce obesity in rats. It is therefore presumed that damage to the nearby noradrenergic bundle or its terminals might be responsible for obesity (Gold, 1973). Moreover, it should be mentioned that tumors are never exactly restricted to one hypothalamic area. For instance, in a patient with hyperphagia and obesity whose VMN was unilaterally lesioned by a hypothalamic astrocytoma, the PVN was also bilaterally involved (Haugh and Markesbery, 1983), which may also have contributed to these symptoms.

If VMN lesions indeed have such a notable effect on the production of episodic rage, emotional instability, hyperphagia with obesity and memory loss, it is remarkable, to say the least, that none of these signs and symptoms have been mentioned following stereotactic destruction of the VMN in patients with "sexual deviations" or in drug addicts. Even in the patient who underwent bilateral destruction of the VMN, the only effect reported was a loss of all interest in sexual activity. The authors explicitly state that psycho-organic disturbances did not occur in any of the patients, although they do not define the exact nature of the disturbances they looked for (Müller et al., 1973). One may thus indeed wonder whether structures in the vicinity of the VMN instead of the VMN itself may be essential for the development of a "ventromedial" hypothalamus syndrome. This possibility is reinforced by the observation of a posttraumatic patient with a lesion in the dorsomedial hypothalamic nucleus who had hyperphagia (Shinoda et al., 1993). Moreover, in the rat, aggression can come from an area below the fornix, just lateral and frontal to the VMN in the hypothalamus. This area almost completely coincides with the intermediate hypothalamic area (Kruk et al., 1998).

Hypothalamic tumors mimicking anorexia nervosa

In cases of anorexia it may be difficult to differentiate between "psychogenic" and "organic" causes. Psychological disturbances without neurological manifestations may be due to occult intracranial tumors masquerading as anorexia nervosa (DeVile et al., 1995). The possibility that anorexia may primarily be a hypothalamic disease is reinforced by a number of case histories of patients who, after they had been diagnosed to suffer from anorexia and sometimes subjected to psychotherapy, were, at autopsy, found to have a tumor in the hypothalamus (for a description of such cases see Swaab, 2003, Chapter 23.2). These cases not only include a number of children that had atypical anorexia, but also adults with a "typical" diagnosis of anorexia nervosa.

Although these case histories show that all the signs and symptoms of anorexia nervosa can be found in patients with a hypothalamic tumor, including the characteristic that the psychodynamic features are the most outstanding, it should be noted that these are very rare causes of anorexia, while the majority of the hypothalamic tumors are not associated with symptoms of anorexia nervosa.

Molecular genetic factors involved in eating and metabolism disorders

Human obesity certainly has an important inherited component. In fact, studies in twins, adoptees and families indicate that 80% of the variance in body mass index is attributable to genetic factors (Rosenbaum et al., 1997). The "obesity gene map 2000" reports on the presence of 47 human cases of obesity caused by single-gene mutation in six different genes, including SIM1, a critical transcription factor for the formation of the supraoptic and paraventricular nucleus (SON and PVN) in mice. In addition, 24 Mendelian disorders exhibiting obesity as one of their clinical manifestations have now been mapped (Pérusse et al., 2001). Yet the genetic factors responsible for most obesity in the general population have remained elusive so far. As far as the single-gene mutations are concerned, obese subjects with a mutation in the gene that encodes for leptin (Montague et al., 1997; Ströbel et al., 1998) or for the leptin receptor (Clément et al., 1998) have been described. The missense leptin mutation described by Ströbel et al. (1998) is associated not only with morbid obesity but also with hypogonadism and primary amenorrhea. The male patient never enters the stage of puberty. The mutation described by Clément et al. (1998) results in a truncated leptin receptor, lacking both the transmembrane and the intracellular domains. In addition to their early onset morbid obesity and lack of pubertal development, patients who are homozygous for this mutation also have reduced secretion of growth hormone, growth retardation and central hypothyroidism. The observations in subjects with mutations in the leptin receptor and leptin itself suggest that leptin not only controls body mass but is also a necessary signal for the initiation of puberty in humans. However, since in a child with congenital leptin deficiency there was no evidence of substantial impairment in basal or total energy expenditure, and her body temperature was normal, leptin may be less central to the regulation of energy expenditure in humans than in mice. Treatment of this 9-year-old child with recombinant leptin led to sustained reduction in weight, predominantly as a result of a loss of fat. This therapeutic response confirms the importance of leptin in the regulation of body weight in humans and establishes an important role for this hormone in the regulation of appetite (Farooqi et al., 1999). Polymorphisms in the leptin receptor gene are associated with increased levels of abdominal fat in postmenopausal overweight women (Wauters et al., 2001).

Another genetic defect was found in a woman with extreme childhood obesity, abnormal glucose homeostasis, hypogonadotropic hypogonadism, hypocortisolism and elevated plasma proinsulin and pro-opiomelanocortin (POMC) concentrations, but a very low insulin level. This disorder seems to be based upon a mutation in the prohormone processing endopeptidase, prohormone convertase 1 (PC1) (Jackson et al., 1997). Severe early onset obesity, adrenal insufficiency and red hair pigmentation were found to be caused by POMC mutations (Krude et al., 1998; Krude and Grüters, 2000). The patients had severe early onset obesity and red hair pigmentation due to mutations truncating the POMC molecule and leading to the complete lack of adrenocorticotropic hormone (ACTH) and α-melanophoric stimulating hormone (α-MSH) (Krude and Grüters, 2000; MacNeil et al., 2002). However, a cryptic trinucleotide repeat polymorphism in exon 3 of POMC that was associated with elevated leptin levels, appeared not to be associated with obesity (Rosmond et al., 2002). Mutations in the MC-4 receptor gene (*MC4R*) seem to be a common cause of monogenic human obesity. Up to 4–6% of severely obese humans have defects of the MC-4 receptor gene. Affected individuals have hyperphagia in childhood, which loses its intensity later in life. These individuals of normal height, demonstrate, according to some studies, binge eating as the major phenotype characteristic. Some patients had cyclothymia or bipolar affective disorder (Cone, 1999; Mergen et al., 2001; Kobayashi et al., 2002; Branson et al., 2003; Farooqi et al., 2003). However, the relatonship between MC4R variants in obese carriers and binge eating phenotype could recently not be confirmed by Hebebrand et al. (2004), who earlier described a patient with both extreme obesity and bulimia nervosa, who has a haploinsufficiency mutation in the MC-4 receptor (Hebebrand et al., 2002). A novel MC-3 receptor mutation has been observed in an obese girl and her father (Lee et al., 2002). However, MC-3 receptor variants are common and generally not considered to explain human morbid obesity (Schalin-Jäntti et al., 2003).

In a number of obese subjects a mutation in the preproghrelin gene was found that corresponds to the last amino acid of ghrelin (Ukkola et al., 2001), but so far there is no firm evidence that sequence variants in the coding region of the ghrelin gene influence body weight (Hinney et al., 2002). However, growth hormone secretogogues such as ghrelin may be important for feeding. When the expression of the receptor for this compound in the arcuate nucleus was blocked, the rats showed lower body weight and less adipose tissue than controls (Shuto et al., 2002).

A glucocorticoid receptor polymorphism is associated with obesity and dysregulation of the hypothalamo–pituitary–adrenal (HPA) axis (Rosmond et al., 2000). A patient with a mutation in the transcription factor steroidogenic factor 1 had a complete sex reversal and developed obesity in late adolescence (Ozisik et al., 2002). An association between a polymorphism of the estrogen receptor beta and bulimia has been reported (Nilsson et al., 2004). Prader–Willi syndrome (PWS) patients, which are characterized by obesity, hypotonia, mental retardation and hypogonadism, usually have a de novo, paternally derived, deletion of the chromosome region 15q11-13. The brain derived neurotrophic factor (BDNF) Met66 variant is strongly associated with all eating disorders, including bulimia (Ribasés et al., 2004).

The heredity of anorexia nervosa is estimated to be around 70%. A meta-analysis showed that the -1438A allele of the *5-HT2A* gene is significantly associated with anorexia nervosa (Gorwood et al., 2003), while also serotonin transporter regulatory region polymophisms are associated with anorexia nervosa (Matsushita et al., 2004). In addition, the BDNFMet66 variant and the BDNF196G/A polymorphism are susceptibility factors for anorexia nervosa (Ribasés et al., 2003, 2004). A single nucleotide polymorphism (SNP) in the agouti-related protein (AGRP), a natural MC-4 receptor agonist, is thought to increase the risk of developing anorexia nervosa (Vink et al., 2001).

Corticosteroids and obesity

In 1912, Harvey Cushing published the first full description of his eponymous syndrome, which results from prolonged exposure of the organism to high levels of glucocorticoids. This syndrome may result from exogenous administration of glucocorticoids, from ACTH excess by a tumor of the pituitary gland, a supra- or extracellular microadenoma (this form is called Cushing's disease), an ectopic ACTH- or corticotropin-releasing hormone (CRH)-secreting tumor such as bronchial carcinoid tumors, medullary thyroid carcinoma, pheochromocytoma or paraganglioma, or from a cortisol-secreting tumor (Magiakou et al., 1997; Murakami et al., 1998; Newell-Price et al., 1999). High levels of corticosteroids may lead to central or visceral obesity (Salehi et al., 2005), hypertension (Kelly et al., 1998) and atypical depression (Dorn et al., 1995; Gold et al., 1995). One-third of the patients receiving corticosteroids experience significant mood disturbances and sleep disruption (Mitchell and O'Keane, 1998).

In an old study of Heinbecker (1944), atrophy of the PVN and SON was described in Cushing's disease, while in some cases a patchy loss of neurons was observed in the SON, PVN, and posterior hypothalamic and mamillary nuclei. These alterations may well be explained by the inhibitory action of the increased corticosteroid levels, on CRH, vasopressin, and thyrotropin releasing hormone (TRH) neurons we observed in the SON and PVN (Erkut et al., 1998; Alkemade et al., 2005; Fig. 5).

Various studies reported the absence of the circadian rhythm characteristics in Cushing's disease (Stewart et al., 1992; Bierwolf et al., 2000). Cushing patients have a loss of normal 24-h blood pressure fluctuations (Piovesan et al., 1990), a disruption of circadian cortisol secretion and elevated cortisol values between 23.00 and 03.00 h. An elevated salivary cortisol late in the evening (e.g. 11 p.m.) suggests the presence of Cushing's syndrome (Raff, 2000). The lack of circadian rhythms in Cushing's disease may also be explained by the inhibitory action of corticosteroids on the SCN (Swaab, 2004; Figs. 6 and 7). That circadian abnormalities (Stewart et al., 1992) are secondary to increased cortisol levels is supported by the observation that patients with Cushing's syndrome, due to excess of exogenous corticosteroids, also lack normal circadian rhythms in other hormones (Biller, 1994; see below), and blood

pressure. Cortisol can increase blood pressure in a dose-dependent fashion (Kelly et al., 1998). CRH itself is thought to suppress food intake via the CRH-2 receptor (Mastorakos and Zapanti, 2004).

Metabolic syndrome-X includes the symptoms insulin resistance, abdominal obesity or visceral obesity with conspicuous similarities with Cushing's syndrome, elevated lipids and blood pressure. The function of the glucocorticoid receptor is abnormal, possibly due to a polymorphism in the first intron of the gene, found in 14% of the Swedish population. The pathogenesis of this syndrome is proposed to start with life events such as psychosocial and socioeconomic handicaps associated with alcohol consumption and smoking, psychiatric traits or mood changes. Perinatal factors may also be involved, preprograming the increased HPA-axis activity. These factors, via the HPA axis, cause elevated cortisol secretion, which is amplified by a deficient feedback inhibition, probably based upon a genetic susceptibility, as has been mentioned earlier. In addition, the sympathetic nervous system is activated (Björntorp and Rosmond, 1999).

Functional imaging and eating/metabolism

Recently functional imaging contributes to our knowledge of the involvement of the hypothalamus in eating and metabolism. Using a new temporal clustering technique for functional MRI (fMRI), Liu et al. (2000) observed two eating-related peaks in neural activity at two different times with distinct localization, i.e. in the "upper-anterior" and "medial" region of the hypothalamus. However, these areas were so far not linked to the microscopic or chemical anatomy of the hypothalamus, although there was a dynamic interaction between these fMRI responses and plasma insulin levels. Following glucose ingestion, a dose-dependent, prolonged decrease of the fMRI signal was observed in the human hypothalamus. This effect was most pronounced in its upper anterior part (Smeets et al., 2005). Satiation produces significant decreases in blood flow in the hypothalamus of obese women (Gautier et al., 2001).

The signs and symptoms of bulimia nervosa were found to be related to the serotonin (5HT)

10

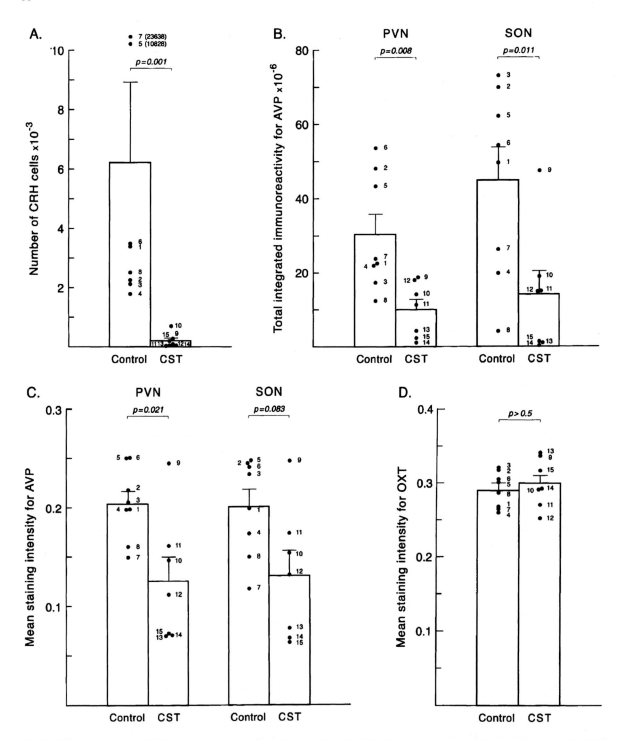

Fig. 5. Estimated number of CRH-immunoreactive cells in the hypothalamic PVN (A), the total integrated immunoreactivity for AVP (B), the mean staining intensity of AVP-immunoreactive cells in the PVN and SON (C), and the mean staining intensity for OXT in the PVN of the controls and the corticosteroid-exposed subjects (D; CST). The numbers of the plotted data refer to the numbers of subjects in the paper. The bars and error lines represent the mean and SEM, and the p values are according to the Mann–Whitney U test. Note that corticosteroids do not only decrease the number of CRH neurons in the PVN, but also the amount of vasopressin staining in the SON and PVN, while OXT stays unaffected (From Erkut et al., 1998; with permission).

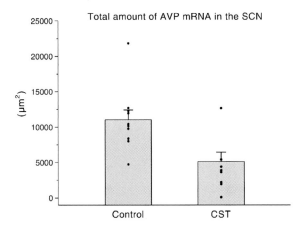

Fig. 6. Estimated total amount of arginine vasopressin (AVP) mRNA in the suprachiasmatic nucleus (SCN; expressed as masked area of silver grains) of the controls and the corticosteroid-exposed subjects (CST). The bars and error lines represent the mean and standard error of the mean (SEM) (From Swaab, 2004; with permission).

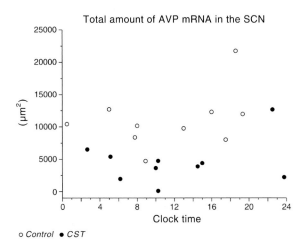

Fig. 7. Day–night fluctuation in the total amount of AVP mRNA of the SCN in controls and in the glucocorticoid-exposed subjects (CST). Note that at any moment of the day the values for CST are lower than those of controls (From Swaab, 2004; with permission).

metabolism. Using single photon emission computed tomography (SPECT), a reduced hypothalamic and thalamic 5-HT transporter availability was found in this disorder. The impaired 5-HT transporter availability was more pronounced with longer duration of the illness (Tauscher et al.,

2001). Reduced 5-HT2A receptor binding may be fundamental to the pathophysiology of anorexia nervosa since this remains after long-term weight restoration (Stamatakis and Hetherington, 2003).

Postmortem studies of the hypothalamus

Apart from the huge problem of the availablity of enough well-documented postmortem material, observations on the hypothalamus using quantitative immunocyto-chemistry or quantitative in situ hybridization reveal excellent functional information on the cellular and circuit level. However, postmortem studies are hampered by a large number of confounding factors that can be distinguished in those occurring before, during or after death. Material of patients with brain diseases should be matched for such factors with appropriate controls or their effect on the measurements should be corrected for, as we did for the effect on storage time on the amount of CRH mRNA in the paraventricular nucleus in depression (Raadsheer et al., 1995). In addition to matched controls that did not die of a neurological or psychiatric disease, samples from related disorders are often useful to control for disease specificity.

Antemortem confounding factors include, e.g., age, sex, use of medicines prior to death, season, clock time of death, lateralization and circulating volume. Factors during dying are, e.g., duration of illness, gravity of illness and agonal state. Postmortem confounding factors concern, e.g., the time between death and fixation or freezing of the tissue, freezing procedures, fixation and storage time (for review, see Swaab, 2003). It is self-evident that a very large control group is necessary to allow the selection of the best matching patients.

Disorders accompanied by disturbances in eating and metabolism

Bulimia nervosa

Bulimia nervosa is characterized by recurrent episodes of binge eating, at least twice a week for a period of 3 months. Large amounts of food are eaten in a discrete period and patients have the feeling that they cannot stop. There is recurrent inappropriate compensatory behavior to prevent weight gain such as vomiting or excessive exercise (DSM IV). Although all the signs and symptoms of this disorder point to a hypothalamic process, no postmortem studies have been performed so far due to a lack of material.

A clear seasonal pattern has been reported in the signs and symptoms of bulimia nervosa. Binge eating behavior, purging and depressed mood were found to be closely associated with the photoperiod in that the symptoms are the most severe in winter and the least severe in summer (Fig. 8). Such seasonal changes in symptoms as seen in bulimia nervosa are not or to a lesser degree present in

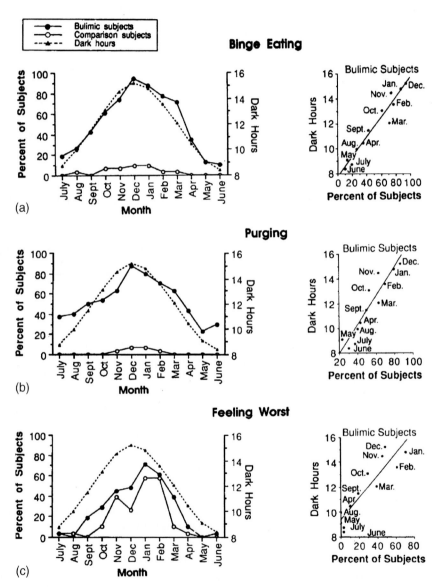

Fig. 8. Seasonal variation in binge eating, purging and feeling worst among 31 bulimic and 31 comparison subjects. Binge eating (a), purging (b), and feeling worst (c) were determined according to the modified seasonal pattern assessment questionnaire. Number of dark hours was defined as 24h minus the average photoperiod for each month (From Blouin et al., 1992; with permission).

anorexia nervosa (Fornari et al., 1994; Lam et al., 1996a; Ghandirian et al., 1999; Yamatsuji et al., 2003). The episodes of symptoms in bulimia were proposed to be related to deminished serotonergic activity (Blouin et al., 1992; Lam et al., 1996a). Seasonal fluctuations have indeed been found in hypothalamic serotonin levels of controls, with a minimum during December and January (Fig. 3). The seasonal fluctuations in activity we observed in the SCN of control patients (Fig. 2) are considered to be the basis for these seasonal fluctuations. The serotonergic system may indeed be involved in the signs and symptoms of bulimia since, using SPECT, a reduced hypothalamic and thalamic 5-HT transporter availability was found in this disorder that was more pronounced with longer duration of the illness (Tauscher et al., 2001). Because the alterations in cortical activity and the elevated concentrations of 5-hydroxyindolacetic acid (5-HIAA) found in the cerebrospinal fluid (CSF) of bulimia nervosa patients remained after recovery, they are proposed to be a trade-related characteristic (Kaye et al., 1998, 2005).

It should be noted though that Posternak and Zimmerman (2002) did not find higher rates of bulimia in winter in an outpatient psychiatric practice in the United States pointing to the presence of subgroups of bulimia patients.

Various observations indicate the presence of a disorder of different hypothalamic systems in subgroups of bulimia patients. The HPA axis is generally found to be hyperactive in bulimia (Licinio et al., 1996; Monteleone et al., 1999; Cotrufo et al., 2000; Neudeck et al., 2001). Bingeing and vomiting in bulimic patients was associated with modest increases in cortisol secretion (Weltzin et al., 1991; Galderisi et al., 2003) and increased dehydroxyepiandrosterone (sulfate) (DHEA(S)) levels (Galderisi et al., 2003), whereas normal-weight bulimic women showed normal circadian ACTH and cortisol variations and levels (Vescovi et al., 1996). A more recent study showed elevated cortisol secretion following exacerbation of bulimic symptoms (Lester et al., 2003).

Prolactin, ACTH, beta-endorphin and melatonin circadian rhythms are disturbed in bulimia nervosa according to some studies (Ferrari et al., 1990; Kaye, 1996; Pacchierotti et al., 2001) and the circadian rhythm of leptin is completely abolished (Balligand et al., 1998). Bulimic women have blunted nocturnal prolactin patterns (Weltzin et al., 1991). Other studies report, however, that the circadian rhythm of melatonin was unaltered in bulimia and anorexia (Brown, 1992), again pointing to the presence of subgroups.

The concept of hypothalamic integration has therapeutic consequences in bulimia. Bulimic patients with worsening of mood and eating symptoms in winter received bright-white light therapy for 2 weeks in a controlled study and were compared to dim light, while in a second open trial, patients received light therapy for 4 weeks. The light treatment was effective for both mood and symptoms of eating disturbances (Lam et al., 1994, 2001). In an other double blind study, giving 3 weeks of bright-light treatment, the binge frequency decreased significantly, but the level of depression did not (Braun et al., 1999), while in a controlled trial of only 1 week of bright light, mood improved in bulimic patients but not the frequency, size or content of binge eating periods (Blouin et al., 1996). Variability in the outcome of light therapy may, at least partly, be explained again by the presence of seasonal and nonseasonal subtypes of bulimia patients (Levitan et al., 1994, 1996).

Binge eating disorder

A new diagnostic concept is binge eating disorder. Like bulimia nervosa, it has binge eating and loss of control as central features, but there is little or no weight control behavior, such as self-induced vomiting and laxative misuse. Some 40% of the binge eating disorder cohort met criteria for obesity in a 5-year follow-up, and this group of patients is highly prevalent (1–30% among often extremely obese subjects seeking weight-loss treatment) (Dingemans et al., 2002; Hsu et al., 2002). The rate of childhood emotional abuse is some 2–3 times more prevalent in this eating disorder than in a normative adult female sample. No other forms of childhood maltreatment, such as physical or sexual abuse, or emotional or physical neglect, were increased in binge eating disorder (Grilo and Masheb, 2002). Binge eating is a major characteristic of subjects with a mutation in the MC4 receptor (Branson et al., 2003). The treatment of choice is currently cognitive behavioral treatment, but interpersonal psychotherapy, self-help and SSRIs also seem effective (Dingemans et al., 2002).

Night eating syndrome

A new, related eating disorder that is different from anorexia nervosa, bulimia nervosa and binge eating, is the night eating syndrome. It is characterized by morning anorexia, evening hyperphagia and insomnia and occurs during periods of stress. Its prevalence has been estimated at 1.5% in the general population and some 27% of severely obese persons. The mood of the night eaters falls during the evening. There are circadian changes, such as an attenuation of the nighttime rise in melatonin and leptin and elevated levels of plasma cortisol (Birketvedt et al., 1999). Nighttime awakenings are far more common among night eaters than among controls and more than half the number of the awakenings are associated with food intake. The typical neuroendocrine characteristics are an attenuation of nocturnal rises in melatonin and leptin and increased diurnal secretion of cortisol. The CRH-induced ACTH and cortisol response are reduced in night eaters (Birketvedt et al., 2002). Observations that light improves the symptoms of nighttime eating syndrome (Friedman et al., 2002) should be further tested in controlled studies.

Major depression

Clinical picture
Major depressive disorder is characterized by a period of at least 14 days with depressed mood or loss of interest or pleasure combined with at least four of the following symptoms: significant weight loss or decrease or increase of appetite; insomnia or hypersomnia; psychomotor agitation or retardation; fatigue or loss of energy; feelings of worthlessness or excessive or inappropriate guilt; diminished ability to think, concentrate, or make decisions; and suicidal tendency (DSM IV). Many

depressed patients also suffer from increased anxiety and decreased libido.

Depression frequently runs an episodic course. Multiple episodes are found in 30–40% of cases and in psychiatric settings recurrence rates of up to almost 90% have been found (Weel-Baumgarten et al., 2000). Then there is seasonal affective disorder (SAD) that is characterized by circannual fluctuations in symptoms and a significant weight gain during winter depression (Levitan et al., 2004). The diurnal variation of depressive state, early morning awakening and seasonal pattern or modulation of onset argues for the involvement of the SCN in the symptomatology, but there are also indications for a disturbed SCN function (see below).

There is a close link between metabolic alterations and changes in mood. Body weight can either decrease or increase. Classically, the melancholic type of depression features anorexia or weight loss. A number of publications points to the relationship between mood disorders and obesity. Children and adolescents with major depression are at increased risk of developing overweight. Patients with bipolar disorder may have elevated rates of overweight, obesity and abdominal obesity, and obese persons seeking weight-loss treatment have elevated rates of depressive and bipolar disorders.

Obesity is associated with depressive disorder in women, and abdominal obesity may be associated with depressive symptoms in females and males. Cortisol elevations are presumed to play a role in these relationships (Brown et al., 2004). Depression is indeed a well-known side effect of glucocorticoids. One-third of the patients receiving glucocorticoids experience significant mood disturbances and sleep disruption. Up to 20% report psychiatric disorders, including depression, mania and psychosis (Mitchell and O'Keane, 1998). Moreover, atypical depression is found in a large proportion of patients with Cushing's disease. Patients with long-term Cushing's syndrome are especially at risk for such psychopathology (Dorn et al., 1995; Gold et al., 1995). The fact that atypical depression is so often seen in Cushing's syndrome indicates that in these patients cortisol causes this type of depression rather than ACTH or CRH. This conclusion is supported by a small study that shows that depression can be treated by ketoconazole, an antiglucocorticoid (Wolkowitz et al., 1999) and by the observation that metyrapone, an inhibitor of the cortisol production, successfully treats depression in Cushing patients (Checkley, 1996). However, it should be noted that, also after correction of hypercortisolism in Cushing's syndrome, atypical depression frequently continues to be present. Suicidal ideation and panic may increase (Dorn et al., 1997). Depression with atypical symptoms in women is significantly more likely to be associated with overweight than depression with typical symptoms. (McElroy et al., 2004).

Among hypertensive subjects with the metabolic syndrome, a condition with increased HPA-axis activity (Björntorp and Rosmond, 1999), the prevalence of depression is greater in women than in men (13.0% vs. 7.3%) (Bonnet et al., 2005). Recently it was found that obesity and metabolic syndrome occur in a circadian *Clock* mutant mouse, suggesting that the circadian network plays an important role in the mammalian energy balance (Turek et al., 2005). Seasonal fluctuations in body weight support this possibility. The typical patient with seasonal affective disorder (SAD) is a premenopausal woman with marked craving for high-carbohydrate/high-fat foods and significant weight gain during winter depression. These patients have a high prevalence of the 7-repeat allele of the dopamine-4 receptor gene (*DRD4*) (Levitan et al., 2004). Patients with SAD show an exaggerated behavioral response to the seasons. Because of the strong fluctuations in weight, SAD has been described as a naturally reversible form of obesity (Rosenthal et al.,1987). Patients with SAD report atypical symptoms of increased appetite, particularly "carbohydrate craving," increased body weight and sleepiness during their winter depression (Kräuchi and Wirz-Justice, 1988). Appetitive symptoms are particularly sensitive to bright-light therapy (Kräuchi et al., 1993).

Furthermore, depressed patients have alterations in their hypothalamo–pituitary–thyroid (HPT)-axis (Musselman and Nemeroff, 1996), as both basal thyroid stimulating hormone (TSH) and thyroxin levels were found to be altered in melancholic and major depressed patients (Maes et al., 1993b), which may also affect metabolism.

Enhanced corticosteroid levels may be responsible for this (Alkemade et al., 2005)

Depressive illness is presumed to result from an interaction between the effects of environmental stress and genetic/developmental predisposition. The HPA axis, a key system in control of the stress response, is considered to be the "final common pathway" for a major part of the depressive symptomatology (Fig. 9), while also subsequent changes in the serotonin system are involved. Indeed, the CRH neurons of the PVN that regulate the HPA axis are strongly activated in depression (Figs. 10 and 11), while in the majority of the depressed patients dexamethasone resistance is prominent (see below). Although the set point of HPA-axis activity and of other central systems is programed by genotype, it can be changed to another level by developmental influences and early negative life events. Long-lasting hyper(re)activity of the CRH neurons, resulting in increased stress responsiveness and reflecting a glucocorticoid resistant state, is commonly seen in depressed individuals (Swaab et al., 2005). Observations in humans further indicate that aversive experiences, both in utero and in the neonatal period, result in sustained HPA-axis activation and in sensitization of emotional and HPA-axis responses to subsequent stress. Maternal stress beginning at infancy and subsequent stress on preschoolers is accompanied by a sensitization of the children's HPA-axis response to subsequent stress exposure. Stressful life events such as bereavement, child abuse, and early maternal separation are also risk factors for depression, anxiety disorder or both. Childhood physical or sexual abuse are important early stressors that may predispose individuals to adult onset depression accompanied by a permanent hyperactivity of the HPA system (Heim and Nemeroff, 2001; Swaab et al., 2005).

In addition, small size at birth is associated with an alteration in set point of the HPA axis and an increased cortisol responsiveness and risk of depression in adulthood (Phillips, 2001; Thompson et al., 2001).

Almost all environmental and genetic risk factors for depression ultimately appear to go together with increased HPA-axis activity in adulthood. On the other hand, when patients or animals

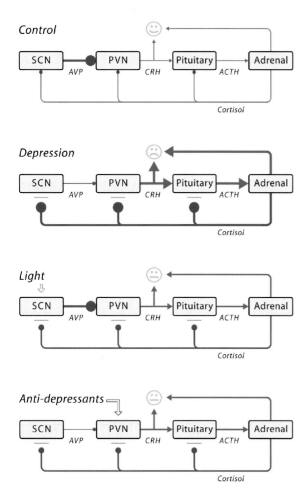

Fig. 9. Depression; schematic illustration of an impaired inter-action between the decreased activity of vasopressin neurons (AVP) in the suprachiasmatic nucleus (SCN) and the increased activity of corticotropin-releasing hormone (CRH) neurons in the paraventricular nucleus (PVN). The hypothalamo–pituitary–adrenal (HPA) system is activated in depression and affects mood, via CRH and cortisol. We found a decreased amount of vasopressin (AVP) mRNA of the SCN in depression. The decreased activity of AVP neurons in the SCN of depressed patients is the basis of the impaired circadian regulation of the HPA system in depression. Moreover, animal data have shown that AVP neurons of the SCN exert an inhibitory influence on CRH neurons in the PVN. Increased levels of circulating glucocorticoids decrease AVP mRNA in the SCN, which will result in smaller inhibition of the CRH neurons. In the light of our data we propose the following hypothesis for the patho-genesis of depression. In depressed patients, stress acting on the HPA system results in a disproportionally high activity of the HPA system because of a deficient cortisol feedback effect due to the presence of glucocorticoid resistance. The glucocorti-coid resistance may either be caused by a polymorphism of

are treated with antidepressants, electroconvulsive therapy, vagus nerve stimulation, or when they show spontaneous remission, the HPA-axis func-tion returns to normal (Nemeroff, 1996; O'Keane et al., 2005).

In addition to these clinical observations, alter-ations in the brain centers that initiate and control the stress response, such as the human hypo-thalamus, have also been reported. In the PVN of patients with major depression or bipolar disorder, CRH, vasopressin and oxytocin neurons are acti-vated (Raadsheer et al., 1994, 1995; Purba et al., 1996).

Moreover, depression is associated with an enhanced pituitary vasopressinergic responsivity (Dinan et al., 1999). Because of their central satiety effect, the activation of oxytocin neurons in depression has been connected to the anorexia and weight loss as reported in the melancholic type of depression (Purba et al., 1996). Interestingly, also the supraoptic nucleus shows enhanced vaso-pressin mRNA production in melancholic depres-sion (Meynen et al., 2006) that results in increased plasma levels of vasopressin (Van Londen et al., 1997, 1998b, 2001). The increased vasopressin production of the SON can at least be partly responsible for the chronic HPA-axis hyperdrive in depression (Von Bardeleben and Holsboer, 1989; Engelmann et al., 2004), acting directly via the portal system or via the systemic circulation on ACTH release in the pituitary (Meynen et al., 2006) and for an enhanced suicide risk (Inder et al., 1997). The possibility that chronically elevated vasopressin levels are involved in the induction of depressive symptomatology is further supported by the case of a 47-year-old man with an esthesioneuroblastoma that was associated with

corticosteroid receptor or by a developmental disorder. Also AVP neurons in the SCN react to the increased cortisol levels and subsequently fail to inhibit sufficiently the CRH neurons in the PVN of depressed patients. Such an impaired negative feed-back mechanism may lead to a further increase in the activity of the HPA system in depression. Both high CRH and cortisol levels contribute to the symptoms of depression. Light therapy activates the SCN, directly inducing an increased synthesis and release of AVP that will inhibit the CRH neurons. (Antidepres-sant medication generally inhibits the activity of CRH neurons in the PVN) (From Swaab, 2004; with permission).

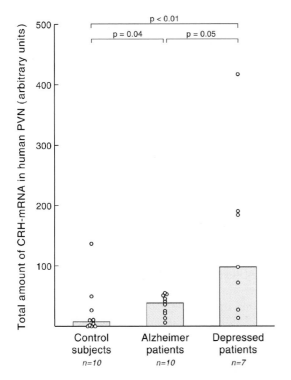

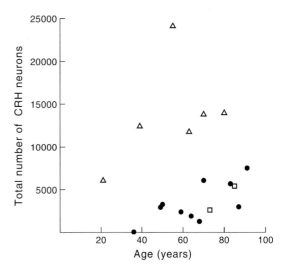

Fig. 11. The total number of CRH-expressing neurons in the PVN of human subjects at different ages. ●, 10 control subjects; Δ, 6 bipolar or major depressed patients; □, 2 "nonmajor depressed" subjects with either an organic mood syndrome or a depressive episode not otherwise specified. Note the high number of neurons expressing CRH in bipolar and major depressed patients (From Raadsheer et al., 1994; with permission).

Fig. 10. Total hybridization signal for human corticotropin-releasing hormone (CRH)-mRNA (arbitrary units) in the para-ventricular nucleus (PVN). Bars indicate median values per patient group. The PVN of the Alzheimer patients ($n = 10$) contained significantly more CRH-mRNA than that of comparison subjects ($n = 10$). The amount of radioactivity in depressed patients ($n = 7$) was significantly higher than in comparison cases and Alzheimer's disease patients (From Raadsheer et al., 1995; with permission).

the onset of a first episode of major depression. The man displayed chronically elevated plasma vasopressin levels due to vasopressin secretion by the tumor. Depressive symptoms markedly improved after surgical resection of the tumor and subsequent normalization of plasma vasopressin levels.

The theory that chronobiological mechanisms play a role in the pathogenetic mechanism of depression is originally based on the classic diurnal variation of depressive state, early morning awakening and seasonal modulation of onset. In addition, this theory is supported by the antidepressant and occasionally mania-inducing effects of manipulations of the sleep–wake cycle and exposure to light. Interestingly, one of the characteristics of jet

lag is exhaustion with mild depression, pointing again to a strong link between affecting disorders and circadian rhythms (Katz et al., 2001). Moreover, a polymorphism in the clock gene *NPAS2* appeared to be associated with SAD (Johansson et al., 2003). The observation that incidence of hospitalization for depression is higher after westbound flights than after eastbound ones, whereas hypomania occurred more frequently after eastbound flights (Wirz-Justice, 1995), also suggests a relationship between circadian phase changes and mood changes. The finding that melatonin may improve not only sleep but also mood (De Vries and Peeters, 1997; Jean-Louis et al., 1998; Lewy et al., 1998; Bellipanni et al., 2001) supports the idea of involvement of the circadian system in mood.

In addition, in a normal population strong seasonal effects in mood are observed. Depression scores are highest in winter and lowest in summer (Harmatz et al., 2000). Some forms of affective illness have a pattern of periodic recurrence, linked to, e.g., hormonal cycles, characteristic sleep disturbances or diurnal or circannual mood fluctuations. The prevalence rate for SAD is some 10% of the depressed patients, with higher rates for people living at higher latitudes (Wicki et al., 1992). Two types of seasonal mood changes have been described in temperate zones, i.e. (1) depression regularly occurring in fall and winter, and (2) depressive episodes in the summer (Neumeister et al., 1997).

A fairly constant finding that demonstrates a disturbance of circadian rhythms in depression is a decrease in circadian amplitude of body temperature, plasma cortisol, plasma corticosterone, noradrenaline, TSH and melatonin. Depressed patients display sleep disturbances and a less rhythmic, more chaotic pattern of cortisol release. Not only are the ACTH and cortisol levels found to be higher in depressed patients, but also the frequency of pulses of these hormones is higher during the evening. Comparison of multiple circadian rhythms during depression and after recovery have suggested that a blunted amplitude is the main chronobiological abnormality (Wirz-Justice, 1995; for review see Swaab, 2004, Chapter 26.4). This may be due to the increased cortisol levels, which inhibit SCN function (Zhou et al., 2001; Swaab, 2004; Figs. 6, 7, 12,13).

The SCN, the clock of the hypothalamus, normally shows strong circadian and circannual variations in neuronal activity (Hofman and Swaab, 1992b,1993; Hofman, 2001, 2003), which are supposed to be related to the circadian and circannual fluctuations in mood and to sleeping disturbances in depression. However, biological rhythms are also to a certain degree disturbed in depression (Van Londen et al., 2001). A disorder of SCN function, as appeared from the increased amount of vasopressin, the decreased amount of vasopressin mRNA in this nucleus (Figs. 12 and 13), and diminished circadian fluctuation of vasopressin mRNA, may not only be the basis of the circadian and sleeping disorders in depression, but may also contribute to hyperactivity of the CRH neurons, since this nucleus extends direct inhibitory projections to the CRH neurons of the PVN (Kalsbeek et al., 1992; Dai et al., 1997; Zhou et al., 2001). One may presume that a basis for disturbed circadian rhythms may be found in the effect of corticosteroids on circadian timing. We have

indeed observed an inhibitory effect of corticosteroids on vasopressin mRNA in the SCN (Swaab, 2004; Figs. 6 and 7). An interesting observation is that the serotonin metabolites in the jugular vein are lowest in winter, and that the turnover of serotonin in the brain rises with increased luminosity (Lambert et al., 2002), suggesting that the SCN may act via the serotonergic system on seasonal mood changes.

Light therapy and the circadian system

A strong argument for a close relationship between the pathogenetic mechanism of depression and the circadian timing system is the effectiveness of light therapy in SAD (Wirz-Justice, 1995; Wileman et al., 2001), in pharmacological treatment-resistant, rapid cycling affective disorders (Kusumi et al., 1995) and in patients with nonseasonal affective disorders (Yamada et al., 1995; Prasko et al., 2002). High intake of sweets in the second half of the day was the best predictor of a rapid and persistent response to light therapy (Kräuchi et al., 1993).

According to some authors, the effect of bright-light therapy on winter depression takes at least 3 weeks before it becomes apparent (Eastman et al., 1998) but would already act after 1 week according to others (Prasko et al., 2002; W.J.G. Hoogendijk, personal communication). After treatment with light, a significantly greater improvement is reported in patients with seasonal depression than in patients with a nonseasonal pattern of depression. However, another study has reported similar effects of light treatment in seasonal and nonseasonal depression and the effects are faster than psychopharmacological treatment (Kripke, 1998). The latter finding has been confirmed by Prasko et al. (2002). Physical exercise is effective in alleviating depressive symptoms, but is much more effective when combined with bright light (Leppämäki et al., 2002).

The mechanisms involved in bright-light treatment are still under investigation. Data from our group indicate a decreased synthesis and transport of vasopressin in the SCN of depressed patients (Zhou et al., 2001; Figs. 12 and 13). Since

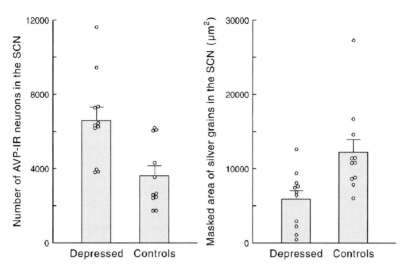

Fig. 12. The number of arginine vasopressin-immunoreactive (AVP-IR) neurons (A) and the mask area of silver grains of the AVP mRNA (B) in the suprachiasmatic nucleus (SCN) in control subjects ($n = 11$) and depressed subjects ($n = 11$). The error bars indicate the SD. Note the change in the balance between the presence of more AVP and less AVP mRNA in depression. There is probably a disorder of the transport of AVP that leads to accumulation of the peptide, in spite of the decreased production rate (From Zhou et al., 2001; with permission).

vasopressin neurons of the SCN inhibit CRH production in the PVN (Kalsbeek et al., 1992; Gomez et al., 1997), this may contribute to the activation of CRH neurons in depression. Light therapy may stimulate the SCN neurons and thus restore the inhibition of the CRH neurons (Fig. 9). Since rapid tryptophan depletion reverses the antidepressant effect of bright-light therapy in patients with SAD, the therapeutic effects of bright light might involve a serotonergic mechanism (Lam et al., 1996b; Neumeister et al., 1997). Measurements of 5-HT metabolites in the jugular vein show that this is indeed the case. The production of 5-HT in the brain rises rapidly with increased luminosity (Lambert et al., 2002).

Narcolepsia

Narcolepsy is a chronic, disabling sleep disorder that affects 1 in 2000 individuals and is characterized by various symptoms, historically bundled into a tetrad-excessive daytime sleepiness, cataplexy, hypnagogic hallucinations and sleep paralysis (Overeem et al., 2001). Daytime hypersomnolence is expressed by the occurrence of irresistible sleep attacks throughout the day. In other words, patients can be awake, but they cannot stay awake. The sleep episodes are commonly superimposed on a more continuous feeling of sleepiness. Cataplexy is a sudden bilateral loss of muscle tone with preserved consciousness, in response to strong emotions such as mirth or laughter (Fig. 14). Most cataplectic attacks are brief, lasting seconds to a few minutes. Hypnagogic hallucinations are unusually vivid dream experiences occurring at sleep onset. These hallucinations can be visual or auditory and are usually frightening. Sleep paralysis is a complete inability to move at sleep onset or awakening. Nowadays, the narcoleptic tetrad is considered incomplete: fragmented nighttime sleep and obesity are important features in many patients as well. The exact cause of obesity is not known. Glucose hypometabolism has been found by positron emission tomography (PET) in the posterior hypothalamus, mediodorsal thalamus and a number of cortical areas (Joo et al., 2004).

Recently it has been established by positioning cloning that an autosomal recessive mutation of the hypocretin (orexin) receptor 2 gene (*HCRT2*) is responsible for the genetic form in a well-established canine model, in Doberman pinchers and Labrador

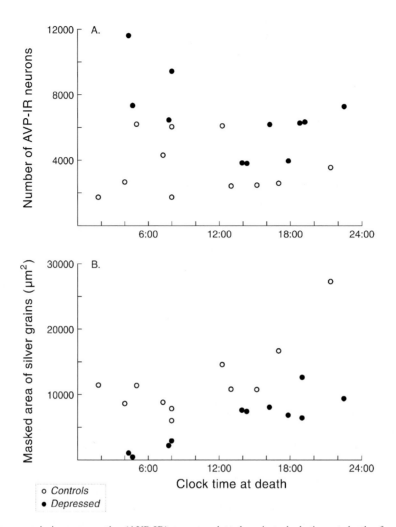

Fig. 13. Number of vasopressin-immunoreactive (AVP-IR) neurons plotted against clock time at death of each individual (11 depressed subjects, and 11 control subjects) (A), area of masked silver grain plotted against clock time of death of each individual (B). The difference between depressed and control subjects is present at different points of the day and there is no overlap between the two groups when you take the clock time at death into account (From Zhou et al., 2001; with permission).

retrievers (Kadotani et al., 1998; Lin et al., 1999). In addition, hypocretin knockout mice exhibit narcolepsy (Chemelli et al., 1999), and modafinil, an antinarcoleptic drug, activates orexin-containing neurons. Hypocretin neurons are localized in the perifornical region in the human brain (Fig. 15). Their number is estimated to be 80,000 neurons (Fronczek et al., 2005). The hypocretin neurons are known to project to brainstem regions linked to motor inhibition as well as to the locus coeruleus (norepinephrine), raphe nucleus (serotonin), laterodorsal tegmental nuclei (acetylcholine) and

ventral tegmentum (dopamine) (Peyron et al., 2000; Van den Pol, 2000; Moore et al., 2001; Thorpy, 2001).

In contrast to animal models, most human cases of narcolepsy are not familial (Siegel, 1999). It is therefore unlikely that a high proportion of human narcoleptics have a mutation or a polymorphism responsible for narcolepsy, either in the *HCRT* gene or in the HCRT-1 or -2 receptors (Ólafsdóttir et al., 2001). Immunocytochemical and in situ hybridization data indicate that there is a substantially (85–95%) reduced number of

Fig. 14. Top: a cataplectic attack in a patient with narcolepsy, here elicited by laughter. Note that the paralysis is not instantly complete; the patient is able to reach out his arms to break the fall. Below: cataplectic attack in a hypocretin-receptor-2 mutated Doberman pincher. The attack was triggered by the excitement from receiving a piece of palatable food. The attack is partial; the most obvious weakness in in the hindlimbs (From Overeem et al., 2002; with permission).

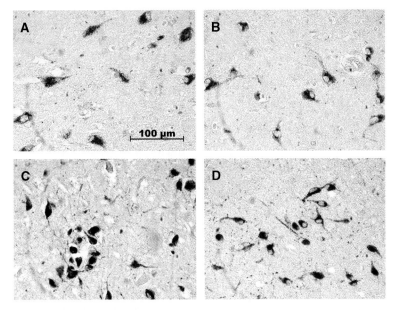

Fig. 15. Examples of staining of hypocretin-IR cell bodies in the lateral hypothalamus of an adult control subject #02-076 (A), an adult Prader–Willi patient #91-058 (B), a control infant #97-153 (C) and a Prader–Willi infant #99-079 (D). There was no significant difference in the intensity of staining and the distribution pattern. Note that the density of cell bodies is higher in the infant subjects, which is accompanied by a smaller volume of the hypothalamic area containing these neurons (From Fronczek et al., with permission).

neurons producing hypocretins in narcoleptics (Peyron et al., 2000; Thannickal et al., 2000; Van den Pol, 2000) and hypocretin-1 levels have been reported to be absent or dramatically decreased in the CSF of the majority of patients suffering from narcolepsy (Nishino et al., 2000, 2001). The detection of increased glial fibrillary acidic protein (GFAP) immunostaining of astrocytes in the perfornical hypocretin area of the brain of narcoleptics (Thannickal et al., 2000; Van de Pol, 2000) argues in favor of some type of neuronal degeneration. However, Peyron et al. (2000) could not confirm the presence of hypothalamic gliosis in narcolepsy. The reason for this discrepancy is not clear at present. Clearly more patients in different phases of the disease should be studied. Some rare polymorphism in the prepro-orexin was claimed to be associated with narcolepsy (Gencik et al., 2001) and a mutated signal sequence of hypocretism in a child with narcolepsy (Peyron et al., 2000).

Hypocretins are not only involved in narcolepsy but also in eating behavior and metabolism. The body mass index (BMI) is increased in narcolepsy patients, indicating altered energy homeostasis (Schuld et al., 2000; Dahmen et al., 2001). Since leptin levels in serum are reduced in narcoleptic patients by more than 50% and increased in CSF, narcolepsy seems to be accompanied by complex alterations in the regulation of food intake and metabolism (Schuld et al., 2000; Nishino et al., 2001; Kok et al., 2002). In narcoleptic patients, hypocretin deficiency is accompanied by a disruption of the circadian distribution of growth hormone-releasing hormone release, in such a way that they secrete about 50% of their growth hormone during the day, whereas controls secrete only 25% of their growth hormone during that period (Overeem et al., 2003). This, and these patients' propensity to fall asleep during the day, supports the notion that the function of the suprachiasmatic nucleus (SCN) is disturbed in this disorder, which may also contribute to obesity (Turek et al., 2005).

Prader–Willi syndrome

Prader–Willi syndrome (PWS), the most common syndromal form of human obesity, is characterized by grossly diminished fetal activity and hypotonia in infancy, mental retardation (mean IQ of 65) or learning disability and a number of hypothalamic symptoms, i.e. feeding problems in infancy, and later insatiable hunger and gross obesity (Fig. 16), hypogonadism and hypogenitalism. PWS occurs in 1 of every 10,000–25,000 births. The majority of PWS cases are sporadic, but familial cases have

been reported (McEntagart et al., 2000). In 70% of the patients a de novo deletion of the paternally inherited chromosome 15q11-13 is present. About 28% of PWS cases are due to maternal uniparental disomy, which would result in a slightly milder phenotype with better cognitive functions. Paternal deletion and maternal uniparental disomy are functionally similar as they both result in the absence of a paternal contribution to the genome in the 15q11-13 region. A third, and the most severe, phenotype with a high incidence of congenital heart disease are the patients with maternal uniparental disomy 15 with mosaic trisomy 15 (Olander et al., 2000). Some PWS candidate genes have been identified. The SNRPN (small nucleoriboprotein-associated polypeptide N) gene is probably part of the putative imprinting center that regulates the expression of several genes in PWS transcriptional domain (Martin et al., 1998a). An intact genomic region and/or transcription of SNRPN exons 2 and 3 seem to play a pivotal role in the manifestations of the clinical phenotype in PWS (Kuslich et al., 1999). However, other human cases tend to exclude SNRPN as the causative gene for PWS genotype (Conroy et al., 1997). In addition, the human necdin gene, *NDN*, which is maternally imprinted and located in PWS chromosomal region, was considered to be a candidate gene (Jay et al., 1997). Although at first necdin-deficient mice did not develop the hypogonadism, infertility or obesity characteristics of PWS (Tsai et al., 1999), later on Necdin mouse mutants were developed that showed hypothalamic and behavioral alterations reminiscent of the human PWS, including a reduction of 90% in oxytocin neurons and of 25% in luteinizing hormone releasing hormone (LHRH)-producing neurons, increased skin-scraping activity (Muscatelli et al., 2000), and a deficiency of respiratory drive (Ren et al., 2003). Others claim that the imprinted genes ZNF-127 and -127 AS may be associated with some of the PWS features (Jong et al., 1999). In addition, there is a small evolutionarily conserved RNA resembling C/D box, small-nucleolar RNA, which is transcribed from *PWCR1*, a novel imprinted gene in the PWS deletion region, which is highly expressed in the brain (De los Santos et al., 2000).

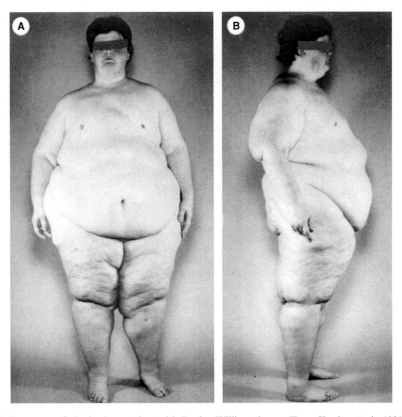

Fig. 16. Characteristic pattern of obesity in a patient with Prader–Willi syndrome (From Kaplan et al., 1991; with permission).

Hypothalamic abnormalities

The major symptoms of this syndrome are seen as the result of hypothalamic disturbances, neuroendocrine and non-neuroendocrine (Swaab, 1997, 2003). This fits in with the experimental data of Keverne et al. (1996), who showed that cells that express only paternal genes accumulate in clusters scattered through the hypothalamus, septum, preoptic area and amygdala, while cells that express only maternal genes accumulate in the cortex and striatum.

Severe fetal hypotonia is often already noticed by the mother during pregnancy; the baby does not seem to move much. Apart from the baby's underactivity, its position in the uterus at the onset of labor is often abnormal (either a transverse, face or breech presentation). These abnormal presentations result in a high percentage of assisted deliveries. In addition, the percentage of asphyctic infants is at least 8 times higher than in the general population. It has often been presumed that the fetal position is caused by hypotonia, the child being too weak to move itself in the correct position. However, there are other congenital disorders in the hypothalamus and pituitary — in which hypotonia is not reported — which are also accompanied by abnormal presentation of the fetus at birth, such as anencephaly and septo-optic dysplasia (De Morsier syndrome). The way the hypothalamus is involved in fetal hypotonia is not known at present. The timing of the moment of birth is often also abnormal; too high a percentage of children with PWS are born either prematurely or too late (Wharton and Bresman, 1989), a phenomenon also found in anencephaly (Honnebier and Swaab, 1973) An abnormality in the PVN of the hypothalamus, which plays a central role in the child's timing of its own birth, may explain these observations (see below).

24

Abnormal function of nerve cells in the hypothalamus containing LHRH is thought to be responsible for decreased levels of sex hormones, resulting in cryptorchism in boys, hypoplastic external genitalia in children of both sexes and delayed or incomplete pubertal development, as well as decreased sexual behavior and insufficient growth during puberty, resulting in short stature (Swaab, 2004).

Short stature and delayed skeletal maturation are the most frequent features of PWS, probably partly due to hypogonadism (see above) and partly due to a growth hormone (GH) deficiency, and are seen in 90% of the PWS patients. GH therapy appeared to have very beneficial effects on body composition and growth velocity. Parents reported that the children were also more alert, had a more stable temperament, were more interested in other children and were easier to handle than before treatment (Lindgren et al., 1997). Growth hormone releasing hormone (GHRH) that is produced in the arcuate nucleus (Fig. 17) was expected to be affected in PWS. However, observations in postmortem material from our group have shown

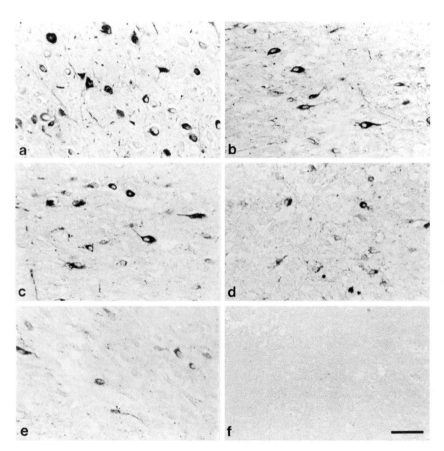

Fig. 17. Growth hormone-releasing hormone Growth hormone releasing hormone (GHRH)-immunoreactive neurons in the infundibular nucleus of 3 controls (left column) and 3 Prader–Willi syndrome (PWS) patients (right column). The top row shows the cases with the highest number of GHRH-immunoreactive neurons, the middle row shows cases with a median GHRH neuron number and the last row represents the cases with the lowest GHRH-neuron number: (a) control case 96-030, (b) PWS case 43830, (c) control case 85-124, (d) PWS case 96-000, (e) control case 80-271 and (f) PWS case 97-049; Scale bar 50 mm. Note that there is a great variability in number of GHRH neurons, both within the control group and in the PWS patient group. Although the PWS patients generally tended to have fewer GHRH neurons and their staining tended to be less intense, this appeared to be due to differences in disease duration and not to PWS per se (cf. Goldstone et al., 2003, preparation by U. Unmehopa).

that the number of GHRH-expressing neurons in this nucleus is not decreased in PWS (Goldstone et al., 2003).

The hypothalamo–pituitary–adrenal and –thyroid axes remain largely intact in PWS, and prolactin and cortisol levels are generally normal (Swaab, 2004).

Leptin is a satiety factor that is produced by fat cells and acts on the infundibular nucleus and other hypothalamic areas in order to inhibit food intake and was, therefore, presumed to be involved in obesity in PWS. Plasma leptin levels are increased in PWS, but this was generally in relation to the increased body mass index (Carlson et al., 1999). In order to see whether an increased activity of neuropeptide-Y (NPY) neurons in the infundibular nucleus might explain the eating disorder in PWS patients, we determined the amount of NPY in the infundibular nucleus immunocytochemically, and NPY mRNA, by means of an image analysis system, in PWS cases, nonsyndromic obese patients and controls. The infundibular nucleus contains NPY cell bodies and an extremely dense network of NPY fibers

that generally do not extend to the most ventral part of the median eminence, which contains the portal capillaries. This indicates that most of the NPY fibers have central projections. NPY immunoreactivity and mRNA are decreased in PWS patients, to the same degree as in the other obese patients (Figs. 18– 20). NPY immunocytochemistry and mRNA increases with longer disease duration (Goldstone et al., 2002; Fig. 21). Apparently the insatiable hunger in PWS is not due to increased NPY expression, as these neurons show a normal reaction to the obese state and disease duration of these patients. No increase was found in the AGRP staining or mRNA in the infundibular nucleus of PWS patients either (Figs. 20 and 21). This peptide, too, is upregulated with disease duration (Goldstone et al., 2002; Fig. 20). AGRP is another peptide that stimulates feeding and is colocalized with NPY, but not with POMC. The decreased NPY and AGRP content of the hypothalamus indicates that the transport of information between the fat cells and the infundibular nucleus to the PVN by the NPY neurons will be largely intact, including

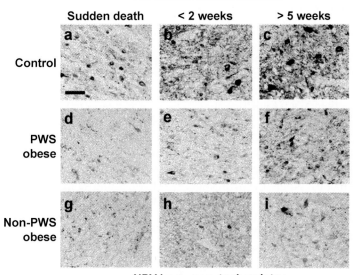

Pre-mortal illness duration

Fig. 18. Hypothalamic neuropeptide-Y (NPY) in human illness and obesity. NPY immunocytochemistry (ICC) staining in the infundibular nucleus of control, Prader–Willi syndrome (PWS) and non-PWS adults, with sudden death, premorbid illness duration of <2 week and >6 week. Note that NPY ICC staining increases with longer periods of illness, but that each illness duration levels are lower in both PWS and non-PWS obese subjects, compared with controls. Bar 50 mm (From Goldstone et al., 2002; with permission).

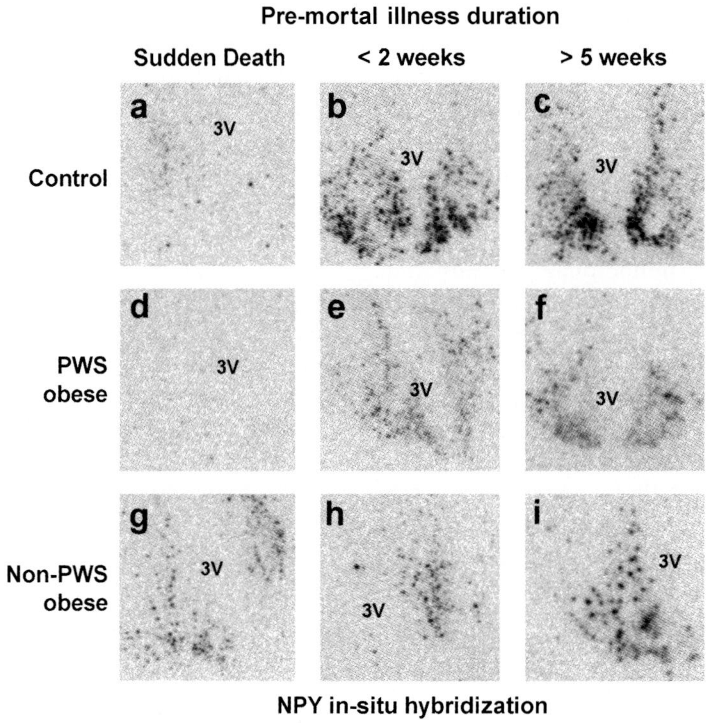

Pre-mortal illness duration

Fig. 19. Hypothalamic neuropeptide-Y (NPY) mRNA in human illness and obesity. Representative autoradiographs of NPY in-situ hybridization in the infundibular nucleus of control, PWS, and non-PWS obese adults, with sudden death, premorbid illness duration of <2 week or >5 week. Note that NPY mRNA expression increases with longer periods of illness, but that at each illness duration levels are lower in PWS or non-PWS obese subjects, compared with controls; 3V, third ventricle (From Goldstone et al., 2002; with permission).

leptin and leptin receptors, and that abnormalities have to be searched for in other peptide systems of the arcuate nucleus or in the area of termination of the NPY fibers, such as the PVN (Goldstone et al., 2002).

The obesity in PWS may be caused by an increased drive to eat as well as by an impaired mechanism of satiation. Both functions are controlled by the hypothalamus. Animal experiments have shown that the parvocellular oxytocin neurons of the hypothalamic PVN are crucial for the regulation of food intake. In the rat, the oxytocin neurons of the PVN project to brainstem nuclei, for example, the nucleus of the solitary tract and the dorsal motor nucleus of the nervus vagus. These connections are held responsible for the satiety effects of oxytocin (Verbalis et al., 1995). Small lesions in the rat PVN cause overeating and obesity (Leibowitz et al., 1981), suggesting that the PVN

usually has an inhibitory effect on eating and body weight. In addition, stimulation of the medial parvocellular subdivision of the rat PVN elicits significant increases in gastric acid secretion (Rogers and Hermann, 1986). Central administration of oxytocin or oxytocin agonists inhibits food intake and gastric motility in rat, whereas these effects are prevented by an oxytocin receptor antagonist (Rogers and Hermann, 1986; Arletti et al., 1989; Benelli et al., 1991; Olson et al., 1991a, 1991b).

We have investigated whether a disorder of the PVN, or, more particularly, of its putative satiety neurons — the oxytocin neurons — might be the basis of the insatiable hunger and obesity in PWS. The thionine-stained volume of the PVN appeared to be 28% smaller in PWS patients and the total cell number of the PVN is 38% lower than in controls (Swaab et al., 1995). Following immunocytochemistry, the immunoreactivities for

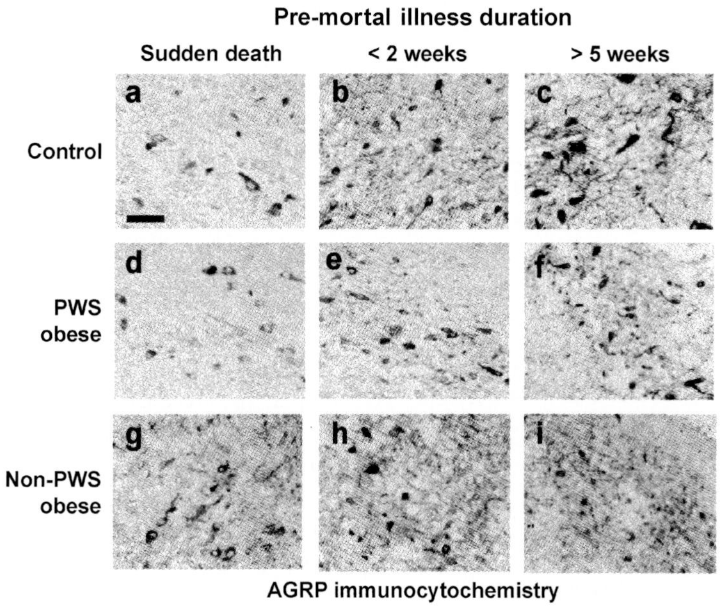

Pre-mortal illness duration

Fig. 20. Hypothalamic agouti-related protein (AGRP) peptide in human illness and obesity. Representative autoradiographs of AGRP ICC staining in the infundibular nucleus of control, PWS, and non-PWS obese adults, with sudden death, premorbid illness duration of <2 week or >5 week. Note that AGRP ICC staining increases with longer periods of illness, and that levels are not increased in PWS or non-PWS obese subjects, compared with controls; Bar 50 mm (From Goldstone et al., 2002; with permission).

oxytocin and vasopressin are decreased in PWS patients (Fig. 22), although the variation within the groups is high. A large and highly significant decrease (42%) in the number of oxytocin-expressing neurons was found in all five PWS patients (Fig. 23). The volume of the PVN containing the oxytocin (OXT) expressing neurons is 54% lower in PWS. The number of vasopressin-expressing neurons in the PVN did not change significantly (Fig. 22). The finding that volume and total cell number and oxytocin cell number was so much lower in PWS patients points to a developmental hypothalamic disorder, and agrees with the hypothesis that oxytocin neurons of the PVN may be good candidates for a physiological role as "satiety neurons" in ingestive behavior, also in the human brain (Swaab et al., 1995).

In conclusion, so far hypothalamic research has revealed an intact NPY/AGRP and GHRH system in PWS syndrome that is inhibited in a normal way by obesity, but the number of oxytocin-expressing neurons in the PVN is clearly diminished.

The hypothalamic hypocretin(orexin) system in PWS

Narcoleptic patients with cataplexy have a general loss of hypocretin (orexin) in the lateral hypothalamus, possibly due to an auto-immune-mediated degeneration of the hypocretin neurons. Obesity is one of the frequent symptoms in narcolepsy (see before). PWS patients have, in addition to hypotonia at birth and in the neonatal period, excessive daytime sleepiness. This was first thought to be due to sleep apnea related to obesity (Richdale et al., 1999). There have been several reports, however, that PWS patients show excessive daytime sleepiness (EDS), sleep onset with REM and in some cases even cataplexy, independent of obesity-related sleep disturbances (Helbing-Zwanenburg et al., 1993; Tobias et al., 2002). Interestingly, there are preliminary studies reporting lower CSF levels of hypocretin in several PWS patients, which suggests that the hypocretin neurons in the lateral hypothalamus are affected

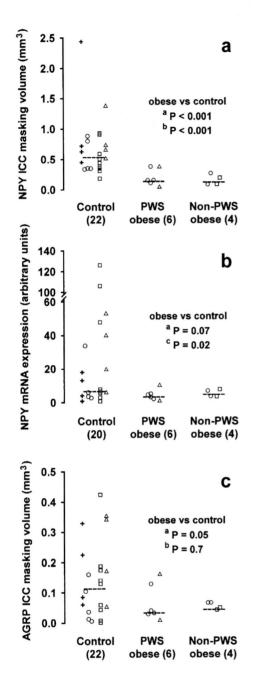

Fig. 21. Hypothalamic neuropeptide-Y (NPY) is decreased and agouti-related peptide (AGRP) is not increased in obesity. (a) NPY ICC staining volume, (b) NPY mRNA expression by ISH and (c) AGRP ICC staining volumes, in the infundibular nucleus/median eminence (INF/ME), in control, Prader–Willi syndrome (PWS) and non-PWS obese subjects. + represents females, ○ hypogonadal females (postmenopausal or PWS), □ intact males, Δ hypogonadal males (castrated controls or PWS).

(Mignot et al., 2002; Nevsimalova et al., 2004). Because of the symptom overlap we estimated the number of hypocretin immunoreactive (IR) neurons in the postmortem hypothalami of 8 PWS adults, 3 PWS infants and 11 matched control subjects using immunocytochemistry and an image analysis system. However, the number of hypocretin-1 IR neurons in postmortem material in PWS patients, i.e. some 80,000 cells, was not different from that in controls (Fronczek et al., 2005; Figs. 24 and 25).

In agreement with the main findings of this paper, we recently measured a normal level of hypocretin-1 in the CSF of one PWS patient (unpublished observation).

In conclusion, neither the hypocretin cell number nor the intensity of staining was different in PWS patients, so that the hypocretin system does not seem to play a major role in the occurrence of childhood hypotonia, obesity, EDS, sleep-onset REM or cataplexy. However, expression levels of hypocretin-1 mRNA and of the receptors for this peptide should still be investigated.

Development of the fetal hypothalamus, birth and programing of metabolism

The coupling of the various hypothalamic systems may not only have its physiological basis in adult adaptive mechanisms (e.g. in the coupling of starvation and anovulation and the fall/winter increase in eating behavior and body weight to accumulate a backup for the winter), but may also already be involved in fetal hypothalamic adaptive mechanisms (e.g. programing of body weight) and in the fetal involvement in the process of birth. The human fetal hypothalamus seems to be quite mature at the moment of birth.

Dashed line represents median for each group. Note that, in obese subjects (PWS and non-PWS) compared with controls, there is a significant reduction in NPY ICC staining and mRNA, but no difference in AGRP ICC staining, when adjusting for significant covariates. P-values: [a]Mann–Whitney test, [b]adjusting for differences in premorbid illness duration by ANCOVA, [c]adjusting for differences in premorbid illness duration and storage time by ANCOVA (From Goldstone et al., 2002; with permission).

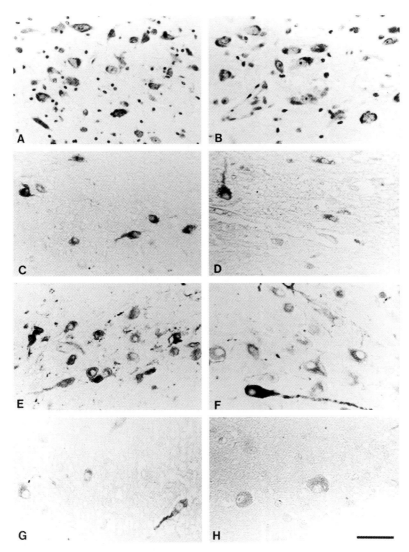

Fig. 22. In thionine-stained sections of the paraventricular nucleus (PVN), no qualitative differences were observed between controls (no. 81255; A) and PWS patients (no. 43830; B). The staining of oxytocin (OXT) (C, D) and vasopressin (AVP) (E, F) was generally lower in PWS patients (no. 43830; D, F) than in controls (no. 81255; C, E). Two PWS patients (no. 1 and 4) had intense and weak OXT staining (no. 93056; G) and only negligible AVP staining (no. 93056; H) in the PVN. Bar 50 mm (From Swaab et al., 1995; with permission).

Development of the human hypothalamus

Koutcherov et al. (2002) have described, in detail, the development of the nuclear organization of the human hypothalamus (Fig. 26). Only minimal signs of nuclear differentiation were found in 9–10 weeks of gestation, but a clear subdivision into three longitudinal zones was found. A well-defined hypo-thalamic sulcus indicates the dorsal hypothalamic boundary and the lens-shaped subthalamic nucleus the lateral hypothalamic border. The tentatively designated posterior hypothalamus was separated from the LHA by fiber bundle 3 (fasciculus 3 of Forel = fibrae hypothalamico-pallidares = fasculus lenticularis) and by the mamillothalamic tract. A cell-sparse SON was also found. In addition, the

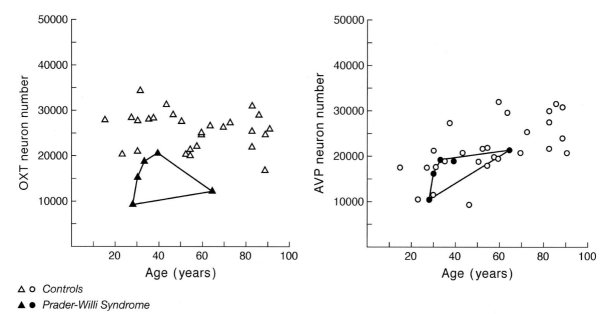

Fig. 23. Number of oxytocin-expressing (OXT) (left panel) and vasopressin-expressing (AVP) (right panel) neurons in the PVN of 27 controls and 5 Prader–Willi syndrome (PWS) patients. The values of the PWS patients are delineated by a minimum convex polygon. Note that the oxytocin neuron number of these patients is about half of that of the controls (left panel), which is not the case for vasopressin (right panel). (From Swaab et al., 1995; with permission).

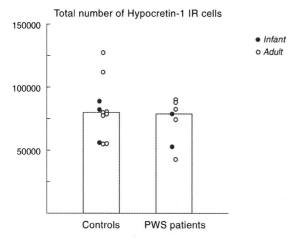

Fig. 24. Median number of hypocretin IR cells in Prader–Willi Syndrome patients (right bar) and controls (left bar). There is no significant difference between the two groups (Mann–Whitney U: $n = 14$, $p = 0.66$) (From Fronczek et al., 2005; with permission).

dorsomedial hypothalamic nucleus (DMN), VMN, the medial preoptic area, the medial mamillary body and the infundibular (= arcuate) nucleus could already be distinguished.

In 11–14 weeks of gestation the fornix became visible and the anlage of the PVN could be distinguished. In this period vasopressin and oxytocin are present (see Swaab, 2003; Chapter 8) and CRH-positive cells and fibers are also found from fetal week 12–16 (Bresson et al., 1987).

In 15–17 weeks of gestation the nucleus tuberalis lateralis, intermediate nucleus (= sexually dimorphic nucleus of the preoptic area = SDN-POA) differentiated. The SDN-POA was embedded in the lateral surface of the teardrop-shaped medial preoptic area. The lateral mamillary body is prominent at 16 weeks of gestation. A supramamillary nucleus was also visible. This nucleus cannot be seen in the adult hypothalamus.

By 18 weeks of gestation the posterior subnucleus of the PVN resembles the postnatal structure. The perifornical area differentiated. This area remains anchored around the fornix, whereas most LHA cells are positively displayed laterally by the successive waves of neurons of the midline and core zones that develop later. The suprachiasmatic nucleus and the retinohypothalamic tract became visible at 23 weeks of gestation. At 21

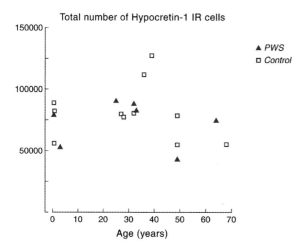

Fig. 25. Correlation between age and total number of hypo-cretin IR cells. Squares indicate controls; triangles indicate Prader–Willi Syndrome patients. The total number of hypocre-tin-1 neurons declines with age. In PWS adults, a correlation between age and total IR cell number was found ($n = 5$, $r = 0.900$, $p = 0.037$) (From Fronczek et al., 2005; with permission).

weeks of gestation the PVN evinced, for the first time, distinct subnuclear subdivisions. Moreover, NPY-positive neurons were present in the infundibular neurons at 21 weeks of gestation. By 24–33 weeks of gestation the fetal hypothalamus has taken on an adult-like appearance. At week 34 to newborn, the nucleus tuberalis lateralis (NTL) and tuberomamillary nucleus can be distinguished.

Body weight is programed during early development as shown by a historical cohort study of 300,000 19-year old men exposed to the Dutch famine of 1944–1945 in Amsterdam during the German occupation and examined at military introduction. During the first half of pregnancy, exposure to famine resulted in a higher obesity rate. This period is presumed to be critical in the development of the hypothalamic centers that regulate food intake and growth (Ravelli et al., 1976). Exposure to famine during the last trimester of pregnancy and first postnatal months resulted in a lower obesity rate. This is a critical period in the development of adipose -tissue cellularity (Ravelli et al., 1976).

The observation that people who had a low birth weight have higher leptin concentrations in adult life than would be expected from their BMI,

agrees with the hunger winter data on the programing of body weight during development. Low birth weight is also accompanied by a higher risk for impaired glucose tolerance and type-2 diabetes in adult life, and a higher prevalence of the metabolic or insulin resistance syndrome. The exact mechanisms involved are not known, but multiple regression data argued against a causative role for leptin in the programing of insulin resistance and glucose intolerance during foetal life (Phillips et al., 1999).

Recent evidence from sheep indicates that not only late gestational undernutrition may program the HPA axis on a higher level, but also periconceptional undernutrition (McMillen et al., 2004). In human, small size at birth leads also to an alteration in set point of the HPA axis and an increased cortisol responsiveness and risk of depression in adulthood (Phillips, 2001; Thompson et al., 2001).

The fetal hypothalamus and the initiation of birth

The concept of the fetal brain playing a crucial role in the start of labor originates from observations in cows and sheep. Gestational length was increased in Guernsey cattle with fetal pituitary aplasia and in cyclop fetuses, due to the poisonous plant *veratrum californicum*, which caused pituitary aplasia or pituitary dystopia. These observations were followed by the classic experiments by C.G. Liggins in sheep in which fetal adrenalectomy, fetal hypophysectomy, infusion of ACTH, and glucocorticoids showed that the fetal HPA axis starts labor in this species (for review, see Swaab, 2004; Chapter 18). Later, a critical experiment was carried out by McDonald and Nathanielsz (1991). They showed that stereotactic destruction of the fetal paraventricular nucleus in sheep was followed by prolonged pregnancy length. Observations on human anencephalics suggest that the human fetal brain, too, is important for the timing of the correct, at term, moment of birth. As a group, however, anencephalics have a shorter gestational length, since anencephalics with hydramnios are born prematurely (Fig. 27; Swaab et al., 1977). Mean gestation length is

32

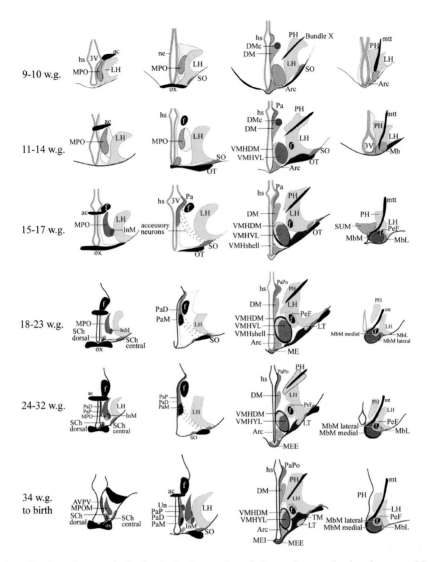

Fig. 26. Organization of major cell groups in the developing human hypothalamus shown at landmark stages of fetal differentiation. The hypothalamus is depicted at four rostrocaudal levels (left to right) for each developmental stage. Gray scale represents hypothalamic structures revealed by cytoarchitecture of the neuroepithelial primordia and transient chemoarchitectonic labeling. Color coding indicates advanced stages of cell groups structural differentiation revealed by chemoarchitecture. Please note that these diagrams are not to scale.

Abbreviations used: 3V, third ventricle; ac, anterior commissure; Arc, arcuate nucleus; AVPV, anteroventral periventricular nucleus; BST, bed nucleus of the stria terminalis; DM, dorsomedial hypothalamic nucleus; DMC, dorsomedial hypothalamic nucleus (compact part); f, fornix; hs, hypothalamic sulcus; IsM, intermediate nucleus (= sexually dimorphic nucleus of the preoptic area); LH, lateral hypothalamic area; LTu, lateral tuberal hypothalamic nucleus; MbL, mamillary body, lateral part; MbM, mamillary body, medial part; MEE, median eminence, external; MEI, median eminence, internal; MPA, medial preoptic area; MPO, medial preoptic nucleus; MPOC, medial preoptic nucleus, central subnucleus; MPOL, medial preoptic nucleus, lateral subnucleus; MPOM, medial preoptic nucleus, medial subnucleus; mt, mtt, mamillothalamic tract; ne, neuronal epithelium; OT, opt, optic tract; ox, optic chiasm; PaD, paraventricular hypothalamic nucleus, dorsal subnucleus; PaM, paraventricular hypothalamic nucleus, magnocellular subnucleus; PaP, paraventricular hypothalamic nucleus, parvicellular subnucleus; PaPo, paraventricular hypothalamic nucleus, posterior subnucleus; PeF, periformical hypothalamic nucleus; PH, posterior hypothalamic area; SCh, suprachiasmatic nucleus; SChD, suprachiasmatic nucleus, dorsal part; SChC, suprachiasmatic nucleus, central part; SO, supraoptic nucleus; SUM, supramamillary nucleus; Un, uncinate nucleus; VMH, ventromedial hypothalamic nucleus; VMHDM, ventromedial hypothalamic nucleus, dorsomedial subnucleus; VMHVL, ventromedial hypothalamic nucleus, ventrolateral subnucleus; VTM, ventral tuberomamillary hypothalamic nucleus (From Koutcherov et al., 2002; with permission).

normal in spontaneously born anencephalics from pregnancies without hydramnios, in contrast to the sheep model. However, a high percentage of prematurely and postmaturely born anencephalic children is found, indicating that the timing of birth is a function of the human fetal brain (Honnebier and Swaab, 1973; Fig. 27). It is presumed that fetal CRH from the paraventricular nucleus

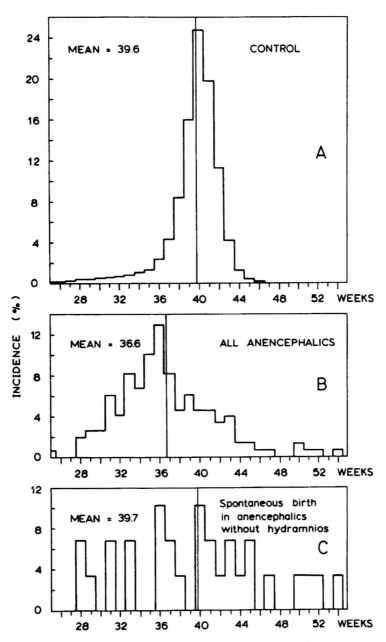

Fig. 27. Frequency distribution of gestation length: (A) for a control group of 49,996 pregnant women; (B) for mothers of all anencephalic fetuses (*n* = 147); (C) for mothers of anencephalic fetuses (*n* = 29) without hydramnios, omitting those who had stillborn fetuses with third-degree maceration, fetuses given intrauterine injections or twins, and those in whom labor was induced (From Swaab et al., 1977; with permission).

plays a key role in the timing mechanism for the initiation of human birth (see below).

In addition, the course of labor is protracted in anencephalic children. In particular, the expulsion stage and birth of the placenta take too long. This suggests that the fetal brain releases compounds such as oxytocin that may speed up delivery (Swaab et al., 1977; see Swaab, 2003, Chapter 8.1). The fact that about 50% of the anencephalics die during the course of labor (Honnebier and Swaab, 1973) points to the importance of intact fetal brain systems for dealing with the stress of birth. The extremely strong release of fetal vasopressin that normally occurs in spontaneous labor and that is absent in anencephalics (Oosterbaan and Swaab, 1987) may play a role, since fetal vasopressin is involved in redistribution of the fetal circulation to those organs that are of vital importance (Iwamoto et al., 1979; Pohjavuori and Fyhrquist, 1980).

Physiological mechanisms involving the same hypothalamic systems as those used to regulate eating and metabolism in the adult are now proposed to be involved in the fetus for the initiation of parturition. Data in sheep show that the fetal HPA axis responds to changes in plasma glucose concentrations within the physiological range and that there may be an increase in the sensitivity of the fetal HPA axis to glucose concentrations during late gestation (McMillen et al., 1995). The decreased levels of fetal glucose, increased levels of cortisol and changes in leptin are presumed to activate NPY neurons in the fetal infundibular nucleus, that activate the fetal HPA axis and thus inducing the cascade that will lead to birth. In the human, NPY is present in the fetal infundibular nucleus from 21 weeks of gestation onwards, and there are projections from this nucleus to the PVN in this stage of pregnancy (Koutcherov et al., 2002). Cortisol not only triggers an increased fetal hypothalamic CRH production, but also placental CRH, which is the main source of this peptide around birth (Mastorakos and Ilias, 2003). In the fetal sheep, glucose concentrations decrease 30% in the second half of gestation and fetal hypothalamic NPY mRNA and NPY increase strongly 7–10 days before delivery. In sheep both intra-fetal infusion of cortisol and fetal undernutrition

stimulate increases in NPY mRNA levels in the hypothalamus before birth. NPY in the PVN results in in the stimulation of CRH and vasopressin (AVP) secretion into the hypophysial portal circulation and thus in an increase of fetal plasma ACTH and glucocorticoids. Corticosteroids in the fetus seem thus to be included in a positive feedback (Warnes et al., 1998) that results in birth.

Birth as the first test of the fetal brain and birth disorders as the first sign of hypothalamic disorders

The fetal brain seems to play an active role in the initiation and course of labor. The idea of an active fetal role of oxytocin neurons in delivery is reinforced by a number of clinical observations. Firstly, human anencephalics do not have a neurohypophysis and have impaired neurohypophysial hormone release (Visser and Swaab, 1979; Oosterbaan and Swaab, 1987). An anencephalics expulsion takes twice as long and the birth of the placenta even takes three times longer, suggesting a role of fetal brain and possibility of neuroendocrine mechanisms in speeding up the course of labor. In addition, the observation that about half of the anencephalics die during the course of labor is a strong indication of the importance of an intact fetal brain to withstand the stress of birth (Honnebier and Swaab, 1973; Swaab et al., 1977). The second observation is derived from children suffering from Prader–Willi syndrome. These children frequently suffer from considerable obstetrical problems (Wharton and Bresman, 1989), and we found that Prader–Willi patients have only 58% of the normal number of oxytocin neurons in adulthood, but a normal number of vasopressin neurons (Swaab et al., 1995). The third argument is based on the frequent perinatal problems found in septo-optic dysplasia (De Morsier syndrome), in which the fetal hypothalamo-neurohypophysial system (HNS) is often damaged (Roessmann et al., 1987). Moreover, prolonged labor and breech delivery have been documented in 50–60% of the idiopathic growth hormone-deficient children, a disorder that seems to be based on congenital hypothalamo–pituitary abnormalities (Maghnie et al., 1991; Swaab, 2003,

2004). Importantly, the causality of the relationship between obstetric complications and neurological or psychiatric diseases such as schizophrenia or autism (Geddes and Lawrie, 1995; Verdoux and Sutter, 2002) might thus be quite the reverse of what is generally thought. How the fetal hypothalamus might play a role in the disorders of fetal presentation, e.g., in PWS (Wharton and Bresman, 1989) is not yet known.

A disturbed labor might thus be the first symptom of a brain disorder, probably even more often than that disturbed labor is the cause of the brain disorder. The same holds for anorexia nervosa. It is remarkable that perinatal factors have an association with the prevalence of anorexia nervosa in later life. Not only was an increased risk for anorexia nervosa found when the patient had had a cephalhematoma (odds ratio (OR) 5.7), but also children with a very preterm birth ($\leqslant$ 32 gestational weeks) had a higher risk (OR 3.2). In very preterm birth, girls who were small for gestational age faced higher risks (OR 5.7) than girls with higher birth weight for gestational age (OR 2.7) (Cnattingius et al., 1999). A perinatal history of delivery complications is even associated with a poor outcome of anorexia later in life (Halmi et al., 1979). Also in children under 7 years with severe feeding and eating difficulties, premature labor and low birth weight were common findings (Douglas and Bryon, 1996). One may therefore wonder whether an abnormal fetal reaction to the diminishment of glucose in gestation is both the cause of premature delivery and the first sign of a hypothalamic disorder in the regulation of eating and metabolism in at least part of the patients with anorexia nervosa and other eating disorders.

Summary and conclusions

The crucial role of the human hypothalamus in eating and metabolic disorders has already been known for a long time from studies on patients with hypothalamic tumors or other lesions causing a spectrum of disorders that run from the ventromedial hypothalamic syndrome with hyperphagia and obesity to anorexia. Such lesions, however, give only information on the level of large brain structures. Recently, functional scanning and molecular genetic studies of mutations and polymorphisms affecting the efficacy of hypothalamic systems have contributed to our knowledge on the human hypothalamus. In addition, postmortem studies using quantitative immunocytochemistry and in situ hybridization are beginning to reveal alterations in the activity of the specific hypothalamic systems known to be involved in eating and metabolism from animal experiments. Valuable detailed functional information can be obtained in this way, as shown in this review, for depression and PWS, both in terms of the cells and circuits that are the basis of the signs and symptoms of the disorder and alterations as a reaction of the brain on the condition. However, the availability of well-characterized postmortem material of patients with eating and other psychiatric disorders is still a major problem.

In disorders of eating and metabolism, there is frequently hypothalamic comorbidity in terms of mood, sleep and rhythm disorders due to the intense hypothalamic interconnectivity and integrative mechanisms. Moreover, these symptoms often occur in an episodic way and generally have a different prevalence for the sexes. Circadian and seasonal fluctuations and sleep disorders may point to the involvement of the SCN in symptom expression. In some conditions a disorder of the SCN itself is involved. In depression, SCN fluctuations are flattened and in "nighttime eating syndrome" a disconnection is present between the circadian control of eating relative to sleep. Disorders of appetite, metabolism and mood may be based on the involvement of, e.g., CRH, cortisol, TRH or orexin. Such disorders may be due to compounds affecting a multitude of hypothalamic centers (e.g. by glucocorticoids in Cushing's syndrome or corticosteroid therapy) affecting metabolism, food intake, mood and sleep or occur as characteristic syndromes, such as bulimia nervosa, nighttime eating syndrome, depression, narcolepsia, or PW or Kleine–Levin syndrome, each with their specific set of signs and symptoms on the foreground and with particular hypothalamic systems involved. For bulimia, with its typical seasonal fluctuations in binge eating, purging and mood, the concept of hypothalamic integration has therapeutical concequences. Light therapy not only improved mood but also the eating disorder. The coupling of the various hypothalamic systems may have its physiological basis not only in mechanisms that were adaptive in adult life in evolution (e.g. coupling of starvation and anovulation, and the fall/winter increase in eating behavior and body weight), but also in adaptive fetal mechanisms in relation to the process of birth. Fetal undernutrition (for human documented in the effects of the Dutch Hunger Winter in Amsterdam, 1944–1945, during the German occupation, but also present in case of placental insufficiency) leads to an adaptive reaction based upon the "fetal expectation" of the presence of a scarcity of food in the environment. This has long-term clinical consequences that are induced by programing of fetal hypothalamic functions. It leads to an increased risk for obesity, hypertension, hyperphagia, hyperinsulinaemia and hyperleptinaemia in the offspring. In addition, the offspring has a reduced locomotor behavior, and are at risk for depression and schizophrenia. A physiological mechanism involving the same hypothalamic system that regulates eating and metabolism in the adult is also proposed to be involved in the fetus for the initiation of parturition. The decreased levels of fetal glucose, increased levels of cortisol and changes in leptin are presumed to activate NPY neurons in the fetal infundibular nucleus, activating the fetal HPA axis and thus inducing the cascade that will lead to birth. Cortisol not only triggers an increased fetal hypothalamic CRH production, but also placental CRH.

Acknowledgment

The author would like to thank Ms. W.T.P. Verweij and Ms. T. Eikelboom for their secretarial support, and Dr. J. Kruisbrink for providing the references.

Appendix: The Dutch Famine 1944–1945

Knowledge on the permanent effects of fetal and neonatal undernutrition on human hypothalamic functions has gained a lot from follow-up studies on subjects who were exposed to the Dutch Famine during the Second World War. Tessa Rosenboom summarized the situation in her PhD thesis (2000) as follows.

"After weeks of heavy fighting following the invasion on the 6th of June 1944," the Allied forces finally broke through German lines. With lightning speed the Allied troops took possession of much of France, Luxemburg and Belgium. By the 4th of September 1944 the Allies had the strategic city of Antwerp in their hands, and on the 14th they entered the Netherlands. Everyone in the Netherlands expected that the German occupation would soon be over. The advance went so quickly that also the commanders of the Allied forces thought it would be only a matter of days before the Germans surrendered. But the advance of the Allies to the north of the Netherlands came to a halt when attempts to get control of the bridge across the river Rhine at Arnhem (operation 'Market Garden') failed.

In order to support the Allied offensive, the Dutch government in exile had called for a strike of the Dutch railways. As a reprisal, the Germans banned all food transports. This embargo on food transports was lifted in early November 1944, when food transport across water was permitted again. By then, it had become impossible to bring in food from the rural east to the urban west of the Netherlands because most canals and waterways were frozen due to the extremely severe winter of 1944–1945, which had started unusually early. Consequently, food stocks in the urban west of the Netherlands ran out rapidly. As a result, the official daily rations for the general adult population — which had decreased gradually from about 1800 kilo calories (kcal) in December 1943 to 1400 kcal in October 1944 — fell abruptly to below 1000 kcal in late November 1944. At the height of the famine from December 1944 to April 1945, the official daily rations varied between 400 and 800 calories. Children younger than 1 year were relatively protected, because their official daily rations never fell below 1000 calories, and the specific nutrient components were always above the standards used by the Oxford Nutritional Survey.

Pregnant and lactating women were entitled to an extra amount of food, but at the peak of the famine these extra supplies could not be provided anymore. In addition to the official rations, food came from church organizations, central kitchens, the black market and foraging trips to the countryside. After the liberation of the Netherlands in early May 1945, the food situation improved swiftly. In June 1945, the rations had risen to more than 2000 kcal.

There was a serious shortage of fuel during the war which caused a gradual decrease and finally a complete shutting down of the production of gas and electricity, and in several places even the water supply had to be cut off, while the authorities were unable to provide fuel for stoves and furnaces in homes. Throughout the winter of 1944–1945 the population had to live without light, without gas, without heat, laundries ceased operating, soap for personal use was unobtainable, and adequate clothing and shoes were lacking in most families. In hospitals, there was serious overcrowding as well as lack of medicines. Above all, hunger dominated all misery.

The famine had a profound effect on the general health of the population. In Amsterdam, the mortality rate in 1945 had more than doubled compared to 1939, and it is very likely that most of this increase in mortality was attributable to malnutrition. But, even during this disastrous famine women conceived and gave birth to babies, and it is in these babies that the effects of maternal malnutrition during different periods of gestation on health in adult life can be studied. Because of its unique experimental characteristics, it is not surprising that people born around the time of the Dutch famine have been studied by many investigators. A number of these studies are referred to in the present review.

References

Alkemade, A., Unmehopa, U.A., Wiersinga, W.M., Swaab, D.F. and Fliers, E. (2005) Glucocorticoids decrease TRH mRNA expression in the paraventricular nucleus (PVN) of the human hypothalamus. J. Clin. Endocrinol. Metab., 90: 323–327.

Arletti, R., Benelli, A. and Bertolini, A. (1989) Influence of oxytocin on feeding behavior in the rat. Peptides, 10: 89–93.

Balligand, J.L., Brichard, S.M., Brichard, V., Desager, J.P. and Lambert, M. (1998) Hypoleptinemia in patients with anorexia nervosa: loss of circadian rhythm and unresponsiveness to short-term refeeding. Eur. J. Endocrinol., 138: 415–420.

Bastrup-Madsen, P. and Greisen, O. (1963) Hypothalamic obesity in acute leukaemia. Acta Haematol., 29: 109–116.

Bellipanni, G., Bianchi, P., Pierpaoli, W., Bulian, D. and Ilyia, E. (2001) Effects of melatonin in perimenopausal and menopausal women: a randomized and placebo controlled study. Exp. Gerontol., 36: 297–310.

Benelli, A., Bertolini, A. and Arletti, R. (1991) Oxytocin-induced inhibition of feeding and drinking: no sexual dimorphism in rats. Neuropeptides, 20: 57–62.

Bierwolf, C., Kern, W., Mölle, M., Born, J. and Fehm, H.L. (2000) Rhythms of pituitary-adrenal activity during sleep in patients with Cushing's disease. Exp. Clin. Endocrinol. Diabetes, 108: 470–479.

Biller, B.M.K. (1994) Pathogenesis of pituitary Cushing's syndrome. Endocrinol. Metab. Clin. N. Am., 23: 547–554.

Birketvedt, G.S., Florholmen, J., Sundsfjord, J., Østerud, B., Dinges, D., Bilker, W. and Stunkard, A. (1999) Behavioral and neuroendocrine characteristics of the night-eating syndrome. JAMA, 282: 657–663.

Birketvedt, G.S., Sundsfjord, J. and Florholmen, J.R. (2002) Hypothalamic-pituitary adrenal axis in the night eating syndrome. Am. J. Physiol., 282: E366–E369.

Björntorp, P. and Rosmond, R. (1999) Hypothalamic origin of the metabolic syndrome X. Ann. N.Y. Acad. Sci., 892: 297–307.

Blouin, A., Blouin, J., Aubin, P., Carter, J., Goldstein, C., Boyer, H. and Perez, E. (1992) Seasonal patterns of bulimia nervosa. Am. J. Psychiatry, 149: 73–81.

Blouin, A.G., Blouin, J.H., Iversen, H., Carter, J., Goldstein, C., Goldfield, G. and Perez, E. (1996) Light therapy in bulimia nervosa: a double-blind, placebo-controlled study. Psychiatry Res., 60: 1–9.

Bonnet, F., Irving, K., Terra, J.-L., Nony, P., Berthezène, F. and Moulin, P. (2005) Depressive symptoms are associated with unhealthy lifestyles in hypertensive patients with the metabolic syndrome. J. Hypertension, 23: 611–617.

Branson, R., Potoczna, N., Kral, J.G., Lentes, K.-U., Hoehe, M.R. and Horber, F.F. (2003) Binge eating as a major phenotype of melanocortin 4 receptor gene mutations. N. Engl. J. Med., 348: 1096–1103.

Braun, D.L., Sunday, S.R., Fornari, V.M. and Halmi, K.A. (1999) Bright light therapy decreases winter binge frequency in women with bulimia nervosa: a double-blind, placebo-controlled study. Comprehensive Psychiatry, 40: 442–448.

Bresson, J.L., Clavequin, M.C., Fellmann, D. and Bugnon, C. (1987) Human corticoliberin hypothalamic neuroglandular system: comparative immunocytochemical study with anti-rat and anti-ovine corticotropin-releasing factor sera in the early stages of development. Brain Res. Dev. Brain Res., 32: 241–246.

Brown, G.M. (1992) Day–Night rhythm disturbance, pineal function and human disease. Horm. Res., 37(Suppl. 3): 105–111.

Brown, S.E., Varghese, F.P. and McEwen, B.S. (2004) Association of depression with medical illness: does cortisol play a role? Biol. Psychiatry, 55: 1–9.

Carlson, M.G., Snead, W.L., Oeser, A.M. and Butler, M.G. (1999) Plasma leptin concentrations in lean and obese human subjects and Prader–Willi syndrome: comparison of RIA and ELISA methods. J. Lab. Clin. Med., 133: 75–80.

Carlsson, A., Svennerholm, L. and Winblad, B. (1980) Seasonal and circadian monoamine variations in human brains examined post mortem. Acta Psychiatr. Scand., 61(Suppl 280): 75–85.

Checkley, S. (1996) The neuroendocrinology of depression and chronic stress. Br. Med. Bull., 52: 597–617.

Chemelli, R.M., Willie, J.T., Sinton, C.M., Elmquist, J.K., Scammell, T., Lee, C., Richardson, J.A., Williams, S.C., Xiong, Y., Kisanuki, Y., Fitch, T.E., Nakazato, M., Hammer, R.E., Saper, C.B. and Yanagisawa, M. (1999) Narcolepsy in orexin knockout mice: molecular genetics of sleep regulation. Cell, 98: 437–451.

Clapham, J.C., Arch, J.R.S. and Tadayyon, M. (2001) Antiobesity drugs: a critical review of current therapies and future opportunities. Pharmacol. Ther., 89: 81–121.

Clément, K., Vaisse, C., Lahlou, N., Cabrol, S., Pelloux, V., Cassuto, D., Gourmelen, M., Dina, C., Chambaz, J., Lacorte, J.-M., Basdevant, A., Bougnères, P., Lebouc, Y., Froguel, P. and Guy-Grand, B. (1998) A mutation in the human leptin receptor gene causes obesity and pituitary dysfunction. Nature, 392: 398–401.

Climo, L.H. (1982) Anorexia nervosa associated with hypothalamic tumor: the search for clinical-pathological correlations. Psychiatr. J. Univ. Ottawa, 7: 20–25.

Cnattingius, S., Hultman, C.M., Dahl, M. and Sparén, P. (1999) Very preterm birth, birth trauma, and the risk of anorexia nervosa among girls. Arch. Gen. Psychiatry, 56: 634–638.

Coffey, R.J. (1989) Hypothalamic and basal forebrain germinoma presenting with amnesia and hyperphagia. Surg. Neurol., 31: 228–233.

Cone, R.D. (1999) The central melanocortin system and energy homeostasis. Trends Endocrinol. Metab., 10: 211–216.

Conroy, J.M., Grebe, T.A., Becker, L.A., Tsuchiya, K., Nicholls, R.D., Buiting, K., Horsthemke, B., Cassidy, S.B. and Schwartz, S. (1997) Balanced translocation 46,XY,t(2;15) (q372;q112) associated with atypical Prader–Willi syndrome. Am. J. Hum. Genet., 61: 388–394.

Cotrufo, P., Monteleone, P., d'Istria, M., Fuschino, A., Serino, I. and Maj, M. (2000) Aggressive behavioral characteristics and endogenous hormones in women with bulimia nervosa. Neuropsychobiology, 42: 58–61.

Dahmen, N., Bierbrauer, J. and Kasten, M. (2001) Increased prevalence of obesity in narcoleptic patients and relatives. Eur. Arch. Psychiatry Clin. Neurosci., 251: 85–89.

Dai, J.P., Swaab, D.F. and Buijs, R.M. (1997) Distribution of vasopressin and vasoactive intestinal polypeptide (VIP) fibers in the human hypthalamus with special emphasis on suprachiasmatic nucleus efferent projections. J. Comp. Neurol., 383: 397–414.

De Castro, J.M. (1991) Seasonal rhythms of human nutrient intake and meal pattern. Physiol. Behav., 50: 243–248.

De los Santos, T., Schweizer, J., Rees, C.A. and Francke, U. (2000) Small evolutionary conserved RNA, resembling C/D box small nucleolar RNA, is trascribed for *PWCR1*, a novel imprinted gene in the Prader–Willi deletion region, which is highly expressed in brain. Am. J. Hum. Genet., 67: 1067–1082.

DeVile, C.J., Sufraz, R., Lask, B.D. and Stanhope, R. (1995) Occult intracranial tumours masquerading as early onset anorexia nervosa. Br. Med. J., 311: 1359–1360.

De Vries, M.W. and Peeters, F.P.M.L. (1997) Melatonin as a therapeutic agent in the treatment of sleep disturbance in depression. J. Nerv. Ment. Dis., 185: 201–202.

Dinan, T.G., Lavelle, E., Scott, L.V., Newell-Price, J., Medbak, S. and Grossman, A.B. (1999) Desmopressin normalizes the blunted adrenocorticotropin response to corticotropin-releasing hormone in melancholic depression: evidence of enhanced vasopressinergic responsivity. J. Clin. Endocrinol. Metab., 84: 2238–2240.

Dingemans, A.E., Bruna, M.J. and Van Furth, E.F. (2002) Binge eating disorder: a review. Int. J. Obesity, 26: 299–307.

Dorn, L.D., Burgess, E.S., Dubbert, B., Simpson, S.-E., Friedman, T., Kling, M., Gold, P.W. and Chrousos, G.P. (1995) Psychopathology in patients with endogenous Cushing's syndrome: 'atypical' or melancholic features. Clin. Endocrinol., 43: 433–442.

Dorn, L.D., Burgess, E.S., Friedman, T.C., Dubbert, B., Gold, P.W. and Chrousos, G.P. (1997) The longitudinal course of psychopathology in Cushing's syndrome after correction of hypercortisolism. J. Clin. Endocrinol. Metab., 82: 912–919.

Douglas, J.E. and Bryon, M. (1996) Interview data on severe behavioural eating difficulties in young children. Arch. Diseases Childhood, 75: 304–308.

Eastman, C.I., Young, M.A., Fogg, L.F., Liu, L. and Meaden, P.M. (1998) Bright light treatment of winter depression. A placebo-controlled trial.. Arch. Gen. Psychiatry, 55: 883–889.

Engelmann, M., Landgraf, R. and Wotjak, C.T. (2004) The hypothalamic-neurohypophysial system regulates the hypothalamic–pituitary–adrenal axis under stress: an old concept revisited. Frontiers Neuroendocrinol., 25: 132–149.

Erkut, Z.A., Pool, C.W. and Swaab, D.F. (1998) Glucocorticoids suppress corticotropin-releasing hormone and vasopressin expression in human hypothalamic neurons. J. Clin. Endocrinol. Metab., 83: 2066–2073.

Farooqi, I.S., Jebb, S.A., Langmack, G., Lawrence, E., Cheetham, C.H., Prentice, A.M., Hughes, I.A., McCamish, M.A. and O'Rahilly, S. (1999) Effects of recombinant leptin therapy in a child with congenital leptin deficiency. New Engl. J. Med., 341: 879–884.

Farooqi, I.S., Keogh, J.M., Yeo, G.S.H., Lank, E.J., Cheetham, T. and O'Rahilly, S. (2003) Clinical spectrum of obesity and mutations in the melanocortin 4 receptor gene. N. Engl. J. Med., 348: 1085–1095.

Ferrari, E., Fraschini, F. and Brambilla, F. (1990) Hormonal circad-ian rhythms in eating disorders. Biol. Psychiatry, 27: 1007–1020.

Flynn, F.G., Cummings, J.L. and Tomiyasu, U. (1988) Altered behavior associated with damage to the ventromedial hypothalamus: a distinctive syndrome. Behav. Neurol., 1: 49–58.

Fornari, V.M., Braun, D.L., Sunday, S.R., Sandberg, D.E., Matthews, M., Chen, L.-L., Mandel, F.S., Halmi, K.A. and Katz, J.L. (1994) Seasonal patterns in eating disorder subgroups. Compr. Psychiatry, 35: 450–456.

Friedman, S., Even, C., Dardennes, R. and Guelfi, J.D. (2002) Light therapy, obesity, and night-eating syndrome. Am. J. Psychiatry, 159: 875–876.

Fronczek, R., Lammers, G.J., Balesar, R., Unmehopa, U.A. and Swaab, D.F. (2005) The number of hypothalamic hypocretin (orexin) neurons is not affected in Prader–Willi Syndrome. J. Clin. Endocrinol. Metab., 90: 5466–5470.

Galderisi, S., Mucci, A., Monteleone, P., Sorrentino, D., Piegari, G. and Maj, M. (2003) Neurocognitive functioning in subjects with eating disorders: the influence of neuroactive steroids. Biol. Psychiatry, 53: 921–927.

Gautier, J.-F., Del Parigi, A., Chen, K., Salbe, A.D., Bandy, D., Pratley, R.E., Ravussin, E., Reiman, E.M. and Tataranni, P.A. (2001) Effect of satiation on brain activity in obese and lean women. Obesity Res., 9: 676–684.

Geddes, J.R. and Lawrie, S.M. (1995) Obstetric complications and schizophrenia: a meta analysis. Br. J. Psychiatry, 167: 786–793.

Gencik, M., Dahmen, N., Wieczorek, S., Kasten, M., Bierbrauer, J., Anghelescu, I., Szegedi, A., Saecker, M. and Epplen, J.T. (2001) A prepro-orexin gene polymorphism is associated with narcolepsy. Neurology, 56: 115–117.

Ghandirian, A.-M., Marini, N., Jabalpurwala, S. and Steiger, H. (1999) Seasonal mood patterns in eating disorders. Gen. Hosp. Psychiatry, 21: 354–359.

Gold, P.W., Licinio, J., Wong, M.L. and Chrousos, G.P. (1995) Corticotropin releasing hormone in the pathophysiology of melancholic and atypical depression and in the mechanism of action of antidepressant drugs. Ann. N.Y. Acad. Sci., 771: 716–729.

Gold, R.M. (1973) Hypothalamic obesity: the myth of the ventromedial nucleus. Science, 182: 488–490.

Goldstone, A.P., Unmehopa, U.A., Bloom, S.R. and Swaab, D.F. (2002) Hypothalamic neuropeptide Y and agouti-related protein are increased in human illness but not in Prader–Willi syndrome and other obese subjects. J. Clin. Endocrinol. Metab., 87: 927–937.

Goldstone, A.P., Unmehopa, U. and Swaab, D.F. (2003) Hypothalamic growth hormone-releasing hormone (GHRH). cell number in human illness, Prader–Willi syndrome and obesity. Clin. Endocrinol., 58: 743–755.

Gomez, F., Chapleur, M., Fernette, B., Burlet, C., Nicolas, J.-P. and Burlet, A. (1997) Arginine vasopressin (AVP) depletion in neurons of the suprachiasmatic nuclei affects the AVP content of the paraventricular neurons and stimulates adrenocorticotrophic hormone release. J. Neurosci. Res., 50: 565–574.

Gorwood, P., Kipman, A. and Foulon, C. (2003) The human genetics of anorexia nervosa. Eur. J. Pharmacol., 480: 163–170.

Grilo, C.M. and Masheb, R.M. (2002) Childhood maltreatment and personality disorders in adult patients with binge eating disorder. Acta Psychiatr. Scand., 106: 183–188.

Hall, J.E., Hildebrandt, D.A. and Kuo, J. (2001) Obesity hypertension: role of leptin and sympathetic nervous system. Am. J. Hypertension, 14(Suppl. 1): 103S–115S.

Halmi, K.A., Goldberg, S.C., Casper, R.C., Eckert, E.D. and Davis, J.M. (1979) Pretreatment predictors of outcome in anorexia nervosa. Br. J. Psychiat., 134: 71–78.

Harmatz, M.G., Well, A.D., Overtree, C.E., Kawamura, K.Y., Rosal, M. and Ockene, I.S. (2000) Seasonal variation of depression and other moods: a longitudinal approach. J. Biol. Rhythms, 15: 344–350.

Haugh, R.M. and Markesbery, W.R. (1983) Hypothalamic astrocytoma. Syndrome of hyperphagia, obesity, and disturbances of behavior and endocrine and autonomic function. Arch. Neurol., 40: 560–563.

Hebebrand, J., Fichter, M., Gerber, G., Görg, T., Hermann, H., Geller, F., Schäfer, H., Remschmidt, H. and Hinney, A. (2002) Genetic predisposition to obesity in bulimia nervosa: a mutation screen of the melanocortin-4 receptor gene. Mol. Psychiatry, 7: 647–651.

Hebebrand, J., Geller, F., Dempfle, A., Heinzel-Gutenbrunner, M., Raab, M., Gerber, G., Wermter, A.-K., Horro, F.F., Blundell, J., Schäfer, H., Remschmidt, H., Herpertz, S. and Hinney, A. (2004) Binge-eating episodes are not characteristic of carriers of melanocortin-4 receptor gene mutations. Mol. Psychiatry, 9: 796–800.

Heim, C. and Nemeroff, C.B. (2001) The role of childhood trauma in the neurobiology of mood and anxiety disorders: preclinical and clinical studies. Biol. Psychiatry, 49: 1023–1039.

Heinbecker, P. (1944) The pathogenesis of Cushing's syndrome. Medicine, 23: 225–247.

Helbing-Zwanenburg, B., Kamphuisen, H.A. and Mourtazaev, M.S. (1993) The origin of excessive daytime sleepiness in the Prader–Willi syndrome. J. Intellect. Disabil. Res., 37: 533–541.

Hinney, A., Hoch, A., Geller, F., Schäfer, H., Siegfried, W., Goldschmidt, H., Remschmidt, H. and Hebebrand, J. (2002) Ghrelin gene: identification of missense variants and a frameshift mutation in extremely obese children and adolescents and healthy normal weight students. J. Clin. Endocrinol. Metab., 87: 2716–2719.

Hofman, M.A. (2001) Seasonal rhythms of neuronal activity in the human biological clock: a mathematical model. Biol. Rhythm Res., 32: 17–34.

Hofman, M.A. (2003) Circadian oscillations of neuropeptide expression in the human biological clock. J. Comp. Physiol. A, 189: 823–832.

Hofman, M.A., Purba, J.S. and Swaab, D.F. (1993) Annual variations in the vasopressin neuron population of the human suprachiasmatic nucleus. Neuroscience, 53: 1103–1112.

Hofman, M.A. and Swaab, D.F. (1992a) The human hypothalamus: comparative morphometry and photoperiodic influences. In: Swaab, D.F., Hofman, M.A., Mirmiran, M., Ravid, R. and Van Leeuwen, F.W. (Eds.) The Human

Hypothalamus in Health and Disease. Progress in Brain Research, Vol. 93. Elsevier, Amsterdam, pp. 133–149.

Hofman, M.A. and Swaab, D.F. (1992b) Seasonal changes in the suprachiasmatic nucleus of man. Neurosci. Lett., 139: 257–260.

Hofman, M.A. and Swaab, D.F. (1993) Diurnal and seasonal rhythms of neuronal activity in the suprachiasmatic nucleus of humans. J. Biol. Rhythms, 8: 283–295.

Honnebier, W.J. and Swaab, D.F. (1973) The influence of anencephaly upon intrauterine growth of fetus and placenta and upon gestation length. J. Obstet. Gynaecol. Br. Cmwlth., 80: 577–588.

Hsu, L.K.G., Mulliken, B., McDonagh, B., Krupa Das, S., Rand, W., Fairburn, C.G., Rolls, B., McCrory, M.A., Saltzman, E., Shikora, S., Dwyer, J. and Roberts, S. (2002) Binge eating disorder in extreme obesity. Int. J. Obesity, 26: 1398–1403.

Inder, W.J., Donald, R.A., Prickett, T.C.R., Frampton, C.M., Sullivan, P.F., Mulder, R.T. and Joyce, P.R. (1997) Arginine vasopressin is associated with hypercortisolemia and suicide attempts in depression. Biol. Psychiatry, 42: 744–747.

Iwamoto, H.S., Rudolph, A.M., Keil, L.C. and Heymann, M.A. (1979) Hemodynamic responses of the sheep fetus to vasopressin infusion. Circ. Res., 44: 430–436.

Jackson, R.S., Creemers, J.W.M., Ohagi, S., Raffin-Sanson, M.-L., Sanders, L., Mantague, C., Hutton, J.C. and O'Rahilly, S. (1997) Obesity and impaired prohormone processing associated with mutations in the human prohormone convertase 1 ((PC1) gene. Nat. Genet., 16: 303–306.

Jay, P., Rougeulle, C., Massacrier, A., Moncla, A., Mattei, M.-G., Malzac, P., Roëckel, N., Taviaux, S., Bergé Lefranc, J.-L., Cau, P., Berta, P., Lalande, M. and Muscatelli, F. (1997) The human necdin gene, NDN, is maternally imprinted and located in the Prader–Willi syndrome chromosomal region. Nat. Genet., 17: 357–361.

Jean-Louis, G., Von Gizycki, H. and Zizi, F. (1998) Melatonin effects on sleep, mood, and cognition in elderly with mild cognitive impairment. J. Pineal Res., 25: 177–183.

Johansson, C., Willeit, M., Smedh, C., Ekholm, J., Paunio, T., Kieseppä, T., Lichtermann, D., Praschak-Rieder, N., Neumeister, A., Nilsson, L.-G., Kasper, S., Peltonen, L., Adolfsson, R., Schalling, M. and Partonen, T. (2003) Circadian clock-related polymorph-isms in seasonal affective disorder and their relevance to diurnal preference. Neuropsychopharmacology, 28: 734–739.

Jong, M.T.C., Gray, T.A., Ji, Y., Glenn, C.C., Saitoh, S., Driscoll, D.J. and Nicholls, R.D. (1999) A novel imprinted gene, encoding a RING zinc-finger protein, and overlapping antisense transcript in the Prader–Willi syndrome critical region. Hum. Mol. Genet., 8: 783–793.

Joo, E.Y., Tae, W.S., Kim, J.H., Kim, B.T. and Hong, S.B. (2004) Glucose hypometabolism of hypothalamus and thalamus in narcolepsy. Ann. Neurol., 56: 437–440.

Kadotani, H., Faraco, J. and Mignot, E. (1998) Genetic studies in the sleep disorder narcolepsy. Genome Res., 8: 427–434.

Kalsbeek, A., Buijs, R.M., Van Heerikhuize, J.J., Arts, M. and Van der Woude, T.P. (1992) Vasopressin-containing neurons of the suprachiasmatic nuclei inhibit corticosterone release. Brain Res., 580: 62–67.

Kaplan, J., Fredrickson, P.A. and Richardson, J.W. (1991) Sleep and breathing in patients with the Prader–Willi syndrome. Mayo Clin. Proc., 66: 1124–1126.

Katz, G., Durst, R., Zislin, Y., Barel, Y. and Knobler, H.Y. (2001) Psychiatric aspects of jet lag: review and hypothesis. Med. Hypotheses, 56: 20–23.

Kaye, W.H. (1996) Neuropeptide abnormalities in anorexia nervosa. Psychiatry Res., 62: 65–74.

Kaye, W.H., Frank, G.K., Bailer, U.F., Henry, S.E., Meltzer, C.C., Price, J.C., Mathis, C.A. and Wagner, A. (2005) Serotonin alterations in anorexia and bulimia nervosa: new insights from imaging studies. Physiol. Behav., 85: 73–81.

Kaye, W., Gendall, K. and Strober, M. (1998) Serotonin neuronal function and selective serotonin reuptake inhibitor treatment in anorexia and bulimia nervosa. Biol. Psychiatry, 44: 825–838.

Kelly, J.J., Mangos, G., Williamson, P.M. and Whitworth, J. A. (1998) Cortisol and hypertension. Clin. Exp. Pharmacol. Physiol. (suppl. 25): S1–S6.

Keverne, E.B., Fundele, R., Narasimha, M., Barton, S.C. and Surani, M.A. (1996) Genomic imprinting and the differential roles of parental genomes in brain development. Brain Res. Dev., 92: 91–100.

Kobayashi, H., Ogawa, Y., Shintani, M., Ebihara, K., Shimodahira, M., Iwakura, T., Hino, M., Ishihara, T., Ikekubo, K., Kurahachi, H. and Nakao, K. (2002) A novel homozygous missense mutation of melanocortin-4 receptor (MC4R) in a Japanese woman with severe obesity. Diabetes, 51: 243–246.

Kok, S.W., Meinders, A.E., Overeem, S., Lammers, G.J., Roelfsema, F., Frolich, M. and Pijl, H. (2002) Reduction of plasma leptin levels and loss of its circadian rhythmicity in hypocretin (orexin)-deficient narcoleptic humans. J. Clin. Endocrinol. Metab., 87: 805–809.

Koutcherov, Y., Mai, J.K., Ashwell, K.W.S. and Paxinos, G. (2002) Organization of human hypothalamus in fetal development. J. Comp. Neurol., 446: 301–324.

Kräuchi, K. and Wirz-Justice, A. (1988) The four seasons: food intake frequency in seasonal affective disorder in the course of a year. Psychiatry Res., 25: 323–338.

Kräuchi, K., Wirz-Justice, A. and Graw, P. (1993) High intake of sweets late in the day predicts a rapid and persistant response to light therapy in winter depression. Psychiatry Res., 46: 107–117.

Kripke, D.F. (1998) Light treatment for nonseasonal depression: speed, efficacy, and combined treatment. J. Affect. Disord., 49: 109–117.

Krude, H., Biebermann, H., Luck, W., Horn, R., Brabant, G. and Grüters, A. (1998) Severe early onset obesity, adrenal insufficiency and red hair pigmentation caused by POMC mutations in humans. Nat. Genet., 19: 155–157.

Krude, H. and Grüters, A. (2000) Implications of proopiomelanocortin (POMC) mutations in humans: the POMC deficiency syndrome. Trends Endocrinol. Metabol., 11: 15–22.

Kruk, M.R., Westphal, K.G.C., Van Erp, A.M.M., Van Asperen, J., Cave, B.K., Slater, E., De Koning, J. and

Haller, J. (1998) The hypothalamus: cross-roads of endocrine and behavioural regulation in grooming and aggression. Neurosci. Biobehav. Rev., 23: 163–177.

Kuslich, C.D., Kobori, J.A., Mohapatra, G., Gregorio-King, C. and Donlon, T.A. (1999) Prader–Willi syndrome is caused by disruption of the *SNRPN* gene. Am. J. Genet., 64: 70–76.

Kusumi, I., Ohmori, T., Kohsaka, M., Ito, M., Honma, H. and Koyama, T. (1995) Chronobiological approach for treatment-resistant rapid cycling affective disorders. Soc. Biol. Psychol., 37: 553–559.

Lam, R.W., Goldner, E.M. and Grewal, A. (1996a) Seasonality of symptoms in anorexia and bulimia nervosa. Int. J. Eat. Disord., 19: 35–44.

Lam, R.W., Goldner, E.M., Solyom, L. and Remick, R.A. (1994) A controlled study of light therapy for bulimia nervosa. Am. J. Psychiatry, 151: 744–750.

Lam, R.W., Lee, S.K., Tam, E.M., Grewal, A. and Yatham, L.N. (2001) An open trial of light therapy for women with seasonal affective disorder and comorbid bulimia nervosa. J. Clin. Psychiatry, 62: 164–168.

Lam, R.W., Zis, A.P., Grewal, A., Delgado, P.L., Charney, D.S. and Krystal, J.H. (1996b) Effects of rapid tryptophan depletion in patients with seasonal affective disorder in remission after light therapy. Arch. Gen. Psychiatry, 53: 41–44.

Lambert, G., Reid, C., Kaye, D., Jennings, G. and Esler, M. (2002) Effect of sunlight and season on serotonin turnover in the brain. Lancet, 360: 1840–1842.

Lee, Y.-S., Kok-Seng Poh, L. and Loke, K.-Y. (2002) A novel melanocortin 3 receptor gene (*MC3R*) mutation associated with severe obesity. J. Clin. Endocrinol. Metab., 87: 1423–1426.

Leibowitz, S.F., Hammer, N.J. and Chang, K. (1981) Hypothalamic paraventricular nucleus lesions produce overeating and obesity in the rat. Physiol. Behav., 27: 1031–1040.

Leppämäki, S., Partonen, T. and Lönnqvist, J. (2002) Bright-light exposure combined with physical exercise elevates mood. J. Affect. Disord., 72: 139–144.

Lester, N.A., Keel, P.K. and Lipson, S.F. (2003) Symptom fluctuation in bulimia nervosa: relation to menstrual-cycle phase and cortisol levels. Psychol. Med., 33: 51–60.

Levitan, R.D., Kaplan, A.S., Levitt, A.J. and Joffe, R.T. (1994) Seasonal fluctuations in mood and eating behavior in bulimia nervosa. Int. J. Eating Disorders, 16: 295–299.

Levitan, R.D., Kaplan, A.S. and Rockert, W. (1996) Characterization of the 'seasonal' bulimic patient. Int. J. Eating Disorders, 19: 187–192.

Levitan, R.D., Masellis, M., Basile, V.S., Lam, R.W., Kaplan, A.S., Davis, C., Muglia, P., Mackenzie, B., Tharmalingam, S., Kennedy, S.H., Macciardi, F. and Kennedy, J.L. (2004) The dopamine-4 receptor gene associated with binge eating and weight gain in women with seasonal affective disorder: an evolutionary perspective. Biol. Psychiatry, 56: 665–669.

Lewy, A.J., Bauer, V.K., Cutler, N.L. and Sack, R.L. (1998) Melatonin treatment of winter depression: a pilot study. Psychiatry Res., 77: 57–61.

Lewy, A.J. and Sack, R.L. (1996) The role of melatonin and light in the human circadian system. Prog. Brain Res., 111: 205–216.

Licinio, J., Wong, M.-L. and Gold, P.W. (1996) The hypothalamic–pituitary–adrenal axis in anorexia nervosa. Psychiatry Res., 62: 75–83.

Lin, L., Faraco, J., Li, R., Kadotani, H., Rogers, W., Lin, X., Qiu, X., De Jong, P.J., Nishino, S. and Mignot, E. (1999) The sleep disorder canine narcolepsy is caused by a mutation in the hypocretin (orexin) receptor 2 gene. Cell, 98: 365–376.

Lindgren, A.C., Hagenäs, L., Müller, J., Blichfeldt, S., Rosenborg, M., Brismar, T. and Ritzén, E.M. (1997) Effects of growth hormone treatment on growth and body composition in Prader–Willi syndrome: a preliminary report. Acta Paediatr. (Suppl), 423: 60–62.

Lino, R.S., Reis, L.C., Reis, M.A., Gobbi, H. and Teixeira, V.P.A. (2000) Hypothalamic neurocysticerosis as a possible cause of obesity. Trans. R. Soc. Trop. Med. Hyg., 94: 294.

Liu, Y., Gao, J.-H., Liu, H.-L. and Fox, P.T. (2000) The temporal response of the brain after eating revealed by functional MRI. Nature, 405: 1058–1062.

Lustig, R.H., Rose, S.R., Burghen, G.A., Velasquez-Mieyer, P., Broome, D.C., Smith, K., Li, H., Hudson, M.M., Heideman, R.L. and Kun, L.E. (1999) Hypothalamic obesity caused by cranial insult in children: altered glucose and insulin dynamics and reversal by a somatostatin agonist. J. Pediatr., 135: 162–168.

MacNeil, D.J., Howard, A.D., Guan, X., Fong, T.M., Nargund, R.P., Bednarek, M.A., Goulet, M.T., Weinberg, D.H., Strack, A.M., Marsh, D.J., Chen, H.Y., Shen, C.P., Chen, A.S., Rosenblum, C.I., MacNeil, T., Tota, M., MacIntyre, E.D. and Van der Ploeg, L.H. (2002) The role of melanocortins in body weight regulation: opportunities for the treatment of obesity. Eur. J. Pharmacol., 440: 141–157.

Maes, M., Meltzer, H.Y., Cosyns, P., Suy, E. and Schotte, C. (1993b) An evaluation of basal hypothalamic–pituitary–thyroid axis function in depression: results of a large-scaled and controlled study. Psychoneuroendocrinology, 18: 607–620.

Maghnie, M., Triulzi, F., Larizza, D., Preti, P., Priora, C., Scotti, G. and Severi, F. (1991) Hypothalamic–pituitary dysfunction in growth hormone-deficient patients with pituitary abnormalities. J. Clin. Endocrinol. Metab., 73: 79–83.

Magiakou, M.A., Mastorakos, G. and Chrousos, G.P. (1997) Cushing syndrome. Differential diagnosis and treatment. In: Wierman, M.E. (Ed.) Diseases of the Pituitary: Diagnosis and Treatment. Contemporary Endocrinology, Vol. 3. Humana Press Inc, Totowa, NJ, pp. 179–202.

Martin, A., State, M., Koenig, K., Schultz, R., Dykens, E.M., Cassidy, S.B. and Leckman, J.F. (1998a) Prader–Willi syndrome. Am. J. Psychiatry, 155: 1265–1273.

Mastorakos, G. and Ilias, I. (2003) Maternal and fetal hypothalamic–pituitary–adrenal axes during pregnancy and postpartum. Ann. N.Y. Acad. Sci., 997: 136–149.

Mastorakos, G. and Zapanti, E. (2004) The hypothalamic–pituitary–adrenal axis in the neuroendocrine regulation of food intake and obesity: the role of corticotropin releasing hormone. Nutr. Neurosci., 7: 271–280.

Matsushita, S., Suzuki, K., Murayama, M., Nishiguchi, N., Hishimoto, A., Takeda, A., Shirakawa, O. and Higuchi, S. (2004) Serotonin transporter regulatory region polymorphism is associated with anorexia nervosa. Am. J. Med. Genet. Part B, 128B: 114–117.

McElroy, S.L., Kotwal, R., Malhotra, S., Nelson, E.B., Keck, P.E. and Nemeroff, C.B. (2004) Are mood disorders and obesity related? A review for the mental health professional. J. Clin. Psychiatry, 65: 634–651.

McEntagart, M.E., Webb, T., Hardy, C. and King, M.D. (2000) Familial Prader–Willi syndrome: case report and a literature review. Clin. Genet., 58: 216–223.

McMillen, I.C., Phillips, I.D., Ross, J.T., Robinson, J.S. and Owens, J.A. (1995) Chronic stress—the key to parturition? Reprod. Fertil. Dev., 7: 499–507.

McMillen, I.C., Schwartz, J., Coulter, C.L. and Edwards, L.J. (2004) Early embryonic environment, the fetal pituitary-adrenal axis and the timing of parturition. Endocrinol. Res., 30: 845–850.

Mergen, M., Mergen, H., Ozata, M., Oner, R. and Oner, C. (2001) A novel melanocortin 4 receptor (MC4R) gene mutation associated with morbid obesity. J. Clin. Endocrinol. Metab., 86: 3448–3451.

Meynen, G., Unmehopa, U.A., van Heerikhuize, J., Hofman, M.A., Swaab, D.F. and Hoogendijk, W.J.G. (2006) Increased arginine vasopressin mRNA expression in the human hypothalamus in depression A Preliminary Report Biol. Psychiatr., in press.

Mitchell, A. and O'Keane, V. (1998) Steroids and depression. Br. Med. J., 316: 244–245.

Montague, C.T., Farooqi, I.S., Whitehead, J.P., Soos, M.A., Rau, H., Wareham, N.J., Sewter, C.P., Digby, J.E., Mohammed, S.N., Hurst, J.A., Cheetham, C.H., Earley, A.R., Barnett, A.H., Prins, J.B. and O'Rahilly, S. (1997) Congenital leptin deficiency is associated with severe early onset obesity in humans. Nature, 387: 903–908.

Monteleone, P., Maes, M., Fabrazzo, M., Tortorella, A., Lin, A., Bosmans, E., Kenis, G. and Maj, M. (1999) Immunoendocrine findings in patients with eating disorders. Neuropsychobiology, 40: 115–120.

Moore, R.Y., Abrahamson, E.A. and Van den Pol, A.N. (2001) The hypocretin neuron system: an arousal system in the human brain. Arch Ital Biol, 139: 195–205.

Müller, D., Roeder, F. and Orthner, H. (1973) Further results of stereotaxis in the human hypothalamus in sexual deviations. First use of this operation in addiction to drugs.. Neurochirurgia, 16: 113–126.

Murakami, N., Furuto-Kato, S., Fujisawa, I., Ohyama, K., Nakao, S., Kuwayama, A. and Kageyama, N. (1998) Supra- and extrasellar pituitary microadenoma as a cause of Cushing's disease. Endocrinol. J., 45: 631–636.

Muscatelli, F., Abrous, D.N., Massacrier, A., Boccaccio, I., Le Moal, M., Cau, P. and Cremer, H. (2000) Disruption of the mouse necdin gene results in hypothalamic and behavioral alteration reminiscent of the human Prader–Willi syndrome. Hum. Mol. Genet., 9: 3101–3110.

Musselman, D.L. and Nemeroff, C.B. (1996) Depression and endocrine disorders: focus on the thyroid and adrenal system. Br. J. Psychiatry, 168: 123–128.

Nemeroff, C.B. (1996) The corticotropin-releasing factor (CRF) hypothesis of depression: new findings and new directions. Mol. Psychiatry, 1: 336–342.

Neudeck, P., Jacoby, G.E. and Florin, I. (2001) Dexamethasone suppression test using saliva cortisol measurement in bulimia nervosa. Physiol. Behav., 72: 93–98.

Neumeister, A., Pirker, W., Willeit, M., Praschak-Rieder, N., Asenbaum, S., Brücke, T. and Kasper, S. (2000) Seasonal variation of availability of serotonin transporter binding sites in healthy female subjects as measured by [^{123}I]-2b-carbomethoxy-3b-(4-iodophenyl)tropane and single photon emission computed tomography. Biol. Psychiatry, 47: 158–160.

Neumeister, A., Praschak-Rieder, N., Besselmann, B., Rao, M.-L., Glück, J. and Kasper, S. (1997) Effects of tryptophan depletion on drug-free patients with seasonal affective disorder during a stable response to bright light therapy. Arch. Gen. Psychiatry, 54: 133–138.

Nevsimalova, S., Vankova, J., Stepanova, I., Seemanova, E., Mignot, E. and Nishino, S. (2004) Hypocretin deficiency in Prader–Willi syndrome. J. Sleep Res., 13(suppl. 1): 526.

Newell-Price, J., Jørgensen, J.O.L. and Grossman, A. (1999) The diagnosis and differential diagnosis of Cushing's syndrome. Horm. Res., 51(Suppl 3): 81–94.

Nilsson, M., Naessén, S., Dahlman, I., Lindén Hirschberg, A., Gustafsson, J.-Å. and Dahlman-Wright, K. (2004) Association of estrogen receptor β gene polymorphisms with bulimic disease in women. Mol. Psychiatry, 9: 28–34.

Nishino, S., Ripley, B., Overeem, S., Nevsimalova, S., Lammers, G.J., Vankova, J., Okun, M., Rogers, W., Brooks, S. and Mignot, E. (2001) Low cerebrospinal fluid hypocretin (orexin) and altered energy homeostasis in human narcolepsy. Ann. Neurol., 50: 381–388.

O'Keane, V., Dinan, T.G., Scott, L. and Corcoran, C. (2005) Changes in hypothalamic–pituitary–adrenal axis measures after vagus nerve stimulation therapy in chronic depression. Biol. Psychiatry, 58: 963–968.

Ólafsdóttir, B.R., Rye, D.B., Scammell, T.E., Matheson, J.K., Stefánsson, K. and Gulcher, J.R. (2001) Polymorphisms in hypocretin/orexin pathway genes and narcolepsy. Neurology, 57: 1896–1899.

Olander, E., Stamberg, J., Steinberg, L. and Wulfsberg, E.A. (2000) Third Prader–Willi syndrome phenotype due to maternal uniparental disomy 15 with mosaic trisomy 15. Am. J. Med. Genet., 93: 215–218.

Olson, B.R., Drutarosky, M.D., Chow, M.S., Hruby, V.J., Stricker, E.M. and Verbalis, J.G. (1991b) Oxytocin and an oxytocin agonist administered centrally decrease food intake in rats. Peptides, 12: 113–118.

Olson, B.R., Drutarosky, M.D., Stricker, E.M. and Verbalis, J.G. (1991a) Brain oxytocin receptor antagonism blunts the effects of anorexigenic treatments in rats: evidence for central oxytocin inhibition of food intake. Endocrinology, 129: 785–791.

Oosterbaan, H.P. and Swaab, D.F. (1987) Circulating neurohypophyseal hormones in anencephalic infants. Am. J. Obstet. Gynecol., 157: 117–119.

Overeem, S., Kok, S.W., Lammers, G.J., Vein, A.A., Frölich, M., Meinders, A.E., Roelfsema, F. and Pijl, H. (2003) Somatotropic axis in hypocretin-deficient narcoleptic humans: altered circadian distribution of GH-secretory events. Am. J. Physiol. Endocrinol. Metab., 284: E641–E647.

Overeem, S., Mignot, E., Van Dijk, G. and Lammers, G.J. (2001) Narcolepsy: clinical features, new pathophysiologic insights, and future perspectives. J. Clin. Neurophysiol., 18: 78–105.

Overeem, S., Van Vliet, J.A., Lammers, G.J., Zitman, F.G., Swaab, D.F. and Ferrari, M.D. (2002) The hypothalamus in episodic brain disorders. Lancet Neurol., 1: 437–444.

Ozisik, G., Achermann, J.C. and Jameson, J.L. (2002) The role of SF1 in adrenal and reproductive function: insight from naturally occurring mutations in humans. Mol. Genet. Metab., 76: 85–91.

Pacchierotti, C., Iapichino, S., Bossini, L., Pieraccini, F. and Castrogiovanni, P. (2001) Melatonin in psychiatric disorders: a review on the melatonin involvement in psychiatry. Front. Neuroendocrinol., 22: 18–32.

Pérusse, L., Chagnon, Y.C., Weisnagel, S.J., Rankinen, T., Snyder, E., Sands, J. and Bouchard, C. (2001) The human obesity map: the 2000 update. Obes. Res., 9: 135–169.

Peyron, C., Faraco, J., Rogers, W., Ripley, B., Overeem, S., Charnay, Y., Nevsimalova, S., Aldrich, M., Reynolds, D., Albin, R., Li, R., Hungs, M., Pedrazzoli, M., Padigaru, M., Kucherlapati, M., Fan, J., Maki, R., Lammers, G.J., Bouras, C., Kucherlapati, R., Nishino, S. and Mignot, E. (2000) A mutation in a case of early onset narcolepsy and a generalized absence of hypocretin peptides in human narcoleptic brains. Nat. Med., 6: 991–997.

Phillips, D.I.W. (2001) Fetal growth and programming of the hypothalamic–pituitary–adrenal axis. Clin. Exp. Pharmacol. Physiol., 28: 967–970.

Phillips, D.I.W., Fall, C.H.D., Cooper, C., Norman, R.J., Robinson, J.S. and Owens, P.C. (1999) Size at birth and plasma leptin concentrations in adult life. Int. J. Obesity, 23: 1025–1029.

Pijl, H. (2003) Reduced dopaminergic tone in hypothalamic neural circuits: expression of a "thrifty" genotype underlying the metabolic syndrome? Eur. J. Pharmacol., 480: 125–131.

Piovesan, A., Panarelli, M., Terzolo, M., Osella, G., Matrella, C., Paccotti, P. and Angeli, A. (1990) 24-Hour profiles of blood pressure and heart rate in Cushing's syndrome: relationship between cortisol and cardiovascular rhythmicities. Chronobiol. Int., 7: 263–265.

Pohjavuori, M. and Fyhrquist, J. (1980) Hemodynamic significance of vasopressin in the newborn infant. J. Pediatr., 97: 462–465.

Polivy, J. and Herman, C.P. (2002) Causes of eating disorders. Ann. Rev. Psychol., 53: 187–213.

Posternak, M.A. and Zimmerman, M. (2002) Lack of association between seasonality and psychopathology in psychiatric outpatients. Psychiat. Res., 112: 187–194.

Prasko, J., Horacek, J., Klaschka, J., Kosova, J., Ondrackova, I. and Sipek, J. (2002) Bright light therapy and/or imipramine for inpatients with recurrent non-seasonal depression. Neuroendocrinol. Lett., 23: 109–113.

Purba, J.S., Hoogendijk, W.J.G., Hofman, M.A. and Swaab, D.F. (1996) Increased number of vasopressin- and oxytocin-expressing neurons in the paraventricular nucleus of the hypothalamus in depression. Arch. Gen. Psychiatry, 53: 137–143.

Raadsheer, F.C., Hoogendijk, W.J.G., Stam, F.C., Tilders, F.J.C. and Swaab, D.F. (1994) Increased numbers of corticotropin-releasing hormone expressing neurons in the hypothalamic paraventricular nucleus of depressed patients. Neuroendocrinology, 60: 436–444.

Raadsheer, F.C., Van Heerikhuize, J.J., Lucassen, P.J., Hoogendijk, W.J.G., Tilders, F.J.H. and Swaab, D.F. (1995) Corticotropin-releasing hormone mRNA levels in the paraventricular nucleus of patients with Alzheimer's disease and depression. Am. J. Psychiatry, 152: 1372–1376.

Raff, H. (2000) Salivary cortisol: a useful measurement in the diagnosis of Cushing's syndrome and the evaluation of the hypothalamic–pituitary–adrenal axis. Endocrinologist, 10: 9–17.

Ravelli, G.-P., Stein, Z.A. and Susser, M.W. (1976) Obesity in young men after famine exposure in utero and early infancy. N. Engl. J. Med., 295: 349–353.

Reeves, A.G. and Plum, F. (1969) Hyperphagia, rage, and dementia accompanying a ventromedial hypothalamic neoplasm. Arch. Neurol., 20: 616–624.

Ren, J., Lee, S., Pagliardini, S., Gérard, M., Stewart, C.L., Greer, J.J. and Wevrick, R. (2003) Absence of Ndn, encoding the Prader–Willi syndrome-deleted gene necdin, results in congenital deficiency of central respiratory drive in neonatal mice. J. Neurosci., 23: 1569–1573.

Ribasés, M., Gratacòs, M., Armengol, L., De Cid, R., Badía, A., Jiménez, L., Solano, R., Vallejo, J., Fernández, F. and Estivill, X. (2003) Met66 in the brain-derived neurotrophic factor (BDNF) precursor is associated with anorexia nervosa restrictive type. Mol. Psychiatry, 8: 745–751.

Ribasés, M., Gratacòs, M., Fernández-Aranda, F., Bellodi, L., Boni, C., Anderluh, M., Cavallini, M.C., Cellini, E., Di Bella, D., Erzegovesi, S., Foulon, C., Gabrovsek, M., Gorwood, P., Hebebrand, J., Hinney, A., Holliday, J., Hu, X., Karwautz, A., Kipman, A., Komel, R., Nacmias, B., Remschmidt, H., Ricca, V., Sorbi, S., Wagner, G., Treasure, J., Collier, D.A. and Estivill, X. (2004) Association of BDNF with anorexia, bulimia and age of onset of weight loss in six European populations. Hum. Mol. Genet., 13: 1205–1212.

Richdale, A.L., Cotton, S. and Hibbit, K. (1999) Sleep and behaviour disturbance in Prader–Willi syndrome: a questionnaire study. J. Intellect. Disabil. Res., 43: 380–392.

Roessmann, U., Velasco, M.E., Small, E.J. and Hori, A. (1987) Neuropathology of septo-optic dysplasia (De Morsier syndrome) with immunohistochemical studies of the hypothalamus and pituitary gland. J. Neuropathol. Exp. Neurol., 46: 597–608.

Rogers, R.C. and Hermann, G.E. (1986) Oxytocin, oxytocin antagonist, TRH, and hypothalamic paraventricular nucleus stimulation effects on gastric motility. Peptides, 8: 505–513.

Rosenbaum, M., Leibel, R.L. and Hirsch, J. (1997) Obesity. N. Engl. J. Med., 337: 396–407.

44

Rosenthal, N.E., Genhart, M., Jacobsen, F.M., Skwerer, R.G. and Wehr, T.A. (1987) Disturbance of appetite and weight regulation in seasonal affective disorder. Ann. N.Y. Acad. Sci., 499: 216–230.

Rosmond, R., Chagnon, Y.C., Holm, G., Chagnon, M., Pérusse, L., Lindell, K., Carlsson, B., Bouchard, C. and Björntorp, P. (2000) A glucocorticoid receptor gene marker is associated with abdominal obesity, leptin, and dysregulation of the hypothalamic–pituitary–adrenal axis. Obesity Res., 8: 211–218.

Rosmond, R., Ukkola, O., Bouchard, C. and Björntorp, P. (2002) Polymorphisms in exon 3 of the proopiomelanocortin gene in relation to serum leptin, salivary cortisol, and obesity in Swedish men. Metabolism, 51: 642–644.

Salehi, M., et al., (2005) Division of Endocrinology and Metabolism, Department of Medicine, Beth Israel Medical Center and Albert Einstein College of Medicine, New York, NY 10003, USA.

Schalin-Jäntti, C., Valli-Jaakola, K., Oksanen, L., Martelin, E., Laitinen, K., Krusius, T., Mustajoki, P., Heikinheimo, M. and Kontula, K. (2003) Melanocortin-3-receptor gene variants in morbid obesity. Int. J. Obesity, 27: 70–74.

Schuld, A., Blum, W.F., Uhr, M., Haack, M., Kraus, T., Holsboer, F. and Pollmächer, T. (2000) Reduced leptin levels in human narcolepsy. Neuroendocrinology, 72: 195–198.

Shinoda, M., Tsugu, A., Oda, S., Masuko, A., Yamaguchi, T., Yamaguchi, T., Tsugane, R. and Sato, O. (1993) Development of akinetic mutism and hyperphagia after left thalamic and right hypothalamic lesions. Child's Nerv. Syst., 9: 243–245.

Shuto, Y., Shibasaki, T., Otagiri, A., Kuriyama, H., Ohata, H., Tamura, H., Kamegai, J., Sugihara, H., Oikawa, S. and Wakabayashi, I. (2002) Hypothalamic growth hormone secretagogue receptor regulates growth hormone secretion, feeding, and adiposity. J. Clin. Invest., 109: 1429–1436.

Siegel, J.M. (1999) Narcolepsy: a key role for hypocretins (orexins). Cell, 98: 409–412.

Smeets, P.A.M., De Graaf, C., Stafleu, A., Van Osch, M.J.P. and Van der Grond, J. (2005) Functional MRI of human hypothalamic responses following glucose ingestion. NeuroImage, 24: 363–368.

Stamatakis, E.A. and Hetherington, M.M. (2003) Neuroimaging in eating disorders. Nutr. Neurosci., 6: 325–334.

Stewart, P.M., Penn, R., Gibson, R., Holder, R., Parton, A., Ratcliffe, J.G. and London, D.R. (1992) Hypothalamic abnormalities in patients with pituitary-dependent Cushing's syndrome. Clin. Endocrinol., 36: 453–458.

Ströbel, A., Issad, T., Camoin, L., Ozata, M. and Strosberg, A.D. (1998) A leptin missense mutation associated with hypogonadism and morbid obesity. Nat. Genet., 18: 213–215.

Swaab, D.F. (1997) Prader–Willi syndrome and the hypothalamus. Acta Paediatr. Suppl., 423: 50–54.

Swaab, D.F. (2003) The human hypothalamus. Basic and clinical aspects. Part I: Nuclei of the hypothalamus. In: Aminoff, M.J., Boller, F. and Swaab, D.F. (Eds.), Handbook of Clinical Neurology. Elsevier, Amsterdam 476pp.

Swaab, D.F. (2004) The human hypothalamus. Basic and clinical aspects. Part II: Neuropathology of the hypothalamus and adjacent brain structures. In: Aminoff, M.J., Boller, F. and Swaab, D.F. (Eds.), Handbook of Clinical Neurology. Elsevier, Amsterdam 596pp.

Swaab, D.F., Bao, A.M. and Lucassen, P.J. (2005) The stress system of the human brain in depression and neurodegeneration. Ageing Res. Rev., 4: 141–194.

Swaab, D.F., Boer, K. and Honnebier, W.J. (1977) The influence of the fetal hypothalamus and pituitary on the onset and course of parturition. In: Knight, J. and O'Connor, M. (Eds.), The Fetus and Birth. Ciba Foundation Symposium 47 Elsevier/North-Holland Biomedical Press, Amsterdam/New York, pp. 379–400.

Swaab, D.F., Purba, J.S. and Hofman, M.A. (1995) Alterations in the hypothalamic paraventricular nucleus and its oxytocin neurons (putative satiety cells) in Prader–Willi syndrome: a study of five cases. J. Clin. Endocrinol. Metab., 80: 573–579.

Tauscher, J., Pirker, W., Willeit, M., De Zwaan, M., Bailer, U., Neumeister, A., Asenbaum, S., Lennkh, C., Praschak-Rieder, N., Brücke, T. and Kasper, S. (2001) [123I]b-CIT and single photon emission computed tomography reveal reduced brain serotonin transporter availability in bulimia nervosa. Biol. Psychiatry, 49: 326–332.

Thannickal, T.C., Moore, R.Y., Nienhuis, R., Ramanathan, L., Gulyani, S., Aldrich, M., Cornford, M. and Siegel, J.M. (2000) Reduced number of hypocretin neurons in human narcolepsy. Neuron, 27: 469–474.

Thompson, C., Syddall, H., Rodin, I., Osmond, C. and Barker, D.J.P. (2001) Birth weight and the risk of depressive disorder in late life. Br. J. Psychiatry, 179: 450–455.

Thorpy, M. (2001) Current concepts in the etiology, diagnosis and treatment of narcolepsy. Sleep Med., 2: 5–17.

Tobias, E.S., Tolmie, J.L. and Stephenson, J.B. (2002) Cataplexy in the Prader–Willi syndrome. Arch. Dis. Child, 87(2): 170.

Tsai, T.-F., Armstrong, D. and Beaudet, A.L. (1999) Necdin-deficient mice do not show lethality or the obesity and infertility of Prader–Willi syndrome. Nat. Gen., 22: 15–16.

Turek, F.W., Joshu, C., Kohsaka, A., Lin, E., Ivanova, G., McDearmon, E., Laposky, A., Losee-Olson, S., Easton, A., Jensen, D.R., Eckel, R.H., Takahashi, J.S. and Bass, J. (2005) Obesity and metabolic syndrome in circadian clock mutant mice. Science, 308: 1043–1045.

Ukkola, O., Ravussin, E., Jacobson, P., Snyder, E.E., Chagnon, M., Sjöström, L. and Bouchard, C. (2001) Mutations in the preproghrelin/ghrelin gene associated with obesity in humans. J. Clin. Endocrinol. Metab., 86: 3996–3999.

Van den Pol, A.N. (2000) Narcolepsy: a neurodegenerative disease of the hypocretin system? Neuron, 27: 415–418.

Van der Ploeg, L.H.T. (2000) Obesity: an epidemic in need of therapeutics. Curr. Opin. Chem. Biol., 4: 452–460.

Van Londen, L., Goekoop, J.G., Kerkhof, G.A., Zwinderman, K.H., Wiegant, V.M. and De Wied, D. (2001) Weak 24-h periodicity of body temperature and increased plasma vasopressin in melancholic depression. Eur. Neuropsychopharmacol., 11: 7–14.

Van Londen, L., Goekoop, J.G., Van Kempen, G.M.J., Frankhuijsen-Sierevogel, A.C., Wiegant, V.M., Van der Velde, E.A. and De Wied, D. (1997) Plasma levels of arginine vasopressin elevated in patients with major depression. Neuropsychopharmacology, 17: 284–292.

Van Londen, L., Kerkhof, G.A., Van denBerg, F., Goekoop, J.G., Zwinderman, K.H., Frankhuijzen-Sierevogel, A.C., Wiegant, V.M. and De Wied, D. (1998b) Plasma arginine vasopressin and motor activity in major depression. Biol. Psychiatry, 43: 196–204.

Van Weel-Baumgarten, E.M., Schers, H.J., Van den Bosch, W.J., Van den Hoogen, H.J. and Zitman, F.G. (2000) Long-term follow-up of depression among patients in the community and in family practice settings. A systematic review., J. Fam. Pract., 49: 1113–1120.

Verbalis, J.G., Blackburn, R.E., Hoffman, G.E. and Stricker, E.M. (1995) Establishing behavioral and physiological functions of central oxytocin: insights from studies of oxytocin and ingestive behaviors. Adv. Exp. Med. Biol., 395: 209–225.

Verdoux, H. and Sutter, A.-L. (2002) Perinatal risk factors for schizophrenia: diagnostic specificity and relationships with maternal psychopathology. Am. J. Med. Genet., 114: 898–905.

Vescovi, P.P., Rastelli, G., Volpi, R., Chiodera, P., Di Gennaro, C. and Coiro, V. (1996) Circadian variations in plasma ACTH, cortisol and b-endorphin levels in normal-weight bulimic women. Neuropsychobiology, 33: 71–75.

Vink, T., Hinney, A., Van Elburg, A.A., Van Goozen, S.H.M., Sandkuijl, L.A., Sinke, R.J., Herpertz-Dahlmann, B.-M., Hebebrand, J., Remschmidt, H., Van Engeland, H. and Adan, R.A.H. (2001) Association between an agouti-related protein gene polymorphism and anorexia nervosa. Mol. Psychiatry, 6: 325–328.

Visser, M. and Swaab, D.F. (1979) Life span changes in the presence of a-melanocyte-stimulating-hormone-containing cells in the human pituitary. J. Dev. Physiol., 1: 161–178.

Von Bardeleben, U. and Holsboer, F. (1989) Corticol response to a combined dexamethasone-human corticotrophin-releasing hormone challenge in patients with depression. J. Neuroendocrinol., 1: 485–488.

Walsh, B.T. and Devlin, M.J. (1998) Eating disorders: progress and problems. Science, 280: 1387–1390.

Wang, L.-N. and Huang, K.W. (1991) Hypothalamic encephalitis with oligodendrocytic glial nodules. Chin. Med. J., 104: 428–431.

Warnes, K.E., Morris, M.J., Symonds, M.E., Phillips, I.D., Clarke, I.J., Owens, J.A. and McMillen, I.C. (1998) Effects of increasing gestation, cortisol and maternal undernutrition on hypothalamic neuropeptide Y expression in the sheep fetus. J. Neuroendocrinol., 10: 51–57.

Wauters, M., Mertens, I., Chagnon, M., Rankinen, T., Considine, R.V., Chagnon, Y.C., Van Gaal, L.F. and Bouchard, C. (2001) Polymorphisms in the leptin receptor gene, body composition and fat distribution in overweight and obese women. Int. J. Obesity, 25: 714–720.

Weltzin, T.E., McConaha, C., McKee, M., Hsu, L.K.G., Perel, J. and Kaye, W.H. (1991) Circadian patterns of cortisol, prolactin, and growth hormonal secretion during bingeing and vomiting in normal weight bulimic patients. Biol. Psychiatry, 30: 37–48.

Wharton, R.H. and Bresman, M.J. (1989) Neonatal respiratory depression and delay in diagnosis in Prader–Willi syndrome. Dev. Med. Child Neurol., 31: 231–236.

Wicki, W., Angst, J. and Merikangas, K.R. (1992) Epidemiology of seasonal depression. Eur. Arch. Psychiatry Clin. Neurosci., 241: 301–306.

Wileman, S.M., Eagles, J.M., Andrew, J.E., Howie, F.L., Cameron, I.M., McCormack, K. and Naji, S.A. (2001) Light therapy for seasonal affective disorder in primary care. Br. J. Psychiatry, 178: 311–316.

Wirz-Justice, A. (1995) Biological rhythms in mood disorders. In: Bloom, F.E. and Kupfer, D.J. (Eds.), Psychopharmacology: the Fourth Generation of Progress. Raven Press, New York, pp. 999–1017.

Wolkowitz, O.M., Reus, V.I., Chan, T., Manfredi, F., Raum, W., Johnson, R. and Canick, J. (1999) Antiglucocorticoid treatment of depression: double-blind ketoconazole. Biol. Psychiatry, 45: 1070–1074.

Yamada, N., Martin-Iverson, M.T., Daimon, K., Tsujimoto, T. and Takahashi, S. (1995) Clinical and chronobiological effects of light therapy on nonseasonal affective disorders. Biol. Psychiatry, 37: 866–873.

Yamatsuji, M., Yamashita, T., Arii, I., Taga, C., Tatara, N. and Fukui, K. (2003) Seasonal variations in eating disorder subtypes in Japan. Int. J. Eat. Disord., 33: 71–77.

Zhou, J.N., Riemersma, R.F., Unmehopa, U.A., Hoogendijk, W.J., Van Heerikhuize, J.J., Hofman, M.A. and Swaab, D.F. (2001) Alterations in arginine vasopressin neurons in the suprachiasmatic nucleus in depression. Arch. Gen. Psychiatry, 58: 655–662.

Kalsbeek, Fliers, Hofman, Swaab, Van Someren & Buijs
Progress in Brain Research, Vol. 153
ISSN 0079-6123

CHAPTER 2

Synaptic plasticity mediating leptin's effect on metabolism

Tamas L. Horvath*

Section of Comparative Medicine, and Departments of Ob./Gyn. & Reproductive Sciences and Neurobiology, Yale University School of Medicine, 375 Congress Ave LSOG 339, New Haven, CT 06520, USA

Introduction

For over a century now, increasingly sophisticated methods have been brought to bear on the problem of brain involvement in the physiology of energy homeostasis and the pathogenesis of obesity. A vast number of experimental observations have been produced and, particularly within the last decade, the combination of novel genetic with sophisticated physiological techniques has allowed for great progress in the identification of metabolic hormones and their relationship to key peptidergic systems in the hypothalamus. While the central integration of afferent signals reflecting acute and chronic energy requirements becomes clearer, the neuronal pathways that actually initiate changes in ingestive behavior or energy expenditure are still largely unknown as is our understanding of the fine signaling modality of central body-weight regulation. It is our discovery during the past 3 years that there is a rapid synaptic remodeling involving hypothalamic peptidergic systems; this observation may shed new light on the mechanism of the central regulation of metabolism and offer explanations as to why, despite the breadth of knowledge gained in the past decades, no cure has been developed for metabolic disorders.

As in all mechanisms controlling physiological processes that are, essentially, of evolutionary character, the brain plays a critical role in the regulation of energy homeostasis. Central nervous circuits assess and integrate peripheral metabolic, endocrine, and neuronal signals reflecting current energy status, to then orchestrate a modulating influence on both behavioral patterns and peripheral metabolism according to acute and chronic requirements (Spiegelman and Flier, 2001).

Information processing is via neurons in the brain. The signal flow within the brain that underlies metabolism regulation is a highly complex process and is based on neuronal interactions. Neurons interact with each other by synapses that are established between axon terminals and dendritic or perikaryal membranes. The information moves from axon terminal to the dendritic or perikaryal membranes. The information transmission in synapses occurs either electrically, chemically via the release of neurotransmitters or modulators from synaptic vesicles of the axon terminal (neurotransmitters and neuromodulators), or by the release of gaseous substances such as nitric oxide or carbon monoxide. In most cases, neurons are capable of signaling by all the three ways, while one mode of transmission might dominate depending on the action potential as well as on substances that may directly signal to the axon terminal from the extracellular space (i.e., other neuromodulators released by nearby axons or

*Corresponding author. Tel.: +1-203-785-4597; Fax: +1-203-785-4747.; E-mail: tamas.horvath@yale.edu

DOI: 10.1016/S0079-6123(06)53002-X

47

available by volume transmission, extracellular anion and cation concentrations, or substances released to the extracellular space by paracrine or endocrine processes). The specificity of signal transduction is insured by the appropriate connectivity within a given network as well as by the availability of receptors at the right sites for released neurotransmitters or neuromodulators as well as for peripheral metabolic hormones. It is reasonable to suggest that neurons and their interplay with each other hold a key to the understanding of metabolism regulation. Thus, the establishment of the connectivity and hierarchical relationship between hypothalamic neuromodulator systems is critical to understanding the blueprint for the hypothalamic regulation of daily energy homeostasis. Much progress has been made in the past decade to accomplish this goal. The schematic illustration in Fig. 1 summarizes some of the critical findings of several laboratories, including our own.

We studied the relationship between key peptidergic circuits and their inputs in the nonhuman primate hypothalamus to establish whether data gained in rodent species correspond to primate anatomy. These studies are still ongoing, but, by and large, it appears that the basic wiring of the hypothalamus in nonhuman primates, and most likely in humans as well, does correspond to that found in either rats or mice. We also aimed to reveal whether alterations in the wiring of hypothalamic metabolic circuits occur in response to the changing metabolic state. Strikingly, our studies conducted in the nonhuman primate hypothalamus revealed a robust and rapid rearrangement of synaptic inputs of orexigenic circuits and their respective interneuronal controllers, in response to short-term fasting. Fasting resulted in a balance of stimulatory inhibitory synapses on orexin and neuropeptide Y (NPY) neurons that favored increased activity of these neurons. These cells have been implicated as key orexigenic neurons in the hypothalamus (Elmquist et al., 1998; Kalra et al., 1999). On the other hand, putative inhibitory interneurons of the same regions (neurons that would inhibit either orexin or NPY neuronal activity), exhibited a synaptic balance during fasting that support neuronal inactivation,

thereby further enhancing the activity levels of orexin and NPY perikarya. These observations raised the intriguing possibility that metabolic signals, leptin in particular, may have an acute effect on synaptic plasticity within the appetite center. The fact that the hypothalamus is not hardwired, i.e., it goes through continuous synaptic reorganization, is not novel. Rapid rearrangement of synapses have been shown to occur in the magnocellular system during changes in water homeostasis (Theodosis et al., 1991; Miyata et al., 1994; Stern and Armstrong, 1998;), the arcuate nucleus–interneuronal system during changes in the gonadal steroid milieu (Garcia-Segura et al., 1986; Parducz et al., 2003) and on the perikarya of luteinizing hormone-releasing hormone neurons during changes in the gonadal steroid milieu (Zsarnovszky et al., 2001) or during changes in photoperiod lengths (Xiong et al., 1997). However, such synaptic plasticity has never been considered as a critical component in the regulation of daily energy homeostasis. Our observations now suggest that synaptic plasticity is a key component in the physiological regulation of energy homeostasis, and that under pathological conditions, the synaptic constellation and its plasticity is impaired.

Leptin is a key metabolic signal associated with the rapid rewiring of hypothalamic pathways

The rapid rearrangement of synapses during fasting in monkeys coincided with diminished circulating levels of leptin (Diano et al., 2003). This, together with the fact that both NPY and orexin neurons express leptin receptors in the nonhuman primate (Horvath et al., 1999), suggested that leptin may be an important contributor to the observed synaptic plasticity during changing metabolic states. We then carried out a series of studies to directly test that proposition. For its critical importance, this endeavor became the central focus of our project.

In order to test our hypothesis, we turned to a rodent model. We did this for two pragmatic reasons: administration or replacement of leptin to fasted nonhuman primates would have been difficult and would have had a number of confounding

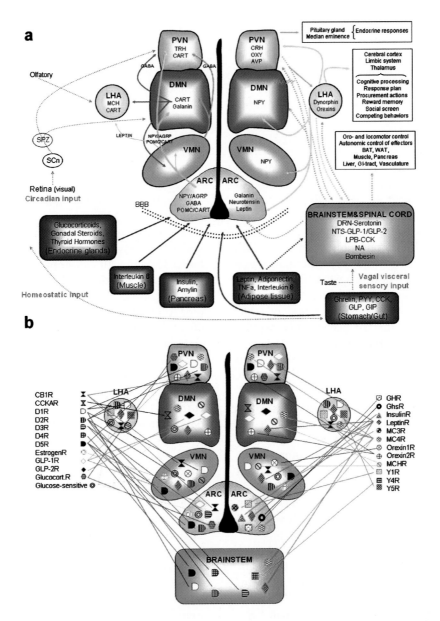

Fig. 1. (a) Schematic overview of the central nervous system circuitry, neuropeptides and neurotransmitters, which are involved in the control of appetite and body fat. It should be noted that this schematic overview is an extremely simplified version. (b) Schematic overview of the distribution of receptors and sensors in the hypothalamus and brainstem that are involved in the control of appetite and body fat. It should be noted that this schematic overview is an extremely simplified version. Cannabinoid receptor-1 (CB1R), cholecystokinin-A receptors (CCKAR), dopamine receptors (1–5) (D1R, D2R, D3R, D4R, and D5R), estrogen receptors (EstrogenR), glucagon-like peptide-1 receptors (GLP-1R), glucagon-like peptide-2 receptors (GLP-2R), glucocorticoid receptors (Glucocort.R), glucose-sensing neurons (glucose-sensitive), growth-hormone receptor (GHR), ghrelin receptor (GhsR), insulin receptors (InsulinR), leptin receptors (LeptinR), melanocortin receptors (MC3R), melanocortin receptor 4 (MC4R), orexin/hypocretin receptor-1 (Orexin1R), orexin/hypocretin receptor-2 (Orexin2R), melanin-concentrating hormone (MCHR), NPY/PYY/PP-receptors 1, 2, 4, and 5) (Y1R, Y2R, Y4R, and Y5R).

factors. For example, there are several other hormones that change during fasting. The use of appropriate controls for these hormones (adrenal, thyroid, gonadal hormones, etc.) would have inflated the number of primates required for the study which would have been unjustifiable either ethically or financially. Instead, we turned our attention to *ob/ob* mice and their wild-type littermates. These mice lack the leptin gene, and, thus, the phenotype of these animals resemble morbid human obesity (Zhang et al., 1994; Campfield et al., 1995; Halaas et al., 1995). Replacement of leptin to *ob/ob* animals rapidly decreases food intake and triggers weight loss (Zhang et al., 1994; Halaas et al., 1995; Campfield et al., 1995). Thus, *ob/ob* mice and their wild-type littermates presented a great model in which we could determine whether the presence of leptin predicts a different wiring pattern of hypothalamic peptidergic circuits. We aimed to analyze two distinct populations of neurons in the arcuate nucleus: one that produces NPY/AgRP and the other that expresses pro-opiomelanocortin (POMC). Arcuate nucleus neurons that coproduce NPY and AgRP (Hahn et al., 1998) are key orexigenic cells, which interact with those local cells that express the POMC-derived peptide, α-melanocyte-stimulating hormone (α-MSH), the most potent anorexigenic peptide, and are considered a *primum movens* of metabolism regulation by the brain (Fan et al., 1997; Barsh and Schwartz, 2002). The leptin receptor, LRb, is coexpressed with both neuronal subtypes (Mercer et al., 1996; Baskin et al., 1999; Lin et al., 2000). Increased NPY/AgRP activity and suppressed POMC tone is thought to underlie feeding and fat deposition. In contrast, increased POMC tone and suppressed NPY/AgRP activity support decreased feeding and lean body mass (Zarjevski et al., 1993; Stephens et al., 1995). In line with this, the *ob/ob* mouse expresses increased NPY and decreased POMC (Schwartz et al., 1997; Mizuno et al., 1998). Thus, the perikarya of these neurons in *ob/ob* mice and their wild-type littermates represented a unique model in which we could test our hypothesis regarding synaptic plasticity and the effect of leptin upon it. While our studies focused on the arcuate nucleus–melanocortin system, it is important to emphasize that energy-metabolism regulation from the brain is organized from various sites, which is not limited to the hypothalamus, but also includes other regions, most notably, the brain stem (Schwartz, 2000; Grill and Kaplan, 2002).

Leptin-deficient *ob/ob* animals have altered synaptology and electrophysiological properties in the arcuate nucleus

We analyzed transgenic animals generated by the laboratory of Dr. Jeffrey Friedman at Rockefeller University in which *tau*-sapphire GFP is expressed under the transcriptional control of the NPY genomic sequence or *tau*-topaz GFP is expressed under the transcriptional control of POMC genomic sequence. We first examined the afferent inputs to POMC and NPY neurons in the Arc of *ob/ob* and wild-type animals utilizing patch-clamp electrophysiology recording in slice preparations. NPY-GFP or POMC-GFP cells were held in the whole-cell voltage-clamp configuration and the number of excitatory and inhibitory postsynaptic currents (EPSC/IPSC) was determined. In wild-type animals, NPY neurons had similar levels of spontaneous excitatory postsynaptic currents (sEPSC) and inhibitory postsynaptic currents (sIPSC). The inputs onto POMC neurons were principally IPSCs. We next compared the afferent inputs onto NPY and POMC neurons in the *ob/ob* mouse relative to those seen in the wild-type mice. The electrophysiological recordings revealed a large shift in the ratio of EPSC to IPSCs onto both the NPY and POMC neurons in the *ob/ob* slices. NPY neurons from *ob/ob* mice showed a significant increase in the frequency of EPSCs combined with a significant decrease in the frequency of IPSCs. Similarly, there was a large shift in the ratio of inputs onto POMC neurons from *ob/ob* mice with a robust increase in the frequency of IPSCs. While there were no alterations in the frequency of EPSCs onto the POMC neurons, there was a marked net increase in inhibitory tone onto these neurons. Thus, in hypothalamic slices from *ob/ob* mice, there were reciprocal effects on NPY and POMC neurons with a marked increase in excitatory tone onto the NPY neurons and a net decrease in excitatory tone onto the POMC neurons. To assess whether the number of synaptic inputs to the NPY and POMC neurons correspond to the above electrophysiological properties, we used stereology to quantify the synaptic density on NPY and POMC perikarya from *ob/ob* and wild-type animals. In *ob/ob* animals, there was a significantly higher total number of synapses on the perikarya of NPY neurons compared to the wild-type littermates. This value was fully accounted for by an increase in the number of excitatory synapses onto these cells that were more numerous than the inhibitory ones in *ob/ob* mice. In contrast, in the wild-type animals, inhibitory synapses onto the NPY neurons were more numerous than the excitatory ones. This altered synaptic profile of NPY cells in the *ob/ob* animals is entirely consistent with their electrophysiological profile. In the POMC cells of wild-type mice, excitatory synapses dominated over the inhibitory contacts, while in *ob/ob* cells, the opposite was observed. In addition, the total number of synapses on the POMC neurons was lower in the *ob/ob* mice compared to the wild-type littermates. Here too, the excitatory/inhibitory synaptic balance is in agreement with the increased inhibitory tone seen on these neurons in slices from the *ob/ob* hypothalami described above.

Leptin induces rapid rewiring of arcuate nucleus-feeding circuits in *ob/ob* mice

A single dose of leptin reduces food intake in *ob/ob* mice within 12 h (Pelleymounter et al., 1995). Groups of *ob/ob* animals were treated with leptin or PBS for 6 h, 48 h, or 12 days. After 6 h of leptin treatment, there was a significant decrease in the number of excitatory inputs and a significant increase in the number of inhibitory inputs onto the NPY neurons of *ob/ob* mice. In contrast, there was a significant increase in the number of excitatory inputs onto the POMC neurons of *ob/ob* mice. The statistical strength in these changes increased after 48 h and 12 days. At 48 h after leptin replacement, we also found that the electrophysiological properties of *ob/ob* NPY/AgRP and POMC neurons shifted toward the wild-type values. Both the electrophysiological and anatomical changes were positively correlated with the changing food intake and body weight gain of the same animals. These data show that leptin has potent and rapid effects on the wiring of key neurons in the hypothalamus. The fact that these synaptic changes that occur prior to changes in feeding behavior and body weight can be detected statistically (albeit these parameters are changing) suggests the possibility that the rapid leptin-induced rewiring of the synaptic inputs to the NPY and POMC cells in *ob/ob* mice may be a prerequisite for some portion of its behavioral effects (Fig. 2).

What is the mechanism of action of leptin in triggering synaptic plasticity?

The molecular basis for leptin-mediated plasticity in the hypothalamus remains elusive. We predict that activation of the long-form leptin receptor is an important step in triggering synaptic plasticity by leptin. Leptin action through the long-form leptin receptor requires the activation of the STAT 3 (Vaisse et al., 1996; Bates et al., 2003; Gao et al., 2004) and/or the PI3 kinase-signaling pathways in order to decrease food intake and increase energy expenditure (Niswender et al., 2001, 2003). Thus, it will be important to test whether leptin-induced synaptic plasticity is impaired in transgenic animals that express the long form of leptin receptor, but lack either the entire STAT 3 gene, the tyrosine residue of the long form of leptin receptor that is critical to the activation of STAT 3. Tyr 1138 of LRb mediates activation of the transcription factor STAT 3 during leptin action, but apparently, activation of STAT 3 is not required for all of leptin's action, including that relating to fertility (Bates et al., 2003). In addition, we will analyze animals in which the STAT 3 and PI3 kinase pathway is selectively altered in POMC neurons. We will analyze the synaptology and electrophysiological properties of hypothalamic POMC neurons. All of the wild-type and transgenic animals will carry the transgene for visualization of POMC cells with the aid of GFP. Additionally, in order to better understand the kinetics of leptin-induced

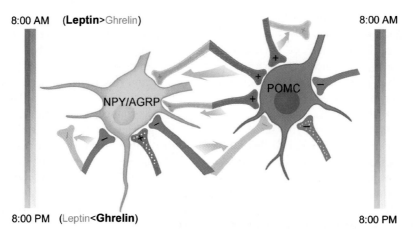

Fig. 2. Perikaryal inputs of NPY/AGRP neurons are dominated by inhibitory (-) connections when circulating leptin levels are high. These connections are rearranged (arrows indicate changes and lighter outlines indicate new locations) when leptin levels diminish (and ghrelin levels increase). Under these circumstances, stimulatory synapses (+) dominate over inhibitory inputs. On the POMC perikarya, the changes occur in the opposite direction of that described for the NPY/AGRP inputs. Some of the inhibitory inputs on the POMC cells are likely to originate from the NPY/AGRP neurons, and some of the stimulatory inputs on both cell types originate in the lateral hypothalamic hypocretin neurons (white dots in + axons). Because rapid synaptic changes were observed in wild-type animals, it is reasonable to propose that synaptic rearrangement of feeding circuits is a continuous phenomenon, which will also occur in a circadian fashion responding to the changing daily metabolic environment.

synaptic plasticity, we are now analyzing multiple time points and dose responses to determine the long-lasting influence of a single leptin injection on synaptology.

Synaptic plasticity may be triggered by leptin, but it is obviously the consequence of transcriptional regulation of genes, the products of which are involved in membrane organization and maintenance, receptor trafficking, and recycling as well as in cell–cell interactions. It will be critical to determine members of these protein families that are involved with leptin-induced synaptic plasticity. While most likely there will be dozens of involved genes, our initiation of these investigations is not without probable candidates. For example, we have been studying synaptic plasticity in the arcuate nucleus in relation to the gonadal steroid, estradiol (Horvath et al., 1995, 1997a, b, 1999; Naftolin et al., 1995, 1996; Leranth et al., 2000; Parducz et al., 2003; Hung et al., 2003), from which, we and others (Theodosis et al., 1991, 1998) have inferred that one of the most abundant cell adhesion molecules in the arcuate nucleus is the polysialic neuronal-cell adhesion molecule (PSA-NCAM). It is reasonable to assume that similar to estradiol-induced synaptic plasticity, leptin's effect

will also alter PSA-NCAM in the arcuate nucleus. We base this prediction on the following observations: (1) leptin receptors are colocalized with estradiol receptors (Diano et al., 1998); (2) estradiol mimics leptin's synaptic effect on melanocortin cells (unpublished data); (3) translocation of PSA-NCAM from the cytosol to the perykaryal membrane is associated with the removal of GABA-ergic synapses (our unpublished observation); and (4) both leptin and estradiol can cause rapid synaptic rearrangement in the arcuate nucleus involving GABA synapses (Parducz et al., 2003).

Another mechanism that is likely to play a role in the regulation of symmetrical synapses in the melanocortin system is phospholipase C beta 1 (PLCβ-1). This signaling pathway is implicated in activity-dependent cortical development, with particular emphasis on symmetrical synaptic connections (Spires et al., 2004). We have shown that this mechanism is also expressed in the adult melanocortin system, and, it is likely to explain the lean phenotype of PLCβ-1-knockout animals (Spires et al., 2005).

Regarding stimulatory synapses, we have recently revealed an important role for *cpg-2*, an activity-regulated gene that encodes a membrane-bound ligand that regulates growth of apposing dendritic and axonal arbors and the maturation of their synapses in excitatory synapse formation of the arcuate nucleus. Previously, we have shown and published (Cottrell et al., 2004) that CPG-2 decreases recycling NMDAR receptors, thus strengthening asymmetrical, stimulatory contacts. We also predicted that the ability of the adult hypothalamus to maintain plasticity must have transcriptional regulators such as homeodomain genes, which are associated with immature, developing brain. Strikingly, we recently revealed adult hypothalamic expression of the *orthopedia* gene product, OTP, a homeodomain gene that is critical for pattern formation of the hypothalamus (Acampora et al., 1999). We found it to be expressed in distinct peptidergic systems, including the NPY/AgRP neurons. We predict that the expression level of OTP will have correlation with leptin and metabolic status.

Note that the aforementioned are only examples of candidate molecules in mediating leptin's effect on arcuate nucleus synaptic plasticity. In order to acquire a broader picture, experiments will be needed to screen the arcuate nucleus of animals with metabolic disorders and wild-type animals for expression patterns of genes belonging to these families of genes and to determine which are affected by leptin signaling. To this end, we are also testing those animals that develop diet-induced obesity.

Are the synaptic organization and electrophysiological properties of the hypothalamic POMC neurons altered in mice with diet-induced obesity?

The observations on nonhuman primates, *ob/ob* and wild-type animals clearly show that leptin's effect in the hypothalamus is associated with synaptic organization of hypothalamic peptidergic systems. If synaptic plasticity is a mandatory element for leptin to influence food intake and energy expenditure, then such plasticity must be impaired during leptin resistance which develops during diet-induced obesity (Frederich et al., 1995; Halaas et al., 1997). Thus, we are testing this hypothesis

by analyzing the electrophysiological properties and synaptology of NPY/AgRP, POMC, orexin, and MCH neurons in GFP transgenic animals at various time points during the course of diet-induced obesity development and in their control littermates.

Relevance to human health

Chronically increased energy intake without a respective increase in energy expenditure leads to obesity and diabetes as well as a variety of life-threatening consequences of diseases such as cancer and cardiovascular diseases (Eckel, 1997; Calle et al., 2003). While it appears intuitively obvious, that in the majority of cases, positive energy balance should be corrected by changes in lifestyle and/or diet, the impressive dynamics of the spreading obesity epidemic (Flegal et al., 2002) certainly suggests that, in modern industrialized civilizations, an efficient and safe pharmacological approach to treat obesity would be useful. Based on our preliminary data, we believe that synaptic plasticity induced by leptin, and its impairment during pathological conditions, is fundamental in the regulation of energy homeostasis.

Acknowledgment

The work described in this review has been supported by the NIH grants DK-060711 and DK-01445.

References

Acampora, D., Postiglione, M.P., Avantaggiato, V., Di Bonito, M., Vaccarino, F.M., Michaud, J. and Simeone, A. (1999) Progressive impairment of developing neuroendocrine cell lineages in the hypothalamus of mice lacking the *Orthopedia* gene. Genes Dev., 13: 2787–2800.

Barsh, G.S. and Schwartz, M.W. (2002) Genetic approaches to studying energy balance: perception and integration. Nat. Rev. Genet., 3: 589–600.

Baskin, D.G., Breininger, J.F. and Schwartz, M.W. (1999) Leptin receptor mRNA identifies a subpopulation of neuropeptide Y neurons activated by fasting in rat hypothalamus. Diabetes, 48: 828–833.

54

Bates, S., Stearns, W., Dundon, T., Schubert, M., Tso, A., Wang, Y., Banks, A., Lavery, H., Haq, A., Maratos-Flier, E., Neel, B.G., Schwartz, M.W. and Myers Jr., M.G. (2003) STAT3 signalling is required for leptin regulation of energy balance but not reproduction. Nature, 421: 856–859.

Calle, E.E., Rodriguez, C., Walker-Thurmond, K. and Thun, M.J. (2003) Overweight, obesity, and mortality from cancer in a prospectively studied cohort of U.S. adults. N. Engl. J. Med., 348: 1625–1638.

Campfield, L.A., Smith, F.J., Guisez, Y., Devos, R. and Burn, P. (1995) Recombinant mouse OB protein: evidence for a peripheral signal linking adiposity and central neural networks. Science, 269: 546–549.

Cottrell, J.R., Borok, E., Horvath, T.L. and Nedivi, E. (2004) CPG2: a brain- and synapse-specific protein that regulates the endocytosis of glutamate receptors. Neuron, 44: 677–690.

Diano, S., Horvath, B., Urbanski, H.F., Sotonyi, P. and Horvath, T.L. (2003) Fasting activates the nonhuman primate hypocretin (orexin) system and its postsynaptic targets. Endocrinology, 144: 3774–3778.

Diano, S., Kalra, S.P., Sakamoto, H. and Horvath, T.L. (1998) Leptin in estrogen receptor-containing neurons of the female rat hypothalamus. Brain Res., 812: 256–259.

Eckel, R.H. (1997) Obesity and heart disease: a statement for healthcare professionals from the Nutrition Committee, American Heart Association. Circulation, 96: 3248–3250.

Elmquist, J.K., Maratos-Flier, E., Saper, C.B. and Flier, J.S. (1998) Unraveling the central nervous system pathways underlying responses to leptin. Nat. Neurosci., 1: 445–450.

Fan, W., Boston, B.A., Kesterson, R.A., Hruby, V.J. and Cone, R.D. (1997) Role of melanocortinergic neurons in feeding and the agouti obesity syndrome. Nature, 385: 165–168.

Flegal, K.M., Carroll, M.D., Ogden, C.L. and Johnson, C.L. (2002) Prevalence and trends in obesity among US adults, 1999–2000. JAMA, 288: 1723–1727.

Frederich, R.C., Hamann, A., Anderson, S., Lollmann, B., Lowell, B.B. and Flier, J.S. (1995) Leptin levels reflect body lipid content in mice: evidence for diet-induced resistance to leptin action. Nat. Med., 1: 1311–1314.

Gao, Q., Wolfgang, M.J., Neschen, S., Morino, K., Horvath, T.L., Shulman, G.I. and Fu, X.-Y. (2004) Disruption of neural STAT3 causes obesity, diabetes, infertility and thermal dysregulation. PNAS, 101: 4661–4666.

Garcia-Segura, L., Baetens, D. and Naftolin, F. (1986) Synaptic remodelling in arcuate nucleus after injection of estradiol valerate in adult female rats. Brain Res., 366: 131–136.

Grill, H.J. and Kaplan, J.M. (2002) The neuroanatomical axis for control of energy balance. Front. Neuroendocrinol., 23: 2–40.

Hahn, T.M., Breininger, J.F., Baskin, D.G. and Schwartz, M.W. (1998) Coexpression of Agrp and NPY in fasting-activated hypothalamic neurons. Nat. Neurosci., 1: 271–272.

Halaas, J.L., Gajiwala, K.S., Maffei, M., Cohen, S.L., Chait, B.T., Rabinowitz, D., Lallone, R.L., Burley, S.K. and Friedman, J.M. (1995) Weight-reducing effects of the plasma protein encoded by the *obese* gene. Science, 269: 543–546.

Halaas, J.L., Boozer, C., Blair-West, J., Fidahusein, N., Denton, D.A. and Friedman, J.M. (1997) Physiological response to long-term peripheral and central leptin infusion in lean and obese mice. Proc. Natl. Acad. Sci. USA, 94: 8878–8883.

Horvath, T.L., Garcia-Segura, L.M. and Naftolin, F. (1997a) Lack of gonadotrophin positive feedback in the male rat predicts lack of estrogen-induced synaptic plasticity in the arcuate nucleus. Neuroendocrinology, 65: 136–141.

Horvath, T.L., Garcia-Segura, L.M. and Naftolin, F. (1997b) Control of gonadotrophin feedback: the role of estrogen-induced hypothalamic synaptic plasticity. Gynec. Endocrin., 11: 139–143.

Horvath, T.L., Leedom, L., Garcia-Segura, L.M., Naftolin, F. (1995) Estrogen-induced hypothalamic synaptic plasticity; implications for the regulation of gonadotrophins. In: Smith, M.S. (Ed.) Current Opinion in Endocrinology and Diabetes, Lippincott Williams & Wilkins (LWW), Vol. 2. pp. 186–190.

Horvath, T.L., Diano, S. and Van den Pol, A.N. (1999) Synaptic interaction between hypocretin (orexin) and NPY cells in the rodent and primate hypothalamus — a novel hypothalamic circuit implicated in metabolic and endocrine regulations. J Neurosci., 19: 1072–1087.

Hung, A.J., Stanbury, M.G., Shanabrough, M., Horvath, T.L., Garcia-Segura, L.M. and Naftolin, F. (2003) Estrogen, synaptic plasticity and hypothalamic reproductive aging. Exp. Gerontol., 38: 53–59.

Kalra, S.P., Xu, B., Dube, M.G., Pu, S., Horvath, T.L. and Kalra, P.S. (1999) Interacting appetite regulating pathways in the hypothalamic regulation of body weight. Endocrinol. Rev., 20: 67–100.

Leranth, C., Sahanbrough, M. and Horvath, T.L. (2000) Hormonal regulation of hippocampal morphology involves subcortical mediation. Neuroscience, 101: 349–356.

Lin, S., Storlien, L.H. and Huang, X. (2000) Leptin receptor, NPY, POMC mRNA expression in the diet-induced obese mouse brain. Brain Res., 875: 89–95.

Mercer, J., Hoggard, N., Williams, L., Lawrence, C., Hannah, L., Morgan, P. and Trayhurn, P. (1996) Coexpression of leptin receptor and preproneuropeptide Y mRNA in arcuate nucleus of mouse hypothalamus. J. Neuroendocrinol., 8: 733–735.

Miyata, S., Nakashima, T. and Kiyohara, T. (1994) Structural dynamics of neural plasticity in the supraoptic nucleus of the rat hypothalamus during dehydration and rehydration. Brain Res. Bull., 34: 169–175.

Mizuno, T.M., Kleopoulos, S.P., Bergen, H.T., Roberts, J.L., Priest, C.A. and Mobbs, C.V. (1998) Hypothalamic pro-opiomelanocortin mRNA is reduced by fasting in *ob/ob* and *db/db* mice, but is stimulated by leptin. Diabetes, 47: 294–297.

Naftolin, F., Mor, G., Horvath, T.L., Luquin, S., Fajer, A.B., Kohen, F. and Garcia-Segura, L.M. (1996) Synaptic remodeling in the arcuate nucleus during the estrus cycle is induced by estrogen and precedes the midcycle gonadotrophin surge. Endocrinology, 137: 5576–5580.

Naftolin, F., Leranth, C., Horvath, T.L. and Garcia-Segura, L.M. (1995) Potential neuronal mechanisms of estrogen actions in synaptogenesis and synaptic plasticity. Cell. Mol. Neurobiol., 16: 213–223.

Niswender, K.D., Morrison, C.D., Clegg, D.J., Olson, R., Baskin, D.G., Myers Jr., M.G., Seeley, R.J. and Schwartz, M.W. (2003) Insulin activation of phosphatidylinositol 3-kinase in the hypothalamic arcuate nucleus: a key mediator of insulin-induced anorexia. Diabetes, 52: 227–231.

Niswender, K., Morton, G.J., Stearns, W.H., Rhodes, C.J., Myers Jr., M.G. and Schwartz, M.W. (2001) Intracellular signalling. Key enzyme in leptin-induced anorexia. Nature, 413: 794–795.

Parducz, A., Zsarnovszky, A., Naftolin, F. and Horvath, T.L. (2003) Estradiol affects axo-somatic contacts of neuroendocrine cells in the arcuate nucleus of adult rats. Neuroscience, 117: 791–794.

Pelleymounter, M.A., Cullen, M.J., Baker, M.B., Hecht, R., Winters, D., Boone, T. and Collins, F. (1995) Effects of the *obese* gene product on body weight regulation in *ob/ob* mice. Science, 269: 540–543.

Schwartz, G.J. (2000) The role of gastrointestinal vagal afferents in the control of food intake: current prospects. Nutrition, 16: 866–873.

Schwartz, M.W., Seeley, R.J., Woods, S.C., Weigle, D.S., Campfield, L.A., Burn, P. and Baskin, D.G. (1997) Leptin increases hypothalamic pro-opiomelanocortin mRNA expression in the rostral arcuate nucleus. Diabetes, 46: 2119–2123.

Spiegelman, B.M. and Flier, J.S. (2001) Obesity and the regulation of energy balance. Cell, 104: 531–543.

Spires, T.L., Molnar, Z., Kind, P.C., Cordery, P.M., Upton, A.L., Blakemore, C. and Hannan, A.J. (2005) Activity-dependent regulation of synapse and dendritic spine morphology in developing barrel cortex requires phospholipase C-β1 signalling. Cereb. Cortex, 15: 385–393.

Stephens, T.W., Bashinski, M., Bristow, P.K., Bue-Valleskey, J.M., Burgett, S.G., Hale, H., Hoffmann, J., Hsiung, H.M., Krauciunas, A., Mackellar, W., Rosteck, P.R., Schoner, B., Smith, D., Tinsley, F.C., Zhang, X.-Y. and Heiman, M. (1995) The role of neuropeptide Y in the antiobesity action of the *obese* gene product. Nature, 377: 530–534.

Stern, J. and Armstrong, W. (1998) Reorganization of the dendritic trees of oxytocin and vasopressin neurons of the rat supraoptic nucleus during lactation. J. Neurosci., 18: 841–853.

Theodosis, D., El Majdoubi, M., Pierre, K. and Poulain, D. (1998) Factors governing activity-dependent structural plasticity of the hypothalamoneurohypophysial system. Cell. Mol. Neurobiol., 18: 285–298.

Theodosis, D., Rougon, G. and Poulain, D. (1991) Retention of embryonic features by an adult neuronal system capable of plasticity: polysialylated neural cell adhesion molecule in the hypothalamo-neurohypophysial system. Proc. Natl. Acad. Sci. USA, 88: 5494–5498.

Vaisse, C., Halaas, J.L., Horvath, C.M., Darnell Jr., J.E., Stoffel, M. and Friedman, J.M. (1996) Leptin activation of Stat3 in the hypothalamus of wild-type and *ob/ob* mice but not *db/db* mice. Nat. Genet., 14: 95–97.

Xiong, J., Karsch, F. and Lehman, M. (1997) Evidence for seasonal plasticity in the gonadotropin-releasing hormone (GnRH) system of the ewe: changes in synaptic inputs onto GnRH neurons. Endocrinology, 138: 1240–1250.

Zarjevski, N., Cusin, I., Vettor, R., Rohner-Jeanrenaud, F. and Jeanrenaud, B. (1993) Chronic intracerebroventricular neuropeptide-Y administration to normal rats mimics hormonal and metabolic changes of obesity. Endocrinology, 133: 1753–1758.

Zhang, Y., Proenca, P., Maffei, M., Barone, M., Leopold, L. and Friedman, J.M. (1994) Positional cloning of the mouse *obese* gene and its human homologue. Nature, 372: 425–432.

Zsarnovszky, A., Horvath, T., Garcia-Segura, L., Horvath, B. and Naftolin, F. (2001) Oestrogen-induced changes in the synaptology of the monkey (*Cercopithecus aethiops*) arcuate nucleus during gonadotropin feedback. J. Neuroendocrinol., 13: 22–28.

Kalsbeek, Fliers, Hofman, Swaab, Van Someren & Buijs
Progress in Brain Research, Vol. 153
ISSN 0079-6123

CHAPTER 3

The hypothalamus, hormones, and hunger: alterations in human obesity and illness

Anthony P. Goldstone*

Imaging Sciences Department, MRC Clinical Sciences Centre, Faculty of Medicine, Imperial College, Hammersmith Hospital Campus, London W12 0NN, UK

Abstract: Obesity is a major global epidemic, with over 300 million obese people worldwide, and nearly 1 billion overweight adults. Being overweight carries significant health risks, reduced quality of life, and impaired socioeconomic success, with profound consequences for health expenditure. The most successful treatment for obesity is gastric bypass surgery, which acts in part by reducing appetite through alterations in gut hormones. Circulating gut hormones, secreted or suppressed after eating food, act in the brain, particularly the hypothalamus, to alter hunger and fullness. Stomach-derived ghrelin increases food intake even in those with anorexia from chronic illness, while pancreatic polypeptide (PP), intestinal peptide YY 3-36 (PYY), oxyntomodulin, and other hormones reduce food intake and appetite. While obese subjects have appropriate reductions in orexigenic ghrelin, other gut-hormone disturbances may contribute to obesity such as reduced anorexigenic PYY and PP. Prader–Willi syndrome (PWS) arises from the loss of paternally inherited genes on chromosome 15q11-13, leading to life-threatening insatiable hunger and obesity from early childhood, through developmental brain, particularly hypothalamic defects. The study of genetically homogenous causes of abnormal-feeding behavior helps our understanding of appetite regulation. PWS subjects have inappropriately elevated plasma ghrelin for their obesity, at least partly explained by preserved insulin sensitivity. It remains unproven if their hyperghrelinemia or other gut-hormone abnormalities contribute to the hyperphagia in PWS, in addition to brain defects. Postmortem human hypothalamic studies and generation of animal models of PWS can also provide insight into the pathophysiology of abnormal-feeding behavior. Changes in orexigenic NPY and AGRP hypothalamic neurons, or anorexigenic oxytocin neurons have been found in illness and PWS. Functional neuroimaging studies, using PET and fMRI, will also allow us to tease apart the hormonal and brain pathways responsible for controlling human appetite, and their defects in obesity.

Keywords: hypothalamus; appetite; obesity; Prader–Willi syndrome; ghrelin; PYY; pancreatic polypeptide; oxyntomodulin

Obesity

Obesity is a major worldwide health issue affecting developed and developing nations. Over 300 million people worldwide are obese, and nearly 1 billion adults are overweight (Deitel, 2002; Kimm and Obarzanek, 2002; Thibault and Rolland-Cachera, 2003). In England in 2003, 60% of adults were overweight, and 23% obese (Department of Health, 2004). Of particular concern is the rise of childhood obesity (Speiser et al., 2005). Over 30% of children in the United States are overweight or obese (Fox, 2003). More than two-thirds of children 10 years

*Tel.: +44-20-8383-1510; Fax: +44-20-8743-5409; E-mail: tony.gold@imperial.ac.uk

DOI: 10.1016/S0079-6123(06)53003-1

and older who are obese will become obese adults (Magarey et al., 2003; Must, 2003).

Being overweight and obese carries significant health risks, including type 2 diabetes mellitus, metabolic syndrome, hypertension, hyperlipidemia, cardiovascular disease, osteoarthritis, sleep apnea, and certain cancers. Obesity leads to increased mortality, reduced quality of life and impaired socioeconomic success for affected individuals, and has huge associated medical costs. Treatment of obesity prevents its medical complications, improves life expectancy and quality of life (Knowler et al., 2002; Avenell et al., 2004).

Increase in obesity is a result of environmental factors with increased dietary caloric and fat intake, and reduced physical activity (Hill and Peters, 1998; Koplan and Dietz, 1999), but underlying genetic predisposing factors are also evident (Perusse et al., 2005). Since food intake may be more important than physical activity for the subsequent development of obesity, unraveling the pathways involved in the control of appetite will be particularly vital for appropriate prevention and treatment strategies (Tataranni et al., 2003; Flier, 2004).

The most successful treatment for obesity is gastric bypass surgery, which appears to achieve at least part of its effect by reducing appetite through alterations in gut hormones (Le Roux and Bloom, 2005); however, this procedure carries significant risk. There is currently only one licensed drug that reduces appetite, sibutramine, for the treatment of obesity (Bray, 1999). There is also interest in the development of 'functional foods' that will modify appetite (Riccardi et al., 2005). However, the development of such anti-obesity therapies is hindered by the absence of reliable quantitative in vivo biomarkers for appetite in humans (De Graaf et al., 2004).

Gut hormones and appetite

Appetite is controlled by a variety of peripheral signals that change in response to food intake or starvation, which act in the brain to alter feelings of hunger and fullness so as to determine meal initiation ('hunger' or 'satiety') and meal termination ('satiation') (Bray, 2000; De Graaf et al., 2004;

Neary et al., 2004a; Badman and Flier, 2005) (Fig. 1). These signals may include a number of ascending neural inputs (e.g., vagus nerve signaling stomach distension), metabolic and hormonal changes, including plasma glucose and insulin, gastrointestinal release of anorexigenic hormones such as peptide YY3-36 (PYY3-36), pancreatic polypeptide (PP), glucagon-like peptide-1 (GLP-1), cholecystokinin (CCK) and oxyntomodulin, suppression of the orexigenic hormone ghrelin, and synergism with the anorexigenic adipocyte hormone, leptin.

Acute infusion of stomach-derived ghrelin increases food intake (Wren et al., 2001; Neary et al., 2004b; Druce et al., 2005; Wynne et al., 2005a), while intestinal PYY3-36, GLP-1 and oxyntomodulin, and PP reduce food intake and appetite in lean humans (Verdich et al., 2001; Batterham et al., 2002, 2003a, b; Cohen et al., 2003; Neary et al., 2005) (Fig. 2). There may also be roles for changes in gastric emptying or digestive secretions, in addition to alterations in appetite, for the effects of gut hormones on food intake and body weight (Allen et al., 1984; Adrian et al., 1985; Naslund et al., 1999; Verdich et al., 2001; Murray et al., 2005).

While absolute fasting plasma levels of orexigenic ghrelin are appropriately reduced in obesity, obese subjects may have a smaller relative decrease postprandially, and may have reduced fasting and/or postprandial secretion of anorexigenic PYY3-36 and PP (Lassmann et al., 1980; Batterham et al., 2003a; Goldstone et al., 2004, 2005; Neary et al., 2004a; Roth et al., 2005; Le Roux et al., 2005b, 2006) (Fig. 3). While these abnormalities may be a consequence of obesity, they may promote the maintenance of excess caloric intake by reducing satiety (Le Roux et al., 2006; Young, 2006).

Unlike leptin, sensitivity to the gastric-emptying, anorexigenic, and orexigenic effects of gut hormones remain intact in simple obesity (Lieverse et al., 1994; Verdich et al., 2001; Batterham et al., 2003a; Druce et al., 2005; Wynne et al., 2005b). Chronic administration of anorexigenic gut hormones or analogs are therefore potential long-term treatments for obesity, which may avoid problems of tachyphylaxis or significant side effects (Small and Bloom, 2004; Wynne et al., 2005b).

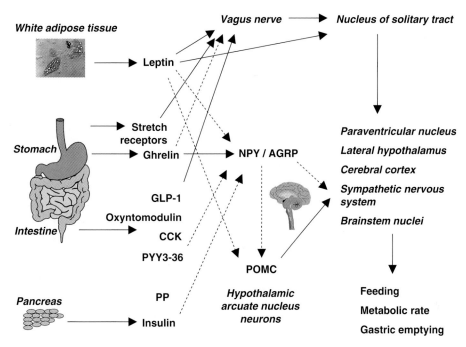

Fig. 1. Circulating gastrointestinal and adipocyte hormones and neural circuits involved in energy homeostasis. A solid line represents a net stimulatory effect and a dashed line represents a net inhibitory effect. Adapted with permission from (Neary et al., 2004a), copyright Blackwell Publishing.

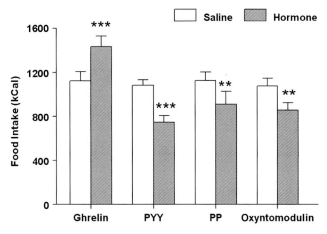

Fig. 2. Gut hormones regulate appetite in humans. Effect on food intake during test meal of acute intravenous infusion of ghrelin (increased 28%), PYY3-36 (decreased 31%), oxyntomodulin (decreased 19%), and pancreatic polypeptide (decreased 22%) in fed (ghrelin) and fasted (other hormones) lean human subjects. Data represent mean $\pm$ SEM: ** $P<0.01$, *** $P<0.001$ vs. saline. Data taken from Wren et al. (2001), Batterham et al. (2003a), Cohen et al. (2003), and Batterham et al. (2003b).

Gut hormones and downstream pathways

Rodent studies have revealed initial targets for the appetite effects of gut hormones as including the vagus nerve, brain stem (such as the nucleus of the solitary tract (NTS), area postrema, and ventral tegmental area), and the hypothalamus. These studies involve such methods as microinjection of

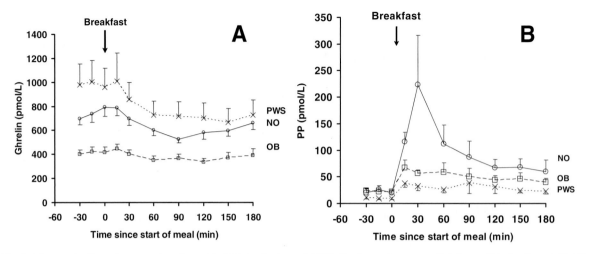

Fig. 3. Plasma ghrelin and pancreatic polypeptide (PP) in obesity and PWS. Fasting and postprandial plasma (A) ghrelin and (B) PP levels following a 522 kcal breakfast in non-obese (○, solid line, NO, $n = 8$), obese subjects (□, dashed line, OB, $n = 9$), and subjects with PWS (x, dotted line, $n = 10$). Data given as mean $\pm$ SEM. The postprandial area under curve (AUC) for ghrelin in OB is less than in both NO and PWS ($P < 0.01$), while the AUC for PP in PWS is less than in both NO ($P < 0.001$) and OB ($P < 0.05$). (A): adapted with permission from Goldstone et al. (2005), copyright The Endocrine Society; (B): from unpublished observations (Goldstone et al.).

peptides, postmortem neuropeptide, and *c-fos* peptide or mRNA expression as a marker of neuronal activation after peripheral hormone administration, brain and vagus nerve localization of hormone receptors, ex vivo and in vivo hormone effects on hypothalamic or vagal nerve electrophysiology, hypothalamic-specific receptor knockout mice, altered effects in lesioned animals, and more recently, manganese-enhanced magnetic resonance imaging (Bray, 2000; Batterham et al., 2002; Sainsbury et al., 2002; Asakawa et al., 2003; Cox and Randich, 2004; Dakin et al., 2004; Korbonits et al., 2004; Kuo et al., 2004; Neary et al., 2004a; Badman and Flier, 2005; Naleid et al., 2005).

Within the hypothalamus, ghrelin activates orexigenic neuropeptide Y (NPY) and agouti-related protein (AGRP) neurons, while inhibiting anorexigenic proopiomelanocortin (POMC) neurons in the arcuate (or infundibular) nucleus, in an opposite manner to anorexigenic leptin (Swaab, 2003a; Korbonits et al., 2004; Neary et al., 2004a; Badman and Flier, 2005) (Fig. 4). These neurons in turn project particularly to the paraventricular nucleus (PVN) and lateral hypothalamus, where neurons containing peptides such as anorexigenic oxytocin, CRH, or TRH and orexigenic melanin-concentrating

hormone (MCH) or orexin are located (Swaab, 2003b, c). Hypothalamic signals for the actions of these hormones include AMP-activated protein kinase (Andersson et al., 2004).

PYY is thought to act on presynaptic Y2 autoreceptors in the arcuate nucleus to reduce the activity of orexigenic NPY neurons that inhibit anorexigenic POMC neurons (Batterham et al., 2002). While PYY may not operate through brain stem pathways, interestingly, the anorexigenic action of PYY is still maintained in animals with disruptions of the melanocortin system (Challis et al., 2004; Halatchev et al., 2004; Martin et al., 2004; Bewick et al., 2005). Oxyntomodulin appears to act through the GLP-1 receptor (Turton et al., 1996; Baggio et al., 2004). There is evidence that PP may inhibit food intake through decreases in orexigenic hypothalamic peptides such as NPY and orexin (Asakawa et al., 2003).

Ascending feeding pathways from the brain stem include GLP-1 neurons in the NTS, and monoaminergic (noradrenaline and dopamine) and serotonin neurons from the locus coeruleus, ventral tegmental area, and dorsal raphe nuclei (Leibowitz and Alexander, 1998; Everitt et al., 1999; Goldstone et al., 2000; Halford et al., 2005; Wellman, 2005). Many of these pathways project

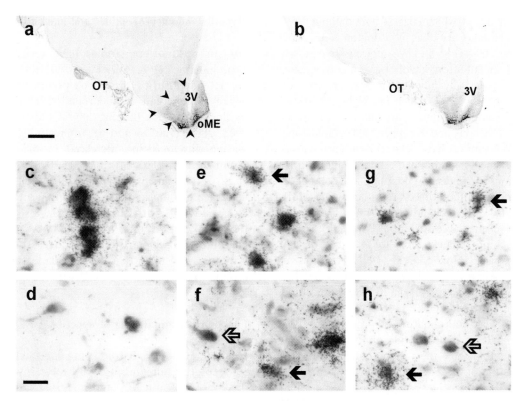

Fig. 4. NPY is colocalized with AGRP, but not POMC, in human hypothalamic neurons. (a, b) Immunocytochemistry (ICC) staining (black) for (a) NPY and (b) AGRP peptide in the human hypothalamus from a control male (#94-118). Note the overlap in the distribution of cell bodies and fibers staining for NPY and AGRP in the infundibular nucleus (INF) and inner layer of the median eminence (ME). The area outlined by arrows indicates the region of the INF/ME used for calculation of the ICC masking area. OT = optic tract, 3 V = third ventricle. Note that there is no NPY or AGRP ICC staining in the outer layer of the median eminence (oME). (c, d) Human infundibular nucleus from a control male (#93-025), double-stained for NPY mRNA (black silver grains, in-situ hybridization (ISH) emulsion autoradiography) with (c) antisense probe or (d) sense probe, and AGRP peptide (brown immunocytochemistry (ICC) staining), with blue thionine counterstaining. (e, f) Infundibular nucleus from a control male (#94-118), double-stained for NPY mRNA with antisense probe (black ISH silver grains), and (e) AGRP or (f) POMC peptides (brown ICC staining). (g, h) Infundibular nucleus from an obese PWS male (#95-104), double-stained for NPY mRNA, and (g) AGRP or (h) POMC peptides. Note that AGRP-peptide-containing cells express NPY mRNA using an antisense probe (c), but that there is no nonspecific ISH signal with the NPY sense probe following AGRP ICC (d). Note that while almost all AGRP-peptide-containing cells express NPY mRNA, some neurons stain only for NPY mRNA (black arrow), in both control (e) and PWS (g) subjects. Note that by contrast, NPY mRNA is not co-localized in POMC neurons in either control (f) or PWS (g) subjects, with cells staining for only NPY mRNA (black arrow) or for only POMC (open arrow). Bar 2 mm in (a) and 20 μm in (d). With permission from Goldstone et al. (2002), copyright The Endocrine Society.

not only to the hypothalamus, but also to the corticolimbic areas such as the nucleus accumbens involved in mood, reward, and arousal.

While there is evidence for involvement of the vagus nerve as a component of the orexigenic effect of ghrelin in humans as in rodents (Date et al., 2002; Le Roux et al., 2005a), the brain and downstream pathways by which these gut hormones alter appetite in humans are unknown.

Prader–Willi syndrome

Prader–Willi syndrome (PWS) is one of the commonest genetic causes of obesity, with a birth incidence of 1 in 29,000 (Whittington et al., 2001). Patients have additional phenotypes that include neonatal hypotonia, hypogonadism, growth hormone deficiency, sleep disturbance, learning difficulties, behavioral problems, and characteristic facial features, many of which suggest hypothalamic dysfunction (Holm et al., 1993; Whittington et al., 2002; Goldstone, 2004) (Fig. 5). PWS subjects have grossly increased appetite from childhood as a consequence of loss of paternally inherited imprinted genes on chromosome 15q11-13, which are highly expressed in the developing brain, particularly, the hypothalamus (Nicholls and Knepper, 2002; Goldstone, 2004). PWS subjects have delayed meal termination, and earlier meal initiation and return of hunger after the previous meal, display hoarding and stealing of food and eat nonfood objects (pica behavior) (Hoffman et al., 1992; Holland et al., 1993; Tan et al., 2004). Given free access to food, PWS subjects will consume approximately three to six times that of control subjects (Holland et al., 1993; Fieldstone et al., 1998). Anorexigenic drugs appear to be ineffective in treating the hyperphagia in PWS (Goldstone, 2004; Shapira et al., 2004).

Postmortem neuroanatomical studies in PWS have looked for neuropeptide abnormalities that may contribute to their hyperphagia and neuroendocrinology, identifying reductions in hypothalamic anorexigenic oxytocin, but normal orexigenic NPY, AGRP and orexin, and GHRH neurons (Swaab et al., 1995; Goldstone et al., 2002, 2003; Goldstone, 2004; Fronczek et al., 2005) (Table 1) (Figs. 4 and 6). The lack of obesity in mouse models of PWS has so far limited their use in the study of hyperphagia in PWS (Chamberlain et al., 2004; Goldstone, 2004).

PWS subjects also have fasting and postprandial elevations in total plasma ghrelin and reductions in plasma PP, but appropriate plasma PYY3-36, for their obesity (Zipf et al., 1981; Goldstone et al., 2004, 2005) (Table 1; Fig. 3). The hyperghrelinemia in PWS appears at least partly explained by preserved insulin sensitivity in turn related to reduced visceral adiposity (Goldstone et al., 2001, 2004, 2005) (Fig. 7). However since infusion of PP has only a minimal effect on food intake in PWS subjects (Zipf et al., 1990; Berntson et al., 1993), and a somatostatin infusion normalizes hyperghrelinaemia in PWS subjects without any acute reduction in appetite (Tan et al., 2004) (Fig. 8), there appears to be a more important role for brain defects than hormonal abnormalities in causing the hyperphagia in PWS, leading to absolute or relative resistance to peripheral anorexigenic signals in PWS.

Hypothalamic neuropeptides and human illness

Human postmortem studies have revealed that hypothalamic NPY, AGRP, and GHRH neurons are activated during prolonged premortem illness which may mediate the neuroendocrine responses to illness (Van den Berghe, 2000; Goldstone et al., 2002, 2003) (Figs. 4 and 6). Interestingly, despite plasma ghrelin being increased in malnutrition (Shimizu et al., 2003; Sturm et al., 2003; Korbonits et al., 2004), patients with anorexia of chronic illness such as renal patients receiving peritoneal dialysis and those with cancer cachexia, also increase food intake in response to acute administration of ghrelin, raising the possibility of novel strategies to improve appetite and nutrition in these vulnerable groups (Neary et al., 2004b; Wynne et al., 2005a).

Functional neuroimaging of appetite

In vivo functional neuroimaging can facilitate the study of the human-brain pathways involved in the control of appetite, and identification of how their dysregulation leads to excess caloric intake in obesity by reducing satiety and increasing hunger (Tataranni and Delparigi, 2003; De Graaf et al., 2004).

Functional magnetic resonance imaging (fMRI) studies have shown a reduction in resting blood oxygen level-dependent (BOLD) signal in the hypothalamus in response to ingestion of oral glucose (Liu et al., 2000; Smeets et al., 2005b) (Fig. 9). Although this change is dose responsive and correlations have been found with plasma insulin levels, the change in BOLD signal can occur very rapidly, and is not found after ingestion of similar sweet or caloric nonglucose loads, even when similar changes in plasma glucose or insulin are induced (Matsuda et al., 1999; Smeets et al., 2005a).

Nonhypothalamic brain regions are also involved in the satiety response in humans, through a variety of hedonic and cognitive processes (Berthoud, 2004; Kishi and Elmquist, 2005). Positron emission tomography (PET) studies have highlighted changes in neuronal activity, as measured by regional resting cerebral blood flow (rCBF), between fasted and fed states, in several brain regions (Tataranni and

(A)

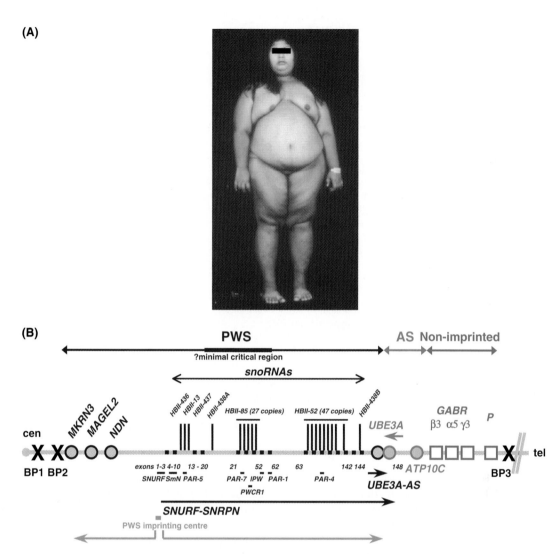

Fig. 5. PWS: from genes to phenotype. (A) A 17-year-old female with PWS. (B) PWS chromosomal region on 15q11-q13 (not to scale) showing the genetic map of the 2 Mb PWS region. Imprinted genes are in blue (paternal allele expressed) and red (maternal allele expressed). Nonimprinted genes are in green. Orange arrows indicate the area of regional imprint control through the imprinting centre at the 5′ end of the bicistronic *SNURF–SNRPN* locus. Vertical bars indicate snoRNA transcripts and horizontal bars, the relative positions of identified exons and other transcripts within the *SNURF–SNRPN* locus. Also indicated are the overlapping sense and antisense transcripts of the Angelman syndrome (AS) gene, *UBE3A*, which is located adjacent to the *PWS* locus. The black crosses indicate common breakpoint (BP) regions for deletions. With permission from Goldstone (2004), copyright Elsevier.

Delparigi, 2003) (Fig. 10). These include the hypothalamus (hunger and satiation), prefrontal cortex (inhibition of inappropriate responses), orbitofrontal cortex (OFC, pleasure, or aversion to food), insula and temporal cortex (dealing with gustatory sensory information), and limbic and paralimbic areas such as amygdala (drive-related and emotional behavior), hippocampus (memory), and parahippocampal gyrus.

fMRI has demonstrated state-dependent regional changes in the reward and arousal responses to food stimuli that are diminished when fed compared to fasted (LaBar et al., 2001; Holsen et al., 2005). In control subjects, viewing food pictures

Table 1. Appetite and neuroendocrine abnormalities in PWS

Appetite
Delayed meal termination
Earlier meal initiation and return of hunger after previous meals
Hoarding and stealing of food
Eating nonfood objects (pica behavior)
Delayed gastric emptying and reduced vomiting

Hormonal and metabolic
Normal plasma leptin relative to obesity
Increased fasting and postprandial plasma ghrelin relative to obesity
Reduced fasting and postprandial plasma pancreatic polypeptide relative to obesity
Reduced fasting and postprandial plasma insulin relative to obesity (preserved insulin sensitivity)
Reduced hypertriglyceridemia relative to obesity
Reduced visceral adiposity relative to obesity
Normal plasma peptide YY relative to obesity
Normal cholecystokinin secretion
Growth hormone deficiency
Hypothalamic hypogonadism (occasionally primary in males)

Hypothalamic
Normal NPY, AGRP, and GHRH neurons in hypothalamic infundibular nucleus
Reduced number of total and oxytocin neurons in hypothalamic PVN
Normal number of vasopressin neurons in PVN
Normal number of orexin neurons in lateral hypothalamus

Functional neuroimaging
Delayed decrease in resting hypothalamic fMRI response to ingestion of oral glucose
Abnormal corticolimbic functional neuroimaging responses to food, consistent with reduced or no brain satiety responses, and greater reward and arousal from food stimuli

(compared to nonfood pictures) leads to an acute increase in fMRI BOLD signal in the insula, inferotemporal cortex (parahippocampal and fusiform gyri), amygdala, parahippocampal, and OFC, when fasted, but not when the studies are repeated after ingestion of a meal. The activation in these cortical and limbic regions when imagining, smelling, tasting, and remembering food stimuli is also suppressed in the fed vs. fasted state (Morris and Dolan, 2001; Hinton et al., 2004; Kringelbach, 2004; Kringelbach and Rolls, 2004). This implies that in the fed state there is less neuronal activation in areas of the brain involved in arousal, reward, and emotion when presented with food stimuli. Viewing pictures of high-calorie foods may elicit different brain-region fMRI responses than viewing low-calorie foods owing to the different motivational salience of such stimuli (Killgore et al., 2003).

The postprandial mediators of these satiety effects seen in functional neuroimaging studies are unknown. Gastric distension is one potential mediator (Stephan et al., 2003), and circulating factors, including glucose and gut hormones, are also prime candidates, particularly acting through connections between these corticolimbic areas and the hypothalamus and brain stem (Berthoud, 2004; Badman and Flier, 2005). For example, animal studies have shown interactions between both leptin and ghrelin with neuropeptides or monoamines that control both feeding and brain reward and motivation circuits (Fulton et al., 2000; Szczypka et al., 2000; Harris et al., 2005; Jo et al., 2005; Naleid et al., 2005).

Defects in brain dopamine motivation and reward circuitry may predispose to human obesity (Wang et al., 2001). There is an increasing interest in the role of addictive behaviors in the development of obesity, particularly with the discovery of the role of endocannabinoids in food intake, leading to the development of CB1-receptor antagonists

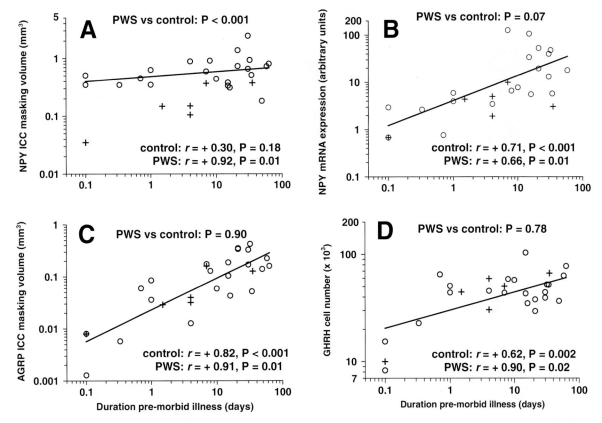

Fig. 6. Hypothalamic NPY, AGRP, and GHRH increase with duration of premorbid illness in control and obese PWS subjects. Relationship between (A) NPY immunocytochemistry (ICC) staining volumes, (B) NPY mRNA expression, (C) AGRP ICC staining volumes, and (D) GHRH cell number in the infundibular nucleus/median eminence, and duration of premorbid illness in control adults (O, solid regression line), and obese PWS adults (+). Note that the y-axes have $\log_{10}$ scales. Note that NPY peptide and mRNA, AGRP peptide, and GHRH cell number increase with illness duration in both control and obese PWS subjects. Note that correcting for illness duration, NPY ICC staining and mRNA expression, but not AGRP ICC staining or GHRH cell number, appear lower in PWS subjects, compared to controls. r represents Pearson correlation coefficient. With permission (A–C) from Goldstone et al. (2002), copyright The Endocrine Society; (D) from Goldstone et al. (2003), copyright Blackwell Publishers.

such as rimonabant, for the treatment of obesity (Di Marzo and Matias, 2005; Jo et al., 2005; Volkow and Wise, 2005).

Functional neuroimaging in obesity

Alterations in the PET rCBF changes in response to satiation have been reported in obese or postobese subjects compared to lean subjects, including the prefrontal cortex, OFC, insula and temporal cortex, hypothalamus, hippocampus, and amygdala (Del Parigi et al., 2002; Del Parigi et al., 2004). Impaired and delayed resting

hypothalamic fMRI responses to ingestion of oral glucose have been reported in obese vs. lean subjects (Matsuda et al., 1999; Liu et al., 2000) (Fig. 9). These defects in brain responses may be appropriate responses to try and limit food intake or could contribute to excess caloric intake and the development of obesity. They may result from alterations in circulating metabolites and hormones, and primary or secondary defects in the brain responses to these peripheral signals. To date, such a hypothesis has only been addressed indirectly by correlation of functional neuroimaging with circulating-hormone levels during these studies (De Graaf et al., 2004).

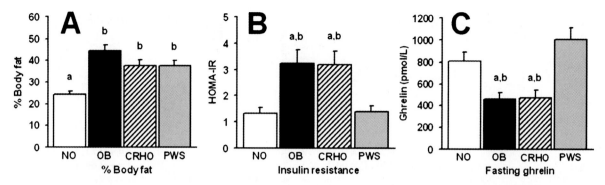

Fig. 7. Increased fasting plasma ghrelin and preserved insulin sensitivity despite obesity in PWS. (A–C) Mean ($\pm$ SEM) values for (A) % body fat, (B) homeostasis model insulin resistance index (HOMA-IR), and (C) fasting plasma ghrelin levels in nonobese (NO, white, $n = 15$) and obese (OB, black, $n = 16$) controls, craniopharyngioma subjects with hypothalamic obesity (CRHO, hatched, $n = 9$) and subjects with PWS (gray, $n = 26$), a: $P < 0.01$ vs. PWS, and b: $P < 0.01$ vs. NO. Despite similar degrees of obesity, subjects with PWS have increased fasting plasma ghrelin and preserved insulin sensitivity, compared to the other two obese groups. With permission from Goldstone et al. (2005), copyright The Endocrine Society.

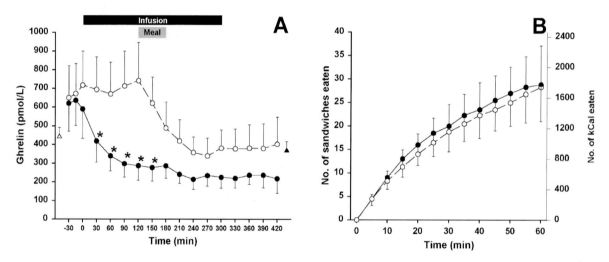

Fig. 8. Effect of somatostatin on plasma ghrelin and acute food intake in PWS. Effect of a somatostatin (250 µg/h: ●, solid line) or saline (○, dashed line) infusion on (A) plasma ghrelin, and (B) cumulative consumption of food (no. of sandwiches or kilocalories eaten) during a 60 min period of free access to food, in four PWS male adults. In (A) the duration of infusion (black bar) and access to food (gray bar) are indicated; Δ represents fasted levels, and ▲ postprandial trough plasma ghrelin levels after a 522 kcal breakfast in five non-PWS similarly obese male adults. Data represents mean $\pm$ SEM. * $P < 0.05$ somatostatin vs. saline at same time point. Note that somatostatin lowers plasma ghrelin but has no significant effect on food intake in PWS subjects. With permission from Tan et al. (2004), copyright The Endocrine Society.

Functional neuroimaging in PWS

Several small studies have demonstrated abnormal functional neuroimaging of appetite in PWS subjects. The maximum time points for changes in the resting fMRI BOLD signal after ingestion of oral glucose are delayed in PWS subjects even longer than in both non-PWS obese and lean subjects, including decreases in fMRI signal in the hypothalamus, OFC, and nucleus accumbens, and increases in the dorsolateral prefrontal cortex and insula (Shapira et al., 2005). Meanwhile, a PET

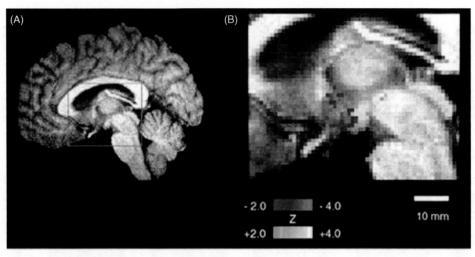

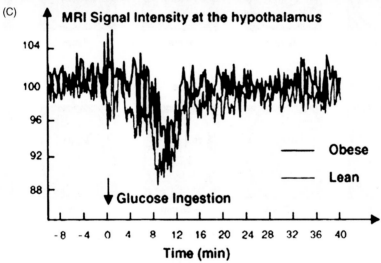

Fig. 9. Hypothalamic functional MRI response to oral glucose ingestion. (A) T1-weighted sagittal anatomical MRI image including the hypothalamic region in an obese subject. (B) fMRI (in color) overlaid on an anatomical MRI image (in gray) depicts the typical inhibitory response observed in the hypothalamus after glucose ingestion in the same obese subject. The color-coded areas represent the areas of statistically significant inhibition ($z < -2.0$; $P < 0.05$) of brain activity after glucose ingestion. After glucose ingestion, two areas with consistent inhibition were observed: the upper anterior hypothalamus (UAH) and the lateral posterior hypothalamus (LPH), which correspond to the paraventricular and ventromedial nuclei, respectively. (C) The time courses of the MRI signal intensities in the LPH area in an obese subject and a lean subject after glucose ingestion. The inhibitory response on fMRI was observed 4 min after the start of oral glucose ingestion and reached a maximum at ~10 min before returning to the baseline after 14 min. With permission from Matsuda et al. (1999), copyright American Diabetes Association.

study found that PWS subjects scanned while choosing food items lacked the normal increases in rCBF in the OFC and inferotemporal cortex seen after ingestion of a meal (Hinton et al., 2004, 2005). Other preliminary reports have also found that PWS patients display reduced or no brain satiety responses, with greater reward and arousal from food stimuli, after ingestion of oral glucose or meals which may underlie their hyperphagia (Holsen et al., 2004; Tataranni et al., 2004; James

68

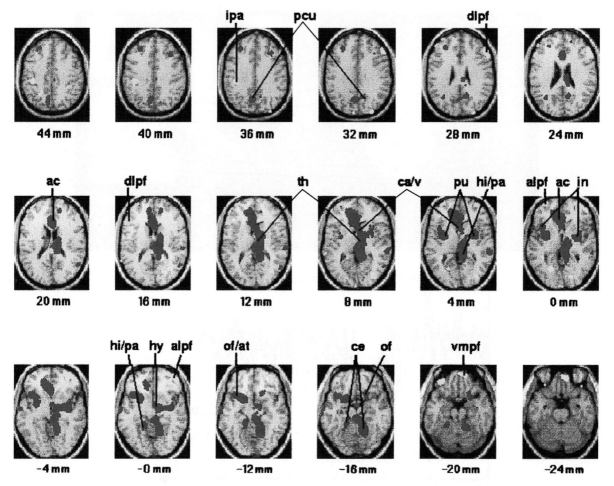

Fig. 10. Positron emission tomography images of brain activation in response to hunger and satiation. Brain regions with significant increase in rCBF in response to hunger are shown in blue; brain regions with significant increase in rCBF in response to satiation are shown in yellow. Images were generated by using PET and MRI data from eleven healthy, lean, male individuals. Color-coded images are superimposed onto an average of the subjects' brain MRIs (grayscale image). Horizontal brain sections correspond to the coordinates of the Talairach and Tournoux brain atlas. The number under each section reflects the distance in millimeters superior (+) or inferior (−) to a horizontal plane between the anterior and posterior commissures. The right hemisphere in each section is on the reader's right. Compared with satiation, hunger was associated with significantly increased rCBF ($P < 0.005$ uncorrected for multiple comparisons) in the hypothalamus (hy), thalamus (th), anterior cingulate (ac), insula/claustrum (in), orbitofrontal and temporal cortex (of/at), hippocampus/parahippocampul gyrus (hi/pa), precuneus (pcu), caudate/ventricle (ca/v), putamen (pu), and cerebellum (ce). Compared with hunger, satiation was associated with increased rCBF ($P < 0.005$ uncorrected for multiple comparisons) in the ventromedial prefrontal cortex (vmpf), dorsolateral prefrontal cortex (dlpf), ventrolateral prefrontal cortex (alpf), and inferior parietal lobule (ipa). With permission from Tataranni et al. (1999), copyright National Academy of Sciences, USA.

et al., 2006). It remains to be determined how these functional defects relate to hormonal abnormalities, developmental hypothalamic, and even cortical neuroanatomical defects in PWS (Goldstone, 2004; Miller et al., 2005).

Functional neuroimaging in the future

Similar functional neuroimaging studies could be used to study the neuroanatomical basis of abnormal feeding behavior in monogenic causes of human obesity, so as to reveal in humans where and how particular genes and signals act in the pathways that regulate appetite, and in time, more common genetic polymorphisms within the appetite pathways as well as assessing or perhaps even predicting the response to anorexigenic drugs or surgery (O'Rahilly et al., 2003; Bell et al., 2005; Le Roux and Bloom, 2005; Spiegel et al., 2005). Such studies will enable a rational approach to the development of treatments for hyperphagia and obesity, both pharmacological and perhaps even neuromodulatory such as deep-brain stimulation to directed areas through chronically implanted microelectrodes (Wallace et al., 2004).

By identifying the brain pathways that drive and inhibit feeding in humans and their interaction with peripheral satiety signals, we can develop brain-imaging biomarkers for appetite modulation so as to more rationally use and combine hormones, their analogs, and drugs in treating obesity, and identify potential novel therapeutic brain targets to reduce appetite.

Acknowledgments

The author wishes to acknowledge the technical support from all his colleagues and collaborators, including those at the Hammersmith Hospital, Imperial College, London; Netherlands Institute for Brain Research, Amsterdam; St. Bartholomew's Hospital, London; University of Florida College of Medicine, USA; and Royal Free Hospital, London. He extends his sincere gratitude to the UK Medical Research Council, the Royal Society of London, the UK and US PWS Associations, The Royal College of Physicians (London), the National Institutes of Health, the Hayward Foundation, and Merck Research Laboratories for financial support and to all the patients, families, carers, and volunteers for their participation in the various research studies.

References

Adrian, T.E., Savage, A.P., Sagor, G.R., Allen, J.M., Bacarese-Hamilton, A.J., Tatemoto, K., Polak, J.M. and Bloom, S.R. (1985) Effect of peptide YY on gastric, pancreatic, and biliary function in humans. Gastroenterology, 89: 494–499.

Allen, J.M., Fitzpatrick, M.L., Yeats, J.C., Darcy, K., Adrian, T.E. and Bloom, S.R. (1984) Effects of peptide YY and neuropeptide Y on gastric emptying in man. Digestion, 30: 255–262.

Andersson, U., Filipsson, K., Abbott, C.R., Woods, A., Smith, K., Bloom, S.R., Carling, D. and Small, C.J. (2004) AMP-activated protein kinase plays a role in the control of food intake. J. Biol. Chem., 279: 12005–12008.

Asakawa, A., Inui, A., Yuzuriha, H., Ueno, N., Katsuura, G., Fujimiya, M., Fujino, M.A., Niijima, A., Meguid, M.M. and Kasuga, M. (2003) Characterization of the effects of pancreatic polypeptide in the regulation of energy balance. Gastroenterology, 124: 1325–1336.

Avenell, A., Broom, J., Brown, T.J., Poobalan, A., Aucott, L., Stearns, S.C., Smith, W.C., Jung, R.T., Campbell, M.K. and Grant, A.M. (2004) Systematic review of the long-term effects and economic consequences of treatments for obesity and implications for health improvement. Health Technol. Assess., 8: 1–182.

Badman, M.K. and Flier, J.S. (2005) The gut and energy balance: visceral allies in the obesity wars. Science, 307: 1909–1914.

Baggio, L.L., Huang, Q., Brown, T.J. and Drucker, D.J. (2004) Oxyntomodulin and glucagon-like peptide-1 differentially regulate murine food intake and energy expenditure. Gastroenterology, 127: 546–558.

Batterham, R.L., Cohen, M.A., Ellis, S.M., Le Roux, C.W., Withers, D.J., Frost, G.S., Ghatei, M.A. and Bloom, S.R. (2003a) Inhibition of food intake in obese subjects by peptide YY3-36. N. Engl. J. Med., 349: 941–948.

Batterham, R.L., Cowley, M.A., Small, C.J., Herzog, H., Cohen, M.A., Dakin, C.L., Wren, A.M., Brynes, A.E., Low, M.J., Ghatei, M.A., Cone, R.D. and Bloom, S.R. (2002) Gut hormone PYY(3-36) physiologically inhibits food intake. Nature, 418: 650–654.

Batterham, R.L., Le Roux, C.W., Cohen, M.A., Park, A.J., Ellis, S.M., Patterson, M., Frost, G.S., Ghatei, M.A. and Bloom, S.R. (2003b) Pancreatic polypeptide reduces appetite and food intake in humans. J. Clin. Endocrinol. Metab., 88: 3989–3992.

Bell, C.G., Walley, A.J. and Froguel, P. (2005) The genetics of human obesity. Nat. Rev. Genet., 6: 221–234.

Berntson, G.G., Zipf, W.B., O'Dorisio, T.M., Hoffman, J.A. and Chance, R.E. (1993) Pancreatic polypeptide infusions reduce food intake in Prader–Willi syndrome. Peptides, 14: 497–503.

Berthoud, H. (2004) Mind versus metabolism in the control of food intake and energy balance. Physiol. Behav., 81: 781–793.

Bewick, G.A., Gardiner, J.V., Dhillo, W.S., Kent, A.S., White, N.E., Webster, Z., Ghatei, M.A. and Bloom, S.R. (2005) Post-embryonic ablation of AgRP neurons in mice leads to a lean, hypophagic phenotype. FASEB J., 19: 1680–1682.

Bray, G.A. (1999) Drug treatment of obesity. Baillieres Best Pract. Res. Clin. Endocrinol. Metab., 13: 131–148.

Bray, G.A. (2000) Afferent signals regulating food intake. Proc. Nutr. Soc., 59: 373–384.

Challis, B.G., Coll, A.P., Yeo, G.S., Pinnock, S.B., Dickson, S.L., Thresher, R.R., Dixon, J., Zahn, D., Rochford, J.J., White, A., Oliver, R.L., Millington, G., Aparicio, S.A., Colledge, W.H., Russ, A.P., Carlton, M.B. and O'Rahilly, S. (2004) Mice lacking pro-opiomelanocortin are sensitive to high-fat feeding but respond normally to the acute anorectic effects of peptide-YY(3-36). Proc. Natl. Acad. Sci. USA, 101: 4695–4700.

Chamberlain, S.J., Johnstone, K.A., DuBose, A.J., Simon, T.A., Bartolomei, M.S., Resnick, J.L. and Brannan, C.I. (2004) Evidence for genetic modifiers of postnatal lethality in PWS-IC deletion mice. Hum. Mol. Genet., 13: 2971–2979.

Cohen, M.A., Ellis, S.M., Le Roux, C.W., Batterham, R.L., Park, A., Patterson, M., Frost, G.S., Ghatei, M.A. and Bloom, S.R. (2003) Oxyntomodulin suppresses appetite and reduces food intake in humans. J. Clin. Endocrinol. Metab., 88: 4696–4701.

Cox, J.E. and Randich, A. (2004) Enhancement of feeding suppression by PYY(3-36) in rats with area postrema ablations. Peptides, 25: 985–989.

Dakin, C.L., Small, C.J., Batterham, R.L., Neary, N.M., Cohen, M.A., Patterson, M., Ghatei, M.A. and Bloom, S.R. (2004) Peripheral oxyntomodulin reduces food intake and body weight gain in rats. Endocrinology, 145: 2687–2695.

Date, Y., Murakami, N., Toshinai, K., Matsukura, S., Niijima, A., Matsuo, H., Kangawa, K. and Nakazato, M. (2002) The role of the gastric afferent vagal nerve in ghrelin-induced feeding and growth hormone secretion in rats. Gastroenterology, 123: 1120–1128.

De Graaf, C., Blom, W.A.M., Smeets, P.A.M., Stafleu, A. and Hendriks, H.F.J. (2004) Biomarkers of satiation and satiety. Am. J. Clin. Nutr., 79: 946–961.

Deitel, M. (2002) The International Obesity Task Force and "globesity". Obes. Surg., 12: 613–614.

Del Parigi, A., Gautier, J.F., Chen, K., Salbe, A.D., Ravussin, E., Reiman, E. and Tataranni, P.A. (2002) Neuroimaging and obesity: mapping the brain responses to hunger and satiation in humans using positron emission tomography. Ann. N.Y. Acad. Sci., 967: 389–397.

Delparigi, A., Chen, K., Salbe, A.D., Hill, J.O., Wing, R.R., Reiman, E.M. and Tataranni, P.A. (2004) Persistence of abnormal neural responses to a meal in postobese individuals. Int. J. Obes. Relat. Metab. Disord., 28: 370–377.

Department of Health (2004) Health Survey for England 2003 www.dh.gov.uk/assetRoot/04/09/89/11/04098911.pdf Vol. 2: 147.

Di Marzo, V. and Matias, I. (2005) Endocannabinoid control of food intake and energy balance. Nat. Neurosci., 8: 585–589.

Druce, M.R., Wren, A.M., Park, A.J., Milton, J.E., Patterson, M., Frost, G., Ghatei, M.A., Small, C. and Bloom, S.R. (2005) Ghrelin increases food intake in obese as well as lean subjects. Int. J. Obes. Relat. Metab. Disord., 29: 1130–1136.

Everitt, B.J., Parkinson, J.A., Olmstead, M.C., Arroyo, M., Robledo, P. and Robbins, T.W. (1999) Associative processes in addiction and reward. The role of amygdala-ventral striatal subsystems. Ann. N.Y. Acad. Sci., 877: 412–438.

Fieldstone, A., Zipf, W.B., Sarter, M.F. and Berntson, G.G. (1998) Food intake in Prader–Willi syndrome and controls with obesity after administration of a benzodiazepine-receptor agonist. Obes. Res., 6: 29–33.

Flier, J.S. (2004) Obesity wars: molecular progress confronts an expanding epidemic. Cell, 116: 337–350.

Fox, R. (2003) Overweight children. Circulation, 108: e9071.

Fronczek, R., Lammers, G.J., Balesar, R., Unmehopa, U.A. and Swaab, D.F. (2005) The number of hypothalamic hypocretin (orexin) neurons is not affected in Prader–Willi syndrome. J. Clin. Endocrinol. Metab., 90: 5466–5470.

Fulton, S., Woodside, B. and Shizgal, P. (2000) Modulation of brain reward circuitry by leptin. Science, 287: 125–128.

Goldstone, A.P. (2004) Prader–Willi syndrome: advances in its genetics, pathophysiology and treatment. Trends Endocrinol. Metab., 15: 12–20.

Goldstone, A.P., Morgan, I., Mercer, J.G., Morgan, D.G., Moar, K.M., Ghatei, M.A. and Bloom, S.R. (2000) Effect of leptin on hypothalamic GLP-1 peptide and brain-stem preproglucagon mRNA. Biochem. Biophys. Res. Commun., 269: 331–335.

Goldstone, A.P., Patterson, M., Kalingag, N., Ghatei, M.A., Brynes, A.E., Bloom, S.R., Grossman, A.B. and Korbonits, M. (2005) Fasting and post-prandial hyperghrelinemia in Prader–Willi syndrome is partially explained by hypoinsulinemia, and is not due to peptide YY 3-36 deficiency or seen in hypothalamic obesity due to craniopharyngioma. J. Clin. Endocrinol. Metab., 90: 2681–2690.

Goldstone, A.P., Thomas, E.L., Brynes, A.E., Bell, J.D., Frost, G., Saeed, N., Hajnal, J.V., Howard, J.K., Holland, A. and Bloom, S.R. (2001) Visceral adipose tissue and metabolic complications of obesity are reduced in Prader–Willi syndrome female adults: evidence for novel influences on body fat distribution. J. Clin. Endocrinol. Metab., 86: 4330–4338.

Goldstone, A.P., Thomas, E.L., Brynes, A.E., Castroman, G., Edwards, R., Ghatei, M.A., Frost, G., Holland, A.J., Grossman, A.B., Korbonits, M., Bloom, S.R. and Bell, J.D. (2004) Elevated fasting plasma ghrelin in Prader–Willi-syndrome adults is not solely explained by their reduced visceral adiposity and insulin resistance. J. Clin. Endocrinol. Metab., 89: 1718–1726.

Goldstone, A.P., Unmehopa, U.A., Bloom, S.R. and Swaab, D.F. (2002) Hypothalamic NPY and agouti-related protein are increased in human illness but not in Prader–Willi syndrome and other obese subjects. J. Clin. Endocrinol. Metab., 87: 927–937.

Goldstone, A.P., Unmehopa, U.A. and Swaab, D.F. (2003) Hypothalamic growth hormone-releasing hormone (GHRH) cell number is increased in human illness, but is not reduced in Prader–Willi syndrome or obesity (erratum in Clin. Endocrinol. 59, 266, 2003). Clin. Endocrinol. (Oxf.), 58: 743–755.

Halatchev, I.G., Ellacott, K.L., Fan, W. and Cone, R.D. (2004) Peptide YY3-36 inhibits food intake in mice through a melanocortin-4 receptor-independent mechanism. Endocrinology, 145: 2585–2590.

Halford, J.C., Harrold, J.A., Lawton, C.L. and Blundell, J.E. (2005) Serotonin (5-HT) drugs: effects on appetite expression and use for the treatment of obesity. Curr. Drug Targets, 6: 201–213.

Harris, G.C., Wimmer, M. and Jones, G.A. (2005) A role for lateral hypothalamic orexin neurons in reward seeking. Nature, 437: 556–559.

Hill, J.O. and Peters, J.C. (1998) Environmental contributions to the obesity epidemic. Science, 280: 1371–1374.

Hinton, E.C., Holland, A.J., Gellatly, M.S., Soni, S., Patterson, M., Ghatei, M.A. and Owen, A.M. (2005) Neural representations of hunger and satiety in Prader–Willi syndrome. Int. J. Obes., Vol. 30, 313–321.

Hinton, E.C., Parkinson, J.A., Holland, A.J., Arana, F.S., Roberts, A.C. and Owen, A.M. (2004) Neural contributions to the motivational control of appetite in humans. Eur. J. Neurosci., 20: 1411–1418.

Hoffman, C.J., Aultman, D. and Pipes, P. (1992) A nutrition survey of and recommendations for individuals with Prader–Willi syndrome who live in group homes. J. Am. Diet. Assoc., 92: 823–830, 833.

Holland, A.J., Treasure, J., Coskeran, P., Dallow, J., Milton, N. and Hillhouse, E. (1993) Measurement of excessive appetite and metabolic changes in Prader–Willi syndrome. Int. J. Obes., 17: 527–532.

Holm, V.A., Cassidy, S.B., Butler, M.G., Hanchett, J.M., Greenswag, L.R., Whitman, B.Y. and Greenberg, F. (1993) Prader–Willi syndrome: consensus diagnostic criteria. Pediatrics, 91: 398–402.

Holsen, L., Zarcone, J., Anderson, M., Young, J., Butler, M.G., Thompson, T. and Savage, C. (2004) Abnormal food motivation in Prader–Willi syndrome: relationship between neural dysfunction and obesity using fMRI. Abstracts 26th USA Prader–Willi Syndrome Conference, Sandusky, Ohio.

Holsen, L.M., Zarcone, J.R., Thompson, T.I., Brooks, W.M., Anderson, M.F., Ahluwalia, J.S., Nollen, N.L. and Savage, C.R. (2005) Neural mechanisms underlying food motivation in children and adolescents. Neuroimage, 27: 669–676.

James, G.A., Miller, J.L., Couch, J., Goldstone, A.P., He, G., Driscoll, D.J. and Liu, Y. (2006) Heightened orbitofrontal BOLD response to food stimuli in Prader–Willi patients. Abstracts 12th Annual Meeting of the Organization for Human Brain Mapping, Florence, Italy.

Jo, Y.H., Chen, Y.J., Chua Jr., S.C., Talmage, D.A. and Role, L.W. (2005) Integration of endocannabinoid and leptin signaling in an appetite-related neural circuit. Neuron, 48: 1055–1066.

Killgore, W.D., Young, A.D., Femia, L.A., Bogorodzki, P., Rogowska, J. and Yurgelun-Todd, D.A. (2003) Cortical and limbic activation during viewing of high- versus low-calorie foods. Neuroimage, 19: 1381–1394.

Kimm, S.Y. and Obarzanek, E. (2002) Childhood obesity: a new pandemic of the new millennium. Pediatrics, 110: 1003–1007.

Kishi, T. and Elmquist, J.K. (2005) Body weight is regulated by the brain: a link between feeding and emotion. Mol. Psychiatry, 10: 132–146.

Knowler, W.C., Barrett-Connor, E., Fowler, S.E., Hamman, R.F., Lachin, J.M., Walker, E.A. and Nathan, D.M. (2002) Reduction in the incidence of type 2 diabetes with lifestyle intervention or metformin. N. Engl. J. Med., 346: 393–403.

Koplan, J.P. and Dietz, W.H. (1999) Caloric imbalance and public health policy. JAMA, 282: 1579–1581.

Korbonits, M., Goldstone, A.P., Gueorguiev, M. and Grossman, A.B. (2004) Ghrelin — a hormone with multiple functions. Front. Neuroendocrinol., 25: 27–68.

Kringelbach, M.L. (2004) Food for thought: hedonic experience beyond homeostasis in the human brain. Neuroscience, 126: 807–819.

Kringelbach, M.L. and Rolls, E.T. (2004) The functional neuroanatomy of the human orbitofrontal cortex: evidence from neuroimaging and neuropsychology. Prog. Neurobiol., 72: 341–372.

Kuo, Y.T., Parkinson, J.R., Herlihy, A.H., So, P.W., Small, C.J., Bloom, S.R. and Bell, J.D. (2004) Imaging appetite in vivo with manganese-enhanced MRI (MEMRI). Abstracts ISMRM 13th Scientific Meeting, Miami Beach, FL, USA.

LaBar, K.S., Gitelman, D.R., Parrish, T.B., Kim, Y.H., Nobre, A.C. and Mesulam, M.M. (2001) Hunger selectively modulates corticolimbic activation to food stimuli in humans. Behav. Neurosci., 115: 493–500.

Lassmann, V., Vague, P., Vialettes, B. and Simon, M.C. (1980) Low plasma levels of pancreatic polypeptide in obesity. Diabetes, 29: 428–430.

Le Roux, C.W., Batterham, R.L., Aylwin, S.J., Patterson, M., Borg, C.M., Wynne, K.J., Kent, A., Vincent, R.P., Gardiner, J., Ghatei, M.A. and Bloom, S.R. (2006) Attenuated peptide YY release in obese subjects is associated with reduced satiety. Endocrinology, 147: 3–8.

Le Roux, C.W. and Bloom, S.R. (2005) Why do patients lose weight after Roux-en-Y gastric bypass? J. Clin. Endocrinol. Metab., 90: 591–592.

Le Roux, C.W., Neary, N.M., Halsey, T.J., Small, C.J., Martinez-Isla, A.M., Ghatei, M.A., Theodorou, N.A. and Bloom, S.R. (2005a) Ghrelin does not stimulate food intake in patients with surgical procedures involving vagotomy. J. Clin. Endocrinol. Metab., 90: 4521–4524.

Le Roux, C.W., Patterson, M., Vincent, R.P., Hunt, C., Ghatei, M.A. and Bloom, S.R. (2005b) Postprandial plasma ghrelin is suppressed proportional to meal calorie content in normal weight but not obese subjects. J. Clin. Endocrinol. Metab., 90: 1068–1071.

Leibowitz, S.F. and Alexander, J.T. (1998) Hypothalamic serotonin in control of eating behavior, meal size, and body weight. Biol. Psychiatry, 44: 851–864.

Lieverse, R.J., Jansen, J.B., Masclee, A.M. and Lamers, C.B. (1994) Satiety effects of cholecystokinin in humans. Gastroenterology, 106: 1451–1454.

Liu, Y., Gao, J.H., Liu, H.L. and Fox, P.T. (2000) The temporal response of the brain after eating revealed by functional MRI. Nature, 405: 1058–1062.

Magarey, A.M., Daniels, L.A., Boulton, T.J. and Cockington, R.A. (2003) Predicting obesity in early adulthood from

72

childhood and parental obesity. Int. J. Obes. Relat. Metab. Disord., 27: 505–513.

Martin, N.M., Small, C.J., Sajedi, A., Patterson, M., Ghatei, M.A. and Bloom, S.R. (2004) Pre-obese and obese agouti mice are sensitive to the anorectic effects of peptide YY(3-36) but resistant to ghrelin. Int. J. Obes. Relat. Metab. Disord., 28: 886–893.

Matsuda, M., Liu, Y., Mahankali, S., Pu, Y., Mahankali, A., Wang, J., DeFronzo, R.A., Fox, P.T. and Gao, J.H. (1999) Altered hypothalamic function in response to glucose ingestion in obese humans. Diabetes, 48: 1801–1806.

Miller, J., Kranzler, J., Hatfield, A., Mueller, O.T., Theriaque, D.W., Goldstone, A.P., Shuster, J.J. and Driscoll, D.J. (2005) Cognitive and behavioral findings in Prader–Willi syndrome and early onset morbid obesity. Abstracts 27th USA Prader–Willi Syndrome Conference, Orlando, FL.

Morris, J.S. and Dolan, R.J. (2001) Involvement of human amygdala and orbitofrontal cortex in hunger-enhanced memory for food stimuli. J. Neurosci., 21: 5304–5310.

Murray, C.D., Martin, N.M., Patterson, M., Taylor, S., Ghatei, M.A., Kamm, M.A., Johnston, C., Bloom, S.R. and Emmanuel, A.V. (2005) Ghrelin enhances gastric emptying in diabetic gastroparesis: a double-blind, placebo-controlled, cross-over study. Gut, 54: 1693–1698.

Must, A. (2003) Does overweight in childhood have an impact on adult health? Nutr. Rev., 61: 139–142.

Naleid, A.M., Grace, M.K., Cummings, D.E. and Levine, A.S. (2005) Ghrelin induces feeding in the mesolimbic reward pathway between the ventral tegmental area and the nucleus accumbens. Peptides, 26: 2274–2279.

Naslund, E., Bogefors, J., Skogar, S., Gryback, P., Jacobsson, H., Holst, J.J. and Hellstrom, P.M. (1999) GLP-1 slows solid gastric emptying and inhibits insulin, glucagon, and PYY release in humans. Am. J. Physiol., 277: R910–R916.

Neary, N.M., Goldstone, A.P. and Bloom, S.R. (2004a) Appetite regulation: from the gut to the hypothalamus. Clin. Endocrinol. (Oxf.), 60: 153–160.

Neary, N.M., Small, C.J., Druce, M.R., Park, A.J., Ellis, S.M., Semjonous, N.M., Dakin, C.L., Filipsson, K., Wang, F., Kent, A.S., Frost, G.S., Ghatei, M.A. and Bloom, S.R. (2005) Peptide YY3-36 and glucagon-like peptide-17-36 inhibit food intake additively. Endocrinology, 146: 5120–5127.

Neary, N.M., Small, C.J., Wren, A.M., Lee, J.L., Druce, M.R., Palmieri, C., Frost, G.S., Ghatei, M.A., Coombes, R.C. and Bloom, S.R. (2004b) Ghrelin increases energy intake in cancer patients with impaired appetite: acute, randomized, placebo-controlled trial. J. Clin. Endocrinol. Metab., 89: 2832–2836.

Nicholls, R.D. and Knepper, J.L. (2002) Genome organization, function, and imprinting in Prader–Willi and Angelman syndromes. Annu. Rev. Genomics Hum. Genet., 2: 153–175.

O'Rahilly, S., Farooqi, I.S., Yeo, G.S. and Challis, B.G. (2003) Minireview: human obesity-lessons from monogenic disorders. Endocrinology, 144: 3757–3764.

Perusse, L., Rankinen, T., Zuberi, A., Chagnon, Y.C., Weisnagel, S.J., Argyropoulos, G., Walts, B., Snyder, E.E. and Bouchard, C. (2005) The human obesity gene map: the 2004 update. Obes. Res., 13: 381–490.

Riccardi, G., Capaldo, B. and Vaccaro, O. (2005) Functional foods in the management of obesity and type 2 diabetes. Curr. Opin. Clin. Nutr. Metab. Care, 8: 630–635.

Roth, C.L., Enriori, P.J., Harz, K., Woelfle, J., Cowley, M.A. and Reinehr, T. (2005) Peptide YY Is a regulator of energy homeostasis in obese children before and after weight loss. J. Clin. Endocrinol. Metab., 90: 6386–6391.

Sainsbury, A., Schwarzer, C., Couzens, M., Fetissov, S., Furtinger, S., Jenkins, A., Cox, H.M., Sperk, G., Hokfelt, T. and Herzog, H. (2002) Important role of hypothalamic Y2 receptors in body weight regulation revealed in conditional knockout mice. Proc. Natl. Acad. Sci. USA, 99: 8938–8943.

Shapira, N.A., Lessig, M.C., He, G.A., James, G.A., Driscoll, D.J. and Liu, Y. (2005) Satiety dysfunction in Prader–Willi syndrome demonstrated by fMRI. J. Neurol. Neurosurg. Psych., 76: 260–262.

Shapira, N.A., Lessig, M.C., Lewis, M.H., Goodman, W.K. and Driscoll, D.J. (2004) Effects of topiramate in adults with Prader–Willi syndrome. Am. J. Ment. Retard., 109: 301–309.

Shimizu, Y., Nagaya, N., Isobe, T., Imazu, M., Okumura, H., Hosoda, H., Kojima, M., Kangawa, K. and Kohno, N. (2003) Increased plasma ghrelin level in lung cancer cachexia. Clin. Cancer Res., 9: 774–778.

Small, C.J. and Bloom, S.R. (2004) Gut hormones as peripheral anti-obesity targets. Curr. Drug Targets CNS Neurol. Disord., 3: 379–388.

Smeets, P.A., De Graaf, C., Stafleu, A., Van Osch, M.J. and Van der Grond, J. (2005a) Functional magnetic resonance imaging of human hypothalamic responses to sweet taste and calories. Am. J. Clin. Nutr., 82: 1011–1016.

Smeets, P.A., De Graaf, C., Stafleu, A., Van Osch, M.J. and Van der Grond, J. (2005b) Functional MRI of human hypothalamic responses following glucose ingestion. Neuroimage, 24: 363–368.

Speiser, P.W., Rudolf, M.C., Anhalt, H., Camacho-Hubner, C., Chiarelli, F., Eliakim, A., Freemark, M., Gruters, A., Hershkovitz, E., Iughetti, L., Krude, H., Latzer, Y., Lustig, R.H., Pescovitz, O.H., Pinhas-Hamiel, O., Rogol, A.D., Shalitin, S., Sultan, C., Stein, D., Vardi, P., Werther, G.A., Zadik, Z., Zuckerman-Levin, N. and Hochberg, Z. (2005) Childhood obesity. J. Clin. Endocrinol. Metab., 90: 1871–1887.

Spiegel, A., Nabel, E., Volkow, N., Landis, S. and Li, T.K. (2005) Obesity on the brain. Nat. Neurosci., 8: 552–553.

Stephan, E., Pardo, J.V., Faris, P.L., Hartman, B.K., Kim, S.W., Ivanov, E.H., Daughters, R.S., Costello, P.A. and Goodale, R.L. (2003) Functional neuroimaging of gastric distention. J. Gastrointest. Surg., 7: 740–749.

Sturm, K., MacIntosh, C.G., Parker, B.A., Wishart, J., Horowitz, M. and Chapman, I.M. (2003) Appetite, food intake, and plasma concentrations of cholecystokinin, ghrelin, and other gastrointestinal hormones in undernourished older women and well-nourished young and older women. J. Clin. Endocrinol. Metab., 88: 3747–3755.

Swaab, D.F. (2003a) Infundibular nucleus (arcuate nucleus), subventricular nucleus and median eminence. In: Aminoff,

M.J., Boller, F. and Swaab, D.F. (Eds.) The Human Hypothalamus: Basic and Clinical Aspects, Part I, Vol. 79 (3rd Series Vol. 1).

Swaab, D.F. (2003b) Lateral hypothalamic area (LHA), including the perifornical area and intermediate hypothalamic area (IHA). In: Aminoff, M.J., Boller, F. and Swaab, D.F. (Eds.) The Human Hypothalamus: Basic and Clinical Aspects, Part I, Vol. 79 (3rd Series Vol. 1).

Swaab, D.F. (2003c) Supraoptic and paraventricular nucleus. In: Aminoff, M.J., Boller, F. and Swaab, D.F. (Eds.) The Human Hypothalamus: Basic and Clinical Aspects, Part I, Vol. 79 (3rd Series Vol. 1).

Swaab, D.F., Purba, J.S. and Hofman, M.A. (1995) Alterations in the hypothalamic paraventricular nucleus and its oxytocin neurons (putative satiety cells) in Prader–Willi syndrome: a study of five cases. J. Clin. Endocrinol. Metab., 80: 573–579.

Szczypka, M.S., Rainey, M.A. and Palmiter, R.D. (2000) Dopamine is required for hyperphagia in Lep(ob/ob) mice. Nat. Genet., 25: 102–104.

Tan, T.M., Vanderpump, M., Khoo, B., Patterson, M., Ghatei, M.A. and Goldstone, A.P. (2004) Somatostatin infusion lowers plasma ghrelin without reducing appetite in adults with Prader–Willi syndrome. J. Clin. Endocrinol. Metab., 89: 4162–4165.

Tataranni, P.A., Gautier, J.F., Chen, K.W., Uecker, A., Bandy, D., Salbe, A.D., Pratley, R.E., Lawson, M., Reiman, E.M. and Ravussin, E. (1999) Neuroanatomical correlates of hunger and satiation in humans using positron emission tomography. Proc. Natl. Acad. Sci. USA, 96: 4569–4574.

Tataranni, P.A., Chen, K., Salbe, A.D., Reiman, E.M. and Del Parigi, A. (2004) Neuroimaging evidence implicates the human hypothalamus in monitoring the relative energy content of a meal. Abstracts 86th Annual Meeting of the American Endocrine Society, New Orleans, OR19-3.

Tataranni, P.A. and Delparigi, A. (2003) Functional neuroimaging: a new generation of human brain studies in obesity research. Obes. Rev., 4: 229–238.

Tataranni, P.A., Harper, I.T., Snitker, S., Del Parigi, A., Vozarova, B., Bunt, J., Bogardus, C. and Ravussin, E. (2003) Body weight gain in free-living Pima Indians: effect of energy intake vs. expenditure. Int. J. Obes. Relat. Metab. Disord., 27: 1578–1583.

Thibault, H. and Rolland-Cachera, M.F. (2003) Prevention strategies of childhood obesity. Arch. Pediatr., 10: 1100–1108.

Turton, M.D., O'Shea, D., Gunn, I., Beak, S.A., Edwards, C.M.B., Meeran, K., Choi, S.J., Taylor, G.M., Heath, M.M., Lambert, P.D., Wilding, J.P.H., Smith, D.M., Ghatei, M.A., Herbert, J. and Bloom, S.R. (1996) A role for glucagon-like peptide 1 in the central regulation of feeding. Nature, 379: 69–72.

Van den Berghe, G. (2000) Novel insights into the neuroendocrinology of critical illness. Eur. J. Endocrinol., 143: 1–13.

Verdich, C., Flint, A., Gutzwiller, J.P., Naslund, E., Beglinger, C., Hellstrom, P.M., Long, S.J., Morgan, L.M., Holst, J.J. and Astrup, A. (2001) A meta-analysis of the effect of glucagon-like peptide-1 (7-36) amide on ad libitum energy intake in humans. J. Clin. Endocrinol. Metab., 86: 4382–4389.

Volkow, N.D. and Wise, R.A. (2005) How can drug addiction help us understand obesity? Nat. Neurosci., 8: 555–560.

Wallace, B.A., Ashkan, K. and Benabid, A.L. (2004) Deep brain stimulation for the treatment of chronic, intractable pain. Neurosurg. Clin. N. Am., 15: 343–357.

Wang, G.J., Volkow, N.D., Logan, J., Pappas, N.R., Wong, C.T., Zhu, W., Netusil, N. and Fowler, J.S. (2001) Brain dopamine and obesity. Lancet, 357: 354–357.

Wellman, P.J. (2005) Modulation of eating by central catecholamine systems. Curr. Drug Targets, 6: 191–199.

Whittington, J., Holland, A., Webb, T., Butler, J., Clarke, D. and Boer, H. (2002) Relationship between clinical and genetic diagnosis of Prader–Willi syndrome. J. Med. Genet., 39: 926–932.

Whittington, J.E., Holland, A.J., Webb, T., Butler, J., Clarke, D. and Boer, H. (2001) Population prevalence and estimated birth incidence and mortality rate for people with Prader–Willi syndrome in one UK Health Region. J. Med. Genet., 38: 792–798.

Wren, A.M., Seal, L.J., Cohen, M.A., Brynes, A.E., Frost, G.S., Murphy, K.G., Dhillo, W.S., Ghatei, M.A. and Bloom, S.R. (2001) Ghrelin enhances appetite and increases food intake in humans. J. Clin. Endocrinol. Metab., 86: 5992–5995.

Wynne, K., Giannitsopoulou, K., Small, C.J., Patterson, M., Frost, G., Ghatei, M.A., Brown, E.A., Bloom, S.R. and Choi, P. (2005a) Subcutaneous ghrelin enhances acute food intake in malnourished patients who receive maintenance peritoneal dialysis: a randomized, placebo-controlled trial. J. Am. Soc. Nephrol., 16: 2111–2118.

Wynne, K., Park, A.J., Small, C.J., Patterson, M., Ellis, S.M., Murphy, K.G., Wren, A.M., Frost, G.S., Meeran, K., Ghatei, M.A. and Bloom, S.R. (2005b) Subcutaneous oxyntomodulin reduces body weight in overweight and obese subjects: a double-blind, randomized, controlled trial. Diabetes, 54: 2390–2395.

Young, A.A. (2006) Obesity: a peptide YY-deficient, but not peptide YY-resistant, state. Endocrinology, 147: 1–2.

Zipf, W.B., O'Dorisio, T.M. and Berntson, G.G. (1990) Short-term infusion of pancreatic polypeptide: effect on children with Prader–Willi syndrome. Am. J. Clin. Nutr., 51: 162–166.

Zipf, W.B., O'Dorisio, T.M., Cataland, S. and Sotos, J. (1981) Blunted pancreatic polypeptide responses in children with obesity of Prader–Willi syndrome. J. Clin. Endocrinol. Metab., 52: 1264–1266.

Kalsbeek, Fliers, Hofman, Swaab, Van Someren & Buijs
Progress in Brain Research, Vol. 153
ISSN 0079-6123

CHAPTER 4

Glucocorticoids, chronic stress, and obesity[☆]

Mary F. Dallman*, Norman C. Pecoraro, Susanne E. La Fleur, James P. Warne, Abigail B. Ginsberg, Susan F. Akana, Kevin C. Laugero, Hani Houshyar, Alison M. Strack, Seema Bhatnagar and Mary E. Bell

University of California at San Francisco, San Francisco, CA 94143-0444, USA

Abstract: Glucocorticoids either inhibit or sensitize stress-induced activity in the hypothalamo-pituitary-adrenal (HPA) axis, depending on time after their administration, the concentration of the steroids, and whether there is a concurrent stressor input. When there are high glucocorticoids together with a chronic stressor, the steroids act in brain in a feed-forward fashion to recruit a stress-response network that biases ongoing autonomic, neuroendocrine, and behavioral outflow as well as responses to novel stressors. We review evidence for the role of glucocorticoids in activating the central stress-response network, and for mediation of this network by corticotropin-releasing factor (CRF). We briefly review the effects of CRF and its receptor antagonists on motor outflows in rodents, and examine the effects of glucocorticoids and CRF on monoaminergic neurons in brain. Corticosteroids stimulate behaviors that are mediated by dopaminergic mesolimbic "reward" pathways, and increase palatable feeding in rats. Moreover, in the absence of corticosteroids, the typical deficits in adrenalectomized rats are normalized by providing sucrose solutions to drink, suggesting that there is, in addition to the feed-forward action of glucocorticoids on brain, also a feedback action that is based on metabolic well being. Finally, we briefly discuss the problems with this network that normally serves to aid in responses to chronic stress, in our current overindulged, and underexercised society.

Both glucocorticoids and stressors come in two flavors that are determined by their duration of action. Glucocorticoids act in several temporal domains through different mechanisms — rapid (nontranscriptional in minutes), intermediate (genomic in hours) and long (nuclear protein–protein in days) (Keller-Wood and Dallman, 1984). Stressors can also be punctate: immediate or potential threats to personal integrity that, once dealt with are also finished with and do not provide pronounced lasting effects; but stressors can also be persistent: chronic threats to the maintenance of mental or bodily integrity that cannot be escaped or controlled, physiologically, behaviorally or psychologically. The sustained stressor may leave prolonged traces when glucocorticoids are sufficiently elevated. Importantly, chronic elevations of glucocorticoids act with different physiological outcomes, depending on whether they are persistently elevated in the presence or absence of concurrent stressors. This is because in the presence of a chronic stressor, a central stress response network is recruited that, through the action of glucocorticoid-induced increases in limbic corticotropin-releasing factor (CRF), biases the brain toward increased use of networks that also result in altered monoaminergic tone. Because of this recruitment, behaviors, autonomic, and neuroendocrine outflows change. With high intensity, or long

☆Preparation of this article and many of the experiments discussed were supported, in part, by NIH grants DK 28172 and DA 16944

*Corresponding author.; E-mail: mary.dallman@ucsf.edu

DOI: 10.1016/S0079-6123(06)53004-3

duration, the interaction between stressors and glucocorticoids on brain may outlast the duration of the stressor, and altered patterns of behavior, autonomic, and neuroendocrine outflow may persist in the absence of ongoing stress. Both acute and chronic glucocorticoid responses to stressors are essential for life; however, the stressor–glucocorticoid-induced plastic effects on brain networks that result from persistent stressors may also have deleterious consequences for the chronically stressed organism.

Here, we will first discuss effects of chronic glucocorticoid treatment on basal activity of the hypothalamo-pituitary-adrenal (HPA) axis, next the rapid actions of stress-induced glucocorticoid secretion on function in the HPA axis, suggesting a potential mechanism, and next, the effects of chronic elevations in glucocorticoids with accompanying stressors on HPA function. We will then describe what we believe is chronic stressor-induced recruitment of the central stress–response network: what the network consists of, and how it is expressed. Finally, we will mention what we believe are the long-term effects of chronic stressors and elevated glucocorticoids on the physiology and behavior of the organism.

Glucocorticoids and function in the HPA axis

Chronic glucocorticoid actions without a concurrent stressor

Probably the best recognized action of glucocorticoids on the HPA axis is strong reduction for the abolition of activity in the system. Chronic treatment with glucocorticoid in the absence of ongoing stress exerts marked feedback effects and is the basis for the well-accepted model of autoregulation in the system. People with diseases, such as severe asthma, or other autoimmune diseases as well as recipients of organ transplants are treated for prolonged periods with exogenous glucocorticoids, and it is well known that they may require additional treatment if they are exposed to additional stress, such as surgery or sepsis, because of the profound, steroid-induced inhibition of their

HPA axis (Graber et al., 1965; Lamberts et al., 1995; Orth, 1995).

In animal experiments, it is clear that exogenous administration of glucocorticoid for several days in the absence of concurrent stress inhibits basal activity in the HPA axis in a very sensitive dose-related fashion, so that in adrenalectomized rats basal CRF and ACTH concentrations are restored to normal when exogenous corticosterone results in mean daily plasma concentrations of $5\,\mu g/dl$ (Akana et al., 1985; Akana and Dallman, 1997). At low doses of corticosterone provided to adrenalectomized rats either through steroid in the drinking water (Wilkinson et al., 1981), or by implanted steroid constant release pellet (Akana et al., 1985, 1992; Levin et al., 1988; Bradbury et al., 1993; Scribner et al., 1993; Strack et al., 1995a; La Fleur et al., 2004), basal activity in the HPA axis decreases as a function of the circulating steroid concentrations in consequence of occupancy of both the high affinity mineralocorticoid receptors (MRs) and the lower affinity glucocorticoid receptors (GRs) (Dallman et al., 1989b; Bradbury et al., 1994). Moreover, at very high daily doses of glucocorticoid for 3 weeks, it requires several days to weeks for basal activity of the HPA axis to return to normal after cessation of the treatment. The sequence of restoration of function in the axis is first, hypothalamic CRF secretion, next, pituitary ACTH secretion, and, finally renewed adrenocortical sensitivity and glucocorticoid secretion (Graber et al., 1965; Nicholson et al., 1984, 1987, 1988; Harbuz et al., 1990).

Chronic glucocorticoid actions on HPA responses to an acute stressor

As with basal activity, HPA axis responses to single acute stressors are also inhibited in proportion to the degree of chronic glucocorticoid treatment; however, the type of stressor also determines the results. Within hours, large doses of the potent synthetic glucocorticoid, dexamethasone entirely inhibit ACTH and corticosterone responses to moderate stimuli, that have steroid-sensitive brain networks afferent to the CRF neuron (Dallman and Yates, 1968); corticosterone treatment inhibits the responses after days (Wilkinson et al., 1981; Scribner et al., 1993). It is likely that the high doses of glucocorticoids exert their major effects on pituitary ACTH synthesis and secretion (Dallman et al., 1985). However, during the intermediate feedback time period that was used after large doses of dexamethasone, the brain components of the HPA axis are not inhibited since it is clear that with sufficiently intense stimuli that are not blocked by steroid-sensitive networks in brain, an HPA response can still be obtained (Dallman and Yates, 1968). At lower daily concentrations of corticosterone there is dose-related inhibition of ACTH and corticosterone responses to acute stressors; and different stimuli can be used to distinguish between glucocorticoid-sensitive and glucocorticoid-insensitive afferent inputs to the CRF neuron (Wilkinson et al., 1981; Watts, 2005).

Differential steroid feedback sensitivity was shown by the finding that the HPA responses of rats exposed to ether vapor is readily and entirely inhibited by pretreatment with glucocorticoids, whereas the response to laparotomy and intestinal manipulation (L+IM) under ether anesthesia is not (Dallman and Yates, 1968; Sato et al., 1975; Wilkinson et al., 1981). Corticosterone responses to both ether alone and sham adrenalectomy are entirely inhibited by pretreatment with dexamethasone, whereas corticosterone responses to L+IM, as well as to hemorrhage still occur (Dallman and Yates, 1968). Figure 1, top, shows changes in CRF bioactivity in the median eminence with time after ether and sham adrenalectomy, or ether and L+IM in intact rats, and Fig. 2 distinguishes between ether alone or ether and L+IM as a function of adrenal number and corticosterone treatment in bilaterally, unilaterally adrenalectomized and intact rats.

The stimulus of L+IM exerts more persistent input to CRF neurons than sham adrenalectomy, with a secondary rise in CRF content in the median eminence that peaks at 80 min (Fig. 1, top). This is supported by the increases in both ACTH and corticosterone that persist for the full 2 h period, in contrast to the responses to sham adrenalectomy which peak within the first 60 min and then are reduced during the second hour of the experiment (Fig. 1). L+IM also demonstrates an

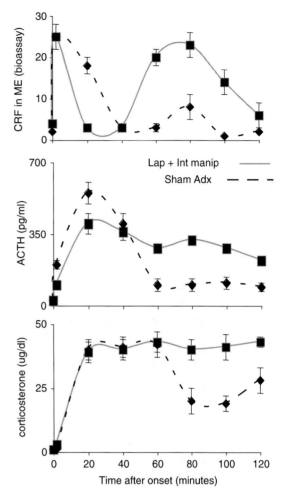

Fig. 1. Differential responses in the hypothalamic-pituitary-adrenal axis to different stressors suggests that some stimuli act through corticosteroid-sensitive inputs to the CRF neurons in the paraventricular hypothalamus (sham adrenalectomy), while others are insensitve to corticosteroid inhibition (laparotomy + intestinal manipulation). Top: bioassayable CRF content in the median eminence. Middle: plasma ACTH. Bottom: plasma corticosterone after imposition of the two stimuli. Note the rapid depletion followed by repletion of CRF and sustained increases in ACTH and corticosterone in rats exposed to laparotomy + intestinal manipulation, compared to the slower depletion and lesser repletion of CRF, and decreased ACTH and corticosterone after 60 min in sham-adrenalectomized rats. Note also that maximal values in the three variables measured are not greater in the rats exposed to laparotomy + intestinal manipulation. Data redrawn from Sato et al. (1975).

entirely different glucocorticoid inhibition regression pattern on ACTH from that observed with ether alone (Fig. 2). These results suggest strongly

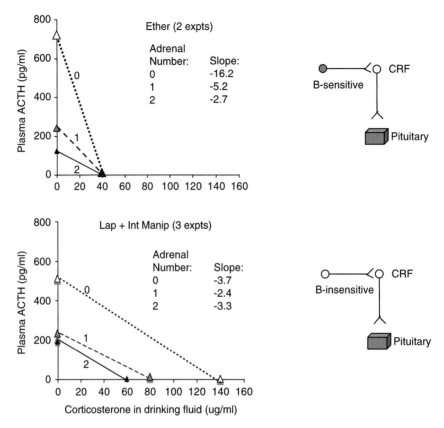

Fig. 2. Differential sensitivity to corticosterone feedback to different stressors in rats that were adrenalectomized (0 adrenals), unilaterally adrenalectomized (1 adrenal) or sham-adrenalectomized (2 adrenals). Fitted regression lines of ACTH on the concentration of corticosterone given to all rats in the drinking water show convergence of the lines at a concentration of about 40 μg/ml corticosterone in rats undergoing ether stress (top), but no convergence, but rather parallel regressions in rats undergoing laparotomy + intestinal manipulation under ether anesthesia (bottom). Models, interpreting the data are shown to the right of each graph, and suggest that ether stress activates CRF secretion through steroid-sensitive pathways, while laparotomy + intestinal manipulation activates CRF secretion through inputs that are not glucocorticoid-sensitive. In both instances, it is assumed that the corticosterone inhibited pituitary ACTH synthesis and secretion to an equal extent. Data redrawn from Wilkinson et al. (1981).

that stimuli evoked by L + IM arrive at the hypothalamus through steroid-insensitive afferent neurons, like afferents from hypovolemia, induced by polyethylene glycol (Watts, 2005). By contrast, the ACTH response to ether alone, converges independently of the adrenal number at ~40 μg/ml corticosterone in the drinking water (Fig. 2, top), suggesting strongly that, in contrast to L + IM (Fig. 2, bottom), this dose of corticosterone entirely inhibits afferent input to CRF neurons from the ether stimulus.

Thus, with acute, single stimuli it is possible to show that some afferent stimulus networks in brain are sensitive to glucocorticoid treatment whereas others are not, although the pituitary corticotrope invariably is inhibited by glucocorticoids (Aguilera, 1994b; Young et al., 1995). Nonetheless, in the presence of chronic steroid treatment and in the absence of concurrent chronic stress, glucocorticoids clearly inhibit both basal and acute stimulus-induced activity in the HPA axis in a dose-related manner.

Summary. In the absence of concurrent stress, prolonged treatment with glucocorticoids reduces activity in the HPA axis, and blunts responses to acute stressors in proportion to the dose of steroid given. However, the sequence of recovery in components of the HPA suggests strongly that there is only a minor effect of the glucocorticoid on the brain, provided that steroid-sensitive brain elements are not involved in transmitting the signal induced by the stimulus, but there is a strong negative effect on ACTH synthesis in the corticotrope. The absence of trophic ACTH stimulation of the adrenal is responsible for prolonged adrenal hyporesponsiveness after prolonged high-dose steroid treatment. The adrenal corticosteroid response to stress is restored to normal only after ACTH secretion returns to normal, or supranormal, levels (Graber et al., 1965).

Acute glucocorticoid actions and acute stressors

In the stressor-naïve, intact person and animal the normal sequence of HPA component responses to acute stimuli is afferent activation of the hypothalamic CRF neuron, secretion of CRF from axons in the median eminence to activate the pituitary corticotrope and, thus, ACTH secretion into the general circulation where it acts at the adrenal cortex to stimulate glucocorticoid secretion.

Systemic glucocorticoids rise within 2–5 min of the stimulus, and act to inhibit the duration of the CRF and ACTH secretory responses to the acute stimulus within minutes (Keller-Wood and Dallman, 1984) in both people and rats (Dallman and Yates, 1969; Dallman et al., 1972; Fehm et al., 1979; Widmaier and Dallman, 1984; Young et al., 1990a, b; Falkerstein et al., 2000; Dallman, 2005). In the absence of the fast feedback action of glucocorticoids, the capacity of the pituitary to secrete ACTH may be exhausted (Dallman et al., 1974; Akana et al., 1986), additionally, excessive amounts of corticosterone would be secreted.

An elegant example of this rapid action of glucocorticoids is shown in sheep exposed to a predator dog, by rapid sampling measurement of CRF secretion from the paraventricular nuclei (PVN) through dialysis (Cook, 2004). In this study, the dog was present for 15 min (Fig. 3). CRF concentrations in the dialysate increased within seconds of exposure, rose to a peak at 5 min and then decayed to a level just above control during the last 10 min in the presence of the dog as plasma cortisol concentrations were rising rapidly in response to the CRF–ACTH stimulus. It is likely that the reduction in CRF secretion that

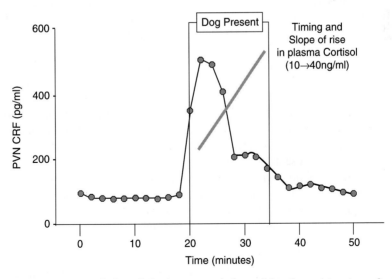

Fig. 3. There is a rapid CRF response over the hypothalamic paraventricular nuclei to the predator stress of a dog in sheep; however, CRF secretion falls pari passu with the rise in endogenous cortisol while the stressor is still present. These results are consistent with the time course and effect of the fast feedback action of glucocorticoids on CRF neurons. Data redrawn, with permission, from Cook (2004) and with permission from Elsevier (2004).

occurred in the presence of the predator was occasioned by the rise in cortisol that occurred between 5 and 15 min after the onset of the stressor.

Electrophysiological studies of hypothalamic slices by Tasker's lab provided a mechanism for the rapid, non-transcriptional feedback effects of glucocorticoids on CRF secretion. Within 3 min of the onset of steroid perfusion, electrical activity of subsequently identified CRF neurons decreased. It was shown that the decrease resulted from glucocorticoid-stimulated secretion of enocannabinoids that then acted on CB-1 receptors on glutamatergic afferent input to the cells (Di et al., 2003).

Although there is much more work to do to test fully whether one of the mechanisms of glucocorticoid fast feedback on the HPA axis is mediated in vivo by endocannabinoids, initial studies suggest that this may be at least the mechanism by which the HPA response to restraint is curbed (Ginsberg et al., unpublished data). The rapid actions of glucocorticoids which curb the duration, but not the peak magnitude, of stimulus-induced ACTH secretion are key to limiting the duration of action of the HPA axis to provide the body with the precise amount of steroid needed to counter the threat, but not to provide overwhelming amounts of glucocorticoids that might provide harmful effects (Munck et al., 1984).

Chronic glucocorticoid actions with a concurrent stressor

Chronic or repeated stressors exert marked effects on transmitters and neurons in most of brain, including the cortex, limbic system, midbrain, and brainstem (see central stress response network, below). Therefore, it should be of no surprise that rather than inhibition, as would be predicted from long-term elevation of glucocorticoids in the absence of a stressor, there is sensitization of activity in the HPA axis in response to a novel stimulus in chronically stressed rats (Dallman and Jones, 1973; Sakellaris and Vernikos-Danellis, 1975; Kant et al., 1983; Armario et al., 1986; Chappell et al., 1986; Fichter and Pirke, 1986; Kant et al., 1987; de Goeij et al., 1991, 1992a, b; Ottenweller

et al., 1992; Sarlis et al., 1992; Anderson et al., 1993; Bartanusz et al., 1993; Dallman, 1993; Harbuz et al., 1993; Aguilera, 1994a; Marti et al., 1994, 1999; Servatius et al., 1994; Chowdrey et al., 1995; Janssens et al., 1995; Akana et al., 1996; Gomez et al., 1996; Albeck et al., 1997; Brown and Sawchenko, 1997; Bonaz and Rivest, 1998; Pittman et al., 1998; Buwalda et al., 1999, 2001; Mizoguchi et al., 2001). Frequently, it is assumed that the sensitized response to a novel stressor represents a failure of glucocorticoid feedback efficacy, because in some instances hippocampal GR content is reduced. However, this seems unlikely. Although the steroid-dose inhibition of the response curve is clearly shifted to the right, as it is between circadian nadir and peak (Dallman et al., 1989a), it actually requires controlled glucocorticoid concentrations in the high (stress), not the control range for the normal sensitized response to novel stress, that is seen in chronically stressed intact rats, to occur in chronically stressed adrenalectomized rats with clamped corticosterone concentrations (Akana and Dallman, 1992, 1997; Scribner et al., 1993; Tanimura et al., 1998; Tanimura and Watts, 1998, 2001; Dallman, 2003).

The fact that HPA sensitization requires corticosterone concentrations above normal replacement levels (i.e. constant values $\geqslant 100$ ng/ml, for at least several days), suggests strongly that elevated corticosteroid concentrations are essential to allow the sensitized responses. Moreover, rather than ineffective glucocorticoid inhibition in brain, it seems to us more likely that the drive to hypothalamic CRF neurons from the rest of brain is increased during chronic stress, and that this is observed as sensitization of HPA responses when new stressors are encountered. As will be detailed below, the effect of persistently elevated glucocorticoids increases amygdalar CRF expression, which is a key component of the central stress response network once it has been recruited by a chronic stressor.

Summary. Sustained treatment with glucocorticoids in the absence of concurrent stress inhibits both basal and acutely stimulated activity in the HPA axis. It seems likely that a major site of this inhibition is at the pituitary, with less central inhibition. The acute corticosteroid response to a stressor also feedsback to inhibit activity in the HPA axis, through non-transcriptionally mediated effects that may, in part, be explained by glucocorticoid-induced endocannabinoid secretion from CRF neurons that reduces afferent input to the neuron through association with CB-1 receptors on presynaptic axons. Fast feedback also occurs at the pituitary to inhibit ACTH secretion (Widmaier and Dallman, 1984; Ginsberg et al., 2005). Finally, in the presence of concurrent stress, HPA responses to novel stimuli are sensitized, only when the tonic mean concentration of glucocorticoids is elevated above the normal daily mean value. This sensitization is a consequence of the recruitment of a chronic stress network that requires elevated corticosterone and is probably initiated by increased limbic CRF activity.

The central stress response network

It now seems unambiguous that, after the action of elevated glucocorticoids and stress, endogenous CRF networks initiate and recruit the activation of the neural networks that promote the changes from normal in behaviors, autonomic neural activity and neuroendocrine responses, that are characteristic of organisms undergoing a period of chronic stress. Extrahypothalamic CRF neuronal cell groups are found throughout the brain: in the central nucleus of the amygdala, Barrington's nucleus, bed nucleus of the stria terminalis, locus coeruleus and olfactory bulb; moreover, there are scattered CRF-synthesizing neurons elsewhere in brain and there is a wide distribution of CRF-immunoreactive cell fibers found throughout the brain (Swanson et al., 1983). Furthermore, there is an equally wide distribution of CRF receptors that is generally, but not always, well matched to CRF-fiber endings (van Pett et al., 2000; Reul and Holsboer, 2002). Thus, in terms of the potential sites of action of CRF, there is an embarrassment of riches. Below we will mention the effects of exogenous CRF and CRF antagonists administered during stressors, and the role of glucocorticoids in activating the central CRF network. Next, we

discuss the probable site and the glucocorticoid-mediated mechanism of the habituation of the CRF network with low-grade, essentially non-threatening repetitious stimuli. Then we will discuss the effect of augmented central CRF secretion on monoaminergic cell groups (Fig. 4).

CRF as the driver of the central stress response network

Since the elucidation of its structure and availability of the peptide, CRF has been suspected of mediating the behavioral and autonomic as well as the HPA responses to stress (Vale et al., 1981; Brown et al., 1982; Fisher et al., 1982). Both treatment of experimental animals with CRF usually given into a cerebral ventricle (icv) either acutely or chronically, or study of responses of genetically manipulated CRF or CRF-receptor knockout mice have consistently suggested that CRF in the brain mimics many of the well-characterized behavioral, autonomic, and neuroendocrine responses to both acute and chronic stressors (Britton et al., 1982; Fisher et al., 1982; Sutton et al., 1982; Ehlers et al., 1983; Krahn et al., 1988; Matsuzaki et al., 1989; Takahashi et al., 1989a, b; Dunn and Berridge, 1990; Koob and Britton, 1990; Rothwell, 1990; Owens and Nemeroff, 1991; Diamant et al., 1992b; Liang et al., 1992; Fisher, 1993; Gray, 1993a; Swiergiel et al., 1993; Buwalda et al., 1997; Denver, 1997; Linthorst et al., 1997; Jones et al., 1998; Arborelius et al., 2000; Shaham et al., 2000; Smagin et al., 2001; Tache et al., 2001; Bale and Vale, 2004; Heinrichs and Koob, 2004). In response to central CRF injections, mice, rats, and monkeys become more aroused, fearful, anxious, active, and aggressive; blood pressure increases as does brown adipose tissue activity, food intake decreases, and intestinal motility is altered. Moreover, learning and memory are affected (Veldhuis and De Wied, 1984; Diamant et al., 1992a; Croiset et al., 2000; Roozendaal et al., 2002; Heinrichs and Koob, 2004). It seems clear that infusion of CRF into brain can produce many of the characteristic effects of stressors and changes in output observed during chronic stress.

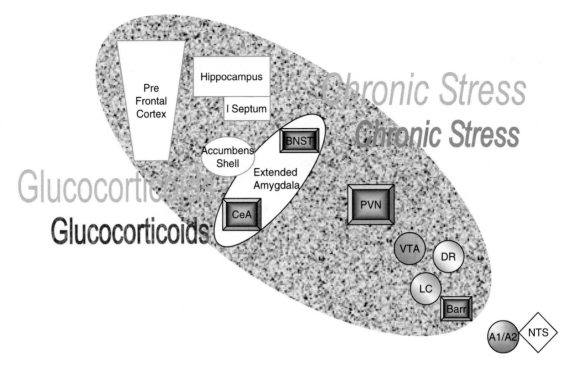

Fig. 4. Glucocorticoids and chronic stress recruit the central brain stress response network. Glucocorticoids acting directly at the amygdala stimulate CRF synthesis and secretion. CRF acts on receptors in structures throughout the brain, and, importantly recruits activity in at least 3 monoaminergic cell groups that innervate essentially the entire forebrain. In consequence, the organism chronically exposed to stress is, through the actions of CRF and monoamines, more capable of adapting to the stressful conditions. Gray picture-frames = cell groups that synthesize CRF: PVN (parvicellular neurons in the paraventricular nuclei. hypothalamus, autonomic and neuroendocrine outflow), CeA (central nucleus of the amygdala, autonomic outflow, forebrain, behavior, monoaminergic cell groups), BNST (bed nuclei of the stria terminalis, autonomic outflow, neuroendocrine, behavior), Barr (Barrington's nucleus, autonomic outflow). Shaded circles = monoaminergic cell groups: VTA (ventral tegmental area:dopamine), DR (dorsal raphe:serotonin), LC (locus coeruleus:norepinephrine), and the A1/A2 medullary noradrenergic cell groups are activated in response to chronic stress, CRF and glucocorticoids. In consequence of recruitment of this network, the stressed animal is more likely to be cautious (fearful), more ready to be diverted from tasks at hand (prefrontal cortex), adopt alternative strategies (lateral septum), enjoy rewards (n. accumbens shell), remember fearful situations (amygdala, hippocampus) better, but possibly not recall them as well (hippocampus). See text.

Stressors ± CRF antagonists

Stressors, particularly processive, psychological stressors that may suggest anticipated disaster, stimulate CRF mRNA and peptide activity in the central nucleus of the amygdala (Gray, 1991; Honkaniemi, 1992; Adamec and McKay, 1993; Kalin et al., 1994; Schulkin et al., 1994; Makino et al., 1995, 1999 Raber et al., 1995; Rivest and Laflamme, 1995; Merali et al., 1998; Palkovits et al., 1998b; Day et al., 1999; Richter and Weiss, 1999; Sajdyk et al., 1999; Palkovits, 2000; Zorrilla et al., 2001; Hand et al., 2002; McNally and Akil, 2002; Houshyar et al., 2003). However, CRF

activity in the amygdala also appears to be sensitive to alterations in metabolic state (Merali et al., 1998, 2004b; Zorilla et al., 2003). Furthermore, using the very mild stress of tail-pinch, which is known to stimulate food intake, a CRF antagonist was shown to block the stress-induced food intake, and CRF administered icv in very low doses stimulated food intake (Samarghandian et al., 2003), suggesting that, like the glucocorticoids (de Kloet et al., 1998), CRF exerts highly dose-dependent effects, and displays an inverted U-shaped dose–response curve.

Many, if not most of the physiological and behavioral effects of acute stressors are blocked by administration of CRFR1 antagonists (Britton et al., 1986a, b; Krahn et al., 1986a, b; Takahashi et al., 1989a; Baldwin et al., 1991; Menzaghi et al., 1991, 1992, 1994; Heinrichs et al., 1992; Strijbos et al., 1992; Korte et al., 1994; Basso et al., 1999; Smagin et al., 1999; Radulovic et al., 1999; Reul and Holsboer, 2002; Samarghandian et al., 2003; Carrasco and Van de Kar, 2003; Bale and Vale, 2004; Greenwood-Van Meerveld et al., 2005). However, less study has been performed on the effects of manipulation of CRF receptors under conditions of chronic stressors except with tests of their effects on paradigms that are thought to cause depression (Cole et al., 1990; Mayer and Fanselow, 2003; Muller et al., 2003; Claes, 2004). Comparison of CRF and CRFR1 expression in brains of suicides and controls has revealed greatly elevated CRF content in hypothalamic PVN (Raadsheer et al., 1994) and prefrontal cortex (PFC), as well as decreased numbers of CRFR1 in PFC (Nemeroff et al., 1988; Merali et al., 2004a). Because stressor-induced alterations in motor outputs are blocked so well, there has been considerable interest in the pursuit of non-peptide CRFR1 antagonists for the potential treatment of human depression and relapse to drug abuse as well as other disorders thought to be mediated by the central or peripheral actions of CRF (Koob, 1992; Schulkin et al., 1994; Griebel et al., 1998; Arvantis et al., 1999; Deak et al., 1999; Holsboer, 1999; Rivier et al., 1999; Arborelius et al., 2000; Habib et al., 2000; Zobel et al., 2000; Koob and Le Moal, 2001; Sarnyai et al., 2001; Griebel et al., 2002; Hsin et al., 2002; Rivier et al., 2002; Merali et al., 2004a). Whether or not the use of CRF receptor antagonists proves to be useful for treatment of human conditions, the compounds have been and certainly will continue to be important for dissection of the role of CRF in mammalian stress responses.

Amygdalar CRF and glucocorticoids

CRF mRNA and peptide concentrations increase in the amygdala after both acute and chronic stressors (Adamec and McKay, 1993; Gray, 1993b; Kalin et al., 1994; Makino et al., 1994b;

1999; Merlo Pich et al., 1994, 1995; Raber et al., 1995; Rivest and Laflamme, 1995; Lee and Davis, 1997; Rodriguez de Fonseca et al., 1997; Feldman and Weidenfeld, 1998; Merali et al., 1998; Richter and Weiss, 1999; Zorrilla et al., 2001; Hand et al., 2002; Cook, 2004). Moreover, when CRF receptor antagonists are infused directly into the amygdala, many behavioral, autonomic, and neuroendocrine responses to acute stressors are reduced, or abolished (Swiergiel et al., 1993; Wiersma et al., 1995; Lee and Davis, 1997; Feldman and Weidenfeld, 1998; Sajdyk et al., 1999; Erb et al., 2001; Bakshi et al., 2002; Roozendaal et al., 2002; Carrasco and Van de Kar, 2003). However, in the absence of concurrent chronic stressors there are only minor effects of repeated injections of CRF into the amygdala on behavior or acutely stimulated HPA activity (Daniels et al., 2004), suggesting that stress-induced activation of endogenous CRF-secreting cells is key to the central stress response network.

Circulating concentrations of glucocorticoids are important to the level of expression of CRF in the amygdala. Adrenalectomy reduces and glucocorticoid treatment restores CRF mRNA in the amygdala dose-dependently (Swanson and Simmons, 1989; Makino et al., 1994a, 1995; Watts and Sanchez-Watts, 1995; Viau et al., 2001). However, the reduction of CRF expression in amygdala by adrenalectomy may result primarily from peripheral metabolic disturbances, since this CRF can be restored to normal in the absence of corticosterone when adrenalectomized rats drink high density sucrose solutions (Laugero et al., 2001a, see below). Nonetheless, it is highly likely that sustained, stress-induced increases in glucocorticoid secretion are in major part responsible for the observed stress-induced increases in amygdalar CRF expression.

The effects of glucocorticoids on amygdalar CRF are exerted directly at the amygdala, and their action on amygdalar CRF is required for sensitized responses to novel stressors to occur in chronically or repeatedly stressed animals. When 30 µg pellets of corticosterone are placed over the central nucleus of the amygdala CRF mRNA expression is increased as is stress-induced anxiety-related behavior (Shepard et al., 2000) and HPA

84

responses to an acute restraint stimulus is also augmented (Shepard et al., 2003). In sheep, exposure to a barking dog stimulates CRF secretion over the amygdala within seconds; after decay of the first, fear-induced peak in CRF, there is a second marked rise in CRF secretion that peaks 30–40 min after the stressor (Cook, 2001, 2002). The secondary rise in amygdala CRF occurs as cortisol concentrations increase in plasma and in the fluid bathing amygdala neurons (Cook, 2001, 2002). Moreover, Cook has shown the secondary rise to be mediated by the action of cortisol at the amygdala, since this is blocked by prior injection of the GR antagonist RU486 (Cook, 2002). Finally, in this remarkable series of experiments on sheep, seven daily inescapable exposures to a dog for 5 min/day resulted in markedly augmented CRF secretion from the amygdala after introduction of the novel stressor of foreleg shock on day 8; the sensitized response to shock was blocked when RU486 was injected into the amygdala 5 min before exposure to the dog on each of the 7 preceding days (Fig. 5). This, again, represents a very rapid action of glucocorticoids. However, there is recent clear immunocytochemical evidence for glucocorticoid receptors embedded in the postsynaptic dendritic membranes of neurons in the lateral amygdala (Johnson et al., 2005), suggesting that the machinery for rapid steroid actions through GR exists in the amygdala, as well as in other brain sites (Liposits and Bohn, 1993).

These results show that elevated glucocorticoids acting at the amygdala are essential for increased CRF expression and secretion from the amygdala, and, that increased amygdalar expression of CRF is tightly coupled to hypersensitivity of the HPA axis to stressors; moreover, it appears that the persistent increase in amygdalar CRF activity requires days to occur and that it does not occur after the response to only a single stressor (Cook, 2002; Laugero et al., 2002), suggesting that a memorial input is also required, at least for the HPA responses.

A critical experiment that coupled the amygdaloid actions of glucocorticoids and amygdala CRF on behavior, HPA activity and autonomic outflow was recently published (Myers et al., 2005). When bilateral implants of corticosterone

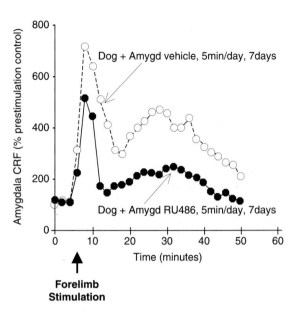

Fig. 5. Repeated inescapable exposure to a dog (5 min/day) for 7 days sensitizes amygdala CRF responses to the novel stressor of forelimb shock (open circles). The sensitization of CRF is blocked by intra-amygdalar administration of the glucocorticoid receptor antagonist, RU486, daily 5 min prior to presenting the dog (closed circles. The animals treated with RU486 had a CRF response to forelimb shock of similar magnitude to that of naïve sheep, or sheep that were allowed to escape from the dog stress (not shown). Data redrawn with permission from Cook (2002) and with permission from Elsevier (2002).

were made but a CRF-R1 antagonist was given just before stress, the receptor antagonist blocked the usual effects of amygdalar corticosterone implants on behavior, HPA activity and autonomic responses (Myers et al., 2005), suggesting strongly that all three motor effects of corticosterone are mediated by the recruited amygdalar CRF. Thus, there is a powerful feed-forward action of corticosterone in brain that may be entirely explained by recruitment of CRF in the amygdala. If so, it would clarify the distinct similarities observed between behavioral, autonomic and endocrine activities of the glucocorticoids and CRF (Holsboer, 1999, 2000) and show a reverberating circuit in which the hypothalamic PVN secretes CRF and stimulates glucocorticoid secretion, which then, in its turn acts to stimulate amygdalar CRF producing the typical motor responses to chronic stressors (Schulkin et al., 2005).

Although responses of amygdalar CRF to chronic stressors and glucocorticoids appear to be essential for many of the other changes that occur with chronic stress (behavioral, autonomic, and neuroendocrine), the stressor-induced trigger to the amygdala to recruit these changes has not been nearly as well explored. Since both the catecholaminergic cell groups in the region of the nucleus of the tractus solitarius (Swanson, 1982; Campeau and Davis, 1995; Myers and Rinaman, 2002; Herman et al., 2003) and the forebrain (Everitt et al., 1999; Hasue and Shammah-Lagnado, 2002; Asan et al., 2005) innervate the amygdala, there is ample structural opportunity for persistent bottom up and top down stimuli to activate the chronic stress network.

However, many low-intensity repeated stressors like restraint, cold, noise, and ethanol provoke habituation rather than sensitization in the HPA axis (reviewed in Dallman and Bhatnagar, 2001). There is very good anatomical and some functional evidence supporting both stressor-receptive (Nauta, 1962; Bohus, 1970; Berendse and Groenewegen, 1990; Bubser and Deutch, 1998; Young and Deutch, 1998; Bhatnagar and Dallman, 1999; Bubser and Deutch, 1999; Smith et al., 2002; Van der Werf et al., 2002; Spencer et al., 2004) and memorial roles (Bohus, 1970; Buchanan et al., 1989; Jones et al., 1989; Berendse and Groenewegen, 1990; Chen and Su, 1990; Miyata et al., 1995; Castro-Alamancos and Connors, 1996; Kinomura et al., 1996; Bubser and Deutch, 1998; Deutch et al., 1998; Bhatnagar and Dallman, 1999; Bubser and Deutch, 1999; Michi et al., 2001; Timofeeva and Richard, 2001; Smith et al., 2002; Van der Werf et al., 2002; Spencer et al., 2004) played by the PVN of the thalamus.

The paraventricular thalamus is a midline cell group that extends throughout the anterior–posterior length of the thalamus, and has been designated the "neuroendocrine thalamus" because of its afferent and efferent neuronal connections (Turner and Herkenham, 1991). It is a cell group that receives polymodal inputs from visceral afferents, sensory afferents, hypothalamus, and cortical regions associated with emotional memory and it has a strong output to Acb, amygdala, and prefrontal cortex.

This long nucleus serves to inhibit responses to new, or repeated stimuli to the HPA axis (McFarland and Haber, 2002; Smith et al., 2002; Jaferi et al., 2003; Spencer et al., 2004), and lesions of this structure augment ACTH responses to both chronic and acute stressors and is sensitive to glucocorticoids (Bhatnagar and Dallman, 1998b; Bhatnagar et al., 2000b; Bhatnagar et al., 2002; Fernandes et al., 2002; Bhatnagar and Vining, 2003; Spencer et al., 2004). Moreover, the posterior paraventricular thalamus innervates all 3 of the areas in the amygdala, the nucleus accumbens and several other limbic structures important for emotional learning and memory (Turner and Herkenham, 1991; Moga et al., 1995; Van der Werf et al., 2002; Pinto et al., 2003; Dong and Swanson, 2004) that showed increased fos-immunoreactivity when rats had been chronically stressed then subjected to novel stress, compared to novel stress alone (Bhatnagar and Dallman, 1998a). When lesions of the paraventricular thalamus were restricted to the posterior part fo the nucleus, there was no effect of these on acute ACTH responses to stress, but the lesions augmented responses to new stimuli in chronically stressed rats, and also blocked the response inhibition that is seen with some repeated stimuli (Bhatnagar and Dallman, 1998c; Bhatnagar et al., 2000b; Bhatnagar et al., 2002; Bhatnagar and Vining, 2003; Bhargava et al., 2004). Given its strong direct input to the central, basolateral and basomedial cell groups of the amygdala, as well as its multimodal afferent innervation, the posterior paraventricular thalamus may be a conduit to a key memorial site, like the prefrontal cortex, and changes in its activity may reflect the possibility that it is one of the sites in brain responsible for the reduction, or habituation of chronic, repeated stress responses through inhibition of CRF cells in the central n. of the amygdala.

Once turned on by chronic stress and high glucocorticoids, increased CRF output from the amygdala (Fig. 5), and probably other sites of CRF cell groups, appears to affect not only autonomic neural outflow, but also, both directly and indirectly influences the activity of many brain monoaminergic cell groups, altering arousal, behavioral responses, and neuroendocrine control (see Fig. 4).

CRF and glucocorticoid on monoaminergic cell groups

Both glucocorticoids and CRF affect neuronal activity in, particularly, the monoaminergic cell groups that massively innervate the forebrain: the serotoninergic cells in the dorsal raphe nuclei cell group, the norepinephrinergic neurons of the locus coeruleus and the dopaminergic cells in the ventral tegmental area. Other effects of CRF and perhaps the glucocorticoids are also exerted at many axonal endings of these monoaminergic cells. The overall effect of the actions of these stress hormones in the presence of stress appears to be to tune the animal for increased alertness, vigilance, rapid learning and taking increased advantage of uncertain situations to find solutions to problems.

Serotonin
Serotoninergic cells in the dorsal raphe nucleus innervate most forebrain structures and modulate ongoing behaviors. Azmitia summarizes the role of serotonin in brain as integrating external and internal signals and maintaining the integrity of the neural target cells for these signals in an optimally responsive state (Azmitia, 1999, 2001). Adrenalectomy and glucocorticoid treatments decrease and increase, respectively, 5HT expression in serotonin neurons of the raphe (Azmitia and McEwen, 1969, 1974), although it is not clear whether these effects of altering glucocorticoids are mediated through their effects on CRF neuronal activity. Both stress and CRF infusions into the dorsal raphe region modulate neuronal activity, serotonin output in the forebrain as well as behaviors in complex ways that appear to be determined by which cells or which CRF receptors are activated (Price et al., 1998; Kirby et al., 2000; Lowry et al., 2000; Price and Lucki, 2001; Valentino et al., 2001a, b; Price et al., 2002; Carrasco and Van de Kar, 2003; Hammack et al., 2003; Oshima et al., 2003; Roche et al., 2003; Thomas et al., 2003; Tan et al., 2004; Asan et al., 2005; Maier and Watkins, 2005; Waselus et al., 2005). For instance, a CRF receptor antagonist infused into the dorsal raphe prior to an initial swim stress blocks the enhanced serotonin secretion in the lateral septum that generally accompanies this stimulus, but an antagonist similarly provided 24 h later enhances serotonin secretion in the lateral septum (Price et al., 2002), actions that are believed to shift swimming behavior from an active to a probably more appropriate passive mode (Valentino and Commons, 2005).

Norepinephrine
The noradrenergic cells in the locus coeruleus (LC) innervate most of the brain, but particularly the forebrain. Firing patterns of cells in the LC have been distinguished along an intensity line by Aston-Jones, from low tonic to nil (observed during sleep) to phasic, in which this activity appears to reinforce specific decisions made by the prefrontal cortex, to high tonic, in which the activity of other cells in forebrain is elevated so that possibly attention will be diverted from the task in hand to other possibilities; this range of activity is summarized as yielding optimal adaptive performance (Aston-Jones and Cohen, 2005). There is major input to the LC from the frontal cortex, and it appears that this input may serve as a positive loop so that the phasic activity of LC cells more easily enables the cortical output. Additionally, CRF fibers from the amygdala innervate the dendritic tree of LC neurons (Van Bockstaele et al., 1998), thus providing the anatomical basis for activated amygdalar CRF neurons to activate, in turn, activity of the LC. Chronic stressors, adrenalectomy, glucocorticoid and CRF treatment all affect activity in noradrenergic cells in the LC, either raising or lowering responsivity. It seems highly likely, as proposed by Chrousos and Gold (1992) that the LC noradrenergic neurons activate the HPA axis, and lesions of the LC decrease HPA responses to acute stress (Ziegler et al., 1999). Furthermore, activity in the LC is reflected by norepinephrine output in the extended amygdala, hippocampus and prefrontal cortex, as well as in downstream structures such as the brainstem A1/A2 noradrenergic cell groups that gather visceral information from inputs to the nucleus of the tractus solitarius and send monosynaptic input to amygdala (Quirarte et al., 1998), bed nucleus of the stria terminalis (Pacak et al., 1995a), prefrontal cortex (Palkovits et al., 1998a), and the paraventricular hypothalamus (Pacak et al., 1995b; Plotsky et al., 1989). Chronic stressors

clearly activate the LC (Valentino et al., 1992, 1983; Melia and Duman, 1991; Smagin et al., 1995; Schulz and Lehnert, 1996; Curtis et al., 1997; Pavcovich and Valentino, 1997; Smagin et al., 1997; Van Bockstaele et al., 1998; Lechner and Valentino, 1999; Okuyama et al., 1999; Van Bockstaele et al., 1999; Kawahara et al., 2000; Bruijnzeel et al., 2001; Jedema et al., 2001; Valentino et al., 2001c; Curtis et al., 2002; Zeng et al., 2003; Dunn et al., 2004), in all probability through the mechanisms suggested above. The role of norepinephrine on stress-induced behavioral responses has recently been critically reviewed (Morilak et al., 2005).

Dopamine

As with the other monoaminergic cell groups, activity of dopaminergic (DA) cells, particularly in the mesocorticolimbic system, are both glucocorticoid- and CRF-sensitive, and are also innervated by CRF fibers from the central nucleus of the amygdala (Hall et al., 2001; Jackson and Moghaddam, 2001; McFarland et al., 2004). The sensitivity of dopamine secretion to glucocorticoids is starkly shown in measurement of DA release from n. accumbens under basal conditions and after injections of cocaine and morphine (Barrot et al., 2001); in adrenalectomized rats under all three conditions DA secretion is low, specifically in the shell but not core region of the nucleus, and restored to normal after glucocorticoid treatment. CRF fibers and endings innervate the ventral tegmental area (VTA), the source of DA to the n. accumbens (Sawchenko and Swanson, 1989), CRF infused into the VTA stimulates DA output in the n. accumbens and CRF-antagonists block stress-induced secretion of VTA CRF on DA output into the shell of the accumbens (Wang et al., 2005a).

The mesocorticolimbic system has enjoyed considerable study because of its role in drug-seeking behaviors and relapse. It seems clear that relapse in addicted rats involves this system and not only glucocorticoids (Goeders and Guerin, 1996; Piazza et al., 1996a, b; Piazza and Le Moal, 1997, 1998; Goeders, 2002), but also CRF (Shaham et al., 1997; Erb et al., 1998; Goeders and Guerin, 2000; Koob and Le Moal, 2001; Sarnyai et al., 2001). Thus, as with the other forebrain-directed monoaminergic

cell groups the activity of DA neurons in the VTA is sensitive to both glucocorticoids and CRF.

Thus it appears that both glucocorticoids and CRF, possibly the former through activation of the latter, increase activity in the monoaminergic systems that nearly globally innervate the forebrain resulting in a differential focus on the tasks at hand, as well as increasing the emotional valence of life at the moment.

Physiological sequelae of activation of glucocorticoid and the central stress response network

Most cells, and all cell types in the body, contain glucocorticoid receptors, and are responsive to these steroids. Therefore, it is not a simple matter to determine in whole animal experiments which of the effects of the glucocorticoids are exerted in the periphery, and which centrally. However, it seems that like many physiological systems, the glucocorticoids have opposite functional effects in the periphery and brain. In the periphery, the steroids are catabolic and reduce energy storage by mobilizing the outlying substrates of amino acids from muscle and free fatty acids and glycerol from fat depots thus supplying energy to the liver for gluconeogenesis. These systemic effects of the glucocorticoids ensure adequate substrate for use of tissues that must use glucose, and an abundant supply of fatty acids for use in heart and other muscle, assuring the immediate energetic capacity to escape stressors (Peters et al., 2004). In brain (Fig. 4), the glucocorticoids act anabolically to augment energy stores by enabling substrate acquisition through their effects on behaviors, autonomic outflow, and the HPA axis. We discuss below anabolic effects of glucocorticoid actions on brain with respect to feeding, reward, learning and memory. In all of these effects, the glucocorticoids appear to foster caloric intake, in opposition to their effects on calorie mobilization and use in the periphery.

Comfort food ingestion

In the absence of stress, glucocorticoids strongly stimulate, in a dose-related fashion, the ingestion of substances that are pleasurable to the animal (Fig. 6), such as saccharin (Bhatnagar et al., 2000a), sucrose (Bell et al., 2000) and lard (La Fleur et al., 2004). Adrenalectomized rats exhibit a 5–10% reduction in food intake, increased sympathetic neural outflow and augmented CRF and ACTH; and increases in corticosterone do not markedly increase chow intake, unless the rats are diabetic. When adrenalectomized rats are made diabetic with streptozotocin, then chow intake increases in proportion to the circulating corticosterone (Fig. 6, left sets of bars (Strack et al., 1995b; La Fleur et al., 2004)), but as insulin is replaced in these rats, chow intake decreases and fat intake increases (La Fleur et al., 2004). Increasing corticosterone in adrenalectomized rats that are not diabetic also increases circulating insulin concentrations; therefore, we have proposed that elevated insulin may act to interfere with the action of corticosterone on chow intake (Dallman et al., 2004; La Fleur et al., 2004; Dallman et al., 2005). When adrenalectomized rats are provided with both chow and lard to eat, as corticosterone concentrations increase, so does lard intake (Fig. 6, middle;

La Fleur et al., 2004). It is notable that with high corticosterone concentrations, both in diabetic adrenalectomized rats eating chow and non-diabetic adrenalectomized rats eating lard, the amount consumed is 50% above that in the control, sham-adrenalectomized rats. It may be that the blunting effect of high saturated fats on insulin action (Brindley et al., 1981; Svedberg et al., 1991; Balkan et al., 1993; Matsuo et al., 1999; Bergman and Ader, 2000; van Dijk et al., 2003) accounts for the marked increase in lard consumption in the high-steroid group that were also hyperinsulinemic. However, in sum, these results also suggest that eating pleasurable calories stimulates other areas of brain in addition to the basic feeding region in the hypothalamus.

Adrenalectomized rats that were not replaced with steroid ate 50–60% as much as controls when the foodstuff contained calories (Fig. 6, middle 3 bars — no insulin, lard, sucrose; Bell et al., 2000) but when given pleasurable, but non-nutritive saccharin (Bhatnagar et al., 2000a), they drank only 4% as much as controls (last set of bars Fig. 6), suggesting that in the absence of steroid, rats may

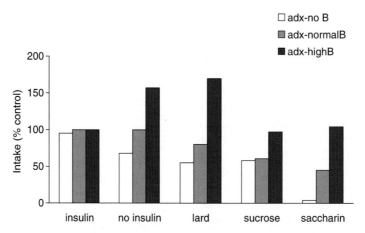

Fig. 6. Corticosterone strongly affects caloric intake, but the food ingested depends on the availability of insulin. Triplets of bars from Left to right: In the presence of insulin, which is stimulated in a dose-related fashion by increasing corticosterone (B), chow intake is minimally affected by B, however, in rats made diabetic with streptozotocin, chow intake is strongly affected by B (Strack et al., 1995b; La Fleur et al., 2004). In the presence of insulin increasing with corticosterone, lard (La Fleur et al., 2004), sucrose (Bell et al., 2002) and saccharin (Bhatnagar et al., 2000a) intake increases in proportion to the steady-state corticosterone concentrations. Note that in the absence of insulin, or presence of lard which damps the efficacy of insulin, the adx-high-B rats ate more than controls. Note also, that when non-caloric saccharin intake was tested, rats drank very little in the absence of corticosterone. Data are redrawn from the cited papers.

eat more because of post-ingestive caloric effects. We studied more completely the effects of voluntary sucrose drinking in adrenalectomized rats.

When adrenalectomized rats were provided with 30% sucrose, but not equally sweet saccharin to drink, all measured adrenalectomy-induced deficits that we measured are restored to normal (Laugero et al., 2001b). Moreover, when asked to choose between these sucrose and saccharin solutions, which sham-adrenalectomized rats like equally well, adrenalectomized rats far prefer the calorically rich sucrose solution (Laugero et al., 2001b). These results suggest strongly that the adrenalectomized rats seek calories to restore their energy balance to normal. Increasing corticosterone in adrenalectomized rats does not, however, increase normal chow intake as it does comfort foods, suggesting that other hormones, such as insulin, may limit the feeding of ordinary, only moderately palatable foods. Centrally administered corticosterone did not take the place of drinking sucrose in restoring adrenalectomized rats to normal, rather central corticosterone worsened the systemic condition and stimulated activity in the HPA axis (Laugero et al., 2002), as might have been anticipated from the effects of glucocorticoid on CRF in the amygdala (discussed above).

Because caloric intake, fat depot weight, brown adipose tissue activity, and expression of CRF mRNA are restored to normal in adrenalectomized rats not treated with corticosterone, but allowed sucrose to drink, we suggested that "comfort foods" result in a metabolic feedback signal that damp brain stress responses (Dallman, 2003; Dallman et al., 2003). Along with previous results that suggested this might be correct (Minor and Saade, 1997; Levin et al., 2000), tests of this hypothesis suggest that the provision of "comfort foods" before and during stressful periods does, indeed, damp HPA responses to stressors (Bell et al., 2002; Pecoraro et al., 2004; La Fleur et al., 2005).

The nucleus accumbens

The fact that "comfort foods" are highly palatable suggests that signals deriving from their ingestion activate cells in the nucleus accumbens (Acb), the "pleasure center". Many of the neural pathways that lead to, and mediate the intake of "comfort foods" have been very well discussed recently (Kelley et al., 2005), with the proposal that understanding of the role of the Acb on food motivation parses into neurochemical and site-specific subsystems that control different modules that regulate calorie acquisition. Three well supported conclusions of that review are that: "GABA output neurons localized exclusively in the Acb shell directly influence hypothalamic effector mechanisms for feeding motor neurons but do not participate in more complicated executive stratagies; enkephalinergic neurons throughout the Acb mediate the hedonic impact of palatable foods; and dopamine transmission in the Acb governs general motoric and arousal processes related to response selection as well as motor learning-related plasticity".

Although we have not found data that specifically examine the effects of glucocorticoids on GABAergic neurons in the Acb, sustained administration of high daily doses of cortisol to macaques for a year stimulates activity of GABAergic neurons and BDNF in hippocampus (McMillan et al., 2004). Moreover, glucocorticoid receptors are reported to reside in membranes of presynaptic terminals, dendrites and synaptic spines as well as cell bodies in the lateral amygdala and 13% of the GABA cells of the lateral amygdala co-express GR (Johnson et al., 2005). In the nucleus of the tractus solitarius, dexamethasone activates GABAergic transmission by a non-transcriptional effect (Wang et al., 2005b). The actions of glucocorticoids at any of these sites (Fig. 4) could alter GABAergic ouflow from the Acb, affecting hypothalamic sites that mediate food intake (Kelley et al., 2005). Enkephalin expressing neurons, particularly in the shell of the Acb are reduced by adrenalectomy and increased by treatment with corticosterone (Ahima et al., 1992; Lucas et al., 1998), therefore, it may be this decrease in enkephalin expression with adrenalectomy, and the corticosteroid dose-related increase in enkephalin that may explain, at least in part, the glucocorticoid dose-related increase in eating palatable sucrose and fat; enkephalin is proposed to augment the hedonic experience of feeding, particularly of highly palatable foods (Kelley et al., 2005). Finally, there is good evidence that glucocorticoids affect activity and transmission in the dopaminergic system in the Acb, particularly in the shell (Rivet et al., 1989; Deroche et al., 1995; Piazza and Le Moal, 1996; Piazza et al., 1996b; Barrot et al., 2001; Czyrak et al., 2003); an action of dopamine in the Acb is proposed to direct behavior toward food and the stimuli that predict it (Kelley et al., 2005). Running, a food-associated activity, is markedly reduced in adrenalectomized rats and is restored by treatment with corticosterone (Leshner, 1971), or dexamethasone (Moberg and Clark, 1976). Moreover, running behavior after administration of morphine or amphetamine is sensitive to circulating glucocorticoids (Martinelli et al., 1994). Another food-associated behavior is schedule-induced polydipsia; this, again is dependent on how much glucocorticoid is in the animal, and is dramatically decreased by adrenalectomy (Cirulli et al., 1994). Moreover, adrenalectomy blocks search behavior when a favored food becomes unavailable and treatment with corticosterone restores this behavior (Pecoraro et al., 2005b). These behaviors may all involve the dopaminergic submodule in the Acb (Kelley et al., 2005). Thus there is good evidence that the glucocorticoids may affect each of the proposed subsystems in the Acb, and that the action of the steroids on these subsystems may result in the integrated steroid-related increased intake of palatable foods in rats.

Learning and memory

Also feeding into the accumbens network, as well as into the rest of brain, are neurons that transmit information about learned and remembered tasks or events (Fig. 5). Again, glucocorticoids, when given into the amygdala or hippocampus, have pronounced effects on learning and memory, although these effects are strongly dependent on when they are provided relative to a stressor (Roozendaal et al., 1996; Roozendaal et al., 1999; Roozendaal et al., 2001a, b; Roozendaal, 2002, 2003; Thompson et al., 2004).

A revealing model of learning and memory is a negative contrast paradigm, in which groups of rats are provided with either 32% sucrose or 4% sucrose, trained to drink it for 5 min/day and then one group is abruptly switched from 32% to 4% sucrose. When c-Fos immunoreactive neurons were measured 60 min after such a switch, to determine structures activated by the change in sucrose concentration, many of the same structures shown in Fig. 4 exhibited fos-ir neurons; the neuronal excitation, reflected by fos-ir in the rats switched from 32% to 4% sucrose, was observed only on the day of the switch, not the next day, when they drank more of the less preferred sweet drink than they had on the previous day (Pecoraro and Dallman, 2005). We organized these fos results using the three pathway brain output model of Swanson (Swanson, 2000; Thompson and Swanson, 2002), and showed specific activation of the food-associated visceromotor pattern-generating cell groups in the hypothalamus. Pecoraro determined that adrenalectomized rats that were treated with a single pulse of corticosterone paired with a taste of 32% sucrose drank more sucrose on the first day of training than rats that had not experienced paired corticosterone-sucrose, and drank more sucrose overall during training than rats that were not previously exposed to paired corticosterone-sucrose, suggesting that corticosterone exerted a memorial effect that persisted through training in the absence of the steroid (Pecoraro et al., 2005a). Finally, the amount of sucrose drunk during training was proportional to the steady-state corticosterone supplied to adrenalectomized rats (Pecoraro et al., 2005b).

Certainly with respect to drug relapse, stressors with associated glucocorticoid secretion appear to tap into similar networks as the drugs themselves, but through a CRF-mediated mechanism involving the amygda and BNST (Shaham et al., 1997; Shaham et al., 1998; Erb and Stewart, 1999; Erb et al., 2001). Stressors result in further pursuit of drugs in rats that have been withdrawn from them for fairly long periods, but through different mechanisms, from giving priming doses of the drug which are known to act specifically at the Acb (Deroche et al., 1997; Piazza et al., 1990, 1991, 1996b; Deroche et al., 1992; Piazza et al., 1993; Deroche et al., 1995; Piazza and Le Moal, 1997, 1998). The relapse-inducing effect of stress and glucocorticoids may also act through their effects on dopamine secretion in the Acb, and it seem likely that much of the glucocorticoid-mediated effect is memorial (Micheau et al., 1981; Zorawski and Killcross, 2003; Di Chiara et al., 2004; LaLumiere et al., 2005).

Summary. Glucocorticoids act on brain in a feed-forward manner to simulate activity in a variety of systems including feeding, learning and memory, sympathetic and HPA outflow. Glucocorticoids stimulate caloric intake. In the absence of insulin, chow intake is increased, but in its presence, only palatable caloric intake is responsive to corticosterone. Glucocorticoids act on neuromodulators regulating output of all 3 behavioral sub-modules of the nucleus accumbens defined by Kelley et al (2005), and it is likely that these effects are important in the observed glucocorticoid-induced intake of palatable foods. Further, glucocorticoids have clear effects on learning and memory, both for palatable foods and for addictive drugs, and it is likely that these effects are mediated through glucocorticoid actions on learning and memory.

Metabolic feedback

Under conditions of elevated glucocorticoids, whether by stressor-induced increases in HPA activity or through treatment with these steroids in the absence of stress, the combination of increased glucocorticoids and insulin rearrange calories stored in the body to favor storage in abdominal fat depots. The prototypical example of this action of the glucocorticoids is the individual with an ACTH-secreting adenoma and Cushing's Disease (Kotake et al., 1996; Bujalska et al., 1997). Adrenalectomy in most genetically obese rodent species reduces body fat content (reviewed in Dallman, 1984; Dallman et al., 1993) and glucocorticoid replacement increases the percentage fat content in proportion to the amount of steroid administered (Strack et al., 1995b), independently of the negative effects of corticosterone on body weight. As reviewed above, sucrose given to adrenalectomized rats to drink can

normalize fat content in these animals and also normalize hypothalamic CRF expression. Either because of an unidentified signal from increased fat content, or quite possibly because of the palatable effects of eating sucrose or fat on the dopaminergic "reward" system, eating comfort foods is associated with a reduction in activity in the HPA axis, in effect, balancing the pronounced feed-forward effects of glucocorticoids on brain.

The downside of the chronic stress response

Although the glucocorticoids appear to make animals and humans more fit to handle stressors, both acute and chronic, through their metabolic actions and multiple effects on the response networks in brain, in our current society there are roadblocks to allowing the normal long-term and life-saving actions to occur (see, e.g., Dallman et al., 2004, 2005). The incidence of chronic social stress is increased, high calorie palatable foods are readily available and the physical effort used to acquire them is decreased (Hemingway and Marmot, 1999; Stansfeld et al., 1999; Wardle and Gibson, 2002; Landbergis et al., 2003; Kunz-Ebrecht et al., 2004; Schlotz et al., 2004). Because of these facts, it seems likely that a long-refined set of mechanisms that have enabled many species to survive, and even thrive under the influence of chronic stress, have not yet adapted to our current climate of easy access to food attainment. In part, this may explain a portion of the current epidemic of obesity that is sweeping the world (Mokdad et al., 2000; Bray et al., 2004; Steinbaum, 2004). Because much of the obesity that occurs, particularly with chronic stress (Rebuffe-Scrive et al., 1990; Rebuffe-Scrive et al., 1992; Fried et al., 1993; Dallman et al., 2005) involves intra-abdominal stores, the prognosis for long-term health is dim (Davy and Hall, 2003; Reaven et al., 2004) without actively removing oneself from the stressful input.

References

Adamec, R.E. and McKay, D. (1993) Amygdala, kindling, anxiety, and corticotrophin-releasing factor (CRF). Physiol. Behav., 54: 423–431.

Aguilera, G. (1994a) Regulation of pituitary ACTH secretion during chronic stress. Front. Neuroendocrinol., 15: 321–350.

Aguilera, G. (1994b) Regulation of piutuitary ACTH secretion during chronic stress. Front. Neuroendocrinol., 15: 321–350.

Ahima, R.S., Garcia, M.M. and Harlan, R.E. (1992) Glucocorticoid regulation of proenkephalin gene expression in the rat forebrain. Molec. Brain Res., 16: 119–127.

Akana, S.F., Cascio, C.S., Du, J.-Z., Levin, N. and Dallman, M.F. (1986) Reset of feedback in the adrenocortical system: an apparent shift in sensitivity of adrenocorticotropin to inhibition by corticosterone between morning and evening. Endocrinology, 119: 2325–2332.

Akana, S.F., Cascio, C.S., Shinsako, J. and Dallman, M.F. (1985) Corticosterone: narrow range required for normal body and thymus weight and ACTH. Am. J. Physiol., 249: R527–R532.

Akana, S.F. and Dallman, M.F. (1992) Feedback and facilitation in the adrenocortical system: unmasking facilitation by partial inhibition of the glucocorticoid response to prior stress. Endocrinology, 131: 57–68.

Akana, S.F. and Dallman, M.F. (1997) Chronic cold in adrenalectomized, corticosterone (B)-treated rats: facilitated corticotropin responses to acute stress emerge as B increases. Endocrinology, 138: 3249–3258.

Akana, S.F., Hanson, E.S., Horsley, C.J., Strack, A.M., Bhatnagar, S., Bradbury, M.J., Milligan, E.D. and Dallman, M.F. (1996) Clamped corticosterone (B) reveals the effect of endogenous B on both facilitated responsivity to acute restraint and metabolic responses to chronic stress. Stress, 1: 33–49.

Akana, S., Scribner, K., Bradbury, M., Strack, A., Walker, C.-D. and Dallman, M. (1992) Feedback sensitivity of the rat hypothalamo-pituitary-adrenal axis and its capacity to adjust to exogenous corticosterone. Endocrinology, 131: 585–594.

Albeck, D.S., McKittrick, C.R., Blanchard, D.C., Blanchard, R.J., Nikulina, J., McEwen, B.S. and Sakai, R.R. (1997) Chronic social stress alters levels of corticotropin-releasing factor and arginine vasopressin mRNA in rat brain. J. Neurosci., 17: 4895–4903.

Anderson, S.M., Kant, G.J. and de Souza, E.B. (1993) Effects of chronic stress on anterior pituitary and brain corticotropin-releasing factor receptors. Pharmacol. Biochem. Behav., 44: 755–761.

Arborelius, L., Skelton, K.H., Thrivikraman, K.V., Plotsky, P.M., Schultz, D.W. and Owens, M.J. (2000) Chronic administration of the selective corticotropin-releasing factor 1 receptor antagonist CP-154-256: behavioral, endocrine and neurochemical effects in the rat. J. Pharmacol. Exp. Therapeut., 294: 588–597.

Armario, A., Lopez-Calderon, A., Jolin, T. and Balasch, J. (1986) Response of anterior pituitary hormones to chronic stress. Neurosci. Biobehav. Rev., 10: 245–250.

Arvantis, A.G., Gilligan, P.J., Chorvat, R.J., Cheeseman, R.S., Christos, T.E., Bakthavatchalam, R., Beck, J.P., Cocuzza, A.J., Hobbs, F.W., Wilde, R.G., Arnold, C., Chidister, D., Curry, M., He, L., Hollis, A., Klaczkiewicz, J., Krenitsky, P.J., Rescinito, J.P., Scholfield, E., Culp, S., De Souza, E.B., Fitzgerald, L., Grigoriadis, D.E., Tam, S.W., Wong, N.,

Huang, S.-M. and Shen, H.L. (1999) Non-peptide corticotropin-releasing hormone antagonists: syntheses and structure-activity relationships of 2-anilinopyrimidines and -triazines. J. Med. Chem., 42: 805–818.

Asan, E., Yilmazer-Hanke, D.M., Eliava, M., Hantsch, M., Lesch, K.P. and Schmidt, A. (2005) The corticotropin-releasing factor (CRF)-system and monoaminergic afferents in the central amygdala: investigations with various mouse strains and comparison with the rat. Neuroscience, 131: 953–967.

Aston-Jones, G. and Cohen, J.D. (2005) An integrative theory of locus coeruleus-norepinephrine function: adaptive gain and optimal performance. Annu. Rev. Neurosci., 28: 403–450.

Azmitia, E.C. (1999) Serotonin neurons, neuroplasticity, and homeostasis of neural tissue. Neuropsychopharmacology, 21: 33S–45S.

Azmitia, E.C. (2001) Modern views on an ancient chemical: serotonin effects on cell proliferation, maturation and apoptosis. Brain Res. Bull., 56: 413–424.

Azmitia, E.C. and McEwen, B.S. (1969) Corticosterone regulation of tryptophan hydroxylase in midbrain of rat. Science, 166: 1274–1276.

Azmitia, E.C. and McEwen, B.S. (1974) Adrenalcortical influence on rat brain tryptophan-hydroxylase activity. Brain Res., 78: 291–302.

Bakshi, V.P., Smith-Roe, S., Newman, S.M., Grigoriadis, D.E. and Kalin, N.H. (2002) Reduction of stress-induced behavior by antagonism of corticotropin-releasing hormone 2 (CRH2) receptors in lateral septum, or CRH1 receptors in amygdala. J. Neurosci., 22: 7926–7935.

Baldwin, H.A., Rassnick, S., Rivier, J., Koob, G.F. and Britton, K.T. (1991) CRF antagonist reverses the "anxiogenic" response to ethanol withdrawal in the rat. Psychopharmacology (Berl.), 103: 227–232.

Bale, T.L. and Vale, W.W. (2004) CRF and CRF receptors: role in stress responsivity and other behaviors. Annu. Rev. Pharmacol. Toxicol., 44: 525–557.

Balkan, B., Strubbe, J.H., Bruggink, J.E. and Steffens, A.B. (1993) Overfeeding-induced obesity in rats: Insulin sensitivity and autonomic regulation of metabolism. Metabolism, 42: 1509–1518.

Barrot, M., Abrous, D.N., Marinelli, M., Rouge-Pont, F., Le Moal, M. and Piazza, P.V. (2001) Influence of glucocorticoids on dopaminergic transmission in the rat dorsolateral striatum. Eur. J. Neurosci., 13: 812–818.

Bartanusz, V., Jezova, D., Bertinin, L.T., Tilders, F.J.H., Aubry, J.-M. and Kiss, J.Z. (1993) Stress-induced increases in vasopressin and corticotropin-releasing factor expression in hypophysiotropic paraventricular neurons. Endocrinology, 132: 895–902.

Basso, A.M., Spina, M.G., RivIer, J., Vale, W. and Koob, G.F. (1999) Corticotropin-releasing factor antagonist attenuates the "anxiogenic-like" effect in the defensive burying paradigm but not in the elevated plus maze following chronic cocaine in rats. Psychopharmacology, 145: 21–30.

Bell, M.E., Bhargava, A., Soriano, L., Laugero, K., Akana, S.F. and Dallman, M.F. (2002) Sucrose and corticosterone interact to modulate behavior, energy balance, autonomic outflow and neuroendocrine responses during chronic cold. J. Neuroendocrinol., 14: 330–342.

Bell, M.E., Bhatnagar, S., Liang, J., Soriano, L., Nagy, T.R. and Dallman, M.F. (2000) Voluntary sucrose ingestion, like corticosterone replacement, prevents the metabolic deficits of adrenalectomy. J. Neuroendocrinol., 12: 461–470.

Berendse, H.W. and Groenewegan, H.J. (1990) Orginization of the thalamostriatal projections in the rat, with special emphasis on the ventral striatum. J. Comp. Neurol., 200: 187–228.

Bergman, R.N. and Ader, M. (2000) Free fatty acids and pathogenesis of type 2 diabetes mellitus. Trends Endocrinol. Metab., 11: 351–356.

Bhargava, A., Dallman, M.F., Pearce, D. and Choi, S. (2004) Long double-stranded RNA-mediated RNA interference as a tool to achieve site-specific silencing of hypothalamic neuropeptides. Brain Res. Brain Res. Protoc., 13: 115–125.

Bhatnagar, S., Bell, M.E., Liang, J., Soriano, L., Nagy, T.R. and Dallman, M.F. (2000a) Corticosterone facilitates saccharin intake in adrenalectomized rats. Does corticosterone increase stimulus salience? J. Neuroendocrinol., 12: 453–460.

Bhatnagar, S. and Dallman, M. (1998a) Neuroanatomical basis for facilitation of hypothalamic-pitutiary-adrenal responses to a novel stressor after chronic stress. Neuroscience, 84: 1025–1039.

Bhatnagar, S. and Dallman, M.F. (1998b) Neuroanatomical basis for facilitation of hypothalamic-pituitary-adrenal responses to a novel stressor after chronic stress. Neuroscience, 84: 1025–1039.

Bhatnagar, S. and Dallman, M.F. (1998c) Neuroanatomical basis for facilitation of the hypothalamo-pituitary-adrenal responses to a novel stress after chronic stress. Neuroscience, 84: 1025–1039.

Bhatnagar, S. and Dallman, M.F. (1999) The paraventricular nucleus of the thalamus alters rhythms in core temperature and energy balance in a state-dependent manner. Brain Res., 851: 66–75.

Bhatnagar, S., Huber, R., Nowak, N. and Trotter, P. (2002) Lesions of the paraventricular thalamus block adaptation of hypothalamic-pituitary-adrenal (HPA) responses to repeated restraint. J. Neuroendocrinol., 14: 403–410.

Bhatnagar, S., Viau, V., Chu, A., Soriano, L., Meijer, O.C. and Dallman, M.F. (2000) A cholecystokinin-mediated pathway to the paraventricular thalamus is recruited in chronically stressed rats and regulates hypothalamic pituitary adrenal function. J. Neurosci., 20: 5564–5573.

Bhatnagar, S. and Vining, C. (2003) Facilitation of hypothalamic-pituitary-adrenal responses to novel stress following repeated social stress using the resident/intruder paradigm. Horm. Behav., 43: 15–165.

Bohus, B. (1970) The medial thalamus and the opposite effect of corticosteroids and adrenocorticotrophic hormone on avoidance extinction in the rat. Acta Physiol. Acad. Scient. Hung., 38: 217–223.

Bonaz, B. and Rivest, S. (1998) Effect of a chronic stress on CRF neuronal activity and expression of its type 1 receptor in the rat brain. Am. J. Physiol., 275: R1438–R1449.

Bradbury, M.J., Akana, S.F. and Dallman, M.F. (1994) Roles of type I and II corticosteroid receptors in regulation of basal activity in the hypothalamo-pituitary-adrenal axis during the diurnal trough and the peak: evidence for a nonadditive effect of combined receptor occupation. Endocrinology, 134: 1286–1296.

Bradbury, M.J., Strack, A.M. and Dallman, M.F. (1993) Lesions of the hippocampal efferent pathway (Fimbria-fornix) do not alter sensitivity of adrenocorticotropin to feedback inhibition by corticosterone in rats. Neuroendocrinology, 58: 396–407.

Bray, G.A., Nielson, S.J. and Popkin, B.M. (2004) Consumption of high-fructose corn syrup in beverages may play a role in the epidemic of obesity. Am. J. Clin. Nutr., 79: 537–543.

Brindley, D.N., Cooling, J., Glenny, H.P., Burditt, S.L. and McKechie, I.S. (1981) Effects of chronic modification of dietary fat and carbohydrate on the insulin, corticosterone and metabolic responses of rats fed acutely with glucose, fructose or ethanol. Biochem. J., 200: 275–284.

Britton, D.R., Koob, G.F., Rivier, J. and Vale, W. (1982) Intraventricular corticotropin-releasing factor enhances behavioral effects of novelty. Life Sci., 31: 363–367.

Britton, K.T., Lee, G., Vale, W., Rivier, J. and Koob, G.F. (1986a) Corticotropin releasing factor (CRF) receptor antagonist blocks activating and 'anxiogenic' actions of CRF in the rat. Brain Res., 369: 303–306.

Britton, K.T., Lee, G., Vale, W., Rivier, J. and Koob, G.F. (1986b) Corticotropin-releasing factor (CRF) receptor antagonist blocks activating and 'anxiogenic' actions of CRF in the rat. Brain Res., 369: 303–306.

Brown, M.R., Fisher, L.A., Spiess, J., Rivier, k., Rivier, J. and Vale, W. (1982) Corticotropin-releasing factor: actions on the sympathetic nervous system and metabolism. Endocrinology, 111: 928–931.

Brown, E.R. and Sawchenko, P.E. (1997) Hypophysiotropic CRF neurons display a sustained immediate-early gene response to chronic stress but not to adrenalectomy. J. Neuroendocrinol., 9: 307–316.

Bruijnzeel, A.W., Stam, R., Compaan, J.C. and Wiegant, V.M. (2001) Stress-induced sensitization of CRH-ir but not P-CREB-ir responsivity in the rat central nervous system. Brain Res., 908: 187–196.

Bubser, M. and Deutch, A.Y. (1998) Thalamic paraventricular nucleus neurons collateralize to innervate the prefrontal cortex and nucleus accumbens. Brain Res., 787: 304–310.

Bubser, M. and Deutch, A.Y. (1999) Stress induces Fos expression in neurons of the thalamic paraventricular nucleus that innervates limbic forebrain sites. Synapse, 32: 13–22.

Buchanan, S.L., Thompson, R.H. and Powell, D.A. (1989) Midline thalamic lesions enhance conditioned bradycardia and the cardiac orienting reflex in rabbits. Psychobiology, 17: 300–306.

Bujalska, I.J., Kumar, S. and Stewart, P.M. (1997) Does central obesity reflect "Cushing's disease of the omentum"? Lancet, 349: 1210–1213.

Buwalda, B., De Boer, S.F., Schmidt, E.D., Felszeghy, K., Nyaka, C., Sgoigo, A., Van der Begt, B.J., Tilders, F.H.J.,

94

Bohus, B. and Koolhaas, J.M. (1999) Long-lasting deficient dexamethasone suppression of hypothalamic-pituitary-adrenocortical activation following peripheral CRF challenge in socially defeated rats. J. Neuroendocrinol., 11: 512–520.

Buwalda, B., de Boer, S.F., Van Kalkeren, A.A. and Koolhaas, J.M. (1997) Physiological and behavioral effects of chronic intracerebroventricular infusion of corticotropin-releasing factor in the rat. Psychoneuroendocrinology, 22: 297–309.

Buwalda, B., Felszehy, K., Horvath, K.M., Nyakas, C., de Boer, S.F., Bohus, B. and Koolhaas, J.M. (2001) Temporal and spatial dynamics of corticosteroid receptor down-regulation in rat brain following social defeat. Physiol. Behav., 72: 349–354.

Campeau, S. and Davis, M. (1995) Involvement of subcortical and cortical afferents to the lateral nucleus of the amygdala in fear conditioning measured with fear-potentiated startle in rats trained concurrently with auditory and visual stimuli. J. Neurosci., 15: 2312–2327.

Carrasco, G.A. and Van de Kar, L.D. (2003) Neuroendocrine pharmacology of stress. Eur. J. Pharmacol., 463: 235–272.

Castro-Alamancos, M.A. and Connors, B.W. (1996) Short-term plasticity of a thalamocortical pathway dynamically modulated by behavioral state. Science, 272: 274–277.

Chappell, P.B., Smith, M.A., Kilts, C.D., Bissette, G., Ritchie, J., Anderson, C. and Nemeroff, C.B. (1986) Alterations in corticotropin-releasing factor-like immunoreactivity in discrete brain regions after acute and chronic stress. J. Neurosci., 6: 2908–2914.

Chen, S. and Su, H.-S. (1990) Afferent connections of the thalamic paraventricular and parataenial nuclei in the rat. A retrograde tracing study with iontophoretic application of fluorogold. Brain Res., 522: 1–6.

Chowdrey, H.S., Larsen, P.J., Harbuz, M.S., Jessop, D.S., Aguilera, G., Eckland, D.J.A. and Lightman, S.L. (1995) Evidence for arginine vasopressin as the primary activator of the HPA axis during adjuvant-induced arthritis. Br. J. Pharmacol., 116: 2417–2424.

Chrousos, G. and Gold, P. (1992) The concepts of stress and stress system disorders. J. Am. Med. Assoc., 267: 1244–1252.

Cirulli, F., van Oers, H., de Kloet, E.R. and Levine, S. (1994) Differential influence of corticosterone and dexamethasone on schedule-induced polydipsia in adrenalectomized rats. Behav. Brain Res., 65: 33–39.

Claes, S.J. (2004) Corticotropin-releasing hormone (CRH) in psychiatry: from stress to psychopathology. Ann. Med., 36: 50–61.

Cole, B.J., Cador, M., Stinus, L., Rivier, J., Vale, W., Koob, G.F. and Le Moal, M. (1990) Central administration of a CRF antagonist blocks the development of stress-induced behavioral sensitization. Brain Res., 512: 343–346.

Cook, C.J. (2001) Measuring of extracellular cortisol and corticotropin-releasing hormone in the amygdala using immunosensor coupled microdialysis. J. Neurosci. Meth., 110: 95–101.

Cook, C.J. (2002) Glucocorticoid feedback increases the sensitivity of the limbic system to stress. Physiol. Behav., 75: 455–464.

Cook, C.J. (2004) Stress induces CRF release in the paraventricular nucleus, and both CRF and GABA release in the amygdala. Physiol. Behav., 82: 751–762.

Croiset, G., Nijsen, M.J.M.A. and Kamphuis, P.J.G.H. (2000) Role of corticotropin-releasing factor, vasopressin and the autonomic nervous system in learning and memory. Eur. J. Pharmacol., 405: 225–234.

Curtis, A.L., Bello, N.T., Connolly, K.R. and Valentino, R.J. (2002) Corticotropin-releasing factor neurones of the central nucleus of the amygdala mediate locus coeruleus activation by cardiovascular stress. J. Neuroendocrinol., 14: 667–682.

Curtis, A.L., Lechner, S.M., Pavcovich, L.A. and Valentino, R.J. (1997) Activation of the locus coeruleus noradrenergic system by intracoerulear microinfusion of corticotropin-releasing factor: effects on discharge rate, cortical norepinephrine levels and cortical electroencephalographic activity. J. Pharmacol. Exp. Therapeut., 281: 163–172.

Czyrak, A., Mackowiak, M., Chocyk, A., Fijal, K. and Wedzony, K. (2003) Role of glucocorticoids in the regulation of dopaminergic neurotransmission. Pol. J. Pharmacol., 55: 667–674.

Dallman, M., Levin, N., Cascio, C., Akana, S., Jacobson, L. and Kuhn, R. (1989a) Pharmacological evidence that the inhibition of diurnal adrenocorticotropin secretion by corticosteroids is mediated via type I corticosterone-preferring receptors. Endocrinology, 124: 2844–2850.

Dallman, M., Pecoraro, N. and laFleur, S. (2005) Chronic stress and comfort foods: self-medication and abdominal obesity. Brain Behav. Immun., 19: 275–282.

Dallman, M.F. (1984) Viewing the ventromedial hypothalamus from the adrenal gland. Am. J. Physiol., 246: R1–R12.

Dallman, M.F. (1993) Stress update. Adaptations of the hypothalamic-pituitary-adrenal axis to chronic stress. Trends Endocrinol. Metab., 4: 62–69.

Dallman, M.F. (2003) A spoon of sugar: feedback signals of energy stores and corticosterone regulate responses to chronic stress. Physiol. Behav., 79: 3–12.

Dallman, M.F. (2005) Fast glucocorticoid actions on brain: back to the future. Front. Neuroendocrinol., 26: 103–108.

Dallman, M.F. and Bhatnagar, S. (2001) Chronic Stress and Energy Balance: Role of the Hypothalamo-Pituitary-Adrenal Axis. Oxford University Press, New York.

Dallman, M.F., DeManincor, D. and Shinsako, J. (1974) Diminishing corticotrope capacity to release ACTH during sustained stimulation: the twenty-four hours after bilateral adrenalectomy in the rat. Endocrinology, 95: 65–73.

Dallman, M.F. and Jones, M.T. (1973) Corticosteroid feedback control of ACTH secretion: effect of stress-induced corticosterone ssecretion on subsequent stress responses in the rat. Endocrinology, 92: 1367–1375.

Dallman, M.F., Jones, M.T., Vernikos-Danellis, J. and Ganong, W.F. (1972) Corticosteroid feedback control of ACTH secretion: rapid effects of bilateral adrenalectomy on plasma ACTH in the rat. Endocrinology, 91: 961–968.

Dallman, M.F., la Fleur, S.E., Pecoraro, N.C., Gomez, F., Houshyar, H. and Akana, S.F. (2004) Minireview: glucocorticoids — food intake, abdominal obesity, and wealthy nations in 2004. Endocrinology, 145: 2633–2638.

Dallman, M.F., Levin, N., Cascio, C.S., Akana, S.F., Jacobson, L. and Kuhn, R.W. (1989b) Pharmacological evidence that the inhibition of diurnal adrenocorticotropin secretion by corticosteroids is mediated via type I corticosterone-preferring receptors. Endocrinology, 124: 2844–2850.

Dallman, M.F., Makara, G.B., Roberts, J.L., Levin, N. and Blum, M. (1985) Corticotrope response to removal of releasing factors and corticosteroids in vivo. Endocrinology, 117: 2190–2197.

Dallman, M.F., Pecoraro, N., Akana, S.F., La Fleur, S.E., Gomez, F., Houshyar, H., Bell, M.E., Bhatnagar, S., Laugero, K.D. and Manalo, S. (2003) Chronic stress and obesity: a new view of "comfort food". Proc. Natl. Acad. Sci. USA, 100: 11696–11701.

Dallman, M.F., Strack, A.M., Akana, S.F., Bradbury, M.J., Hanson, E.S., Scribner, K.A. and Smith, M. (1993) Feast and famine: critical role of glucocorticoids with insulin in daily energy flow. Front. Neuroendocrinol., 14: 303–347.

Dallman, M.F. and Yates, F.E. (1968) Anatomical and functional mapping of central neural input and feedback pathways of the adrenocortical system. Mem. Soc. Endocrinol. (Lond.), 17: 39–72.

Dallman, M.F. and Yates, F.E. (1969) Dynamic asymmetries in the corticosteroid feedback path and distribution-metabolism-binding elements of the adrenocortical system. Ann. N.Y. Acad. Sci., 156: 696–821.

Daniels, W.M.U., Richter, L. and Stein, D.J. (2004) The effects of repeated intra-amygdala CRF injections on rat behavior and HPA axis function after stress. Metab. Brain Dis., 19: 15–23.

Davy, K.P. and Hall, J.E. (2003) Obesity and hypertension: two epidemics or one? Am. J. Physiol., 286: R803–R813.

Day, H.E.W., Curran, E.J., Watson Jr., S.J. and Akil, H. (1999) Distinct neurochemical populatons in the rat central nucleus of the amygdala and bed nucleus of the stria terminalis: evidence for their activation by interleukin-1B. J. Comp. Neurol., 413: 113–128.

de Goeij, D.C.E., Binnekade, R. and Tilders, F.J.H. (1992a) Chronic stress enhances vasopressin but not corticotropin releasing factor secretion during hypoglycemia. Am. J. Physiol., 263: E394–E399.

de Goeij, D.C.E., Dijkstra, H. and Tilders, F.J.H. (1992b) Chronic psychosocial stress enhances vasopressin, but not corticotropin-releasing factor in the external zone of the median eminence of male rats: relationship to subordinate status. Endocrinology, 131: 847–893.

de Goeij, D.C.E., Kvetnansky, R., Whitnall, M.H., Jezova, D., Berkenbosch, F. and Tilders, F.J.H. (1991) Repeated stress-induced activation of corticotropin-releasing factor neurons enhances vasopressin stores and colocalization with corticotorpin-releasing factor in the median eminence. Neuroendocrinology, 53: 150–159.

de Kloet, E.R., Vreugdenhil, E., Oitzl, M. and Joels, M. (1998) Brain corticosteroid receptor balance in health and disease. Endocr. Rev., 19: 269–301.

Deak, T., Nguyen, K.T., Ehrlich, A.L., Watkins, L.R., Spencer, R.L., Maier, S.F., Licinio, J., Wong, M.-L., Chrousos, G.P., Webster, E. and Gold, P.W. (1999) The impact of the nonpeptide corticotropin-releasing hormone antagonist antalarmin on behavioral and endocrine responses to stress. Endocrinology, 140: 79–86.

Denver, R.J. (1997) Environmental stress as a developmental cue: corticotropin-releasing hormone is a promiximate mediator of adaptive phenotypic plasticity in amphibian metamorphosis. Horm. Behav., 31: 169–179.

Deroche, V., Marinelli, M., Le Moal, M. and Piazza, P.V. (1997) Glucocorticoids and behavioral effects of psychostimulants. II: cocaine intravenous self-administration and reinstatement depend on glucocorticoid levels. J. Pharmacol. Exp. Ther., 281: 1401–1407.

Deroche, V., Marinelli, M., Maccari, S., Le Moal, M., Simon, H. and Piazza, P.V. (1995) Stress-induced sensitization and glucocorticoids. I. Sensitization of dopamine-dependent locomotor effects of amphetamine and morphine depends on stress-induced corticosterone secretion. J. Neurosci., 15: 7181–7188.

Deroche, V., Piazza, P.V., Casolini, P., Maccari, S., Le Moal, M. and Simon, H. (1992) Stress-induced sensitization to amphetamine and morphine psychomotor effects depend on stress-induced corticosterone secretion. Brain Res., 598: 343–348.

Deutch, A.Y., Bubser, M. and Young, C.D. (1998) Psychostimulant-induced Fos protein expression in the thalamic paraventricular nucleus. J. Neurosci., 18: 10680–10687.

Di Chiara, G., Bassareo, V., Fenu, S., De Luca, M.A., Spina, L., Cadoni, C., Acquas, E., Carboni, E., Valentini, V. and Lecca, D. (2004) Dopamine and drug addiction: the nucleus accumbens shell connection. Neuropharmacology, 47(Suppl. 1): 227–241.

Di, S., Malcher-Lopes, R., Halmos, K.C. and Tasker, J.G. (2003) Nongenomic glucocorticoid inhibition via endocannabinoid release in the hypothalamus: a fast feedback mechanism. J. Neurosci., 23: 4850–4857.

Diamant, M., Croiset, G. and de Wied, D. (1992a) The effect of corticotropin-releasing factor (CRF) on autonomic and behavioral responses during shock-prod burying test in rats. Peptides, 13: 1149–1158.

Diamant, M., Kashtanov, S.I., Fodor, M. and de Wied, D. (1992b) Corticotropin-releasing factor induces differential behavioral and cardiovascular effects after intracerebroventricular and lateral hypothalamic/perifornical injections in rats. Neuroendocrinology, 56: 750–760.

Dong, H.-W. and Swanson, L.W. (2004) Projections from the bed nuclei of the stria terminalis, posterior division: implications for cerebral hemisphere regulation of defensive and reproductive behaviors. J. Comp. Neurol., 471: 396–433.

Dunn, A.J. and Berridge, C.W. (1990) Physiological and behavioral responses to corticotropin releasing factor administration: is CRF a mediator of anxiety or stress responses? Brain Res. Rev., 15: 1–100.

Dunn, A.J., Swiergiel, A.H. and Palamarchouk, V. (2004) Brain circuits involved in corticotropin-releasing factor-norepinephrine interactions during stress. Ann. N.Y. Acad. Sci., 1018: 25–34.

Ehlers, C.L., Henriksen, S.J., Wang, M., Rivier, J., Vale, W. and Bloom, F.E. (1983) Corticotropin releasing factor produces increases in brain excitability and convulsive seizures in rats. Brain Res., 278: 332–336.

Erb, S., Salmaso, N., Rodaros, D. and Stewart, J. (2001) A role for the CRF-containing pathway from central neuclus of the amygdala to bed nucleus of the stria terminalis in the stress-induced reinstatement of cocaine seeking in rats. Psychopharmacology, 158: 360–365.

Erb, S., Shaham, Y. and Stewart, J. (1998) The role of corticotropin-releasing factor and corticosterone in stress- and cocaine-induced relapse to cocaine seeking in rats. J. Neurosci., 18: 5529–5536.

Erb, S. and Stewart, J. (1999) A role for the bed nucleus of the stria terminalis, but not the amygdala, in the effects of corticotropin-releasing factor on stress-induced reinstatement of cocaine seeking. J. Neurosci., 19: RC351–RC356.

Everitt, B.J., Parkinson, J.A., Olmstead, M.C., Arroyo, M., Robledo, P. and Robbins, T.W. (1999) Associative processes in addiction and reward. The role of amygdala- ventral striatal subsystems. Ann. N.Y. Acad. Sci., 877: 412–438.

Falkerstein, E., Tillmann, H.-C., Christ, M., Feuring, M. and Wehling, M. (2000) Multiple actions of steroid hormones-a focus on rapid, nongenomic effects. Pharm. Rev., 52: 513–555.

Fehm, H.L., Voigt, K., Kummer, G. and Pfeiffer, E.F. (1979) Positive rate-sensitive corticosteroid feedback mechanism of ACTH secretion in Cushing's disease. J. Clin. Invest., 64: 102–108.

Feldman, S. and Weidenfeld, J. (1998) The excitatory effects of the amygdala on hypothalamo-pituitary-adrenocortical responses are mediated by hypothalamic norepinephrine, serotonin and CRF-41. Brain Res. Bull., 45: 389–393.

Fernandes, G.A., Perks, P., Cox, N.K., Lightman, S.L., Ingram, C.D. and Shanks, N. (2002) Habituation and cross-sensitization of stress-induced hypothalamic-pituitary-adrenal activity: effect of lesions in the paraventricular nucleus of the thalamus or bed nuclei of the stria terminalis. J. Neuroendocrinol., 14: 593–602.

Fichter, M.M. and Pirke, K.M. (1986) Effect of experimental and pathological weight loss upon the hypothalamic-pituitary-adrenal axis. Psychoneuroendocrinology, 11: 295–305.

Fisher, L.A. (1993) Central actions of corticotropin-releasing factor on autonomic nervous activity and cardiovascular function. Ciba Foundation Symposium: Corticotropin-releasing factor, 172: 243–253.

Fisher, L.A., Rivier, J., Rivier, C., Spiess, J., Vale, W. and Brown, M.R. (1982) Corticotropin-releasing factor: effects on the autonomic nervous system and visceral systems. Endocrinology, 110: 2222–2224.

Fried, S.K., Russell, C.D., Grauso, N.L. and Brolin, R.E. (1993) Lipoprotein lipase regulation by insulin and glucocorticoid in subcutaneous and omental adipose tissues of obese women and men. J. Clin. Invest., 92: 2191–2198.

Ginsberg, A., Frank, M.G., Francis, A.B., Rubin, B.R., O'Connor, K.A. and Spencer, R.L. (2005) Specific and Time-Dependent effects of Glucocorticoid Receptor Agonist RU28362 on Stress-Induced Gene Expression in the Pituitary. J. Neuroendocrinol, 18: 129–138.

Goeders, N.E. (2002) The HPA axis and cocaine reinforcement. Psychoneuroendocrinology, 27: 13–33.

Goeders, N.E. and Guerin, G.F. (1996) Effects of surgical and pharmacological adrenalectomy on the initiation and maintenance of intravenous cocaine self-administration in rats. Brain Res., 722: 145–152.

Goeders, N.E. and Guerin, G.F. (2000) Effects of the CRH receptor antagonist CP[154-526] on intravenous cocaine self-administration in rats. Neuropsychopharmacology, 23: 577–586.

Gomez, F., Lahmame, A., de Kloet, E.R. and Armario, A. (1996) Hypothalamic-pituitary-adrenal response to chronic stress in five inbred rat strains: differential responses are mainly located at the adrenocortical level. Neuroendocrinology, 63: 327–337.

Graber, A.L., Ney, R.L., Nicholson, W.E., Island, D.P. and Liddle, G.W. (1965) Natural history of pituitary-adrenal recovery following long-term suppression with corticosteroids. J. Clin. Endo. Met., 25: 11–16.

Gray, T. S. (1991) Limbic pathways and neurotransmitters as mediators of autonomic and neuroendocrine responses to stress. In: Stress: neurobiology and neuroendocrinology., pp. 73–89. Eds. M. R. Brown, G. F. Koob and C. Rivier (Eds.). CRC Press, Boca Raton, pp. 73–89.

Gray, T.S. (1993a) Amydgaloid CRF pathways. Role in autonomic, neuroendocrine, and behavioral responses to stress. In: Tache, Y.C. and Rivier, Y. (Eds.) Corticotropin-Releasing Factor and Cytokines: Role in the Stress Response, Vol. 697. New York Academy of Sciences, New York, pp. 53–60.

Gray, T.S. (1993b) Amygdaloid CRF pathways. Role in autonomic, neuroendocrine, and behavioral responses to stress. Ann. N.Y. Acad. Sci., 697: 53–60.

Greenwood-Van Meerveld, B., Johnson, A.C., Cochrane, S., Schulkin, J. and Myers, D.A. (2005) Corticotropin-releasing factor 1 receptor-mediated mechanisms inhibit colonic hypersensitivity in rats. Neurogastroenterol. Motil., 17: 415–422.

Griebel, G., Perrault, G. and Sanger, D.J. (1998) Characterization of the behavioral profile of the non-peptide CRF receptor antagonist CP-154, 526 in anxiety models in rodents. Psychopharmacology, 138: 55–66.

Griebel, G., Simiand, J., Steinberg, R., Jung, M., Gully, D., Roger, P., geslin, M., Scatton, B., Maffrand, J.-P. and Soubrie, P. (2002) 4-(2-Chloro-4-methoxy-5-methylphenyl)-N-[(1S)-2-cyclopropyl-1-(3fluoro-4-methylphenyl)ethyl]5-methyl-N-(2-propanyl)-1,3thiazol-2-amine hydrochloride (SSR 125543A): a potent and selective corticotropin-releasing factor1 receptor antagonist: II. Characterization in rodent models of stress-related disorders. J Pharmacol. Exp. Ther., 301: 333–345.

Habib, K.E., Weld, K.P., Rice, K.C., Pushkas, J., Champoux, M., Listwak, S., Webster, E.L., Atkinson, A.J., Schulkin, J., Contoreggi, C., Chrousos, G.P., McCann, S.M., Suomi, S.J., Higley, J.D. and Gold, P.W. (2000) Oral administration of a corticotropin-releasing hormone receptor antagonist significantly attenuates behavioral, neuroendocrine and autonomic responses to stress in primates. Proc. Natl. Acad. Sci. USA, 97: 6079–6084.

Hall, J., Parkinson, J.A., Connor, T.M., Dickinson, A. and Everitt, B.J. (2001) Involvement of the central nucleus of the amygdala and nucleus accumbens core in mediating Pavlovian influences on instrumental behaviour. Eur. J. Neurosci., 13: 1984–1992.

Hammack, S.E., Pepin, J.L., DesMarteau, J.S., Watkins, L.R. and Maier, S.F. (2003) Low doses of corticotropin-releasing hormone injected into the dorsal raphe nucleus block the behavioral consequences of uncontrollable stress. Behav. Brain Res., 147: 55–64.

Hand, G.A., Hewitt, C.B., Fulk, L.J., Stock, H.S., Carson, J.A., Davis, J.M. and Wilson, M.A. (2002) Differential release of corticotropin-releasing hormone (CRH) in the amygdala during different types of stressors. Brain Res., 949: 122–130.

Harbuz, M.S., Nicholson, S.A., Gillham, B. and Lightman, S.L. (1990) Stress-responsiveness of hypothalamic corticotrophin-releasing factor and pituitary proopiomelanocortin mRNAs following high-dose glucocorticoid treatment and withdrawal in the rat. J. Endocrinol., 127: 407–415.

Harbuz, M.S., Rees, R.G. and Lightman, S.L. (1993) HPA axis responses to acute stress and adrenalectomy during adjuvant-induced arthritis in the rat. Am. J. Physiol., 264: R179–R185.

Hasue, R.H. and Shammah-Lagnado, S.J. (2002) Origin of the dopaminergic innervation of the central extended amygdala and accumbens shell: a combined retrograde tracing and immunohistochemical study in the rat. J. Comp. Neurol., 454: 15–33.

Heinrichs, S.C. and Koob, G.F. (2004) Corticotropin-releasing factor in brain: a role in activation, arousal and affect regulation. J. Pharmacol. Exp. Ther., 311: 427–440.

Heinrichs, S.C., Merlo Pich, E., Miczek, K.A., Britton, K.T. and Koob, G.F. (1992) Corticotropin-releasing factor antagonist reduces emotionality in socially deprived rats via direct neurotropic action. Brain Res., 581: 190–197.

Hemingway, H. and Marmot, M. (1999) Psychosocial factors in the aetiology and prognosis of coronary heart disease: systematic review of prospective cohort studies. BMJ, 318: 1460–1467.

Herman, J.P., Figueiredo, H.F., Mueller, N.K., Ulrich-Lai, Y., Ostrander, M.M., Choi, D.C. and Culinan, W.E. (2003) Central mechanisms of stress integration: hierarchical circuitry controlling hypothalamo-pituitary-adrenocortical responsiveness. Front. Neuroendocrinol., 24: 151–180.

Holsboer, F. (1999) The rationale for corticotropin-releasing hormone receptor (CRH-R) antagonists to treat depression and anxiety. J. Psychiat. Res., 33: 181–214.

Holsboer, F. (2000) The corticosteroid receptor hypothesis of depression. Neuropsychopharmacology, 23: 477–501.

Honkaniemi, J. (1992) Colocalization of peptide- and tyrosine hydroxylase-like immunoreactivities with Fos-immunoreactive neurons in rat central amygdaloid nuclei after immobilization stress. Brain Res., 598: 107–113.

Houshyar, H., Gomez, F., Manalo, S., Bhargava, A. and Dallman, M.F. (2003) Intermittent morphine administration induces dependence and is a chronic stressor in rats. Neuropsychopharmacology, 28: 1960–1972.

Hsin, L.-W., Tian, X., Webster, E.L., Coop, A., Caldwell, T.M., Jacobson, A.E., Chrousos, G.P., Gold, P.W., Habib, K.E., Ayala, A., Eckelman, W.C., Contoreggi, C. and Rice, K.C. (2002) CRHR1 receptor binding and lipophilicity of pyrrolopyrimidines, potential nonpeptide corticotropin-releasing hormone type 1 receptor antagonists. Bioorg. Med. Chem., 10: 175–183.

Jackson, M.E. and Moghaddam, B. (2001) Amygdala regulation of nucleus accumbens dopamine output is governed by the prefrontal cortex. J. Neurosci., 21: 676–681.

Jaferi, A., Nowak, N. and Bhatnagar, S. (2003) Negative feedback functions in chronically stressed rats: role of the posterior paraventricular thalamus. Physiol. Behav., 78: 365–373.

Janssens, C.J.J.G., Helmond, F.A., Loyens, L.W.S., Schouten, W.G.P. and Wiegant, V.M. (1995) Chronic stress increases the opioid-mediated inhibition of the pituitary-adrenocortical response to acute stress in pigs. Endocrinology, 136: 1468–1473.

Jedema, H.P., Finlay, J.M., Sved, A.F. and Grace, A.A. (2001) Chronic cold exposure potentites CRF-evoked increases in electrophysiologic activity of locus coeruleus neurons. Biol. Psychiat., 49: 351–359.

Johnson, L., Farb, C., Morrison, J.H., McEwen, B.S. and Ledoux, J.E. (2005) Localization of glucocorticoid receptors at postsynaptic membranes in the lateral amygdala. Neuroscience, 136: 289–299.

Jones, M.W., Kilpatrick, I.C. and Phillipson, O.T. (1989) Regulation of dopamine function in the nucleus accumbens of the rat by the thalamic paraventricular nucleus and adjacent midline nuclei. Exp. Brain Res., 76: 572–580.

Jones, D.N.C., Kortekaas, R., Slade, P.D., Middlemiss, D.N. and Hagen, J.J. (1998) The behavioral effects of corticotropin-releasing factor-related peptides in rats. Psychopharmacology, 138: 124–132.

Kalin, N.H., Takahashi, L.K. and Chen, F.L. (1994) Restraint stress increases corticotropin-releasing factor mRNA content in the amygdala and paraventricular nucleus. Brain Res., 656: 182–186.

Kant, G.J., Bunnell, B.N., Mougey, E.H., Pennington, L.L. and Meyerhoff, J.L. (1983) Effects of repeated stress on pituitary cyclic AMP and plasma prolactin, corticosterone and growth hormone in male rats. Pharmacol. Biochem. Behav., 22: 967–971.

Kant, G.J., Leu, J.R., Anderson, S.M. and Mougey, E.H. (1987) Effects of chronic stress on plasma corticosterone, ACTH and prolactin. Physiol. Behav., 40: 775–779.

Kawahara, H., Kawahara, Y. and Westerink, B.H.C. (2000) The role of afferents to the locus coeruleus in the handling stress-induced increase in the release of noradrenaline in the medial prefrontal cortex: a dual-probe microdialysis study. Eur. J. Pharmacol., 387: 279–286.

Keller-Wood, M.E. and Dallman, M.F. (1984) Corticosteroid inhibition of ACTH secretion. Endocr. Rev., 5: 1–24.

Kelley, A.E., Baldo, B.A., Pratt, W.E. and Will, M.J. (2005) Corticostriatal-hypothalamic circuitry and food motivation: integration of energy, action and reward. Physiol. Behav., 86: 773–795.

98

Kinomura, S., Larsson, J., Bulyas, B. and Roland, P.E. (1996) Activation by attention of the human reticular formation and thalamic intralaminar nuclei. Science, 271: 512–515.

Kirby, L.G., Rice, K.C. and Valentino, R.J. (2000) Effects of corticotropin-releasing factor on neuronal activity in the serotonergic dorsal raphe nucleus. Neuropsychopharmacology, 22: 148–162.

Koob, G.F. (1992) Neural mechanisms of drug reinforcement. Ann. N.Y. Acad. Sci., 654: 171–191.

Koob, G.F. and Britton, K.T. (1990) Behavioral Effects of Corticotropin-Releasing Factor, Chap. 17. CRC Press, Boca-Raton.

Koob, G.F. and Le Moal, M. (2001) Drug addiction, dysregulation of reward, and allostasis. Neuropsychopharmacology, 24: 97–129.

Korte, S.M., Korte-Bouws, G.A., Bohus, B. and Koob, G.F. (1994) Effect of corticotropin-releasing factor antagonist on behavioral and neuroendocrine responses during exposure to defensive burying paradigm in rats. Physiol. Behav., 56: 115–120.

Kotake, M., Nakai, A., Mokuno, T., Oda, N., Sawai, Y., Itoh, Y., Shimazaki, K., Kato, R., Hayakawa, N., Uchikawa, A., Oiso, Y., Hirooka, Y., Mitsuma, T., Itoh, M. and Nagasaka, A. (1996) Short stature due to growth hormone deficiency associated with Cushing's disease and ulcerative colitis. Horm. Metab. Res., 28: 565–569.

Krahn, D.D., Gosnell, B.A., Grace, M. and Levine, A.S. (1986a) CRF antagonist partially reverses CRF- and stress-induced behavioral effects. Brain Res. Bull., 17: 285–289.

Krahn, D.D., Gosnell, B.A., Grace, M. and Levine, A.S. (1986b) CRF antagonist partially reverses CRF- and stress-induced effects on feeding. Brain Res. Bull., 17: 285–289.

Krahn, D.D., Gosnell, B.A., Levine, A.S. and Morley, J.E. (1988) Behavioral effects of corticotropin-releasing factor: localization and characterization of central effects. Brain Res., 443: 63–69.

Kunz-Ebrecht, S.R., Kirschbaum, C., Marmot, M. and Steptoe, A. (2004) Differences in cortisol awakening response on work days and weekends in women and men from Whitehall II cohort. Psychoneuroendocrinology, 29: 516–528.

La Fleur, S.E., Akana, S.F., Manalo, S. and Dallman, M.F. (2004) Interaction between corticosterone and insulin in obesity: regulation of lard intake and fat stores. Endocrinology, 145: 2174–2185.

La Fleur, S.E., Houshyar, H., Roy, M. and Dallman, M.F. (2005) Choice of lard, but not total lard calories, damps ACTH responses to restraint. Endocrinology, 146: 2193–2199.

LaLumiere, R.T., Nawar, E.M. and McGaugh, J.L. (2005) Modulation of memory consolidation by the basolateral amygdala or nucleus accumbens shell requires concurrent dopamine receptor activation in both brain regions. Learn. Mem., 12: 296–301.

Lamberts, S.W.J., Bruning, H.A. and de Jong, F.H. (1995) Corticosteroid therapy in severe illness. N. Engl. J. Med., 337: 1285–1292.

Landbergis, P.A., Schnall, P.L., Pickering, T.G., Warren, K. and Schwartz, J.E. (2003) Life-course exposure to job strain and ambulatory blood pressure in men. Am. J. Epidemiol., 157: 998–1006.

Laugero, K.D., Bell, M.E., Bhatnagar, S., Soriano, L. and Dallman, M.F. (2001a) Sucrose ingestion normalizes central expresion of corticotropin-releasing factor mRNA and energy balance in adrenalectomized rats: a glucocorticoid-metabolic-brain axis? Endocrinology, 142: 2796–2804.

Laugero, K.D., Bell, M.E., Bhatnagar, S., Soriano, L. and Dallman, M.F. (2001b) Sucrose ingestion normalizes central expression of corticotropin-releasing-factor messenger ribonucleic acid and energy balance in adrenalectomized rats: a glucocorticoid-metabolic-brain axis? Endocrinology, 142: 2796–2804.

Laugero, K.D., Gomez, F., Manalo, S. and Dallman, M.F. (2002) Corticosterone infused intracerebroventricularly inhibits energy storage and stimulates the hypothalamo-pituitary axis in adrenalectomized rats drinking sucrose. Endocrinology, 143: 4552–4562.

Lechner, S.M. and Valentino, R.J. (1999) Glucocorticoid receptor-immunoreactivity in corticotropin-releasing factor afferents to locus coeruleus. Brain Res., 816: 17–28.

Lee, Y. and Davis, M. (1997) Role of the hippocampus, the bed nucleus of the stria terminalis, and the amygdala in th excitatory effect of corticotropin-releasing hormone on the acoustic startle reflex. J. Neurosci., 17: 6434–6446.

Leshner, A.I. (1971) The adrenals and the regulatory nature of running wheel activity. Physiol. Behav., 6: 551–558.

Levin, B.E., Richard, D., Michel, C. and Servatius, R. (2000) Differential stress responsivity in diet-induced obese and resistant rats. Am. J. Physiol., 279: R1357–R1364.

Levin, N., Shinsako, J. and Dallman, M.F. (1988) Corticosterone acts on the brain to inhibit adrenalectomy-induced adrenocorticotropin secretion. Endocrinology, 122: 694–704.

Liang, K.C., Melia, K.R., Miserendino, M.J., Falls, W.A., Campeau, S. and Davis, M. (1992) Corticotropin-releasing factor: long-lasting facilitation of the acoustic startle reflex. J. Neurosci., 12: 2303–2312.

Linthorst, A.C., Falachskamm, C., Hopkins, S.J., Hoadley, M.E., Labeur, M.S., Holsboer, F. and Reul, J.M. (1997) Long-term intracerebroventricular infusion of corticotropin-releasing hormone alters neuroendocrine, neurochemical, autonomic, behavioral and cytokine responses to systemic cytokine challenge. J. Neurosci., 17: 4448–4460.

Liposits, Z. and Bohn, M.C. (1993) Association of glucocorticoid receptor immunoreactivity with cell membrane and transport vesicles in hippocampal and hypothalamic neurons of the rat. J. Neurosci. Res., 35: 14–19.

Lowry, C.A., Rodda, J.E., Lightman, S.L. and Ingraham, C.D. (2000) Corticotropin-releasing factor increases in vitro firing rates of serotoninergic neurons in the rat dorsal raphe nucleus: evidence for activation of a topograhically organized mesolimbic serotoninergic system. J. Neurosci., 20: 7728–7736.

Lucas, L.R., Pompei, P., Ono, J. and McEwen, B.S. (1998) Effects of adrenal steroids on basal ganglia neuropeptide mRNA and tyrosine hydroxylase radioimmunoreactive levels in the adrenalectomized rat. J. Neurochem., 71: 833–843.

Maier, S.F. and Watkins, L.R. (2005) Stressor controllability and learned helplessness: the roles of the dorsal raphe nucleus, serotonin, and corticotropin-releasing factor. Neurosci. Biobehav. Rev., 29: 829–841.

Makino, S., Gold, P.W. and Schulkin, J. (1994a) Corticosterone effects on corticotropin-releasing hormone mRNA in the central nucleus of the amygdala and the parvocellular region of the paraventricular nucleus of the hypothalamus. Brain Res., 640: 105–112.

Makino, S., Gold, P.W. and Schulkin, J. (1994b) Effects of corticosterone on CRH mRNA and content in bed nucleus of the stria terminalis; comparison with the effects in the bed nucleus of the amygdala and the paraventricular nucleus of the hypothalamus. Brain Res., 657: 141–149.

Makino, S., Schulkin, J., M.A., S., Pacak, K., Palkovits, M. and Gold, P.W. (1995) Regulation of corticotropin-releasing hormone receptor messenger ribonucleic acid in the rat brain and pituitary by glucocorticoids and stress. Endocrinology, 136: 4517–4525.

Makino, S., Shibasaki, T., Yamauchi, N., Nishioka, T., Mimoto, T., Wakabayashi, I., Gold, P.W. and Hashimoto, K. (1999) Psychological stress increased corticotropin-releasing hormone mRNA and content in the central nucleus of amygdala but not in the hypothalamic paraventricular nuclei of the rat. Brain Res., 850: 136–143.

Marti, O., Gavalda, A., Gomez, F. and Armario, A. (1994) Direct evidence for chronic stress-induced facilitation of the adrenocorticotropin response to a novel acute stressor. Neuroendocrinology, 60: 1–7.

Marti, O., Harbuz, M.S., Andres, R., Lightman, S.L. and Armario, A. (1999) Activation of the hypothalamic-pituitary axis in adrenalectomized rats: potentiation by chronic stress. Brain Res., 831: 1–7.

Martinelli, M., Piazza, P.V., Deroshe, V., Maccari, S., Le Moal, M. and Simon, H. (1994) Corticosterone circadian secretion differentially facilitates dopamine-mediated psychomotor effect of cocaine and morphine. J. Neurosci., 14: 2724–2731.

Matsuo, T., Iswashita, S., Komuro, M. and Suzuki, M. (1999) Effects of high fat diet intake on glucose uptake in central and peripheral tissues of non-obese rats. J. Nutr. Soc. Vitaminol., 45: 667–673.

Matsuzaki, I., Takamatsu, Y. and Moroji, T. (1989) The effects of introcerebroventricularly injected corticotropin-releasing factor (CRF) on the central nervous system: behavioral and biochemical studies. Neuropeptides, 13: 147–155.

Mayer, E.A. and Fanselow, M.S. (2003) Dissecting the components of the central response to stress. Nat. Neurosci., 6: 1011–1012.

McFarland, K., Davidge, S.B., Lapish, C.C. and Kalivas, P.W. (2004) Limbic and motor circuitry underlying footshock-induced reinstatement of cocaine-seeking behavior. J. Neurosci., 24: 1551–1560.

McFarland, N.R. and Haber, S.N. (2002) Thalamic relay nuclei of the basal ganglia for both reciprocal and nonreciprocal cortical connections, linking multiple frontal cortical areas. J. Neurosci., 22: 8117–8132.

McMillan, P.G., Wilkinson, C.W., Greenup, L., Raskind, M.A., Peskind, E.R. and Leverenz, J.B. (2004) Chronic cortisol exposure promotes the development of a GABAergic phenotype in the primate hippocampus. J. Neurochem., 91: 843–851.

McNally, G.P. and Akil, H. (2002) Role of corticotropin-releasing hormone in the amygdala and bed nucleus of the stria terminalis in the behavioral, pain modulatory, and endocrine consequences of opiate withdrawal. Neuroscience, 12: 605–617.

Melia, K. and Duman, R. (1991) Involvement of corticotropin-releasing factor in chronic stress regulation of the brain noradrenergic system. Proc. Natl. Acad. Sci. USA, 88: 8382–8386.

Menzaghi, F., Burlet, A., Chapleur, M., Nicolas, J.-P. and Burlet, C. (1992) Alteration of pituitary-adrenal responses to adrenalectomy by the immunological targeting of CRF neurons. Neurosci. Lett., 135: 49–52.

Menzaghi, F., Heinrichs, S.C., Merlo-Pich, E., Tilders, F.J.H. and Koob, G.F. (1994) Involvement of hypothalamic corticotropin-releasing factor neurons in behavioral responses to novelty in rats. Neurosci. Lett., 168: 139–142.

Menzaghi, F., Heinrichs, S.C., Pich, E.M., Weiss, F. and Koob, G.F. (1993) The role of limbic and hypothalamic corticotropin-releasing factor in behavioral responses to stress. In: Tache, Y. and Rivier, C. (Eds.) Corticotropin-Releasing Factor and Cytokines: Role in the Stress Response, Vol. 697. New York Academy of Sciences, New York, pp. 142–154.

Merali, Z., Du, L., Hrdina, P., Palkovits, M., Faludi, G., Poulter, M.O. and Anisman, H. (2004a) Dysregulation in the suicide brain: mRNA expression of corticotropin-releasing hormone receptors and GABAa receptor subunits in frontal cortical brain region. J. Neurosci., 24: 1478–1485.

Merali, Z., Khan, S., Michaud, D.S., Shippy, S.A. and Anisman, H. (2004b) Does amygdaloid corticotropin-releasing hormone (CRH) mediate anxiety-like behaviors? Dissociation of anxiogenic effects and CRH release. J. Neuroendocrinol., 20: 229–239.

Merali, Z., McIntosh, J., Kent, P., Michaud, D. and Anisman, H. (1998) Aversive and appetitive events evoke the release of corticotropin-releasing hormone and bombesin-like peptides at the central nucleus of the amygdala. J. Neurosci., 18: 4758–4766.

Merlo Pich, E., Koob, G.F., Vale, W. and Weiss, F. (1994) Release of corticotropin releasing factor (CRF) from the amygdala of ethanol-dependent rats measured with microdialysis. Alcohol Clin. Exp. Res., 18: 522.

Merlo Pich, E., Lorang, M.T., Yeganeh, M., Rodriguez de Fonseca, F., Raber, J., Koob, G.F. and Weiss, F. (1995) Increase of extracellular corticotropin-releasing factor-like immunoreactivity levels in the amygdala of awake rats during restraint stress and ethanol withdrawal as measured by microdialysis. J. Neurosci., 15: 5439–5447.

Micheau, J., Destrade, C. and Soumireu-Mourat, B. (1981) Intraventricular corticosterone injection facilitates memory of an appetitive discriminative task in mice. Behav. Neural Biol., 31: 100–104.

Michi, T., Jocic, M., Heinemann, A., Schuligoi, R. and Holzer, P. (2001) Vagal afferent signaling of a gastric mucosal acid insult to medullary, pontine, thalamic, hypothalamic and limbic but not cortical nuclei of the rat brain. Pain, 92: 19–27.

Minor, T.R. and Saade, S. (1997) Postsstress glucose mitigates behavioral impairment in rats in the "learned helplessness" model of psychopathology. Biol. Psychiat., 42: 324–334.

Miyata, S., Ishiyama, M., Shido, O., Nakashima, T., Sibata, M. and Kiyohara, T. (1995) Central mechanism of neural activation with cold acclimation of rats using Fos immunohistochemistry. Neurosci. Res., 22: 209–218.

Mizoguchi, K., Yuzurihara, M., Ishige, A., Sasaki, H., Chui, D.H. and Tabira, T. (2001) Chronic stress differentially regulates glucocorticoid negative feedback response in rats. Psychoneuroendocrinology, 26: 443–459.

Moberg, G.P. and Clark, C.R. (1976) Effect of adrenalectomy and dexamethasone treatment on circadian running in the rat. Physiol. Behav., 4: 617–619.

Moga, M.M., Weis, R.P. and Moore, R.Y. (1995) Efferent projections of the paraventricular thalamic nucleus of the rat. J. Comp. Neurol., 359: 221–238.

Mokdad, A.H., Serdula, M.K., Dietz, W.H., Bowman, B.A., Marks, J.S. and Koplan, J.P. (2000) The continuing epidemic of obesity in the United States. JAMA, 284: 1650–1651.

Morilak, D.A., Garera, G., Echevarria, D.J., Garcia, A.S., Hernandez, A., Ma, S. and Petre, C.O. (2005) Role of brain norepinephrine in the behavioral response to stress. Prog. Neuro-Psychopharmacol. & Biol. Psychol., 29: 1214–1224.

Muller, M.B., Zimmermann, S., Sillaber, I., Hagemeyer, T.P., Deussing, J.M., Timpl, P., Kormann, M.S.D., Droste, S.K., Kuhn, R., Reul, J.M.H.M., Holsboer, F. and Wurst, W. (2003) Limbic corticotropin-releasing hormone receptor 1 mediates anxiety-related behavior and hormonal adaptation to stress. Nat. Neurosci., 6: 1100–1107.

Munck, A., Guyre, P.M. and Holbrook, N.J. (1984) Physiological functions of glucocorticoids in stress and their relation to pharmacological actions. Endocr. Rev., 5: 25–44.

Myers, D.A., Gibson, M., Schulkin, J. and Greenwood Van-Meerveld, B. (2005) Corticosterone implants to the amygdala and type 1 CRH receptor regulation: effects on behavior and colonic sensitivity. Behav. Brain. Res., 161: 39–44.

Myers, E.A. and Rinaman, L. (2002) Viscerosensory activation of noradrenergic input to the amygdala. Physiol. Behav., 77: 723–729.

Nauta, W. (1962) Neural associations of the amygdaloid complex in the monkey. Brain, 85: 505–520.

Nemeroff, C.B., Owens, M.J., Bissette, G., Andorn, A.C. and Stanley, M. (1988) Reduced corticotropin-releasing factor biding sites in the frontal cortex of suicide victims. Arch. Gen. Psychiatr., 45: 577–579.

Nicholson, S.A., Campbell, E.A., Gillham, B. and Jones, M.T. (1987) Recovery of the components of the hypothalamo-pituitary-adrenocortical axis in the rat after chronic treatment with prednisolone. J. Endocrinol., 113: 239–247.

Nicholson, S., Campbell, E., Torrellas, A., Beckford, U., Altaher, R., Sandford, R., Scraggs, R., Gillham, B. and Jones, M. (1984) Recovery of the hypothalamo-pituitary-adrenocortical axis in the rat after long-term dexamethasone treatment. Neuroendocrinology, 39: 343–349.

Nicholson, S.A., Gillham, B. and Jones, M.T. (1988) Influence of prolonged glucocorticoid treatment on intracellular mechanisms involved in ACTH secretion in the rat. J. Mol. Endocrinol., 1: 202–212.

Okuyama, S., Chaki, S., Kawashima, N., Suzuki, Y., Ogawa, S.-I., Nakazato, A., Kumagai, T., Okubo, T. and Tomisawa, K. (1999) Receptor binding, behavioral and electrophysiological profiles of nonpeptide corticotropin-releasing factor subtype 1 receptor antagonists CRA1000 and CRA 1001. J. Pharmacol. Exp. Ther., 289: 926–935.

Orth, D.N. (1995) Cushing's syndrome. N. Engl. J. Med., 332: 791–803.

Oshima, A., Flachskamm, C., Reul, J.M.H.M., Holsboer, F. and Linthorst, A.C. (2003) Altered serotoninergic neurotransmission but normal hypothalamic-pituitary-adrenocortical axis activity in mice chronically treated with corticotropin-releasing hormone receptor type I agonist NBI 30775. Neuropsychopharmacology, 28: 2148–2159.

Ottenweller, J.E., Servatius, R.J., Tapp, W.N., Drastal, S.D., Bergen, M.T. and Natelson, B.H. (1992) A chronic stress state in rats: effects of repeated stress on basal corticosterone and behavior. Physiol. Behav., 51: 689–698.

Owens, M.J. and Nemeroff, C.B. (1991) Physiology and pharmacology of corticotropin-releasing factor. Pharmacol. Rev., 43: 425–473.

Pacak, K., McCarty, R., Palkovits, M., Kopin, I.J. and Goldstein, D.S. (1995a) Effects of immobilization on in vivo release of norepinephrine in the bed nucleus of the stria terminalis in conscious rats. Brain Res., 688: 242–246.

Pacak, K., Palkovits, M., Kopin, I.J. and Goldstein, D.S. (1995b) Stress-induced norepinephrine release in hypothalamic paraventricular nucleus and pituitary-adrenocortical and sympathoadrenal activity: in vivo microdialysis studies. Front. Neuroendocrinol., 16: 89–150.

Palkovits, M. (2000) Stress-induced expression of co-localized neuropeptides in hypothalamic and amygdaloid neurons. Eur. J. Pharmacol., 405: 161–166.

Palkovits, M., Baffi, J., Toth, Z.E. and Pacak, K. (1998a) Brain catecholamine systems in stress. Adv. Pharmacol., 42: 572–575.

Palkovits, M., Young III, W.S., Kovacs, K., Toth, T. and Makara, G.B. (1998b) Alterations in corticotropin-releasing hormone gene expression of central amygdaloid neurons following long-term paraventricular lesions and adrenalectomy. Neuroscience, 85: 135–147.

Pavcovich, L.A. and Valentino, R.J. (1997) Regulation of a putative neurotransmitter effect of corticotropin-releasing factor: effects of adrenalectomy. J. Neurosci., 17: 401–408.

Pecoraro, N. and Dallman, M. (2005) c-Fos after incentive shifts: expectancy, incredulity, and recovery. Behav. Neurosci., 119: 366–387.

Pecoraro, N., Gomez, F., la Fleur, S., Roy, M. and Dallman, M.F. (2005a) Single, but not multiple pairings of sucrose and corticosterone enhance memory for sucrose drinking and amplify remote reward relativity effects. Neurobiol. Learn. Mem., 83: 188–195.

Pecoraro, N., Reyes, F., Gomez, F., Bhargava, A. and Dallman, M.F. (2004) Chronic stress promotes palatable feeding, which reduces signs of stress: feedforward and feedback effects of chronic stress. Endocrinology, 145: 3754–3762.

Pecoraro, N.C., Roy, M. and Dallman, M.F. (2005b) Glucocorticoids dose-dependently remodel energy stores and amplify incentive reality effects. Psychoneuroendocrinology, 30: 815–825.

Peters, A., Schweiger, U., Pellerin, L., Hubold, C., Oltmanns, K.M., Conrad, M., Schultes, B., Born, J. and Fehm, H.L. (2004) The selfish brain: competition for energy resources. Neurosci. Biobehav. Rev., 28: 143–180.

Piazza, P.V., Barrot, M., Rouge-Pont, F., Marinelli, M., Maccari, S., Abrous, D.N., Simon, H. and Le Moal, M. (1996a) Suppression of glucocorticoid secretion and antipsychotic drugs have similar effects on the mesolimbic dopaminergic transmission. Proc. Natl. Acad. Sci. USA, 93: 15445–15450.

Piazza, P.V., Deminiere, J.M., le Moal, M. and Simon, H. (1990) Stress- and pharmacologically induced behavioral sensitization increases vulnerability to acquisition of amphetamine self-administration. Brain Res., 514: 22–26.

Piazza, P.V., Deroche, V., Deminiere, J.-M., Maccari, S., Le Moal, M. and Simon, H. (1993) Corticosterone in the range of stress-induced levels possesses reinforcing properties: implications for sensation-seeking behaviors. Proc. Natl. Acad. Sci. USA, 90: 11738–11742.

Piazza, P.V. and Le Moal, M.L. (1996) Pathophysiological basis of vulnerability to drug abuse: role of an interaction between stress, glucocorticoids, and dopaminergic neurons. Annu. Rev. Pharmacol. Toxicol., 36: 359–378.

Piazza, P.V. and Le Moal, M. (1997) Glucocorticoids as a biological substrate of reward: physiological and pathophysiological implications. Brain Res. Rev., 25: 359–372.

Piazza, P.V. and Le Moal, M. (1998) The role of stress in drug self-administration. TiPS, 19: 67–74.

Piazza, P.V., Maccari, S., Deminiere, J.M., Le Moal, M., Mormede, P. and Simon, H. (1991) Corticosterone levels determine individual vulnerability to amphetamine self administration. Proc. Natl. Acad. Sci. USA, 88: 2088–2092.

Piazza, P.V., Rouge-Pont, F., Derche, V., Maccari, S., Simon, H. and Le Moal, M. (1996b) Glucocorticoids have state-dependent stimulant effects on the mesencephalic dopaminergic transmission. Proc. Natl. Acad. Sci. USA, 93: 8716–8720.

Pinto, A., Jankowski, M. and Sesack, S.R. (2003) Projections from the paraventricular nucleus of the thalamus to the rat prefrontal cortex and nucleus accumbens shell: ultrastructural characteristics and spatial relationships with dopamine afferents. J. Comp. Neurol., 459: 142–155.

Pittman, D.L., Ottenweller, J.E. and Natelson, B.H. (1998) Plasma corticosterone levels during repeated presentation of two intensities of restraint stress: chronic stress and habituation. Physiol. Behav., 43: 47–55.

Plotsky, P.M., Cunningham Jr., E.T. and Widmaier, E.P. (1989) Catecholaminergic modulation of corticotropin-releasing factor and adrenocorticotropin secretion. Endocr. Rev., 10: 437–458.

Price, M.L., Curtis, A.L., Kirby, L.G., Valentino, R.J. and Lucki, I. (1998) Effects of corticotropin-releasing factor on brain serotoninergic activity. Neuropsychopharmacology, 18: 492–508.

Price, M.L., Kirby, L.G., Valentino, R.J. and Lucki, I. (2002) Evidence for corticotropin-releasing factor regulation of serotonin in the lateral septum during acute swim stress: adaptation produced by repeated swimming. Psychopharmacology, 162: 406–414.

Price, M.L. and Lucki, I. (2001) Regulation of serotonin release in the lateral septum and striatum by corticotropin-releasing factor. J. Neurosci., 21: 2833–2841.

Quirarte, G.L., Galvez, R., Roozendaal, B. and McGaugh, J.L. (1998) Norepinephrine release in the amygdala in response to footshock and opioid peptidergic drugs. Brain Res., 808: 134–140.

Raadsheer, F.C., Hoogendijk, W.J.G., Stam, F.C., Tilders, F.H.J. and Swaab, D.F. (1994) Increased numbers of corticotropin-releasing hormone expressing neurons in the hypothalamic paraventricular nucleus of depressed patients. Neuroendocrinology, 60: 433–436.

Raber, J., Koob, G.F. and Bloom, F.E. (1995) Interleukin-2 (Il-2) induces corticotropin-releasing factor (CRF) release from the amygdala and involves a nitric oxide-mediated signalling; comparison with the hypothalamic response. J. Pharmacol. Exp. Ther., 272: 815–824.

Radulovic, J., Ruhmann, A., Liepold, T. and Spiess, J. (1999) Modulation of learning and anxiety by corticotripin-releasing factor (CRF) and stress: differential roles of CRF receptors 1 and 2. J. Neurosci., 19: 5016–5025.

Reaven, G., Abbasi, F. and McLaughlin, T. (2004) Obesity, insulin resistance, and cardiovascular disease. Recent Prog. Horm. Res., 59: 207–223.

Rebuffe-Scrive, M., Bronnegard, M., Nilsson, A., Eldh, J., Gustafsson, J.-A. and Bjorntorp, P. (1990) Steroid hormone receptors in human adipose tissues. J. Clin. Endocrinol. Metabol., 71: 1215–1219.

Rebuffe-Scrive, M., Walsh, U.A., McEwen, B. and Rodin, J. (1992) Effect of chronic stress and exogenous glucocorticoids on regional fat distribution and metabolism. Physiol. Behav., 52: 583–590.

Reul, J.M.H.M. and Holsboer, F. (2002) Corticotropin-releasing factor receptors 1 and 2 in anxiety and depression. Curr. Opin. Pharm., 2: 23–33.

Richter, R.M. and Weiss, F. (1999) In vivo CRF release in rat amygdala is increased during cocaine withdrawal in self-administering rats. Synapse, 32: 254–261.

Rivest, S. and Laflamme, N. (1995) Neuronal activity and neuropeptide gene transcription in the brains of immune-challenged rats. J. Neuroendocrinol., 7: 501–525.

Rivet, J.M., Stinus, L., LeMoal, M. and Mormede, P. (1989) Behavioral sensitization to amphetamine is dependent on corticosteroid receptor activation. Brain Res., 498: 149–153.

Rivier, J., Gulyas, J., Kirby, D., Low, W.,]Perrin, M.H., Kunitake, K.S., DiGruccio, M., Vaughan, J., Reubi, J.C., Waser, B., Koerber, S.C., Martinez, V., Wang, L., Tache, Y. and Vale, W. (2002) Potent and long-acting corticotropin releasing factor (CRF) receptor 2 selective peptide competitive antagonists. J. Med. Chem., 45: 4737–4747.

Rivier, J.E., Kirby, D.A., Lahricht, S.L., Corrigan, a., Vale, W.W. and Rivier, C.L. (1999) Constrained corticotropin releasing factor (CRF) antagonists (Astressin analogs) with long duration of acti9n in the rat. J. Med. Chem., 42: 3175–3182.

Roche, M., Commons, K.G., Peoples, A. and Valentino, R.J. (2003) Circuitry underlying regulation of the serotonin system by swim stress. J. Neurosci., 23: 970–977.

Rodriguez de Fonseca, F., Carrera, M.R., Navarro, M., Koob, G.F. and Weiss, F. (1997) Activation of corticotropin-releasing factor in the limbic system during cannabinoid withdrawal. Science, 276: 2050–2054.

Roozendaal, B. (2002) Stress and memory: opposing effects of glucocorticoids on memory consolidation and memory retrieval. Neurobiol. Learn. Mem., 78: 578–595.

Roozendaal, B. (2003) Systems mediating acute glucocorticoid effects on memory consolidation and retrieval. Prog. Neuropsychopharmacol. Biol. Psychiat., 27: 1213–1223.

Roozendaal, B., Bohus, B. and McGaugh, J.L. (1996) Dose-dependent suppression of adrenocortical activity with metyrapone: effects on emotion and memory. Psychoneuroendocrinology, 21: 681–693.

Roozendaal, B., Brunson, K.L., Holloway, B.L., McGaugh, J.L. and Baram, T.Z. (2002) Involvement of stress-released corticotropin-releasing hormone in the basolateral amygdala in regulating memory consolidation. PNAS, 99: 13908–13913.

Roozendaal, B., de Quervqin, E.-F., Ferry, B., Setalo, G. and McGaugh, J.L. (2001a) Basolateral amygdala-nucleus accumbens interactions in mediating glucocorticoid enhancement of memory consolidation. J. Neurosci., 21: 2518–2525.

Roozendaal, B., Phillips, R.G., Power, A.E., Brooke, S.M., Sapolsky, R.M. and McGaugh, J.L. (2001b) Memory retrieval impairment induced by hippocampal CA3 lesions is blocked by adrenocortical suppression. Nat. Neurosci., 4: 1169–1171.

Roozendaal, B., Williams, C.L. and McGaugh, J.L. (1999) glucocorticoid receptor activation in the rat nucleus of the solitary tract facilitates memory consolidation: involvement of the basolateral amygdala. Eur. J. Neurosci., 11: 1317–1323.

Rothwell, N.J. (1990) Central effects of CRF on metabolism and energy balance. Neurosci. Biobehav. Rev., 14: 263–271.

Sajdyk, T.J., Schober, D.A., Gehlert, D.R. and Shekhar, A. (1999) Role of corticotropin-releasing factor and urocortin within the basolateral amygdala of rats in anxiety and panic attacks. Behav. Brain Res., 100: 207–215.

Sakellaris, P.C. and Vernikos-Danellis, J. (1975) Increased rate of response of the pituitary-adrenal system in rats adapted to chronic stress. Endocrinology, 97: 597–602.

Samarghandian, S., Ohata, N., Yamauchi, N. and Shibasaki, T. (2003) Corticotropin-releasing factor as well as opioid and dopamine are involved in tail-pinch-induced food intake of rats. Neuroscience, 116: 519–524.

Sarlis, N.J., Chowdrey, H.S., Stephanou, A. and Lightman, S.L. (1992) Chronic activation of the hypothalamo-pituitary-adrenal axis and loss of circadian rhythm during adjuvant-induced arthritis in the rat. Endocrinology, 130: 1775–1779.

Sarnyai, B., Shaham, Y. and Heinrichs, S.C. (2001) The role of corticotropin-releasing factor in drug addiction. Pharm. Rev., 53: 209–243.

Sato, T., Sato, M., Shinsako, J. and Dallman, M.F. (1975) Corticosterone induced changes in hypothalamic corticotropin releasing factor (CRF) content after stress. Endocrinology, 97: 265–274.

Sawchenko, P.E. and Swanson, L.W. (1989) Organization of CRF immunoreactive cells and fibers in the rat brain: immunohistochemical studies. In: Nemeroff, C.B. and de Souza, E.B. (Eds.), CRC Critical Reviews in Corticotropin-Releasing Factor: Basic and Clinical Studies of a Neuropeptide. CRC Press, Boca Raton, pp. 29–51.

Schlotz, W., Hellhammer, J., Schulz, P. and Stone, A.A. (2004) Perceived work overload and chronic worrying predict weekend-weekday differences in the cortisol awakening response. Psychosom. Med., 66: 207–214.

Schulkin, J., McEwen, B.S. and Gold, P.W. (1994) Allostasis, amygdala, and anticipatory angst. Neurosci. Biobehav. Rev., 18: 385–396.

Schulkin, J., Morgan, M.A. and Rosen, J.B. (2005) A neuroendocrine mechanism for sustaining fear. Trends Neurosci., 28: 629–635.

Schulz, C. and Lehnert, H. (1996) Activation of noradrenergic neurons in the locus coeruleus by corticotropin-releasing factor. Neuroendocrinology, 63: 454–458.

Scribner, K.A., Akana, S.F., Walker, C.-D. and Dallman, M.F. (1993) Streptozotocin-diabetic rats exhibit facilitated adrenocorticotropin responses to acute stress, but normal sensitivity to feedback by corticosteroids. Endocrinology, 133: 2667–2674.

Servatius, R.J., Ottenweller, J.E., Bergen, M.T., Soldan, S. and Natelson, B.H. (1994) Persistent stress-induced sensitization of adrenocortical and startle responses. Physiol. Behav., 56: 945–954.

Shaham, Y., Erb, S., Leung, S., Buczek, Y. and Stewart, J. (1998) CP-154,526, a selective, non-peptide antagonist of the corticotropin-releasing factor1 receptor attenuates stress-induced relapse to drug seeking in cocaine- and heroin-trained rats. Psychopharmacology (Berl.), 137: 184–190.

Shaham, Y., Erb, S. and Stewart, J. (2000) Stress-induced relapse to heroin and cocaine seeking in rats: a review. Brain Res. Rev., 33: 13–33.

Shaham, Y., Funk, D., Erb, S., Brown, T.J., Walker, C.-D. and Stewart, J. (1997) Corticotropin-releasing factor, but not corticosterone, is involved in stress-induced relapse to heroin seeking in rats. J. Neurosci., 17: 2605–2614.

Shepard, J.D., Barron, K.W. and Myers, D.A. (2000) Corticosterone delivery to the amygdala increases corticotropin-releasing factor mRNA in the central amygdaloid nucleus and anxiety-like behavior. Brain Res., 861: 288–295.

Shepard, J.D., Barron, K.W. and Myers, D.A. (2003) Stereotaxic localization of corticosterone to the amygdala enhances hypothalamo-pituitary-adrenal responses to behavioral stress. Brain Res., 963: 203–213.

Smagin, G.N., Heinrichs, S.C. and Dunn, A.J. (2001) The role of CRH in behavioral responses to stress. Peptides, 22: 713–724.

Smagin, G.N., Howell, L.A., Redmann Jr., S., Ryan, D.H. and Harris, R.B. (1999) Prevention of stress-induced weight loss by third ventricle CRF receptor antagonist. Am. J. Physiol., 276: R1461–R1468.

Smagin, G.N., Swiergiel, A.H. and Dunn, A.J. (1995) Corticotropin-releasing factor administered into the locus coeruleus, but not the parabrachial nucleus, stimulates norepinephrine release in the prefrontal cortex. Brain Res. Bull., 36: 71–76.

Smagin, G.N., Zhou, J., Harris, R.B.S. and Ryan, D.H. (1997) CRF receptor antagonist attenuates immobilization stress-induced norepinephrine release in the prefrontal cortex of rats. Brain Res. Bull., 42: 431–434.

Smith, D.M., Freeman Jr., J.M., Nicholson, D. and Gabriel, M. (2002) Limbic thalamic lesions, appetitively motivated discrimination learning and training-induced neuronal activity in rabbits. J. Neurosci., 22: 8212–8221.

Spencer, S.J., Fox, J.C. and Day, T.A. (2004) Thalamic paraventricular nucleus lesions facilitate central amygdala neuronal responses to acute psychological stress. Brain Res., 997: 234–237.

Stansfeld, S.A., Fuhrer, R., Shipley, M.J. and Marmot, M.G. (1999) Work characteristics predict psychiatric disorder: prospective results from the Whitehall II study. Occup. Environ. Med., 56: 302–307.

Steinbaum, S.R. (2004) The metabolic syndrome: an emerging health epidemic in women. Prog. Cardiovasc. Dis., 46: 321–336.

Strack, A.M., Horsley, C.J., Sebastian, R.J., Akana, S.F. and Dallman, M.F. (1995a) Glucocorticoids and insulin: complex interaction on brown adipose tissue. Am. J. Physiol., 268: R1209–R1216.

Strack, A.M., Sebastian, R.J., Schwartz, M.W. and Dallman, M.F. (1995b) Glucocorticoids and insulin: reciprocal signals for energy balance. Am. J. Physiol., 268: R142–R149.

Strijbos, P.J., Hardwick, A.J., Relton, J.K., Carey, F. and Rothwell, N.J. (1992) Inhibition of central actions of cytokines on fever and thermogenesis by lipocortin-1 involves CRF. Am. J. Physiol., 263: E632–E636.

Sutton, R.E., Koob, G.F., Le Moal, M., Rivier, J. and Vale, W. (1982) Corticotropin-releasing factor (CRF) produces behavioral activation in rats. Nature, 297: 331–333.

Svedberg, J., Pjorntorp, P., Lonnroth, P. and Smith, U. (1991) Prevention of inhibitiory effect of free fatty acids on insulin binding and action in isolated rat hepatocytes by etomixir. Diabetes, 40: 783–786.

Swanson, L.W. (1982) The projections of the ventral tegmental area and adjacent regions: a combined fluorescent retrograde tracer and immunofluorescence study in the rat. Brain Res. Bull., 9: 321–353.

Swanson, L.W. (2000) Cerebral hemisphere regulation of motivated behavior. Brain Res., 886: 113–164.

Swanson, L.W., Sawchenko, P.E., Rivier, J. and Vale, W.W. (1983) Organization of ovine corticotropin-releasing factor immunoreactive cells and fibers in the rat brain: an immunohistochemical study. Neuroendocrinology, 36: 165–186.

Swanson, L.W. and Simmons, D.M. (1989) Differential steroid hormone and neural influences on peptide mRNA levels in CRH cells of the paraventricular nucleus: a hybridization histochemical study in the rat. J. Comp. Neurol., 285: 413–435.

Swiergiel, A.H., Takahashi, L.K. and Kalin, N.H. (1993) Attenuation of stress-induced behavior by antagonism of corticotropin-releasing factor receptors in the central amygdala of the rat. Brain Res., 623: 229–234.

Tache, Y., Martinez, V., Million, M. and Wang, L. (2001) Stress and the gastrointestinal tract III. Stress-related alterations of gut motor function: role of brain corticotropin-releasing factor receptors. Am. J. Physiol., 280: G173–G177.

Takahashi, L.K., Kalin, N.H., Vanden Burgt, J.A. and Sherman, J.E. (1989a) Corticotropin-releasing factor modulates defensive-withdrawal and exploratory behavior in rats. Behav. Neurosci., 103: 648–654.

Takahashi, L.K., Kalin, N.H., Venden Burgt, J.A. and Sherman, J.A. (1989b) Corticotropin-releasing factor modulates defensive-withdrawal and exploratory behavior in rats. Behav. Neurosci., 103: 648–654.

Tan, H., Zhong, P. and Yan, Z. (2004) Corticotropin-releasing factor and acute stress prolongs serotoninergic regulation of GABA transmission in prefrontal cortical neurons. J. Neurosci., 24: 5000–5008.

Tanimura, S.M., Sanchez-Watts, G. and Watts, A.G. (1998) Peptide gene activation, secretion, and steroid feedback during stimulation of rat neuroendocrine corticotropin-releasing hormone neurons. Endocrinology, 139: 3822–3829.

Tanimura, S.M. and Watts, A.G. (1998) Corticosterone can facilitate as well as inhibit corticotropin-releasing hormone gene expression in the rat hypothalamic paraventricular nucleus. Endocrinology, 139: 3830–3836.

Tanimura, S.M. and Watts, A.G. (2001) Corticosterone modulation of ACTH secretogogue gene expression in the paraventricular nucleus. Peptides, 22: 775–783.

Thomas, E., Pernar, L., Lucki, I. and Valentino, R.J. (2003) Corticotropin-releasing factor in the dorsal raphe nucleus regulates activity of lateral septal neurons. Brain Res., 960: 201–208.

Thompson, B.L., Erickson, K., Schulkin, J. and Rosen, J.B. (2004) Corticosterone facilitates retention of contextually conditioned fear and increases CRH mRNA expression in the amygdala. Behav. Brain Res., 149: 209–215.

Thompson, R.H. and Swanson, L.W. (2002) Structural characterization of a hypothalamic visceromotor pattern generator network. Brain Res. Rev., 41: 153–202.

Timofeeva, E. and Richard, D. (2001) Activation of the central nervous system in obese Zucker rats during food deprivation. J. Comp. Neurol., 441: 71–89.

Turner, B.H. and Herkenham, M. (1991) Thalamoamygdaloid projections in the rat: a test of the amygdala's role in sensory processing. J. Comp. Neurol., 313: 295–325.

Vale, W., Spiess, J., Rivier, C. and Rivier, J. (1981) Characterization of a 41-residue ovine hypothalamic peptide that stimulates secretion of corticotropin and beta-endorphin. Science, 213: 1394–1397.

Valentino, R.J. and Commons, K.G. (2005) Peptides that fine-tune the serotonin system. Neuropeptides, 39: 1–8.

Valentino, R.J., Foote, S.L. and Aston-Jones, G. (1983) Corticotropin-releasing factor activates noradrenergic neurons of the locus coeruleus. Brain Res., 270: 363–367.

Valentino, R.J., Louterman, L. and Van Bockstaele, E.J. (2001a) Evidence for regional heterogeneity in corticotropin-releasing factor interactions in the dorsal raphe nuclei. J. Comp. Neurol., 435: 450–463.

Valentino, R.J., Louterman, L. and Van Bockstaele, E.J. (2001b) Evidence for regional heterogeneity in corticotropin-releasing factor interactions in the dorsal raphe nuclei. J. Comp. Neurol., 435: 450–463.

Valentino, R.J., Page, M., Van Bockstaele, E.J. and Aston-Jones, G. (1992) Corticotropin-releasing factor innervation of the locus coeruleus region: distribution and sources of input. Neuroscience, 48: 689–705.

Valentino, R.J., Rudoy, C., Saunders, A., Liu, X.-B. and Van Bockstaele, E.J. (2001c) Corticotropin-releasing factor is preferentially colocalized with excitatory rather than inhibitory amino acids in axon terminals in the peri-locus coeruleus retion. Neuroscience, 106: 375–384.

Van Bockstaele, E.J., Colago, E.E. and Valentino, R.J. (1998) Amygdaloid corticotropin-releasing factor targets locus coeruleus dendrites: substrate for the co-ordination of emotional and cognitive limbs of the stress response. J. Neuroendocrinol., 10: 743–757.

Van Bockstaele, E.J., Peoples, J. and Valentino, R.J. (1999) Anatomic basis for differential regulation of the rostrolateral peri-locus coeruleus region by limbic afferents. Biol. Psychiatr., 46: 1352–1363.

Van der Werf, Y.D., Witter, M.P. and Groenewegen, H.J. (2002) The intralaminar and midline nuclei of the thalamus: anatomical and functional evidence for participation in processes of arousal and awareness. Brain Res. Rev., 39: 107–140.

van Dijk, G., de Vires, K., Behntham, L., Nyakas, C., Buwalda, B. and Scheurink, A.J.W. (2003) Neuroendocrinology of insulin resistance: metabolic and endocrine aspects of adiposity. Eur. J. Pharmacol., 480: 31–42.

van Pett, K., Viau, V., Bittencourt, J.C., Chan, R.K.W., LI, H.-Y., Arias, C.M., Prins, G.S., Perrin, M., Vale, W. and Sawchenko, P.E. (2000) Distribution of mRNAs encoding CRF receptors in brain and pituitary of rat and mouse. J. Comp. Neurol., 428: 191–212.

Veldhuis, H.D. and De Wied, D. (1984) Differential behavioral actions of corticotropin-releasing factor (CRF). Pharmacol. Biochem. Behav., 21: 707–713.

Viau, V., Soriano, L. and Dallman, M.F. (2001) Androgens alter corticotropin-releasing hormone and vasopressin mRNA within forebrain sites known to regulate activity in the hypothalamic-pituitary-adrenal axis. J. Neuroendocrinol., 13: 442–452.

Wang, L.-L., Ou, C.-C. and Chan, J.Y.H. (2005b) Receptor-independent activation of GABAergic neurotransmission and receptor-dependent non transcriptional activation of phosphatidylinositol 3-kinase/protein kinase Akt pathway in short-term cardiovascular actions of dexamethasone at the nucleus tractus solitarii of the rat. Mol. Pharmacol., 67: 489–498.

Wang, B., Shaham, Y., Zitzman, D., Azari, S., Wise, R.A. and You, Z.-B. (2005a) Cocaine experience establishes control of midbrain glutamate and dopamine by corticotropin-releasing factor: a role in stress-induced relapse to drug seeking. J. Neurosci., 25: 5389–5396.

Wardle, J. and Gibson, E.L. (2002) Impact of stress on diet: processes and implications. In: Stansfeld, S.A. and Marmot, M.G. (Eds.), Stress and the Heart: Psychosocial Pathways to Coronary Heart Disease. BMJ Books, London, pp. 124–149.

Waselus, M., Valentino, R.J. and Van Bocksdale, E.J. (2005) Ultrastructural evidence for a role of g-aminobutyric acid in mediating the effects of corticotropin-releasing factor on the rat dorsal raphe serotonin system. J. Comp. Neurol., 482: 155–165.

Watts, A.G. (2005) Glucocorticoid regulation of peptide genes in neuroendocrine CRH neurons: a complexity beyond negative feedback. Front Neuroendocrinol., 26: 109–130.

Watts, A.G. and Sanchez-Watts, G. (1995) Region-specific regulation of neuropeptide mRNAs in rat limbic forebrain neurones by aldosterone and corticosterone. J. Physiol., 484: 721–736.

Widmaier, E.P. and Dallman, M.F. (1984) The effects of corticotropin-releasing factor on adrenocorticotropin secretion from perifused pituitaries in vitro: rapid inhibition by glucocorticoids. Endocrinology, 115: 2368–2374.

Wiersma, A., Baauw, A.D., Bohus, B. and Koolhaas, J.M. (1995) Behavioural activation produced by CRH but not alpha-helical CRH (CRH-receptor antagonist) when microinfused into the central neuclus of the amygdala under stress-free conditions. Psychoneuroendocrinology, 20: 423–432.

Wilkinson, C.W., Engeland, W.C., Shinsako, J. and Dallman, M.F. (1981) Nonsteroidal adrenal feedback demarcates two types of pathways to CRF-ACTH release. Am. J. Physiol., 240: E136–E145.

Young, E.A., Akana, S.F. and Dallman, M.F. (1990a) Decreased sensitivity to glucocorticoid fast feedback in chronically stressed rats. Neuroendocrinology, 51: 536–542.

Young, C. and Deutch, A.Y. (1998) The effects of thalamic paraventricular nucleus lesions on cocaine-induced locomotor activity and sensitization. Pharmacol. Biochem. Behav., 60: 753–758.

Young, E.A., Kwak, S.P. and Kottak, J. (1995) Negative feedback regulation following administration of chronic exogenous corticosterone. J. Neuroendocrinol., 7: 37–45.

Young, E.A., Spencer, R.L. and McEwen, B.S. (1990b) Changes at multiple levels of the hypothalamo-pituitary adrenal axis following repeated electrically induced seizures. Psychoneuroendocrinology, 15: 165–172.

Zeng, J., Kitayama, I., Yoshizato, H., Zhang, K. and Okazaki, Y. (2003) Increased expression of corticotropin-releasing factor receptor mRNA in the locus coeruleus of stress-induced rat model of depression. Life Sci., 73: 1131–1139.

Ziegler, D.R., Cass, W.A. and Herman, J.P. (1999) Excitatory influence of the locus coeruleus in hypothalamic-pituitary-adrenocortical axis responses to stress. J. Neuroendocrinol., 11: 361–369.

Zobel, A.W., Nickel, T., Kunzel, H.E., Ackl, N., Sonntag, A., Ising, M. and Holsboer, F. (2000) Effects of the high-affinity corticotropin-releasing hormone receptor 1 antagonist R121919 in major depression: the first 20 patients treated. J. Psychiat. Res., 34: 171–181.

Zorawski, M. and Killcross, S. (2003) Glucocorticoid receptor agonist enhances pavlovian appetitive conditioning but disrupts outcome-specific associations. Behav. Neurosci., 117: 1453–1457.

Zorilla, E.P., Tache, Y. and Koob, G.F. (2003) Nibbling at CRF receptor control of feeding and gastrocolonic motility. TIPS, 24: 421–427.

Zorrilla, E.P., Valdez, G.H. and Weiss, F. (2001) Changes in levels of regional CRF-like immunoreactivity and plasma corticosterone during protracted drug withdrawal in dependent rats. Psychopharmacology, 158: 374–381.

Kalsbeek, Fliers, Hofman, Swaab, Van Someren & Buijs
Progress in Brain Research, Vol. 153
ISSN 0079-6123

CHAPTER 5

Design and synthesis of (ant)-agonists that alter appetite and adiposity

Lex H.T. Van der Ploeg[1,*], Akio Kanatani[2], Douglas MacNeil[3], Tung Ming Fong[3], Alison Strack[4], Ravi Nargund[5] and Xiao-Ming Guan[3]

[1]*Merck Research Laboratories, Boston, MA, USA*
[2]*Tsukuba Research Institute, Banyu, Japan*
[3]*Department of Metabolic Disorders, Merck Research Laboratories, Rahway, NJ, US*
[4]*Department of Pharmacology, Merck Research Laboratories, Rahway, NJ, USA*
[5]*Department of Medicinal Chemistry, Merck Research Laboratories, Rahway, NJ, USA*

Abstract: Over the past decade, hypothalamic circuits have been described that impact energy homeostasis in rodents and humans. Our drug development efforts for the treatment of obesity and the metabolic syndrome have largely focused on selected genetic and/or pharmacologically validated pathways. The translation of these pathways into therapeutics for the treatment of obesity will find its first clinical successes over the coming decade. Initial efforts have focused on gaining a better understanding of the relevance of rodent pharmacological and genetic observations for the development of therapeutics for the treatment of human obesity. We pursue pathways defined by the expression of the ghrelin receptor, melanin-concentrating hormone receptors, melanocortin receptors, cannabinoid receptors and neuropeptide Y1 and Y5 receptors. In this review, we will discuss drug development efforts for the treatment of obesity, focused on selective melanocortin 4 receptor agonists and neuropeptide Y1 and Y5 receptor antagonists. These drug development efforts required an in-depth understanding of cell-based observations which drive the development of compound structure-activity relationships. These include understanding of receptor function in selected cell-based backgrounds and early evaluation and validation of ex vivo observations in appropriate in vivo models. In order to develop selective and safe anti-obesity drugs, diverse approaches are needed to increase the likelihood of clinical success, including: (i) developing a detailed understanding of the predictive value of rodent pathways for treatment of human disease; (ii) knowledge of the exact location of targeted receptor subtypes for the clinical indication under study in order to derive a suitable compound profile; (iii) predictive measures of in vivo and/or ex vivo receptor occupancy required to bring about a desired physiological effect; (iv) predictive parameters that outline that the drug-derived effects are safe and mechanism-based; and (v) the refinement of selected compound classes, aimed at their clinical use.

Melanocortin receptors

The melanocortin system is well documented to play an important role in energy balance. Both mutations in the *POMC* (proopiomelanocortin)

gene and the *MC4R* gene have been identified that cause obesity in humans and rodents (Huszar et al., 1997; Farooqi and O'Rahilly, 2005). The *POMC* gene encodes several peptides that can activate the five melanocortin receptor subtypes. These peptides include α-, β-, γ-melanocyte-stimulating hormone (α-MSH, β-MSH, γ-MSH) and adrenocorticotropin (ACTH). Of the five receptor

*Corresponding author. Tel.: +1-617-992-2017; Fax: +1-617-992-2408; E-mail: lex_van_der_ploeg@merck.com

108

subtypes, MC3R and MC4R are primarily expressed in the CNS and have been considered to mediate most of the effects of melanocortin on food intake and energy expenditure (Hadley et al., 1996; MacNeil et al., 2002; Gantz and Fong, 2003; Cone, 2005). Of particular interest is the presence of an endogenous melanocortin receptor antagonist (AGRP or agouti-related protein) that is produced by neurons in the arcuate nucleus and released at various CNS sites (Fong et al., 1997; Ollmann et al., 1997; Shutter et al., 1997; Tota et al., 1999). The presence of an endogenous antagonist of MC4R may present additional challenges in drug development, as it is anticipated that MC4R agonism is required for efficacy.

Early pharmacological studies revealed that non-selective MC agonists such as Melanotan-II (MT-II) reduce food intake and body weight in rodents (Fan et al., 1997). By combining pharmacological and genetic tools, Chen et al. (2000b) showed that MT-II administration reduced food intake and increased metabolic rate only in the wild-type mice but not in the *Mc4r−/−* mice, thereby demonstrating that the acute efficacy of MT-II is predominantly mediated by MC4R. While MC3R may not play as dominant a role as MC4R, studies with *Mc3r−/−* suggested that the MC3R may serve a subtle role in controlling energy balance (Butler et al., 2000; Chen et al., 2000a).

Selective melanocortin receptor agonists

Based on the overwhelming genetic validation of the MC4R as a potential anti-obesity target, we proceeded to develop small molecule MC4R agonists. With known non-selective peptide agonists such as MT-II, the challenges would be to reduce the size and increase selectivity, while maintaining potency and improving brain penetration, and Pharmacokinetics (PK) properties while assuring an exquisite safety profile. Previous Alanine scanning studies identified the HFRW motif in α-MSH and MT-II as the most important element for agonist activity. Through a combination of screening and medicinal chemistry efforts, we identified THIQ and other small molecule MC4R agonists (Van der Ploeg et al., 2002; Sebhat et al., 2003). Modeling of these small molecules and peptides showed that the small molecule agonists mimic the His6, D-Phe7 and Trp9 residues in MT-II or α-MSH without mimicking the Arg8 side chain of MT-II.

As shown in Fig. 1, these small molecule agonists are potent and selective for MC4R. When developing in vitro assays for a drug-discovery program, one often faces the challenge of using appropriate in vitro assays that are relevant to in vivo physiology. For example, the possibility of species-dependent pharmacology dictates that

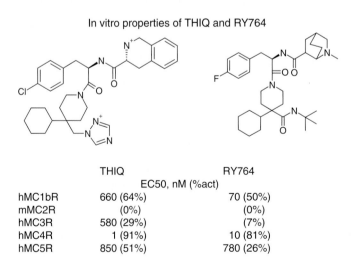

In vitro properties of THIQ and RY764

	THIQ	RY764
	EC50, nM (%act)	
hMC1bR	660 (64%)	70 (50%)
mMC2R	(0%)	(0%)
hMC3R	580 (29%)	(7%)
hMC4R	1 (91%)	10 (81%)
hMC5R	850 (51%)	780 (26%)

Fig. 1. Structure and biochemical characterization of selective MC4R agonists at the human (h) and mouse (m) melanocortin receptors (MC-Rs). EC50s are shown in nM and % agonism at the receptor is given in parenthesis.

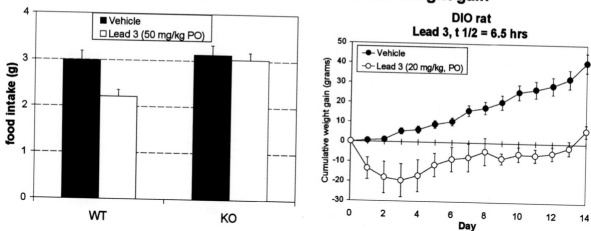

Fig. 2. MC4R agonist lead 3-dosed orally in wt and MC4R−/− mice selectively inhibits overnight cumulative food intake. Chronic treatment of diet-induced obese (DIO) rats, leads to reduction in adiposity. V, vehicle; lead 3, MC4R agonist lead 3.

both human and non-human receptor activity and selectivity be evaluated. For MC4R agonists discovered at Merck, most have very similar binding affinity and activation potency for both human and rat MC4R, unlike for instance, certain NK1R antagonists that show species-dependent pharmacology. In addition, the potential effect of receptor reserve on GPCR agonist efficacy dictates that recombinant cell lines have to be validated so that in vitro EC50 can be shown to be consistent with ex vivo and in vivo EC50s. In the case of MC4R agonists, slice electrophysiology of MC4R responsive neurons revealed EC50 values comparably to those obtained in cell line derived data, thus validating the recombinant cell lines used (Fong and Van der Ploeg, 2000), and giving early insights into anticipated in vivo efficacy.

In vivo efficacy of melanocortin receptor agonists

Various compounds in the Merck series have shown in vivo efficacy in inhibiting food intake and causing weight loss in DIO mice and rats (Ye et al., 2005). Most of the weight loss can be accounted by the loss of fat mass. As food intake reduction and weight loss sometimes can be due to structure-based toxicity, it is desirable to ascertain that MC4R agonists indeed cause weight loss

mediated by selective agonism of the MC4R. To address this issue, we combined pharmacology with genetics and were able to show that the MC4R agonists inhibit food intake in an MC4R-dependent and hence mechanism-based manner as demonstrated by their lack of efficacy in *Mc4r* knockout (KO) mice (Fig. 2).

Investigation of the site of anorectic action of melanocortin receptor agonists

While a critical role of the CNS MC4Rs in mediating effects on energy metabolism is widely accepted, the potential sites of action outside the brain are poorly understood. To better understand the degree of brain penetration required for melanocortin-induced anorectic effects, we developed an ex vivo brain MC4R occupancy assay using ^{125}I-NDPαMSH binding in the superior colliculi as a surrogate readout (the superior colliculus in rats is a region where MC4R accounts for most of ^{125}I-NDPαMSH binding sites). Fig. 3 shows examples of brain MC4R occupancy of several MC4R agonists following their systemic administration. With the exception of lead 8, most compounds exhibited only modest to minimal MC4R occupancy at the superior colliculus when administered at doses efficacious for reducing food intake. Table 1 summarizes

110

Ex Vivo Brain Receptor Occupancy of Melanocortin Compounds

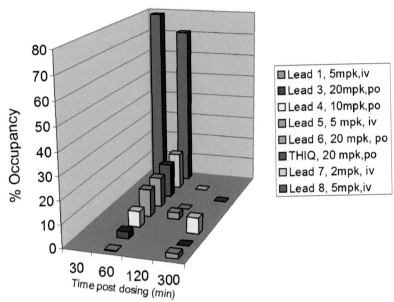

• Receptor occupancy results are in general agreement with brain penetration results

Fig. 3. Brain MC4R occupancy of MC4R agonists following systemic administration in Sprague-Dawley rats. Receptor occupancy was measured ex vivo by quantitative autoradiography of MC4R binding in the superior colliculi with ^{125}I-NDPαMSH as the radioligand. Percent receptor occupancy was calculated by the reciprocal of percent maximal binding as determined from vehicle-treated animals. Data are presented as the average of at least three independent determinations, for eight melanocortin receptor agonists.

Table 1. Correlation of brain/plasma ratio with brain MC4R occupancy by MC4R agonists in rats

Compounds	Brain/plasma ratio	Brain receptor occupancy (%)
Lead 5	0.07	12
THIQ	0.11	15[a]
Lead 7	0.07	16
Lead 8	0.97	71

Note: Brain/plasma ratios were determined at 15 min post injection. Receptor occupancy was measured at 1 h after dosing. Data are means from three independent determinations.
[a] All compounds were administered intravenously with one exception of oral dosing.

the measurements of MC4R agonist concentrations in brain homogenates, showing that the receptor occupancy data and physical measurements of compound concentrations nicely match. We conclude

that high MC4R occupancy is not required to elicit a profound effect on appetite and metabolism.

To further investigate the site of action of MC4R agonists, we conducted a series of studies using a cyclic heptapeptide (MT-II) (Trivedi et al., 2003). MT-II is a non-selective melanocortin receptor agonist with high affinity to MC1R, 3R, 4R, and 5R (Hruby et al., 1995), but its effect on energy balance is primarily mediated via MC4R (Marsh et al., 1999; Chen et al., 2000b). Intravenous (IV) injection of MT-II resulted in profound inhibition of food intake comparable to that observed following intracerebroventricular administration in rats. Unlike central administration, however, IV injection of MT-II did not result in a detectable induction of c-Fos like immunoreactivity in brain regions important for energy regulation such as the paraventricular nucleus of the hypothalamus (PVN). In addition, the amount of

MT-II in the brain as measured directly by LC-MS-MS following IV injection was low (30–90 nM) but higher than the EC50 of MT-II. Furthermore, brain autoradiography following IV injection of ^{125}I-MT-II failed to detect any significant radioactivity in the brain parenchyma (please note that iodo-MT-II is comparably potent as MT-II in inhibiting feeding when given IV). Instead, the radioactivity was intensely detected in a group of circumventricular organs (CVOs), such as the subfornical organ, median eminence, area postrema and choroid plexus, which are anatomically situated outside of the blood–brain barrier (BBB) (Trivedi et al., 2003). Taken together, these results support the notion that, besides MC4R located in the brain parenchyma, the activation of MC4R in brain regions in close proximity to the CVOs or sites outside of the BBB, including CVOs or other peripheral systems may also contribute to effects on appetite and metabolism.

Substantial progress has been made in the discovery of selective small molecule MC4R agonists, which are efficacious in reducing food intake and body weight in rodents. The availability of these pharmacological tools has provided excellent rodent validation, correlating pharmacological efficacy with genetic evidence for the involvement of MC4R in controlling energy balance. The site of action of these agonists is likely to include two separate compartments with MCRs at the CVOs and in the CNS. The next challenge will be to further develop some of these molecules and test these in clinical trials in order to determine whether MC4R agonists would be effective anti-obesity agents.

Development of NPY antagonists and their anorectic effects

Neuropeptide Y (NPY), a 36-amino acid polypeptide identified in 1982, is a member of pancreatic polypeptide family, which consists of NPY, peptide YY (PYY) and pancreatic polypeptide (PP) (Tatemoto and Mutt, 1980; Tatemoto et al., 1982; Clark et al., 1984; Stanley and Leibowitz, 1984). Although several physiological functions of NPY have been reported thus far, its strong orexigenic effects attracted a great deal of attention.

The presence of at least six distinct subtypes of NPY receptors has been described (Blomqvist and Herzog, 1997). However, the Y3 receptor has not been cloned and has only been biochemically characterized. It is assumed that the Y3 receptor might be identical to the Y2 receptor (Clark et al., 1984). As for y6 receptor, the expression of functional y6 receptor has not yet been detected (Gregor et al., 1996). Therefore, current investigations are focusing on four subtypes of NPY receptors, the Y1, Y2, Y4 and Y5 receptors. Among these, the Y1 and Y5 receptors are of particular interest in feeding regulation. However, recent investigation shows that Y2 and Y4 receptors are also involved in feeding regulation as anorectic receptors, indicating that NPY-mediated feeding regulation could be controlled in a complex manner (Batterham et al., 2002, 2003a, b; Asakawa et al., 2003). In pharmaceutical companies, development of Y2 and Y4 selective ligands has only just started. In this article, we will discuss roles of Y1 and Y5 receptors applying by highly selective antagonists including compound validation using receptor-deficient mice and receptor-occupancy studies.

Y1 antagonists

To address roles of Y1 receptors in feeding regulation, we developed potent and selective Y1 antagonists and successfully developed several selective compounds (Kanatani et al., 2001). Simultaneous intracerebroventriclular (ICV) injection of NPY and Compound A significantly inhibited NPY-induced food intake in rats in a dose-dependent manner. In addition, ICV NPY pre-dosing followed by IV injection of Compound A also inhibited ICV NPY-induced food intake (Kanatani et al., 2001). These results show that exogenous NPY-induced feeding involves activation of the Y1 receptor.

Second, we evaluated effects of Compound A on spontaneous food intake in mice. IV injection of Compound A before the beginning of the dark cycle significantly and dose-dependently reduced food intake in lean C57BL mice. The same effects were also observed in genetically obese *db/db* mice, which are known to express increased levels of

CNS NPY levels. We conclude that the Y1 receptor is a key receptor in NPY-related physiological feeding. Interestingly, anorectic efficacy of Compound A was more potent in obese *db/db* mice as compared with lean C57BL mice, suggesting that the increased Y1 signaling may participate in hyperphagia observed in *db/db* mice (Kanatani et al., 2001).

We evaluated whether anorectic effects of small molecules result from non-specific effects due to toxicity or other off-target activity. These issues specifically concerned us as such non-specific effects easily cause feeding reduction. We confirmed mechanism-based effects of Compound A using NPY receptor-deficient mice. ICV injection of NPY caused significant increases in food intake in WT and *Npy5r* KO mice, and their feeding was significantly inhibited by Compound A. In *Npy1r* KO mice, NPY-induced food intake was marginal when compared with those of WT and *Npy5r* KO mice. However, NPY-induced food intake in *Npy1r* KO mice was still significantly increased when compared with that of vehicle-treated mice indicating that other NPY receptors contribute to the behavioral effects (Kanatani et al., 2001). As anticipated for a selective effect of compound A, NPY-induced and spontaneous food intake in *Npy1r* KO mice was not affected by Compound A. Therefore, anorectic effects of Compound A are caused by the inhibition of the Y1 receptor. Hence, we believe that the Y1 receptor is a likely feeding receptor.

We subsequently evaluated whether Y1 receptors are involved in the development of obesity as a feeding receptor using Zucker fatty rats. We employed oral administration of compounds, because IV or IP administration is often more stressful and can easily induce non-specific body weight changes in chronic studies. To increase oral bioavailability, we developed the selective Y1 antagonist Compound B, which shows a higher oral bioavailability than Compound A (Kanatani et al., 1999). Oral administration of Compound B at 100 mg/kg for 2 weeks significantly suppressed the daily food intake and body weight gain in Zucker fatty rats (Fig. 4; Ishihara et al., 2002) accompanied with a reduction of fat cell size and plasma corticosterone levels (Ishihara et al., 2002). Despite the fact that food intake gradually returned to the control level, the cumulative amount of food ingested during the treatment period remained significantly less when compared to that of the vehicle controls (Fig. 4; Ishihara et al., 2002). These findings show that the Y1 receptor plays a definitive role as a feeding receptor in the development of obesity in rodents.

Of Y1 antagonists reported thus far, Pfizer/Neurogen's NGD95-1 reached the clinical trial stage as an anti-obese agent. Although details including its structure were not disclosed, clinical trials with

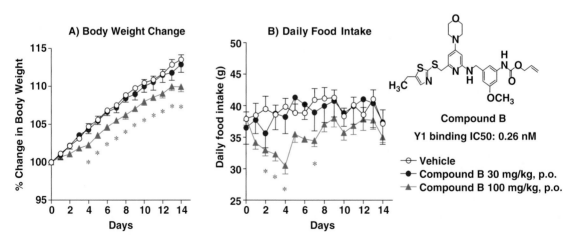

Fig. 4. Effect of the Y1 antagonist Compound B, on body weight change (A) and daily food intake (B) in Zucker fatty rats. Values are mean $\pm$ s.e. mean. *$P < 0.05$ vs. vehicle-treated group ($N = 6$, repeated measurement ANOVA followed by Bonferroni test).

NGD95-1 were suspended (apparently due to the induction of liver enzymes; Neurogen halts obesity trial (SCRIP, 1997, 25, 2294). Thus, the potential of Y1 antagonists for the treatment of obesity in humans awaits the application of an effective Y1 antagonist (Ishihara et al., 2002). The anorectic effects of Y1 antagonists in laboratory animals were confirmed by several groups (Hipskind et al., 1997; Wieland et al., 1998; Carpino et al., 2001; Poindexter et al., 2002). A common issue of current Y1 antagonists is their high molecular weight. Y1 antagonists developed thus far generally carry bulky moieties to achieve high affinity and selectivity, resulting in loss of stability while exhibiting undesirably low brain-penetrability.

Y5 antagonists

Thus far, the function of Y1 receptors as orexigenic receptors has been broadly accepted based on several lines of evidence. However, in the early 1990s, evidence emerged to outline that the feeding receptor of NPY may be a Y1-like receptor, and not the typical Y1 receptor. The discovery in 1996 of the Y5 receptor therefore attracted significant interest and this receptor was proposed to be a dominant NPY-dependent feeding receptor (Gerald et al., 1996). To investigate a role for the Y5 receptor in feeding regulation, we started the development of selective small molecule Y5 antagonists and generated Compound C (Kanatani et al., 2000). As compared with Y1 antagonists described before, the intrinsic potency of Compound C is relatively weak. However, the compound shows high metabolic stability and good bioavailability. Several experiments were performed to evaluate the impact of compound C on food intake and body composition, using its oral administration in rodents.

After oral dosing of Compound C at 10 mg/kg, Y5 agonist-derived, bovine pancreatic polypeptide (bPP)-induced food intake was significantly suppressed. At time points where food intake was measured, brain levels of Compound C were 100-fold higher than the IC_{50} value of Compound C (as determined by the calcium flux cell-based functional assays). We subsequently tested the effects of Compound C on the less selective, NPY ligand in NPY-induced food intake studies. Surprisingly, Compound C had no effect on food intake induced by ICV-injected NPY. The latter data indicated that the effects of ICV NPY are not likely to be mediated through the Y5 receptor in rodents. We also evaluate the effects of Compound C on spontaneous food intake in *db/db* mice and Zucker fatty rats, in which Y1 antagonists significantly inhibited food intake. In these experiments, Compound C did not show any inhibitory effects on food intake in these obese rodent models (Kanatani et al., 2000). Next, we employed Y5 selective peptide agonist in order to further elucidate the role of the Y5 receptor in energy homeostasis.

D-Trp34NPY selectively binds the Y5 receptor over the Y1 receptor (>20-fold). The critical difference between bPP and D-Trp34NPY includes enhanced selectivity of D-Trp34NPY over Y2 and Y4 receptors. ICV-infusion of D-Trp34NPY in mice for 2 weeks resulted in significant body weight gain with hyperphagia. Furthermore, the increased food intake and body weight gain could be completely suppressed by the oral administration of the highly selective Y5 antagonist, Compound C (Mashiko et al., 2003). We concluded that D-Tpr34NPY-induced hyperphagia and body weight gain are largely caused by the activation of Y5 receptors. We hypothesized that there might be selected types of obesity in which Y5 receptors might be contributing to appetite and metabolism. Subsequently, we identified that diet-induced obese mice are sensitive to Y5 antagonists (Ishihara et al., 2006).

Oral administration of the Y5R antagonist Compound C to mice eating a moderately high fat diet (MHF; Compound C at 30 and 100 mg/kg for 2 weeks) significantly suppressed the diet-induced body weight gain in a dose-dependent manner and reduced the calorie intake of the MHF diet by 7.6 and 10.0% at 30 and 100 mg/kg, respectively when compared to the vehicle controls (Fig. 5; Ishihara et al., 2006). To examine whether the anti-obesity effects of the Y5 antagonist are mechanism-based, we administered the Y5 antagonist to *Npy5r* KO (Y5 KO) in a crossover study with three dosing phases (14–19 days each arm).

114

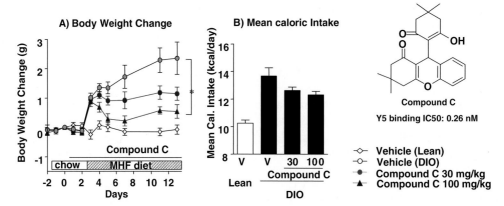

Fig. 5. Oral treatment with the Y5R antagonist, Compound C inhibited the body weight gain in mice-fed MHF diet. (A) Body weight change. (B) Mean caloric intake. Veh., vehicle. Data are expressed means ± SE (n = 8–9). *P<0.05 (compared with vehicle).

The Y5 antagonist (100 mg/kg) suppressed the body weight gain in the wild-type mice comparably to the effects described in the prior study. By contrast, the Y5 antagonist had no significant effect on body weight gain in *Npy5r* KO mice throughout the experiment (Ishihara et al., 2006). These results showed that the anti-obesity and anorectic effects of the Y5 antagonist are mechanism-based. Therefore, the Y5 receptor plays a critical and specific role in the development of diet-induced obesity in rodents.

Although it was previously reported that Synaptic's Y5 antagonist potently suppressed food intake and body weight gain in genetically obese rodents (Mashiko et al., 2003), recent investigations showed that these effects were also observed in *Npy5* KO mice. Possible targets through which this effect was brought about may have included monoaminergic receptors or transporters (Criscione et al., 1998; Criscione, 2001). A second distinct and structurally diverse Y5 antagonist was subsequently reported to be a potent anorectic ligand (Della Zuana et al., 2001). The functions of Y5 receptor as an orexigenic receptor require further characterization to elucidate the mechanisms through which these compounds exert effects on appetite and metabolism.

Y5 binding and receptor occupancy

To better understand Y5 receptor function, we established a brain receptor-occupancy assay for

the Y5 receptor. We developed for ex vivo receptor occupancy assays two [35S]-labeled Y5 antagonists (Guan et al., unpublished data). Both radioligands are highly potent (K_d values range from 0.2 to 0.4 nM in rat brain tissues) and selective for the Y5 receptor. In vitro receptor autoradiography with these [35S]-compounds revealed specific high affinity binding in a number of brain regions including the hippocampus and striatum. The specificity of binding was further confirmed by demonstrating the complete lack of equivalent bindings in the *Npy5r*−/− mouse brain. Using these radioligands, we performed ex vivo receptor occupancy assays, which enabled us to evaluate the degree of drug exposure in the rodent brain following the systemic administration of cold therapeutically active Y5 antagonist leads. We were thus able to establish a correlation between brain exposure and in vivo efficacy in a number of animal models. These receptor occupancy assays allowed us to conclude that a lack of efficacy of Y5 antagonists in certain animal models did not result from a lack of Y5 receptor occupancy.

Both Y1 and Y5 receptors appear to be receptors that can mediate the effects of NPY. We conclude that the Y1 receptor is a relevant receptor that impacts feeding behavior under physiologically relevant conditions in rodents. The Y5 receptor is also shown to be effective though its efficacy was restricted to selective feeding conditions.

Melanocortin receptor agonists interact with NPY receptor responsive neurons

The NPY and melanocortin neurotransmitter pathways are closely linked anatomically. The peptidergic NPY and melanocortin neurons are expressed in hypothalamic nuclei and the arcuate nucleus, and have highly overlapping projections. Projection sites for neurons expressing brain Y1 and MC4R receptors also overlap. These areas include the PVN and the central amygdalar nucleus (Kishi et al., 2005). While similar colocalzation studies have not been done with Y5 and MC4R, Y5 receptors are observed in MC4R expressing nuclei in humans, including the arcuate and PVN in humans (Jacques et al., 1998). Applying electrophysiological techniques MT-II has been shown to increase the inhibitory GABA synaptic transmission and NPY inhibits GABA transmission in the PVN of rat hypothalamus (Cowley et al., 1999), consistent with the PVN as an area that may integrate these two systems.

Previously, we demonstrated that the melanocortin pan-agonist, MT-II, can suppress feeding evoked by NPY (Murphy et al., 1998). More recently, we examined the actions of MT-II in light of agonism by NPY itself or bPP, a Y4 and Y5 agonist. hNPY or bPP, both increase food intake when administered centrally to Sprague-Dawley rats (Fig. 6). These increases in food intake are attenuated by MT-II administration. Feeding evoked by hNPY, an agonist at Y1, Y2 and Y5 receptors, is entirely blocked by MT-II, with MT-II only partially suppresses bPP (Y4 and Y5 agonist)-evoked feeding, suggesting a greater role for the interaction of the melanocortin system with the Y1 and/or Y2 receptors than with Y4 or Y5 in short-term food intake studies (Fig. 6). Further studies of the potential interaction of NPY, melanocortin and other related CNS systems modulating energy homeostasis are likely to provide insights into the potential for combination therapeutics for the treatment of obesity.

Summary and conclusions

Over the past decade, the development of therapeutics for the treatment of obesity has gained momentum. Several hypothalamic circuits have been identified that serve key roles in controlling appetite and metabolism. The application of selective ligands for receptors modulating these pathways in diverse rodent and primate species will allow the determination of their relevance for humans. The

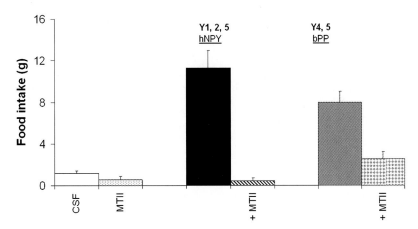

Fig. 6. Suppression of NPY agonism evoked food intake by the melanocortin agonist, MT-II in the Sprague Dawley rat. Central administration of the 5 µg hNPY (Y1, Y2 and Y5 agonist)-induced feeding is entirely blocked by MT-II (4 µg), while 5-µg bPP (Y4 and Y5 agonist)-mediated feeding is only partially blocked by MT-II co-administration. Values are mean ± standard error.

Y5 receptor may take a distinct position in this family of appetite control pathways as its efficacy appears selective for particular feeding paradigms.

Once data have accumulated that allow the comparison of efficacy in preclinical test species with clinical observations, rodent models can be further refined to help predict efficacy in humans.

The interactions of melanocortin and NPY receptor feeding pathways may enable the testing of different combination therapeutic regimens in order to overcome the inevitable compensatory pathways that minimize the efficacy of drugs aimed at the control of body weight.

Abbreviations

ACTH	adrenocorticotropic hormone
bPP	bovine pancreatic polypeptide
CNS	central nervous system
CVO	circumventricular organ
ICV	intracerebroventricular
NPY	neuropeptide Y
MCR	melanocortin receptor
MSH	melanocyte-stimulating hormone
MT-II	melanotan II
POMC	proopiomelanocortin
PYY	peptide YY
Y1	neuropeptide Y1 receptor
Y5	neuropeptide Y5 receptor

Acknowledgments

We thank the numerous MRL researchers that are the driving force of some of the research described in this review, for their input and motivation and we thank Nancee Jenne and Rhonda Wozniak for their expert help in completing this manuscript.

References

Asakawa, A., Inui, A., Yuzuriha, H., Ueno, N., Katsuura, G., Fujimiya, M., Fujino, M.A., Niijima, A., Meguid, M.M. and Kasuga, M. (2003) Characterization of the effects of pancreatic polypeptide in the regulation of energy balance. Gastroenterology, 124: 1325–1336.

Batterham, R.L., Cohen, M.A., Ellis, S.M., LeRoux, C.W., Withers, D.J., Frost, G.S., Ghatei, M.A. and Bloom, S.R. (2003b) Inhibition of food intake in obese subjects by peptide YY3-36. N. Engl. J. Med., 349: 941–948.

Batterham, R.L., Cowley, M.A., Small, C.J., Herzog, H., Cohen, M.A., Dakin, C.L., Wren, A.M., Brynes, A.E., Low, M.J., Ghatei, M.A., Cone, R.D. and Bloom, S.R. (2002) Gut hormone PYY(3-36) physiologically inhibits food intake. Nature, 418: 650–654.

Batterham, R.L., Le Roux, C.W., Cohen, M.A., Park, A.J., Ellis, S.M., Patterson, M., Frost, G.S., Ghatei, M.A. and Bloom, S.R. (2003a) Pancreatic polypeptide reduces appetite and food intake in humans. J. Clin. Endocrinol. Metab., 88: 3989–3992.

Blomqvist, A.G. and Herzog, H. (1997) Y-receptor subtypes: how many more? Trends Neurosci., 20: 294–298.

Butler, A.A., Kesterson, R.A., Khong, K., Cullen, M.J., Pelleymounter, M.A., Dekoning, J., Baetscher, M. and Cone, R.D. (2000) A unique metabolic syndrome causes obesity in the melanocortin-3 receptor-deficient mouse. Endocrinology, 141: 3518–3521.

Carpino, P. A., Griffith, D. A., Maurer, T. S. et al. (2001) Characterization of CP-671906, a potent NPY-Y1 antagonist. Abstract O.43, 6th Int. NPY Conf., Sydney, April 22–26.

Chen, A.S., Marsh, D.J., Trumbauer, M.E., Frazier, E.G., Guan, X.M., Yu, H., Rosenblum, C.I., Vongs, A., Feng, Y., Cao, L., Metzger, J.M., Strack, A.M., Camacho, R.E., Mellin, T.N., Nunes, C.N., Min, W., Fisher, J., Gopal-Truter, S., MacIntyre, D.E., Chen, H.Y. and Van der Ploeg, L.H. (2000a) Inactivation of the mouse melanocortin-3 receptor results in increased fat mass and reduced lean body mass. Nat. Genet., 26: 97–102.

Chen, A.S., Metzger, J.M., Trumbauer, M.E., Guan, X.M., Yu, H., Frazier, E.G., Marsh, D.J., Forrest, M.J., Gopal-Truter, S., Fisher, J., Camacho, R.E., Strack, A.M., Mellin, T.N., MacIntyre, D.E., Chen, H. and Van der Ploeg, L.H.T. (2000b) Role of the melanocortin-4 receptor in metabolic rate and food intake in mice. Transgenic Res., 9: 145–154.

Clark, J.T., Kalra, P.S., Crowley, W.R. and Kalra, S.P. (1984) Neuropeptide Y and human pancreatic polypeptide stimulate feeding behavior in rats. Endocrinology, 115: 4277–4429.

Cone, R.D. (2005) Anatomy and regulation of the central melanocortin system. Nat. Neurosci., 8: 571–578.

Cowley, M.A., Pronchuk, N., Fan, W., Dinulescu, D.M., Colmers, W.F. and Cone, R.D. (1999) Integration of NPY, AGRP and melanocortin signals in the hypothalamic paraventricular nucleus. Neuron, 24: 155–168.

Criscione, L. (2001). Effects of the the Y5 receptor antagonist CGP71683 on food intake in male Y5 receptor knockout mice. Abstract O46, 6th Int. NPY Conf. Sydney, Australia, April 22–26.

Criscione, L., Rigollier, P., Batzl-Hartmann, C., Rüeger, H., Stricker-Krongrad, A., Wyss, P., Brunner, L., Whitebread, S., Yamaguchi, Y., Gerald, C., Heurich, R.O., Walker, M.W., Chiesi, M., Schilling, W., Hofbauer, K.G. and Levens, N. (1998) Food intake in free-feeding and energy-deprived lean rats is mediated by the neuropeptide Y5 receptor. J. Clin. Invest., 102: 2136–2145.

Della Zuana, O., Sadlo, M., Germain, M., Feletou, M., Charmorro, S., Tisserand, F., de Montrion, C., Boivin, J.F., Duhault, J., Boutin, J.A. and Levens, N. (2001) Reduced food intake in response to CGP 71683 may be due to mechanisms other than NPY Y5 receptor blockade. Int. J. Obes., 25: 84–94.

Fan, W., Boston, B.A., Kesterson, R.A., Hruby, V.J. and Cone, R.D. (1997) Role of melanocortinergic neurons in feeding and the agouti obesity syndrome. Nature, 385: 165–168.

Farooqi, I.S. and O'Rahilly, S. (2005) Monogenic obesity in humans. Annu. Rev. Med., 56: 443–458.

Fong, T.M., Mao, C., MacNeil, T., Kalyani, R., Smith, T., Weinberg, D.H., Tota, M. and Van der Ploeg, L.H.T. (1997) ART (protein product of agouti-related transcript) as an antagonist of MC3 and MC4 receptors. Biochem. Biophys. Res. Comm., 237: 629–631.

Fong, T.M. and Van der Ploeg, L.H. (2000) A melanocortin agonist reduces neuronal firing rate in rat hypothalamic slices. Neurosci. Lett., 283: 5–8.

Gantz, I. and Fong, T.M. (2003) The melanocortin system. Am. J. Physiol. Endocrinol. Metab., 284: E468–E474.

Gerald, C., Walker, M.W., Criscione, L., Gustafson, E.L., Batzl-Hartman, C., Smith, K.E., Vaysse, P., Durkin, M.M., Laz, T.M., Linemeyer, D.L., Schaffhauser, A.O., Whitebread, S., Hofbauer, K.G., Taber, R.I., Branchek, T.A. and Weinshank, R.I. (1996) A receptor subtype involved in neuropeptide-Y-induced food intake. Nature, 382: 168–171.

Gregor, P., Feng, Y., DeCarr, L.B., Cornfield, L.J. and MaCaleb, M.L. (1996) Molecular characterization of a second mouse pancreatic polypeptide receptor and its inactivated human homologue. J. Biol. Chem., 271: 27776–27781.

Hadley, M.E., Hruby, V.J., Jiang, J., Sharma, S.D., Fink, J.L., Haskell-Luevano, C., Bentley, D.L., Al-Obeid, A. and Sawyer, T.K. (1996) Melanocortin receptors: identification and characterization by melanotropic peptide agonists and antagonists. Pigment Cell Res., 9: 213–234.

Hipskind, P.A., Lobb, K.L., Nixon, J.A., Britton, T.C., Bruns, R.F., Catlow, J., Dieckman-McGinty, D.K., Gackenheimer, S.L., Gitter, B.D., Iyengar, S., Schober, D.A., Simmons, R.M.A., Swanson, S., Zarrinmayeh, H., Zimmerman, D.M. and Gehlert, D.R. (1997) Potent and selective 1,2,3-trisubstituted indole NPY Y-1 antagonists. J. Med. Chem., 40: 3712–3714.

Hruby, V.J., Lu, D., Sharma, S.D., Castrucci, L., Kesterson, R.A., al-Obeidi, F.A., Hadley, M.E. and Cone, R.D. (1995) Cyclic lactam alpha-melanotropin analogues of Ac-Nle4-cyclo[Asp5, D-Phe7,Lys10] alpha-melanocyte-stimulating hormone-(4-10)-NH2 with bulky aromatic amino acids at position 7 show high antagonist potency and selectivity at specific melanocortin receptors. J. Med. Chem., 38: 3454–3461.

Huszar, D., Lynch, C.A., Fair-Huntress, V., Dunmore, J.H., Fang, Q., Berkemeier, L.R., Gu, W., Kesterson, R.A., Boston, B.A., Cone, R.D., Smith, F.J., Campfield, L.A., Burn, P. and Lee, F. (1997) Targeted disruption of the melanocortin-4 receptor results in obesity in mice. Cell, 88: 131–141.

Ishihara, A., Kanatani, A., Mashiko, S., Tanaka, T., Hidaka, M., Gomori, A., Iwaasa, H., Murai, N., Egashira, S., Murai, T., Mitobe, Y., Matsushita, H., Okamota, O., Sato, N., Jitsuoka, M., Fukuroda, T., Ohe, T., Guan, X., MacNeil, D., Van der Ploeg, L. H. T., Nishikibe, M., Ishii, Y., Ihara, M. and Fukami, T. (2006) A neuropeptide Y Y5 antagonist selectively ameliorates body weight gain and associated parameters in diet-induced obese mice. Proc. Natl. Acad. Sci., submitted.

Ishihara, A., Kanatani, A., Okada, M., Hidaka, M., Tanaka, T., Mashiko, S., Gomori, A., Kanno, T., Hata, M., Kanesaka, M., Tominaga, T., Sato, N., Kobayashi, M., Murai, T., Watanabe, K., Ishii, Y., Fukuroda, T., Fukami, T. and Ihara, M. (2002) Blockade of body weight gain and plasma corticosterone levels in Zucker fatty rats using an orally active neuropeptide Y Y1 antagonist. Br. J. Pharmacol., 136: 341–346.

Jacques, D., Tong, Y., Shen, S.H. and Quirion, R. (1998) Discrete distribution of the neuropeptide Y Y5 receptor gene in the human brain: an in situ hybridization study. Mol. Brain Res., 61: 100–107.

Kanatani, A., Hata, M., Mashiko, S., Ishihara, A., Okamoto, O., Haga, Y., Ohe, T., Kanno, T., Murai, N., Ishii, Y., Fukuroda, T., Fukami, T. and Ihara, M. (2001) A typical Y1 receptor regulates feeding behaviors: effects of a potent and selective Y1 antagonist, J-115814. Mol. Pharmacol., 59: 501–505.

Kanatani, A., Ishihara, A., Iwaasa, H., Nakamura, K., Okamoto, O., Hidaka, M., Ito, J., Fukuroda, T., MacNeil, D.J., Van der Ploeg, L.H.T., Ishii, Y., Okabe, T., Fukami, T. and Ihara, M. (2000) L-152, 804: orally active and selective neuropeptide Y Y5 receptor antagonist. Biochem. Biophys. Res. Commun., 27: 169–173.

Kanatani, A., Kanno, T., Ishihara, A., Hata, M., Sakuraba, A., Tanaka, T., Tsuchiya, Y., Mase, T., Fukuroda, T., Fukami, T. and Ihara, M. (1999) The novel neuropeptide Y Y1 antagonist J-104870: a potent feeding suppressant with oral bioavailability. Biochem. Biophys. Res. Commun., 266: 88–91.

Kishi, T., Aschkenasi, C.J., Choi, B.J., Lopez, M.E., Lee, C.E., Liu, H., Hollenberg, A.N., Friedman, J.M. and Elmquist, J.K. (2005) Neuropeptide Y Y1 receptor mRNA in rodent brain: distribution and colocalization with melanocortin-4 receptor. J. Comp. Neurol., 482: 217–243.

MacNeil, D.J., Howard, A.D., Guan, X., Fong, T.M., Nargund, R.P., Bednarek, M.A., Goulet, M.T., Weinberg, D.H., Strack, A.M., Marsh, D.J., Chen, H.Y., Shen, C.P., Chen, A.S., Rosenblum, C.I., MacNeil, T., Tota, M., MacIntyre, E.D. and Van der Ploeg, L.H.T. (2002) The role of melanocortins in body weight regulation: opportunities for the treatment of obesity. Eur. J. Pharmacol., 450: 93–109.

Marsh, D.J., Hollopeter, G., Huszar, D., Laufer, R., Yagaloff, K.A., Fisher, S.L., Burn, P. and Palmiter, R.D. (1999) Response of melanocortin-4 receptor-deficient mice to anorectic and orexigenic peptides. Nat. Genet, 21: 119–122.

Mashiko, S., Ishihara, A., Iwaasa, H., Sano, H., Oda, Z., Ito, J., Yumoto, M., Okawa, M., Suzuki, J., Fukuroda, T., Jitsuoka, M., Morin, N.R., MacNeil, D.J., Van der Ploeg, L.H.T., Ihara, M., Fukami, T. and Kanatani, A. (2003)

118

Characterization of neuropeptide Y (NPY) Y5 receptor-mediated obesity in mice: chronic intracerebroventricular infusion of D-Trp34NPY. Endocrinology, 144: 1793–1801.

Murphy, B., Nunes, C., Ronan, J., Harper, C., Beall, M., Hanaway, M., Firhurst, A.M., Van der Ploeg, L.H., MacIntyre, D.E. and Mellin, T.N. (1998) Melanocortin mediated inhibition of feeding behavior in rats. Neuropeptides, 32: 491–497.

Ollmann, M.M., Wilson, B.D., Yang, Y.-K., Kerns, J.A., Chen, Y., Gantz, I. and Barsh, G.S. (1997) Antagonism of central melanocortin receptors in vitro and in vivo by agouti-related protein. Science, 278: 135–138.

Poindexter, G.S., Bruce, M.A., LeBoulluec, K.L., Monkovic, I., Martin, S.W., Parker, E.M., Iben, L.G., McGovern, R.T., Ortiz, A.A., Stanley, J.A., Mattson, G.K., Kozlowski, M., Arcuri, M. and Antal-Zimanyi, I. (2002) Dihydropyridine neuropeptide Y Y_1 receptor antagonists. Bioorg. Med. Chem. Lett., 12: 379–382.

Sebhat, I., Ye, Z., Bednarek, M., Weinberg, D., Nargund, R. and Fong, T.M. (2003) Melanocortin-4 receptor agonists and antagonists: chemistry and potential therapeutic utilities. Ann. Rep. Med. Chem., 38: 31–40.

Shutter, J.R., Gramham, M., Kinsey, A.C., Scully, S., Luthy, R. and Stark, K. (1997) Hypothalamic expression of ART, a novel gene related to agouti, is up-regulated in obese and diabetic mutant mice. Gene Dev., 11: 593–602.

Stanley, B.G. and Leibowitz, S.F. (1984) Neuropeptide Y: stimulation of feeding and drinking by injection into the paraventricular nucleus. Life Sci., 35: 2635–2642.

Tatemoto, K., Carlquist, M. and Mutt, V. (1982) Neuropeptide Y: a novel brain peptide with structural similarities to peptide YY and pancreatic polypeptide. Nature, 296: 659–660.

Tatemoto, K. and Mutt, V. (1980) Isolation of two novel, candidate hormones using a chemical method for finding naturally occurring polypeptides. Nature, 285: 417–418.

Tota, M.R., Smith, T.S., Mao, C., MacNeil, T., Mosley, R.T., Van der Ploeg, L.H.T. and Fong, T.M. (1999) Molecular interaction of agouti protein and agouti-related protein with human melanocortin receptors. Biochemistry, 38: 897–904.

Trivedi, P., Jiang, M., Tamvakopoulos, C., Shen, X., Yu, H., Mock, S., Fenyk-Melody, J., Van der Ploeg, L.H. and Guan, X.M. (2003) Exploring the site of anorectic action of peripherally administered synthetic melanocortin peptide MT-II in rats. Brain Res., 977: 221–230.

Van der Ploeg, L.H.T., Martin, W.J., Howard, A.D., Nargund, R.P., Austin, C.P., Guan, X., Drisko, J., Cashen, D., Sebhat, I., Patchett, A.A., Figueroa, D.J., DiLella, A.G., Connolly, D.H., Weinberg, D.H., Tan, C.T., Palyha, O.C., Pong, S., MacNeil, T., Rosenblum, C., Vongs, A., Tang, R., Yu, H., Sailer, A.W., Fong, T.M., Huang, C., Tota, M., Chang, R.S., Stearns, R., Tamvakopoulos, T., Christ, G., Drazen, D.L., Spar, b.D., Nelson, R.J. and MacIntyre, D.E. (2002) A role for the melanocortin 4 receptor in sexual function. Proc. Natl. Acad. Sci. USA, 99: 11381–11386.

Wieland, H.A., Engel, W., Eberlein, W., Rudolf, K. and Doods, H.N. (1998) Subtype selectivity of the novel nonpeptide neuropeptide Y Y1 receptor antagonist BIBO 3304 and its effect on feeding in rodents. Br. J. Pharmacol., 125: 549–555.

Ye, Z., Guo, L., Barakat, K.J., Pollard, P.G., Palucki, B.L., Sebhat, I.K., Bakshi, R.K., Tang, R., Kalyani, R.N., Vongs, A., Chen, A.S., Chen, H.Y., Rosenblum, C.I., MacNeil, T., Weinberg, D.H., Peng, Q., Tamvakopoulos, C., Miller, R.R., Stearns, R.A., Cashen, D.E., Martin, W.J., Metzger, J.M., Strack, A.M., MacIntyre, D.E., Van der Ploeg, L.H., Patchett, A.A., Wyvratt, M.J. and Nargund, R.P. (2005) Discovery and activity of (O1R,4S,6R)-N-[(1R)-2-[4-cyclohexyl-4-[[(1,1-dimethylethyl)amino]carbonyl]-1-piperidinyl]-1-[(4-fluorophenyl)methyl]-2-oxoethyl]-2-methyl-2-azabicyclo[2.2.2]octane-6- carboxamide (3, RY764), a potent and selective melanocortin subtype-4 receptor agonist. Bioorg. Med. Chem. Lett., 15: 3501–3505.

Kalsbeek, Fliers, Hofman, Swaab, Van Someren & Buijs
Progress in Brain Research, Vol. 153
ISSN 0079-6123

CHAPTER 6

Monogenic human obesity syndromes

I.S. Farooqi[*]

Departments of Medicine & Clinical Biochemistry, University of Cambridge, Addenbrooke's Hospital, Cambridge, UK

Abstract: Over the past decade we have witnessed a major increase in the scale of scientific activity devoted to the study of energy balance and obesity. This explosion of interest has, to a large extent, been driven by the identification of genes responsible for murine obesity syndromes, and the novel physiological pathways revealed by those genetic discoveries. Others and we have also recently identified several single gene defects causing severe human obesity. Many of these defects have been in molecules identical or similar to those identified as a cause of obesity in rodents. I will review the human monogenic obesity syndromes that have been characterised to date and discuss how far such observations support the physiological role of these molecules in the regulation of human body weight and neuroendocrine function.

Introduction

The concept that mammalian body fat mass is homeostatically regulated has its underpinning in experimental science going back over 50 years. Thus, the adipostatic theory of Kennedy, which emerged in the 1950s, was based on his own observations of the responses of rodents to perturbations of food intake (Kennedy, 1953) together with the hypothalamic lesioning studies of Hetherington (Hetherington and Ranson, 1940) and Anand (Anand and Brobeck, 1951) and the parabiosis experiments of Hervey (1959). The subsequent emergence of several murine genetic models of obesity (Bray and York, 1971), and their study in parabiosis experiments by Coleman (1973) led to the consolidation of the concept that a circulating factor might be involved in the mediation of energy homeostasis. However, it was not until the 1990s when the precise molecular basis for the *agouti, ob/ob, db/db* and *fat/fat* mouse emerged, that the molecular components of an energy balance regulatory network began to be pieced together (Leibel et al., 1997). The use of gene targeting technology has gone on to demonstrate the

critical roles of certain other key molecules such as the melanocortin 4 receptor (MC4R) (Huszar et al., 1997) and melanin-concentrating hormone (MCH) (Shimada et al., 1998; Chen et al., 2002) in that network.

A critical question raised by these discoveries is the extent to which these regulatory pathways are operating in the control of human body weight. Over the past few years, a number of novel monogenic disorders causing human obesity have emerged (Barsh et al., 2000). In many cases, the mutations are found in components of the regulatory pathways identified in rodents. The importance of these human studies is several fold. First, they have established for the first time that humans can become obese due to a simple inherited defect. Second, it has been notable that in all cases the principal effect of the genetic mutation has been to disrupt mechanisms regulating food intake. Third, some defects, although rare, are amenable to rational therapy.

Congenital leptin deficiency

In 1997, we reported two severely obese cousins from a highly consanguineous family of Pakistani

---

*Corresponding author. E-mail: fd219@cam.ac.uk

DOI: 10.1016/S0079-6123(06)53006-7

origin (Montague et al., 1997). Both children had undetectable levels of serum leptin and were found to be homozygous for a frameshift mutation in the *ob* gene (ΔG133), which resulted in a truncated protein that was not secreted (Montague et al., 1997; Rau et al., 1999). We have since identified several further affected individuals (Farooqi et al., 2002) (and unpublished observations) who are also homozygous for the same mutation in the leptin gene. All the families are of Pakistani origin but not known to be related over five generations. A large Turkish family who carry a homozygous missense mutation have also been described (Strobel et al., 1998). All subjects in these families are characterised by severe early-onset obesity and intense hyperphagia (Farooqi et al., 1999, 2002; Ozata et al., 1999). Hyperinsulinaemia and an advanced bone-age are also common features (Farooqi et al., 1999, 2002). Some of the Turkish subjects are adults with hypogonadotropic hypogonadism (Ozata et al., 1999). Although normal pubertal development did not occur, there was some evidence of a delayed but spontaneous pubertal development in one person (Ozata et al., 1999).

We demonstrated that children with leptin deficiency had profound abnormalities of T cell number and function (Farooqi et al., 2002), consistent with high rates of childhood infection and a high reported rate of childhood mortality from infection in obese Turkish subjects (Ozata et al., 1999).

Most of these phenotypes closely parallel those seen in murine leptin deficiency. However, there are some phenotypes where the parallels between human and mouse are not clear-cut. Thus, while *ob/ob* mice are stunted (Dubuc and Carlisle, 1988), it appears that growth retardation is not a feature of human leptin deficiency (Farooqi et al., 1999, 2002), although abnormalities of dynamic growth hormone secretion have been reported in one human subject (Ozata et al., 1999). The *ob/ob* mice have marked activation of the hypothalamic pituitary adrenal axis with very elevated corticosterone levels (Dubuc, 1977). In humans, abnormalities of cortisol secretion are, if present at all, much more subtle (Farooqi et al., 2002). The contribution of reduced energy expenditure to the obesity of the *ob/ob* mouse is reasonably well established (Trayhurn et al., 1977).

In leptin-deficient humans we found no detectable changes in resting or free-living energy expenditure (Farooqi et al., 2002), although it was not possible to examine how such systems adapted to stressors such as cold. Ozata et al. (1999) reported abnormalities of sympathetic nerve function in leptin-deficient humans consistent with defects in the efferent sympathetic limb of thermogenesis.

Response to leptin therapy

Recently, we reported the dramatic and beneficial effects of daily subcutaneous injections of leptin reducing body weight and fat mass in three congenitally leptin-deficient children (Farooqi et al., 2002). We have recently commenced therapy in the other two children and seen comparably beneficial results (personal observations). All children showed a response to initial leptin doses designed to produce plasma leptin levels at only 10% of those predicted by height and weight (i.e. approximately 0.01 mg/kg of lean body mass) (Farooqi et al., 2002). The most dramatic example of leptin's effects was with a 3-year-old boy, severely disabled by gross obesity (wt 42 kg), who now weighs 32 kg (75th centile for weight) after 48 months of leptin therapy.

The major effect of leptin was on appetite with normalisation of hyperphagia. Leptin therapy reduced energy intake during an 18 MJ *adlibitum* test meal by up to 84% (5 MJ ingested pre-treatment vs. 0.8 MJ post-treatment in the child with the greatest response) (Farooqi et al., 2002). We were unable to demonstrate a major effect of leptin on basal metabolic rate (BMR) or free-living energy expenditure (Farooqi et al., 2002), but, as weight loss by other means is associated with a decrease in BMR (Rosenbaum et al., 2002), the fact that energy expenditure did not fall in our leptin-deficient subjects is notable.

The administration of leptin permitted progression of appropriately timed pubertal development in the single child of appropriate age and did not cause the early onset of puberty in the younger children (Farooqi et al., 2002). Free thyroxine and TSH levels, although in the normal range before treatment, had consistently increased at the earliest post-treatment time point and subsequently stabilised at this elevated level (Farooqi et al., 2002). These findings are consistent with evidence from animal models that leptin influences TRH release from the hypothalamus (Legradi et al., 1997; Nillni et al., 2000; Harris et al., 2001) and from studies illustrating the effect of leptin deficiency on TSH pulsatility in humans (Mantzoros et al., 2001).

Throughout the trial of leptin administration, weight loss continued in all subjects, albeit with refractory periods which were overcome by the increases in leptin dose (Farooqi et al., 2002). The families in the UK harbour a mutation that leads to a prematurely truncated form of leptin, and thus wild-type leptin is a novel antigen to them. Thus, all subjects developed antileptin antibodies after ~6 weeks of leptin therapy, which interfered with interpretation of serum leptin levels and in some cases were capable of neutralizing leptin in a bioassay (Farooqi et al., 2002). These antibodies are the likely cause of refractory periods occurring during therapy. The fluctuating nature of the antibodies probably reflects the complicating factor that leptin deficiency is itself an immuno-deficient state (Lord et al., 1998; Matarese, 2000) and administration of leptin leads to a change from the secretion of predominantly Th2 to Th1 cytokines, which may directly influence antibody production. Thus far, we have been able to regain control of weight loss by increasing the dose of leptin.

Is there a heterozygous phenotype?

The major question with respect to the potential therapeutic use of leptin in more common forms of obesity relates to the shape of the leptin dose–response curve. We have clearly shown that at the lower end of plasma leptin levels, raising leptin levels from undetectable to detectable has profound effects on appetite and weight (Farooqi et al., 2002). Heymsfield et al. (1999) administered supraphysiological doses (0.1–0.3 mg/kg body weight) of leptin to obese subjects for 28 weeks. On average, subjects lost significant weight, but the extent of weight loss and the variability between subjects has led many to conclude that the leptin resistance of common obesity cannot be usefully overcome by leptin supplementation, at least when administered peripherally.

However, on scientific rather than pragmatic grounds, it is of interest that there was a significant effect on weight, suggesting that plasma leptin can continue to have a dose/response effect on energy homeostasis across a wide plasma concentration range. To test this hypothesis, we studied the heterozygous relatives of our leptin-deficient subjects. Serum leptin levels in the heterozygous subjects were found to be significantly lower than expected for % body fat and they had a higher prevalence of obesity than seen in a control population of similar age, sex and ethnicity (Farooqi et al., 2001). Additionally, % body fat was higher than predicted from their height and weight in the heterozygous subjects compared to control subjects of the same ethnicity (Farooqi et al., 2001). These findings closely parallel those in heterozygous ob– and db/– mice (Coleman, 1979; Chung et al., 1998). These data provide further support for the possibility that leptin can produce a graded response in terms of body composition across a broad range of plasma concentrations.

All heterozygous subjects had normal thyroid function and appropriate gonadotropins, normal development of secondary sexual characteristics, normal menstrual cycles and fertility suggesting that low leptin levels are sufficient to preserve these functions (Farooqi et al., 2001). This is consistent with the data of Ioffe and colleagues who demonstrated that several of the neuroendocrine features associated with leptin deficiency were abolished in low level leptin transgenic mice, which were fertile with normal corticosterone levels (Ioffe et al., 1998).

Our findings in the heterozygous individuals have some potential implications for the treatment of common forms of obesity. While serum leptin concentrations correlate positively with fat mass, there is considerable inter-individual variation at any particular fat mass. Leptin is inappropriately low in some obese individuals and the relative hypoleptinemia in these subjects may be actively contributing to their obesity and may be responsive to leptin therapy (Ravussin et al., 1997). Heymsfield et al. (1999) found no relationship between baseline plasma leptin levels and therapeutic response; however, study subjects were not pre-selected for relative hypoleptinemia. A therapeutic trial in a subgroup of subjects selected for disproportionately low circulating leptin levels would be of great interest.

Leptin receptor deficiency

A mutation in the leptin receptor has been reported in one consanguineous family with three affected subjects (Clement et al., 1998). Affected individuals were found to be homozygous for a mutation that truncates the receptor before the transmembrane domain. The mutant receptor ectodomain is shed from cells and circulates bound to leptin. The phenotype has similarities to that of leptin deficiency. Leptin receptor-deficient subjects were also of normal birthweight but exhibited rapid weight gain in the first few months of life, with severe hyperphagia and aggressive behaviour when food was denied (Clement et al., 1998). Basal temperature and resting metabolic rate were normal, cortisol levels were in the normal range and all individuals were normoglycaemic with mildly elevated plasma insulins similar to leptin-deficient subjects.

POMC

Two unrelated obese German children have been reported with homozygous or compound heterozygous mutations in POMC (pro-opiomelanocortin) (Krude et al., 1998). Both children were hyperphagic and developed early onset obesity presumably due to impaired melanocortin signalling in the hypothalamus (Krude et al., 1998). Presentation was in neonatal life with adrenal crisis due to isolated ACTH deficiency (POMC is a precursor of ACTH in the pituitary). The children had pale skin and red hair due to the lack of MSH function at melanocortin 1 receptors in the skin (Krude et al., 1998). Three further subjects with homozygous or compound heterozygous complete loss of function mutations of the POMC gene have been described (Krude and Gruters, 2000). Recently, a number of groups have identified a heterozygous missense mutation (Arg236Gly) in POMC that disrupts the dibasic amino acid processing site between β-MSH and β-endorphin (Echwald et al., 1999; Del Giudice et al., 2001; Challis et al., 2002). This results in an aberrant β-MSH/β-endorphin fusion peptide which binds to MC4R with an affinity identical to that of α- and β-MSH but has a markedly reduced ability to activate the receptor (Challis et al., 2002). Therefore, this cleavage site mutation in POMC may confer susceptibility to obesity through a novel molecular mechanism. Mutations affecting this processing site have been reported in obese children from several different populations, and therefore may be a relatively common contributor to early onset obesity.

Prohormone convertase 1 deficiency

Further evidence for the role of the melanocortin system in the regulation of body weight in humans comes from the description of a 47-year-old woman with severe childhood obesity, abnormal glucose homeostasis, very low plasma insulin but with elevated levels of proinsulin, hypogonadotropic hypogonadism and hypocortisolaemia associated with increased levels of POMC (Jackson et al., 1997). She was found to be a compound heterozygote for mutations in prohormone convertase 1 (PC1), which cleaves prohormones at pairs of basic amino acids, leaving C-terminal basic residues that are excised by carboxypeptidase E (CPE) (Jackson et al., 1997). Although the inability to cleave POMC is a likely mechanism for obesity in these patients, PC1 cleaves a number of other neuropeptides in the hypothalamus including glucagon-like-peptide 1, which may influence feeding behaviour. The phenotype of these subjects is very similar to that seen in the CPE-deficient fat/fat mouse (Naggert et al., 1995) implicating this part of the pathway may be important in the control of body weight in humans. To date, however, no humans with CPE defects have been described.

Human MC4R deficiency

Of the five known melanocortin receptors, the MC4R has been most closely linked to control energy balance in rodents (Yeo et al., 2000). Mice homozygous for a deleted MC4 receptor become severely obese; heterozygotes have body weights intermediate between wild-type and homozygote null animals (Huszar et al., 1997). In 1998, two groups reported heterozygous mutations in the MC4 receptor in humans which were associated with dominantly inherited obesity (Vaisse et al., 1998; Yeo et al., 1998). Since then, heterozygous mutations in MC4R have been reported in obese humans from various ethnic groups (Hinney et al., 1999; Farooqi et al., 2000; Vaisse et al., 2000).

We have studied over 500 severely obese probands and found that approximately 5–6% have pathogenic MC4R mutations that are non-conservative in nature, not found in control subjects from the background population and co-segregate with obesity in families (Farooqi et al., 2003). MC4R deficiency represents the commonest known monogenic cause of human obesity. Some studies have observed a lower prevalence and this may be explained by the differing prevalence in certain ethnic groups although it is more likely to reflect the later onset and reduced severity of obesity of the subjects in these studies (Jacobson et al., 2002). While we found a 100% penetrance of early onset obesity in heterozygous probands, others have described obligate carriers who were not obese (Vaisse et al., 2000). Given the large number of potential influences on body weight, it is perhaps not surprising that both genetic and environmental modifiers will have important effects in some pedigrees.

We have now studied over 100 MC4R mutant carriers in our Clinical Research Facility. Alongside the increase in fat mass, MC4R mutant subjects also have an increase in lean mass that is not seen in leptin deficiency (Farooqi et al., 2003). Linear growth of these subjects is striking with affected children having a height standard deviation score (SDS) of $+2$ compared to population standards (mean height SDS of other obese children in our cohort $= +0.5$) (Farooqi et al., 2003). MC4R-deficient subjects also have higher levels of fasting insulin than age, sex and BMI SDS-matched children (Farooqi et al., 2003). The accelerated linear growth and the disproportionate early hyperinsulinaemia are consistent with observations in the MC4R KO mouse (Fan et al., 2000).

Affected subjects are objectively hyperphagic, but this is not as severe as that seen with leptin deficiency (Farooqi et al., 2003). Of particular note is the finding that the severity of receptor dysfunction seen in in vitro assays can predict the amount of food ingested at a test meal by the subject harbouring that particular mutation. One notable feature of this syndrome is that the severity of many of the phenotypic features appears to partially ameliorate with time.

We have studied in detail the signalling properties of many of these mutant receptors and this information should help to advance the understanding of structure/function relationships within the receptor (Yeo et al., 2003). Importantly, we have been unable to demonstrate evidence for dominant negativity associated with these mutants, which suggests that MC4R mutations are more likely to result in a phenotype through haploinsufficiency (Yeo et al., 2003).

MC4R mutations appear to be the commonest monogenic cause of obesity thus far described in humans. The maintenance of this reasonably high disease frequency is likely to be partly due to the fact that obesity is expressed in heterozygotes and that there is no evidence of any apparent effect of the mutations on reproductive function.

Summary

Several monogenic forms of human obesity have now been identified by searching for mutations homologous to those causing obesity in mice. Although such monogenic obesity syndromes are rare, the successful use of murine models to study human obesity indicates that substantial homology exists across mammalian species in the functional organisation of the weight regulatory system. More importantly, the identification of molecules that control food intake has generated new targets for drug development in the treatment of obesity and related disorders. These considerations indicate that an expanded ability to diagnose the pathophysiological basis of human obesity will have direct applications to its treatment. A more detailed understanding of the molecular pathogenesis of human obesity may ultimately guide treatment of affected individuals.

References

Anand, B.K. and Brobeck, J.R. (1951) Hypothalamic control of food intake in rats and cats. Yale J. Biol. Med., 24: 123–146.

Barsh, G.S., Farooqi, I.S. and O'Rahilly, S. (2000) Genetics of body-weight regulation. Nature, 404: 644–651.

Bray, G.A. and York, D.A. (1971) Genetically transmitted obesity in rodents. Physiol. Rev., 51: 598–646.

Challis, B.G., Pritchard, L.E., Creemers, J.W., Delplanque, J., Keogh, J.M., Luan, J., Wareham, N.J., Yeo, G.S., Bhattacharyya, S., Froguel, P., White, A., Farooqi, I.S. and

O'Rahilly, S. (2002) A missense mutation disrupting a dibasic prohormone processing site in pro-opiomelanocortin (POMC) increases susceptibility to early onset obesity through a novel molecular mechanism. Hum. Mol. Genet., 11: 1997–2004.

Chen, Y., Hu, C., Hsu, C.K., Zhang, Q., Bi, C., Asnicar, M., Hsiung, H.M., Fox, N., Slieker, L.J., Yang, D.D., Heiman, M.L. and Shi, Y. (2002) Targeted disruption of the melanin-concentrating hormone receptor-1 results in hyperphagia and resistance to diet-induced obesity. Endocrinology, 143: 2469–2477.

Chung, W.K., Belfi, K., Chua, M., Wiley, J., Mackintosh, R., Nicolson, M., Boozer, C.N. and Leibel, R.L. (1998) Heterozygosity for Lep(ob) or Lepr(db) affects body composition and leptin homeostasis in adult mice. Am. J. Physiol., 274: R985–R990.

Clement, K., Vaisse, C., Lahlou, N., Cabrol, S., Pelloux, V., Cassuto, D., Gourmelen, M., Dina, C., Chambaz, J., Lacorte, J.M., Basdevant, A., Bougneres, P., Lebouc, Y., Froguel, P. and Guy-Grand, B. (1998) A mutation in the human leptin receptor gene causes obesity and pituitary dysfunction. Nature, 392: 398–401.

Coleman, D.L. (1973) Effects of parabiosis of obese with diabetes and normal mice. Diabetologia, 9: 294–298.

Coleman, D.L. (1979) Obesity genes: beneficial effects in heterozygous mice. Science, 203: 663–665.

Del Giudice, E.M., Santoro, N., et al. (2001) Molecular screening of the proopiomelanocortin (POMC) gene in Italian obese children: report of three new mutations. Int. J. Obes. Relat. Metab. Disord., 25: 61–67.

Dubuc, P.U. (1977) Basal corticosterone levels of young og/ob mice. Horm. Metab. Res., 9: 95–97.

Dubuc, P.U. and Carlisle, H.J. (1988) Food restriction normalizes somatic growth and diabetes in adrenalectomized ob/ob mice. Am. J. Physiol., 255: R787–R793.

Echwald, S.M., Sorensen, T.I., Andersen, T., Tybjaerg-Hansen, A., Clausen, J.O. and Pedersen, O. (1999) Mutational analysis of the proopiomelanocortin gene in Caucasians with early onset obesity. Int. J. Obes. Relat. Metab. Disord., 23: 293–298.

Fan, W., Dinulescu, D.M., Butler, A.A., Zhou, J., Marks, D.L. and Cone, R.D. (2000) The central melanocortin system can directly regulate serum insulin levels. Endocrinology, 141: 3072–3079.

Farooqi, I.S., Jebb, S.A., Langmack, G., Lawrence, E., Cheetham, C.H., Prentice, A.M., Hughes, I.A., McCamish, M.A. and O'Rahilly, S. (1999) Effects of recombinant leptin therapy in a child with congenital leptin deficiency. N. Engl. J. Med., 341: 879–884.

Farooqi, I.S., Keogh, J.M., Kamath, S., Jones, S., Gibson, W.T., Trussell, R., Jebb, S.A., Lip, G.Y. and O'Rahilly, S. (2001) Partial leptin deficiency and human adiposity. Nature, 414: 34–35.

Farooqi, I.S., Keogh, J.M., Yeo, G.S., Lank, E.J., Cheetham, T. and O'Rahilly, S. (2003) Clinical spectrum of obesity and mutations in the melanocortin 4 receptor gene. N. Engl. J. Med., 348: 1085–1095.

Farooqi, I.S., Matarese, G., Lord, G.M., Keogh, J.M., Lawrence, E., Agwu, C., Sanna, V., Jebb, S.A., Perna, F., Fontana, S., Lechler, R.I., DePaoli, A.M. and O'Rahilly, S. (2002) Beneficial effects of leptin on obesity, T cell hyporesponsiveness, and neuroendocrine/metabolic dysfunction of human congenital leptin deficiency. J. Clin. Invest., 110: 1093–1103.

Farooqi, I.S., Yeo, G.S., Keogh, J.M., Aminian, S., Jebb, S.A., Butler, G., Cheetham, T. and O'Rahilly, S. (2000) Dominant and recessive inheritance of morbid obesity associated with melanocortin 4 receptor deficiency. J. Clin. Invest., 106: 271–279.

Harris, M., Aschkenasi, C., Elias, C.F., Chandrankunnel, A., Nillni, E.A., Bjoorbaek, C., Elmquist, J.K., Flier, J.S. and Hollenberg, A.N. (2001) Transcriptional regulation of the thyrotropin-releasing hormone gene by leptin and melanocortin signaling. J. Clin. Invest., 107: 111–120.

Hervey, G.R. (1959) The effects of lesions in the hypothalamus in parabiotic rats. J. Physiol. London, 145: 336.

Hetherington, A.W. and Ranson, S.W. (1940) Hypothalamic lesions and adiposity in the rat. Anat. Rec., 78: 149–172.

Heymsfield, S.B., Greenberg, A.S., Fujioka, K., Dixon, R.M., Kushner, R., Hunt, T., Lubina, J.A., Patane, J., Self, B., Hunt, P. and McCamish, M. (1999) Recombinant leptin for weight loss in obese and lean adults: a randomized, controlled, dose-escalation trial. JAMA, 282: 1568–1575.

Hinney, A., Schmidt, A., Nottebom, K., Heibult, O., Becker, I., Ziegler, A., Gerber, G., Sina, M., Gorg, T., Mayer, H., Siegfried, W., Fichter, M., Remschmidt, H. and Hebebrand, J. (1999) Several mutations in the melanocortin-4 receptor gene including a nonsense and a frameshift mutation associated with dominantly inherited obesity in humans. J. Clin. Endocrinol. Metab., 84: 1483–1486.

Huszar, D., Lynch, C.A., Fairchild-Huntress, V., Dunmore, J.H., Fang, Q., Berkemeier, L.R., Gu, W., Kesterson, R.A., Boston, B.A., Cone, R.D., Smith, F.J., Campfield, L.A., Burn, P. and Lee, F. (1997) Targeted disruption of the melanocortin-4 receptor results in obesity in mice. Cell, 88: 131–141.

Ioffe, E., Moon, B., Connolly, E. and Friedman, J.M. (1998) Abnormal regulation of the leptin gene in the pathogenesis of obesity. Proc. Natl. Acad. Sci. USA, 95: 11852–11857.

Jackson, R.S., Creemers, J.W., Ohagi, S., Raffin-Sanson, M.L., Sanders, L., Montague, C.T., Hutton, J.C. and O'Rahilly, S. (1997) Obesity and impaired prohormone processing associated with mutations in the human prohormone convertase 1 gene. Nat. Genet., 16: 303–306.

Jacobson, P., Ukkola, O., Rankinen, T., Snyder, E.E., Leon, A.S., Rao, D.C., Skinner, J.S., Wilmore, J.H., Lonn, L., Cowan Jr., G.S., Sjostrom, L. and Bouchard, C. (2002) Melanocortin 4 receptor sequence variations are seldom a cause of human obesity: the Swedish Obese Subjects, the HERITAGE Family Study, and a Memphis cohort. J. Clin. Endocrinol. Metab., 87: 4442–4446.

Kennedy, G.C. (1953) The role of depot fat in the hypothalamic control of food intake in the rat. Proc. R. Soc. London, 140: 578–596.

Krude, H., Biebermann, H., Luck, W., Horn, R., Brabant, G. and Gruters, A. (1998) Severe early onset obesity, adrenal insufficiency and red hair pigmentation caused by POMC mutations in humans. Nat. Genet., 19: 155–157.

Krude, H. and Gruters, A. (2000) Implications of pro-opiomelanocortin (POMC) mutations in humans: the POMC deficiency syndrome. Trends Endocrinol. Metab., 11: 15–22.

Legradi, G., Emerson, C.H., Ahima, R.S., Flier, J.S. and Lechan, R.M. (1997) Leptin prevents fasting-induced suppression of prothyrotropin-releasing hormone messenger ribonucleic acid in neurons of the hypothalamic paraventricular nucleus. Endocrinology, 138: 2569–2576.

Leibel, R.L., Chung, W.K. and Chua Jr., S.C. (1997) The molecular genetics of rodent single gene obesities. J. Biol. Chem., 272: 31937–31940.

Lord, G.M., Matarese, G., Howard, J.K., Baker, R.J., Bloom, S.R. and Lechler, R.I. (1998) Leptin modulates the T-cell immune response and reverses starvation-induced immunosuppression. Nature, 394: 897–901.

Mantzoros, C.S., Ozata, M., Negrao, A.B., Suchard, M.A., Ziotopoulou, M., Caglayan, S., Elashoff, R.M., Cogswell, R.J., Negro, P., Liberty, V., Wong, M.L., Veldhuis, J., Ozdemir, I.C., Gold, P.W., Flier, J.S. and Licinio, J. (2001) Synchronicity of frequently sampled thyrotropin (TSH) and leptin concentrations in healthy adults and leptin-deficient subjects: evidence for possible partial TSH regulation by leptin in humans. J. Clin. Endocrinol. Metab., 86: 3284–3291.

Matarese, G. (2000) Leptin and the immune system: how nutritional status influences the immune response. Eur. Cytokine Netw., 11: 7–14.

Montague, C.T., Farooqi, I.S., Whitehead, J.P., Soos, M.A., Rau, H., Wareham, N.J., Sewter, C.P., Digby, J.E., Mohammed, S.N., Hurst, J.A., Cheetham, C.H., Earley, A.R., Barnett, A.H., Prins, J.B. and O'Rahilly, S. (1997) Congenital leptin deficiency is associated with severe early onset obesity in humans. Nature, 387: 903–908.

Naggert, J.K., Fricker, L.D., Varlamov, O., Nishina, P.M., Rouille, Y., Steiner, D.F., Carroll, R.J., paigen, B.J. and Leiter, E.H. (1995) Hyperproinsulinaemia in obese fat/fat mice associated with a carboxypeptidase E mutation which reduces enzyme activity. Nat. Genet., 10: 135–142.

Nillni, E.A., Vaslet, C., Harris, M., Hollenberg, A., Bjorbak, C. and Flier, J.S. (2000) Leptin regulates prothyrotropin-releasing hormone biosynthesis. Evidence for direct and indirect pathways. J. Biol. Chem., 275: 36124–36133.

Ozata, M., Ozdemir, I.C. and Licinio, J. (1999) Human leptin deficiency caused by a missense mutation: multiple endocrine defects, decreased sympathetic tone, and immune system dysfunction indicate new targets for leptin action, greater central than peripheral resistance to the effects of leptin, and spontaneous correction of leptin-mediated defects. J. Clin. Endocrinol. Metab., 84: 3686–3695.

Rau, H., Reaves, B.J., O'Rahilly, S. and Whitehead, J.P. (1999) Truncated human leptin (delta133) associated with extreme obesity undergoes proteasomal degradation after defective intracellular transport. Endocrinology, 140: 1718–1723.

Ravussin, E., Pratley, R.E., Maffei, M., Wang, H., Friedman, J.M., Bennett, P.H. and Bogardus, C. (1997) Relatively low plasma leptin concentrations precede weight gain in Pima Indians. Nat. Med., 3: 238–240.

Rosenbaum, M., Murphy, E.M., Heymsfield, S.B., Matthews, D.E. and Leibel, R.L. (2002) Low dose leptin administration reverses effects of sustained weight-reduction on energy expenditure and circulating concentrations of thyroid hormones. J. Clin. Endocrinol. Metab., 87: 2391.

Shimada, M., Tritos, N.A., Lowell, B.B., Flier, J.S. and Maratos-Flier, E. (1998) Mice lacking melanin-concentrating hormone are hypophagic and lean. Nature, 396: 670–674.

Strobel, A., Issad, T., Camoin, L., Ozata, M. and Strosberg, A.D. (1998) A leptin missense mutation associated with hypogonadism and morbid obesity. Nat. Genet., 18: 213–215.

Trayhurn, P., Thurlby, P.L. and James, W.P.T. (1977) Thermogenic defect in pre-obese ob/ob mice. Nature, 266: 60–62.

Vaisse, C., Clement, K., Durand, E., Hercberg, S., Guy-Grand, B. and Froguel, P. (2000) Melanocortin-4 receptor mutations are a frequent and heterogeneous cause of morbid obesity. J. Clin. Invest., 106: 253–262.

Vaisse, C., Clement, K., Guy-Grand, B. and Froguel, P. (1998) A frameshift mutation in human MC4R is associated with a dominant form of obesity. Nat. Genet., 20: 113–114.

Yeo, G.S., Farooqi, I.S., Aminian, S., Halsall, D.J., Stanhope, R.G. and O'Rahilly, S. (1998) A frameshift mutation in MC4R associated with dominantly inherited human obesity. Nat. Genet., 20: 111–112.

Yeo, G.S., Farooqi, I.S., Challis, B.G., Jackson, R.S. and O'Rahilly, S. (2000) The role of melanocortin signalling in the control of body weight: evidence from human and murine genetic models. QJM, 93: 7–14.

Yeo, G.S., Lank, E.J., Farooqi, I.S., Keogh, J., Challis, B.G. and O'Rahilly, S. (2003) Mutations in the human melanocortin-4 receptor gene associated with severe familial obesity disrupts receptor function through multiple molecular mechanisms. Hum. Mol. Genet., 12: 561–574.

Hypothalamic Integration of Blood-borne Signals

Kalsbeek, Fliers, Hofman, Swaab, Van Someren & Buijs
Progress in Brain Research, Vol. 153
ISSN 0079-6123

CHAPTER 7

The selfish brain: competition for energy resources

H.L. Fehm*, W. Kern and A. Peters

Medizinische Klinik I, Universität Lübeck, Ratzeburger Allee 160, D-23538 Lübeck, Germany

Abstract: Although the brain constitutes only 2% of the body mass, its metabolism accounts for 50% of total body glucose utilization. This delicate situation is aggravated by the fact that the brain depends on glucose as energy substrate. Thus, the contour of a major problem becomes evident: how can the brain maintain constant fluxes of large amounts of glucose to itself in the presence of powerful competitors as fat and muscle tissue.

Activity of cortical neurons generates an "energy on demand" signal which eventually mediates the uptake of glucose from brain capillaries. Because energy stores in the circulation (equivalent to ca. 5 g glucose) are also limited, a second signal is required termed "energy on request"; this signal is responsible for the activation of allocation processes. The term "allocation" refers to the activation of the "behavior control column" by an input from the hippocampus–amygdala system. As far as eating behavior is concerned the behavior control column consists of the ventral medial hypothalamus (VMH) and periventricular nucleus (PVN). The PVN represents the central nucleus of the brain's stress systems, the hypothalamus–pituitary–adrenal (HPA) axis and the sympathetic nervous system (SNS). Activation of the sympatico-adrenal system inhibits glucose uptake by peripheral tissues by inhibiting insulin release and inducing insulin resistance and increases hepatic glucose production. With an inadequate "energy on request" signal neuroglucopenia would be the consequence. A decrease in brain glucose can activate glucose-sensitive neurons in the lateral hypothalamus (LH) with the release of orexigenic peptides which stimulate food intake. If the energy supply of the brain depends on activation of the LH rather than on increased allocation to the brain, an increase in body weight is evitable. An increase in fat mass will generate feedback signals as leptin and insulin, which activate the arcuate nucleus. Activation of arcuate nucleus in turn will stimulate the activity of the PVN in a way similar to the activation by the hippocampus–amydala system.

The activity of PVN is influenced by the hippocampal outflow which in turn is the consequence of a balance of low-affinity and high-affinity glucocorticoid receptors. This set-point can permanently be displaced by extreme stress situations, by starvation, exercise, hormones, drugs or by endocrine-disrupting chemicals. Disorders in the "energy on request" process will influence the allocation of energy and in so doing alter the body mass of the organism. In this "selfish brain theory" the neocortex and the limbic system play a central role in the pathogenesis of diseases, such as anorexia nervosa, obesity and diabetes mellitus type II.

From these considerations it appears that the primary disturbance in obesity is a displacement of the hippocampal set-point of the system. The resulting permanent activation of the feedback system must result in a likewise permanent activation of the sympatico-adrenal system, which induces insulin resistance, hypertension and the other components of the metabolic syndrome. Available therapies for treatment of

*Corresponding author. Tel.: +49-451-500-2305; Fax: +49-451-500-3339; E-mail: fehm@uni-luebeck.de

DOI: 10.1016/S0079-6123(06)53007-9

129

the metabolic syndrome (blockade of α- and β-adrenergic receptors, insulin and insulin secretagogues) interfere with mechanisms, which must be considered compensatory. This explains why these therapies are disappointing in the long run. New therapeutic strategies based on the "selfish brain theory" will be discussed.

Introduction

We are witnessing a world-wide surge in the incidence of obesity and type 2 diabetes. There is evidence that the increase in longevity that has been observed over the last 150 years might be arrested by obesity and its major consequence, i.e. the so-called metabolic syndrome. The necessity to understand the pathophysiology of obesity and type 2 diabetes in more detail has never been as urgent as it is now. Fortunately, over the last 10 years there has also been dramatic progress in understanding the brain's fundamental role in energy homeostasis and body fat mass regulation. This development started with the discovery of leptin, a hormone produced in fat tissue which communicates the status of body energy stores to the central nervous system (CNS) (Zhang et al., 1994). Well before the discovery of leptin, insulin was known to exert similar effects; however, at that time these findings were widely neglected (Woods et al., 1984). The discovery of leptin also led to a reappraisal of the fundamental role of cortisol in the regulation of body weight and energy homeostasis (Jeanrenaud and Rohner-Jeanrenaud, 2000). More recently, a number of peptides have been described that are released from the intestinal tract and are able to induce sensations of satiety (Schwartz et al., 2000). All these paved the way for elucidating the role of the CNS in regulating body weight and energy homeostasis. In the wake of these findings, a number of orexigenic and anorexigenic neuropeptides have been identified that are produced in hypothalamic neurons and which integrate peripheral and CNS signals to control food intake (Fig. 1). Given the large number of factors and hormones, which participate in this regulation, it has become a major task to understand how these factors interact and define their relative importance. In an attempt to build a foundation to deal with this problem the "selfish brain theory" (Peters et al., 2004) has been developed, which aims to define the fundamental role of the brain in regulating food intake and energy homeostasis, and which emphasizes the brain's primacy in the control of energy fluxes.

Maintenance of glucose fluxes to the brain

Although the brain constitutes only 2% of the body mass, its metabolism accounts for 50% of total body glucose utilization, i.e. it takes up approximately 100 g of glucose each day (Owen et al., 1967). In comparison to other organs the brain is the most energy demanding. This precarious situation is aggravated by the fact that the brain depends on glucose as its energy substrate; in contrast to muscle and fat tissue it preferentially utilizes glucose instead of other energy substrates such as fatty acids or amino acids. This organ, which uses the most energy of all organs, prefers to run on the fuel that is stored the worst. As such, all the trappings of a major problem begin to emerge: how can the brain maintain constant fluxes of large amounts of glucose to itself during periods of fasting and in the presence of powerful competitors such as the fat and muscle tissue?

The initiating event is the energy consumption of cortical neurons; energy homeostasis of cortical neurons has been studied in depth by Pierre Magistretti and Luc Pellerin (Pellerin and Magistretti, 1994; Magistretti et al., 1999). The astrocyte plays a central role in a process which has been termed "energy on demand". Upon stimulation, excitatory cortical neurons release glutamate into the synaptic cleft. The activity of glutamate is terminated by an efficient glutamate uptake system into astrocytes. Glutamate induces a series of events within the astrocyte which results in the uptake of one molecule of glucose for each glutamate molecule. Foot processes of the astrocytes surround the brain capillary vessels and are equipped with glucose transporters (GLUT 1) for the uptake of

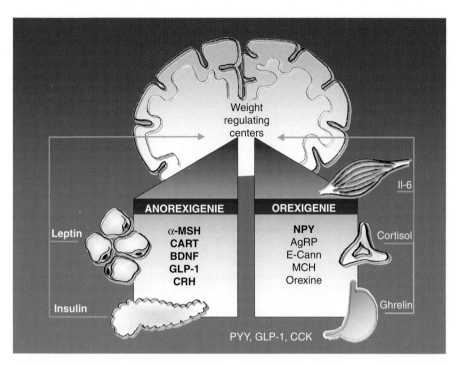

Fig. 1. Elements of energy homeostasis. MSH, melanocyte-stimulating hormone; CART, cocaine–amphetamine-related transcript; BDNF, brain-derived neurotrophic factor; GLP-1, glucagon-like peptide; CRH, corticotrophin-releasing hormone; NPY, neuropeptide Y; AgRP, Agouti-related peptide; E-Cann, endocannabinoids; MCH, melanin-concentrating hormone; PYY, peptide YY; CCK, cholecystokinin; Il-6, interleukine 6.

glucose. However, the total amount of glucose present in the circulation is limited; given a blood volume of 5 l with a glucose concentration of 100 mg/dl, the amount of circulating glucose that is available at a given time can be as low as 5 g. This is in striking contrast to the estimated amount consumed over 24 h, i.e. 100 g glucose. To ensure an adequate supply the brain must prevent or impair glucose fluxes to peripheral tissues, and stimulate glucose production by the liver (glycolysis and gluconeogenesis). These actions are mediated by the "behavior control column" as described by Larry Swanson (Swanson, 2000). The behavior control column receives major input from the hippocampus and the amygdala. As far as eating behavior is concerned, this column consists of the ventral medial hypothalamus (VMH) and the periventricular nucleus (PVN). The hippocampus in turn receives projections from all parts of the neocortex, which enables this structure to organize an adequate activation of the behavior control

column. This part of the behavior control column is more or less identical to what is known as the brain's "stress system" in other contexts. Activation of the PVN entails activation of the hypothalamus–pituitary–adrenal (HPA) axis and at the same time activation of the sympathetic nervous system (SNS). The activity of the sympatico-adrenal system can easily be monitored by generally available methods (measuring plasma ACTH/cortisol levels or plasma norepinephrine concentrations or more directly by means of microneurography among other techniques; Low, 2003). Activation of the PVN will eventually suppress insulin secretion, induce insulin resistance and stimulate hepatic glycolysis and gluconeogenesis. Together, these events will increase glucose fluxes to the brain. We have coined the term "energy on request" to describe this series of events (Fig. 2) (Peters et al., 2004).

Should the hippocampal "energy on request" signal be inadequate, a deficiency in brain glucose

132

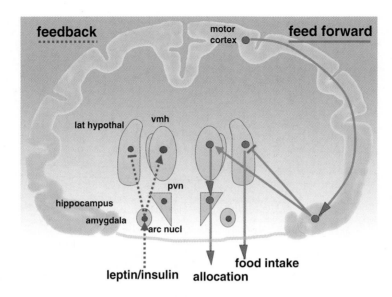

Fig. 2. The "energy on request" principle.

concentration would result, i.e. a neuroglucopenia. This is where the lateral hypothalamus (LH) comes into play. Glucose-sensitive neurons are present in the LH. These neurons respond to a decrease in brain glucose by increasing their firing rate in a manner presumably mediated by Na/K ATPase (Levin et al., 1999). Thus, orexigenic peptides produced in the LH such as orexin A and B and melanin concentrating hormone (MCH) are released to stimulate appetite and eating behavior. Overall, the brain has two main mechanisms at its disposal to ensure an adequate glucose supply: activation of allocation behavior by increasing the activity of PVN, and stimulation of eating behavior via LH.

If the brain is forced to utilize stimulation of eating behavior because allocation is inadequate, total energy intake and the proportion ending up in peripheral tissues will increase. Over time, this will result in an increase in fat stores and as such body weight. The feedback signals, leptin and insulin are activated by the increasing fat stores. These adipostatic signals are recognized by and processed within the arcuate nucleus (Spanswick et al., 1997; Spanswick et al., 2000), an area that among other things responds by releasing melanocyte stimulating hormone (α-MSH). The central melanocortin

system is perhaps the best-characterized neuronal pathway involved in the regulation of energy homeostasis, and there are several excellent reviews on this topic (Seeley et al., 2004).

Given the fact that the arcuate nucleus mediates feedback signals generated by increased fat stores it is unlikely that defective arcuate nucleus functions represent the primary disturbance in obesity; the primary disturbance is to be expected rather in the feed-forward pathway of energy regulation, i.e. in the functions of the hippocampus.

The role of the hippocampus/amygdala system in energy homeostasis

The hippocampus and amygdala are well known for their role in the formation of new memories (Bliss and Collingridge, 1993). Excitatory synaptic transmission in these brain regions depends on glutamate. This neurotransmitter is released into the synapse by the presynaptic neurons and binds to receptors on the postsynaptic neuronal membrane (Fig. 3). There are two types of glutamate receptors: AMPA receptors, which regulate basal synaptic transmission, and NMDA receptors, which modulate a specific type of synaptic

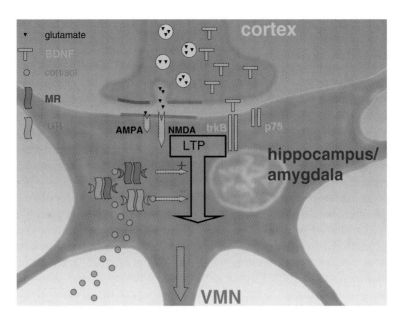

Fig. 3. The role of the hippocampus/amygdala system in energy homeostasis.

plasticity called long-term potentiation (LTP) (Shi et al., 2001). LTP may form the cellular basis of learning and memory. Among many other molecules that regulate LTP is a neurotrophin called brain-derived neurotrophic factor (BDNF) (Barde et al., 1982). Kovalchuk et al. demonstrated that BDNF in concert with glutamate induces LTP by binding to receptors on the postsynaptic neuronal membrane (Kovalchuk et al., 2002). Again, there are two types of receptors for BDNF: TrkB and p75. Binding of BDNF to TrkB receptors is essential for LTP induction in the hippocampus (Ikegaya et al., 2002).

Apart from receptors for glutamate and BDNF, neurons in the hippocampus and amygdala are also equipped with glucocorticoid receptors (GRs) and mineralocorticoid receptors (MRs). Indeed, these receptors are most abundant within these brain sites (McEwen et al., 1968). The endogenous ligand for both of these receptors is cortisol, where MR represents the high-affinity and GR the low-affinity receptor for cortisol. The role of these receptors in the regulation of LTP is well known (Pavlides et al., 1995; Pavlides and McEwen, 1999).

What is the evidence that all such molecules present in hippocampal and amygdala neurons are

involved in the regulation of glucose fluxes and energy homeostasis?

The fundamental role of MR's and GR's is evident from clinical observations in patients with a deficiency (Addison's disease) or an excess of cortisol (Cushing's disease). Addison's disease is inevitably accompanied by a weight reduction, in many patients of a dramatic nature, while cortisol excess of all causes is accompanied by an increase in body fat, especially visceral fat. In experimental animals the critical role of glucocorticoids in everyday energy flow has been discussed by Dallman et al. (1993). In humans, polymorphisms of the GR gene and their association with metabolic parameters and body composition have been studied (van Rossum and Lamberts, 2004). The polymorphism N363S has been associated with increased sensitivity to glucocorticoids and increased BMI. In contrast, the ER 22/23 EK polymorphism has been associated with a relative resistance for glucocorticoid and a healthy metabolic profile, although BMI was not changed significantly. Finally, bilateral lesions of the amygdala in animal experiments regularly lead to an excessive increase in body weight (King et al., 1996, 2003; Rollins and King, 2000).

The importance of hippocampal and amygdala functions in the regulation of metabolic processes is underlined by the fact that the brain is capable of establishing unconscious memories of metabolic events: after a hypoglycemic episode, the response to a second, otherwise identical, episode is different. Diabetic patients with repeated hypoglycemic episodes are prone to "hypoglycemia unawareness" (Gerich et al., 1991). This learned behavior makes these patients unable to recognize hypoglycemia at a time (Smith and Amiel, 2002) when they are still able to prevent a more severe hypoglycemia by ingestion of carbohydrates (Strachan et al., 2004). It appears that the hypothalamus readily adapts to conditions of hypoglycemia (Fruehwald-Schultes et al., 1999). Interestingly, this adaptation of counter regulatory reactions pertains not to hypoglycemia-induced feelings of hunger (Schultes et al., 2003). In our model, hunger is generated in the LH and in contrast to the VMH the LH is not activated by the hippocampus.

The role of NMDA receptors in generating the endocrine response to hypoglycemia has been studied by P. Molina et al. (28). They demonstrated that blockade of NMDA receptors in dogs profoundly attenuated the response of the HPA axis to neuroglucopenia (Molina and Abumrad, 2001).

More recently, severe obesity in an 8-year-old male could be traced to a de novo mutation affecting human TrkB (Yeo et al., 2004). This is the first observation in humans, which points to the role of TrkB and LTP in the regulation of body weight in humans.

Perhaps the most convicting evidence, however, comes from anatomic studies showing that the major input to the behavior control column responsible for eating behavior comes from the hippocampus and the amygdala (Swanson, 2000).

The balance between food intake and glucose allocation

The cortico-hypothalamic circuits described above create two main outputs, i.e. control of eating behavior and allocation behavior. Given the high-energy requirements of cortical areas it is evident that any change in food intake must be accompanied by a corresponding change in allocation activity. The reciprocal relationship between the need for food intake and the need for allocation to supply the brain with sufficient fuel can be described using a simple hyperbolic curve (Fig. 4). Since the activity of the allocation behavior (VMH) is influenced by massive input from the hippocampus/amygdala system (Swanson, 2000), food intake must be adjusted correspondingly. The target of the regulatory system is not a constant energy intake level or a constant body weight, but rather a constant energy content within the brain. Of course, in a healthy subject the hippocampal outflow is such that energy expenditure and energy uptake are equal, a situation which results in a constant body weight. This situation represents the set-point of the system: the energy needs of the brain are satisfied and peripheral energy stores are stable.

A displacement of the set-point to the right means less food intake but greater activity of allocation mechanisms i.e. activation of SNS and HPA-axis. With time, this must result in a weight reduction which will be accelerated by further displacement to the right. At the same time, activation of the SNS and HPA-axis is a prerequisite for the development of insulin resistance and hypertension. Conversely, a displacement of the set-point to the left results inevitably in a weight increase (increased food intake, whereby a larger proportion goes to the peripheral stores) accompanied by increased insulin sensitivity and decreased sympathetic outflow.

How does this system behave in a situation with excessive calorie intake? An increase in food intake allows for a decrease in allocation activity: the stress systems can relax. However, the inevitable consequence of this displacement of the balance point to the left is an increase in body weight. This increase represents a burden to the regulatory system which implies that the balance point moves upwards to curves characteristic for higher body weights. At the same time, feedback mechanisms are activated via increased leptin and insulin levels. These feedback signals reduce eating behavior and activate the SNS. Both actions must be tuned in such a way that the brain's glucose needs are not

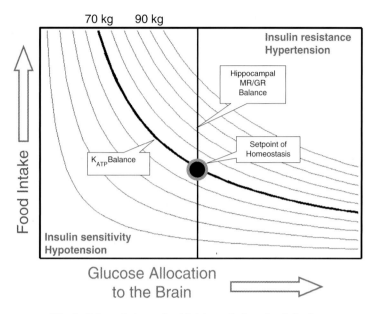

Fig. 4. Balance between food intake and allocation behavior.

jeopardized. This brings the balance point back to the vicinity of the original set-point, albeit upwardly displaced. Whenever excessive calorie intake ceases, the feedback actions of leptin and insulin displace the balance point to the right of the original set-point into an area characterized by weight reduction. With decreasing body weight, the feedback effects become weaker until the original set-points and the balance point are identical. The same course of events — albeit in the opposite direction — will occur upon calorie restriction. It is well known that weight reduction by dieting or fasting is only temporarily effective and eventually results in the starting weight being returned to.

Obesity and diabetes mellitus type 2 as a brain disease?

A constant weight reduction can be achieved only with a displacement of the set-point to the right, while obesity can only occur with a displacement to the left.

What are the consequences of a permanent displacement of the set-point to the left? The inevitable increase in body weight will activate the feedback mechanisms. Even in this situation

a stable situation will result, however, with increased body weight and a slight but permanent activation of the stress systems and their malign sequelae, i.e. the metabolic syndrome. We therefore postulate that the development of the metabolic syndrome starts with a decrease in outflow of the hippocampus/amygdala system which entails a displacement of the set-point to the left in Fig. 4. In this situation there is increased insulin sensitivity but increasing body weight. With time there is an increase in leptin and insulin levels activating the VMH and the arcuate nucleus. The increased activity of the VMH and the arcuate nucleus can compensate for the decreased hippocampal outflow. At some time during further development the balance point approaches the vicinity of the original set-point, at which time the increase in body weight will stop before being replaced by a reduction in body weight, namely when the feedback effects of leptin/insulin shift the balance point to the right. At this time, the stress systems are activated so that insulin resistance and hypertension occur. It is well known from epidemiological observations of untreated patients with type 2 diabetes that the increase in body weight is replaced by a decrease upon manifestation of the diabetes mellitus (Looker et al., 2001). With this way of looking at things, the metabolic syndrome represents the organism's attempt to compensate for the original displacement of the set-point.

In the light of this concept, current strategies for treating type 2 diabetes appear to be of somewhat doubtful effectiveness. A cornerstone of current treatment is strict glucose control. In the United Kingdom Prospective Diabetes Study (UKPDS) and other studies it was shown that microvascular complications were significantly and consistently reduced in the more aggressively controlled patient groups (UKPDS Study Group, 1998). In spite of these advantages of intensive glycemic control, diabetes-related mortality and all-cause mortality did not differ between intensive and conventional groups (UKPDS Study Group, 1998). Regarding the adverse effects of intensive glycemic control, the UKPDS revealed that despite regular supervision by a dietician, weight gain was significantly higher in the intensive group than it was in the conventional group, and that patients assigned insulin had a greater gain in weight than those assigned oral agents. The progression of the disease as evidenced by the slope of the increase in HbA1c levels was not influenced by intensive treatment. Thus, the results of the UKPDS demonstrate that strict glycemic control represents a problematic approach for treating diabetic patients: it promotes weight gain when obesity appears to play a fundamental role in the pathophysiology of the disease.

The second cornerstone of treatment of diabetic patients is control of hypertension. Hypertension is common in diabetic patients and 39% of newly diagnosed patients are already hypertensive. Control of elevated blood pressure can be achieved in these patients using a range of antihypertensive compounds. One of the oldest drug families in this respect are the β-blockers. Again, one of the adverse effects of β-blockers is weight gain; at least some of these drugs may in fact impair glycemic control.

In the "selfish brain theory," insulin resistance and an activated sympathetic nervous system represent counter regulatory mechanisms that aim to compensate for a displacement of the set-point. This theory predicts that interference with these compensatory mechanisms by overcoming insulin resistance and/or by blocking the sympathetic nervous system will worsen rather than improve the situation. The prevailing enthusiasm in commentaries about the UKPDS usually suppresses these sobering study results, although some experts have been more critical (Kopelman and Hitman, 1998). There would appear to be an urgent need for treatment options based on a better understanding of the fundamental defect in obesity and diabetes mellitus type 2.

Future treatment strategies for obesity and diabetes mellitus type 2

An ideal therapeutic strategy for treating type 2 diabetes would be to correct the displacement of the metabolic set-point. Such correction is theoretically feasible since set-points are defined by the activity of neurons in the hippocampus/amygdala, and this activity is prone to plasticity. Metabolic set-points can be "learned" and "relearned" similar to other types of memory, and mechanisms which control memory consolidation and long-term potentiation are also related to metabolic set-points. In such a case a single, one-time intervention would suffice to ensure a life-long treatment for a disease wherever a set-point displacement is the primary disturbance. However, such a therapeutic approach is not yet available.

A more realistic alternative might be a permanent stimulation of the activity of the arcuate nucleus. As was outlined earlier, the arcuate nucleus is capable of compensating for a disturbance in the outflow of the hippocampus/amygdala system. The activity of the arcuate nucleus can be stimulated by insulin and leptin, and the increased activity of the arcuate nucleus can be mimicked by administration of α-MSH. The problem is that these peptides cannot cross the blood–brain barrier and therefore have no access to receptors located beyond this barrier. We have recently shown in humans that the intranasal administration of peptides such as insulin and α-MSH allows direct access to the cerebrospinal fluid compartment within 30 min in a manner that bypasses uptake into the blood stream (Born et al., 2002). By exploiting these methods we were able to study the effects of insulin and α-MSH analogues on CNS function and the control of body weight in humans.

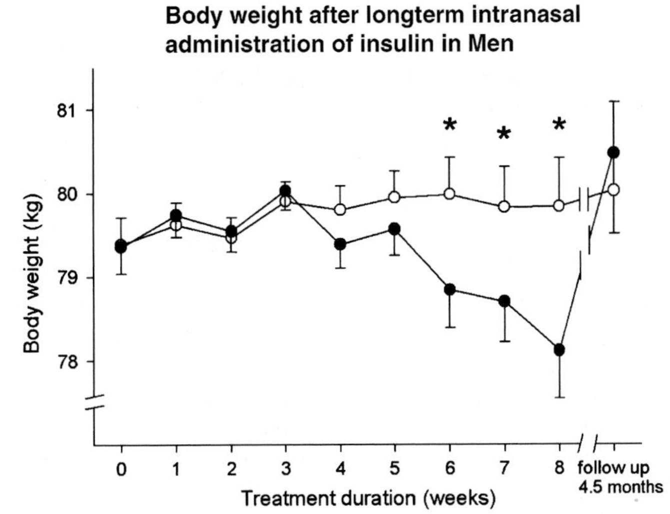

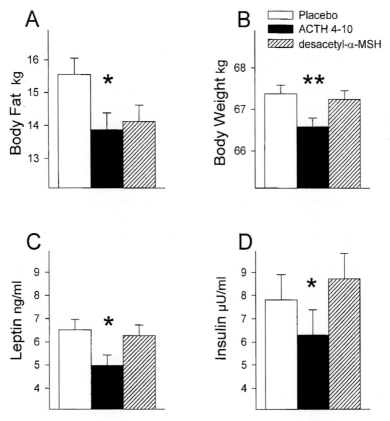

**Body weight after longterm intranasal
administration of insulin in Men**

Fig. 5. Effects of intranasal insulin on body weight in human male subjects.

Fig. 6. Effects of intranasal α-MSH$_{4-10}$ on body weight in human subjects.

(a) *Insulin*. 40 IU insulin were administered intranasally four times a day: in the morning, at noon, in the evening and before going to bed (Fig. 5). Twenty healthy, normal weight subjects were treated over a period of 8 weeks. The insulin-treated men lost an average of 1.28 kg body weight and 1.38 kg body fat, while their waist girth decreased by 1.63 cm. Plasma leptin levels dropped by 27%. In contrast, the insulin-treated women did not lose any body fat and in fact gained 1.04 kg in body weight due to an increase in extracellular water. This sexual dimorphism was also observed in animal experiments with intracerebro-venticular insulin infusion(Clegg et al., 2003).

(b) *α-MSH*. Insulin acts on energy homeostasis by stimulating neurons in the arcuate nucleus to release α-MSH. Intranasal administration of α-MSH or α-MSH analogues should therefore yield similar results to intranasal administration of insulin. Again, healthy young subjects who were unaware that they were participating in a study on body weight were treated with α-MSH_{4-10} intranasally over a period of 6 weeks (Fig. 6) (Fehm et al., 2001). α-MSH_{4-10} reduced body fat by 1.68 kg and body weight by 0.79 kg. Concurrently, plasma leptin levels were decreased by 24% and insulin levels by 20% (Hallschmid et al., 2004b).

The results with intranasal administration of insulin and α-MSH underline the notion that already existing knowledge about the role of the arcuate nucleus in the control of energy homeostasis can be used to design new strategies for treating obesity and type 2 diabetes mellitus in human subjects (Fehm et al., 2001; Hallschmid et al., 2004a). However, much progress still needs to be made. At present we do not know whether the weight-reducing effects of insulin and α-MSH can be observed in overweight subjects. In fact, preliminary results show that overweight subjects are resistant to these effects, since they are also resistant to the peripheral effects of insulin and leptin.

Conclusions

Given the extraordinary requirements of the brain with regard to the amount of energy and the types of fuel, it is clear that the brain must be capable of controlling energy fluxes within the body. The mechanisms exerting this control are described in the "selfish brain theory," which emphasizes the brain's primacy in the allocation of energy fluxes to the brain. The main players, i.e. the HPA-axis and the SNS, are well characterized, although they have usually been viewed in different contexts.

According to the "selfish brain theory," obesity and type II diabetes represent the sequelae of a dysfunction in the hippocampus/amygdala system. The important question that remains to be answered is the cause underlying the displacement of the hippocampal/amygdala set-point. The current surge in the incidence of obesity and diabetes mellitus started some 20 years ago. It is therefore clear that the initiating event is an environmental factor rather than a genetic disturbance. Despite this, the vast majority of research concerning obesity concerns its genetic and molecular biological basis.

With the recent advances in obesity research, especially the discovery of the orexigenic and anorexigenic peptides of the hypothalamus, there is a general agreement that obesity is a brain disease. Unfortunately, most research has been committed to studying the hypothalamus and especially the arcuate nucleus. Much of the current obesity research has therefore addressed the feedback-regulation exerted by insulin and leptin via the arcuate nucleus and fascinating insights have certainly been gained over the last years. However, a primary disturbance in the feedback-regulation will be silent unless the regulatory system is challenged. Focusing on the hypothalamic integration of energy metabolism might therefore be inadequate to understand the complex pathophysiology of obesity and diabetes. Future research must include extrahypothalamic sites, especially the hippocampus and amygdala, and certainly also other cortical areas as proposed by the "selfish brain theory".

References

Barde, Y.A., Edgarm, D. and Thoenen, H. (1982) Purification of a new neurotrophic factor from mammalian brain. EMBO J., 1: 549–553.

Bliss, T.V. and Collingridge, G.L. (1993) A synaptic model of memory: long-term potentiation in the hippocampus. Nature, 361: 31–39.

Born, J., Lange, T., Kern, W., McGregor, G.P., Bickel, U. and Fehm, H.L. (2002) Sniffing neuropeptides: a transnasal approach to the human brain. Nat. Neurosci., 5: 514–516.

Clegg, D.J., Riedy, C.A., Smith, K.A., Benoit, S.C. and Woods, S.C. (2003) Differential sensitivity to central leptin and insulin in male and female rats. Diabetes, 52: 682–687.

Dallman, M.F., Strack, A.M., Akana, S.F., Bradbury, M.J., Hanson, E.S., Scribner, K.A. and Smith, M. (1993) Feast and famine: critical role of glucocorticoids with insulin in daily energy flow. Front. Neuroendocrinol., 14: 303–347.

Fehm, H.L., Smolnik, R., Kern, W., McGregor, G.P., Bickel, U. and Born, J. (2001) The melanocortin melanocyte-stimulating hormone/adrenocorticotropin (4–10) decreases body fat in humans. J. Clin. Endocrinol. Metab., 86: 1144–1148.

Fruehwald-Schultes, B., Kern, W., Deininger, E., Wellhoener, P., Kerner, W., Born, J., Fehm, H.L. and Peters, A. (1999) Protective effect of insulin against hypoglycemia-associated counterregulatory failure. J. Clin. Endocrinol. Metab., 84: 1551–1557.

Gerich, J.E., Mokan, M., Veneman, T., Korytkowski, M. and Mitrakou, A. (1991) Hypoglycemia unawareness. Endocr. Rev., 12: 356–371.

Hallschmid, M., Benedict, C., Born, J., Fehm, H.L. and Kern, W. (2004a) Manipulating central nervous mechanisms of food intake and body weight regulation by intranasal administration of neuropeptides in man. Physiol. Behav., 83: 55–64.

Hallschmid, M., Benedict, C., Schultes, B., Fehm, H.L., Born, J. and Kern, W. (2004b) Intranasal insulin reduces body fat in men but not in women. Diabetes, 53: 3024–3029.

Ikegaya, Y., Ishizaka, Y. and Matsuki, N. (2002) BDNF attenuates hippocampal LTD via activation of phospholipase C: implications for a vertical shift in the frequency-response curve of synaptic plasticity. Eur. J. Neurosci., 16: 145–148.

Jeanrenaud, B. and Rohner-Jeanrenaud, F. (2000) CNS-periphery relationships and body weight homeostasis: influence of the glucocorticoid status. Int. J. Obes. Relat. Metab. Disord., 24(Suppl 2): S74–S76.

King, B.M., Cook, J.T. and Dallman, M.F. (1996) Hyperinsulinemia in rats with obesity-inducing amygdaloid lesions. Am. J. Physiol., 271: R1156–R1159.

King, B.M., Cook, J.T., Rossiter, K.N. and Rollins, B.L. (2003) Obesity-inducing amygdala lesions: examination of anterograde degeneration and retrograde transport. Am. J. Physiol. Regul. Integr. Comp. Physiol., 284: R965–R982.

Kopelman, P.G. and Hitman, G.A. (1998) Diabetes. Exploding type II. Lancet, 352(Suppl 4): SIV5.

Kovalchuk, Y., Hanse, E., Kafitz, K.W. and Konnerth, A. (2002) Postsynaptic induction of BDNF-mediated long-term potentiation. Science, 295: 1729–1734.

Levin, B.E., Dunn-Meynell, A. and Routh, V.H. (1999) Brain glucose sensing and body energy homeostasis: role in obesity and diabetes. Am. J. Physiol., 276: R1223–R1231.

Looker, H.C., Knowler, W.C. and Hanson, R.L. (2001) Changes in BMI and weight before and after the development of type 2 diabetes. Diabetes Care, 24: 1917–1922.

Low, P.A. (2003) Testing the autonomic nervous system. Semin. Neurol., 23: 407–421.

Magistretti, P.J., Pellerin, L., Rothman, D.L. and Shulman, R.G. (1999) Energy on demand. Science, 283: 496–497.

McEwen, B.S., Weiss, J.M. and Schwartz, L.S. (1968) Selective retention of corticosterone by limbic structures in rat brain. Nature, 220: 911–912.

Molina, P.E. and Abumrad, N.N. (2001) Contribution of excitatory amino acids to hypoglycemic counter-regulation. Brain Res., 899: 201–208.

Owen, O.E., Morgan, A.P., Kemp, H.G., Sullivan, J.M., Herrera, M.G. and Cahill Jr., G.F. (1967) Brain metabolism during fasting. J. Clin. Invest., 46: 1589–1595.

Pavlides, C. and McEwen, B.S. (1999) Effects of mineralocorticoid and glucocorticoid receptors on long-term potentiation in the CA3 hippocampal field. Brain Res., 851: 204–214.

Pavlides, C., Watanabe, Y., Magarinos, A.M. and McEwen, B.S. (1995) Opposing roles of type I and type II adrenal steroid receptors in hippocampal long-term potentiation. Neuroscience, 68: 387–394.

Pellerin, L. and Magistretti, P.J. (1994) Glutamate uptake into astrocytes stimulates aerobic glycolysis: a mechanism coupling neuronal activity to glucose utilization. Proc. Natl. Acad. Sci. USA, 91: 10625–10629.

Peters, A., Schweiger, U., Pellerin, L., Hubold, C., Oltmanns, K.M., Conrad, M., Schultes, B., Born, J. and Fehm, H.L. (2004) The selfish brain: competition for energy resources. Neurosci. Biobehav. Rev., 28: 143–180.

Rollins, B.L. and King, B.M. (2000) Amygdala-lesion obesity: what is the role of the various amygdaloid nuclei? Am. J. Physiol. Regul. Integr. Comp. Physiol., 279: R1348–R1356.

Schultes, B., Oltmanns, K.M., Kern, W., Fehm, H.L., Born, J. and Peters, A. (2003) Modulation of hunger by plasma glucose and metformin. J. Clin. Endocrinol. Metab., 88: 1133–1141.

Schwartz, M.W., Woods, S.C., Porte Jr., D., Seeley, R.J. and Baskin, D.G. (2000) Central nervous system control of food intake. Nature, 404: 661–671.

Seeley, R.J., Drazen, D.L. and Clegg, D.J. (2004) The critical role of the melanocortin system in the control of energy balance. Annu. Rev. Nutr., 24: 133–149.

Shi, S., Hayashi, Y., Esteban, J.A. and Malinow, R. (2001) Subunit-specific rules governing AMPA receptor trafficking to synapses in hippocampal pyramidal neurons. Cell, 105: 331–343.

Smith, D. and Amiel, S.A. (2002) Hypoglycaemia unawareness and the brain. Diabetologia, 45: 949–958.

Spanswick, D., Smith, M.A., Groppi, V.E., Logan, S.D. and Ashford, M.L. (1997) Leptin inhibits hypothalamic neurons by activation of ATP-sensitive potassium channels. Nature, 390: 521–525.

Spanswick, D., Smith, M.A., Mirshamsi, S., Routh, V.H. and Ashford, M.L. (2000) Insulin activates ATP-sensitive K + channels in hypothalamic neurons of lean, but not obese rats. Nat. Neurosci., 3: 757–758.

140

Strachan, M.W., Ewing, F.M., Frier, B.M., Harper, A. and Deary, I.J. (2004) Food cravings during acute hypoglycaemia in adults with Type 1 diabetes. Physiol. Behav., 80: 675–682.

Swanson, L.W. (2000) Cerebral hemisphere regulation of motivated behavior. Brain Res., 886: 113–164.

UKPDS Study Group. (1998) Intensive blood-glucose control with sulphonylureas or insulin compared with conventional treatment and risk of complications in patients with type 2 diabetes (UKPDS 33). Lancet, 352: 837–853.

van Rossum, E.F. and Lamberts, S.W. (2004) Polymorphisms in the glucocorticoid receptor gene and their associations with metabolic parameters and body composition. Recent Prog. Horm. Res., 59: 333–357.

Woods, S.C., Stein, L.J., McKay, L.D. and Porte Jr., D. (1984) Suppression of food intake by intravenous nutrients and insulin in the baboon. Am. J. Physiol., 247: R393–R401.

Yeo, G.S., Connie Hung, C.C., Rochford, J., Keogh, J., Gray, J., Sivaramakrishnan, S., O'Rahilly, S. and Farooqi, I.S. (2004) A de novo mutation affecting human TrkB associated with severe obesity and developmental delay. Nat. Neurosci., 7: 1187–1189.

Zhang, Y., Proenca, R., Maffei, M., Barone, M., Leopold, L. and Friedman, J.M. (1994) Positional cloning of the mouse obese gene and its human homologue. Nature, 372: 425–432.

Kalsbeek, Fliers, Hofman, Swaab, Van Someren & Buijs
Progress in Brain Research, Vol. 153
ISSN 0079-6123

CHAPTER 8

Integration of metabolic stimuli in the hypothalamic arcuate nucleus

M. van den Top* and D. Spanswick

Division of Clinical Sciences, Warwick Medical School, The University of Warwick, Coventry CV4 7AL, UK

Abstract: Integration of peripheral and central anabolic and catabolic inputs within the hypothalamic arcuate nucleus (ARC) is believed to be central to the maintenance of energy balance. In order to perform this complex task, neurons in the ARC express receptors for all major humoral and central transmitters involved in the maintenance of energy homeostasis. The integration of these inputs occurs at the cellular and circuit level and the resulting electrical output forms the origins for the activation of feeding and energy balance-related networks. Here, we discuss the role that active intrinsic membrane conductances, K_{ATP} channels and intracellular second messenger systems play in the integration of metabolic stimuli at the cellular level in the ARC. We conclude that the research into the integration of hunger and satiety signals in the ARC has made substantial progress in the last decade, but we are far from unraveling the complex neuronal networks involved in the maintenance of energy homeostasis. The diverse range of inputs, neuronal integrative properties, targets, output signals and how these signals relate to the physiological output provides us with a colossal challenge for years to come. However, to battle the current obesity epidemic, target-specific drugs need to be developed for which the knowledge of neuronal pathways involved in the maintenance of energy homeostasis will be crucial.

Keywords: arcuate nucleus; active membrane conductances; ATP-sensitive potassium channels; leptin; glucose; PI3K

Energy homeostasis is maintained by balancing food intake with energy expenditure resulting in consistent levels of stored energy in the form of fat. In order to maintain this balance the central nervous system (CNS) needs to integrate information originating from the periphery that signals energy status at the level of the central metabolic drive. Thus, a complex feedback system has evolved that utilizes humoral and neuronal signaling to inform the CNS of the whole body metabolic status. Hormones such as insulin, leptin, ghrelin and Peptide YY_{3-36} (PYY_{3-36}) are secreted from the periphery in quantities related to metabolic status and vagal

pathways relay information from the gut back to the CNS, through the brainstem (for review see Broberger, 2005). Subsequently, neurons and their associated circuits within function-specific areas of the autonomic subdivision of the CNS sense, process and formulate anabolic or catabolic outputs. These outputs, as part of a complex neuronal network, drive the appropriate changes relative to the energy status of the organism and as such maintain energy homeostasis. Aspects of the CNS therefore function as key integrative centers responsible for information processing and the coordination of appropriate output. The arcuate nucleus (ARC) of the hypothalamus is such an integrative center.

*Corresponding author. Tel.: +44 2476 572538; Fax: +44 2476 523701; E-mail: Marco.van-den-Top@warwick.ac.uk

The best described anabolic neurons residing in the ARC use Neuropeptide Y and agouti-related protein (NPY/AgRP) as neurotransmitters/neuromodulators (Broberger et al., 1998; Morton and Schwartz, 2001). Proopiomelanocortin (POMC)-expressing neurons, which can potentially cleave a range of neurotransmitters, are by far the best described catabolic neurons (Cone, 2005). Together, these two populations of neurons form parallel and opposing pathways that work together to maintain energy balance. However, a number of different neuronal networks originating within the ARC are involved in the maintenance of energy balance (Howard et al., 2000; Gottsch et al., 2004), although an in-depth understanding of the functioning of these networks is presently lacking. Within the ARC, the complexity of the neuronal networks controlling energy balance is reflected by the inputs, outputs and integrative properties of individual ARC neurons. At the cellular level, the physiological properties of ARC neurons and cellular mechanisms by which they integrate the multitude of orexigenic and anorexigenic signals remains poorly understood. A number of studies have shown that gene-expression characteristics of ARC neurons change in response to altered energy status or hormones, such as leptin and insulin (Schwartz et al., 1997; Kristensen et al., 1998; Mizuno et al., 1998; Mizuno and Mobbs, 1999; Howard et al., 2000; Jureus et al., 2000; Shimokawa et al., 2003). The changes in the expression levels of neuronal neuropeptidergic transmitters under various metabolic regimens have been shown to be a reliable tool to predict the overt function of these neurons. However, how these changes in metabolic status translate into appropriate modulation of electrical output and neurotransmitter release onto target neurons is unclear.

The electrical activity generated in ARC neurons depends on the neurons' intrinsic electrical properties, the type of input, the spatial organization and functional relationship between these properties and the relationship of the individual neuron with other component neurons forming the function-specific circuits. Taken together these neuronal properties are crucial for a neuron and its associated circuit to function as an integrative center for metabolic stimuli. The complexity of this integrative process and the resulting output signal is reflected in the diversity of neuronal components controlling this process and the extent of the input and output pathways involved. A detailed review of the efferent and afferent pathways associated with the ARC is outside the remit of this chapter (for comprehensive reviews of this subject, see Chronwall, 1985; Broberger and Hokfelt, 2001) and thus we will highlight only a few facets of these pathways to illustrate the complexity of the integration processes performed by ARC neurons at the single cell level.

To illustrate the complexity of the inputs to the ARC we will touch on the inputs originating from lateral hypothalamus (LH) and circulating factors. Such inputs utilize a combination of transmitters including fast neurotransmitters (gamma aminobutyric acid (GABA) and glutamate), neuropeptides (including orexin, melanin-concentrating hormone (MCH), Dynorphin-A, cocaine- and amphetamine-regulated transcript (CART) and in the case of circulating factors, hormones such as leptin, insulin, PYY_{3-36} and ghrelin (Elias et al., 1998; Peyron et al., 1998; Broberger, 1999; Date et al., 1999; Horvath et al., 1999; Chou et al., 2001; Elias et al., 2001; Hewson et al., 2002; Rosin et al., 2003). The general circulation contains a range of messengers that modulate the central metabolic drive through an action on neurons in the CNS including leptin, insulin, ghrelin and glucose (for review, see Woods et al., 1996; Elmquist et al., 1998; Horvath et al., 2001; Levin et al., 2004). The ARC is ideally situated and equipped to sense these humoral factors as it is located in a part of the brain with an impaired blood–brain barrier and neurons in the ARC express some of the highest levels in the CNS of receptors for circulating factors, including the anorexigens leptin (Ob-Rb) and insulin, and the orexigen ghrelin (growth hormone secretagogue receptor) (van Houten et al., 1980; Mercer et al., 1996; Willesen et al., 1999; Shuto et al., 2002). The humoral factors are all released in quantities relative to the nutritional and peripheral metabolic status, whereas in the CNS neurotransmitters are differentially released following the integration of the peripheral signals within the central drive. The inputs to the ARC of central origin comprise a multitude of transmitters and originate from a range of sources as outlined above. These transmitters are commonly colocalized, for example orexin colocalizes with glutamate and dynorphin-A, while CART colocalizes with MCH (Broberger, 1999; Chou et al., 2001; Rosin et al., 2003). The LH is classically referred to as a feeding center, which is highlighted by the fact that the majority of neuropeptides expressed in this hypothalamic nucleus are orexigenic. However, CART is an anorexigenic neuropeptide which is coexpressed with the orexigenic neuropeptide MCH (Qu et al., 1996; Kristensen et al., 1998; Shimada et al., 1998; Broberger, 1999). The functional significance of this coexpression of peptides with apparently opposing functions is presently unclear. The output of ARC neurons is distributed to functionally related areas in the CNS including: other hypothalamic nuclei such as the LH, paraventricular nucleus (PVN) and dorsomedial nucleus (DMN) with which the ARC reciprocally interacts; the brainstem and spinal cord (Bai et al., 1985; Baker and Herkenham, 1995; Beck, 2001). In addition to the NPY/AgRP and POMC-derived peptides, other energy balance-related neurons originating here utilize galanin-like peptide (GALP), neuromedin U and CART that colocalizes with POMC. (Chronwall, 1985; Baker and Herkenham, 1995; Broberger et al., 1998; Legradi and Lechan, 1999; DeFalco et al., 2001) These neuropeptides are coexpressed with fast neurotransmitters GABA and glutamate as well as other neuropeptides, thus adding to the complexity, diversity and computational capability of these neurons and their associated circuits (Horvath et al., 1997; Elias et al., 1998; Hahn et al., 1998; Cowley et al., 2001; Collin et al., 2003; Hentges et al., 2004). How the differential release of neurotransmitters in the CNS relates to the dynamics of the neuronal activity of the neurons expressing them is poorly understood. For example, in neurosecretory neurons the release of neuropeptides has been shown to be dependent on the pattern of neuronal activity rather than the rate of activity (Dutton and Dyball, 1979; Bicknell and Leng, 1981; Poulain and Wakerley, 1982). In order to understand the role ARC neurons play in coordinating the central response to maintain energy balance, a greater appreciation of the functional organization of these neurons and their circuits, the dynamics of their output and

input and the cellular signal transduction mechanisms by which these neurons integrate metabolic signals is required. In this chapter, we will focus on the integrative properties of ARC neurons at a membrane level and signal transduction level.

Electrophysiological properties of ARC neurons

The function, integration and computational capability of a neuron depends on its architecture and passive and active membrane properties, all of which contribute to its ability to instigate, modulate and set limits of sensitivity to detection of external stimuli. We have studied the electrophysiological and morphological properties of large number of ARC neurons maintained in rat hypothalamic slice preparations in vitro. ARC neurons have cell body sizes ranging from 10 to 30 μm with one to five fine neuronal projections. These neurons express a diversity of membrane resistances or input conductances ranging from 420 to 4667 MΩ suggesting heterogeneity within the population. Similarly, ARC neurons express a range of active membrane conductances of which several are readily identifiable in the current–voltage relationships (Fig. 1) These active conductances are differentially expressed, again indicating functional heterogeneity. In the absence of active membrane conductances negative current injections induced membrane potential responses directly proportional to the amount of current injected. However, the activation of certain active conductances results in nonlinear relationships between current and voltage and/or modulates the return to resting membrane potential following hyperpolarization. Differentially expressed active conductances observed in ARC neurons include anomalous inward rectification (I_{AN}), hyperpolarization-activated non-selective cation conductances (I_H), transient outward rectification (I_A) and low voltage-activated T-type calcium conductances (I_T) (Fig. 1).

In some ARC neurons, the decrease in input resistance at more negative membrane potentials was instantaneous and non-inactivating and indicated by a decrease in the slope of the plot of the current–voltage relationship (Fig. 1Ai). The conductance resembles those described in other

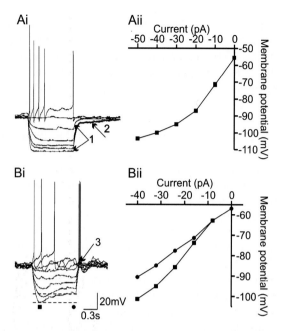

Ai

Aii

Current (pA)
-50 -40 -30 -20 -10 0

Membrane potential (mV)
-50
-60
-70
-80
-90
-100
-110

Bi

Bii

Current (pA)
-40 -30 -20 -10 0

Membrane potential (mV)
-60
-70
-80
-90
-100

20mV
0.3s

Fig. 1. Active conductances in ARC neurons. Ai: Current–voltage (*I–V*) relationship of an ARC neuron expressing anomalous inward rectifying and A-like potassium conductances. The plot shows superimposed traces from a continuous current clamp recording in which negative rectangular-wave current injections of constant increment induced a non-linear decrease in input resistance at more negative membrane potentials as a result of the activation of I_{AN}. Arrow (1) indicates the decrease in the peak amplitude of membrane response at negative holding potentials (downward arrow) relative to the responses seen near resting membrane potential (upward arrow). Arrow (2) indicates the delayed return to baseline following negative current injection as a result of the activation of an A-like conductance. Aii: Plot of the *I–V* relationship obtained from the neuron shown in Ai. The reduced slope of the plot at more negative holding potentials is the result of the activation of I_{AN}. Bi: *I–V* relationship of an ARC neuron expressing I_H and I_T. Note the sag in the membrane response during negative current injection as a result of the activation of I_H. Rebound depolarizations were induced following the release of the holding current due to the activation of an I_T and the slow inactivation of I_H (Arrow (1)). Bii: Plot of the current–voltage relationship of the neuron shown in Bi.

neurons of the CNS including hypothalamic nuclei following activation of the so-called "anomalous inward rectifier," which comprises potassium channels that effectively allow passage of potassium ions against its concentration gradient (Constanti and Galvan, 1983; Tasker and Dudek, 1991). The function of I_{AN} is unclear, but as it acts to prevent cells becoming "too" hyperpolarized,

this conductance has been suggested to maintain the membrane potential of neurons in a functional range through its activation at negative membrane potentials. At the break of negative current injection a subpopulation of ARC neurons show a rebound excitation. The properties of the rebound depolarization are similar to those described resulting from the activation of I_T (Llinas and Yarom, 1981; Llinas and Jahnsen, 1982). I_T are implicated in the generation of burst-like firing patterns in neurons of the CNS (Huguenard, 1996). Moreover, I_T can modulate the properties of synaptic inputs in the form of both inhibitory post-synaptic potentials (IPSPs) and excitatory post-synaptic potentials (EPSPs) (Huguenard and Prince, 1994; Gillessen and Alzheimer, 1997). Another conductance differentially expressed in ARC neurons is a time- and voltage-dependent inward rectification. The characteristics of the inward rectification resemble the features of I_H (Halliwell and Adams, 1982; Pape, 1996). I_H is a non-selective cation conductance involved in the generation of pacemaker activity, as described in the thalamus (Pape, 1996), and for its ability to modulate synaptic inputs (Magee, 1998, 1999). Transient outward rectifying conductances are a further feature of a subpopulation of ARC neurons. The conductance, observed at the break of the membrane response to negative current injection in ARC neurons as a delayed return to rest of the membrane potential, resembles the activation of A-like potassium conductances (Fig. 1; Connor and Stevens, 1971; Rogawski, 1985). Indeed, A-like conductances have been identified in a number of other hypothalamic nuclei and mouse ARC neurons (Bourque, 1988; Bouskila and Dudek, 1995; Luther et al., 2000; Burdakov et al., 2003). I_A channels are classically involved in the repolarizing phase of the action potential and are thus important in the regulation of repetitive firing (Rogawski, 1985). Mouse and most rat Arc neurons express an A-like conductance similar to those classically reported for these currents (Burdakov et al., 2003 and our unpublished observations).

The type, properties and location of active conductances expressed together with the location, size and kinetics of the synaptic input determine the final integrative properties of ARC neurons. In this respect the subset of ARC NPY/AgRP-expressing neurons in the ARC that contain an orexigen-sensitive pacemaker which we recently identified are of interest (van den Top et al., 2004). These neurons generated a burst-firing pattern in the presence of the orexigens, orexin and ghrelin (Fig. 2) and are inhibited by the catabolic hormone leptin (Fig. 5A). Within these neurons the active

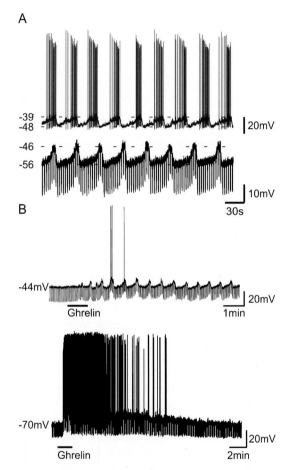

Fig. 2. Pacemaker activity in the ARC. A: Trace showing typical burst firing as observed in orexigen-sensitive NPY/AgRP pacemaker neuron (top trace). Underlying oscillations in membrane potential were revealed by injection of negative current (bottom traces). B: Continuous current clamp recording showing burst firing induced in ARC pacemaker neurons in the presence of ghrelin. Subpopulations of ghrelin-responsive ARC neurons respond differentially in the presence of this orexigenic neuropeptide. The bottom trace shows an ARC neuron that is depolarized by ghrelin resulting in action potential firing in a repetitive manner.

conductances I_A and I_T play a key role in the generation and properties of the membrane potential oscillations underlying burst firing. Thus, blocking I_T prevents the generation of membrane potential oscillations, while blocking I_A increases the frequency of these events. The I_A underlying the pacing of membrane potential oscillations has unusual properties, compared to other ARC neurons, in that these conductances can be activated around and more negative than rest and show an unusually long inactivation. At present it is unclear what underlies these differential properties of I_A in pacemaker neurons compared to similar conductances in other neurons activated at more depolarized membrane potentials. Functionally, these neurons may form the origin of the central drive to feed although it remains to be established where these neurons project to and what transmitters are released through the generated burst firing. However, both orexin and ghrelin are believed to mediate their orexigenic effects through a central mechanism including the direct activation of NPY/AgRP neurons and subsequent downstream activation of NPY receptors (Dube et al., 2000; Jain et al., 2000; Yamanaka et al., 2000). Moreover, orexin neurons impinge directly onto ARC NPY neurons and ARC neurons, including pacemaker neurons, express orexin and ghrelin receptors (Horvath et al., 1999; Marcus et al., 2001; Chen et al., 2004; van den Top et al., 2004). Therefore, conditional pacemaker activity generated by NPY/AgRP neurons in response to orexigens may underpin the release of NPY and AgRP. Furthermore, the ability to modulate this activity through external metabolic and neurotransmitter signals provides a substrate to differentially regulate release thus adding to the level of functional plasticity and computational power of the system.

Modulation of synaptic inputs by active conductances

ARC neurons receive inputs from both GABAergic and glutamatergic neurons observed as IPSPs and EPSPs, respectively the origins of which remain to be determined (Fig. 3; Kiss et al., 2005).

146

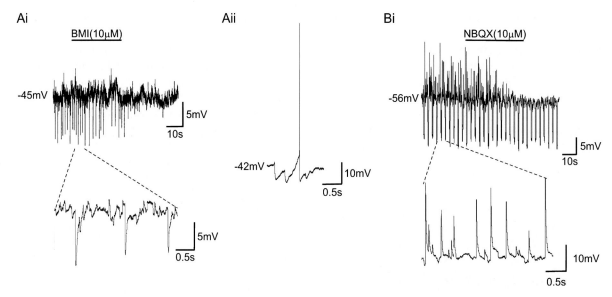

Fig. 3. Chemical synaptic inputs to ARC neurons. Ai: Continuous current clamp recording showing spontaneous inhibitory post-synaptic potentials (IPSPs, downward deflection) and the subsequent block of these events in the presence of bicuculline methiodide (BMI), a blocker of $GABA_A$ receptors. A section of the trace is shown under the trace on an expanded time scale. Aii: Continuous current clamp recording in which the summation of IPSPs results in the induction of a rebound excitation and the discharge of an action potential. Bi: Trace from a current clamp recording showing spontaneous excitatory post-synaptic potentials (EPSPs, upward deflections) and the subsequent block of these events in the presence of NBQX, a blocker of non-NMDA glutamate receptors. The downward deflections are membrane responses to repetitive negative rectangular wave current injections.

EPSPs are blocked by the application of 6-nitro-7-sulphamoylbenzo(f)-quinoxaline-2,3-dione (NBQX) suggesting them to be resulting from the activation of AMPA-type glutamate receptors (Fig. 3B). The IPSPs can be blocked with bicuculline methiodide (BMI), a $GABA_A$ receptor blocker, and IPSPs are generally assumed to reduce neuronal activity (Fig. 3Ai). However, functional significance of such inputs can only be realized when the intrinsic integrative properties of neurons are taken into account (Huguenard, 1996; Pape, 1996; Magee, 1998; van den Top et al., 2004). For example, as reported in other tissues, single or summated IPSPs can trigger rebound excitations in ARC and other neurons (Fig. 3Aii, Whyment et al., 2004). These rebound excitations have been reported to be mediated through the activation of I_H and/or I_T following the IPSP-induced transient membrane hyperpolarization (Huguenard and Prince, 1994).

Recently, leptin was shown to rapidly modulate the synaptic inputs to ARC neurons in that leptin increases the number of inhibitory synapses onto NPY/AgRP neurons, while reducing the excitatory inputs to POMC neurons (Pinto et al., 2004). However, to what extent this apparent synaptic plasticity plays in information processing in the ARC or its overall contribution to information processing is difficult to reconcile without understanding the degree of "hard-wiring" in the ARC pathways and the contribution of more conventional means of synaptic plasticity, such as pre- and post-synaptic modulation of inputs. Thus, precisely how neurons retract synaptic inputs and then rewire them in a functionally relevant manner to the post-synaptic cell to allow for integration by intrinsic conductances within the cell needs to be explored further. Indeed, in ghrelin excited and thus presumed orexigenic neurons in the ARC of Zucker *fa/fa* rats, which lack a functional long form of the Ob-Rb leptin receptor, we observe ghrelin-induced IPSPs and EPSPs similar to the ones observed in Zucker lean rats (Fig. 4; Phillips et al., 1996). These findings imply that the reported fast rewiring of the ARC does not occur in all neuronal populations and circuits of the ARC. As

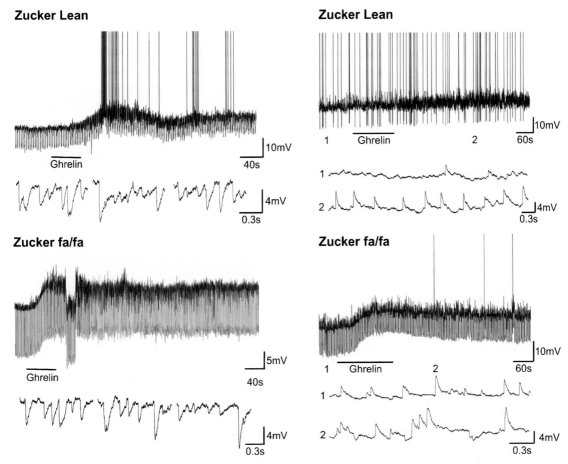

Fig. 4. Synaptic hardwiring in the ARC. Left: Continuous current clamp recordings showing induction of an excitation in conjunction with an increase in IPSPs in Zucker obese (*fa/fa*) and lean rats, respectively. The traces shown on an expanded time base and scale were taken at the peak of the response. Right: Trace showing the induction of EPSPs in Zucker lean and obese rats, respectively. Traces shown on an expanded time and voltage base are taken prior to (Top) and in the presence of ghrelin (Bottom). These findings suggest that the synaptic connections formed in the presence and absence of central leptin signaling are similar.

NPY and POMC neurons are not homogeneous neuronal populations, the discrepancy between our findings and those reported previously might partially lie in the fact that we have recorded from different neuronal subpopulations (Horvath et al., 1997; Collin et al., 2003; Hentges et al., 2004; our unpublished observations).

K_{ATP} channels and central integration of humoral factors

So far, we have discussed the effects that membrane components of ARC neurons play in the integration of inputs. However, intracellular signaling pathways are involved in this process too. The CNS needs to integrate peripheral inputs signaling the energy reserve status with central and peripheral short-term satiety/hunger inputs to maintain energy homeostasis. Classic examples of humoral factors that feedback information concerning the energy status are insulin, leptin and glucose. Insulin and leptin receptors are highly expressed in the ARC making this nucleus a prime target for the integration of these humoral factors into the central metabolic drive and maintain energy balance. The signal transduction pathways activated by insulin and leptin include common components as

148

observed in the periphery i.e. both involve the activation of phosphoinositide 3-kinase (PI3K) (Niswender and Schwartz, 2003). Thus, if leptin and insulin impinge on the same target they may be capable of modulating their respective functions as a result of the cross talk between their intracellular-signaling cascades. This could, for example explain central leptin and insulin resistance observed in the obese (Munzberg and Myers, 2005).

In the hypothalamus, the cross talk between the leptin and insulin intracellular-signaling pathways was originally investigated utilizing electrophysiological-recording techniques (Spanswick et al., 2000). Previously, Spanswick et al. (1997) had reported the activation of ATP-sensitive potassium channels (K_{ATP}) by leptin in glucose responsive neurons (GR-neurons) located in the ARC and the ventromedial nucleus of the hypothalamus (VMN). Subsequently, insulin was found to inhibit GR-neurons through a PI3K-dependent activation of K_{ATP} channels. Moreover, these two circulating anorectic hormones were shown to impinge on the same ARC neurons in a mutually exclusive manner. Finally, in line with previous findings in which insulin was shown to be incapable of reducing food intake, insulin was found unable to activate K_{ATP} in Zucker *fa/fa* rats (Ikeda et al., 1986; Spanswick et al., 2000).

Recently, the interest in both (i) the role of PI3K and (ii) the role that K_{ATP} channels play in the hypothalamic control of energy homeostasis has gained momentum.

(i) Studies investigating the role of PI3K in the leptin- and insulin-induced anorectic response that this enzyme is necessary for their anorectic response (Niswender et al., 2001, 2003). However, as discussed above the maintenance of energy balance involves subpopulations of neurons in the ARC of which neurons containing POMC (catabolic) and NPY/AgRP (anabolic) form parallel and apposing pathways that are differentially modulated by leptin. Thus, leptin/insulin inhibit the expression of NPY/AgRP, while they activate the expression of POMC in ARC neurons. Similarly, at the level of the cellular membrane leptin differentially modulates the neuronal activity of ARC POMC and NPY/AgRP neurons (Fig. 5A) (Cowley et al., 2001; van den Top et al., 2004). In

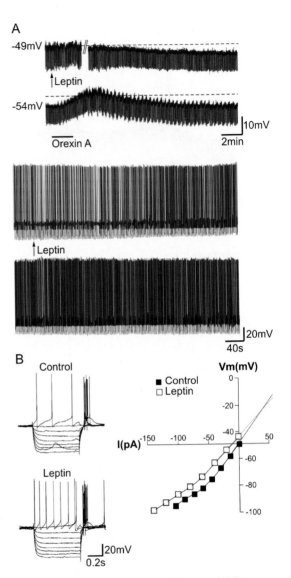

Fig. 5. Opposing effects of leptin on ARC neurons. A: (Top) Continuous current clamp recording showing the slow leptin-induced (50 nM) membrane potential hyperpolarization associated with a reduction in membrane resistance. In ARC pacemaker neurons, leptin induced a membrane potential hyperpolarization. Subsequent orexin application induced pacemaker activity. (Bottom) Continuous current clamp recording showing a leptin-induced (50 nM) excitation associated with a reduction in membrane resistance. B: Leptin excites ARC neurons through the activation of a non-selective cation channel. Superimposed plots of I–V relationships obtained in the presence and absence of leptin. Plot of these I–V relationships reveals an extrapolated reversal potential for the leptin-induced conductance of ~ −25 mV.

agreement with previous experiments performed on POMC neurons in mice, we also found in rat that the leptin-induced excitation is induced through the activation of a non-selective cation channel (Fig. 5B; Cowley et al., 2001). Both NPY/ AgRP and POMC neurons express leptin and insulin receptors thus raising the question, how identical receptors can impose opposite neuronal responses. Moreover, in both these neuronal populations leptin and insulin application results in a modulation of the PI3K activation level (Xu et al., 2005). Unfortunately, at present it is unclear how PI3K differentially modulates neuronal activity in NPY and POMC neurons. However, recently it has been shown that this enzyme mediates leptin-induced inhibition of NPY and AgRP levels in the ARC. Logically, the neuronal activity will follow/ induce the changes in neuropeptide expression levels in order to achieve release, thus suggesting that the inhibition of NPY/AgRP neurons is likely to be mediated through an effect mediated via PI3K.

(ii) Studies investigating the role the mediobasal hypothalamus in the control of hepatic glucose production have revealed a key role for K_{ATP} channels (Obici and Rossetti, 2003). Neurons sensing circulating levels of insulin and long-chain fatty acids are located in the mediobasal hypothalamus, which utilize K_{ATP} channels to modulate their neuronal activity in order to control hepatic glucose production (Obici et al., 2002a, b, c; Lam et al., 2005; Pocai et al., 2005). These findings, in combination with our previous results regarding the action of leptin and insulin in the ARC and VMN, imply K_{ATP} channels to be a common neuronal target for humoral factors in the CNS. However, although there are binding sites for fatty acids on K_{ATP} channels, it remains to be shown that this results in the modulation of these potassium channels in the ARC (Branstrom et al., 1997). Moreover, channels are widely expressed in the CNS including the hypothalamic ARC and do not form a functionally homogeneous population. In the ARC, electrophysiological studies have shown that diazoxide-sensitive K_{ATP} channels are expressed in the majority of neurons including neurons that express NPY/AgRP and POMC, respectively (Fig. 6; Ibrahim et al., 2003; Thomzig et al., 2005; our unpublished observations). Thus,

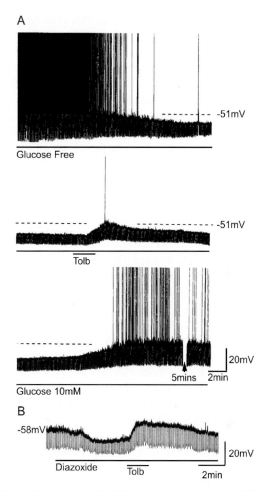

Fig. 6. The majority of ARC neurons express functional K_{ATP} channels. A: Continuous recording in current clamp (top three traces) showing the induction of a membrane hyperpolarization associated with a reduction in membrane resistance following a reduction of the extracellular glucose concentration from 10 to 0 mM (glucose free). The downward deflections are the membrane potential responses to repetitive negative rectangular wave current injections utilized to monitor input resistance. Note that the membrane response is gradually reduced as the membrane hyperpolarization progresses, indicating a reduction in membrane resistance. The application of the K_{ATP} channel blocker tolbutamide (100 μM) or the reintroduction of 10 mM glucose reverses the glucose free-induced response. This response was observed in 70% of the ARC neurons tested and included orexigen-sensitive NPY/AgRP pacemaker neurons. B: Current clamp recording performed in the presence of the sodium channel blocker tetrodotoxin (TTX) to electrically isolate the neuron, showing that diazoxide (500 μM) directly induced hyperpolarization of ARC neurons through the activation of K_{ATP} channels. Note the reduction in input resistance associated with this response. The application of tolbutamide (200 μM) reverses the effects of diazoxide.

central diazoxide application as reported recently to elucidate the role of central K_{ATP} channels in the control of hepatic glucose production is likely to result in the inhibition of a number of functionally distinct neuronal populations in the ARC. In order to have a coordinated output from the ARC-opposing pathways like the NPY/AgRP and POMC, neurons need to be differentially modulated. However, how this is achieved remains to be clarified.

In addition to representing a functional endpoint for leptin and insulin signaling, the K_{ATP} channel has classically been linked to glucose-sensing both in the periphery and in the CNS (Levin et al., 2004; Ashcroft, 2005). Centrally, the VMN has historically attracted the most attention but recent studies indicate a potential role for the ARC. The compromised blood–brain barrier here coupled with the widespread expression of K_{ATP} channels render this nucleus ideally situated and equipped to detect changes in circulating glucose levels. A recent study suggested that ARC neurons can detect physiological changes in glucose levels via a K_{ATP}-dependent mechanism (Wang et al., 2004). However, the inability to identify the functional phenotype of the neurons responding, given the widespread distribution of the channels, renders the observations difficult to interpret in a functional context. In order to elucidate the role of K_{ATP} channels expressed in ARC neurons in the maintenance of energy homeostasis, we need to establish how these channels are differentially activated under physiologically relevant conditions.

Future perspectives

Since the discovery of leptin, the level of research activity and developments in our understanding of the central control of energy balance has expanded at an incredible rate. Unfortunately, so has our appreciation of the level of complexity we are dealing with. While our knowledge concerning the neurotransmitters and neurohormonal signals involved, their receptors and signal transduction mechanisms has improved considerably our understanding of the functional organization and operation of the neuronal components and their

circuits through which these systems operation has lagged behind.

Electrophysiological studies in the ARC have described the effects of many feeding-related ligands on the electrical activity of ARC neurons including NPY, orexin, leptin, insulin, glucose, cholecystokinin and opioids (Kelly et al., 1990; Spanswick et al., 1997, 2000; Kinney et al., 1998; Cowley et al., 2001; Burdakov and Ashcroft, 2002; Burdakov et al., 2003; Ibrahim et al., 2003; Poulain et al., 2003; Roseberry et al., 2004; Wang et al., 2004). More recently, the development of transgenic mice that selectively express green fluorescent protein in POMC or NPY-containing neurons and single cell reverse transcriptase-polymerase chain reaction (RT-PCR) techniques has provided us with powerful tools to study functionally different neuronal populations (Cowley et al., 2001; Ibrahim et al., 2003; van den Top et al., 2004; Acuna-Goycolea et al., 2005; Takahashi and Cone, 2005). These latter advances will undoubtedly assist in the quest to understand the fundamental properties of ARC neurons and associated circuits controlling energy balance and the cellular- and molecular-signaling mechanisms by which these neurons coordinate a central response to perturbations in energy balance. However, significant investment is required to understand the operation of these neurons, their organization and operation in function-specific circuits and how these component parts detect, process, integrate and formulate appropriate output to perturbations in energy balance. The introduction of mathematical-modeling techniques and the development of predictive models will also greatly assist in the evolution of our understanding of the central control of energy balance and how disruption or dysfunction to this system manifests as obesity.

Abbreviations

AgRP	agouti-related peptide
AMPA	α-amino-3-hydroxy-5-methyl-4-isoxazole proprionic acid
ARC	arcuate nucleus
ATP	adenosine triphosphate
BMI	bicuculline methiodide

CART	cocaine- and amphetamine-regulated transcript
CNS	central nervous system
DMN	dorsomedial nucleus of the hypothalamus
EPSPs	excitatory post-synaptic potentials
GABA	gamma aminobutyric acid
GALP	galanin-like peptide
GR-neurons	glucose responsive neurons
I_A	transient outward rectification
I_{AN}	anomalous inward rectification
I_H	hyperpolarization-activated non-selective cation channel
I_T	low voltage-activated T-type calcium conductance
IPSPs	inhibitory post-synaptic potentials
K_{ATP}	ATP-sensitive potassium channels
LH	lateral hypothalamus
MCH	melanin concentrating hormone
α-MSH	α-melanocyte stimulating hormone
NBQX	6-nitro-7-sulphamoylbenzo(f)-quinoxaline-2,3-dione
NMU	neuromedin U
NPY	neuropeptide Y
Ob-Rb	long-form of the leptin receptor
PI3K	phosphoinositide 3-kinase
POMC	proopiomelanocortin
PVN	paraventricular nucleus of the hypothalamus
PYY_{3-36}	peptide YY_{3-36}
RT-PCR	reverse transcriptase-polymerase chain reaction
TTX	tetrodotoxin
VMN	ventromedial nucleus of the hypothalamus

Acknowledgments

The financial support from the Biotechnology and Biological Sciences Research Council (BBSRC), the Foundation for Prader-Willi Research and British Heart Foundation (BHF) is gratefully acknowledged.

References

Acuna-Goycolea, C., Tamamaki, N., Yanagawa, Y., Obata, K. and van den Pol, A.N. (2005) Mechanisms of neuropeptide Y, peptide YY, and pancreatic polypeptide inhibition of identified green fluorescent protein-expressing GABA neurons in the hypothalamic neuroendocrine arcuate nucleus. J. Neurosci., 25: 7406–7419.

Ashcroft, F.M. (2005) ATP-sensitive potassium channelopathies: focus on insulin secretion. J. Clin. Invest., 115: 2047–2058.

Bai, F.L., Yamano, M., Shiotani, Y., Emson, P.C., Smith, A.D., Powell, J.F. and Tohyama, M. (1985) An arcuato-paraventricular and -dorsomedial hypothalamic neuropeptide Y-containing system which lacks noradrenaline in the rat. Brain Res., 331: 172–175.

Baker, R.A. and Herkenham, M. (1995) Arcuate nucleus neurons that project to the hypothalamic paraventricular nucleus: neuropeptidergic identity and consequences of adrenalectomy on mRNA levels in the rat. J. Comp. Neurol., 358: 518–530.

Beck, B. (2001) KOs and organisation of peptidergic-feeding behavior mechanisms. Neurosci. Biobehav. Rev., 25: 143–158.

Bicknell, R.J. and Leng, G. (1981) Relative efficiency of neural firing patterns for vasopressin release in vitro. Neuroendocrinology, 33: 295–299.

Bourque, C.W. (1988) Transient calcium-dependent potassium current in magnocellular neurosecretory cells of the rat supraoptic nucleus. J. Physiol., 397: 331–347.

Bouskila, Y. and Dudek, F.E. (1995) A rapidly activating type of outward rectifier K+ current and A-current in rat suprachiasmatic nucleus neurones. J. Physiol., 488(Part 2): 339–350.

Branstrom, R., Corkey, B.E., Berggren, P.O. and Larsson, O. (1997) Evidence for a unique long chain acyl-CoA ester-binding site on the ATP-regulated potassium channel in mouse pancreatic beta cells. J. Biol. Chem., 272: 17390–17394.

Broberger, C. (1999) Hypothalamic cocaine- and amphetamine-regulated transcript (CART) neurons: histochemical relationship to thyrotropin-releasing hormone, melanin-concentrating hormone, orexin/hypocretin and neuropeptide Y. Brain Res., 848: 101–113.

Broberger, C. (2005) Brain regulation of food intake and appetite: molecules and networks. J. Intern. Med., 258: 301–327.

Broberger, C. and Hokfelt, T. (2001) Hypothalamic and vagal neuropeptide circuitries regulating food intake. Physiol. Behav., 74: 669–682.

Broberger, C., Johansen, J., Johansson, C., Schalling, M. and Hokfelt, T. (1998) The neuropeptide Y/agouti gene-related protein (AGRP) brain circuitry in normal, anorectic, and monosodium glutamate-treated mice. Proc. Natl. Acad. Sci. USA, 95: 15043–15048.

Burdakov, D. and Ashcroft, F.M. (2002) Cholecystokinin tunes firing of an electrically distinct subset of arcuate nucleus

152

neurons by activating A-type potassium channels. J. Neurosci., 22: 6380–6387.

Burdakov, D., Liss, B. and Ashcroft, F.M. (2003) Orexin excites GABAergic neurons of the arcuate nucleus by activating the sodium–calcium exchanger. J. Neurosci., 23: 4951–4957.

Chen, H.Y., Trumbauer, M.E., Chen, A.S., Weingarth, D.T., Adams, J.R., Frazier, E.G., Shen, Z., Marsh, D.J., Feighner, S.D., Guan, X.M., Ye, Z., Nargund, R.P., Smith, R.G., Van der Ploeg, L.H., Howard, A.D., MacNeil, D.J. and Qian, S. (2004) Orexigenic action of peripheral ghrelin is mediated by neuropeptide Y and agouti-related protein. Endocrinology, 145: 2607–2612.

Chou, T.C., Lee, C.E., Lu, J., Elmquist, J.K., Hara, J., Willie, J.T., Beuckmann, C.T., Chemelli, R.M., Sakurai, T., Yanagisawa, M., Saper, C.B. and Scammell, T.E. (2001) Orexin (hypocretin) neurons contain dynorphin. J. Neurosci., 21: RC168.

Chronwall, B.M. (1985) Anatomy and physiology of the neuroendocrine arcuate nucleus. Peptides, 6(Suppl 2): 1–11.

Collin, M., Backberg, M., Ovesjo, M.L., Fisone, G., Edwards, R.H., Fujiyama, F. and Meister, B. (2003) Plasma membrane and vesicular glutamate transporter mRNAs/proteins in hypothalamic neurons that regulate body weight. Eur. J. Neurosci., 18: 1265–1278.

Cone, R.D. (2005) Anatomy and regulation of the central melanocortin system. Nat. Neurosci., 8: 571–578.

Connor, J.A. and Stevens, C.F. (1971) Voltage clamp studies of a transient outward membrane current in gastropod neural somata. J. Physiol., 213: 21–30.

Constanti, A. and Galvan, M. (1983) Fast inward-rectifying current accounts for anomalous rectification in olfactory cortex neurones. J. Physiol., 335: 153–178.

Cowley, M.A., Smart, J.L., Rubinstein, M., Cerdan, M.G., Diano, S., Horvath, T.L., Cone, R.D. and Low, M.J. (2001) Leptin activates anorexigenic POMC neurons through a neural network in the arcuate nucleus. Nature, 411: 480–484.

Date, Y., Ueta, Y., Yamashita, H., Yamaguchi, H., Matsukura, S., Kangawa, K., Sakurai, T., Yanagisawa, M. and Nakazato, M. (1999) Orexins, orexigenic hypothalamic peptides, interact with autonomic, neuroendocrine and neuroregulatory systems. Proc. Natl. Acad. Sci. USA, 96: 748–753.

DeFalco, J., Tomishima, M., Liu, H., Zhao, C., Cai, X., Marth, J.D., Enquist, L. and Friedman, J.M. (2001) Virus-assisted mapping of neural inputs to a feeding center in the hypothalamus. Science, 291: 2608–2613.

Dube, M.G., Horvath, T.L., Kalra, P.S. and Kalra, S.P. (2000) Evidence of NPY Y5 receptor involvement in food intake elicited by orexin A in sated rats. Peptides, 21: 1557–1560.

Dutton, A. and Dyball, R.E. (1979) Phasic firing enhances vasopressin release from the rat neurohypophysis. J. Physiol., 290: 433–440.

Elias, C.F., Lee, C., Kelly, J., Aschkenasi, C., Ahima, R.S., Couceyro, P.R., Kuhar, M.J., Saper, C.B. and Elmquist, J.K. (1998) Leptin activates hypothalamic CART neurons projecting to the spinal cord. Neuron, 21: 1375–1385.

Elias, C.F., Lee, C.E., Kelly, J.F., Ahima, R.S., Kuhar, M., Saper, C.B. and Elmquist, J.K. (2001) Characterization of

CART neurons in the rat and human hypothalamus. J. Comp. Neurol., 432: 1–19.

Elmquist, J.K., Maratos-Flier, E., Saper, C.B. and Flier, J.S. (1998) Unraveling the central nervous system pathways underlying responses to leptin. Nat. Neurosci., 1: 445–450.

Gillessen, T. and Alzheimer, C. (1997) Amplification of EPSPs by low Ni(2+)- and amiloride-sensitive Ca2+ channels in apical dendrites of rat CA1 pyramidal neurons. J. Neurophysiol., 77: 1639–1643.

Gottsch, M.L., Clifton, D.K. and Steiner, R.A. (2004) Galanin-like peptide as a link in the integration of metabolism and reproduction. Trends Endocrinol. Metab., 15: 215–221.

Hahn, T.M., Breininger, J.F., Baskin, D.G. and Schwartz, M.W. (1998) Coexpression of Agrp and NPY in fasting-activated hypothalamic neurons. Nat. Neurosci., 1: 271–272.

Halliwell, J.V. and Adams, P.R. (1982) Voltage-clamp analysis of muscarinic excitation in hippocampal neurons. Brain Res., 250: 71–92.

Hentges, S.T., Nishiyama, M., Overstreet, L.S., Stenzel-Poore, M., Williams, J.T. and Low, M.J. (2004) GABA release from proopiomelanocortin neurons. J. Neurosci., 24: 1578–1583.

Hewson, A.K., Tung, L.Y., Connell, D.W., Tookman, L. and Dickson, S.L. (2002) The rat arcuate nucleus integrates peripheral signals provided by leptin, insulin, and a ghrelin mimetic. Diabetes, 51: 3412–3419.

Horvath, T.L., Bechmann, I., Naftolin, F., Kalra, S.P. and Leranth, C. (1997) Heterogeneity in the neuropeptide Y-containing neurons of the rat arcuate nucleus: GABAergic and non-GABAergic subpopulations. Brain Res., 756: 283–286.

Horvath, T.L., Diano, S., Sotonyi, P., Heiman, M. and Tschop, M. (2001) Minireview: ghrelin and the regulation of energy balance — a hypothalamic perspective. Endocrinology, 142: 4163–4169.

Horvath, T.L., Diano, S. and van den Pol, A.N. (1999) Synaptic interaction between hypocretin (orexin) and neuropeptide Y cells in the rodent and primate hypothalamus: a novel circuit implicated in metabolic and endocrine regulations. J. Neurosci., 19: 1072–1087.

Howard, A.D., Wang, R., Pong, S.S., Mellin, T.N., Strack, A., Guan, X.M., Zeng, Z., Williams Jr., D.L., Feighner, S.D., Nunes, C.N., Murphy, B., Stair, J.N., Yu, H., Jiang, Q., Clements, M.K., Tan, C.P., McKee, K.K., Hreniuk, D.L., McDonald, T.P., Lynch, K.R., Evans, J.F., Austin, C.P., Caskey, C.T., Van der Ploeg, L.H. and Liu, Q. (2000) Identification of receptors for neuromedin U and its role in feeding. Nature, 406: 70–74.

Huguenard, J.R. (1996) Low-threshold calcium currents in central nervous system neurons. Annu. Rev. Physiol., 58: 329–348.

Huguenard, J.R. and Prince, D.A. (1994) Intrathalamic rhythmicity studied in vitro: nominal T-current modulation causes robust antioscillatory effects. J. Neurosci., 14: 5485–5502.

Ibrahim, N., Bosch, M.A., Smart, J.L., Qiu, J., Rubinstein, M., Ronnekleiv, O.K., Low, M.J. and Kelly, M.J. (2003) Hypothalamic proopiomelanocortin neurons are glucose responsive and express K(ATP) channels. Endocrinology, 144: 1331–1340.

Ikeda, H., West, D.B., Pustek, J.J., Figlewicz, D.P., Greenwood, M.R., Porte Jr., D. and Woods, S.C. (1986) Intraventricular insulin reduces food intake and body weight of lean but not obese Zucker rats. Appetite, 7: 381–386.

Jain, M.R., Horvath, T.L., Kalra, P.S. and Kalra, S.P. (2000) Evidence that NPY Y1 receptors are involved in stimulation of feeding by orexins (hypocretins) in sated rats. Regul. Peptides, 87: 19–24.

Jureus, A., Cunningham, M.J., McClain, M.E., Clifton, D.K. and Steiner, R.A. (2000) Galanin-like peptide (GALP) is a target for regulation by leptin in the hypothalamus of the rat. Endocrinology, 141: 2703–2706.

Kelly, M.J., Loose, M.D. and Ronnekleiv, O.K. (1990) Opioids hyperpolarize beta-endorphin neurons via mu-receptor activation of a potassium conductance. Neuroendocrinology, 52: 268–275.

Kinney, G.A., Emmerson, P.J. and Miller, R.J. (1998) Galanin receptor-mediated inhibition of glutamate release in the arcuate nucleus of the hypothalamus. J. Neurosci., 18: 3489–3500.

Kiss, J., Csaba, Z., Csaki, A. and Halasz, B. (2005) Glutamatergic innervation of neuropeptide Y and pro-opiomelanocortin-containing neurons in the hypothalamic arcuate nucleus of the rat. Eur. J. Neurosci., 21: 2111–2119.

Kristensen, P., Judge, M.E., Thim, L., Ribel, U., Christjansen, K.N., Wulff, B.S., Clausen, J.T., Jensen, P.B., Madsen, O.D., Vrang, N., Larsen, P.J. and Hastrup, S. (1998) Hypothalamic CART is a new anorectic peptide regulated by leptin. Nature, 393: 72–76.

Lam, T.K., Schwartz, G.J. and Rossetti, L. (2005) Hypothalamic sensing of fatty acids. Nat. Neurosci., 8: 579–584.

Legradi, G. and Lechan, R.M. (1999) Agouti-related protein containing nerve terminals innervate thyrotropin-releasing hormone neurons in the hypothalamic paraventricular nucleus. Endocrinology, 140: 3643–3652.

Levin, B.E., Routh, V.H., Kang, L., Sanders, N.M. and Dunn-Meynell, A.A. (2004) Neuronal glucosensing: what do we know after 50 years? Diabetes, 53: 2521–2528.

Llinas, R. and Jahnsen, H. (1982) Electrophysiology of mammalian thalamic neurones in vitro. Nature, 297: 406–408.

Llinas, R. and Yarom, Y. (1981) Electrophysiology of mammalian inferior olivary neurones in vitro. Different types of voltage-dependent ionic conductances. J. Physiol., 315: 549–567.

Luther, J.A., Halmos, K.C. and Tasker, J.G. (2000) A slow transient potassium current expressed in a subset of neurosecretory neurons of the hypothalamic paraventricular nucleus. J. Neurophysiol., 84: 1814–1825.

Magee, J.C. (1998) Dendritic hyperpolarization-activated currents modify the integrative properties of hippocampal CA1 pyramidal neurons. J. Neurosci., 18: 7613–7624.

Magee, J.C. (1999) Dendritic Ih normalizes temporal summation in hippocampal CA1 neurons. Nat. Neurosci., 2: 508–514.

Marcus, J.N., Aschkenasi, C.J., Lee, C.E., Chemelli, R.M., Saper, C.B., Yanagisawa, M. and Elmquist, J.K. (2001) Differential expression of orexin receptors 1 and 2 in the rat brain. J. Comp. Neurol., 435: 6–25.

Mercer, J.G., Hoggard, N., Williams, L.M., Lawrence, C.B., Hannah, L.T. and Trayhurn, P. (1996) Localization of leptin receptor mRNA and the long form splice variant (Ob-Rb) in mouse hypothalamus and adjacent brain regions by in situ hybridization. FEBS Lett., 387: 113–116.

Mizuno, T.M., Kleopoulos, S.P., Bergen, H.T., Roberts, J.L., Priest, C.A. and Mobbs, C.V. (1998) Hypothalamic pro-opiomelanocortin mRNA is reduced by fasting and [corrected] in ob/ob and db/db mice, but is stimulated by leptin. Diabetes, 47: 294–297.

Mizuno, T.M. and Mobbs, C.V. (1999) Hypothalamic agouti-related protein messenger ribonucleic acid is inhibited by leptin and stimulated by fasting. Endocrinology, 140: 814–817.

Morton, G.J. and Schwartz, M.W. (2001) The NPY/AgRP neuron and energy homeostasis. Int. J. Obes. Relat. Metab. Disord., 25(Suppl 5): S56–S62.

Munzberg, H. and Myers, M.G. (2005) Molecular and anatomical determinants of central leptin resistance. Nat. Neurosci., 8: 566–570.

Niswender, K.D., Morrison, C.D., Clegg, D.J., Olson, R., Baskin, D.G., Myers Jr., M.G., Seeley, R.J. and Schwartz, M.W. (2003) Insulin activation of phosphatidylinositol 3-kinase in the hypothalamic arcuate nucleus: a key mediator of insulin-induced anorexia. Diabetes, 52: 227–231.

Niswender, K.D., Morton, G.J., Stearns, W.H., Rhodes, C.J., Myers Jr., M.G. and Schwartz, M.W. (2001) Intracellular signalling. Key enzyme in leptin-induced anorexia. Nature, 413: 794–795.

Niswender, K.D. and Schwartz, M.W. (2003) Insulin and leptin revisited: adiposity signals with overlapping physiological and intracellular signaling capabilities. Front. Neuroendocrinol., 24: 1–10.

Obici, S., Feng, Z., Karkanias, G., Baskin, D.G. and Rossetti, L. (2002) Decreasing hypothalamic insulin receptors causes hyperphagia and insulin resistance in rats. Nat. Neurosci., 5: 566–572.

Obici, S., Feng, Z., Morgan, K., Stein, D., Karkanias, G. and Rossetti, L. (2002) Central administration of oleic acid inhibits glucose production and food intake. Diabetes, 51: 271–275.

Obici, S. and Rossetti, L. (2003) Minireview: nutrient sensing and the regulation of insulin action and energy balance. Endocrinology, 144: 5172–5178.

Obici, S., Zhang, B.B., Karkanias, G. and Rossetti, L. (2002) Hypothalamic insulin signaling is required for inhibition of glucose production. Nat. Med., 8: 1376–1382.

Pape, H.C. (1996) Queer current and pacemaker: the hyperpolarization-activated cation current in neurons. Annu. Rev. Physiol., 58: 299–327.

Peyron, C., Tighe, D.K., van den Pol, A.N., de Lecea, L., Heller, H.C., Sutcliffe, J.G. and Kilduff, T.S. (1998) Neurons containing hypocretin (orexin) project to multiple neuronal systems. J. Neurosci., 18: 9996–10015.

Phillips, M.S., Liu, Q., Hammond, H.A., Dugan, V., Hey, P.J., Caskey, C.J. and Hess, J.F. (1996) Leptin receptor missense mutation in the fatty Zucker rat. Nat. Genet., 13: 18–19.

154

Pinto, S., Roseberry, A.G., Liu, H., Diano, S., Shanabrough, M., Cai, X., Friedman, J.M. and Horvath, T.L. (2004) Rapid rewiring of arcuate nucleus feeding circuits by leptin. Science, 304: 110–115.

Pocai, A., Lam, T.K., Gutierrez-Juarez, R., Obici, S., Schwartz, G.J., Bryan, J., Aguilar-Bryan, L. and Rossetti, L. (2005) Hypothalamic K(ATP) channels control hepatic glucose production. Nature, 434: 1026–1031.

Poulain, D.A. and Wakerley, J.B. (1982) Electrophysiology of hypothalamic magnocellular neurones secreting oxytocin and vasopressin. Neuroscience, 7: 773–808.

Poulain, P., Decrocq, N. and Mitchell, V. (2003) Direct inhibitory action of galanin on hypothalamic arcuate nucleus neurones expressing galanin receptor Gal-r1 mRNA. Neuroendocrinology, 78: 105–117.

Qu, D., Ludwig, D.S., Gammeltoft, S., Piper, M., Pelleymounter, M.A., Cullen, M.J., Mathes, W.F., Przypek, R., Kanarek, R. and Maratos-Flier, E. (1996) A role for melanin-concentrating hormone in the central regulation of feeding behaviour. Nature, 380: 243–247.

Rogawski, M.A. (1985) The A-current: how ubiquitous a feature of excitable cells is it? Trends Neurosci., 8: 214–219.

Roseberry, A.G., Liu, H., Jackson, A.C., Cai, X. and Friedman, J.M. (2004) Neuropeptide Y-mediated inhibition of proopiomelanocortin neurons in the arcuate nucleus shows enhanced desensitization in ob/ob mice. Neuron, 41: 711–722.

Rosin, D.L., Weston, M.C., Sevigny, C.P., Stornetta, R.L. and Guyenet, P.G. (2003) Hypothalamic orexin (hypocretin) neurons express vesicular glutamate transporters VGLUT1 or VGLUT2. J. Comp. Neurol., 465: 593–603.

Schwartz, M.W., Seeley, R.J., Woods, S.C., Weigle, D.S., Campfield, L.A., Burn, P. and Baskin, D.G. (1997) Leptin increases hypothalamic pro-opiomelanocortin mRNA expression in the rostral arcuate nucleus. Diabetes, 46: 2119–2123.

Shimada, M., Tritos, N.A., Lowell, B.B., Flier, J.S. and Maratos-Flier, E. (1998) Mice lacking melanin-concentrating hormone are hypophagic and lean. Nature, 396: 670–674.

Shimokawa, I., Fukuyama, T., Yanagihara-Outa, K., Tomita, M., Komatsu, T., Higami, Y., Tuchiya, T., Chiba, T. and Yamaza, Y. (2003) Effects of caloric restriction on gene expression in the arcuate nucleus. Neurobiol. Aging, 24: 117–123.

Shuto, Y., Shibasaki, T., Otagiri, A., Kuriyama, H., Ohata, H., Tamura, H., Kamegai, J., Sugihara, H., Oikawa, S. and Wakabayashi, I. (2002) Hypothalamic growth hormone secretagogue receptor regulates growth hormone secretion, feeding, and adiposity. J. Clin. Invest., 109: 1429–1436.

Spanswick, D., Smith, M.A., Groppi, V.E., Logan, S.D. and Ashford, M.L. (1997) Leptin inhibits hypothalamic neurons by activation of ATP-sensitive potassium channels. Nature, 390: 521–525.

Spanswick, D., Smith, M.A., Mirshamsi, S., Routh, V.H. and Ashford, M.L. (2000) Insulin activates ATP-sensitive K+ channels in hypothalamic neurons of lean, but not obese rats. Nat. Neurosci., 3: 757–758.

Takahashi, K.A. and Cone, R.D. (2005) Fasting induces a large, leptin-dependent increase in the intrinsic action potential frequency of orexigenic arcuate nucleus neuropeptide Y/ Agouti-related protein neurons. Endocrinology, 146: 1043–1047.

Tasker, J.G. and Dudek, F.E. (1991) Electrophysiological properties of neurones in the region of the paraventricular nucleus in slices of rat hypothalamus. J. Physiol., 434: 271–293.

Thomzig, A., Laube, G., Pruss, H. and Veh, R.W. (2005) Pore-forming subunits of K-ATP channels, Kir6.1 and Kir6.2, display prominent differences in regional and cellular distribution in the rat brain. J. Comp. Neurol., 484: 313–330.

van den Top, M., Lee, K., Whyment, A.D., Blanks, A.M. and Spanswick, D. (2004) Orexigen-sensitive NPY/AgRP pacemaker neurons in the hypothalamic arcuate nucleus. Nat. Neurosci., 7: 493–494.

van Houten, M., Posner, B.I., Kopriwa, B.M. and Brawer, J.R. (1980) Insulin-binding sites localized to nerve terminals in rat median eminence and arcuate nucleus. Science, 207: 1081–1083.

Wang, R., Liu, X., Hentges, S.T., Dunn-Meynell, A.A., Levin, B.E., Wang, W. and Routh, V.H. (2004) The regulation of glucose-excited neurons in the hypothalamic arcuate nucleus by glucose and feeding-relevant peptides. Diabetes, 53: 1959–1965.

Whyment, A.D., Wilson, J.M., Renaud, L.P. and Spanswick, D. (2004) Activation and integration of bilateral GABA-mediated synaptic inputs in neonatal rat sympathetic preganglionic neurones in vitro. J. Physiol., 555: 189–203.

Willesen, M.G., Kristensen, P. and Romer, J. (1999) Co-localization of growth hormone secretagogue receptor and NPY mRNA in the arcuate nucleus of the rat. Neuroendocrinology, 70: 306–316.

Woods, S.C., Chavez, M., Park, C.R., Riedy, C., Kaiyala, K., Richardson, R.D., Figlewicz, D.P., Schwartz, M.W., Porte Jr., D. and Seeley, R.J. (1996) The evaluation of insulin as a metabolic signal influencing behavior via the brain. Neurosci. Biobehav. Rev., 20: 139–144.

Xu, A.W., Kaelin, C.B., Takeda, K., Akira, S., Schwartz, M.W. and Barsh, G.S. (2005) PI3K integrates the action of insulin and leptin on hypothalamic neurons. J. Clin. Invest., 115: 951–958.

Yamanaka, A., Kunii, K., Nambu, T., Tsujino, N., Sakai, A., Matsuzaki, I., Miwa, Y., Goto, K. and Sakurai, T. (2000) Orexin-induced food intake involves neuropeptide Y pathway. Brain Res., 859: 404–409.

Kalsbeek, Fliers, Hofman, Swaab, Van Someren & Buijs
Progress in Brain Research, Vol. 153
ISSN 0079-6123

CHAPTER 9

Adipokines that link obesity and diabetes to the hypothalamus

Rexford S. Ahima[1,2,*], Yong Qi[1] and Neel S. Singhal[2]

[1]Department of Medicine, Division of Endocrinology, Diabetes and Metabolism, University of Pennsylvania
School of Medicine, Philadelphia, PA, USA
[2]Neuroscience Graduate Group, Philadelphia, PA, USA

Abstract: Adipose tissue plays a crucial role in energy homeostasis not only in storing triglyceride, but also responding to nutrient, neural, and hormonal signals, and producing factors which control feeding, thermogenesis, immune and neuroendocrine function, and glucose and lipid metabolism. Adipose tissue secretes leptin, steroid hormones, adiponectin, inflammatory cytokines, resistin, complement factors, and vasoactive peptides. The endocrine function of adipose tissue is typified by leptin. An increase in leptin signals satiety to neuronal targets in the hypothalamus. Leptin activates Janus-activating kinase2 (Jak2) and STAT 3, resulting in stimulation of anorexigenic peptides, e.g., α-MSH and CART, and inhibition of orexigenic peptides, e.g., NPY and AGRP. The reduction in leptin levels during fasting stimulates appetite, decreases thermogenesis, thyroid and reproductive hormones, and increases glucocorticoids. Leptin also stimulates fatty acid oxidation, insulin release, and peripheral insulin action. These effects involve regulation of PI-3 kinase, PTP-1B, suppressor of cytokine signaling-3 (SOCS-3), and AMP-activated protein kinase in the brain and peripheral organs. There is emerging evidence that leptin, adiponectin, and resistin act through overlapping pathways. Understanding the signal transduction of adipocyte hormones will provide novel insights on the pathogenesis and treatment of obesity, diabetes, and various metabolic disorders.

Keywords: Adipose; Adipokine; Leptin; Adiponectin; Resistin; Hypothalamus; Neuropeptide

Adipose tissue

Obesity has attained epidemic levels in developed countries and has been increasing worldwide (Kopelman, 2000; Hedley et al., 2004). This trend has serious health consequences, since obesity is associated with increased risk of diabetes, cardiovascular disease, obstructive sleep apnea, nonalcoholic fatty liver disease, malignancies, gall bladder disease, arthritis, subfertility, and various complications (Kopelman, 2000). Obesity also increases mortality (Kopelman, 2000; Flegal et al., 2005). Interest in the pathophysiology of obesity and metabolic abnormalities has focused on the biology of adipose tissue (Flier, 2004).

Adipose tissue is composed of lipid-filled cells (adipocytes) surrounded by a connective tissue matrix containing a rich blood supply, nerves, stromal fibroblasts, adipocyte precursors, and immune cells. In addition to providing a vast storage capacity for triglyceride, adipose tissue secretes factors with diverse effects on metabolism (Flier, 2004; Table 1). Leptin, adiponectin, and proinflammatory cytokines act through well-defined receptors on distant target tissues, thus making

*Corresponding author. Tel.: +1-215-573-1872; Fax: +1-215-573-5809; E-mail: ahima@mail.med.upenn.edu

DOI: 10.1016/S0079-6123(06)53009-2

Table 1. Factors produced by adipose tissue

Secreted factors	Receptors	Metabolic enzymes and transporters
Adipokines	*Membrane*	*Lipid metabolism*
Leptin	Insulin	Lipoprotein lipase
Adiponectin	Glucagon	Apolipoprotein E
Resistin	Thyroid-stimulating hormone	Cholesterol ester transfer protein
Tumor necrosis factor-α	Growth hormone	Adipocyte fatty-acid binding protein
Interleukin (IL)-6	Angiotensin II gastrin/	(aP2)
Adipsin	cholecystokinin B	CD36
Acylation-stimulating protein	TNF-α	
Angiotensinogen	IL-6	
Fasting-induced adipose factor	Leptin	
Plasminogen activator inhibitor-1		
Tissue factor	*Nuclear*	*Steroid metabolism*
Monocyte chemoattractant protein-1		
Tranforming growth factor-β	PPARγ	Aromatase
Visfatin	Glucocorticoid	11β Hydroxysteroid dehydrogenase-1
Retinol-binding protein-4	Thyroid	17β Hydroxysteroid dehydrogenase
	Estrogen	(17βHSD)
	Progesterone	
Steroid hormone conversion (11βHSD-1)	Androgen	*Glucose metabolism*
	Vitamin D	
Cortisone → cortisol	Nuclear factor (NF)-KB	GLUT4
(Aromatase)		Phosphatidylinositol 3-kinase
Testosterone → estradiol		Protein kinase B (Akt)
Androstenedione → estrone		Protein kinase λ/ζ
(17βHSD)		Glycogen synthase kinase a
Androstenedione → testosterone		
Estrone → estradiol		*Others*
		Inducible nitric oxide synthase

adipose tissue a bona fide endocrine organ (Flier, 2004; Ahima, 2005). Adipose-secreted factors called "adipokines" also act through paracrine and autocrine mechanisms to control metabolism, immunity, cardiovascular function, and various physiological systems (Flier, 2004). In turn, adipocytes respond to nutrients, amines, peptides, and steroid hormones (Flier, 2004; Table 1). This review will present an overview of major adipocyte hormones, in particular how leptin controls energy balance, hormone levels, and glucose by targeting the hypothalamus.

Kennedy (1953) first proposed that a circulating factor derived from adipose tissue signaled to the brain to match food intake with energy expenditure, with the goal of maintaining a constant body weight. The idea of an "adipostat" gained support

from parabiosis (cross-circulation) experiments (Hervey, 1959; Parameswaran et al., 1977). Parabiosis between rats made obese by ventromedial hypothalamic (VMH)-lesion and normal (nonlesioned) rats resulted in inhibition of feeding in the normal rat, while the obese partner gained weight. In contrast, obese (VMH-lesioned) rats remained hyperphagic and gained weight when parabiosed with each other. These results suggested that the VMH or its connections detected a blood-borne signal related to energy stores (Hervey, 1959). The discovery of recessive mutations, *obese* (*ob*) and *diabetes* (*db*), both of which led to hyperphagia and severe obesity lent further support to the existence of hormones linking energy stores in adipose tissue to the brain (Ingalls et al., 1950; Coleman and Hummel, 1969; Coleman, 1973).

Based on parabiosis studies in normal and obese mice, Coleman showed that the *ob* locus encoded a circulating satiety factor, while the *db* locus was responsible for the action of the ob product (Coleman and Hummel, 1969; Coleman, 1973). Decades later, these seminal findings were confirmed by the cloning of genes for leptin and its receptor (Tartaglia et al., 1995; Zhang et al., 1995; Chen et al., 1996).

Leptin

The product of the *lep* gene, leptin, is expressed mainly by adipocytes, although low levels have been detected in the gastric fundus, intestine, placenta, skeletal muscle, mammary epithelium, and the brain (Friedman and Halaas, 1998). Leptin has a relative mass of 16 kDa, a structure similar to cytokines and is highly conserved in mammals (Zhang et al., 1995). The levels of leptin in adipose tissue and plasma are related to fat mass and the status of energy balance (Friedman and Halaas, 1998). Leptin is higher in obesity and increases several hours after overfeeding (Frederich et al., 1995a, b; Considine et al., 1996). Conversely, leptin decreases rapidly during fasting and weight loss (Frederich et al., 1995a b; Ahima et al., 1996; Kolaczynski et al., 1996b). The precise mechanisms linking nutrition to leptin are unclear, although insulin and glucose are closely related to changes in leptin during feeding and fasting, and insulin directly stimulates leptin (Ahima et al., 1996; Kolaczynski et al., 1996a; Barr et al., 1997). Even after the menopause, leptin is higher in women than in men (Ahima and Flier, 2000). The gender difference is due to inhibition of leptin by androgens, stimulation by estrogen, and higher leptin expression in subcutaneous adipose tissue. Chronic glucocorticoid exposure, acute infection, and inflammatory cytokines increase leptin (Masuzaki et al., 1997; Ahima and Flier, 2000). On the other hand, exposure to cold, β3 adrenergic stimulation, growth hormone, thyroid hormone, melatonin, and smoking decrease leptin (Ahima and Flier, 2000). These changes in leptin occur at the levels of synthesis or secretion.

At the time of its discovery, it was proposed that leptin acted as a negative-feedback signal from adipose tissue to the brain, leading to appetite suppression and reduction in weight and fat (Halaas et al., 1995; Pelleymounter et al., 1995; Campfield et al., 1995, 1996). This idea was based on the metabolic features of rodents and humans with deficiency of leptin or leptin receptor (Montague et al., 1997; Clement et al., 1998; Friedman and Halaas, 1998). Leptin-deficient humans and rodents manifest voracious feeding, early-onset obesity, insulin resistance, immunosuppression, and a variety of hormonal abnormalities, in particular hypothalamic hypogonadism. Moreover, $Lep^{ob/ob}$ and $Lepr^{db/db}$ mice and $Lepr^{fa/fa}$ rats have reduced thermogenesis and increased corticosterone levels (Lee et al., 1997; Friedman and Halaas, 1998). As predicted, systemic and especially intracerebroventricular (i.c.v.) leptin treatment corrected these abnormalities in congenital leptin deficiency (Halaas et al., 1997; Farooqi et al., 2002). However, the initial excitement about the use of leptin in obesity therapy was soon tempered by studies showing that leptin was increased, rather than reduced, in most obese humans and rodents (Frederich et al., 1995a, b; Considine et al., 1996). The failure of high endogenous leptin to prevent obesity suggested "leptin resistance". This was confirmed in studies showing reduced sensitivity of normal and diet-induced obese rodents and humans to leptin treatment (Heymsfield et al., 1999; El-Haschimi et al., 2000; Hukshorn et al., 2000, 2002; Westerterp-Plantenga et al., 2001; Takahashi et al., 2002). Leptin resistance has been associated with reduced brain uptake, induction of suppressors of cytokine signaling (SOCS)-3, and alterations in hypothalamic neuropeptides (Levin and Dunn-Meynell, 1997; Bergen et al., 1999; El-Haschimi et al., 2000; Takahashi et al., 2002; Flier, 2004).

In contrast to leptin resistance in obesity, the decrease in leptin during fasting triggers robust metabolic and neuroendocrine responses and increased sensitivity to leptin treatment (Ahima et al. 1996, 1999b). In rodents, low leptin during fasting mediates reductions in thyroid, growth, and reproductive hormones, increased ACTH and corticosterone, immunosuppression, hyperphagia, and reduction in thermogenesis (Ahima et al., 1996, 1999b). In humans, low leptin during fasting and chronic energy deprivation has been linked to hypothalamic hypogonadism and central inhibition of thyroid and growth hormones (Rosenbaum et al., 2002; Chan et al., 2003; Welt et al., 2004). As in congenital leptin deficiency and lipodystrophy, physiologic leptin replacement during fasting reverses the metabolic and hormonal abnormalities in humans (Oral et al., 2002; Rosenbaum et al., 2002; Chan et al., 2003; Welt et al., 2004; Musso et al., 2005; Javor et al., 2005a, b). We have proposed that leptin's role is skewed toward responding to energy deficiency instead of excess (Ahima et al., 1996, 1999b; Flier, 2004). This function is likely to have evolved as a protection against the threat of starvation (Flier, 2004).

Leptin acutely increases glucose metabolism by enhancing glucose turnover and uptake in peripheral tissues, without decreasing hepatic glycogen content (Kamohara et al., 1997). Similar effects are observed after intravenous or i.c.v. leptin treatment (Kamohara et al., 1997). Insulin and leptin have distinct central effects on glucose fluxes. Insulin administered i.c.v. inhibits gluconeogenesis and glucose production (Obici et al., 2001). In contrast, leptin stimulates gluconeogenesis, decreases glycogenolysis, and does not affect glucose production (Gutierrez-Juarez et al., 2004). Leptin administered i.c.v. or intravenously increases activities of glucose-6-phosphatase (G-6-Pase) and phosphoenolpyruvate carboxykinase (PEPCK) (Gutierrez-Juarez et al., 2004). Central melanocortin blockade by the melanocortin (MC)3 and 4 receptor antagonist, SHU9119, abolishes the systemic and central actions of leptin and insulin on glucose fluxes, and thus demonstrates a common central pathway for these hormones (Obici et al., 2001; Gutierrez-Juarez et al., 2004).

Leptin decreases glucose-stimulated insulin secretion. Under hyperglycemic clamp (11 mmol/L), i.c.v. leptin infusion decreased insulin secretion by 50% (Muzumdar et al., 2004). This effect was replicated by a 3-fold increase in plasma leptin following intravenous leptin infusion (Muzumdar et al., 2004). SHU9119 administered i.c.v. or intravenously prevented the effect of leptin on insulin secretion (Muzumdar et al., 2004).

Leptin modulates the reward circuitry for feeding and taste perception, is involved in brain development and maturation of hypothalamic feeding circuits, vascular function, autonomic function, hematopoiesis, angiogenesis, bone development, and wound healing (Ahima and Flier, 2000; Fulton et al., 2000; Kawai et al., 2000). As with energy balance and neuroendocrine regulation, these diverse actions of leptin occur through the leptin receptor (LR) (Tartaglia, 1997). LR is a member of the cytokine receptor class I superfamily, containing an extracellular ligand-binding domain, a transmembrane domain, and cytoplasmic signaling domain. Five leptin receptor isoforms, LRa–LRe derived from alternate splicing of *lepr* transcript have been identified, but current evidence indicates that most of the actions of leptin are signaled through the long receptor LRb (Vaisse et al., 1996; Tartaglia, 1997). As will be discussed in detail later, LRb activates a member of the Janus kinase family, Janus activating kinase2 (Jak2), resulting in activation of STAT-3 and transcription of neuropeptides and other leptin target genes. Leptin is a large protein (16 kDa) secreted mainly by adipose tissue; thus, it is inferred that leptin enters the brain through a specific transport mechanism (Ahima et al., 2001). Understanding the blood-to-brain leptin transport is important not only for the study of biology of leptin, but also for the study of other polypeptide hormones that act in the brain.

Blood–brain leptin transport

The blood–brain barrier (BBB) comprises of a modification of the vascular bed to prevent unrestricted transfer of substances between the blood and the extracellular fluid of the brain. While most tissues have capillary endothelium separated by intercellular spaces and transendothelial channels to permit transport of large molecules, capillary endothelial cells in the brain are joined by tight junctions and devoid of transendothelial channels. Transport of molecules across the BBB is directly proportional to lipid solubility and inversely proportional to molecular weight. Thus, drugs enter the brain by virtue of size (< 400) or lipophilicity (Pardridge, 1981). In contrast, nutrients and polypeptides typically cross the BBB via saturable transport systems (Pardridge, 1981).

Banks et al. (1996) first reported that leptin entered the rodent brain through a specific saturable mechanism distinct from insulin. Leptin is present in the brain in an intact form in rodents, but the nature of the BBB transporter is unknown. Since the short leptin receptor LRa, which is abundant in brain microvessels binds and transports ^{125}I-leptin in polarized epithelium, it was proposed that LRa and other short receptors served as BBB leptin transporters (Devos et al., 1996; Golden et al., 1997; Boado et al., 1998; Hileman et al., 2000, 2002). This idea was tested in obese Koletsky rats lacking all membrane leptin receptors (Banks et al., 2002). The entry of intravenously injected leptin into brain was attenuated in Koletsky rats compared with the controls, but the concentrations of leptin in the CSF and brain leptin perfusion were both normal in Koletsky rats (Banks et al., 2002). Furthermore, neither the absence of leptin in $Lep^{ob/ob}$ nor long receptor in $Lepr^{db/db}$ mice affected brain leptin transport (Maness et al., 2000). Together, these results demonstrated that the BBB leptin transporter was not encoded by *lepr*.

In normal rodents, the entry of leptin into the brain is partially saturated over a wide range of plasma leptin concentrations, from low levels during fasting to high levels in overfeeding and obesity (Banks et al., 2000). The highest leptin uptake is in the hippocampus and hypothalamus, and the lowest is in the frontal cortex, in contrast to insulin, which is highly concentrated in the olfactory bulb, cerebral cortex, and hypothalamus (Banks et al., 2000). Brain leptin transport decreases rapidly during fasting and increases with rising plasma leptin during feeding (Banks et al., 2000). Although rodents prone to obesity have normal BBB leptin transport while on chow diet, a switch to high-fat diet decreases leptin transport across the BBB (Banks et al., 2002; Levin et al., 2004). This occurs despite an increase in LRa levels in brain microvessels (Boado et al., 1998). The reduction in brain leptin transport in diet-induced obesity is reversible and partly attributed to dietary fat or elevation of plasma triglyceride (Banks et al., 1999, 2003, 2004). Aging is also associated with reduced BBB leptin transport, which may contribute to excessive fat accumulation and abnormal glucose and lipid metabolism (Banks and Farrell, 2003). Brain leptin transport is decreased by glucocorticoid deficiency and lipopolysaccharide and increased by epinephrine via α-1 receptor (Banks et al., 2003). Together, these results demonstrate that the BBB is subject to regulation by nutritional, hormonal, and environmental factors (Kastin and Pan, 2000).

The circumventricular organs (CVOs), including the organum vasculosum of the lamina terminalis, median eminence, subfornical organ, and area postrema, have been suggested as conduits for entry of leptin into brain. CVOs have fenestrated capillaries that are restricted by tight junctions in the ependymal lining. ^{125}I-leptin accumulates rapidly in the arcuate nucleus (Arc) located dorsal to the median eminence and the choroid plexus after intravenous injection in mice (Maness et al., 1998). In contrast, ^{125}I-leptin accumulates in the choroid plexus and periventricular region following i.c.v. injection, but there is limited transport of leptin into the brain parenchyma (Maness et al., 1998). On the other hand, leptin has been detected in the hypothalamus after intrathecal injection in baboons, demonstrating transport across the CVO (McCarthy et al., 2002).

Leptin binds specifically to human brain endothelium (Golden et al., 1997). Human CSF leptin is a thousand times lower than plasma leptin (Schwartz et al., 1996). The concentration of leptin in CSF is strongly related to plasma leptin and body-mass index, suggesting that brain leptin transport in humans is proportional to body fat (Schwartz et al., 1996). The CSF/plasma leptin ratio is lower in obese than in lean individuals, but whether this indicates reduced efficiency of brain leptin transport is unclear (Schwartz et al., 1996). Paradoxically, the CSF/plasma leptin ratio is significantly elevated in anorexia nervosa and reverts to normal before body weight is restored during refeeding (Mantzoros et al., 1997). It has been suggested that this relative increase in CNS leptin may be involved in alteration of hunger perception and body image in anorexia nervosa (Mantzoros et al., 1997).

Central neuronal circuits for leptin

Leptin acts on specific neurons in the hypothalamus and brain stem (Ahima et al., 2001;

Flier, 2004). The long leptin receptor LRb is enriched in the arcuate, dorsomedial, ventromedial, and ventral premamillary hypothalamic nuclei. Moderate expression of LRb is present in the periventricular region and posterior hypothalamic nucleus, and low LRb levels are detected in the paraventricular nucleus (PVN) and lateral hypothalamic area (LHA). LRb is also present in several areas of the brain stem, including the nucleus tractus solitarius (NTS), lateral parabrachial nucleus (LPB), and motor and sensory nuclei not normally associated with energy balance (Grill and Kaplan, 2002). LRb is present in the same neurons as STAT3 and neuropeptide targets of leptin (Ahima et al., 2001). Neuropeptide Y (NPY) and Agouti-related protein (AGRP) are coexpressed in the medial Arc and regulate melanin–concentrating hormone (MCH) and orexin neurons in the LHA as well as thyrotropin-releasing hormone (TRH), corticotropin-releasing hormone (CRH), and oxytocin neurons in the PVN (Ahima et al., 2001). NPY acts through Y1 and Y5 receptors to stimulate feeding, decrease thermogenesis, and increase weight. In contrast, α-melanocyte-stimulating hormone (α-MSH) (derived from proopiomelanocortin (POMC)) and cocaine and amphetamine-regulated transcript (CART) are produced in the lateral subdivision of the Arc and decrease feeding and weight (Ahima et al., 2001; Flier, 2004). AGRP antagonizes the actions of α-MSH at melanocortin (MC)-3 and 4 receptors in the PVN and perifornical area. Leptin directly inhibits NPY and AGRP, indirectly inhibits MCH and orexins, and stimulates POMC and CART (Fig. 1). The net effect is to decrease appetite, increase thermogenesis, stimulate fatty acid oxidation, and decrease body weight (Ahima et al., 2001).

The hypothalamic actions of leptin have been confirmed using lesion and genetic methods. Ablation of the Arc disrupts the effect of leptin on TRH (Legradi et al., 1998), and ability of leptin to decrease feeding and weight (Dube et al., 1999). Importantly, specific deletion of LRb in neurons produced an obese phenotype similar to $Lepr^{db/db}$ mice (Cohen et al., 2001). Conversely, viral-mediated expression of LRb in the Arc, but not in the lateral hypothalamus reversed obesity in Koletsky

rats (Morton et al., 2003). Loss of NPY partially reversed hyperphagia, thermoregulatory defect, obesity, diabetes, and neuroendocrine abnormalities in $Lep^{ob/ob}$ mice (Erickson et al., 1996). MCH deficiency partially reversed obesity in $Lep^{ob/ob}$ mice (Segal-Lieberman et al., 2003). POMC deficiency also attenuated the hypothalamic action of leptin to inhibit feeding (Balthasar et al., 2004).

The PVN is uniquely positioned to transduce the actions of leptin during periods of changing energy availability (Elias et al., 2000; Ahima et al., 2001; Flier, 2004; Fig. 1). Apart from controlling feeding, the PVN is the source of preganglionic autonomic projections that mediate thermogenesis, insulin secretion, and gut motility (Ahima et al., 2001; Flier, 2004). Leptin regulates the anterior pituitary via PVN peptides, e.g., CRH, TRH, growth hormone-releasing hormone (GHRH), and somatostatin (Ahima et al., 2001; Fig. 1). The dorsomedial nucleus (DMN) and subparaventricular zone respond directly to leptin, and project to the PVN to integrate the circadian regulation of hormones and feeding (Elmquist et al., 1998; Chou et al., 2003; Fig. 1). The VMN has been implicated in glucose regulation (Borg et al., 1994), but it is doubtful whether leptin acts directly in this nucleus since the level of LRb is low (Elmquist et al., 1998; Elias et al., 2000; Ahima et al., 2001). As mentioned earlier, leptin inhibits feeding partly by reducing MCH and orexins in the LHA. MCH and orexins are produced by distinct neurons in LHA which project to the cerebral cortex and limbic areas, and thus may provide a channel for transducing the effect of leptin and other blood-borne signals to higher CNS centers (Elias et al., 1999; Ahima et al., 2001).

Interestingly, Kreier et al. (2002) have demonstrated that parasympathetic innervation of white adipose tissue is critical for insulin-mediated glucose and free fatty acid uptake, indicating that vagal input to adipose tissue serves an anabolic role. Parasympathetic denervated animals also exhibited reductions in white adipose tissue leptin mRNA. Utilizing a retrograde transneuronal tracer, parasympathetic afferents originating in white adipose tissue were found to transynaptically project to the PVN and LHA, suggesting that in addition to mediating many of leptin's effects,

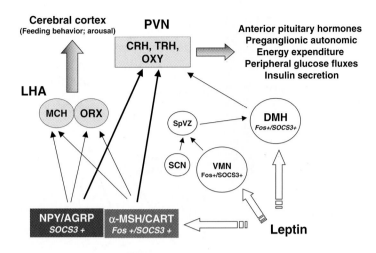

Fig. 1. Neuronal targets of leptin in the hypothalamus. Leptin acts directly in the Arc to increase POMC and CART and reduce NPY and AGRP. Arcuate neurons project to the LHA and PVN, where leptin decreases feeding and controls preganglionic autonomic outflow and anterior pituitary hormones. Leptin also acts through the DMN and VMN to integrate feeding with circadian rhythms and glucose homeostasis. Hypothalamic nuclei demonstrating strong induction of Fos and SOCS-3 in response to leptin are labeled as such. SpVZ —subparaventricular zone; SCN — suprachiasmatic nucleus.

hypothalamic input can also regulate leptin synthesis (Kreier et al., 2002).

Leptin signal transduction in hypothalamus

Leptin activates the Jak-STAT pathway (Tartaglia, 1997; Myers, 2004). Jak associates constitutively with conserved box 1 and 2 motifs in the intracellular domain of LRb. Binding of leptin to the extracellular domain of LRb results in autophosphorylation of Jak2, phosphorylation of tyrosine residues on the cytoplasmic domain of LRb, and activation of downstream transcription factors, named STATs. In rodents, LRb has three conserved tyrosine residues: Y985, Y1077, and Y1138 in the intracellular domain. Leptin phosphorylates Y1138 and recruits STAT3 via its SH2 domain. Tyrosyl-phosphorylated STAT3 undergoes homodimerization, is translocated into the nucleus, and regulates transcription of neuropeptides and other target genes of leptin (Fig. 2). The critical role Y1138 in the signal transduction of leptin was demonstrated by replacing this amino acid residue with serine (Bates et al., 2003, 2004).

Homozygous $Lepr^{S1138}$ mutation disrupted the ability of LRb to phosphorylate and activate STAT3. $Lepr^{S1138}$ homozygous mice became hyperphagic, had reduced energy expenditure, and developed early onset obesity, similar to $Lepr^{db/db}$ mice (Bates et al., 2003, 2004). However, $Lepr^{S1138}$ did not affect sexual maturation, fertility, or body length, all of which are attenuated in $Lepr^{db/db}$ mice. $Lepr^{S1138}$ were also less prone diabetes than $Lepr^{db/db}$ mice (Bates et al., 2003). NPY was increased in the hypothalamus of $Lepr^{db/db}$ mice, but not $Lepr^{S1138}$ homozygotes; however, POMC was increased in both (Bates et al., 2003). Together, these results demonstrate that STAT3 is important for leptin's ability to regulate feeding, thermogenesis, and weight, but is not required for leptin's effects on reproduction, growth, glucose, or expression of NPY level in the hypothalamus (Bates et al., 2003, 2004). These results have been contradicted by another study, in which neuron-specific deletion produced an obese phenotype identical to $Lepr^{db/db}$ mice (Gao et al., 2004). In contrast to $Lepr^{S1138}$ homozygotes, neuron-specific STAT3 ablation resulted in hyperphagia, impaired thermoregulation, elevated corticosterone, reduced linear

162

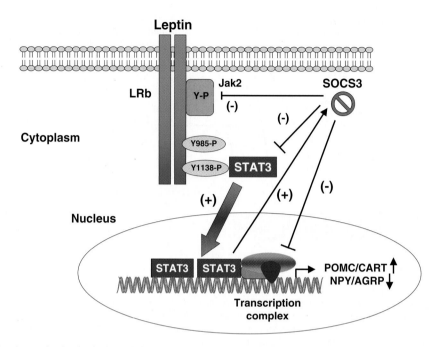

Fig. 2. Leptin signal transduction in the hypothalamus. Activation of LRb by leptin leads to increased activity of Jak2 associated with the proximal intracellular region of LRb. Jak2 phosphorylates tyrosine residues Y985 and Y1138 on LRb. STAT3 binds to pY1138 on LRb, is phosphorylated on Y705 by Jak2, dimerizes, and is translocated into the nucleus to inhibit the expression of NPY and AGRP and increase POMC and CART. SOCS-3 is induced by STAT 3 pathway and inhibits leptin signaling via LRb.

growth, infertility, insulin resistance, diabetes, severe steatosis, and hyperlipidemia (Bates et al., 2003, 2004; Gao et al., 2004).

Leptin regulates insulin targets, e.g., insulin receptor substrate-1 (IRS-1) and 2, MAP kinase, ERK, Akt, and PI3-kinase (Morton et al., 2003, 2005; Niswender and Schwartz, 2004; Fig. 3). Leptin stimulates IRS2-mediated activation of PI3 kinase in hypothalamus, which is related to inhibition of feeding and weight loss (Niswender and Schwartz, 2004). Conversely, inhibition of PI3 kinase activity prevents leptin's ability to phosphorylate and activate STAT3 and inhibit feeding (Niswender et al., 2001; Zhao et al., 2002). Insulin activates PI3-K and Jak2 in the hypothalamus, and this has been associated with the anorexigenic action of insulin in the brain (Niswender and Schwartz, 2004; Fig. 3). The overlap between leptin and insulin signaling in the hypothalamus provides a molecular basis for how these hormones that are related to energy stores in adipose tissue may converge to integrate metabolism (Fig. 3).

Leptin induces the suppressor of cytokine signaling-3 (SOCS-3), a member of a family of proteins which inhibits Jak-STAT signaling (Flier, 2004; Fig. 2). Earlier studies showed that leptin acted in the hypothalamus to increase in SOCS-3, decrease STAT3 phosphorylation, and reduce feeding and weight (Bjorbaek et al., 1998; El-Haschimi, 2000). As predicted, deletion of SOCS-3 especially in neurons enhanced leptin-induced hypothalamic STAT3 phosphorylation, increased POMC level in the hypothalamus, and reduced food intake and body weight (Howard et al., 2004; Mori et al., 2004). The increase in leptin sensitivity in SOCS-3-deficiency enhanced insulin sensitivity (Mori et al., 2004). Protein-tyrosine phosphastase (PTP)-1B terminates leptin signaling by binding and dephosphorylating Jak2 (Flier, 2004). PTP-1B is present in the same neurons as STAT3 and neuropeptide targets of leptin in the hypothalamus. In support of its role as a negative regulator of leptin, deficiency of PTP-1B in mice increased tyrosyl phosphorylation of STAT3 in the hypothalamus

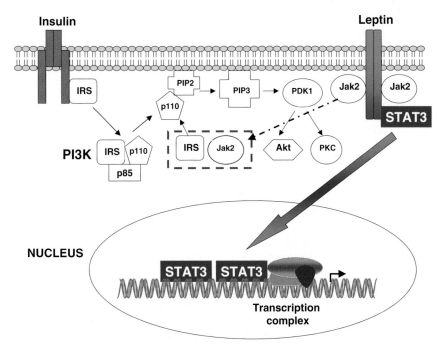

Fig. 3. Crosstalk between leptin and insulin in the hypothalamus. Insulin binds to its receptor and activates an intrinsic tyrosine kinase, leading to phosphorylation of the intracellular domain of the insulin receptor. IRS proteins bind to the phosphorylated residues on the insulin receptor, become activated by tyrosine phosphorylation, and in turn, activate PI3 kinase. PI3-K catalyzes the phosphorylation of phosphatidylinositol (4,5)-bisphosphate (PIP2) to phosphatidylinositol (3,4,5)-trisphosphate (PIP3), leading to the activation of various downstream targets, including 3-phosphoinositide-dependent kinase-1 (PDK1), glycogen synthase kinase-3 (GSK3), and protein kinase B (PKB/Akt). Leptin binds to LRb, which dimerizes and recruits Jak2, resulting in activation of STAT 3. Insulin and leptin both activate PI3-K, possibly through Jak2, thus providing a mechanism for regulation of common target genes.

and leptin sensitivity, and resistance to obesity (Zabolotny et al., 2002).

Effects of leptin on neurotransmission and synaptic plasticity

Leptin has rapid effects on neurotransmission and neuropeptide secretion that cannot be explained by a transcriptional mechanism. Soon after its discovery, it was noted that leptin rapidly inhibited NPY secretion from hypothalamic explants (Stephens et al., 1995). Subsequently, Cone and Cowley demonstrated a direct depolarization of POMC neurons in the Arc when leptin was applied to hypothalamic slices (Cowley et al., 2001). Leptin activates POMC neurons partly by decreasing the inhibitory tone of GABA in the Arc, resulting in a net increase in α-MSH (Cowley et al., 2001).

In contrast, leptin hyperpolarizes and inactivates NPY neurons in the Arc (Cowley et al., 2001).

In normal mice, fasting induces a rapid increase in action potential frequency in NPY/AGRP neurons (Takahashi and Cone, 2005). $Lep^{ob/ob}$ and $Lepr^{db/db}$ mice have higher action potential frequency in NPY/AGRP neurons, but there is no further increase during fasting (Takahashi and Cone, 2005). Leptin treatment reduces the action potential frequency in fasted and $Lep^{ob/ob}$ mice, demonstrating a critical role of leptin in neurotransmission (Takahashi and Cone, 2005). Rapid electrophysiological actions of leptin have been demonstrated in the supraoptic nucleus, vagal afferents in the gut, and glucose-sensitive neurons in the brain (Spanswick et al., 1997; Gaige et al., 2002; Honda et al., 2002). Leptin-deficient $Lep^{ob/ob}$ mice differ from wild-type mice by having greater excitatory than inhibitory synapses in NPY and POMC neurons (Pinto et al., 2004). This difference is reversed by leptin treatment and precedes the inhibition of feeding (Pinto et al., 2004). Together, these results suggest that leptin controls feeding through synaptic plasticity in the hypothalamus.

We first proposed a role of leptin in brain development based on earlier reports of smaller brains in $Lep^{ob/ob}$ and $Lepr^{db/db}$ mice (Sena et al., 1985; Ahima et al., 1999a). This abnormality is associated with severe reductions in glial and synaptic proteins (Ahima et al., 1999a). Leptin normalizes brain weight and partially restores the levels of neuronal proteins in $Lep^{ob/ob}$ mice (Ahima et al., 1999a; Steppan and Swick, 1999). Importantly, leptin treatment also reverses structural deficits in the brains of genetically leptin-deficient patients (Matochik et al., 2005). Neurotrophic actions of leptin have been demonstrated in vivo and in vitro. $Lep^{ob/ob}$ mice manifest a delay in maturation of the Arc-PVN projection, which is corrected by leptin treatment during the postnatal period (Bouret et al., 2004). Leptin prevents neuronal apoptosis (Russo et al., 2004). A neonatal leptin surge has been implicated in obesity and insulin resistance during adulthood (Yura et al., 2005). These results in conjunction with the close correlation between the ontogeny of hypothalamic neuropeptides and timing of leptin's effects on thermogenesis and feeding, suggest that leptin may be involved in obesity by controlling neuronal structure and function (Ahima et al., 1999b; Ahima and Hileman, 2000; Yura et al., 2005).

Effects of leptin on metabolic enzymes

Recent studies have focused on effects of adipocyte hormones on enzymes normally associated with metabolism in peripheral tissues (Winder and Hardie, 1999; Kahn et al., 2005). In the fed state, fatty acid synthase (FAS) catalyzes the condensation of acetyl-CoA and malonyl-CoA to generate long-chain fatty acids. Elevated levels of malonyl-CoA allosterically inhibit carnitine palmitoylacyltransferase (CPT-1), an integral membrane protein that catalyzes the esterification of long-chain fatty

acyl-CoAs to L-carnitine, and mediates the transport of acyl moieties from the cytosol into mitochondria to undergo fatty acid oxidation. Conversely, during fasting, when energy stores are low, malonyl-CoA is reduced, releasing the inhibition of CPT-1, and allowing fatty acids to enter mitochondria to be oxidized for energy. The regulation of FAS and CPT-1 is coupled to AMP-activated protein kinase (AMPK), a heterotrimeric protein kinase present in most mammalian cells, including neurons (Kahn et al., 2005). AMPK is composed of a catalytic α subunit and two regulatory β and γ subunits. AMPK is activated by metabolic stressors, exercise, and hypoxia that deplete cellular ATP, leading to increased AMP-to-ATP ratio and phosphorylation and activation of AMPK. AMPK phosphorylates and inactivates biosynthetic pathways, e.g. acetyl-CoA carboxylase (ACC), thereby preventing further depletion of ATP (Kahn et al., 2005). Additionally, AMPK stimulates glucose uptake, glycolysis, and fatty acid oxidation, in an attempt to restore cellular ATP levels (Kahn et al., 2005).

While leptin is thought to act mostly through the Jak-STAT pathway, some of the rapid actions on lipid and glucose metabolism are mediated through AMPK (Kahn et al., 2005). Leptin selectively stimulates the phosphorylation and activation of the α-subunit of AMPK in skeletal muscle through a biphasic action involving a direct transient activation of AMPK action in muscle, followed by sustained activation lasting several hours (Minokoshi et al., 2002). The longer action of leptin is mediated through the a hypothalamic-sympathetic stimulation of α-adrenergic receptors in muscle (Minokoshi et al., 2002). Leptin suppresses ACC2 activity in muscle in parallel with activation of AMPK, thereby increasing fatty acid oxidation and preventing triglyceride accumulation (Minokoshi et al., 2002). Leptin-induced AMPK also inhibits lipogenesis in the liver (Kahn et al., 2005). Collectively, these antisteatotic actions of leptin improve insulin sensitivity (Kahn et al., 2005).

Leptin inhibits AMPK activity in the Arc and PVN, in parallel with its satiety and weight-reducing actions (Minokoshi et al., 2004). These processes require an intact melanocortin pathway, since leptin's ability to activate AMPK and inhibit feeding are both abolished in MC4R knockout mice (Minokoshi et al., 2004). As with leptin, insulin, glucose, and feeding inhibit AMPK activity in the hypothalamus (Minokoshi et al., 2004). In contrast, fasting and orexigenic peptides, e.g., ghrelin and AGRP activate AMPK (Minokoshi et al., 2004). Thus, the regulation of AMPK activity is coupled to energy stores through adiposity-related hormones and hypothalamic neuropeptides. Efforts are underway to determine whether AMPK is involved in the pathogenesis of obesity and disorders of glucose and lipid metabolism.

Adiponectin

Adiponectin was identified by various laboratories as a protein secreted exclusively by differentiated adipocytes (Kadowaki and Yamauchi, 2005). Adiponectin has a relative mass of 30 kDa and a primary structure that contains an amino-terminal signal sequence, a variable region, a collagen-like tail, and carboxy-terminal globular head domain. Adiponectin shares strong sequence homology with complement factor C1q and types VIII and X collagen. The tertiary structure of the globular head resembles TNFα. Low- and high-molecular weight adiponectin complexes have been identified in plasma. The high-molecular weight adiponectin mediates the biological activity of adiponectin in mammalian systems (Waki et al., 2003; Pajvani et al., 2004). Both the total and high-molecular weight adiponectin are increased by thiazolidinediones and related to the insulin-sensitizing action of these drugs (Pajvani et al., 2004). Conversely, adiponectin is decreased in obesity and inversely related to insulin resistance, hyperlipidemia, and atherosclerosis (Pajvani et al., 2004; Fisher et al., 2005; Kadowaki and Yamauchi, 2005). As predicted, ablation of the adiponectin gene resulted in insulin resistance, glucose intolerance and dyslipidemia, and increased susceptibility to vascular injury and atherosclerosis in diet-induced obese and $Lep^{ob/ob}$ mice (Kubota et al., 2002; Maeda et al., 2002; Yamauchi et al., 2003a, b). Adiponectin treatment reversed these abnormalties (Kubota et al., 2002; Maeda et al., 2002).

Adiponectin receptors (AdipoR) 1 and 2 contain seven transmembrane domains, but are distinct from G protein-coupled receptors (Yamauchi et al., 2003a). Binding of adiponectin to AdipoR phosphorylates AMPK in muscle and liver (Yamauchi et al., 2002). AdipoR1 is highly expressed in muscle and functions as a high-affinity receptor for globular adiponectin and a low affinity receptor for full-length adiponectin. AdipoR2 is abundant in liver and has intermediate affinity receptor for globular and full-length adiponectin. Both receptors are abundant in the hypothalamus and various regions of the brain (Ahima, unpublished results; Yamauchi et al., 2004). In muscle, adiponectin treatment activates AMPK and increases fatty acid oxidation and glucose transport (Tomas et al., 2002; Yamauchi et al., 2003a, b). In liver, AMPK activation by adiponectin inhibits PEPCK, glucose-6-phosphatase, and hepatic glucose production (Yamauchi et al., 2003a, b). Importantly, the action of adiponectin is abolished by the expression of dominant-negative AMPK, establishing a functional role of AMPK (Yamauchi et al., 2003a, b).

Adiponectin increased energy expenditure and fatty acid oxidation when administered peripherally in mice on a high fat diet (Tomas et al., 2002). Adiponectin reduced food intake and decreased body weight when expressed chronically in muscle of diet-induced obese rats (Shklyaev et al., 2003). Given the similarities between the metabolic actions of adiponectin and leptin, we hypothesized that adiponectin acts centrally to regulate energy balance (Qi et al., 2004). We detected an increase in adiponectin in CSF after intravenous injection, suggesting brain transport (Qi et al., 2004). In contrast to leptin, adiponectin decreased body weight and fat by increasing energy expenditure, but did not affect food intake. The full-length and globular adiponectin and a mutant adiponectin with cysteine at position 39 replaced with serine were all effective, whereas the collagenous tail was not (Qi et al., 2004). $Lep^{ob/ob}$ mice have reduced endogenous adiponectin level, and increased sensitivity to i.c.v. and peripheral adiponectin treatment, which stimulated fatty acid oxidation and decreased glucose (Qi et al., 2004). Adiponectin potentiated the thermogenic, lipolytic, and glucose-lowering actions of leptin (Qi et al., 2004). Leptin and adiponectin both increased CRH in the hypothalamus, but in contrast to leptin, adiponectin did not affect NPY, AGRP, CART or POMC (Qi et al., 2004). Adiponectin induced immuno-staining of Fos protein mostly in PVN, while leptin induced Fos immunostaining in PVN as well as the arcuate, ventromedial, dorsomedial, posterior mammillary nucleus, NTS, LPB, and several nuclei (Elias et al., 2000; Qi et al., 2004). Agouti A^y/a mice are insensitive to leptin and adiponectin, indicating that both hormones require MC3/4 receptors for their central actions on energy balance, glucose, and lipids (Qi et al., 2004; Fig. 4).

These findings are contrary to another study which did not observe a significant effect of central adiponectin treatment (Masaki et al., 2003). In contrast to our study, the latter administered bacterially expressed adiponectin in A^y/a mice, which we found not to be responsive to adiponectin or leptin (Masaki et al., 2003; Qi et al., 2004). Crucial questions include how adiponectin is transported across the BBB, what form of adiponectin reaches the hypothalamus, and the precise signaling pathways in the brain. It is possible that adiponectin acts through AMPK in the hypothalamus to increase thermogenesis, fatty acid oxidation, and insulin sensitivity in peripheral tissues (Fig. 4).

Resistin

Resistin was discovered during screening for proteins by inhibited by thiazolidinediones (Steppan et al., 2001a). Resistin has a relative mass of 12 kDa and belongs to a unique family of cysteine-rich C-terminal domain proteins called resistin-like molecules (RELMs) (Steppan et al., 2001b). Rodent studies showed an increase in resistin protein levels in obese rodents and a positive correlation with insulin resistance and glucose (Steppan et al., 2001a; Rajala et al., 2004). Resistin treatment inhibited insulin action by suppressing hepatic glucose production (Banerjee et al., 2004; Patel et al.,

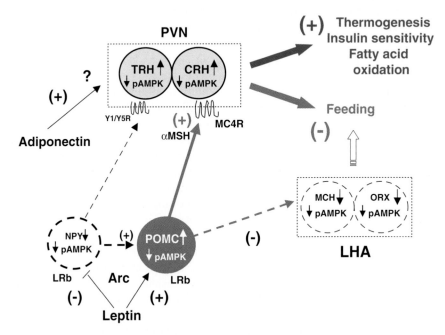

Fig. 4. Regulation of AMPK activity in the hypothalamus by leptin. Leptin's ability to inhibit feeding and decrease weight involves inhibition of AMPK in the Arc and PVN. Leptin-mediated suppression of AMPK activity inhibits NPY and AGRP, and increases α-MSH (derived from POMC). The MC4R is necessary for leptin to inhibit AMPK and reduce feeding. Adiponectin potentiates leptin's effect to stimulate thermogenesis, fatty acid oxidation, and insulin sensitivity. Both hormones require central melanocortin signaling for their actions. It is possible that adiponectin also interacts with leptin through AMPK to influence metabolism.

2004), while transgenic overexpression of resistin in liver or muscle increased insulin resistance and glucose (Rangwala et al., 2004; Satoh et al., 2004). Conversely, ablation of *rstn* gene or reduction in resistin protein using RNAi decreased glucose by suppressing hepatic gluconeogenesis (Banerjee et al., 2004; Muse et al., 2004). The resistin-mediated effects on insulin sensitivity appear to involve AMPK (Banerjee et al., 2004; Muse et al., 2004; Satoh et al., 2004). Moreover, resistin induces SOCS-3, and this may mediate insulin resistance (Steppan et al., 2005).

As with leptin, resistin decreases during fasting and increases in response to feeding (Ahima et al., 1996; Kim et al., 2001; Steppan et al., 2001a; Rajala et al., 2004). The nutritional regulation of resistin is mediated through insulin and glucose (Kim et al., 2001; Rajala et al., 2004). Administration of NPY i.c.v. increases resistin (Yuzuriha et al., 2003). On the other hand, resistin inhibits feeding, albeit transiently, when injected i.c.v. (Tovar et al., 2005). The latter is associated with induction of Fos immunostaining in the Arc (Tovar et al., 2005). Furthermore, resistin is expressed de novo in the hypothalamus and hippocampus, suggesting a paracrine action (Brunetti et al., 2004). Resistin stimulates norepinephrine and dopamine release from hypothalamic synaptosomes, further supporting a central action (Brunetti et al., 2004).

Resistin inhibits adipogenesis in vitro (Kim et al., 2001). Importantly, the latter is supported by in vivo studies in which expression of a dominant negative resistin prevented endogenous resistin action and enhanced adipogenesis (Kim et al., 2004). As expected, these resistin-deficient mice were more insulin sensitive despite the excess adiposity (Kim et al., 2004). Resistin acts centrally to increase insulin secretion (Park et al., 2005a). Administration of resistin i.c.v. increases hepatic glucose production, whereas neutralization of endogenous resistin decreases glucose production (Muse et al., 2005).

Despite these well-characterized actions in rodents, the biology of resistin in humans remains controversial (Steppan and Lazar, 2004; Kusminski et al., 2005). In contrast to rodents, resistin is expressed by mononuclear cells in the stromovascular compartment of adipose tissue (Savage et al., 2001). Human resistin shares only 64% homology with murine resistin, unlike the close homologies between murine and human leptin and adiponectin (Friedman and Halaas, 1998; Steppan & Lazar, 2004; Kadowaki and Yamauchi, 2005). Some epidemiological studies have failed to show a consistent association between resistin levels in adipose tissue and circulation, and adiposity or insulin resistance (Lee et al., 2003; Barb et al., 2005). Nonetheless, increased resistin levels have been associated with obesity, insulin resistance, and inflammation in other studies (Degawa-Yamauchi et al., 2003; Lehrke et al., 2004; Vozarova de Courten et al., 2004; Reilly et al., 2005).

Proinflammatory cytokines

TNFα is produced by adipocytes and stromovascular cells, increased in obese rodents and humans and related to abnormal glucose and lipid metabolism (Moller, 2000). The transmembrane 26 kDa TNF protein is produced in adipose tissue and cleaved into a 17 kDa biologically active (soluble) form which acts on type I and type II receptors. TNFα induces insulin resistance in rodents, whereas the loss of TNFα or its receptors lowers lipid levels and increases insulin sensitivity in rodents (Moller, 2000). However, neutralization of TNFα did not improve insulin action in humans, raising doubts about the clinical significance of the rodent studies (Ofei et al., 1996). TNFα and various cytokines are induced in the hypothalamus and other brain areas by lipopolysaccharide and tumors and related to anorexia and cachexia (Tracey and Cerami, 1992; Ilyin et al., 1998; Chance et al., 2003).

IL-6 circulates as glycosylated proteins, 22–27 kDa in size, and binds to a receptor homologous to the leptin receptor (Tartaglia, 1997). IL-6 genotype polymorphisms and increased plasma concentrations are related to obesity, insulin resistance, and cardiovascular morbidity (Berthier et al., 2003; Wolford et al., 2003; Park et al., 2005b). IL-6 deficiency in mice results in obesity, and elevated glucose and lipids, which are reversible by IL-6 treatment (Wallenius et al., 2002).

168

In the latter model, IL-6 administered i.c.v. increased energy expenditure (Wallenius et al., 2002). Although IL-6 has been suggested as a paracrine factor that regulates peripheral glucose metabolism in humans (Bastard et al., 2002), its role as a central regulator of glucose and energy homeostasis is unclear.

Conclusion

Our understanding of the interactions between adipose tissue and the brain has benefited from classic hypothalamic lesion and physiological studies and recent advances in molecular genetics. Leptin provides important lessons for the systematic characterization of adipokines, especially how they are linked to nutrition, transported to targets in the brain and peripheral organs, and act through specific signaling pathways involved in energy and glucose homeostasis. Ultimately, studies in rodents and other models have to be translated to humans in order to gain better insights on the physiology of adipokines and their roles in disease. Such studies could provide novel markers of obesity-related diseases and potential therapeutic targets.

Acknowledgment

This work was supported by grants from the National Institutes of Health (NIDDK).

References

Ahima, R.S. (2005) Central actions of adipocyte hormones. Trends Endocrinol. Metab., 16: 307–313.

Ahima, R.S., Bjorbaek, C., Osei, S. and Flier, J.S. (1999a) Regulation of neuronal and glial proteins by leptin: implications for brain development. Endocrinology, 140: 2755–2762.

Ahima, R.S. and Flier, J.S. (2000) Leptin. Annu. Rev. Physiol., 62: 413–437.

Ahima, R.S. and Hileman, S.M. (2000) Postnatal regulation of hypothalamic neuropeptide expression by leptin: implications for energy balance and body weight regulation. Regul. Peptides, 92: 1–7.

Ahima, R.S., Kelly, J., Elmquist, J.K. and Flier, J.S. (1999b) Distinct physiologic and neuronal responses to decreased leptin and mild hyperleptinemia. Endocrinology, 140: 4923–4931.

Ahima, R.S., Prabakaran, D., Mantzoros, C., Qu, D., Lowell, B., Maratos-Flier, E. and Flier, J.S. (1996) Role of leptin in the neuroendocrine response to fasting. Nature, 382: 250–252.

Ahima, R.S., Saper, C.B., Flier, J.S. and Elmquist, J.K. (2001) Leptin regulation of neuroendocrine systems. Front. Neuroendocrinol., 21: 263–307.

Balthasar, N., Coppari, R., McMinn, J., Liu, S.M., Lee, C.E., Tang, V., Kenny, C.D., McGovern, R.A., Chua Jr., S.C., Elmquist, J.K. and Lowell, B.B. (2004) Leptin receptor signaling in POMC neurons is required for normal body weight homeostasis. Neuron, 42: 983–991.

Banerjee, R.R., Rangwala, S.M., Shapiro, J.S., Rich, A.S., Rhoades, B., Qi, Y., Wang, J., Rajala, M.W., Pocai, A., Scherer, P.E., Steppan, C.M., Ahima, R.S., Obici, S., Rossetti, L. and Lazar, M.A. (2004) Regulation of fasted blood glucose by resistin. Science, 303: 1195–1198.

Banks, W.A. and Farrell, C.L. (2003) Impaired transport of leptin across the blood–brain barrier in obesity is acquired and reversible. Am. J. Physiol. Endocrinol. Metab., 285: E10–E15.

Banks, W.A., Clever, C.M. and Farrell, C.L. (2000) Partial saturation and regional variation in the blood-to-brain transport of leptin in normal weight mice. Am. J. Physiol. Endocrinol. Metab., 278: E1158–E1165.

Banks, W.A., Coon, A.B., Robinson, S.M., Moinuddin, A., Shultz, J.M., Nakaoke, R. and Morley, J.E. (2004) Triglycerides induce leptin resistance at the blood–brain barrier. Diabetes, 53: 1253–1260.

Banks, W.A., DiPalma, C.R. and Farrell, C.L. (1999) Impaired transport of leptin across the blood–brain barrier in obesity. Peptides, 20: 1341–1345.

Banks, W.A., Kastin, A.J., Huang, W., Jaspan, J.B. and Maness, L.M. (1996) Leptin enters the brain by a saturable system independent of insulin. Peptides, 17: 305–311.

Banks, W.A., Niehoff, M.L., Martin, D. and Farrell, C.L. (2002) Leptin transport across the blood–brain barrier of the Koletsky rat is not mediated by a product of the leptin receptor gene. Brain Res., 950: 130–136.

Barb, D., Wadhwa, S.G., Kratzsch, J., Gavrila, A., Chan, J.L., Williams, C.J., Karchmer, A.W. and Mantzoros, C.S. (2005) Circulating resistin levels are not associated with fat redistribution, insulin resistance, or metabolic profile in patients with the highly active antiretroviral therapy-induced metabolic syndrome. J. Clin. Endocrinol. Metab., 90: 5324–5328.

Barr, V.A., Malide, D., Zarnowski, M.J., Taylor, S.I. and Cushman, S.W. (1997) Insulin stimulates both leptin secretion and production by rat white adipose tissue. Endocrinology, 138: 4463–4472.

Bastard, J.P., Maachi, M., Van Nhieu, J.T., Jardel, C., Bruckert, E., Grimaldi, A., Robert, J.J., Capeau, J. and Hainque, B. (2002) Adipose tissue IL-6 content correlates with resistance to insulin activation of glucose uptake both in vivo and in vitro. J. Clin. Endocrinol. Metab., 87: 2084–2089.

Bates, S.H., Dundon, T.A., Seifert, M., Carlson, M., Maratos-Flier, E. and Myers Jr., M.G. (2004) LRb-STAT3 signaling is required for the neuroendocrine regulation of energy expenditure by leptin. Diabetes, 53: 3067–3073.

Bates, S.H., Stearns, W.H., Dundon, T.A., Schubert, M., Tso, A.W., Wang, Y., Banks, A.S., Lavery, H.J., Haq, A.K., Maratos-Flier, E., Neel, B.G., Schwartz, M.W. and Myers Jr, M.G. (2003) STAT3 signalling is required for leptin regulation of energy balance but not reproduction. Nature, 421: 856–859.

Bergen, H.T., Mizuno, T., Taylor, J. and Mobbs, C.V. (1999) Resistance to diet-induced obesity is associated with increased proopiomelanocortin mRNA and decreased neuropeptide Y mRNA in the hypothalamus. Brain Res., 851: 198–203.

Berthier, M.T., Paradis, A.M., Tchernof, A., Bergeron, J., Prud'homme, D., Despres, J.P. and Vohl, M.C. (2003) The interleukin 6-174 G/C polymorphism is associated with indices of obesity in men. J. Hum. Genet., 48: 14–19.

Bjorbaek, C., Elmquist, J.K., Frantz, J.D., Shoelson, S.E. and Flier, J.S. (1998) Identification of SOCS-3 as a potential mediator of central leptin resistance. Mol. Cell., 1: 619–625.

Boado, R.J., Golden, P.L., Levin, N. and Pardridge, W.M. (1998) Up-regulation of blood–brain barrier short-form leptin receptor gene products in rats fed a high-fat diet. J. Neurochem., 71: 1761–1764.

Borg, W.P., During, M.J., Sherwin, R.S., Borg, M.A., Brines, M.L. and Shulman, G.I. (1994) Ventromedial hypothalamic lesions in rats suppress counterregulatory responses to hypoglycemia. J. Clin. Invest., 93: 1677–1682.

Bouret, S.G., Draper, S.J. and Simerly, R.B. (2004) Trophic action of leptin on hypothalamic neurons that regulate feeding. Science, 304: 108–110.

Brunetti, L., Orlando, G., Recinella, L., Michelotto, B., Ferrante, C. and Vacca, M. (2004) Resistin, but not adiponectin, inhibits dopamine and norepinephrine release in the hypothalamus. Eur. J. Pharmacol., 493: 41–44.

Campfield, L.A., Smith, F.J., Guisez, Y., Devos, R. and Burn, P. (1995) Recombinant mouse OB protein: evidence for a peripheral signal linking adiposity and central neural networks. Science, 269: 546–549.

Campfield, L.A., Smith, F.J. and Burn, P. (1996) The OB protein (leptin) pathway–a link between adipose tissue mass and central neural networks. Horm. Metab. Res., 28: 619–632.

Chan, J.L., Heist, K., DePaoli, A.M., Veldhuis, J.D. and Mantzoros, C.S. (2003) The role of falling leptin levels in the neuroendocrine and metabolic adaptation to short-term starvation in healthy men. J. Clin. Invest., 111: 1409–1421.

Chance, W.T., Sheriff, S., Dayal, R. and Balasubramaniam, A. (2003) Refractory hypothalamic α-mSH satiety and AGRP feeding systems in rats bearing MCA sarcomas. Peptides, 24: 1909–1919.

Chen, H., Charlat, O., Tartaglia, L.A., Woolf, E.A., Weng, X., Ellis, S.J., Lakey, N.D., Culpepper, J., Moore, K.J., Breitbart, R.E., Duyk, G.M., Tepper, R.I. and Morgenstern, J.P. (1996) Evidence that the *diabetes* gene encodes the leptin receptor: identification of a mutation in the leptin receptor gene in *db/db* mice. Cell, 84: 491–495.

Chou, T.C., Scammell, T.E., Gooley, J.J., Gaus, S.E., Saper, C.B. and Lu, J. (2003) Critical role of dorsomedial hypothalamic nucleus in a wide range of behavioral circadian rhythms. J. Neurosci., 23: 10691–10702.

Clement, K., Vaisse, C., Lahlou, N., Cabrol, S., Pelloux, V., Cassuto, D., Gourmelen, M., Dina, C., Chambaz, J., Lacorte, J.M., Basdevant, A., Bougneres, P., Lebouc, Y., Froguel, P. and Guy-Grand, B. (1998) A mutation in the human leptin receptor gene causes obesity and pituitary dysfunction. Nature, 392: 398–401.

Cohen, P., Zhao, C., Cai, X., Montez, J.M., Rohani, S.C., Feinstein, P., Mombaerts, P. and Friedman, J.M. (2001) Selective deletion of leptin receptor in neurons leads to obesity. J. Clin. Invest., 108: 1113–1121.

Coleman, D.L. (1973) Effects of parabiosis of obese with diabetes and normal mice. Diabetologia, 9: 294–298.

Coleman, D.L. and Hummel, K.P. (1969) Effects of parabiosis of normal with genetically diabetic mice. Am. J. Physiol., 217: 1298–1304.

Considine, R.V., Sinha, M.K., Heiman, M.L., Kriauciunas, A., Stephens, T.W., Nyce, M.R., Ohannesian, J.P., Marco, C.C., McKee, L.J., Bauer, T.L. and Caro, J.F. (1996) Serum immunoreactive-leptin concentrations in normal-weight and obese humans. N. Engl. J. Med., 334: 292–295.

Cowley, M.A., Smart, J.L., Rubinstein, M., Cerdan, M.G., Diano, S., Horvath, T.L., Cone, R.D. and Low, M.J. (2001) Leptin activates anorexigenic POMC neurons through a neural network in the arcuate nucleus. Nature, 41: 480–484.

Degawa-Yamauchi, M., Bovenkerk, J.E., Juliar, B.E., Watson, W., Kerr, K., Jones, R., Zhu, Q. and Considine, R.V. (2003) Serum resistin (FIZZ3) protein is increased in obese humans. J. Clin. Endocrinol. Metab., 88: 5452–5455.

Devos, R., Richards, J.G., Campfield, L.A., Tartaglia, L.A., Guisez, Y., Van der Heyden, J., Travernier, J., Plaetinck, G. and Burn, P. (1996) OB protein binds specifically to the choroid plexus of mice and rats. Proc. Natl. Acad. Sci. USA, 93: 5668–5673.

Dube, M.G., Xu, B., Kalra, P.S., Sninsky, C.A. and Kalra, S.P. (1999) Disruption in neuropeptide Y and leptin signaling in obese ventromedial hypothalamic-lesioned rats. Brain Res., 816: 38–46.

El-Haschimi, K., Pierroz, D.D., Hileman, S.M., Bjorbaek, C. and Flier, J.S. (2000) Two defects contribute to hypothalamic leptin resistance in mice with diet-induced obesity. J. Clin. Invest., 105: 1827–1832.

Elias, C.F., Aschkenasi, C., Lee, C., Kelly, J., Ahima, R.S., Bjorbaek, C., Flier, J.S., Saper, C.B. and Elmquist, J.K. (1999) Leptin differentially regulates NPY and POMC neurons projecting to the lateral hypothalamic area. Neuron, 23: 775–786.

Elias, C.F., Kelly, J.F., Lee, C.E., Ahima, R.S., Drucker, D.J., Saper, C.B. and Elmquist, J.K. (2000) Chemical characterization of leptin-activated neurons in the rat brain. J. Comp. Neurol., 423: 261–281.

Elmquist, J.K., Ahima, R.S., Elias, C.F., Flier, J.S. and Saper, C.B. (1998) Leptin activates distinct projections from the dorsomedial and ventromedial hypothalamic nuclei. Proc. Natl. Acad. Sci. USA, 95: 741–746.

Erickson, J.C., Hollopeter, G. and Palmiter, R.D. (1996) Attenuation of the obesity syndrome of *ob/ob* mice by the loss of neuropeptide Y. Science, 274: 1704–1707.

Farooqi, I.S., Matarese, G., Lord, G.M., Keogh, J.M., Lawrence, E., Agwu, C., Sanna, V., Jebb, S.A., Perna, F., Fontana, S., Lechler, R.I., DePaoli, A.M. and O'Rahilly, S. (2002) Beneficial effects of leptin on obesity, T cell hyporesponsiveness, and neuroendocrine/metabolic dysfunction of human congenital leptin deficiency. J. Clin. Invest., 110: 1093–1103.

Fisher, F.F., Trujillo, M.E., Hanif, W., Barnett, A.H., McTernan, P.G., Scherer, P.E. and Kumar, S. (2005) Serum high molecular weight complex of adiponectin correlates better with glucose tolerance than total serum adiponectin in Indo-Asian males. Diabetologia, 48: 1084–1087.

Flegal, K.M., Graubard, B.I., Williamson, D.F. and Gail, M.H. (2005) Excess deaths associated with underweight, overweight, and obesity. JAMA, 293: 1861–1867.

Flier, J.S. (2004) Obesity wars: molecular progress confronts an expanding epidemic. Cell, 116: 337–350.

Frederich, R.C., Hamann, A., Anderson, S., Lollmann, B., Lowell, B.B. and Flier, J.S. (1995a) Leptin levels reflect body lipid content in mice: evidence for diet-induced resistance to leptin action. Nat. Med., 1: 1311–1314.

Frederich, R.C., Lollmann, B., Hamann, A., Napolitano-Rosen, A., Kahn, B.B., Lowell, B.B. and Flier, J.S. (1995b) Expression of ob mRNA and its encoded protein in rodents. Impact of nutrition and obesity. J. Clin. Invest., 96: 1658–1663.

Friedman, J.M. and Halaas, J.L. (1998) Leptin and the regulation of body weight in mammals. Nature, 395: 763–770.

Fulton, S., Woodside, B. and Shizgal, P. (2000) Modulation of brain reward circuitry by leptin. Science, 287: 125–128 Erratum in: Science 287,1931.

Gaige, S., Abysique, A. and Bouvier, M. (2002) Effects of leptin on cat intestinal vagal mechanoreceptors. J. Physiol., 543: 679–689.

Gaige, S., Abysique, A. and Bouvier, M. (2002) Effects of leptin on cat intestinal vagal mechanoreceptors. J. Physiol., 555: 297–310.

Gao, Q., Wolfgang, M.J., Neschen, S., Morino, K., Horvath, T.L., Shulman, G.I. and Fu, X.Y. (2004) Disruption of neural signal transducer and activator of transcription 3 causes obesity, diabetes, infertility, and thermal dysregulation. Proc. Natl. Acad. Sci. USA, 101: 4661–4666.

Golden, P.L., Maccagnan, T.J. and Pardridge, W.M. (1997) Human blood–brain barrier leptin receptor. Binding and endocytosis in isolated human brain microvessels. J. Clin. Invest., 99: 14–18.

Grill, H.J. and Kaplan, J.M. (2002) The neuroanatomical axis for control of energy balance. Front. Neuroendocrinol., 23: 2–40.

Gutierrez-Juarez, R., Obici, S. and Rossetti, L. (2004) Melanocortin-independent effects of leptin on hepatic glucose fluxes. J. Biol. Chem., 279: 49704–49715.

Halaas, J.L., Boozer, C., Blair-West, J., Fidahusein, N., Denton, D.A. and Friedman, J.M. (1997) Physiological response to long-term peripheral and central leptin infusion in lean and obese mice. Proc. Natl. Acad. Sci. USA, 94: 8878–8883.

Halaas, J.L., Gajiwala, K.S., Maffei, M., Cohen, S.L., Chait, B.T., Rabinowitz, D., Lallone, R.L., Burley, S.K. and Friedman, J.M. (1995) Weight-reducing effects of the plasma protein encoded by the obese gene. Science, 269: 543–546.

Hedley, A.A., Ogden, C.L., Johnson, C.L., Carroll, M.D., Curtin, L.R. and Flegal, K.M. (2004) Prevalence of overweight and obesity among US children, adolescents and adults, 1999–2002. JAMA, 291: 2847–2850.

Hervey, G.R. (1959) The effects of lesions in the hypothalamus in parabiotic rats. J. Physiol., 145: 336–352.

Heymsfield, S.B., Greenberg, A.S., Fujioka, K., Dixon, R.M., Kushner, R., Hunt, T., Lubina, J.A., Patane, J., Self, B., Hunt, P. and McCamish, M. (1999) Recombinant leptin for weight loss in obese and lean adults: a randomized, controlled, dose-escalation trial. JAMA, 282: 1568–1575.

Hileman, S.M., Pierroz, D.D., Masuzaki, H., Bjorbaek, C., El-Haschimi, K., Banks, W.A. and Flier, J.S. (2002) Characterization of short isoforms of the leptin receptor in rat cerebral microvessels and of brain uptake of leptin in mouse models of obesity. Endocrinology, 143: 775–783.

Hileman, S.M., Tornoe, J., Flier, J.S. and Bjorbaek, C. (2000) Transcellular transport of leptin by the short leptin receptor isoform ObRa in Madin–Darby canine kidney cells. Endocrinology, 141: 1955–1961.

Honda, K., Narita, K., Murata, T. and Higuchi, T. (2002) Leptin affects the electrical activity of neurons in the hypothalamic supraoptic nucleus. Brain Res. Bull., 57: 721–725.

Howard, J.K., Cave, B.J., Oksanen, L.J., Tzameli, I., Bjorbaek, C. and Flier, J.S. (2004) Enhanced leptin sensitivity and attenuation of diet-induced obesity in mice with haploinsufficiency of SOCS-3. Nat. Med., 10: 734–738.

Hukshorn, C.J., Saris, W.H., Westerterp-Plantenga, M.S., Farid, A.R., Smith, F.J. and Campfield, L.A. (2000) Weekly subcutaneous pegylated recombinant native human leptin (PEG-OB) administration in obese men. J. Clin. Endocrinol. Metab., 85: 4003–4009.

Hukshorn, C.J., Van Dielen, F.M., Buurman, W.A., Westerterp-Plantenga, M.S., Campfield, L.A. and Saris, W.H. (2002) The effect of pegylated recombinant human leptin (PEG-OB) on weight loss and inflammatory status in obese subjects. Int. J. Obes. Relat. Metab. Disord., 26: 504–509.

Ilyin, S.E., Gayle, D., Flynn, M.C. and Plata-Salaman, C.R. (1998) Interleukin-1β system (ligand, receptor type I, receptor accessory protein and receptor antagonist), TNF-α, TGF-β1 and neuropeptide Y mRNAs in specific brain regions during bacterial LPS-induced anorexia. Brain Res. Bull., 45: 507–515.

Ingalls, G.M., Dickie, M.M. and Snell, G.D. (1950) Obese, a new mutation in the house mouse. J. Hered., 41: 317–378.

Javor, E.D., Cochran, E.K., Musso, C., Young, J.R., Depaoli, A.M. and Gorden, P. (2005a) Long-term efficacy of leptin replacement in patients with generalized lipodystrophy. Diabetes, 54: 1994–2002.

Javor, E.D., Ghany, M.G., Cochran, E.K., Oral, E.A., DePaoli, A.M., Premkumar, A., Kleiner, D.E. and Gorden, P. (2005b) Leptin reverses nonalcoholic steatohepatitis in patients with severe lipodystrophy. Hepatology, 41: 753–760.

Kadowaki, T. and Yamauchi, T. (2005) Adiponectin and adiponectin receptors. Endocrinol. Rev., 26: 439–451.

Kahn, B.B., Alquier, T., Carling, D. and Hardie, D.G. (2005) AMP-activated protein kinase: ancient energy gauge provides clues to modern understanding of metabolism. Cell Metab., 1: 15–25.

Kamohara, S., Burcelin, R., Halaas, J.L., Friedman, J.M. and Charron, M.J. (1997) Acute stimulation of glucose metabolism in mice by leptin treatment. Nature, 389: 374–377.

Kastin, A.J. and Pan, W. (2000) Dynamic regulation of leptin entry into brain by the blood–brain barrier. Regul. Peptides, 25(92): 37–43.

Kawai, K., Sugimoto, K., Nakashima, K., Miura, H. and Ninomiya, Y. (2000) Leptin as a modulator of sweet taste sensitivities in mice. Proc. Natl. Acad. Sci. USA, 97: 11044–11049.

Kennedy, G.C. (1953) The role of depot fat in the hypothalamic control of food intake in the rat. Proc. R. Soc. Lond. B. Biol. Sci., 140: 578–596.

Kim, K.H., Lee, K., Moon, Y.S. and Sul, H.S. (2001) A cysteine-rich adipose tissue-specific secretory factor inhibits adipocyte differentiation. J. Biol. Chem., 276: 11252–11256.

Kim, K.H., Zhao, L., Moon, Y., Kang, C. and Sul, H.S. (2004) Dominant inhibitory adipocyte-specific secretory factor (ADSF)/resistin enhances adipogenesis and improves insulin sensitivity. Proc. Natl. Acad. Sci. USA, 101: 6780–6785.

Kolaczynski, J.W., Nyce, M.R., Considine, R.V., Boden, G., Nolan, J.J., Henry, R., Mudaliar, S.R., Olefsky, J. and Caro, J.F. (1996a) Acute and chronic effects of insulin on leptin production in humans: studies in vivo and in vitro. Diabetes, 45: 699–701.

Kolaczynski, J.W., Ohannesian, J.P., Considine, R.V., Marco, C.C. and Caro, J.F. (1996b) Response of leptin to short-term and prolonged overfeeding in humans. J. Clin. Endocrinol. Metab., 81: 4162–4165.

Kopelman, P.G. (2000) Obesity as a medical problem. Nature, 404: 635–643.

Kreier, F., Fliers, E., Voshol, P.J., Van Eden, C.G., Havekes, L.M., Kalsbeek, A., Van Heijningen, C.L., Sluiter, A.A., Mettenleiter, T.C., Romijn, J.A., Sauerwein, H.P. and Buijs, R.M. (2002) Selective parasympathetic innervation of subcutaneous and intra-abdominal fat — functional implications. J. Clin. Invest., 110: 1243–1250.

Kubota, N., Terauchi, Y., Yamauchi, T., Kubota, T., Moroi, M., Matsui, J., Eto, K., Yamashita, T., Kamon, J., Satoh, H., Yano, W., Froguel, P., Nagai, R., Kimura, S., Kadowaki, T. and Noda, T. (2002) Disruption of adiponectin causes insulin resistance and neointimal formation. J. Biol. Chem., 277: 25863–25866.

Kusminski, C.M., McTernan, P.G. and Kumar, S. (2005) Role of resistin in obesity, insulin resistance and Type II diabetes. Clin. Sci. (Lond)., 109: 243–256.

Lee, G., Li, C., Montez, J., Halaas, J., Darvishzadeh, J. and Friedman, J.M. (1997) Leptin receptor mutations in 129 db3J/db3J mice and NIH facp/facp rats. Mamm. Genome, 8: 445–447.

Lee, J.H., Chan, J.L., Yiannakouris, N., Kontogianni, M., Estrada, E., Seip, R., Orlova, C. and Mantzoros, C.S. (2003) Circulating resistin levels are not associated with obesity or insulin resistance in humans and are not regulated by fasting or leptin administration: cross-sectional and interventional studies in normal, insulin-resistant, and diabetic subjects. J. Clin. Endocrinol. Metab., 88: 4848–4856.

Legradi, G., Emerson, C.H., Ahima, R.S., Rand, W.M., Flier, J.S. and Lechan, R.M. (1998) Arcuate nucleus ablation prevents fasting-induced suppression of ProTRH mRNA in the hypothalamic paraventricular nucleus. Neuroendocrinology, 68: 89–97.

Lehrke, M., Reilly, M.P., Millington, S.C., Iqbal, N., Rader, D.J. and Lazar, M.A. (2004) An inflammatory cascade leading to hyperresistinemia in humans. PLoS Med., 1(2): e45.

Levin, B.E. and Dunn-Meynell, A.A. (1997) Dysregulation of arcuate nucleus preproneuropeptide Y mRNA in diet-induced obese rats. Am. J. Physiol., 272: R1365–R1370.

Levin, B.E., Dunn-Meynell, A.A. and Banks, W.A. (2004) Obesity-prone rats have normal blood–brain barrier transport but defective central leptin signaling before obesity onset. Am. J. Physiol. Regul. Integr. Comp. Physiol., 286: R143–R150.

Maeda, N., Shimomura, I., Kishida, K., Nishizawa, H., Matsuda, M., Nagaretani, H., Furuyama, N., Kondo, H., Takahashi, M., Arita, Y., Komuro, R., Ouchi, N., Kihara, S., Tochino, Y., Okutomi, K., Horie, M., Takeda, S., Aoyama, T., Funahashi T. and Matsuzawa, Y. (2002) Diet-induced insulin resistance in mice lacking adiponectin/ACRP30. Nat. Med., 8: 731–737.

Maness, L.M., Kastin, A.J., Farrell, C.L. and Banks, W.A. (1998) Fate of leptin after intracerebroventricular injection into the mouse brain. Endocrinology, 139: 4556–4562.

Maness, L.M., Banks, W.A. and Kastin, A.J. (2000) Persistence of blood-to-brain transport of leptin in obese leptin-deficient and leptin receptor-deficient mice. Brain Res., 873: 165–167.

Mantzoros, C., Flier, J.S., Lesem, M.D., Brewerton, T.D. and Jimerson, D.C. (1997) Cerebrospinal fluid leptin in anorexia nervosa: correlation with nutritional status and potential role in resistance to weight gain. J. Clin. Endocrinol. Metab., 82: 1845–1851.

Masaki, T., Chiba, S., Yasuda, T., Tsubone, T., Kakuma, T., Shimomura, I., Funahashi, T., Matsuzawa, Y. and Yoshimatsu, H. (2003) Peripheral, but not central, administration of adiponectin reduces visceral adiposity and upregulates the expression of uncoupling protein in agouti yellow (Ay/a) obese mice. Diabetes, 52: 2266–2273.

Masuzaki, H., Ogawa, Y., Hosoda, K., Miyawaki, T., Hanaoka, I., Hiraoka, J., Yasuno, A., Nishimura, H., Yoshimasa, Y., Nishi, S. and Nakao, K. (1997) Glucocorticoid regulation of leptin synthesis and secretion in humans: elevated plasma leptin levels in Cushing's syndrome. J. Clin. Endocrinol. Metab., 82: 2542–2547.

Matochik, J.A., London, E.D., Yildiz, B.O., Ozata, M., Caglayan, S., DePaoli, A.M., Wong, M.L. and Licinio, J. (2005) Effect of leptin replacement on brain structure in genetically leptin-deficient adults. J. Clin. Endocrinol. Metab., 90: 2851–2854.

172

McCarthy, T.J., Banks, W.A., Farrell, C.L., Adamu, S., Derdeyn, C.P., Snyder, A.Z., Laforest, R., Litzinger, D.C., Martin, D., LeBel, C.P. and Welch, M.J. (2002) Positron emission tomography shows that intrathecal leptin reaches the hypothalamus in baboons. J. Pharmacol. Exp. Ther., 301: 878–883.

Minokoshi, Y., Alquier, T., Furukawa, N., Kim, Y.B., Lee, A., Xue, B., Mu, J., Foufelle, F., Ferre, P., Birnbaum, M.J., Stuck, B.J. and Kahn, B.B. (2004) AMP-kinase regulates food intake by responding to hormonal and nutrient signals in the hypothalamus. Nature, 428: 569–574.

Minokoshi, Y., Kim, Y.B., Peroni, O.D., Fryer, L.G., Muller, C., Carling, D. and Kahn, B.B. (2002) Leptin stimulates fatty-acid oxidation by activating AMP-activated protein kinase. Nature, 415: 339–343.

Moller, D.E. (2000) Potential role of TNF-α in the pathogenesis of insulin resistance and Type 2 diabetes. Trends Endocrinol. Metab., 11: 212–217.

Montague, C.T., Farooqi, I.S., Whitehead, J.P., Soos, M.A., Rau, H., Wareham, N.J., Sewter, C.P., Digby, J.E., Mohammed, S.N., Hurst, J.A., Cheetham, C.H., Earley, A.R., Barnett, A.H., Prins, J.B. and O'Rahilly, S. (1997) Congenital leptin deficiency is associated with severe early-onset obesity in humans. Nature, 387: 903–908.

Mori, H., Hanada, R., Hanada, T., Aki, D., Mashima, R., Nishinakamura, H., Torisu, T., Chien, K.R., Yasukawa, H. and Yoshimura, A. (2004) SOCS-3 deficiency in the brain elevates leptin sensitivity and confers resistance to diet-induced obesity. Nat. Med., 10: 739–743.

Morton, G.J., Blevins, J.E., Williams, D.L., Niswender, K.D., Gelling, R.W., Rhodes, C.J., Baskin, D.G. and Schwartz, M.W. (2005) Leptin action in the forebrain regulates the hindbrain response to satiety signals. J. Clin. Invest., 115: 703–710.

Morton, G.J., Niswender, K.D., Rhodes, C.J., Myers Jr., M.G., Blevins, J.E., Baskin, D.G. and Schwartz, M.W. (2003) Arcuate nucleus-specific leptin receptor gene therapy attenuates the obesity phenotype of Koletsky [fa(k)/fa(k)] rats. Endocrinology, 144: 2016–2024.

Muse, E.D., Lam, T.K.T., Scherer, P.E. and Rossetti, L. (2005) Central administration of recombinant resistin induces hepatic insulin resistance. American Diabetes Association abstract 19-OR.

Muse, E.D., Obici, S., Bhanot, S., Monia, B.P., McKay, R.A., Rajala, M.W., Scherer, P.E. and Rossetti, L. (2004) Role of resistin in diet-induced hepatic insulin resistance. J. Clin. Invest., 114: 232–239.

Musso, C., Cochran, E., Javor, E., Young, J., Depaoli, A.M. and Gorden, P. (2005) The long-term effect of recombinant methionyl human leptin therapy on hyperandrogenism and menstrual function in female and pituitary function in male and female hypoleptinemic lipodystrophic patients. Metabolism, 54: 255–263.

Muzumdar, R., Ma, X., Yang, X., Atzmon, G., Bernstein, J., Karkanias, G. and Barzilai, N. (2004) Decrease in glucose-stimulated insulin secretion with aging is independent of insulin action. Diabetes, 53: 441–446.

Myers Jr., M.G. (2004) Leptin receptor signaling and the regulation of mammalian physiology. Recent Prog. Horm. Res., 59: 287–304.

Niswender, K.D., Morton, G.J., Stearns, W.H., Rhodes, C.J., Myers Jr., M.G. and Schwartz, M.W. (2001) Intracellular signaling. Key enzyme in leptin-induced anorexia. Nature, 413: 794–795.

Niswender, K.D. and Schwartz, M.W. (2004) Insulin and its evolving partnership with leptin in the hypothalamic control of energy homeostasis. Trends Endocrinol. Metab., 15: 362–369.

Obici, S., Feng, Z., Tan, J., Liu, L., Karkanias, G. and Rossetti, L. (2001) Central melanocortin receptors regulate insulin action. J. Clin. Invest., 108: 1079–1085.

Ofei, F., Hurel, S., Newkirk, J., Sopwith, M. and Taylor, R. (1996) Effects of an engineered human anti-TNF-α antibody (CDP571) on insulin sensitivity and glycemic control in patients with NIDDM. Diabetes, 45: 881–885.

Oral, E.A., Ruiz, E., Andewelt, A., Sebring, N., Wagner, A.J., Depaoli, A.M. and Gorden, P. (2002) Effect of leptin replacement on pituitary hormone regulation in patients with severe lipodystrophy. J. Clin. Endocrinol. Metab., 87: 3110–3117.

Pajvani, U.B., Hawkins, M., Combs, T.P., Rajala, M.W., Doebber, T., Berger, J.P., Wagner, J.A., Wu, M., Knopps, A., Xiang, A.H., Utzschneider, K.M., Kahn, S.E., Olefsky, J.M., Buchanan, T.A. and Scherer, P.E. (2004) Complex distribution, not absolute amount of adiponectin, correlates with thiazolidinedione-mediated improvement in insulin sensitivity. J. Biol. Chem., 279: 12152–12162.

Parameswaran, S.V., Steffens, A.B., Hervey, G.R. and De Ruiter, L. (1977) Involvement of a humoral factor in regulation of body weight in parabiotic rats. Am. J. Physiol., 232: R150–R157.

Pardridge, W.M. (1981) Transport of nutrients and hormones through the blood–brain barrier. Diabetologia, 20(Suppl): 246–254.

Park, S., Park, C.H., Jang, J.S. and Choi, S.B. (2005a) Chronic leptin and resistin effects on insulin resistance and insulin secretion in pancreatectomized rats. American Diabetes Association abstract no. 1515-P.

Park, H.S., Park, J.Y. and Yu, R. (2005b) Relationship of obesity and visceral adiposity with serum concentrations of CRP, TNF-α and IL-6. Diabetes Res. Clin. Pract., 69: 29–35.

Patel, S.D., Rajala, M.W., Rossetti, L., Scherer, P.E. and Shapiro, L. (2004) Disulfide-dependent multimeric assembly of resistin family hormones. Science, 304: 1154–1158.

Pelleymounter, M.A., Cullen, M.J., Baker, M.B., Hecht, R., Winters, D., Boone, T. and Collins, F. (1995) Effects of the obese gene product on body weight regulation in ob/ob mice. Science, 269: 540–543.

Pinto, S., Roseberry, A.G., Liu, H., Diano, S., Shanabrough, M., Cai, X., Friedman, J.M. and Horvath, T.L. (2004) Rapid rewiring of arcuate nucleus feeding circuits by leptin. Science, 304: 110–115.

Qi, Y., Takahashi, N., Hileman, S.M., Patel, H.R., Berg, A.H., Pajvani, U.B., Scherer, P.E. and Ahima, R.S. (2004) Adiponectin acts in the brain to decrease body weight. Nat. Med., 10: 524–529 Erratum in: Nat. Med. 10, 2004, 649.

Rajala, M.W., Qi, Y., Patel, H.R., Takahashi, N., Banerjee, R., Pajvani, U.B., Sinha, M.K., Gingerich, R.L., Scherer, P.E. and Ahima, R.S. (2004) Regulation of resistin expression and circulating levels in obesity, diabetes, and fasting. Diabetes, 53: 1671–1679.

Rangwala, S.M., Rich, A.S., Rhoades, B., Shapiro, J.S., Obici, S., Rossetti, L. and Lazar, M.A. (2004) Abnormal glucose homeostasis due to chronic hyperresistinemia. Diabetes, 53: 1937–1941.

Reilly, M.P., Lehrke, M., Wolfe, M.L., Rohatgi, A., Lazar, M.A. and Rader, D.J. (2005) Resistin is an inflammatory marker of atherosclerosis in humans. Circulation, 111: 932–939.

Rosenbaum, M., Murphy, E.M., Heymsfield, S.B., Matthews, D.E. and Leibel, R.L. (2002) Low dose leptin administration reverses effects of sustained weight-reduction on energy expenditure and circulating concentrations of thyroid hormones. J. Clin. Endocrinol. Metab., 87: 2391–2394.

Russo, V.C., Metaxas, S., Kobayashi, K., Harris, M. and Werther, G.A. (2004) Antiapoptotic effects of leptin in human neuroblastoma cells. Endocrinology, 145: 4103–4112.

Satoh, H., Nguyen, M.T., Miles, P.D., Imamura, T., Usui, I. and Olefsky, J.M. (2004) Adenovirus-mediated chronic "hyper-resistinemia" leads to in vivo insulin resistance in normal rats. J. Clin. Invest., 114: 224–231.

Savage, D.B., Sewter, C.P., Klenk, E.S., Segal, D.G., Vidal-Puig, A., Considine, R.V. and O'Rahilly, S. (2001) Resistin/Fizz3 expression in relation to obesity and peroxisome proliferator-activated receptor-γ action in humans. Diabetes, 50: 2199–2202.

Schwartz, M.W., Peskind, E., Raskind, M., Boyko, E.J. and Porte Jr., D. (1996) Cerebrospinal fluid leptin levels: relationship to plasma levels and to adiposity in humans. Nat. Med., 2: 589–593.

Segal-Lieberman, G., Bradley, R.L., Kokkotou, E., Carlson, M., Trombly, D.J., Wang, X., Bates, S., Myers Jr., M.G., Flier, J.S. and Maratos-Flier, E. (2003) Melanin-concentrating hormone is a critical mediator of the leptin-deficient phenotype. Proc. Natl. Acad. Sci. USA, 100: 10085–10090.

Sena, A., Sarlieve, L.L. and Rebel, G. (1985) Brain myelin of genetically obese mice. J. Neurol. Sci., 68: 233–243.

Shklyaev, S., Aslanidi, G., Tennant, M., Prima, V., Kohlbrenner, E., Kroutov, V., Campbell-Thompson, M., Crawford, J., Shek, E.W., Scarpace, P.J. and Zolotukhin, S. (2003) Sustained peripheral expression of transgene adiponectin offsets the development of diet-induced obesity in rats. Proc. Natl. Acad. Sci. USA, 100: 14217–14222.

Spanswick, D., Smith, M.A., Groppi, V.E., Logan, S.D. and Ashford, M.L. (1997) Leptin inhibits hypothalamic neurons by activation of ATP-sensitive potassium channels. Nature, 390: 521–525.

Stephens, T.W., Basinski, M., Bristow, P.K., Bue-Valleskey, J.M., Burgett, S.G., Craft, L., Hale, J., Hoffmann, J., Hsiung, H.M., Kriauciunas, A., Mackellar, W., Rosteck, P.R., Schoner, B., Smith, D., Tinsley, F.C., Zhang, X.Y. and Heiman, M. (1995) The role of neuropeptide Y in the anti-obesity action of the *obese* gene product. Nature, 377: 530–532.

Steppan, C.M., Bailey, S.T., Bhat, S., Brown, E.J., Banerjee, R.R., Wright, C.M., Patel, H.R., Ahima, R.S. and Lazar, M.A. (2001a) The hormone resistin links obesity to diabetes. Nature, 409: 307–312.

Steppan, C.M., Brown, E.J., Wright, C.M., Bhat, S., Banerjee, R.R., Dai, C.Y., Enders, G.H., Silberg, D.G., Wen, X., Wu, G.D. and Lazar, M.A. (2001b) A family of tissue-specific resistin-like molecules. Proc. Natl. Acad. Sci. USA, 98: 502–506.

Steppan, C.M. and Lazar, M.A. (2004) The current biology of resistin. J. Intern. Med., 255: 439–447.

Steppan, C.M. and Swick, A.G. (1999) A role for leptin in brain development. Biochem. Biophys. Res. Commun., 256: 600–602.

Steppan, C.M., Wang, J., Whiteman, E.L., Birnbaum, M.J. and Lazar, M.A. (2005) Activation of SOCS-3 by resistin. Mol. Cell Biol., 25: 1569–1575.

Takahashi, K.A. and Cone, R.D. (2005) Fasting induces a large, leptin-dependent increase in the intrinsic action potential frequency of orexigenic arcuate nucleus NPY/AgRP neurons. Endocrinology, 146: 1043–1047.

Takahashi, N., Patel, H.R., Qi, Y., Dushay, J. and Ahima, R.S. (2002) Divergent effects of leptin in mice susceptible or resistant to obesity. Horm. Metab. Res., 34: 691–697.

Tartaglia, L.A. (1997) The leptin receptor. J. Biol. Chem., 272: 6093–6096.

Tartaglia, L.A., Dembski, M., Weng, X., Deng, N., Culpepper, J., Devos, R., Richards, G.J., Campfield, L.A., Clark, F.T., Deeds, J., Muir, C., Sanker, S., Moriarty, A., Moore, K.J., Smutko, J.S., Mays, G.G., Wool, E.A., Monroe, C.A. and Tepper, R.I. (1995) Identification and expression cloning of a leptin receptor, OB-R. Cell, 83: 1263–1271.

Tomas, E., Tsao, T.S., Saha, A.K., Murrey, H.E., Zhang, C.C., Itani, S.I., Lodish, H.F. and Ruderman, N.B. (2002) Enhanced muscle fat oxidation and glucose transport by ACRP30 globular domain: acetyl-CoA carboxylase inhibition and AMP-activated protein kinase activation. Proc. Natl. Acad. Sci. USA, 99: 16309–16313.

Tovar, S., Nogueiras, R., Tung, L.Y., Castaneda, T.R., Vazquez, M.J., Morris, A., Williams, L.M., Dickson, S.L. and Dieguez, C. (2005) Central administration of resistin promotes short-term satiety in rats. Eur. J. Endocrinol., 153: R1–R5.

Tracey, K.J. and Cerami, A. (1992) Tumor necrosis factor and regulation of metabolism in infection: role of systemic vs. tissue levels. Proc. Soc. Exp. Biol. Med., 200: 233–239.

Vaisse, C., Halaas, J.L., Horvath, C.M., Darnell Jr., J.E., Stoffel, M. and Friedman, J.M. (1996) Leptin activation of STAT 3 in the hypothalamus of wild-type and *ob/ob* mice but not *db/db* mice. Nat. Genet., 14: 95–97.

Vozarova de Courten, B., Degawa-Yamauchi, M., Considine, R.V. and Tataranni, P.A. (2004) High serum resistin is associated with an increase in adiposity but not a worsening of insulin resistance in Pima Indians. Diabetes, 53: 1279–1284.

Waki, H., Yamauchi, T., Kamon, J., Ito, Y., Uchida, S., Kita, S., Hara, K., Hada, Y., Vasseur, F., Froguel, P., Kimura, S., Nagai, R. and Kadowaki, T. (2003) Impaired multimerization of human adiponectin mutants associated with diabetes. Molecular structure and multimer formation of adiponectin. J. Biol. Chem., 278: 40352–40363.

174

Wallenius, V., Wallenius, K., Ahren, B., Rudling, M., Carlsten, H., Dickson, S.L., Ohlsson, C. and Jansson, J.O. (2002) Interleukin-6-deficient mice develop mature-onset obesity. Nat. Med., 8: 75–79.

Welt, C.K., Chan, J.L., Bullen, J., Murphy, R., Smith, P., DePaoli, A.M., Karalis, A. and Mantzoros, C.S. (2004) Recombinant human leptin in women with hypothalamic amenorrhea. N. Engl. J. Med., 351: 987–997.

Westerterp-Plantenga, M.S., Saris, W.H., Hukshorn, C.J. and Campfield, L.A. (2001) Effects of weekly administration of pegylated recombinant human OB protein on appetite profile and energy metabolism in obese men. Am. J. Clin. Nutr., 74: 426–434.

Winder, W.W. and Hardie, D.G. (1999) AMP-activated protein kinase, a metabolic master switch: possible roles in Type 2 diabetes. Am. J. Physiol., 277: E1–E10.

Wolford, J.K., Colligan, P.B., Gruber, J.D. and Bogardus, C. (2003) Variants in the interleukin-6-receptor gene are associated with obesity in Pima Indians. Mol. Genet. Metab., 80: 338–343.

Yamauchi, T., Kamon, J., Ito, Y., Tsuchida, A., Yokomizo, T., Kita, S., Sugiyama, T., Miyagishi, M., Hara, K., Tsunoda, M., Murakami, K., Ohteki, T., Uchida, S., Takekawa, S., Waki, H., Tsuno, N.H., Shibata, Y., Terauchi, Y., Froguel, P., Tobe, K., Koyasu, S., Taira, K., Kitamura, T., Shimizu, T., Nagai, R. and Kadowaki, T. (2003a) Cloning of adiponectin receptors that mediate antidiabetic metabolic effects. Nature, 423: 762–769 Erratum in: Nature 431, 2004, 1123.

Yamauchi, T., Kamon, J., Minokoshi, Y., Ito, Y., Waki, H., Uchida, S., Yamashita, S., Noda, M., Kita, S., Ueki, K., Eto, K., Akanuma, Y., Froguel, P., Foufelle, F., Ferre, P., Carling, D., Kimura, S., Nagai, R., Kahn, B.B. and Kadowaki, T. (2002) Adiponectin stimulates glucose utilization and fatty-acid oxidation by activating AMP-activated protein kinase. Nat. Med., 8: 1288–1295.

Yamauchi, T., Kamon, J., Waki, H., Imai, Y., Shimozawa, N., Hioki, K., Uchida, S., Ito, Y., Takakuwa, K., Matsui, J., Takata, M., Eto, K., Terauchi, Y., Komeda, K., Tsunoda, M., Murakami, K., Ohnishi, Y., Naitoh, T., Yamamura, K., Ueyama, Y., Froguel, P., Kimura, S., Nagai, R. and Kadowaki, T. (2003b) Globular adiponectin protected ob/ob mice from diabetes and ApoE-deficient mice from atherosclerosis. J. Biol. Chem., 278: 2461–2468.

Yura, S., Itoh, H., Sagawa, N., Yamamoto, H., Masuzaki, H., Nakao, K., Kawamura, M., Takemura, M., Kakui, K., Ogawa, Y. and Fuji, S. (2005) Role of premature leptin surge in obesity resulting from intrauterine undernutrition. Cell Metab., 1: 371–378.

Yuzuriha, H., Inui, A., Goto, K., Asakawa, A., Fujimiya, M. and Kasuga, M. (2003) Intracerebroventricular administration of NPY stimulates resistin gene expression in mice. Int. J. Mol. Med., 11: 675–676.

Zabolotny, J.M., Bence-Hanulec, K.K., Stricker-Krongrad, A., Haj, F., Wang, Y., Minokoshi, Y., Kim, Y.B., Elmquist, J.K., Tartaglia, L.A., Kahn, B.B. and Neel, B.G. (2002) PTP1B regulates leptin signal transduction in vivo. Dev. Cell, 2: 489–495.

Zhang, Y., Proenca, R., Maffei, M., Barone, M., Leopold, L. and Friedman, J.M. (1994) Positional cloning of the mouse obese gene and its human homologue. Nature, 372: 425–432 Erratum in: Nature 374, 1995, 479.

Zhao, A.Z., Huan, J.N., Gupta, S., Pal, R. and Sahu, A. (2002) A phosphatidylinositol 3-kinase phosphodiesterase 3B-cyclic AMP pathway in hypothalamic action of leptin on feeding. Nat. Neurosci., 5: 727–728.

Hypothalamic Control of Bone and Thyroid Metabolism

Kalsbeek, Fliers, Hofman, Swaab, Van Someren & Buijs
Progress in Brain Research, Vol. 153
ISSN 0079-6123

CHAPTER 10

The circadian modulation of leptin-controlled bone formation

Loning Fu[1,2,♯], Millan S. Patel[1,2,♯] and Gerard Karsenty[1,2,3,*]

[1]*Department of Molecular and Human Genetics, Baylor College of Medicine, Houston, TX, USA*
[2]*Bone Disease Program of Texas, Baylor College of Medicine, Houston, TX, USA*
[3]*Children's Nutrition Research Center, Baylor College of Medicine, Houston, TX, USA*

Abstract: Mice with circadian gene *Period* and *Cryptochrome* mutations develop high bone mass early in life. Such a phenotype is accompanied by an increase in osteoblast numbers in mutant bone and cannot be corrected by leptin intracerebroventricular infusion. Thus, the molecular clock plays a key role in leptin-mediated sympathetic regulation of bone formation. Indeed, we found that leptin-dependent sympathetic signaling induces the expression of AP1 and circadian genes in bone and in osteoblasts with similar kinetics, and these two pathways play opposite roles in controlling *c-myc* expression. Mutations in the *Period 1* and *2* genes result in uncontrolled *c-myc* signaling, overexpression of *G1 cyclins*, and increased osteoblast proliferation and bone-formation parameters. These results indicate that the role of leptin-dependent sympathetic signaling in bone formation is achieved through regulating two antagonistic pathways in osteoblasts.

Keywords: bone remodeling; leptin control of bone mass; the molecular clock

Introduction

In the life of all vertebrates, bone mass is controlled by a dynamic process called bone remodeling, which is a fine balance between bone formation and resorption. In bone formation, osteoblasts make new bone by deposition of extracellular matrix. In bone resorption, osteoclasts break down the preexisting mineralized bone. Recent studies have demonstrated that osteoblasts and osteoclasts also cross talk: osteoblasts could control the differentiation of osteoclasts (Rodan and Martin, 2000). Loss of balance between bone formation and resorption could lead to high bone mass or osteoporosis. However, although osteoporosis is the most prevalent degenerative disease in developed countries, the molecular mechanisms of bone remodeling are still poorly understood (Riggs and Melton, 1986; Chien and Karsenty, 2005).

Bone remodeling is controlled both locally and systemically. The local regulation is controlled by factors such as insulin-like growth factors, RANK ligand, and osteoprotegerin secreted by osteoblasts or osteoclasts. The systemic regulation is controlled by hormones or neural input such as sex steroids, parathyroid hormone, leptin, and sympathetic signaling. Among the many hormones that are involved in bone remodeling, leptin is unique in that it affects both bone formation and resorption profoundly. Our previous studies have demonstrated that leptin, by acting on hypothalamic neurons, could indirectly control sympathetic signaling that controls bone formation directly by regulating osteoblast numbers, and bone resorption indirectly by regulating

*Corresponding author.; E-mail: karsenty@bcm.tmc.edu
♯These authors contributed equally to this work.

DOI: 10.1016/S0079-6123(06)53010-9

osteoclast differentiation (Ducy et al., 2000; Takeda et al., 2002; Elefteriou et al., 2005).

Recent studies have shown that the serum level of leptin follows the circadian oscillation pattern. In addition, the synthesis of type I collagen and osteocalcin, the two main biosynthetic products of osteoblasts, also show diurnal variation in vivo. Thus, the circadian clock may participate in bone remodeling by either controlling the serum level of leptin or by mediating the leptin-dependent sympathetic signaling in osteoblasts (Simmons and Nichols, 1966; Gundberg et al., 1985).

The mammalian circadian clock is composed of a master clock in the suprachiasmatic nucleus of hypothalamus and peripheral clocks in most tissues. Both clocks are operated by the same set of the circadian genes (Reppert and Weaver, 2002). Among these genes, Clock and Bmal1 encode basic-helix-loop-helix (bHLH)-PAS transcription factors that heterodimerize to stimulate other core circadian genes such as Period (Per1, 2, and 3), Cryptochrome (Cry1 and 2), and Rev-Erbα. After being synthesized in the cytoplasm, the PER and CRY proteins form a hetero-multimeric complex with the circadian regulator casein kinase Iε and translocate into the nucleus to suppress the activity of BMAL1/CLOCK heterodimers, whereas REV-ERBα inhibits Bmal1 transcription through a retinoic acid-related orphan receptor response element in the Bmal1 promoter. These interlinked feedback loops of circadian genes result in a 24 h rhythmic activity of the molecular clock. The molecular clock also targets noncircadian genes. Most key cellular processes, therefore, display circadian rhythms as well (Fu and Lee, 2003).

To test the role of the molecular clock in bone remodeling, we studied mice lacking Per1 and 2, or Cry1 and 2 (Vitaterna et al., 1999; Zheng et al., 2001) and found that these mice display a high bone mass (HBM) phenotype early in life owing to increased bone formation. The HBM in these mice occurs in the presence of high level of serum leptin and cannot be corrected by intracerebroventricular (i.c.v.) leptin infusion that results in a decrease in bone mass in wild-type (wt) mice. Thus, the circadian genes mediate leptin-dependent sympathetic signaling in osteoblasts. We found that the sympathetic signaling activates AP1 and clock genes simultaneously in osteoblasts and in bone, and these two pathways play opposite roles in controlling c-myc expression. Mutation in the circadian genes results in uncontrolled AP1-c-myc signaling, overexpression of G1 cyclins, hyperplastic growth of osteoblasts, and abnormal bone formation. Thus, the circadian modulation of osteoblast proliferation in leptin-dependent sympathetic signaling has important physiological relevance.

High bone mass in circadian gene-mutant mice

Mice lacking Per1 and Per2 genes ($Per1^{-/-}$; $Per2^{-/-}$), Per1 gene, and the Per2 PAS domain ($Per1^{-/-}$;$Per2^{m/m}$), or Cry1 and Cry2 genes ($Cry1^{-/-}$;$Cry2^{-/-}$) showed a significant increase in bone mass (HBM) in both vertebrate and long bones as determined by histology and micro-computed tomography (Figs. 1A and B). The HBM was first observed at 6 weeks of age in these mice and worsened over time. A similar HBM shown by mice deficient in different circadian genes suggests that dysfunction of the molecular clock leads to HBM.

To identify which arm of bone remodeling was affected by the loss of clock function, we examined urinary level of deoxypyridinoline (DpD) cross-links, a byproduct of collagen degradation, and found that the DpD levels were similar in wt and ($Per1^{-/-}$;$Per2^{m/m}$) mice. Therefore, bone resorption was not overtly affected in these mice at young age (data not shown). In contrast, there was a significant increase in the number of osteoblasts and in bone-formation parameters (mineral apposition rate and bone-formation rate) in ($Per1^{-/-}$;$Per2^{m/m}$), ($Per1^{-/-}$;$Per2^{-/-}$), and ($Cry1^{-/-}$;$Cry2^{-/-}$) mice and in $Bmal1^{-/-}$ mice that did not show HBM (Fig. 1C). The absence of HBM in $Bmal1^{-/-}$ mice may be explained, at least in part, by their hypogonadism, a condition enhancing bone resorption (data not shown) (Takeda and Karsenty, 2001).

Bone has a peripheral clock

The similar osteoblast abnormalities observed in different clock gene-deficient mouse models suggested that loss of function in the circadian clock could lead to abnormal bone formation. Therefore, bone, like other tissues, may have a peripheral clock.

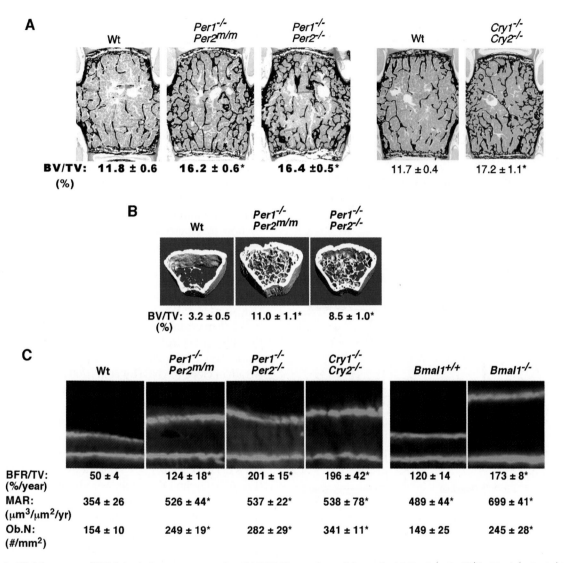

Fig. 1. High bone mass (HBM) in clock gene-mutant mice. (A) HBM in vertebrae of 6-month-old ($Per1^{-/-}$;$Per2^{m/m}$), ($Per1^{-/-}$;$Per2^{-/-}$), and ($Cry1^{-/-}$;$Cry2^{-/-}$) mice; (B) HBM in distal femora of 6-month-old ($Per1^{-/-}$;$Per2^{m/m}$) and ($Per1^{-/-}$;$Per2^{-/-}$) mice; and (C) elevated bone-formation parameters (BFR, bone-formation rate; MAR, mineral apposition rate) and osteoblast number (ObN) in wt and clock gene-mutant mice at 6 months of age. $Bmal1^{-/-}$ and control mice were studied at 2 months of age before mobility became hampered in the $Bmal1^{-/-}$ mice. Asterisks indicate statistically significant differences.

Consistent with this hypothesis, we found that all core circadian genes studied such as *Per1*, *Per2*, *Cry1*, *Bmal1*, and *Clock*, showed circadian expression in wt bones and their expression was decreased and found to be arrhythmic in ($Per1^{-/-}$;$Per2^{m/m}$) bones. Interestingly, the expression of *Npas2*, whose deletion does not affect bone mass, was not detected in either wt or ($Per1^{-/-}$;$Per2^{m/m}$) bones (Fig. 2).

Increased osteoblast proliferation contributes to HBM in ($Per1^{-/-}$;$Per2^{m/m}$) mice

A common feature observed among different clock gene-mutant mice was a large increase in osteoblast number in bone (Fig. 1D), suggesting that loss of function in the molecular clock may result in uncontrolled osteoblast proliferation that contributed

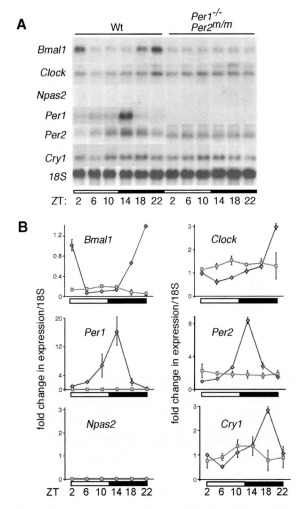

Fig. 2. Bone has a peripheral clock. (A) Northern blot analysis of core circadian gene expression in wt and ($Per1^{-/-}$;$Per2^{m/m}$) bones over a 24 h period. The mutant $Per2$ mRNA is seen as a faster-migrating band owing to an 87-amino acid in-frame deletion. (B) PhosphorImager quantification of three independent Northern blot analyses. After normalizing to 18S RNA, expression levels for each gene at ZT2 was arbitrarily set as 1. Error bars indicate standard deviation ($\pm$SD).

to HBM. Indeed, we found that the mitotic index of osteoblasts followed circadian rhythm in wt bone and was high at all times studied in ($Per1^{-/-}$; $Per2^{m/m}$) bone (Fig. 3A). In addition, ($Per1^{-/-}$;$Per2^{m/m}$) osteoblasts also grew faster than wt controls in vitro (Fig. 3B).

To study the mechanism resulting in deregulated ($Per1^{-/-}$;$Per2^{m/m}$) osteoblast proliferation, we first synchronized wt and ($Per1^{-/-}$;$Per2^{m/m}$) osteoblasts

at the G1 phase of cell cycle by serum starvation and then re-fed the cells with fresh medium to initiate cell proliferation. When the cell cycle profiles of these cells were examined by flow cytometry, we found that 16 h after re-feeding, synchronized ($Per1^{-/-}$; $Per2^{m/m}$) osteoblasts had a two-fold increase in G2 population, whereas wt osteoblasts only showed an increase in S but not in G2 population (Fig. 3C). We then synchronized wt and ($Per1^{-/-}$;$Per2^{m/m}$) osteoblasts in S phase using hydroxyurea and returned them to normal growth conditions, pulse labeled them with BrdU at various times in the subsequent 26 h period and determined the ratio of S-phase cells by flow cytometry. We found that the two S-phase peaks were separated by 24.0 ± 0.2 h in wt osteoblasts, but only by 21.0 ± 1.0 h in ($Per1^{-/-}$;$Per2^{m/m}$) osteoblasts ($p<0.05$) (Fig. 3D). Thus, ($Per1^{-/-}$; $Per2^{m/m}$) osteoblasts had a shorter G1 phase leading to a shorter cell cycle period.

We then examined the expression of *G1 cyclins*, as they play a key role in controlling the G1-phase duration (Fu et al., 2004) and are controlled by the circadian clock in vivo (Fu et al., 2002; Matsuo et al., 2003). We found that the expression of *cyclins D1* and *E* was significantly elevated in ($Per1^{-/-}$; $Per2^{m/m}$) osteoblasts (Fig. 4A). In addition, the expression of cyclin D1 showed circadian variation in wt bone, but was arrhythmic and elevated at most times in ($Per1^{-/-}$;$Per2^{m/m}$) bone during a 24 h period (Fig. 4B). Cyclin D1 protein was also more abundant in ($Per1^{-/-}$;$Per2^{m/m}$) osteoblasts than in wt controls (Fig. 4C). Taken together, these results demonstrated that loss of function in *Per1* and *2* resulted in elevated *G1 cyclin* expression, leading to shortening of cell cycle and increased osteoblast proliferation.

We then studied whether *cyclin D1* was a direct target of the molecular clock. We found that BMAL1 and CLOCK heterodimers did not affect *cyclin D1* promoter activity in DNA cotransfection assays. Instead, they inhibited the promoter activity of *c-myc*, a master regulator of *cyclin D1* (Fig. 4D). In EMSA, wt but not $Bmal1^{-/-}$ osteoblast nuclear extracts could bind to the proximal E box present in *c-myc* P1 promoter. However, no clear binding to the E-box sequence in *cyclin D1* promoter was detected in the same experiment (Fig. 4E and data not shown). Consistent with these observations, *c-myc* was overexpressed in ($Per1^{-/-}$;$Per2^{m/m}$) bone at

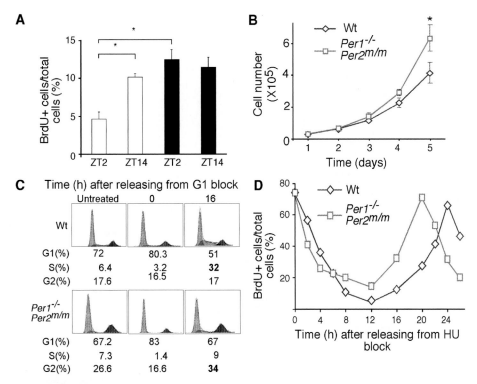

Fig. 3. Increased proliferation of ($Per1^{-/-}$;$Per2^{m/m}$) osteoblasts in vivo and in vitro. (A) Mitotic index of calvarial osteoblasts shows circadian variation in 6-week-old wt mice (clear bars), but was high at all times when studied in ($Per1^{-/-}$;$Per2^{m/m}$) mice (filled bars). Error bars indicate standard error of means ($\pm$ SEM). (B) Growth curves of primary wt and ($Per1^{-/-}$;$Per2^{m/m}$) osteoblasts over 5 days, showing a shorter doubling time for the mutant osteoblasts ($\pm$ SD). (C) Cell cycle profiles of wt and ($Per1^{-/-}$;$Per2^{m/m}$) osteoblasts under normal growth conditions and at 0 and 16 h after releasing from serum starvation. Note the increase in G2 population of ($Per1^{-/-}$;$Per2^{m/m}$) osteoblasts at 16 h after releasing G1 arrest owing to the synchronized cells progressing into G2/M phase, but not G2/M arrest. (D) The proportion of S-phase cells, determined by BrdU pulse-chasing and flow cytometry, in wt and ($Per1^{-/-}$;$Per2^{m/m}$) osteoblasts during a 26 h period after releasing from hydroxyurea block. The time interval between the two S-phase peaks was shorter in ($Per1^{-/-}$;$Per2^{m/m}$) osteoblasts than in wt controls.

most time points studied (Fig. 4F) and both c-myc mRNA and protein were overexpressed in ($Per1^{-/-}$; $Per2^{m/m}$) osteoblasts (Fig. 4G data not shown). Thus, the molecular clock may control *G1 cyclins* indirectly via controlling *c-myc* expression.

The central control of bone formation

We have previously demonstrated that leptin regulates bone formation via controlling sympathetic signaling. The ($Per1^{-/-}$;$Per2^{m/m}$) mice shared the similar HBM phenotypes with $Adr\beta2^{-/-}$ and ob/ob mice, i.e., increased osteoblast numbers and increased bone-formation parameters (Ducy et al., 2000; Takeda et al., 2002; Elefteriou et al., 2004). Because

Adrβ2, a G-coupled protein, is an adrenergic receptor expressed in osteoblasts, we hypothesized that the circadian clock may either control or mediate the leptin-dependent sympathetic regulation of bone formation.

To test this hypothesis, we first studied the serum level of leptin, and found that it followed circadian oscillation pattern in wt mice and was elevated at all time points during the 24 h period in ($Per1^{-/-}$; $Per2^{m/m}$) mice (Fig. 5A). Urinary levels of epinephrine and norepinepherine, and the expression of *Uncoupling Protein-1* (*Ucp1*) in brown fat tissue, which is a direct target of sympathetic tone (Scarpace and Matheny, 1998), all showed circadian variation in wt mice, and were significantly elevated at ZT22 for *Ucp1* mRNA and norepinephrine, and at ZT2 for

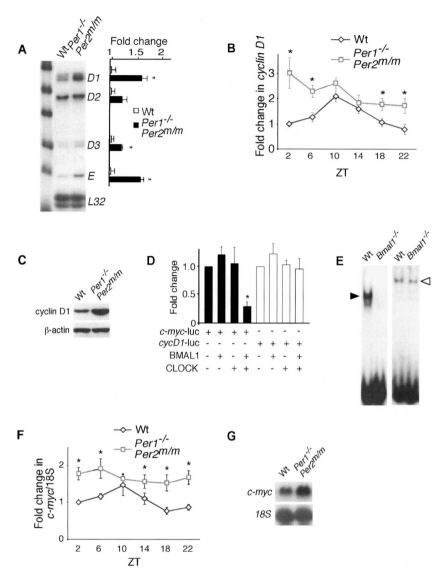

Fig. 4. Deregulation of *c-myc* signaling in (*Per1⁻/⁻;Per2^{m/m}*) osteoblasts. (A) RNase protection assays (RPA) showing upregulation of *G1 cyclins* in (*Per1⁻/⁻;Per2^{m/m}*) osteoblasts. (B) PhosphorImager quantification of three independent Northern blots for *cyclin D1* expression in wt and (*Per1⁻/⁻;Per2^{m/m}*) bones during a 24 h period (±SD). (C) Western blot analysis showing overexpression of cyclin D1 protein in (*Per1⁻/⁻;Per2^{m/m}*) osteoblasts. (D). BMAL1/CLOCK heterodimers inhibit *c-myc* promoter, but not *cyclin D1* promoter in DNA cotransfection assays (±SD). (E) EMSA: a protein–DNA complex (filled arrowhead) formed when wt but not *Bmal1⁻/⁻* nuclear extracts were incubated with the *c-myc* −510 E box (left panel), in contrast to normal Sp1 site binding by both extracts (hollow arrowhead). (F) PhosphorImager quantification of three independent Northern blots for *c-myc* expression in wt and (*Per1⁻/⁻;Per2^{m/m}*) bones during a 24 h period (±SD). (G) Northern blot shows increased *c-myc* expression in (*Per1⁻/⁻;Per2^{m/m}*) osteoblasts.

epinephrine in (*Per1⁻/⁻;Per2^{m/m}*) mice (Fig. 5B–D). These results indicate that leptin-dependent sympathetic signaling is deregulated in clock gene-mutant mice. However, elevated serum level of leptin and sympathetic tone did not result in lowering bone

mass in (*Per1⁻/⁻;Per2^{m/m}*) mice as previously observed in wt mice (Ducy et al., 2000; Takeda et al., 2002; Elefteriou et al., 2004). Instead, increase in leptin level by i.c.v. leptin infusion further increased bone mass in (*Per1⁻/⁻;Per2^{m/m}*) mice (Fig. 5E). In

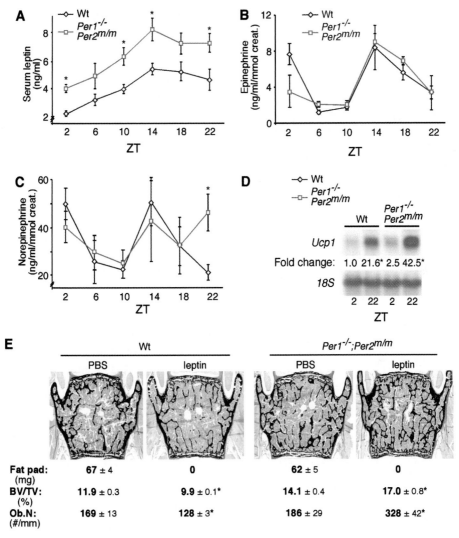

Fig. 5. Central control of bone formation. (A) Serum level of leptin (ng/ml) is elevated in ($Per1^{-/-}$;$Per2^{m/m}$) mice during a 24 h period ($n = 15$/time point) ($\pm$SEM). (B, C) Urinary catecholamines (ng/ml/mmol creat.) show circadian variation and were elevated in ($Per1^{-/-}$;$Per2^{m/m}$) mice at ZT2 for epinephrine and at ZT22 for norepinepherine ($\pm$SEM). (D) $Ucp1$ expression in brown fat was elevated at ZT22 in ($Per1^{-/-}$;$Per2^{m/m}$) mice. (E). Leptin infusion (i.c.v.) further increases bone mass and osteoblast number in ($Per1^{-/-}$;$Per2^{m/m}$) mice. Asterisks indicate statistically significant differences.

addition, mice with osteoblast-specific inactivation of $Per2$ on a $Per1^{-/-}$ background (mice lacking $Per1$ alone did not have bone phenotypes) showed similar HBM as ($Per1^{-/-}$;$Per2^{m/m}$) mice, as measured by increase in osteoblast numbers and bone-formation parameters (data not shown). Thus, apart from generating the circadian rhythm in leptin signaling, the molecular clock may also play a key role in mediating leptin-dependent sympathetic signaling in osteoblasts.

The peripheral clock is a target of sympathetic signaling in osteoblasts

To test whether the clock genes are direct targets of sympathetic signaling in osteoblasts, we treated wt and ($Per1^{-/-}$;$Per2^{m/m}$) osteoblasts with isoproterenol, which mimics sympathetic signaling in vivo. We found that both $Per1$ and $Per2$ are the direct targets of CREB activation (Fig. 6A). In

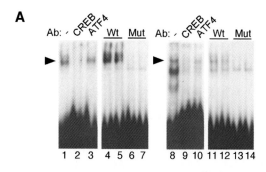

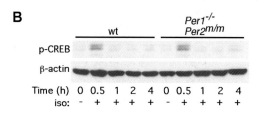

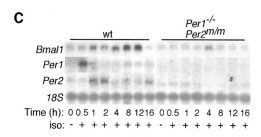

Fig. 6. Clock genes respond to sympathetic signaling in osteo-blasts. (A) EMSA: CREB antibody (lane 2), non preimmune serum (lane 1), or ATF4 antibody (lane 3) prevented formation of a protein–DNA complex (filled arrow) on the *Per1* CRE site. A protein–DNA complex formed when nuclear extracts from wt (lane 4) or from *Atf4*$^{-/-}$ (lane 5) osteoblasts were incubated with a wt *Per1* CRE, but not with a mutant *Per1* CRE (lanes 6 and 7). Similar results were obtained using the *Per2* CRE site (lanes 8–14). (B) Western blot showing that mutations in *Per1* and 2 had no effect on CREB phosphorylation in response to sympathetic signaling in osteoblasts. (C). The induction of *Per1, 2*, and *Baml1* genes in wt, but not in (*Per1*$^{-/-}$;*Per2*$^{m/m}$) osteoblasts by isoproterenol treatment (10 μM).

addition, although isoproterenol treatment acti-vated transcription factor CREB equally well in wt and (*Per1*$^{-/-}$;*Per2*$^{m/m}$) osteoblasts (Fig. 6B), the expression of *Per1* and 2, and *Bmal1* was only in-duced in wt osteoblasts, but not in mutant osteo-blasts (Fig. 6C). Thus, the molecular clock directly responds to sympathetic signaling in osteoblasts

and mutations in *Per1* and 2 abolished such re-sponse.

Sympathetic signaling activates cell cycle clock via AP1 genes in osteoblasts

Since i.c.v. infusion of leptin paradoxically in-creased osteoblast number in (*Per1*$^{-/-}$;*Per2*$^{m/m}$), we hypothesized that the peripheral clock may play a critical role in controlling osteoblast proliferation in response to sympathetic signaling. To test this hypothesis, we studied the expression of G1 cyclins in wt and (*Per1*$^{-/-}$;*Per2*$^{m/m}$) osteoblasts after isop-roterenol treatment. We found that 2 h after isop-roterenol treatment, mRNA levels of cyclins D1 and E were decreased in wt osteoblasts, but were further increased in (*Per1*$^{-/-}$;*Per2*$^{m/m}$) osteoblasts (Fig. 7A). Therefore, in the absence of a functional peripheral clock, sympathetic signaling could result in uncontrolled osteoblast proliferation.

To identify the molecular pathway that links the sympathetic signaling and cell cycle clock in os-teoblasts, we studied the expression of transcrip-tion factors that affect osteoblast differentiation and/or proliferation, and found that among all the genes examined, the expression of *c-fos* showed the most pronounced increase in (*Per1*$^{-/-}$;*Per2*$^{m/m}$) osteoblasts (Fig. 7B). In addition, most members of AP-1 gene family were induced by isoproterenol treatment in wt osteoblasts at the mRNA level and all of them were induced at the protein level (Fig. 7C data not shown).

While no canonical AP-1 binding site could be found in the *cyclin D1* gene, there was a perfect AP-1 binding site in both human and mouse *c-myc* pro-moters. In DNA cotransfection assays, AP1 tran-scription factors can activate a reporter construct driven by wt *myc*-AP1 sequences, but not by mutant *myc*-AP1 sequences (Fig. 7D). In DNA-binding as-says, a DNA–protein complex formed on *myc*-AP1 site in wt osteoblast nuclear extracts, but not in *c-fos*-mutant nuclear extracts (Fig. 7E). We also found that *c-myc* expression was significantly decreased in *c-fos*$^{-/-}$ osteoblasts (Fig. 7F). Taken together, these results indicate that *c-myc* is a direct target of AP1 transcription factors in response to sympathetic signaling in osteoblasts. Because *c-myc* is a master

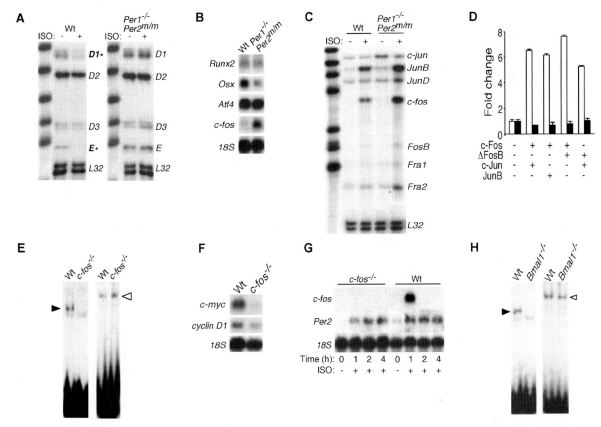

Fig. 7. AP1 genes activate cell cycle clock in response to sympathetic signaling. (A) RPAs showing isoproterenol treatment (10 μM) decreased *G1 cyclin* expression in wt (left panel), but not in (*Per1*$^{-/-}$;*Per2*$^{m/m}$) osteoblasts (right panel). Asterisks indicate statistically significant differences. (B) Expression of transcription factors affecting osteoblast differentiation and proliferation. (C) RPAs showing that AP-1 family members are induced by isoproterenol treatment (10 μM) in wt osteoblasts and most of them are overinduced in (*Per1*$^{-/-}$;*Per2*$^{m/m}$) osteoblasts. (D) AP-1 transcription factors activate a luciferase reporter construct containing four copies of a wt (clear bars), but not mutant (filled bars) *c-myc* AP-1 site (±SD). (E) EMSA: a protein–DNA complex (filled arrowhead) formed when wt, but not *c-fos*$^{-/-}$ osteoblast nuclear extracts were incubated with the *c-myc* AP-1 site, in contrast to normal Sp1 site binding by both extracts (hollow arrowhead). (F) Northern blot showing decreased expression of *c-myc* in *c-fos*$^{-/-}$ osteoblasts. (G) Loss of *c-fos* function does not affect the induction of *Per2* by isoproterenol treatment (10 μM) in osteoblasts. (H) EMSA: a protein–DNA complex (filled arrowhead) formed when wt, but not *Bmal1*$^{-/-}$ nuclear extracts were incubated with the *c-fos* –454 E box, in contrast to normal Sp1 site binding by both extracts (hollow arrowhead).

regulator of *G1 cyclins*, the sympathetic signaling in osteoblasts, therefore, could activate the cell cycle clock.

We also found that although mutation in *c-fos* did not affect the response of *Per2* to isopreterenol treatment (Fig. 7G), mutation in *Per1* and *2* resulted in overexpression of AP1 in response to sympathetic signaling (Fig. 7C). In EMSA, wt but not *Bmal1*$^{-/-}$ osteoblast nuclear extracts bound to the E box at −454 in the mouse *c-fos* promoter (Fig. 7G). Thus, the AP1 and clock-signaling

pathways also cross talk in osteoblasts and the molecular clock play a dominant role in mediating sympathetic signaling in these cells because it downregulates both *c-myc* and AP1 expression.

Leptin-dependent sympathetic signaling controls AP1 and clock genes in vivo

To confirm that the activation of the peripheral clock and AP1 genes by leptin-dependent

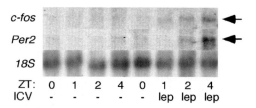

Fig. 8. AP1 and clock genes are regulated by leptin-dependent sympathetic signaling in vivo. Induction of *c-fos* and *Per2* in bone following a single i.c.v. injection of leptin into *ob/ob* mice.

sympathetic signaling indeed happened in vivo, we transiently infused leptin in the third ventricle of *ob/ob* mice, which lacked leptin and responded to i.c.v. leptin infusion with dramatic phenotypic, cellular, and molecular consequences (Zhang et al., 1994), and studied AP1 and clock gene expression in bone. We found that *Per2* and *c-fos* expression was induced following acute injection of leptin with similar kinetics in *ob/ob* bone (Fig. 8), indicating that leptin-dependent sympathetic signaling activates two antagonistic pathways with similar kinetics in osteoblast to control cell proliferation.

Discussion

Our study indicates that the molecular clock plays a key role in leptin-mediated sympathetic regulation of bone formation. It also demonstrates that the sympathetic signaling controls cell proliferation using two antagonistic pathways: a clock-mediated pathway to restrict cell proliferation and an AP1-mediated pathway to promote cell proliferation. Thus, as in the case of leptin-dependent regulation of bone resorption, leptin also uses dual mechanisms to regulate bone formation (Elefteriou et al., 2005).

It has been shown previously that the serum level of leptin follows diurnal variation (Kalsbeek et al., 2001). We found that mutation in *Per1* and *2* resulted in increased serum leptin level throughout the 24 h period. However, instead of having low bone mass (Elefteriou et al., 2004), these mice showed HBM early in life, indicating that the clock genes are required for mediating the leptin-dependent sympathetic signaling in osteoblasts. The evidences supporting this hypothesis include: (1) i.c.v. leptin infusion paradoxically increases bone mass in ($Per1^{-/-};Per2^{m/m}$) mice; (2) the

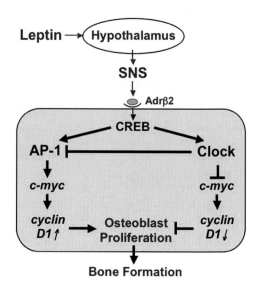

Fig. 9. Current model of the leptin-dependent sympathetic regulation of bone formation. Through CREB, sympathetic signaling activates the peripheral clock and AP-1 genes in osteoblasts. The clock genes inhibit osteoblast proliferation by inhibiting *c-myc*, which leads to *cyclin D1* downregulation. Counteracting these effects, AP-1 genes promote osteoblast proliferation by upregulating *c-myc* and thereby *cyclin D1*. The clock genes exert a dominant influence in part because they also inhibit AP-1 gene expression.

clock genes are regulated by sympathetic signaling in osteoblasts; (3) mice with an osteoblast-specific deletion of *Per2* on a *Per1^{-/-}* background showed that HBM resulted from increased bone formation; and (4) transient infusion of leptin in the third ventricle of the brain induced *Per* expression in the bone of *ob/ob* mice.

The role of the circadian clock in leptin-controlled bone formation is studied at the molecular and cellular levels. Based on these studies, we propose a model to explain how the circadian clock acts in leptin-dependent sympathetic control of bone formation. In this model, the molecular clock responds to sympathetic signaling in osteoblasts to indirectly decrease *cyclin D1* expression by directly inhibiting the expression of *c-myc*, a *cyclin D1* activator that is highly expressed in osteoblasts (Resnitzky and Reed, 1995; Perez-Roger et al., 1999). However, at the same time when the peripheral clock is activated, the sympathetic signaling also activates AP1 genes. Because AP1 transcription factors can directly activate *c-myc*, the sympathetic signaling could lead to osteoblast proliferation as well. In ($Per1^{-/-}$;$Per2^{m/m}$) osteoblasts, sympathetic signaling only activates AP1 genes, but not the peripheral clock, resulting in uncontrolled AP1-c-myc signaling and overexpression of $G1$ $cyclins$. Because the molecular clock also downregulates AP1 in response to sympathetic signaling, there is a functional hierarchy between the clock and AP1 genes, and the clock-mediated pathway plays the dominant role in response to sympathetic signaling. Therefore, increase in the sympathetic tone by i.c.v. leptin only results in a further increase in osteoblast proliferation and bone formation (Fig. 9).

It has been shown recently that c-$fos^{-/-}$ mice have lower cortical osteoblast numbers (Grigoriadis, et al., 1995), $JunB^{\Delta/\Delta}$ mice display osteopenia (Kenner et al., 2004) and mice overexpressing $fra1$ or $\Delta fosB$ have increased osteoblast proliferation (Jochum et al., 2000; Sabatakos et al., 2000). Our results further demonstrate that AP-1 genes play a key role in osteoblast proliferation and bone formation.

We should emphasize, however, that our model does not exclude the possibility that the molecular clock controls osteoblast proliferation through additional mechanisms. It has been shown recently that the molecular clock plays a role in liver regeneration, by controlling the $Wee1$-mediated G2/M checkpoint (Matsuo et al., 2003). Although under normal growth conditions, we did not observe an abnormal activation of G2/M checkpoint in ($Per1^{-/-}$;$Per2^{m/m}$) osteoblasts, nor a differential expression of Wee1 protein in wt and ($Per1^{-/-}$; $Per2^{m/m}$) osteoblasts (data not shown), the possibility remains that $Wee1$ may regulate osteoblast proliferation under other physiological conditions.

A feature progressively emerging from the molecular dissection of leptin's regulation of bone mass is that its mediators are not involved in a significant manner in mediating the other cardinal functions of leptin such as control of appetite and reproduction in unchallenged animals. This specificity of action is surprising in the case of clock genes and AP-1 genes that are not osteoblast-specific genes. The narrow specificity of action of these mediators, the ability of leptin to regulate powerfully both aspects of bone remodeling, along with the fact that it first appears during evolution in vertebrates underscore the importance of leptin's regulation of bone mass and raise the prospect that this may have been the ancestral function of this hormone.

Abbreviations

AP-1	activator protein 1
BFR	bone-formation rate
Cry	cryptochrome
DpD	deoxypyridinoline
HBM	high bone mass
i.c.v.	intracerebroventricular
iso	isoproterenol
MAR	mineral apposition rate
ob	osteoblast
Per	period
Ucp1	Uncoupling Protein-1
wt	wild type

Acknowledgments

We are indebted to Drs. C.C. Lee, R. Van Gelder, S. Reppert, D. Weaver, S.L. McKnight, and P. Ducy for providing $Per2$-deficient, Cry-deficient, $Bmal1$-deficient, and $Npas2$-deficient mice and for critical reading of the manuscript. Gerard Karsenty thanks Dr. S. Reppert for judicious suggestions during the course of this study. This work was supported by the National Institutes of Health (G.K., L.F.) and a Canadian Institute for Health Research Fellowship (M.S.P.).

References

Chien, K.R. and Karsenty, G.. (2005) Longevity and lineages: toward the integrative biology of degenerative diseases in heart, muscle, and bone. Cell, 120: 533–544.

Ducy, P., Amling, M., Takeda, S., Priemel, M., Schilling, A.F., Beil, F.T., Shen, J., Vinson, C., Rueger, J.M. and Karsenty, G.. (2000) Leptin inhibits bone formation through a hypothalamic relay: a central control of bone mass. Cell, 100: 197–207.

Elefteriou, F., Ahn, J.D., Takeda, S., Starbuck, M., Yang, X., Liu, X., Kondo, H., Richards, W.G., Bannon, T.W., Noda, M., et al. (2005) Leptin regulation of bone resorption by the

188

sympathetic nervous system and CART. Nature, 434: 514–520.

Elefteriou, F., Takeda, S., Ebihara, K., Magre, J., Patano, N., Kim, C.A., Ogawa, Y., Liu, X., Ware, S.M., Craigen, W.J., et al. (2004) Serum leptin level is a regulator of bone mass. Proc. Natl. Acad. Sci. USA, 101: 3258–3263.

Fu, L. and Lee, C.C.. (2003) The circadian clock: pacemaker and tumour suppressor. Nat. Rev. Cancer, 3: 350–361.

Fu, L., Pelicano, H., Liu, J., Huang, P. and Lee, C.. (2002) The circadian gene *Period2* plays an important role in tumor suppression and DNA damage response in vivo. Cell, 111: 41–50.

Fu, M., Wang, C., Li, Z., Sakamaki, T. and Pestell, R.G.. (2004) Minireview: Cyclin D1: normal and abnormal functions. Endocrinology, 145: 5439–5447.

Grigoriadis, A.E., Wang, Z.Q. and Wagner, E.F.. (1995) Fos and bone cell development: lessons from a nuclear oncogene. Trends Genet., 11: 436–441.

Gundberg, C.M., Markowitz, M.E., Mizruchi, M. and Rosen, J.F.. (1985) Osteocalcin in human serum: a circadian rhythm. J. Clin. Endocrinol. Metab., 60: 736–739.

Jochum, W., David, J.P., Elliott, C., Wutz, A., Plenk Jr., H., Matsuo, K. and Wagner, E.F.. (2000) Increased bone formation and osteosclerosis in mice overexpressing the transcription factor Fra-1. Nat. Med., 6: 980–984.

Kalsbeek, A., Fliers, E., Romijn, J.A., La Fleur, S.E., Wortel, J., Bakker, O., Endert, E. and Buijs, R.M.. (2001) The suprachiasmatic nucleus generates the diurnal changes in plasma leptin levels. Endocrinology, 142: 2677–2685.

Kenner, L., Hoebertz, A., Beil, T., Keon, N., Karreth, F., Eferl, R., Scheuch, H., Szremska, A., Amling, M., Schorpp-Kistner, M., et al. (2004) Mice lacking JunB are osteopenic due to cell-autonomous osteoblast and osteoclast defects. J. Cell Biol., 164: 613–623.

Matsuo, T., Yamaguchi, S., Mitsui, S., Emi, A., Shimoda, F. and Okamura, H.. (2003) Control of mechanism of the circadian clock for timing of cell division in vivo. Science, 302: 255–259.

Perez-Roger, I., Kim, S.H., Griffiths, B., Sewing, A. and Land, H.. (1999) Cyclins D1 and D2 mediate myc-induced prolif-

eration via sequestration of p27(Kip1) and p21(Cip1). Embo J., 18: 5310–5320.

Reppert, S.M. and Weaver, D.R.. (2002) Coordination of circadian timing in mammals. Nature, 418: 935–941.

Resnitzky, D. and Reed, S.I.. (1995) Different roles for cyclins D1 and E in regulation of the G1-to-S transition. Mol. Cell Biol., 15: 3463–3469.

Riggs, B.L. and Melton 3rd., L.J.. (1986) Involutional osteoporosis. N. Engl. J. Med., 314: 1676–1686.

Sabatakos, G., Sims, N.A., Chen, J., Aoki, K., Kelz, M.B., Amling, M., Bouali, Y., Mukhopadhyay, K., Ford, K., Nestler, E.J. and Baron, R.. (2000) Overexpression of delta FosB transcription factor(s) increases bone formation and inhibits adipogenesis. Nat. Med., 6: 985–990.

Scarpace, P.J. and Matheny, M.. (1998) Leptin induction of *UCP1* gene expression is dependent on sympathetic innervation. Am. J. Physiol., 275: E259–E264.

Simmons, D.J. and Nichols Jr., G.. (1966) Diurnal periodicity in the metabolic activity of bone tissue. Am. J. Physiol., 210: 411–418.

Takeda, S., Elefteriou, F., Levasseur, R., Liu, X., Zhao, L., Parker, K.L., Armstrong, D., Ducy, P. and Karsenty, G.. (2002) Leptin regulates bone formation via the sympathetic nervous system. Cell, 111: 305–317.

Takeda, S. and Karsenty, G.. (2001) Central control of bone formation. J. Bone Mineral Metab., 19: 195–198.

Vitaterna, M.H., Selby, C.P., Todo, T., Niwa, H., Thompson, C., Fruechte, E.M., Hitomi, K., Thresher, R.J., Ishikawa, T., Miyazaki, J., et al. (1999) Differential regulation of mammalian period genes and circadian rhythmicity by *cryptochromes 1* and *2*. Proc. Natl. Acad. Sci. USA, 96: 12114–12119.

Zhang, Y., Proenca, R., Maffei, M., Barone, M., Leopold, L. and Friedman, J.M.. (1994) Positional cloning of the mouse *obese* gene and its human homologue. Nature, 372: 425–432.

Zheng, B., Albrecht, U., Kaasik, K., Sage, M., Lu, W., Vaishnav, S., Li, Q., Sun, Z.S., Eichele, G., Bradley, A. and Lee, C.C.. (2001) Nonredundant roles of the *mPer1* and *mPer2* genes in the mammalian circadian clock. Cell, 105: 683–694.

Kalsbeek, Fliers, Hofman, Swaab, Van Someren & Buijs
Progress in Brain Research, Vol. 153
ISSN 0079-6123

CHAPTER 11

Hypothalamic thyroid hormone feedback in health and disease

Eric Fliers[1,*], Anneke Alkemade[1,2], Wilmar M. Wiersinga[1] and Dick F. Swaab[2]

[1]*Department of Endocrinology and Metabolism, Academic Medical Center, University of Amsterdam,*
Amsterdam, The Netherlands
[2]*Netherlands Institute for Brain Research, Amsterdam, The Netherlands*

Abstract: The role of the human hypothalamus in the neuroendocrine response to illness has only recently begun to be explored. Extensive changes in the hypothalamus–pituitary–thyroid (HPT) axis occur within the framework of critical illness. The best-documented change in the HPT axis is a decrease in serum concentrations of the biologically active thyroid hormone triiodothyronine (T3). From studies in post-mortem human hypothalamus it appeared that low serum T3 and thyrotropin (TSH) during illness (non-thyroidal illness, NTI) are paralleled by decreased thyrotropin-releasing hormone (TRH)mRNA expression in the hypothalamic paraventricular nucleus (PVN), pointing to a major alteration in HPT axis setpoint regulation. A strong decrease in TRHmRNA expression is also present in the PVN of patients with major depression as well as in glucocorticoid-treated patients. By inference, hypercortisolism in hospitalized patients with severe depression or in critical illness may induce down-regulation of the HPT axis at the level of the hypothalamus. In order to start defining the determinants and mechanisms of these setpoint changes in various clinical conditions, it is important to note that an increasing number of hypothalamic proteins appears to be involved in central thyroid hormone metabolism. In recent studies, we have investigated the distribution and expression of thyroid hormone receptor (TR) isoforms, type 2 and type 3 deiodinase (D2 and D3), and the thyroid hormone transporter monocarboxylate transporter 8 (MCT8) in the human hypothalamus by a combination of immunocytochemistry, mRNA in situ hybridization and enzyme activity assays. Both D2 and D3 enzyme activities are detectable in the mediobasal hypothalamus. D2 immunoreactivity is prominent in glial cells of the infundibular nucleus/median eminence region and in tanycytes lining the third ventricle. Combined D2, D3, MCT8 or TR immunocytochemistry and TRHmRNA in situ hybridization indicates that D3, MCT8 and TRs are all expressed by TRH neurons in the PVN, whereas D2 is not. Taken together, these results suggest that the prohormone thyroxine (T4) is taken up in glial cells that convert T4 into the biologically active T3 via the enzyme D2; T3 is subsequently transported to TRH producing neurons in the PVN where it may bind to TRs and/or may be degraded into inactive iodothyronines by D3. This model for thyroid hormone action in the human hypothalamus awaits confirmation in future experimental studies.

Keywords: TRH; TSH; thyroid hormone; hypothalamus; PVN; depression; critical illness; nonthyroidal illness

*Corresponding author. Tel.: +31-20-566-6071; Fax: +31-20-691-7682; E-mail: e.fliers@amc.uva.nl

DOI: 10.1016/S0079-6123(06)53011-0
189

Introduction

The tripeptide thyrotropin-releasing hormone (TRH) was the first hypothalamic hormone to be isolated and structurally characterized (Bøler et al., 1969; Burgus et al., 1969). Immunocytochemical studies in the rat hypothalamus in the 1970s and 1980s demonstrated TRH neurons in a number of hypothalamic nuclei. In the mean time, a key role for TRH neurons in the paraventricular nucleus of the hypothalamus (PVN) in the neuroendocrine regulation of thyroid hormone was revealed. Serum thyroid hormone levels and TRHmRNA expression in the PVN show an inverse relationshipduring experimentally induced hypo- and hyperthyroidism in rats (Segerson et al., 1987) representing the central part of a classical endocrine negative feedback loop. The rat PVN consists of magnocellular neurons containing, e.g., vasopressin and oxytocin in the lateral portions, and of a parvocellular part in the more medial portions of the nucleus. Only TRH synthesizing neurons in the medial and periventricular parvocellular subdivisions of the PVN project to the median eminence (ME), in keeping with observations in experimental models of hypothyroidism showing increased TRHmRNA only in these subdivisions of the PVN (see Lechan and Fekete, 2006). These so-called hypophysiotropic TRH neurons project to the ME and terminate in its external zone, where TRH is released into the portal capillaries for transport to the anterior pituitary and regulation of TSH release. Hypophysiotropic TRH neurons in the PVN receive monosynaptic projections from leptin-responsive neurons in the arcuate nucleus (ARC) containing either alpha-melanocyte-stimulating hormone (α-MSH) and cocaine and amphetamine-regulated transcript (CART), or neuropeptide Y (NPY) and agouti-related protein (AGRP). These projections play a key role in the resetting of the thyroid axis during food deprivation resulting in reduced TRHmRNA expression in the PVN, which contributes to lower serum concentrations of thyroid hormone in the rat (for review see Lechan and Fekete, 2006). Endotoxin administration, which has been used as a model for acute infection, also reduces TRHmRNA in the PVN, possibly via in-

creased activity of the enzyme type 2 deiodinase (D2) in tanycytes stimulating local production of triiodothyronine (T3). In addition to this neuroendocrine role of TRH in the parvocellular part of the PVN, TRH in additional (hypothalamic) areas is involved in the regulation of body temperature, appetite, digestion and locomotor activity.

In the light of these multiple key functions of TRH in the regulation of energy homeostasis, relatively little attention has been paid to TRH neurons in the human hypothalamus. In this chapter, we review a number of recent studies on the functional neuroanatomy of thyroid hormone feedback in the human hypothalamus in health and disease.

Thyroid hormone feedback in the human hypothalamus

TRH neurons in the human hypothalamus

In the 1990s, it became clear that TRH containing neurons and fibers are present throughout the human hypothalamus, including some TRH neurons in the hypothalamic gray. Immunocytochemical studies showed that the human PVN contains many spindle-shaped and spheric multipolar parvocellular TRH neurons, especially in its dorsocaudal portion, while only a small number of magnocellular neurons express TRH. Both the suprachiasmatic nucleus (SCN), which is the circadian pacemaker of the brain acting as a biological clock, and the sexually dimorphic nucleus (SDN) of the human hypothalamus contain a small number of bipolar TRH cells (Fliers et al., 1994). The exact efferent projections of hypothalamic TRH containing neurons in the human brain are unknown, but dense TRH fiber networks, e.g., in the tuberomammillary nucleus (TMN) and perifornical area, suggest an important role for nonhypophysiotropic TRH neurons in the human brain as demonstrated earlier in the rat. The distribution of TRH neurons shown by immunocytochemistry was confirmed by a subsequent study using mRNA in situ hybridization (Guldenaar et al., 1996) reporting numerous TRHmRNA containing cells in the medial region of the dorsocaudal portion of the PVN. In addition to the PVN and SCN, TRH mRNA-labeled neurons appeared to be present in the perifornical area where dense TRH fiber networks may mask immunocytochemical detection of TRH neurons.

TR expression

The TR isoforms α1, α2, β1 and β2 are expressed in a large number of rat brain areas, including the hypothalamus. Prominent TR expression has been reported in the PVN, the ARC and median eminence of the adult rat, notably of TRβ2 (Bradley et al., 1989; Cook et al., 1992; Lechan et al., 1994). A pivotal role for TRβ2 in thyroid hormone negative feedback on TRH neurons in the PVN was suggested by studies in TR isoform-specific knockout mice (Abel et al., 2001), although TRH neurons in the rat PVN express all four TR isoforms (Lechan et al., 1994). Using a set of polyclonal antisera raised against the specific isoforms of TR, we studied the distribution of TR isoform expression in the human hypothalamus (Alkemade et al., 2005c). Although TRs have been assumed to be largely confined to the nuclear compartment, we found predominantly cytoplasmic localization of TR staining with less marked nuclear staining in most subjects studied. Earlier in vitro studies (Hager et al., 2000) had shown that TRs may shuttle rapidly between the nuclear and cytoplasmic compartment. Such translocation of TR may be ligand dependent (Zhu et al., 1998) and require protein interactions with various cofactors (Baumann et al., 2001). Cytoplasmic localization, in addition to nuclear localization, appears to be a general phenomenon in the human brain for steroid hormone receptors (Hager et al., 2000; Kruijver et al., 2002; Maruvada et al., 2003). Thus, given the extensive specificity tests performed with the TR antisera including negative Western blots in TR isoform-specific knockout mice, an artefact seems unlikely as an explanation for the observed cytoplasmic TR staining.

The general distribution of TR expression in the human hypothalamus is represented schematically in Fig. 1. Both parvocellular and magnocellular neurons in the human PVN express all four isoforms studied (Fig 2). In addition, magnocellular

neurons in the supraoptic nucleus (SON) show TR immunoreactivity. Since this neuronal population contains vasopressin in the human hypothalamus (Fliers et al., 1985), this may represent an anatomical substrate for interactions between thyroid hormone status and osmoregulation (e.g., Skowsky and Kikuchi, 1978). Most prominent TR staining is present in the infundibular nucleus (IFN), which is the human homologue of the rat ARC. The distribution of TR expression in the human hypothalamus appears to be in agreement with the rat, where prominent TR isoform staining in TRH expressing cells in the PVN and in the ARC had been reported earlier. In an attempt to identify TR expressing neurons in the IFN we performed combined immunocytochemistry for TRs and in situ hybridization for NPY and pro-opiomelanocortin (POMC)mRNA as candidate neurons for colocalization on the basis of observations in the rat (Legradi and Lechan, 1998). Only sparse colocalization was observed, suggesting that neurons other than those expressing NPY and POMC are involved in thyroid hormone feedback in the IFN. Combined TR immunostaining with TRHmRNA in situ hybridization showed that part of the TRHmRNA expressing neurons in the human PVN express TRs (Alkemade et al., 2005a). In contrast to the rat, the human SCN appeared to contain no TR expressing neurons, which may reflect a difference in expression levels, or an interspecies difference.

Expression of deiodinases and MCT8

Although the thyroid gland mainly secretes thyroxine (T4), most thyroid hormone actions are mediated via binding of the biologically more active thyroid hormone T3 to its receptor. The biologic activity of thyroid hormone in target cells is, therefore, determined in part by the intracellular concentration of T3 which depends on the activity of the iodothyronine deiodinases. In the brain, D2 is responsible for deiodination of T4 into T3, whereas type 3 deiodinase (D3) inactivates thyroid hormone by converting T3 into the biologically inactive T2 and T4 into reverse T3 (rT3) (for review see Bianco et al., 2002). MCT8 has recently been identified in humans as a thyroid hormone transporter with a

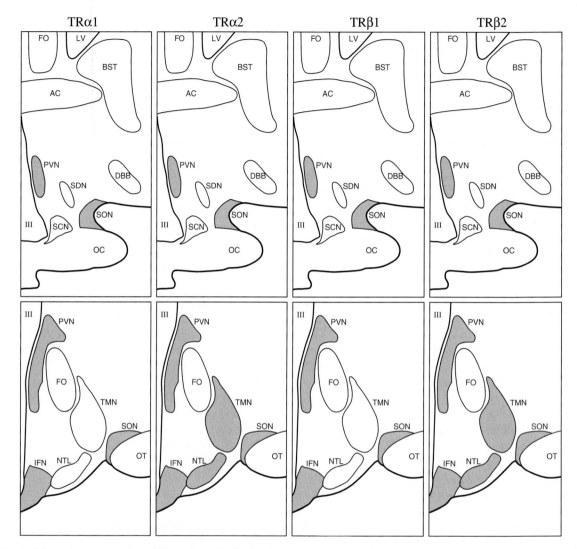

Fig. 1. Schematic representation of TR isoform distribution in the human hypothalamus (upper panels: rostral level; lower panels: caudal level). Gray areas represent TR expression. III, third ventricle; AC, anterior commissure; BST, bed nucleus of the stria terminalis; DBB, diagonal band of Broca; FO, fornix; LV, lateral ventricle; NTL, nucleus tuberalis lateralis; OC, optic chiasm; OT, optic tract; SCN, suprachiasmatic nucleus; SDN, sexually dimorphic nucleus; SON, supraoptic nucleus; TMN, tuberomammillary nucleus. Reproduced with permission from Alkemade et al. (2005c).

preference for T3 as the ligand. It appears to play a crucial role in thyroid hormone metabolism in the CNS by providing cells expressing deiodinases with thyroid hormone (Friesema et al., 2003; Heuer et al., 2005). The functional importance of MCT8 is evident from observations in a number of patients with severe psychomotor retardation who carry mutations or deletions in the MCT8 gene (Friesema et al., 2004). Additional transporters capable of thyroid

hormone transport have been identified. Among these, the organic anion transporter OATP1C1 has a high affinity for T4 and may be critical for the transport of T4 over the blood–brain barrier (BBB) in view of its expression in capillaries throughout the brain (for review see Jansen et al., 2005).

A number of recent studies have reported the expression and distribution of D2, D3 and MCT8 in the human hypothalamus using a combination of

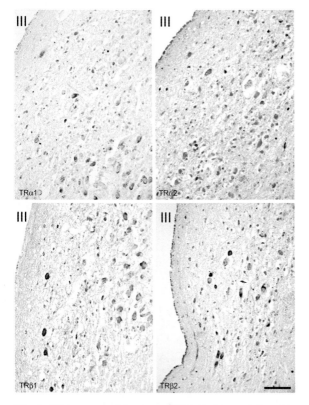

TRα1 · TRα2

TRβ1 · TRβ2

Fig. 2. TR immunostaining in the PVN. Positive staining with antisera against all four TR isoforms is shown. Staining intensity varies between isoforms. For all isoforms, staining is mainly cytoplasmic. Scale bar, 100 μm. Reproduced with permission from Alkemade et al. (2005c).

immunocytochemistry, in situ hybridization and enzyme activity assays. D2 and D3 enzyme activities are detectable in pituitary and hypothalamic tissue samples obtained during autopsy (Alkemade et al., 2005a). The topographic distribution of D2 immunoreactivity is schematically represented in Fig 3. Prominent D2 immunoreactivity is present in cells throughout the ependymal layer of the third ventricle (Fig 4, Alkemade et al., 2005a). D2 staining is also evident in glial cells within the IFN/ME region and in blood vessel walls, extending beyond the endothelial cells into the smooth muscle layer. In addition to glial staining, the IFN shows occasional weak neuronal staining. The distribution pattern of D2 immunostaining is generally in agreement with earlier studies in rats in which subsequent studies using electron microscopy have identified D2 immunoreactive cells as astrocytes, including tanycytes (Tu et al., 1997;

Diano et al., 2003). Tanycytes are cells lining the third ventricle establishing contacts with blood vessels of the ME, and they have been proposed to be involved in providing T3 to the CNS following T4 uptake from the cerebrospinal fluid (CSF) (Tu et al., 1997). D2 expression in rats was reported to be confined to the ventral part of the lining of the third ventricle, but is extended over the entire height of the lining of the third ventricle in the human hypothalamus, which may represent a different expression level or, alternatively, an interspecies difference.

The distribution of D3 immunoreactivity throughout the hypothalamus is schematically presented in Fig 5 (Alkemade et al., 2005a). The distribution of hypothalamic D3 expression clearly differs from that of D2, showing intensely staining neurons in the IFN, PVN and SON (Fig 6). Less intense neuronal D3 staining is present in the SCN, tuberolateral nucleus (NTL), TMN around the mammilary bodies and — sporadically — in the perifornical area. In contrast to D2, D3 is expressed exclusively in neurons showing a strong distribution overlap with TR, suggesting that D3 is expressed in T3-responsive neurons to terminate T3 action.

MCT8 is expressed by neurons in the PVN, SON and IFN (Fig 7). Surprisingly, the perifornical area and lateral hypothalamus (LHA) express MCT8 most prominently (for a schematic representation of MCT8 expression see Fig 8). In the rat, D2 activity was reported in the LHA (Peeters et al., 2001), but in the human LHA no D2 and only sporadic D3 expression was reported. The LHA in the rat is innervated by leptin-sensitive neurons from the ARC and is implicated in regulating food intake and body weight (Sakurai et al., 1998; Elias et al., 2001; Peeters et al., 2001). An experimental study in rats reported that T3 stimulates food intake via the hypothalamic ventromedial nucleus (Kong et al., 2004) The presence of dense TRH fiber networks in the human perifornical area (Fliers et al., 1994) in combination with marked MCT8 and sparse D3 immunoreactivity (Alkemade et al., 2005a) may suggest a neuroanatomical basis for effects of thyroid hormone on feeding behavior in humans as well. Using vimentin and NeuN as markers for tanycytes and neurons, respectively, the expression of D2 by glial cells and of D3 by neurons was confirmed, whereas MCT8 was present in both cell types (Fig 9).

194

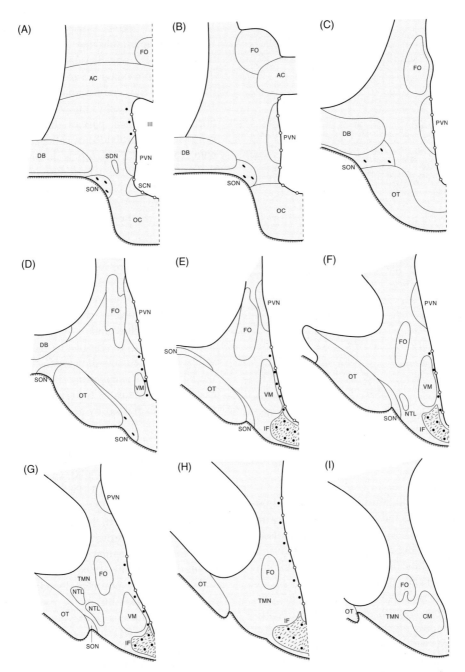

Fig. 3. Schematic illustration of the distribution of D2 immunoreactivity in the human hypothalamus. AC, anterior commissure; CM, corpus mammillare; DB, diagonal band of Broca; FO, fornix; NTL, lateral tuberal nucleus; OC, optic chiasm; OT, optic tract; SDN, sexually dimorphic nucleus; VM, ventromedial nucleus. Both open and closed circles represent glial cells. Reproduced with permission from Alkemade et al. (2005a).

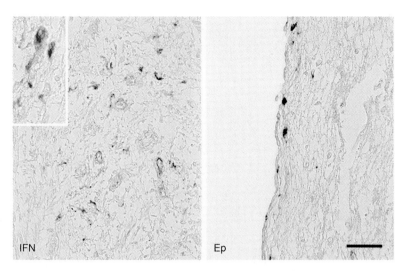

Fig. 4. D2 immunostaining in the IFN and the ependymal layer of the third ventricle (Ep). The inset shows a high-power magnification of D2 expression in the IFN of the same section. D2 is mainly expressed in small glial cells within the IFN and is also present in the ependymal lining of the third ventricle. Scale bars, 100 μm (IFN and Ep) and 25 μm (inset). Reproduced with permission from Alkemade et al. (2005a).

Three models for thyroid hormone feedback in the human hypothalamus

The classical scheme for feedback of thyroid hormone on TRH neurons of the PVN has focused on direct effects of thyroid hormone on TRH neurons. Indeed, unilateral stereotaxic implants of T3 within the anterior hypothalamus was reported to induce a marked reduction of proTRHmRNA in the PVN of rats (Dyess et al., 1988). Based on above-mentioned data, T4 might indeed be taken up locally within the PVN from the blood by astrocytes. This process would require active T4 transport, e.g., by the organic anion transporter OATP1C1 which has been proposed to serve as a high-affinity T4 transporter in the human brain (for review see Jansen et al., 2005). In the rat, OATP1C1 is localized in brain capillaries and may be particularly important for transport of T4 across the BBB (see Jansen et al., 2005). Following uptake, T4 may then be converted to T3 by D2 in astrocytes and subsequently transported to TRH neurons by MCT8 (Fig. 10A). Co-expression of D3, MCT8 and TRs with TRH in the PVN certainly supports the existence of this route for thyroid hormone feedback action in the human hypothalamus.

Feedback following thyroid hormone uptake from the CSF represents an alternative possibility (Fig. 10B). This route has been proposed earlier by other investigators based on neuroanatomical studies in rats (Guadano-Ferraz et al., 1997; Tu et al., 1997; Diano et al., 2003; Lechan and Fekete, 2004). In this model, thyroid hormone is taken up from the CSF in the third ventricle and transported by tanycytes to neurons in the arcuate nucleus that project to TRH cells in the PVN. T3 action in IFN neurons may alter firing pattern, thereby affecting TRH gene expression in PVN neurons via monosynaptic pathways demonstrated earlier by neuroanatomical-tracing studies in the rat. Alternatively, T3 may be transported from the IFN to TRH neurons in the PVN by anterograde axonal transport as proposed in the locus coeruleus by Gordon et al (1999). Our observations in the human hypothalamus support uptake from the CSF at the level of the IFN as a possible route for thyroid hormone feedback on TRH neurons on the basis of immunostaining of D2-expressing glial cells along the lining of the basal part of the third ventricle, and MCT8, TR and D3 expression by neurons in the IFN. In addition, uptake of T4 from the third ventricle at the level of the PVN followed by transport to TRH cells in the PVN appears to be a possibility in view of D2-positive glial cells along the ependyma of the third ventricle in close approximation of the PVN.

196

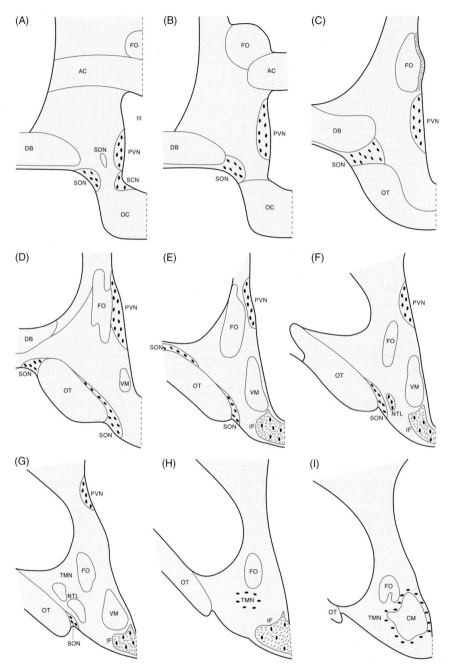

Fig. 5. Schematic illustration of the distribution of D3 immunoreactivity in the human hypothalamus. AC, anterior commissure; CM, corpus mammillare; DB, diagonal band of Broca; FO, fornix; OC, optic chiasm; OT, optic tract; SDN, sexually dimorphic nucleus; VM, ventromedial nucleus. Reproduced with permission from Alkemade et al. (2005a).

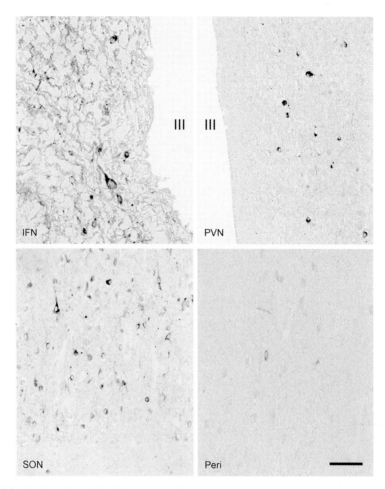

Fig. 6. D3 immunostaining in the IFN, PVN, SON and perifornical area (Peri). D3 is exclusively expressed in neurons. Scale bar, 100 μm. Reproduced with permission from Alkemade et al. (2005a).

Finally, thyroid hormone may have direct access from the circulation to the IFN in view of the absence of the BBB in this part of the mediobasal hypothalamus (Fig. 10C). That cells in the ARC sense intravascular thyroid hormone concentrations is supported by increased D2 activity and mRNA in the ARC/ME after induction of hypothyroidism in rats (Riskind et al., 1987; Tu et al., 1997). In addition, TRβ2 is predominantly expressed in rat ARC neurons that may be able to sense T3 produced by glial cells expressing D2. NPY, POMC and AGRP containing neurons from the ARC project to TRH neurons in the PVN (Legradi and Lechan, 1998, 1999). The finding of D2 in the human IFN in conjunction with earlier observation of TRβ2 expression in the same area (Alkemade et al., 2005c) is suggestive of a similar pathway in the human hypothalamus. If indeed thyroid hormone may act via the vascular compartment–IFN–PVN route in the human hypothalamus remains, however, hypothetical at this stage.

Major depression and glucocorticoid treatment

Major depressive disorder has been associated with changes in the HPT axis and the hypothalamic–pituitary–adrenal (HPA) axis. In the HPT axis a decrease in serum TSH, a blunted TSH response to TRH and an increase in serum free T_4 (FT_4) have

198

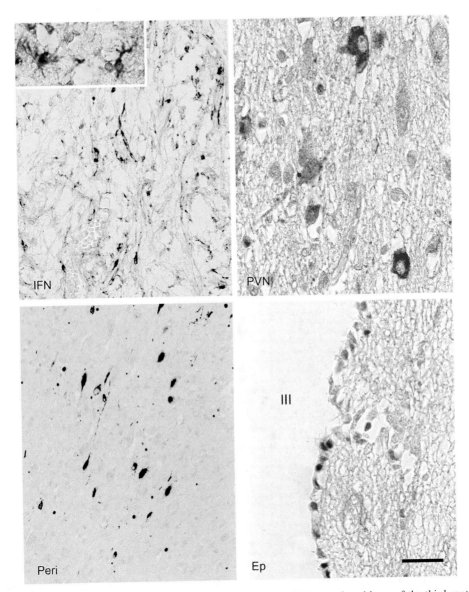

Fig. 7. MCT8 immunoreactivity in the IFN, PVN, perifornical area (Peri) and the ependymal layer of the third ventricle (Ep). The inset shows a high-power magnification of MCT8 expression in the IF of the same section. MCT8 is prominent in both glial cells and neurons. Scale bar, 100 μm (IFN and Peri) and 25 μm (PVN, Ep and inset). Reproduced with permission from Alkemade et al. (2005a).

been reported in inpatients by various authors (for review see Jackson, 1998; Kirkegaard and Faber, 1998) as well as an increased prevalence of subclinical hypothyroidism and thyroid peroxidase (TPO) antibodies (Nemeroff et al., 1985; Pop et al., 1998). A recent controlled study in a large number of untreated outpatients, however, reported slightly higher serum TSH (although still within the reference range)

and unchanged FT4 in major depression (Brouwer et al., 2005), leading to the recognition of the in- vs. outpatient status of depressed patients as a possible determinant of HPT axis changes. Hypercortisolism in depression has been reported in many studies as reflected by elevated mean 24-h serum cortisol concentrations and increased 24 h urinary excretion of cortisol. In the dexamethasone-suppression test

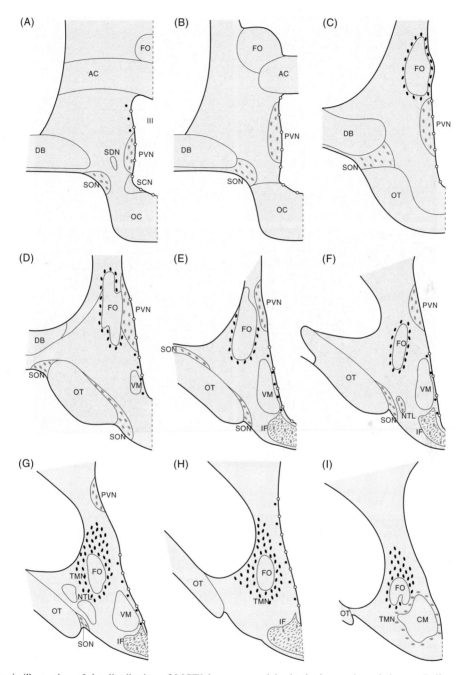

Fig. 8. Schematic illustration of the distribution of MCT8 immunoreactivity in the human hypothalamus. *Different shades of gray* represent staining intensity. AC, anterior commissure; CM, corpus mammillare; DB, diagonal band of Broca; FO, fornix; OC, optic chiasm; OT, optic tract; SDN, sexually dimorphic nucleus; VM, ventromedial nucleus. Both open and closed circles represent glial cells. Reproduced with permission from Alkemade et al. (2005a).

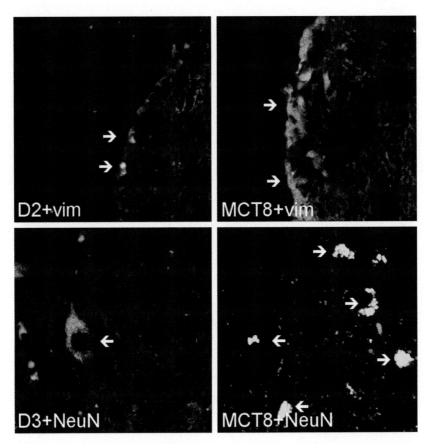

Fig. 9. Colocalization of D2 and MCT8 immunoreactivity in the ependymal layer of the third ventricle with vimentin and colocalization of D3 and MCT8 immunoreactivity with NeuN. Reproduced with permission from Alkemade et al. (2005a).

(DST), serum cortisol and ACTH concentrations are inadequately suppressed in some 20–50% of patients (for review see Holsboer and Barden, 1996; Gold and Chrousos, 2002).

In view of the often blunted TSH response to TRH and the absent nocturnal TSH surge reported in depression (Bartalena et al., 1990) alterations in hypothalamic TRH may be involved in the pathogenesis of HPT axis changes in major depression. This assumption was further supported by one study showing beneficial effects of intrathecal administration of TRH to patients with treatment-resistant depression (Marangell et al., 1997). Comparison of TRHmRNA expression in the PVN of patients with a long history of depression ($n = 7$) to age- and sex-matched controls by quantitative in situ hybridization using image analysis revealed a strong decrease in TRHmRNA expression in depression

(median 24.5 vs. 56.8 au, Mann-Whitney U $p < 0.05$; (Alkemade et al., 2003) (Fig. 11). That this decrease may be the result of hypercortisolism was supported by a subsequent study in glucocorticoid- vs. nonglucocorticoid-treated patients without a psychiatric diagnosis showing decreased TRHmRNA expression in the glucocorticoid-treated patients (Alkemade et al., 2005b).

Hyperthyroidism

We had the unique opportunity to study TRH, deiodinase and MCT8 expression in the hypothalamus of a patient who died acutely from massive pulmonary embolism in whom hyperthyroidism had been documented biochemically just before death (serum FT4 44.2 pmol/L, $n = 10$–23; TSH <0.01 mU/L, $n = 0.4$–4.0). This patient was found

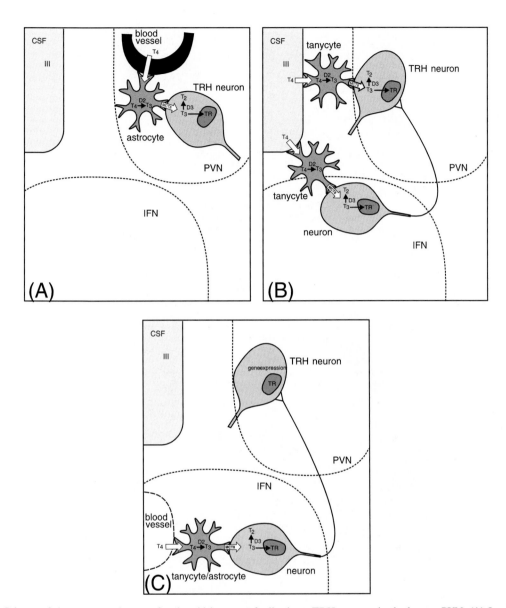

Fig. 10. Scheme of three proposed routes for thyroid hormone feedback on TRH neurons in the human PVN. (A) Local thyroid hormone uptake from the vascular compartment within the PVN, (B) thyroid hormone uptake from the CSF in the third ventricle followed by transport to TRH neurons in the PVN or IFN neurons projecting to TRH neurons in the PVN, and (C) thyroid hormone sensing in the IFN of the mediobasal hypothalamus by neurons projecting to TRH neurons in the PVN. (A) T4 is taken up locally from the vascular compartment within the PVN by astrocytes expressing D2. In these astrocytes, T4 is converted to T3 and transported to TRH neurons by MCT8. In the TRH neuron, T3 may bind to the TR and exert a negative feedback on TRH gene expression. T3 can be inactivated by D3 present in the TRH neuron. (B) T4 is taken up from the CSF by tanycytes expressing D2. In these cells, T4 is converted to T3 and subsequently transported by MCT8 to TRH neurons of the PVN or to neurons in the IFN. T3 may bind to TRs expressed in these neurons and may (subsequently) be degraded by D3. TR binding in the PVN can alter TRH gene expression directly. In addition, TR binding in the IFN may result in altered firing of neurons projecting to TRH neurons in the PVN. Alternatively, T3 may travel from IFN neurons to TRH neurons in the PVN through axonal transport. (C) D2 expressing astrocytes or tanycytes sense T4 in the IFN/ME area, in which the BBB is absent and these cells convert T4 into T3. T3 is transported by MCT8 to IFN neurons. T3 may bind locally to the TR, altering gene expression of IFN neurons, or may be inactivated by D3. T3 may thus result in altered firing pattern of IFN neurons projecting to TRH neurons in the PVN. Alternatively, T3 may travel to TRH neurons in the PVN through axonal transport. Reproduced with permission from Alkemade et al. (2005a).

202

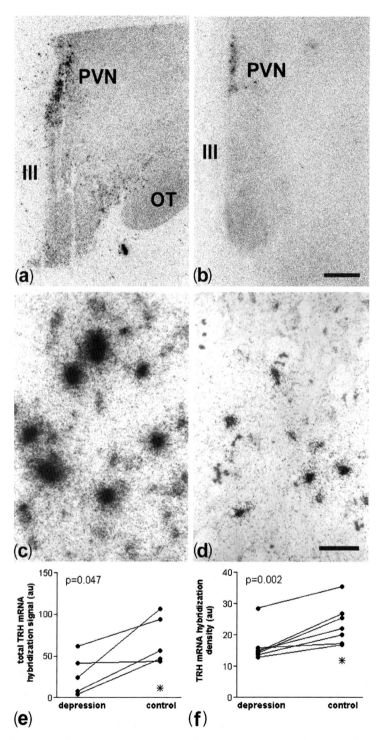

Fig. 11. (a,b) Macroscopic photographs of film autoradiograms of representative sections of a control subject and weaker hybridization signal in a subject with major depression, respectively. (c,d) TRH mRNA containing cells in the PVN of the same subjects. Bars represent 2 mm (a,b) and 50 mm (c,d). OT, optic tract. (e,f) TRH mRNA hybridization signals in depressed subjects and controls. Matched pairs are interconnected with lines. Each dot represents one subject. Note, strongly decreased TRH hybridization signal in subjects with depression. Asterisk represents a subject with primary hyperthyroidism. Reproduced with permission from Alkemade et al. (2003).

to have extremely low TRHmRNA expression in the PVN (Alkemade et al., 2003), in keeping with negative feedback action of thyroid hormone on TRH gene expression in the human PVN. Furthermore, the hypothalamus showed rather weak D2 immunostaining and unremarkable D3 and MCT8 staining. Studies in rats have shown somewhat increased D3 expression in the SON during hyperthyroidism, while extrahypothalamic areas show a more pronounced response to altered thyroid status. Furthermore, hypothyroidism induces up-regulation of hypothalamic D2 enzyme activity and mRNA expression (Riskind et al., 1987; Tu et al., 1997) with opposite changes in hyperthyroidism (Cowley et al., 2003). Therefore, our observations in the patient with hyperthyroidism (i.e., markedly decreased TRHmRNA expression in the PVN, low D2 and unchanged D3 immunoreactivity) are suggestive of similar regulation of hypothalamic TRH and deiodinase by thyroid hormone in human and rat hypothalamus. For obvious reasons, it is unlikely that these preliminary data on effects of hyperthyroidism in the human hypothalamus will be extended in the future.

Nonthyroidal illness

During illness, serum T3 levels decrease without giving rise to higher TSH levels, a phenomenon which is known as nonthyroidal illness (NTI) (Wiersinga, 2005). Although serum TSH remains within the normal range, the nocturnal TSH surge may be absent (Romijn and Wiersinga, 1990). The magnitude of the drop in serum T3 within 24 h after the onset of a severe physical stress, such as surgery or trauma reflects the severity of the stressor. The acute changes in the thyroid axis have been interpreted teleologically as an attempt to reduce energy expenditure, at least when they occur in the framework of starvation. Whether this also holds for other acute stress conditions is unknown (for review see Van den Berghe, 2000). In patients treated in intensive care units (ICU) for prolonged periods, a dramatic decrease in TSH pulsatility occurs. Specifically, a loss of TSH pulse amplitude occurs in this setting in direct relation to low-serum levels of thyroid hormone (Van den Berghe et al., 1999). Continuous iv administration of TRH, in particular when administered together

with the growth hormone (GH) secretagogue GH releasing peptide-2 (GHRP-2), to ICU patients with prolonged critical illness restores TSH pulsatility as well as physiological levels of thyroid hormones (Van den Berghe et al., 1998). These observations have been replicated in a rabbit model for acute and prolonged critical illness (Weekers et al., 2002).

The pathogenesis of NTI is incompletely understood at present but the paradoxically low TSH in the setting of decreased serum T3 points to a hypothalamic factor. In patients whose serum concentrations of TSH, T3 and T4 were assessed in a serum sample taken <24 h before in-hospital death, a positive correlation was reported between total TRHmRNA hybridization signal in the PVN on the one hand and serum TSH and T3 on the other hand, but not between TRHmRNA and serum T4 (Fliers et al., 1997). In addition, patients who die after severe illness have less than half of the concentrations of tissue T3 in the hypothalamus and pituitary compared to patients who die acutely from trauma (Arem et al., 1993). The combination of low hypothalamic T3 and low TRH expression in the PVN implies a major change in hypothalamic thyroid feedback regulation during critical illness, and suggests that the drop in thyroid hormone production in chronic severe illness has — at least in part — a hypothalamic origin (for review see Peeters et al., 2006).

The mediobasal hypothalamus probably plays a role in the resetting of thyroid hormone feedback in NTI induced by lipopolysaccharide injection, which is an animal model for acute infection, since a dramatic increase of local D2 expression occurs both in rats and mice in this setting (Boelen et al., 2004; Fekete et al., 2004). In humans, a positive correlation between total NPY immunoreactivity in the IFN and total TRHmRNA expression in the PVN was reported in patients with biochemically documented NTI suggesting a role for decreased NPY input from the IFN in the resetting of thyroid hormone feedback on hypothalamic TRH cells in NTI (Fliers et al., 2001).

It is unknown at present whether up-regulation of D2 expression in the human mediobasal hypothalamus occurs in the setting of NTI. Another theoretical possibility leading to relatively high hypothalamic T3 concentrations in critical illness is down-regulation of D3 in the PVN although lower hypothalamic T3 concentrations in patients who died after prolonged illness (Arem et al., 1993) do not support this. Regulation of MCT8 expression in TRH neurons of the human PVN and perhaps of additional thyroid hormone transporters such as OATP1C1 during illness represent additional theoretical determinants of thyroid hormone feedback regulation. The same holds true for differential regulation of hypothalamic TR isoforms. In this context, it is of interest that TR isoform expression is differentially regulated by thyroid hormone status in different brain regions including the PVN (Clerget-Froidevaux et al., 2004). These possibilities will need to be explored in future studies.

Conclusion

The distribution and expression of TR isoforms, D2 and D3, and the thyroid hormone transporter MCT8 have recently been reported in the human hypothalamus. Both D2 and D3 enzyme activities are detectable post-mortem in the human mediobasal hypothalamus. Glial cells of the IFN/ME region and tanycytes lining the third ventricle show D2 immunoreactivity. D3, MCT8 and TRs are all expressed by TRH neurons in the PVN, whereas D2 is not. These observations suggest that the prohormone T4 is taken up in hypothalamic glial cells that convert T4 into the biologically active T3, which is subsequently transported to TRH producing neurons in the PVN. In these neurons, T3 may bind to TRs and/or may be degraded into inactive iodothyronines. This model for thyroid hormone action in the human hypothalamus is to be validated in future experimental studies. Set-point changes in the human HPT axis as observed in clinically euthyroid patients during critical illness and in depression, involving paradoxically low hypothalamic TRH mRNA expression, may be mediated at all these molecular levels.

Abbreviations

AC	anterior commissure
AGRP	agouti-related protein
ARC	arcuate nucleus
BBB	blood–brain barrier
CART	cocaine- and amphetamine-regulated transcript
CM	corpus mamillare
D2	type 2 deiodinase
D3	type 3 deiodinase
DB	diagonal band of Broca
Ep	ependymal layer
FO	fornix
HPT axis	hypothalamus–pituitary–thyroid axis
IFN	infundibular nucleus
LHA	lateral hypothalamus
MCT8	monocarboxylate transporter 8
ME	median eminence
αMSH	alpha-melanocyte-stimulating hormone
NPY	neuropeptide Y
NTI	nonthyroidal illness
NTL	tuberolateral nucleus
OC	optic chiasm
OT	optic tract
POMC	pro-opiomelanocortin
PVN	paraventricular nucleus
SCN	suprachiasmatic nucleus
SDN	sexually dimorphic nucleus
SON	supraoptic nucleus
T4	thyroxine
T3	triiodothyronine
TMN	tuberomammilary nucleus
TSH	thyrotropin
TR	thyroid hormone receptor
TRH	thyrotropin-releasing hormone

Acknowlegdments

Brain material was obtained from the Netherlands Brain Bank (coordinator: Dr. R. Ravid). We are indebted to Unga Unmehopa and to Bart Fisser for technical assistance. This work was financially supported by NWO-MW (grant #903-40-201, The Brain Foundation of The Netherlands and the Ludgardine Bouwman Foundation).

References

Abel, E.D., Ahima, R.S., Boers, M.E., Elmquist, J.K. and Wondisford, F.E. (2001) Critical role for thyroid hormone receptor beta2 in the regulation of paraventricular thyrotropin-releasing hormone neurons. J. Clin. Invest., 107(8): 1017–1023.

Alkemade, A., Friesema, E.C., Unmehopa, U.A., Fabriek, B.O., Kuiper, G.G., Leonard, J.L., Wiersinga, W.M., Swaab, D.F., Visser, T.J. and Fliers, E. (2005a) Neuroanatomical pathways for thyroid hormone feedback in the human hypothalamus. J. Clin. Endocrinol. Metab., 90(7): 4322–4334.

Alkemade, A., Unmehopa, U.A., Brouwer, J.P., Hoogendijk, W.J., Wiersinga, W.M., Swaab, D.F. and Fliers, E. (2003) Decreased thyrotropin-releasing hormone gene expression in the hypothalamic paraventricular nucleus of patients with major depression. Mol. Psychiatry, 8(10): 838–839.

Alkemade, A., Unmehopa, U.A., Wiersinga, W.M., Swaab, D.F. and Fliers, E. (2005b) Glucocorticoids decrease thyrotropin-releasing hormone messenger ribonucleic acid expression in the paraventricular nucleus of the human hypothalamus. J. Clin. Endocrinol. Metab., 90(1): 323–327.

Alkemade, A., Vuijst, C.L., Unmehopa, U.A., Bakker, O., Vennstrom, B., Wiersinga, W.M., Swaab, D.F. and Fliers, E. (2005c) Thyroid hormone receptor expression in the human hypothalamus and anterior pituitary. J. Clin. Endocrinol. Metab., 90(2): 904–912.

Arem, R., Wiener, G.J., Kaplan, S.G., Kim, H.S., Reichlin, S. and Kaplan, M.M. (1993) Reduced tissue thyroid hormone levels in fatal illness. Metabolism, 42(9): 1102–1108.

Bartalena, L., Placidi, G.F., Martino, E., Falcone, M., Pellegrini, L., Dell'Osso, L., Pacchiarotti, A. and Pinchera, A. (1990) Nocturnal serum thyrotropin (TSH) surge and the TSH response to TSH-releasing hormone: dissociated behavior in untreated depressives. J. Clin. Endocrinol. Metab., 71(3): 650–655.

Baumann, C.T., Maruvada, P., Hager, G.L. and Yen, P.M. (2001) Nuclear cytoplasmic shuttling by thyroid hormone receptors. Multiple protein interactions are required for nuclear retention. J. Biol. Chem., 276(14): 11237–11245.

Bianco, A.C., Salvatore, D., Gereben, B., Berry, M.J. and Larsen, P.R. (2002) Biochemistry, cellular and molecular biology, and physiological roles of the iodothyronine selenodeiodinases. Endocr. Rev., 23(1): 38–89.

Boelen, A., Kwakkel, J., Thijssen-Timmer, D.C., Alkemade, A., Fliers, E. and Wiersinga, W.M. (2004) Simultaneous changes in central and peripheral components of the hypothalamus–pituitary–thyroid axis in lipopolysaccharide-induced acute illness in mice. J. Endocrinol., 182(2): 315–323.

Bøler, J., Enzmann, F., Folkers, K., Bowers, C.Y. and Schally, A.V. (1969) The identity of chemical and hormonal properties of the thyrotropin-releasing hormone and pyroglutamyl-histidyl-proline amide. Biochem. Biophys. Res. Commun., 37: 705–710.

Bradley, D.J., Young III, W.S. and Weinberger, C. (1989) Differential expression of alpha and beta thyroid hormone receptor genes in rat brain and pituitary. Proc. Natl. Acad. Sci. USA, 86(18): 7250–7254.

Brouwer, J.P., Appelhof, B.C., Hoogendijk, W.J., Huyser, J., Endert, E., Zuketto, C., Schene, A.H., Tijssen, J.G., Van Dyck, R., Wiersinga, W.M. and Fliers, E. (2005) Thyroid and

adrenal axis in major depression: a controlled study in outpatients. Eur. J. Endocrinol., 152(2): 185–191.

Burgus, R.T., Dunn, T., Desiderio, D. and Guillemin, R. (1969) Structure moelculaire du facteur hypothalamique hypophysiotrope TRF d'origine ovine: mise en evidence par spectrometre de masse de la sequence PCA-His-Pro-NH2. CR Acad. Sci. (Paris), 269: 1870–1873.

Clerget-Froidevaux, M.S., Seugnet, I. and Demeneix, B.A. (2004) Thyroid status co-regulates thyroid hormone receptor and co-modulator genes specifically in the hypothalamus. FEBS Lett., 569(1–3): 341–345.

Cook, C.B., Kakucska, I., Lechan, R.M. and Koenig, R.J. (1992) Expression of thyroid hormone receptor beta 2 in rat hypothalamus. Endocrinology, 130(2): 1077–1079.

Cowley, M.A., Smith, R.G., Diano, S., Tschop, M., Pronchuk, N., Grove, K.L., Strasburger, C.J., Bidlingmaier, M., Esterman, M., Heiman, M.L., Garcia-Segura, L.M., Nillni, E.A., Mendez, P., Low, M.J., Sotonyi, P., Friedman, J.M., Liu, H., Pinto, S., Colmers, W.F., Cone, R.D. and Horvath, T.L. (2003) The distribution and mechanism of action of ghrelin in the CNS demonstrates a novel hypothalamic circuit regulating energy homeostasis. Neuron, 37(4): 649–661.

Diano, S., Leonard, J.L., Meli, R., Esposito, E. and Schiavo, L. (2003) Hypothalamic type II iodothyronine deiodinase: a light and electron microscopic study. Brain Res., 976(1): 130–134.

Dyess, E.M., Segerson, T.P., Liposits, Z., Paull, W.K., Kaplan, M.M., Wu, P., Jackson, I.M. and Lechan, R.M. (1988) Triiodothyronine exerts direct cell-specific regulation of thyrotropin-releasing hormone gene expression in the hypothalamic paraventricular nucleus. Endocrinology, 123(5): 2291–2297.

Elias, C.F., Lee, C.E., Kelly, J.F., Ahima, R.S., Kuhar, M., Saper, C.B. and Elmquist, J.K. (2001) Characterization of CART neurons in the rat and human hypothalamus. J. Comp. Neurol., 432(1): 1–19.

Fekete, C., Gereben, B., Doleschall, M., Harney, J.W., Dora, J.M., Bianco, A.C., Sarkar, S., Liposits, Z., Rand, W., Emerson, C., Kacskovics, I., Larsen, P.R. and Lechan, R.M. (2004) Lipopolysaccharide induces type 2 iodothyronine deiodinase in the mediobasal hypothalamus: implications for the nonthyroidal illness syndrome. Endocrinology, 145(4): 1649–1655.

Fliers, E., Guldenaar, S.E., Wiersinga, W.M. and Swaab, D.F. (1997) Decreased hypothalamic thyrotropin-releasing hormone gene expression in patients with nonthyroidal illness. J. Clin. Endocrinol. Metab., 82(12): 4032–4036.

Fliers, E., Noppen, N.W., Wiersinga, W.M., Visser, T.J. and Swaab, D.F. (1994) Distribution of thyrotropin-releasing hormone (TRH)-containing cells and fibers in the human hypothalamus. J. Comp. Neurol., 350(2): 311–323.

Fliers, E., Swaab, D.F., Pool, C.W. and Verwer, R.W. (1985) The vasopressin and oxytocin neurons in the human supraoptic and paraventricular nucleus; changes with aging and in senile dementia. Brain Res., 342(1): 45–53.

Fliers, E., Unmehopa, U.A., Manniesing, S., Vuijst, C.L., Wiersinga, W.M. and Swaab, D.F. (2001) Decreased neuropeptide Y (NPY) expression in the infundibular nucleus of patients with nonthyroidal illness. Peptides, 22(3): 459–465.

Friesema, E.C., Ganguly, S., Abdalla, A., Manning Fox, J.E., Halestrap, A.P. and Visser, T.J. (2003) Identification of monocarboxylate transporter 8 as a specific thyroid hormone transporter. J. Biol. Chem., 278(41): 40128–40135.

Friesema, E.C., Grueters, A., Biebermann, H., Krude, H., von Moers, A., Reeser, M., Barrett, T.G., Mancilla, E.E., Svensson, J., Kester, M.H., Kuiper, G.G., Balkassmi, S., Uitterlinden, A.G., Koehrle, J., Rodien, P., Halestrap, A.P. and Visser, T.J. (2004) Association between mutations in a thyroid hormone transporter and severe X-linked psychomotor retardation. Lancet, 364(9443): 1435–1437.

Gold, P.W. and Chrousos, G.P. (2002) Organization of the stress system and its dysregulation in melancholic and atypical depression: high vs low CRH/NE status. Mol. Psychiatry, 7: 254–275.

Gordon, J.T., Kaminski, D.M., Rozanov, C.B. and Dratman, M.B. (1999) Evidence that 3,3′,5-triiodothyronine is concentrated in and delivered from the locus coeruleus to its noradrenergic targets via anterograde axonal transport. Neuroscience, 93(3): 943–954.

Guadano-Ferraz, A., Obregon, M.J., St Germain, D.L. and Bernal, J. (1997) The type 2 iodothyronine deiodinase is expressed primarily in glial cells in the neonatal rat brain. Proc. Natl. Acad. Sci. USA, 94(19): 10391–10396.

Guldenaar, S.E., Veldkamp, B., Bakker, O., Wiersinga, W.M., Swaab, D.F. and Fliers, E. (1996) Thyrotropin-releasing hormone gene expression in the human hypothalamus. Brain Res., 743(1–2): 93–101.

Hager, G.L., Lim, C.S., Elbi, C. and Baumann, C.T. (2000) Trafficking of nuclear receptors in living cells. J. Steroid Biochem. Mol. Biol., 74(5): 249–254.

Heuer, H., Maier, M.K., Iden, S., Mittag, J., Friesema, E.C., Visser, T.J. and Bauer, K. (2005) The monocarboxylate transporter 8 linked to human psychomotor retardation is highly expressed in thyroid hormone-sensitive neuron populations. Endocrinology, 146(4): 1701–1706.

Holsboer, F. and Barden, N. (1996) Antidepressants and hypothalamic–pituitary–adrenocortical regulation. Endocr. Rev., 17(2): 187–205.

Jackson, I.M. (1998) The thyroid axis and depression. Thyroid, 8(10): 951–956.

Jansen, J., Friesema, E.C., Milici, C. and Visser, T.J. (2005) Thyroid hormone transporters in health and disease. Thyroid, 15(8): 757–768.

Kirkegaard, C. and Faber, J. (1998) The role of thyroid hormones in depression. Eur. J. Endocrinol., 138(1): 1–9.

Kong, W.M., Martin, N.M., Smith, K.L., Gardiner, J.V., Connoley, I.P., Stephens, D.A., Dhillo, W.S., Ghatei, M.A., Small, C.J. and Bloom, S.R. (2004) Triiodothyronine stimulates food intake via the hypothalamic ventromedial nucleus independent of changes in energy expenditure. Endocrinology, 145(11): 5252–5258.

Kruijver, F.P., Balesar, R., Espila, A.M., Unmehopa, U.A. and Swaab, D.F. (2002) Estrogen receptor-alpha distribution in the human hypothalamus in relation to sex and endocrine status. J. Comp. Neurol., 454: 115–139.

Lechan, R.M. and Fekete, C. (2004) Feedback regulation of thyrotropin-releasing hormone (TRH): mechanisms for the non-thyroidal illness syndrome. J. Endocrinol. Invest., 27(Suppl 6): 105–119.

Lechan, R.M. and Fekete, C. (2006) The TRH neuron: a hypothalamic integrator of energy metabolism. Progress in Brain Research, Vol. 153. Elsevier, Amsterdam, in press.

Lechan, R.M., Qi, Y., Jackson, I.M. and Mahdavi, V. (1994) Identification of thyroid hormone receptor isoforms in thyrotropin-releasing hormone neurons of the hypothalamic paraventricular nucleus. Endocrinology, 135(1): 92–100.

Legradi, G. and Lechan, R.M. (1998) The arcuate nucleus is the major source for neuropeptide Y-innervation of thyrotropin-releasing hormone neurons in the hypothalamic paraventricular nucleus. Endocrinology, 139(7): 3262–3270.

Legradi, G. and Lechan, R.M. (1999) Agouti-related protein containing nerve terminals innervate thyrotropin-releasing hormone neurons in the hypothalamic paraventricular nucleus. Endocrinology, 140(8): 3643–3652.

Marangell, L.B., George, M.S., Callahan, A.M., Ketter, T.A., Pazzaglia, P.J., L'Herrou, T.A., Leverich, G.S. and Post, R.M. (1997) Effects of intrathecal thyrotropin-releasing hormone (protirelin) in refractory depressed patients. Arch. Gen. Psychiatry, 54(3): 214–222.

Maruvada, P., Baumann, C.T., Hager, G.L. and Yen, P.M. (2003) Dynamic shuttling and intranuclear mobility of nuclear hormone receptors. J. Biol. Chem., 278(14): 12425–12432.

Nemeroff, C.B., Simon, J.S., Haggerty Jr., J.J. and Evans, D.L. (1985) Antithyroid antibodies in depressed patients. Am. J. Psychiatry, 142(7): 840–843.

Peeters, R.P., Debaveye, Y., Fliers, E. and Visser, T.J. (2006) Changes within the thyroid axis during the course of critical illness. Crit. Care Clin., 22: 41–55.

Peeters, R., Fekete, C., Goncalves, C., Legradi, G., Tu, H.M., Harney, J.W., Bianco, A.C., Lechan, R.M. and Larsen, P.R. (2001) Regional physiological adaptation of the central nervous system deiodinases to iodine deficiency. Am. J. Physiol. Endocrinol. Metab., 281(1): E5461.

Pop, V.J., Maartens, L.H., Leusink, G., van Son, M.J., Knottnerus, A.A., Ward, A.M., Metcalfe, R. and Weetman, A.P. (1998) Are autoimmune thyroid dysfunction and depression related? J. Clin. Endocrinol. Metab., 83(9): 3194–3197.

Riskind, P.N., Kolodny, J.M. and Larsen, P.R. (1987) The regional hypothalamic distribution of type II 5′-monodeiodinase in euthyroid and hypothyroid rats. Brain Res., 420(1): 194–198.

Romijn, J.A. and Wiersinga, W.M. (1990) Decreased nocturnal surge of thyrotropin in nonthyroidal illness. J. Clin. Endocrinol. Metab., 70(1): 35–42.

Sakurai, T., Amemiya, A., Ishii, M., Matsuzaki, I., Chemelli, R.M., Tanaka, H., Williams, S.C., Richardson, J.A., Kozlowski, G.P., Wilson, S., Arch, J.R., Buckingham, R.E., Haynes, A.C., Carr, S.A., Annan, R.S., McNulty, D.E., Liu, W.S., Terrett, J.A., Elshourbagy, N.A., Bergsma, D.J. and

Yanagisawa, M. (1998) Orexins and orexin receptors: a family of hypothalamic neuropeptides and G protein-coupled receptors that regulate feeding behavior. Cell, 92(4): 573–585.

Segerson, T.P., Kauer, J., Wolfe, H.C., Mobtaker, H., Wu, P., Jackson, I.M. and Lechan, R.M. (1987) Thyroid hormone regulates TRH biosynthesis in the paraventricular nucleus of the rat hypothalamus. Science, 238(4823): 78–80.

Skowsky, R.W. and Kikuchi, T.A. (1978) The role of vasopressin in the impaired water excretion of myxedema. Am. J. Med., 64: 613–621.

Tu, H.M., Kim, S.W., Salvatore, D., Bartha, T., Legradi, G., Larsen, P.R. and Lechan, R.M. (1997) Regional distribution of type 2 thyroxine deiodinase messenger ribonucleic acid in rat hypothalamus and pituitary and its regulation by thyroid hormone. Endocrinology, 138(8): 3359–3368.

Van den Berghe, G. (2000) Novel insights into the neuroendocrinology of critical illness. Eur. J. Endocrinol., 143(1): 1–13.

Van den Berghe, G., de Zegher, F., Baxter, R.C., Veldhuis, J.D., Wouters, P., Schetz, M., Verwaest, C., Van, d.V., Lauwers, P., Bouillon, R. and Bowers, C.Y. (1998) Neuroendocrinology of prolonged critical illness: effects of exogenous thyrotropin-releasing hormone and its combination with growth hormone secretagogues. J. Clin. Endocrinol. Metab., 83(2): 309–319.

Van den Berghe, G., Wouters, P., Bowers, C.Y., de Zegher, F., Bouillon, R. and Veldhuis, J.D. (1999) Growth hormone-releasing peptide-2 infusion synchronizes growth hormone, thyrotrophin and prolactin release in prolonged critical illness. Eur. J. Endocrinol., 140(1): 17–22.

Weekers, F., Van Herck, E., Coopmans, W., Michalaki, M., Bowers, C.Y., Veldhuis, J.D. and Van den, B.G. (2002) A novel in vivo rabbit model of hypercatabolic critical illness reveals a biphasic neuroendocrine stress response. Endocrinology, 143(3): 764–774.

Wiersinga, W.M. (2005) Nonthyroidal illness. In: Braverman, L.E. and Utiger, R.D. (Eds.), The Thyroid (9th Ed). Philadelphia, Lippincot, Williams & Wilkins, pp. 247–263.

Zhu, X.G., Hanover, J.A., Hager, G.L. and Cheng, S.Y. (1998) Hormone-induced translocation of thyroid hormone receptors in living cells visualized using a receptor green fluorescent protein chimera. J. Biol. Chem., 273(42): 27058–27063.

Kalsbeek, Fliers, Hofman, Swaab, Van Someren & Buijs
Progress in Brain Research, Vol. 153
ISSN 0079-6123

CHAPTER 12

The TRH neuron: a hypothalamic integrator of energy metabolism[☆]

Ronald M. Lechan[1,2,*] and Csaba Fekete[1,3]

[1]Tupper Research Institute and Department of Medicine, Division of Endocrinology, Diabetes and Metabolism, Box 268, Tufts-New England Medical Center, Boston, MA 02111, USA
[2]Department of Neuroscience, Tufts University School of Medicine, Boston, MA 02111, USA
[3]Department of Endocrine Neurobiology, Institute of Experimental Medicine, Hungarian Academy of Sciences, Budapest 1083, Hungary

Abstract: Thyrotropin-releasing hormone (TRH) has an important role in the regulation of energy homeostasis not only through effects on thyroid function orchestrated through hypophysiotropic neurons in the hypothalamic paraventricular nucleus (PVN), but also through central effects on feeding behavior, thermogenesis, locomotor activation and autonomic regulation. Hypophysiotropic TRH neurons are located in the medial and periventricular parvocellular subdivisions of the PVN and receive direct monosynaptic projections from two, separate, populations of leptin-responsive neurons in the hypothalamic arcuate nucleus containing either α-melanocyte-stimulating hormone (α-MSH) and cocaine- and amphetamine-regulated transcript (CART), peptides that promote weight loss and increase energy expenditure, or neuropeptide Y (NPY) and agouti-related protein (AGRP), peptides that promote weight gain and reduce energy expenditure. During fasting, the reduction in TRH mRNA in hypophysiotropic neurons mediated by suppression of α-MSH/CART simultaneously with an increase in NPY/AGRP gene expression in arcuate nucleus neurons contributes to the fall in circulating thyroid hormone levels, presumably by increasing the sensitivity of the TRH gene to negative feedback inhibition by thyroid hormone. Endotoxin administration, however, has the paradoxical effect of increasing circulating levels of leptin and melanocortin signaling and CART gene expression in arcuate nucleus neurons, but inhibiting TRH gene expression in hypophysiotropic neurons. This may be explained by an overriding inhibitory effect of endotoxin to increase type 2 iodothyroine deiodinase (D2) in a population of specialized glial cells, tanycytes, located in the base and infralateral walls of the third ventricle. By increasing the conversion of T4 into T3, tanycytes may increase local tissue concenetrations of thyroid hormone, and thereby induce a state of local tissue hyperthyroidism in the region of hypophysisotrophic TRH neurons. Other regions of the brain may also serve as metabolic sensors for hypophysiostropic TRH neurons including the ventrolateral medulla and dorsomedial nucleus of the hypothalamus that have direct monosynaptic projections to the PVN. TRH also exerts a number of effects within the central nervous system that may contribute to the regulation of energy homeostasis. Included are an increase in core body temperature mediated through neurons in the anterior hypothalamic-preoptic area that coordinate a variety of autonomic responses; arousal and locomotor activation through cholinergic and dopaminergic mechanisms on the septum and nucleus accumbens, respectively; and regulation of the cephalic phase of digestion. While

[☆]This work was supported by Grants NIH DK-37021 and OTKA T046492.

*Corresponding author. Tel.: +1-617-636-8517; Fax: +1-617-636-4719; E-mail: rlechan@tufts-nemc.org

DOI: 10.1016/S0079-6123(06)53012-2

the latter responses are largely mediated through cholinergic mechanisms via TRH neurons in the brainstem medullary raphe and dorsal motor nucleus of the vagus, effects of TRH on autonomic loci in the hypothalamic PVN may also be important. Contrary to the actions of T3 to increase appetite, TRH has central effects to reduce food intake in normal, fasting and stressed animals. The precise locus where TRH mediates this response is unknown. However, evidence that an anatomically separate population of nonhypophysiotropic TRH neurons in the anterior parvocellular subdivision of the PVN is integrated into the leptin regulatory control system by the same arcuate nucleus neuronal populations that innervate hypophysiotropic TRH neurons, raises the possibility that anterior parvocellular TRH neurons may be involved, possibly through interactions with the limbic nervous system.

Thyrotropin-releasing hormone (TRH), a tripeptide amide (pGlu-His-ProNH$_2$), was originally isolated and characterized on the basis of its action as a releasing hormone involved in the regulation of the hypothalamic-pituitary-thyroid (HPT) axis (Reichlin, 1989), but is now known to subserve a wide variety of other biologic functions (Table 1). As a result, it is becoming increasingly clear that TRH has an important role in the regulation of energy homeostasis not only through its hypophysiotropic actions on anterior pituitary thyrotrophs to regulate circulating levels of thyroid hormone, but also through direct effects exerted within the central nervous system, itself. While TRH-synthesizing neurons in the medial and periventricular parvocellular subdivisions of the hypothalamic paraventricular nucleus (PVN) are critical to a number of the metabolic actions of TRH, under specific physiological circumstances separate populations of TRH neurons in the hypothalamus may also participate together with or independently of hypophysiotropic TRH neurons. The major actions of TRH in the

hypothalamus to regulate energy homeostasis will be discussed below.

Role of hypophysiotropic TRH neurons in energy homeostasis

Organization of the hypothalamic-pituitary-thyroid axis

The origin of TRH-containing neurons that regulate anterior pituitary TSH secretion is the hypothalamic (PVN), a triangular, midline nuclear group symmetrically located on either side of the dorsal portion of the third ventricle (Fig. 1). The PVN is composed of two major parts; magnocellular neurons involved in vasopressin and oxytocin secretion located in more lateral portions of the nucleus and parvocellular neurons located in more medial portions of the nucleus. The parvocellular component has a number of subcompartments including anterior, medial, periventricle, ventral,

Table 1. Nonhypophysiotropic effects of TRH

Analgesia	Increased gastric emptying
Anticonvulsant activity	Increased gastrointestinal motility
Arousal	Increased hepatic blood flow
Cerebral vasodilation	Increased insulin secretion
Diaphoresis	Increased locomotor activity
Facilitation of memory	Increased respiration
Facilitation of motoneuron excitability	Inhibition of food intake
Hypertension	Neurotrophic effects on spinal cord
Improved memory	Motoneurons
Increased blood pressure	Peripheral vasoconstriction
Increased body temperature	Tachycardia
Increased gastric acid secretion	Tremor
Increased gastro-duodenal blood Flow	

dorsal and lateral parvocellular subdivisions (Swanson and Kuypers, 1980). TRH-synthesizing neurons are found primarily in the anterior, medial and periventricular parvocellular subdivisions (Fig. 2 A, B), and to a lesser extent in the dorsal subnuclei (Lechan and Hollenberg, 2003). However, only TRH neurons in medial and periventricular parvocellular subdivisions project to the median eminence, and thereby are functionally distinct from TRH neurons in the anterior and dorsal parvocellular subdivisions. Support for this view is apparent in experimental models of hypothyroidism in which only TRH neurons in the medial and periventricular parvocellular subdivisions

show marked increases in TRH mRNA (Segerson et al., 1987; Lechan and Kakucska, 1992). In addition, cocaine- and amphetamine-regulated transcript (CART) is co-expressed with TRH only in medial and periventricular parvocellular TRH neurons (Fekete et al., 2000c).

Axons from hypophysiotropic TRH neurons in the PVN project to the hypothalamic median eminence, one of the several so-called "circumventricular organs" in the mammalian brain that lie outside of the blood–brain barrier (Lechan, 2003). Here, they terminate primarily in the external zone of the median eminence (Fig. 2C), allowing access of TRH to the fenestrated capillaries of the portal system for direct transport to the anterior pituitary, where TRH regulates the release of TSH. TSH then stimulates the release of thyroid hormones, thyroxine (T4) and triiodothyronine (T3), from the thyroid gland into the peripheral circulation that in turn, inhibit hypothalamic hypophysiotropic TRH neurons and anterior pituitary thyrotrophs to complete what is recognized as a classic example of a negative feedback loop system (Lechan and Hollenberg, 2003).

Thyroid hormone and energy expenditure

Thyroid hormone has an important role in the regulation of energy expenditure primarily through effects on both obligatory thermogenesis (energy expenditure necessary to sustain basal homeostatic functions) and adaptive thermogenesis (additional heat produced in response to triggering signals to

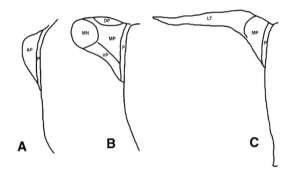

Fig. 1. Schematic of the hypothalamic paraventricular nucleus showing major subdivisions. (A) anterior, (B) mid and (C) caudal levels. AP, anterior parvocellular subdivision; DP, dorsal parvocellular subdivision; MN, magnocellular division; MP, medial paravocellular subdivision; P, periventricular parvocellular subdivision, VP, ventral parvocellular subdivision; LT, lateral parvocellular subdivision. (Modified from Sarkar et al. (2004), with permission from Elsevier.)

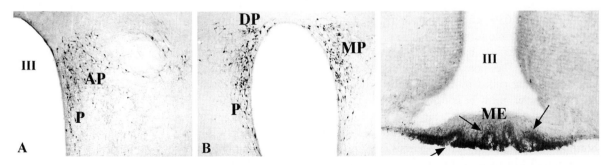

Fig. 2. Immunocytochemical delineation of TRH in the hypothalamic paraventricular nucleus (PVN) and median eminence (ME). (A and B) Neurons containing TRH are present in the anterior (AP), medial (MP), periventricular (P) and dorsal (DP) parvocellular subdivisions. Axons from TRH neurons in the MP and P terminate in the ME shown in (C). Note that axons terminals in the external zone of the ME are in close apposition to the portal capillaries (arrows). III, third ventricle.

sustain core temperature) (Silva, 1995, 2001, 2003; Bianco et al., 2002). Indeed, basal metabolic rate can be reduced by as much as 30% in the absence of thyroid hormone, and adaptive thermogenesis in cold exposed animals is also markedly impaired (Silva, 2003). Conversely, hyperthyroidism increases metabolic rate and accelerates metabolic processes, affecting thermogenesis by increasing ATP consumption and reducing the thermodynamic efficiency of ATP synthesis (Silva, 2003). Although the molecular mechanisms are not precisely known, thyroid hormone has a potent action on uncoupling proteins (Gong et al., 1997; De Lange et al., 2001; Lebon et al., 2001; Ribeiro et al., 2001; Queiroz et al., 2004) which function by short-circuiting the proton gradient across the inner mitochondrial membrane, uncoupling fuel oxidation from the synthesis of ATP to generate heat (Jezek, 2002). While UCP1 has an essential role in thyroid hormone-mediated adaptive thermogenesis in brown adipose tissue (Enerback et al., 1997; Ribeiro et al., 2001; Bianco et al., 2002), however, the contribution of uncoupling proteins to thermogenesis by thyroid hormone in other tissues remains controversial (Collin et al., 2005). Thyroid hormone markedly upregulates UCP3 in skeletal muscle and heart (Boss et al., 1997; Gong et al., 1997; de Lange et al., 2001; Queiroz et al., 2004), increases muscle tricarboxylic acid cycle fluxes without increasing the synthesis of ATP (Lebon et al., 2001) and is permissive for methamphetamine-induced hyperthermia in rats (Sprague et al., 2004). These observations suggest that UCP3 might be a molecular determinant in the regulation of resting metabolic rate by T3 by promoting mitochondrial energy uncoupling in muscle. Nevertheless, UCP3 knockout mice (KO) treated with thyroid hormone show a similar increase in resting metabolic rate as wild-type animals (Gong et al., 2000). The possibility that compensatory mechanisms are called into play in the absence of UCP3, however, must be considered as an explanation for this negative data.

In addition to effects on heat generation, thyroid hormone also has effects on lipogenesis and appetite regulation. T3 increases expression of genes coding for lipogenic enzymes such as malic enzyme, glucose 6-phosphate dehydrogenase and acetyl-coenzyme A carboxylase (Goodridge, 1978; Miksicek and Towle,

1982; Bianco and Silva, 1987; Carvalho et al., 1993; Bianco et al., 1998) that use fatty acids derived from adipose tissue as the primary source of substrate (Oppenheimer et al., 1991), and is well known to increase food intake (Oppenheimer et al., 1991). The latter may be mediated centrally by increased ATP utilization in the mediobasal hypothalamus, perhaps as a result of increased Na/K ATPase activity (Luo and MacLean, 2003), and/or direct effects on the hypothalamic ventromedial nucleus (VMN) (Kong et al., 2004). Kong et al. (2004) have demonstrated that T3 increases immediate early gene activation in the VMN and induces a fourfold increase in food intake when injected directly into the VMN. T3 also inhibits leptin synthesis both in primary cultures of white and brown adipose tissue (Medina-Gomez et al., 2004) and in vivo, following systemic administration of thyrotoxic doses to rats Ishii et al., 2003), suggesting an alternative mechanism by which T3 could increase appetite. Evidence for reduced circulating leptin levels in human subjects with thyrotoxicosis, however, has not been found (Corbetta et al., 1997; Diekman et al., 1998; Nakamura et al., 2000). Presumably, the effects of T3 on lipogenesis and appetite are compensatory, preventing obligatory and/or adaptive thermogenesis from depleting fat stores and to counterbalance increased energy expenditure when T3 levels are elevated (Silva, 2003).

Mechanisms for modulation of the set point for feedback regulation by thyroid hormone on hypophysiotropic TRH neurons

While maintaining normal circulating levels of thyroid hormone is an essential function of HPT axis, under certain circumstances such as cold exposure, fasting or infection, circulating levels of thyroid hormone can be increased or decreased, presumably as a homeostatic mechanism to increase or reduce metabolic rate, respectively (Lechan and Fekete, 2004). These changes are regulated at several levels of the HPT axis (De Groot, 1999) but include resetting of the setpoint for feedback inhibition of hypophysiotropic TRH by thyroid hormone. This is accomplished primarily through the effects of afferent inputs to TRH-producing neurons in the hypothalamic PVN that arise from at least three regions

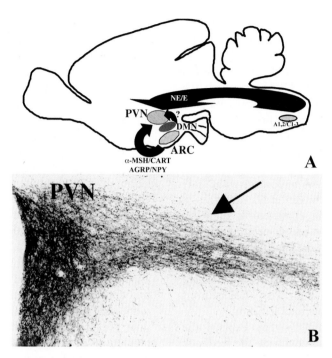

Fig. 3. (A) Sagittal drawing of the rat brain showing three major sources of innervation of TRH neurons in the hypothalamic paraventricular nucleus (PVN): the hypothalamic arcuate nucleus (ARC) that carries neuropeptides α-MSH, CART, AGRP and NPY to the PVN, brainstem A1,2/C1-3 cell groups that carry the catecholamines norepinephrine (NE) and epinenphrine (E) to the PVN, and the hypothalamic dorsomedial nucleus (DMN) for which the neurotransmitters/neuropeptide mediators to the PVN are unknown. (B) Neurons arising in each of the three regions innervate the PVN by traversing or joining fibers in the medial forebrain bundle (arrow). (Modified from Lechan and Hollenberg (2003), with permission from Elsevier.)

of the brain, the A1,2/C1-3 nuclear groups in the medulla of the brainstem, the hypothalamic arcuate nucleus and the hypothalamic dorsomedial nucleus (Fig. 3A). Under certain circumstances, however, the local production of T3 from T4 in the mediobasal hypothalamus by a specialized cell type that lines the third ventricle may also contribute to the regulation of hypophysiotropic TRH neurons (Lechan and Fekete, in press).

Role of brainstem catecholamine and CART projections

Axon terminals arising from neurons in the medulla are thought to contribute approximately 20% of all synapses on TRH neurons in the PVN (Shioda et al., 1986). These axons establish mostly asymmetric synapses with both perikarya and dendrites of TRH neurons, suggesting an excitatory function (Liposits, 1993). The trajectory of these axons to the PVN is to course rostrally in the dorsal and ventral

noradrenergic bundles, ultimately to enter the medial forebrain bundle in the hypothalamus before terminating in the PVN (Fig. 3B). This pathway may be of importance to increase adaptive thermogenesis by acutely increasing circulating thyroid hormone levels in response to cold exposure in some animal species by raising the threshold for feedback inhibition of hypophysiotropic TRH to thyroid hormone. Cold exposure is associated with a rapid increase in the transcription of the TRH gene in hypophysiotropic neurons of the PVN (Zoeller et al., 1990; Uribe et al., 1993), resulting in increased secretion of TRH into the portal capillaries from axon terminals in the median eminence (Rondeel et al., 1991) that can be abolished with TRH antiserum (Szabo and Frohman, 1977; Mori et al., 1978) or by ablation of the PVN (Ishikawa et al., 1984). Catecholamines would appear essential to this response as cold exposure increases the hypothalamic concentration of epinephrine and norepinephrine as shown by push–pull perfusion (Rondeel et al., 1991), and depletion of central catecholamines abolishes cold-induced activation of the HPT axis (Onaya and Hashizume, 1976; Annunziato et al., 1977; Schettini et al., 1979; Arancibia et al., 1989). Furthermore, the rise in circulating thyroid hormone levels with cold exposure is not observed within the first 10 days after birth in the rat when the hypothalamic catecholamine innervation is still immature (Ignar and Kuhn, 1988). As CART is co-contained in approximately 50% of catecholamine axons in contact with TRH neurons in the PVN (Wittmann et al., 2004a), and CART increases TRH gene expression in the PVN (Fekete et al., 2000c) and monoamine accumulation in several regions of the brain (Vaarmann and Kask, 2001), CART may also contribute to activation of the HPT axis during cold exposure either through direct effects on hypophysisotrophic TRH neurons or by potentiating catecholamine secretion.

The molecular mechanisms that explain the ability of catecholamines to override the feedback effects of thyroid hormone on TRH gene expression are not known, but may involve the phosphorylation of the transcription factor, CREB, and its nuclear translocation after binding of catecholamines to G protein-coupled $\alpha 2$ receptors on TRH neurons. The TRH gene is composed of three exons that are well conserved among all animal species studied including the human gene (Lechan and Hollenberg, 2003). The promoter region of the TRH gene is located 5' to exon 1 and is presumed to be the major locus where regulation of the gene takes place. A region termed Site 4 (TGACCTCA), located between -52 and -60 in the human gene and conserved in the mouse and rat promoters (Fig. 4), is important for basal expression of the TRH promoter (Hollenberg et al., 1995). Site 4 also controls regulated expression of the TRH gene by acting as a multifunctional binding site for the phosphorylated form of CREB and thyroid hormone receptors (Hollenberg et al., 1995; Stevenin and Lee, 1995; Satoh et al., 1996). The phosphorylation of CREB following activation of the cAMP signaling cascade is likely a key mechanism for positive regulation of TRH gene expression. When CREB is phosphorylated, phosphoCREB (PCREB) recruits co-activator CREB-binding protein (CBP) (Mayr and Montminy,

2001) to allow transcriptional activation of the TRH gene. Mutation or deletion of Site 4 can prevent activation of the TRH promoter by substances that activate the cAMP signaling cascade, such as forskolin (Harris et al., 2001). In contrast, the binding of thyroid hormone to its receptor on the TRH gene inhibits transcriptional activation (Segerson et al., 1987), primarily mediated by the thyroid hormone receptor beta 2 isoform (TRβ2) (Abel et al., 2001). Thus, mice with targeted deletion of TRβ2 show no significant increase in TRH mRNA concentration in response to PTU-induced hypothyroidism or a decrease in TRH mRNA in response to the exogenous administration of T3 (Abel et al., 2001). As the binding of CREB and thyroid hormone receptors to Site 4 appear to be mutually exclusive, it is conceivable that competition between CREB and thyroid hormone receptors for binding to this site is responsible for establishing the setpoint for feedback regulation of TRH gene expression by thyroid hormone, and for regulation of TRH gene transcription. By increasing CREB phosphorylation in TRH neurons and thereby the availability of PCREB to bind Site 4 in the TRH promoter, the inhibitory feedback of thyroid hormone bound to its receptor on the TRH gene would be reduced due to increased competition by elevated intranuclear concentrations of PCREB. As schematically illustrated in Fig. 4B, the end result would be an elevation in the setpoint for feedback inhibition of the TRH gene by thyroid hormone. The ensuing increase in TRH secreted to the anterior pituitary gland would have the effect of increasing TSH secretion and ultimately, circulating levels of thyroid hormone.

Role of arcuate nucleus α-MSH/CART and AGRP/NPY projections

The hypothalamic arcuate nucleus is the predominant source for afferent inputs to TRH neurons in the PVN, projecting primarily ipsilateral to the PVN and entering the PVN together with the catecholamine/CART axon trajectory from the brainstem (Lechan and Hollenberg, 2003). Physiologically, the interaction of the arcuate nucleus inputs on hypophysiotropic TRH neurons may be the primary mechanism for the development of central

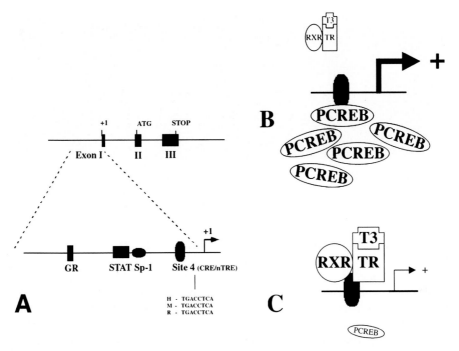

Fig. 4. (A) Genomic and promoter structure of TRH. The murine, rat and human TRH genes are composed of three exons and two inntrons. The coding sequence for the precursor protein is present on exons 2 and 3. As depicted, the TRH promoter region precedes the transcription start site in exon 1. The proximal 250 bp sequences of the human, mouse and rat promoters are similar and share the indicated transcription factor binding sites. The location of the CREB binding site (Site 4) and sequences in human (H), mouse (M) and rat (R) are shown. (B and C) Hypothesized schematic representation of the interaction between PCREB and the thyroid hormone receptor at Site 4. (B) illustrates that in the presence of abundant PCREB, there may be less availability for binding of the thyroid hormone receptor/T3 complex, hence, an increase in TRH gene transcription. When PCREB concentrations fall as shown in (C), increased binding of the thyroid hormone receptor/T3 complex reduces TRH gene transcription. (Modified from Lechan and Hollenberg (2003), with permission from Elsevier.)

hypothyroidism associated with fasting, orchestrated by the circulating protein, leptin (Ahima et al., 1996; Legradi et al., 1997, 1998; Lechan and Fekete, 2004). During nutrient insufficiency, as occurs with fasting or starvation, leptin secretion is reduced resulting in increased appetite, energy conservation and a shift to a neuroendocrine profile that facilitates metabolic adaptation (Wiersinga and Boelen, 1996; De Groot, 1999). Under these circumstances, there is a fall in thyroid hormone levels in the peripheral blood, but a seemingly paradoxical reduction of TRH mRNA in the PVN (Fig. 5A,B), reduced secretion of TRH into the portal blood, and low or inappropriately normal plasma TSH rather than the anticipated increase in all of these parameters as seen in primary hypothyroidism (Reichlin and Glaser, 1958; Harris et al., 1978; Hugues et al., 1988; Blake et al., 1991; Rondeel

et al., 1992; Van Haasteren et al., 1995; Legradi et al., 1997). Thus, the normal feedback mechanism for the regulation of hypophysiotropic TRH neurons is overridden and a state of central hypothyroidism is induced, commonly referred to as "nonthyroidal illness" or the "sick euthyroid syndrome" in man (Wiersinga and Boelen, 1996; De Groot, 1999). If leptin is administered either systemically or centrally to fasting animals (Ahima et al., 1996; Legradi et al., 1997), the reduction in circulating levels of thyroid hormone, TSH and hypophysiotropic TRH mRNA in the PVN is prevented (Fig. 5C). Furthermore, pharmacologic ablation of the arcuate nucleus, which effectively removes most of the afferent input from arcuate nucleus neurons to hypophysisotrophic TRH neurons in the PVN (Broberger et al., 1998; Legradi et al., 1998), not only abolishes the response

216

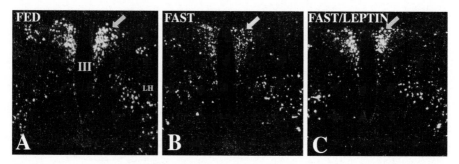

Fig. 5. In situ hybridization autoradiographs of proTRH mRNA in the PVN of (A) normal fed and (B) fasting animals. Note the marked reduction in hybridization signal by fasting. (C) ProTRH mRNA levels are restored to normal in a fasting animals administered leptin. (From Legradi et al. (1997), with permission from The Endocrine Society.)

of the HPT axis to fasting, but its response to the exogenous administration of leptin is lost as well (Ahima et al., 1996; Legradi et al., 1998).

The input from the arcuate–PVN pathway to TRH neurons includes two separate groups of neurons with opposing functions that either co-synthesize the 13-amino acid peptide, alpha-melanocyte stimulating hormone (α-MSH), and CART (Vrang et al., 1999), or the 108-amino acid homologue of agouti, agouti-related protein (AGRP), a potent, selective endogenous antagonist of α-MSH at melanocortin receptors 3 and 4 (Ollmann et al., 1997), and neuropeptide Y (NPY) (Broberger et al., 1998; Hahn et al., 1998). The organization of these neuronal populations in the arcuate nucleus has been well delineated in the rat (Dube et al., 1978; Hahn et al., 1998) with α-MSH/CART-producing neurons located more laterally in the arcuate nucleus (Fig. 6A), whereas AGRP/NPY neurons are more medially located (Fig. 6C). Both neuronal populations have similar terminal projection fields in the PVN (Broberger et al., 1998; Bagnol et al., 1999; Haskell-Luevano et al., 1999; Wilson et al., 1999), and are particularly pronounced in the anterior, medial and periventricular parvocellular subdivisions where TRH neurons reside (Fig. 6B,D). The presence of distinct populations of α-MSH- and AGRP-producing neurons in the arcuate nucleus of the human brain and their projections to TRH neurons in the PVN (Mihaly et al., 2000), suggests the evolutionary importance of this neuroregulatory system. Both melanocortin 3 and 4 receptors and NPY Y1 and Y5 receptors are present in the PVN (Mountjoy et al., 1994; Broberger et al., 1999; Kishi et al., 2003;

Wolak et al., 2003; Coppola et al., 2004), further establishing the significance of arcuate nucleus melanocortin and NPY signaling systems in the regulation of neurons in the PVN.

Practically all TRH neurons within the boundaries of the PVN show evidence of close apposition by AGRP and NPY nerve terminals on their cell bodies and proximal dendrites (Legradi and Lechan, 1998, 1999). In particular, hypophysiotropic TRH neurons in the medial and periventricular parvocellular subdivisions are heavily inundated by dense network of AGRP- and NPY-containing fibers, which by ultrastructural analysis, establish primarily symmetric axosomatic and axodendritic synapses (Legradi and Lechan, 1998, 1999). In keeping with the observation that AGRP is synthesized exclusively in the hypothalamic arcuate nucleus, pharmacologic ablation of the arcuate nucleus with monosodium glutamate results in near complete loss of AGRP fibers in the PVN (Broberger et al., 1998; Legradi and Lechan, 1999) and an approximately 80% reduction in the number NPY fibers contacting TRH neurons (Legradi and Lechan, 1998). The remaining NPY fibers in contact with TRH neurons have been shown to arise from the brainstem adrenergic neurons (Wittmann et al., 2002).

Similarly, all TRH neurons in the PVN are innervated by CART and approximately 70% of TRH neurons in the anterior and periventricular parvocellular subdivisions are innervated by α-MSH (Fekete et al., 2000a,c), but only 34% in the medial parvocellular subdivision are juxtaposed by fibers containing α-MSH (Fekete et al., 2000a). The

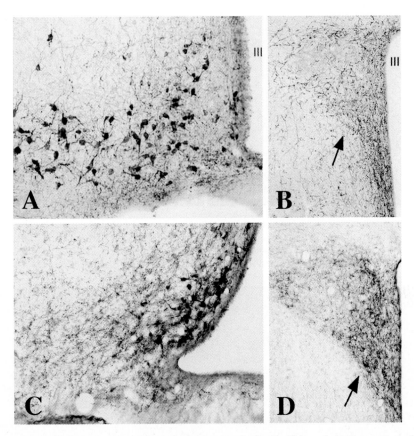

Fig. 6. Location of (A, B) α-MSH- and (C, D) AGRP-producing neurons in the (A, C) hypothalamic arcuate nucleus and (B, D) their axon terminals in the PVN (arrows). Note that α-MSH and AGRP-containing neurons are located in distinct populations of neurons with AGRP-containing neurons organized in more medial portions of the arcuate nucleus. Both neuronal populations have overlapping projections to the PVN. III, third ventricle.

explanation for this discrepancy is that more than 60% of CART fibers in association with TRH neurons in the PVN derive from the brainstem catecholamine cell groups (Fekete et al., 2005a; Wittmann et al., 2004a), whereas all fibers containing α-MSH derive from the arcuate nucleus (Eskay et al., 1979; Sawchenko et al., 1982; Joseph and Michael, 1988; Fekete et al., 2000a). Nevertheless, all α-MSH-containing axon terminals in association with TRH neurons in the PVN co-express CART (Wittmann et al., 2004a) and all TRH neurons in the PVN innervated by α-MSH/CART fibers are dually innervated by AGRP/NPY fibers (Fekete et al., 2000a), providing morphologic evidence to suggest an interaction between these peptides to regulate the HPT axis (Fig. 7).

Both α-MSH (Fekete et al., 2000a; Kim et al., 2000, 2002b; Nillni et al., 2000) and CART (Fekete et al., 2000c; Stanley et al., 2001) have activating effects on hypophysiotropic TRH neurons (Fig. 8). When administered intracerebroventricularly, α-MSH and CART have potent effects to restore the fasting-induced suppression of TRH mRNA in hypophysiotropic neurons to levels found in ad lib feeding animals (Fekete et al., 2000a,c). In addition, in vitro data indicate that both α-MSH and CART not only increase TRH gene expression, but also effectively increase TRH release from hypothalamic slices and hypothalamic explants in culture (Fekete et al., 2000c; Kim et al., 2000; Nillni et al., 2000). The activating mechanism for α-MSH on the TRH promoter is likely

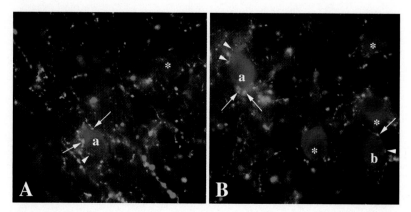

Fig. 7. Triple-labeling fluorescent immunocytochemistry showing dual innervation of (a) periventricular and (b) medial parvocellular subdivision TRH neurons (blue) in the PVN by axon terminals containing α-MSH (red, arrowheads) and AGRP (green, arrows). Note that all TRH neurons contacted by axon terminals containing AGRP also are contacted by several axon terminals containing α-MSH, whereas other TRH neurons (asterisks) are only innervated by AGRP. (From Fekete et al. (2000a), with permission.)

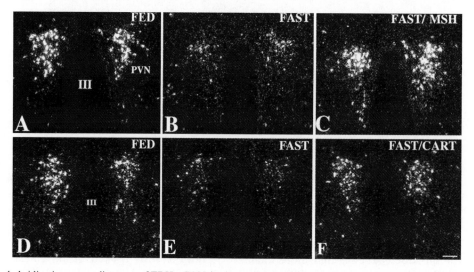

Fig. 8. In situ hybridization autoradiograms of TRH mRNA in the medial and periventricular parvocellular subdivision of the PVN of (A) fed, (B) fasted, and (C) fasted animals receiving α-MSH icv, and (D) fed, (E) fasted and (F) fasted animals receiving CART icv. Fasted animals receiving α-MSH or CART show a marked increase in TRH mRNA that is similar to that of their respective fed controls. III, third ventricle. (Modified from Fekete et al. (2000a, c), with permission.)

mediated by CREB. All melanocortin receptors are coupled in a stimulatory fashion to cAMP (Mountjoy et al., 1992), and following the intraventricular administration of α-MSH, there is a marked increase in the number of TRH neurons in the PVN that contain PCREB in their nucleus from approximately 3% to greater than 80% (Sarkar et al., 2002). In addition, studies by Harris et al. (2001) have demonstrated that in a cell culture system, activation of the TRH gene by α-MSH can be re-duced by more than 50% by mutating the CREB binding site in the TRH promoter and abolished by expressing a dominant inhibitor of CREB that cannot be phosphorylated. In contrast, the mechanism by which CART activates TRH neurons in the PVN remains unknown. The possibility that phosphorylation of CREB is involved in CART signaling is made unlikely by the observation that the intracerebroventricular administration of CART does not increase PCREB immunostaining

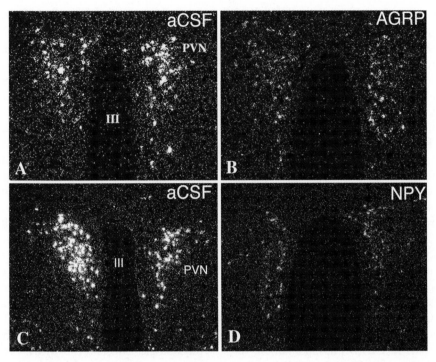

Fig. 9. Darkfield illumination photomicrographs of proTRH mRNA in the medial parvocellular subdivision of the hypothalamic paraventricular nucleus (PVN) in (A, C) artificial CSF-treated control animals, and animals treated with either (B) AGRP or (D) NPY. Note marked reduction in silver grains over neurons of the PVN in the AGRP and NPY-treated groups. III, third ventricle. (Modified from Fekete et al. (2002a, b), with permission from The Endocrine Society.)

in the nucleus of TRH neurons, whereas a marked increase is observed in adjacent CRH neurons (Sarkar and Lechan, 2003).

AGRP and NPY have potent inhibitory effects on hypophysiotropic TRH neurons (Fig. 9). Either substance when administered intracerebroventricularly to ad lib feeding animals creates a state of central hypothyroidism that closely compares to that observed in fasting animals (Fekete et al., 2001, 2002b). Following AGRP administration, there is more than a 60% reduction in TRH mRNA in hypophysiotropic neurons and approximately 50% reduction in circulating thyroid hormone levels, whereas NPY results in an approximately 55% reduction in TRH mRNA and even more profound reduction in circulating thyroid hormone levels (Fekete et al., 2001, 2002b).

The inhibitory effects of AGRP on TRH gene expression is a result of antagonizing the activating effects of α-MSH at melanocortin receptors and/or by suppressing constitutively active melanocortin receptors by functioning as an inverse agonist (Haskell-Luevano and Monck, 2001; Nijenhuis et al., 2001; Adan and Kas, 2003). While the possibility has been suggested that AGRP might also bind to other than melanocortin receptors (Hagan et al., 2000; Kim et al., 2002a), this does not appear to pertain to hypophysiotropic TRH neurons as AGRP has no inhibitory effects on the HPT axis when administered to the MC4R KO mouse (Fekete et al., 2004b). The inhibitory effects of NPY on TRH gene expression are mediated through both Y1 and Y5 receptors (Fekete et al., 2002a). As both receptors are coupled to G proteins that inhibit cAMP (Michel et al., 1998), the presumed mechanism for inhibition is a reduction in PCREB in the nucleus of TRH neurons. Thus, animals pretreated with NPY immediately before α-MSH administration show a 40% reduction in the percentage of TRH neurons containing PCREB compared to α-MSH-treated animals pretreated with artificial CSF (Sarkar and Lechan, 2003). Monosynaptic associations between NPY and α-MSH-producing neurons within the arcuate nucleus may also contribute to inhibition of hypophysiotropic TRH by reducing neuronal activity of α-MSH-producing neurons (Cowley et al., 2001).

Interactions between the activating effects of arcuate α-MSH/CART projections and inhibitory AGRP/NPY fiber projections to hypophysiotropic TRH neurons in the PVN are presumably important to establish the setpoint at which feedback inhibition of hypophysiotropic TRH gene expression by thyroid hormone occurs. Both neuronal populations of arcuate nucleus neurons express leptin receptors and are inversely regulated by circulating leptin levels (Ahima et al., 2000). Thus, during fasting when circulating leptin levels decline (Ahima, 2000), there is simultaneous inhibition of α-MSH and CART production and increase in AGRP and NPY production in arcuate nucleus neurons (Ahima, 2000; Ahima et al., 2000). Under these circumstances, the end result for hypophysiotropic TRH neurons in the PVN would be a decline in CREB phosphorylation, reducing the availability of PCREB for binding to Site 4 in the TRH promoter, schematically illustrated in Fig. 4C. Therefore, the inhibitory feedback effects of thyroid hormone bound to its receptor on the TRH promoter may become enhanced due to reduced competition by lower intranuclear concentrations of PCREB, essentially lowering the setpoint for feedback inhibition of the TRH gene by thyroid hormone. By reducing circulating thyroid hormone levels and hence, obligatory thermogenesis, this mechanism is an important adaptive response in mammalian species to reduce energy expenditure until food once again becomes available.

Dorsomedial nucleus projections
In addition to the PVN, the arcuate nucleus neurons also projects to the hypothalamic dorsomedial nucleus (Bellinger and Bernardis, 2002) heavily innervating this nucleus with both α-MSH- and AGRP-containing axons (Jacobowitz and O'Donohue, 1978; Legradi and Lechan, 1999). Furthermore, when focal injections of the anterogradely transported marker substance, *Phaseolus vulgaris*-leucoagglutinin (PHA-L), are made into the dorsomedial nucleus, a dense projection pathway to the PVN is elucidated that inundates greater than 90% of

TRH-containing neurons and establishes synaptic contacts (Mihaly et al., 2001). At least two, distinct subdivisions of the dorsomedial nucleus provide input to the parvocellular subdivisions of the PVN and are under direct regulation by the melanocortin signaling system. These include neurons in the ventral subdivision of the dorsomedial nucleus where approximately 65% of the total number of retrogradely labeled cells in the dorsomedial nucleus reside after injections into the PVN, and neurons in the dorsal subdivision, where most of the remaining retrogradely labeled cells reside (Singru et al., 2005). Of the retrogradely labeled neurons, approximately 61% in the ventral subdivision and 40% in the dorsal subdivision receive contacts by α-MSH-containing axon terminals (Fig. 10).

The functional significance of two, separate, major regions of the dorsomedial nucleus mediating the actions of the melanocortin signaling on the PVN are uncertain, but might indicate that distinct subregions of the DMN subserve different physiologic effects of α-MSH. This concept is based on observations by Mihaly et al. (2001) that when PHA-L was focally injected into the dorsal subregion of the DMN, nearly 90% of PHA-L-containing axon varicosities were juxtaposed to TRH neurons in the hypothalamic PVN. In contrast, substantially fewer TRH neurons in the PVN were juxtaposed to PHA-L-containing axon terminals when PHA-L was injected into the ventral subregion of the DMN. In addition to direct effects of α-MSH on TRH neurons in the PVN, therefore, α-MSH may also exert indirect effects through a multisynaptic pathway that involves the dorsal subdivision neurons of the DMN.

Infundibular tanycytes and modulation of hypophysiotropic TRH
While arcuate nucleus-derived peptides contribute to suppression of the HPT axis associated with fasting, this mechanism does not explain suppression of the HPT axis observed following endotoxin administration (Kondo et al., 1997). As opposed to reduced melanocortin signaling associated with fasting, substantial evidence suggests increased melanocortin signaling following bacterial lipopolysaccharide (LPS) administration (Marks et al., 2001; Sergeyev et al., 2001; Wisse et al.,

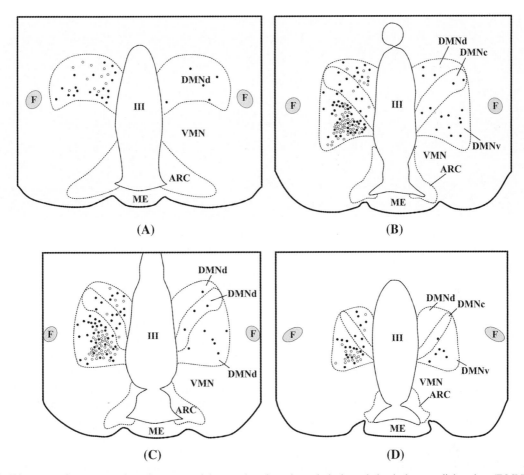

Fig. 10. Diagrammatic representation of rostro-caudal coronal sections through the hypothalamic dorsomedial nucleus (DMN) (A–D) showing organization of the retrogradely labeled cells (circles) in the dorsal subdivision (DMNd), compact zone (DMNc) and ventral subdivision (DMNv) following injection of CtB into the hypothalamic paraventricular nucleus. Open circles represent CtB cells in the DMN contacted by α-MSH-containing axons. ARC, arcuate nucleus; F, fornix; ME, median eminence; VMN, ventromedial nucleus; III, third ventricle. (From Singru et al. (2005), with permission from Elsevier.)

2003). In addition, endotoxin administration increases CART gene expression (Sergeyev et al., 2001), does not increase NPY or AGRP expression (Sergeyev et al., 2001), increases norepinephrine release (Francis et al., 2000) and elevates circulating levels of leptin (Grunfeld et al., 1996). Furthermore, ascending brainstem pathways exert a net stimulatory effect on hypophysiotropic TRH neurons after LPS administration (Fekete et al., 2005b). Collectively, these responses would predict an increase in TRH gene expression in hypophysiotropic neurons rather than a suppression as observed (Kondo et al., 1997; Legradi et al.,

1998). Presumably, therefore, a more potent, overriding inhibitory influence mediated by forebrain mechanisms other than that described above for fasting become activated during infection, superceding any stimulatory action of leptin, α-MSH, CART and/or catecholamines on the HPT axis.

The observation that tanycytes, a specialized type of elongated ependymal cell of glial origin that lines the floor and ventrolateral walls of the third ventricle between the rostral and caudal limits of the hypothalamic median eminence, express the highest concentration of type 2 iodothyronine deiodinase (D2) and D2 mRNA in the brain (Guadano-Ferraz

222

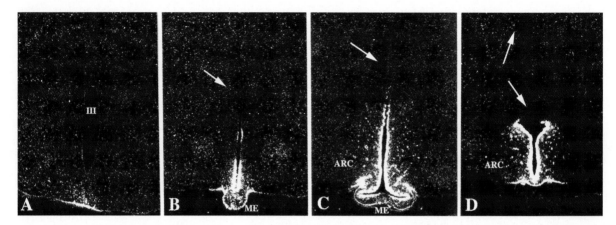

Fig. 11. Rostral-caudal distribution of type 2 iodothyronine deiodinase (D2) mRNA in the mediobasal hypothalamus (darkfield illumination). In the most rostral section (A), hybridization is present in the external zone of the median eminence (ME) but is absent from walls of the third ventricle (III). In more caudal sections (B–D), intense hybridization is seen over cells lining the floor and infralateral walls of the third ventricle and in the arcuate nucleus (ARC) and ME. Hybridization is absent in the dorsal portions and roof of third ventricular wall in B–D (arrows). (Modified from Tu et al. (1997), with permission from The Endocrine Society.)

et al., 1997; Tu et al., 1997; Diano et al., 2003), has led to the hypothesis that tanycytes may have an integral role in the regulation of thyroid function. D2 is one of the three known iodothyronine deiodinases (Croteau et al., 1995, 1996; Bianco et al., 2002) and the predominant 5'-iodothyronine deiodinase in the brain involved in the intracellular generation of the biologically active form of thyroid hormone, tri-iodothyronine (T3), from its less potent precursor, thyroxine (T4) (Bianco et al., 2002). Since greater than 75% of nuclear T3 in the brain is derived from local conversion of T4 into T3 (Bianco et al., 2002), D2 has an essential role in mediating the effects of circulating thyroid hormone on the CNS.

By in situ hybridization histochemistry, D2 mRNA is found in the tanycyte cell bodies lining the floor and infralateral walls of the third ventricle, abruptly ceasing 1/2 to 2/3 up the third ventricular wall (Fig. 11), but also extending into its cytoplasmic projections that surround capillaries in the arcuate nucleus and into the median eminence (Guadano-Ferraz et al., 1997; Tu et al., 1997; Fekete et al., 2000b). A similar distribution of D2 mRNA is also present in the human hypothalamus (Alkemade et al., 2005b).

The location of tanycytes at the interface of the blood–brain barrier and CSF–brain barrier, the well described endocytotic potential of these cells and

their high concentration of D2 mRNA and D2 enzymatic activity, raises the possibility that tanycytes could extract T4 from the bloodstream from end processes terminating on portal capillaries or capillaries in the arcuate nucleus, or from the CSF via apical specializations after T4 has traversed the choroid plexus (Spector and Levy, 1975; Schreiber et al., 1990, 1993). This possibility is made even stronger by the recent discovery that the T4 transporter, monocarboxylate transporter 8 (MCT8), is highly expressed in tanycytes in the mouse, rat and human brain (Alkemade et al., 2005a; Heuer et al., 2005). As the PVN contains little, if any, D2 activity or D2 mRNA (Riskind et al., 1987; Guadano-Ferraz et al., 1997; Tu et al., 1997), and thereby is incapable of intracellular conversion of T4 into biologically active T3, hypophysiotropic TRH neurons in the PVN are dependent upon exogenous sources of T3. Presumably, after converting T4 into T3, tanycytes could release T3 into the CSF, providing a source of T3 to hypophysiotropic TRH neurons in the hypothalamic PVN by volume transmission (Agnati et al., 1995). Alternatively, tanycytes could release T3 into the median eminence for uptake and retrograde transport by axon terminals of hypophysiotropic TRH neurons, and/or into the arcuate nucleus neurons to influence neuronal populations that have known projections to TRH neurons in the PVN (see above). Indeed, in the absence of T4, nearly twice the normal circulating levels of T3 are required to restore TRH mRNA to euthyroid levels in hypothyroid animals Kakucska et al., 1992), indicating that the inhibitory action of thyroid hormone on hypophysiotropic TRH is dependent upon the conversion of T4 into T3 within the CNS.

Recent studies by Fekete et al. (2004a) have shown that the systemic administration of LPS increases D2 activity in the rat mediobasal hypothalamus approximately 400% compared to control animals (Fig. 12A). Similar observations have been made by Boelen et al. (2004) in the mouse hypothalamus. In contrast to the cerebral cortex and anterior pituitary, changes in D2 activity in the medial basal hypothalamus are not secondary to the LPS-induced fall in circulating thyroid hormone levels (Fekete et al., 2005a). Thyroid hormone replacement in thyroidectomized rats treated with endotoxin still show a fourfold increase in D2 in the mediobasal hypothalamus, whereas D2 activity in the cortex and anterior pituitary is suppressed (Fig. 12B).

The possibility that endotoxin exerts its effects on tanycytes by activating cytokines such as tumor necrosis factor (TNF), is suggested by the increased expression of TRF type I (p55) receptors in the median eminence following endotoxin administration in a distribution reminiscent of tanycytes (Nadeau and Rivest, 1999). In addition, TNF is a potent inducer of the activation and translocation of NF-κB into the nucleus (Rivest et al., 2000), and the promoter of the D2 gene (dio2) contains multiple putative NF-κB binding sites (Fekete et al., 2004). Furthermore, co-expression of p65 (RelA), a required component of the activated NF-κB heterodimer, together with a 6.5 kb human dio2 5′ flanking region CAT construct in a HEK-293 cell line, leads to an approximately 50-fold increase in the transcriptional activity of the hdio2 promoter that can be abolished by truncation of the promoter (Fekete et al., 2004a). NF-κB has also been shown to be activated in the median eminence and wall of the third ventricle following LPS administration (Nadeau and Rivest, 1999), although the precise cell types have not yet been identified.

It is conceivable, therefore, that endotoxin results in tissue-specific D2-mediated thyrotoxicosis

in the mediobasal hypothalamus caused by increased T4 to T3 conversion by tanycytes, overriding any effects that might activate the TRH gene. Thus, the increase in mediobasal hypothalamic T3 may suppress the synthesis of TRH in hypophysiotropic neurons and may also inhibit TSH secretion if T3 is released into the portal capillary system for conveyance to the anterior pituitary. A similar mechanism has been proposed to contribute to the mechanism whereby fasting suppresses hypophysiotropic TRH, although as opposed to endotoxin administration, fasting is associated with only a small but significant 1.6-fold increase in D2 activity in the mediobasal hypothalamus (Diano et al., 1998; Coppola et al., 2005).

Role of nonhypophysiotropic TRH neurons in energy homeostasis

TRH and appetite regulation

In addition to indirect effects of TRH on appetite regulation by controlling circulating levels of thyroid hormone, TRH also has separate and diverse actions on appetite regulation through effects exerted directly within the central nervous system. Central administration of TRH or TRH analogues consistently reduces food intake and the time spent interacting with food in all animals models studied (Vijayan and McCann, 1977; Vogel et al., 1979; Morley, 1980; Suzuki et al., 1982; Horita, 1998; Steward et al., 2003). This includes a reduction in the intake of food in normal ad lib feeding animals, animals that have been subjected to a fast and then reintroduced to food, as well as in models of stress-induced eating. The effect is not catecholamine-dependent as brain catecholamine depletion with 6-hydroxydopamine does not block the action of TRH to reduce food ingestion (Vogel et al., 1979).

The precise locus of the TRH effect is not known but direct injection of TRH into the medial hypothalamus and nucleus accumbens, regions that contain TRH axon terminals, can induce anorexia (Suzuki et al., 1982). Along these lines, it is of interest to note that similar to hypophysiotropic

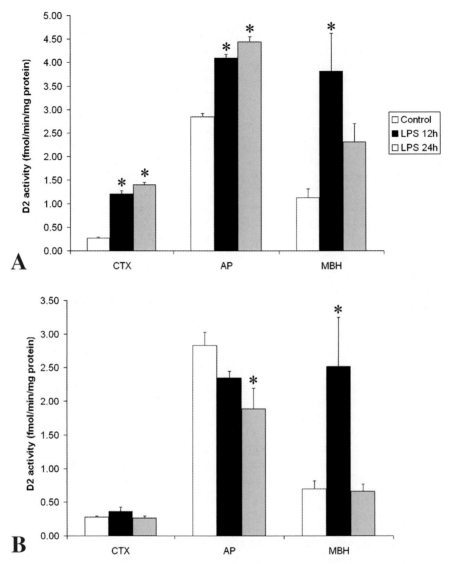

Fig. 12. Effects of LPS administration on D2 activity in the cerebral cortex (CTX); anterior pituitary (AP); and mediobasal hypothalamus (MBH). (A) In intact animals, D2 activity in the cortex and anterior pituitary show a significant increase 12 and 24 h after LPS administration. Peak activity in the MBH occurred 12 h after treatment. (B) In T4 replaced thyroidectomized animals, LPS has no effect on D2 activity in the CTX, an inhibitory effect in the AP, but a persistent stimulatory effect in the MBH. *$P < 0.05$ compared with saline controls. (From Fekete et al. (2005a), with permission from Elsevier.)

TRH-synthesizing neurons in the PVN, non-hypophysiotropic TRH neurons in the anterior parvocellular subdivision of the PVN (Ishikawa et al., 1988; Kawano et al., 1991) are also closely integrated into the arcuate nucleus regulatory control system. Thus, over 70% of anterior parvocellular subdivision TRH neurons are innervated by axon terminals containing α-MSH, and nearly 100%

are densely innervated by axons containing CART, NPY and AGRP (Fekete et al., 2000a, c; Legradi and Lechan, 1998, 1999). In addition, similar to hypophysiotropic TRH neurons, the central administration of α-MSH induces the phosphorylation of CREB in the nucleus of anterior parvocellular TRH neurons (Sarkar et al., 2002), indicating the functional importance of melanocortin signaling in these neurons. Furthermore, the anterior parvocellular subdivision of the PVN including TRH neurons, but not medial or periventricular parvocellular subdivisions, is specifically targeted for innervation by axons containing galanin-like peptide (GALP) (Wittmann et al., 2004b) and calcitonin gene-related peptide (CGRP) (personal observations). GALP arises exclusively from leptin-regulated neurons in the hypothalamic arcuate nucleus (Takatsu et al., 2001) and when injected into the brain, alters food intake, decreases body weight and increases both core temperature and uncoupling protein mRNA in brown adipose tissue (Hansen et al., 2003). CGRP also inhibits food intake when injected into the CSF or directly into the PVN (Dhillo et al., 2003). Thus, anterior parvocellular TRH neurons may have a role in the regulation of appetite and satiety as part of an integrated response together with other leptin-mediated actions through arcuate nucleus neurons. Further support for the anterior parvocellular subdivision in the regulation of appetite is given by Randich et al. (2004), showing that intestinal infusions of long-chain fatty acids increases single-unit recordings of anterior parvocellular neurons, suggesting that anterior PVN neurons contribute to suppression of food intake by intestinal administration of lipids.

Surprisingly, very little is known about the anterior parvocellular subdivision of the PVN either with respect to its projection fields, afferent input or function. Studies using retrogradely transported marker proteins from autonomic centers in the brainstem or the hypothalamic median eminence show little tracer accumulation in anterior parvocellular neurons (Swanson and Kuypers, 1980; Sawchenko and Swanson, 1983). However, large injections of the anterogradely transported marker protein, PHA-L, into the PVN that extend to the anterior parvocellular subdivision show terminal fields in the amygdala, bed nucleus of the stria terminals and nucleus accumbens (Luiten et al., 1985; Roeling et al., 1993), raising the possibility that anterior parvocellular TRH neurons may be involved in limbic functions such as feeding behavior. Extensive studies by Kelley (2004) have demonstrated that the nucleus accumbens and central nucleus of the amygdala are involved in the control of food intake by functioning as part of a critical link between cortical circuits and hypothalamic/brainstem circuits. The amygdala is densely innervated by TRH-containing axons (Lechan et al., 1987) and has an important role in control of food consumption by learned cues (Petrovich and Gallagher, 2003; Kelley, 2004). If the source of TRH neurons in the amygdala is the PVN, TRH neurons may mediate the effect of peripheral satiety signals to the amygdala where it could be integrated with visual and olfactory signals.

TRH and thermogenesis

In addition to the effects of TRH on thermogenesis mediated through the HPT axis, TRH may also participate in the central control of thermoregulation through effects independent of elevations in circulating thyroid hormone levels. Without affecting circulating T3 levels, intracerebroventricular administration of TRH increases rectal temperature and temperature of brown adipose tissue that can be prevented by pretreatment with antibodies to the type 1 TRH receptor (Shintani et al., 2005). As bilateral denervation of the sympathetic nerves innervating brown adipose tissue or the administration of β-adrenergic antagonists markedly attenuate the thermogenic response of centrally administered TRH (Chi and Lin, 1983; Shintani et al., 2005), the effects of TRH are presumably mediated through autonomic centers in the brain. This action may be centered in the hypothalamic dorsomedial nucleus and/or anterior hypothalamic-preoptic area (Boschi and Rips, 1981; Griffiths et al., 1988; Shintani et al., 2005), which are closely integrated into the autonomic nervous system. The anterior hypothalamic-preoptic area is well recognized as a primary site in the CNS for thermoregulation, which through multisynaptic pathways to the brainstem and spinal cord (Nagashima et al., 2000), coordinate a variety of responses that increase heat production by both shivering and nonshivering thermogenesis, including activation of UCP1 in brown adipose tissue and redirection of blood from cutaneous to deep vascular beds (Lechan and Toni, 2003). It is of note, therefore, that when TRH is focally injected into the anterior hypothalamic-preoptic area, it decreases the activity of warm-sensitive neurons and increases the activity of cold-sensitive neurons (Hori et al., 1988), which would have the effect to increase heat production (Lechan and Toni, 2003). TRH has also been reported to cause cutaneous vasoconstriction (Huang et al., 1992), increase pancreatic blood flow (Goto et al., 2004) and induce shivering (Lahti et al., 1983), all contributing to mechanisms whereby the autonomic nervous system could elevate body temperature.

TRH and arousal and locomotor activation

Spontaneous locomotor activity is becoming increasingly recognized as contributing to the regulation of energy homeostasis (Castaneda et al., 2005). Thus, even seemingly trivial activities such as fidgeting may play an important role in weight regulation (Levine et al., 2000). An expanding list of neuroactive substances may be involved in locomotor activity. For example, ghrelin has a suppressive effect on spontaneous locomotor activity (Castaneda et al., 2005), whereas orexin has a potent activating effect (Siegel, 2004).

TRH has long been recognized to be involved in arousal and locomotor activity (see Horita (1998) and Nillni and Sevarino, 1999; for review). Either systemic or central administration of TRH increases wake time (Arnold et al., 1991; Nishino et al., 1997), can arouse animals from drug- or alcohol-induced narcosis (Breese et al., 1975; Glue et al., 1992; French et al., 1993) and decrease the time of spontaneous movements in unconscious animals following head injury (Tanaka et al., 1992). In addition, one of the authors (RML) has observed dramatic awakening by the systemic administration of TRH in a young man with somnolence and extensive hypothalamic involvement by Langerhans Cell Histiocytosis (unpublished). TRH neurons in the preoptic area of the hypothalamus and bed nucleus of the striae terminalis and their axonal projections to the septal nuclei may be responsible for these actions of TRH (Ishikawa et al., 1986). One of the proposed mechanisms whereby TRH exerts this alerting action may be through the release of acetylcholine from pathways that connect the septum to the hippocampus via the septohippocampal pathway. The analeptic effects of TRH can be abolished by atropine or scopolamine (Breese et al., 1975; Horita et al., 1976) and setpohippocampal lesions

prevent TRH from antagonizing pentobarbital-induced narcosis (Kalivas et al., 1981). In addition, by microdialysis, TRH strongly increases ACh release in the hippocampus (Giovannini et al., 1991). Other neurotransmitters, including dopamine and norepinephrine, may also be important (Reichlin, 1986; Horita et al., 1991).

TRH also has profound effects on locomotor activity including both gross motor movements such as forward locomotion and rearing, and fine movements such as head swaying, grooming and chewing (Wei et al., 1975; Ervin et al., 1981; Lin et al., 1987), and potentiates d-amphetamine- and cocaine-induced locomotor activity (Collu et al., 1992; Przegalinski et al., 2004). As the effects of TRH on locomotor activity can be replicated when focally injected into the nucleus accumbens and blocked with dopamine antagonists (Miyamoto and Nagawa, 1977), evidence would suggest that the effects of TRH on locomotor activity are mediated through the mesolimbic dopamine system. Injections of TRH into the ventromedial hypothalamus (Lin et al., 1987) and ventral tegmental area (Kalivas et al., 1987), however, also can elicit locomotion, indicating that more than one region in the brain may be involved.

TRH and autonomic regulation

In addition to the effects of TRH on thermoregulation, mediated by the autonomic nervous system (see TRH and Thermogenesis), TRH also has a number of other well-described effects on the autonomic nervous system that contribute to the cephalic phase of digestion. These include increased gastric acid secretion, gastric motility, gastric emptying, gastro-duodenal and hepatic blood flow and pancreatic exocrine and endocrine secretion (Tache et al., 1980; LaHann and Horita, 1982; Maeda-Hagiwara and Tache, 1987; Garrick et al., 1994; Okumura et al., 1995; Yang et al., 2002; Yoneda et al., 2003). While there is considerable evidence to suggest that these actions are mediated primarily by a vagal, cholinergic-dependent mechanism via TRH neurons in the brainstem medullary raphe and dorsal motor nucleus of the vagus (Tache et al., 1993; Garrick et al., 1992; Garrick et al., 1994; Okumura et al., 1995; Martinez et al., 2002;

Yang et al., 2002; Yoneda et al., 2003), hypothalamic TRH neurons may also participate. In particular, microinjection of TRH into the PVN has been shown to increase gastric contractility (Morrow et al., 1994). As this response can be abolished by vagotomy (Morrow et al., 1994), it would appear to be mediated by central vagal mechanisms.

The hypothalamic PVN has an important role in the regulation of visceral responses through projections from neurons in the dorsal and ventral parvocellular subdivisions to brainstem vegetative centers (Saper et al., 1976; Swanson and Kuypers, 1980; Swanson and Sawchenko, 1980; Swanson et al., 1980; Luiten et al., 1985). TRH-producing neurons are found in the dorsal parvocellular subdivision of the PVN (Lechan and Jackson, 1982), and thereby could contribute to descending projections to autonomic centers in the brainstem. It is intriguing to speculate, however, that hypophysiotropic TRH neurons in the medial and periventricular parvocellular subdivisions of the PVN might contribute to autonomic responses by either sending axonal collaterals to brainstem or by interacting with autonomic centers in the adjacent parvocellular autonomic subdivisions of the PVN through dendritic associations. This mechanism would have the advantage of coordinating the effects of TRH on energy homeostasis through effects on the HPT axis and gastrointestinal system. Precedent for axon collateralization arising from the PVN to innervate autonomic centers in the brainstem and spinal cord have been recently described (Shafton et al., 1998; Pyner and Coote, 2000).

Conclusions

TRH has multiple functions in the regulation of energy homeostasis. Most importantly, it has a critical role in establishing circulating levels of thyroid hormone under normal conditions and during special circumstances such as cold exposure, fasting and infection when changes in thyroid status are required for adaptation. TRH also exerts a number of effects on energy homeostasis directly within the brain. Some of these actions such as an increase in gut motility and endocrine and exocrine secretion would suggest a role for TRH in a coordinated effort to mediate the cephalic phase of digestion. Others, such as an increase in core body temperature and locomotion and decrease in appetite would suggest a role for TRH to increase energy expenditure. Whether these diverse actions of TRH are part of an integrated effort to regulate energy balance or simply independent effects that utilize a common neuropeptide for specific physiological and/or pathophysiological conditions, however, are unknown. Elucidation of circumstances during which each of the observed central actions of TRH come into play and the origin of the TRH neuronal populations responsible for these actions will require further investigation.

References

Abel, E.D., Ahima, R.S., Boers, M.E., Elmquist, J.K. and Wondisford, F.E. (2001) Critical role for thyroid hormone receptor beta2 in the regulation of paraventricular thyrotropin-releasing hormone neurons. J. Clin. Invest., 107: 1017–1023.

Adan, R.A. and Kas, M.J. (2003) Inverse agonism gains weight. Trends Pharmacol. Sci., 24: 315–321.

Agnati, L.F., Zoli, M., Stromberg, I. and Fuxe, K. (1995) Intercellular communication in the brain: wiring versus volume transmission. Neuroscience, 69: 711–726.

Ahima, R.S. (2000) Leptin and the neuroendocrinology of fasting. Front. Horm. Res., 26: 42–56.

Ahima, R.S., Prabakaran, D., Mantzoros, C., Qu, D., Lowell, B., Maratos-Flier, E. and Flier, J.S. (1996) Role of leptin in the neuroendocrine response to fasting. Nature, 382: 250–252.

Ahima, R.S., Saper, C.B., Flier, J.S. and Elmquist, J.K. (2000) Leptin regulation of neuroendocrine systems. Front. Neuroendocrinol., 21: 263–307.

Alkemade, A., Friesema, E.C., Unmehopa, U.A., Fabriek, B.O., Kuiper, G.G., Leonard, J.L., Wiersinga, W.M., Swaab, D.F., Visser, T.J. and Fliers, E. (2005a) Neuroanatomical pathways for thyroid hormone feedback in the human hypothalamus. J. Clin. Endocrinol. Metab., 90: 4322–4334.

Alkemade, A., Vuijst, C.L., Unmehopa, U.A., Bakker, O., Vennstrom, B., Wiersinga, W.M., Swaab, D.F. and Fliers, E. (2005b) Thyroid hormone receptor expression in the human hypothalamus and anterior pituitary. J. Clin. Endocrinol. Metab., 90: 904–912.

Annunziato, L., Di Renzo, G., Lombardi, G., Scopacasa, F., Schettini, G., Preziosi, P. and Scapagnini, U. (1977) The role of central noradrenergic neurons in the control of thyrotropin secretion in the rat. Endocrinology, 100: 738–744.

Arancibia, S., Tapia-Arancibia, L., Astier, H. and Assenmacher, I. (1989) Physiological evidence for alpha 1-adrenergic facilitatory control of the cold-induced TRH release in the rat, obtained by push-pull cannulation of the median eminence. Neurosci. Lett., 100: 169–174.

Arnold, R., Klingberg, F. and Schaker, W. (1991) Systemically applied thyrotropin-releasing hormone (TRH) modifies spontaneous behaviour of rats. Biomed. Biochim. Acta, 50: 1217–1224.

Bagnol, D., Lu, X.Y., Kaelin, C.B., Day, H.E., Ollmann, M., Gantz, I., Akil, H., Barsh, G.S. and Watson, S.J. (1999) Anatomy of an endogenous antagonist: relationship between Agouti-related protein and proopiomelanocortin in brain. J. Neurosci., 19: RC26.

Bellinger, L.L. and Bernardis, L.L. (2002) The dorsomedial hypothalamic nucleus and its role in ingestive behavior and body weight regulation: lessons learned from lesioning studies. Physiol. Behav., 76: 431–442.

Bianco, A.C., Carvalho, S.D., Carvalho, C.R., Rabelo, R. and Moriscot, A.S. (1998) Thyroxine 5'-deiodination mediates norepinephrine-induced lipogenesis in dispersed brown adipocytes. Endocrinology, 139: 571–578.

Bianco, A.C., Salvatore, D., Gereben, B., Berry, M.J. and Larsen, P.R. (2002) Biochemistry, cellular and molecular biology, and physiological roles of the iodothyronine selenodeiodinases. Endocr. Rev., 23: 38–89.

Bianco, A.C. and Silva, J.E. (1987) Optimal response of key enzymes and uncoupling protein to cold in BAT depends on local T3 generation. Am. J. Physiol., 253: E255–E263.

Blake, N.G., Eckland, D.J., Foster, O.J. and Lightman, S.L. (1991) Inhibition of hypothalamic thyrotropin-releasing hormone messenger ribonucleic acid during food deprivation. Endocrinology, 129: 2714–2718.

Boelen, A., Kwakkel, J., Thijssen-Timmer, D.C., Alkemade, A., Fliers, E. and Wiersinga, W.M. (2004) Simultaneous changes in central and peripheral components of the hypothalamus-pituitary-thyroid axis in lipopolysaccharide-induced acute illness in mice. J. Endocrinol., 182: 315–323.

Boschi, G. and Rips, R. (1981) Effects of thyrotropin releasing hormone injections into different loci of rat brain on core temperature. Neurosci. Lett., 23: 93–98.

Boss, O., Samec, S., Paoloni-Giacobino, A., Rossier, C., Dulloo, A., Seydoux, J., Muzzin, P. and Giacobino, J.P. (1997) Uncoupling protein-3: a new member of the mitochondrial carrier family with tissue-specific expression. FEBS Lett., 408: 39–42.

Breese, G.R., Cott, J.M., Cooper, B.R., Prange Jr., A.J., Lipton, M.A. and Plotnikoff, N.P. (1975) Effects of thyrotropin-releasing hormone (TRH) on the actions of pentobarbital and other centrally acting drugs. J. Pharmacol. Exp. Ther., 193: 11–22.

Broberger, C., Johansen, J., Johansson, C., Schalling, M. and Hokfelt, T. (1998) The neuropeptide Y/agouti gene-related protein (AGRP) brain circuitry in normal, anorectic, and monosodium glutamate-treated mice. Proc. Natl. Acad. Sci. USA, 95: 15043–15048.

Broberger, C., Visser, T.J., Kuhar, M.J. and Hokfelt, T. (1999) Neuropeptide Y innervation and neuropeptide-Y-Y1-receptor-expressing neurons in the paraventricular hypothalamic nucleus of the mouse. Neuroendocrinology, 70: 295–305.

Carvalho, S.D., Negrao, N. and Bianco, A.C. (1993) Hormonal regulation of malic enzyme and glucose-6-phosphate dehydrogenase in brown adipose tissue. Am. J. Physiol., 264: E874–E881.

Castaneda, T.R., Jurgens, H., Wiedmer, P., Pfluger, P., Diano, S., Horvath, T.L., Tang-Christensen, M. and Tschop, M.H. (2005) Obesity and the neuroendocrine control of energy homeostasis: the role of spontaneous locomotor activity. J. Nutr., 135: 1314–1319.

Chi, M.L. and Lin, M.T. (1983) Involvement of adrenergic receptor mechanisms within hypothalamus in the fever induced by amphetamine and thyrotropin-releasing hormone in the rat. J. Neural Transm., 58: 213–222.

Collin, A., Cassy, S., Buyse, J., Decuypere, E. and Damon, M. (2005) Potential involvement of mammalian and avian uncoupling proteins in the thermogenic effect of thyroid hormones. Domest. Anim. Endocrinol., 29: 78–87.

Collu, M., D'Aquila, P.S., Gessa, G.L. and Serra, G. (1992) TRH activates mesolimbic dopamine system: behavioural evidence. Behav. Pharmacol., 3: 639–641.

Coppola, A., Meli, R. and Diano, S. (2005) Inverse shift in circulating corticosterone and leptin levels elevates hypothalamic deiodinase type 2 in fasted rats. Endocrinology, 146: 2827–2833.

Coppola, J.D., Horwitz, B.A., Hamilton, J. and McDonald, R.B. (2004) Expression of NPY Y1 and Y5 receptors in the hypothalamic paraventricular nucleus of aged Fischer 344 rats. Am. J. Physiol. Regul. Integr. Comp. Physiol., 287: R69–R75.

Corbetta, S., Englaro, P., Giambona, S., Persani, L., Blum, W.F. and Beck-Peccoz, P. (1997) Lack of effects of circulating thyroid hormone levels on serum leptin concentrations. Eur. J. Endocrinol., 137: 659–663.

Cowley, M.A., Smart, J.L., Rubinstein, M., Cerdan, M.G., Diano, S., Horvath, T.L., Cone, R.D. and Low, M.J. (2001) Leptin activates anorexigenic POMC neurons through a neural network in the arcuate nucleus. Nature, 411: 480–484.

Croteau, W., Davey, J.C., Galton, V.A. and St Germain, D.L. (1996) Cloning of the mammalian type II iodothyronine deiodinase. A selenoprotein differentially expressed and regulated in human and rat brain and other tissues. J. Clin. Invest., 98: 405–417.

Croteau, W., Whittemore, S.L., Schneider, M.J. and St Germain, D.L. (1995) Cloning and expression of a cDNA for a mammalian type III iodothyronine deiodinase. J. Biol. Chem., 270: 16569–16575.

De Groot, L.J. (1999) Dangerous dogmas in medicine: the nonthyroidal illness syndrome. J. Clin. Endocrinol. Metab., 84: 151–164.

De Lange, P., Lanni, A., Beneduce, L., Moreno, M., Lombardi, A., Silvestri, E. and Goglia, F. (2001) Uncoupling protein-3 is a molecular determinant for the regulation of resting metabolic rate by thyroid hormone. Endocrinology, 142: 3414–3420.

Dhillo, W.S., Small, C.J., Jethwa, P.H., Russell, S.H., Gardiner, J.V., Bewick, G.A., Seth, A., Murphy, K.G., Ghatei, M.A. and Bloom, S.R. (2003) Paraventricular nucleus administration of calcitonin gene-related peptide inhibits food intake and stimulates the hypothalamo-pituitary-adrenal axis. Endocrinology, 144: 1420–1425.

Diano, S., Leonard, J.L., Meli, R., Esposito, E. and Schiavo, L. (2003) Hypothalamic type II iodothyronine deiodinase: a light and electron microscopic study. Brain Res., 976: 130–134.

Diano, S., Naftolin, F., Goglia, F. and Horvath, T.L. (1998) Fasting-induced increase in type II iodothyronine deiodinase activity and messenger ribonucleic acid levels is not reversed by thyroxine in the rat hypothalamus. Endocrinology, 139: 2879–2884.

Diekman, M.J., Romijn, J.A., Endert, E., Sauerwein, H. and Wiersinga, W.M. (1998) Thyroid hormones modulate serum leptin levels: observations in thyrotoxic and hypothyroid women. Thyroid, 8: 1081–1086.

Dube, D., Lissitzky, J.C., Leclerc, R. and Pelletier, G. (1978) Localization of alpha-melanocyte-stimulating hormone in rat brain and pituitary. Endocrinology, 102: 1283–1291.

Enerback, S., Jacobsson, A., Simpson, E.M., Guerra, C., Yamashita, H., Harper, M.E. and Kozak, L.P. (1997) Mice lacking mitochondrial uncoupling protein are cold-sensitive but not obese. Nature, 387: 90–94.

Ervin, G.N., Schmitz, S.A., Nemeroff, C.B. and Prange Jr., A.J. (1981) Thyrotropin-releasing hormone and amphetamine produce different patterns of behavioral excitation in rats. Eur. J. Pharmacol., 72: 35–43.

Eskay, R.L., Giraud, P., Oliver, C. and Brown-Stein, M.J. (1979) Distribution of alpha-melanocyte-stimulating hormone in the rat brain: evidence that alpha-MSH-containing cells in the arcuate region send projections to extrahypothalamic areas. Brain Res., 178: 55–67.

Fekete, C., Gereben, B., Doleschall, M., Harney, J.W., Dora, J.M., Bianco, A.C., Sarkar, S., Liposits, Z., Rand, W., Emerson, C., Kacskovics, I., Larsen, P.R. and Lechan, R.M. (2004a) Lipopolysaccharide induces type 2 iodothyronine deiodinase in the mediobasal hypothalamus: implications for the nonthyroidal illness syndrome. Endocrinology, 145: 1649–1655.

Fekete, C., Kelly, J., Mihaly, E., Sarkar, S., Rand, W.M., Legradi, G., Emerson, C.H. and Lechan, R.M. (2001) Neuropeptide Y has a central inhibitory action on the hypothalamic-pituitary-thyroid axis. Endocrinology, 142: 2606–2613.

Fekete, C., Legradi, G., Mihaly, E., Huang, Q.H., Tatro, J.B., Rand, W.M., Emerson, C.H. and Lechan, R.M. (2000a) Alpha-melanocyte-stimulating hormone is contained in nerve terminals innervating thyrotropin-releasing hormone-synthesizing neurons in the hypothalamic paraventricular nucleus and prevents fasting-induced suppression of prothyrotropin-releasing hormone gene expression. J. Neurosci., 20: 1550–1558.

Fekete, C., Marks, D.L., Sarkar, S., Emerson, C.H., Rand, W.M., Cone, R.D. and Lechan, R.M. (2004b) Effect of agouti-related protein (Agrp) in regulation of the hypothalamic-pituitary-thyroid (HPT) axis in the Mc4-R Ko mouse. Endocrinology, 145: 4816–4821.

Fekete, C., Mihaly, E., Herscovici, S., Salas, J., Tu, H., Larsen, P.R. and Lechan, R.M. (2000b) DARPP-32 and CREB are present in type 2 iodothyronine deiodinase-producing tanycytes: implications for the regulation of type 2 deiodinase activity. Brain Res., 862: 154–161.

Fekete, C., Mihaly, E., Luo, L.G., Kelly, J., Clausen, J.T., Mao, Q., Rand, W.M., Moss, L.G., Kuhar, M., Emerson, C.H., Jackson, I.M. and Lechan, R.M. (2000c) Association of cocaine- and amphetamine-regulated transcript-immunoreactive elements with thyrotropin-releasing hormone-synthesizing neurons in the hypothalamic paraventricular nucleus and its role in the regulation of the hypothalamic-pituitary-thyroid axis during fasting. J. Neurosci., 20: 9224–9234.

Fekete, C., Sarkar, S., Christoffolete, M.A., Emerson, C.H., Bianco, A.C. and Lechan, R.M. (2005a) Bacterial lipopolysaccharide (LPS)-induced type 2 iodothyronine deiodinase (D2) activation in the mediobasal hypothalamus (MBH) is independent of the LPS-induced fall in serum thyroid hormone levels. Brain Res., 1056: 97–99.

Fekete, C., Sarkar, S. and Lechan, R.M. (2005) Relative contribution of brainstem afferents to the cocaine- and amphetamine-regulated transcript (CART) innervation of thyrotropin-releasing hormone synthesizing neurons in the hypothalamic paraventricular nucleus (PVN). Brain Res., 1032: 171–175.

Fekete, C., Sarkar, S., Rand, W.M., Harney, J.W., Emerson, C.H., Bianco, A.C., Beck-Sickinger, A. and Lechan, R.M. (2002a) Neuropeptide Y1 and Y5 receptors mediate the effects of neuropeptide Y on the hypothalamic-pituitary-thyroid axis. Endocrinology, 143: 4513–4519.

Fekete, C., Sarkar, S., Rand, W.M., Harney, J.W., Emerson, C.H., Bianco, A.C. and Lechan, R.M. (2002b) Agouti-related protein (AGRP) has a central inhibitory action on the hypothalamic-pituitary-thyroid (HPT) axis; comparisons between the effect of AGRP and neuropeptide Y on energy homeostasis and the HPT axis. Endocrinology, 143: 3846–3853.

Fekete, C., Singru, P.S., Sarkar, S., Rand, W.M. and Lechan, R.M. (2005) Ascending brainstem pathways are not involved in lipopolysaccharide-induced suppression of thyrotropin-releasing hormone gene expression in the hypothalamic paraventricular nucleus. Endocrinology, 146: 1357–1363.

Francis, J., MohanKumar, S.M. and MohanKumar, P.S. (2000) Correlations of norepinephrine release in the paraventricular nucleus with plasma corticosterone and leptin after systemic lipopolysaccharide: blockade by soluble IL-1 receptor. Brain Res., 867: 180–187.

French, T.A., Masserano, J.M. and Weiner, N. (1993) Influence of thyrotropin-releasing hormone and catecholaminergic interactions on CNS ethanol sensitivity. Alcohol Clin. Exp. Res., 17: 99–106.

Garrick, T., Prince, M., Yang, H., Ohning, G. and Tache, Y. (1994) Raphe pallidus stimulation increases gastric contractility via TRH projections to the dorsal vagal complex in rats. Brain Res., 636: 343–347.

Garrick, T., Yang, H., Trauner, M., Livingston, E. and Tache, Y. (1992) Thyrotropin-releasing hormone analog injected into the raphe pallidus and obscurus increases gastric contractility in rats. Eur. J. Pharmacol., 223: 75–81.

230

Giovannini, M.G., Casamenti, F., Nistri, A., Paoli, F. and Pepeu, G. (1991) Effect of thyrotropin releasing hormone (TRH) on acetylcholine release from different brain areas investigated by microdialysis. Br. J. Pharmacol., 102: 363–368.

Glue, P., Bailey, J., Wilson, S., Hudson, A. and Nutt, D.J. (1992) Thyrotropin-releasing hormone selectively reverses lorazepam-induced sedation but not slowing of saccadic eye movements. Life Sci., 50: PL25–PL30.

Gong, D.W., He, Y., Karas, M. and Reitman, M. (1997) Uncoupling protein-3 is a mediator of thermogenesis regulated by thyroid hormone, beta3-adrenergic agonists, and leptin. J. Biol. Chem., 272: 24129–24132.

Gong, D.W., Monemdjou, S., Gavrilova, O., Leon, L.R., Marcus-Samuels, B., Chou, C.J., Everett, C., Kozak, L.P., Li, C., Deng, C., Harper, M.E. and Reitman, M.L. (2000) Lack of obesity and normal response to fasting and thyroid hormone in mice lacking uncoupling protein-3. J. Biol. Chem., 275: 16251–16257.

Goodridge, A.G. (1978) Regulation of malic enzyme synthesis by thyroid hormone and glucagon: inhibitor and kinetic experiments. Mol. Cell Endocrinol., 11: 19–29.

Goto, M., Yoneda, M., Nakamura, K., Terano, A. and Haneda, M. (2004) Effect of central thyrotropin-releasing hormone on pancreatic blood flow in rats. Regul. Pept., 121: 57–63.

Griffiths, E.C., Rothwell, N.J. and Stock, M.J. (1988) Thermogenic effects of thyrotrophin-releasing hormone and its analogues in the rat. Experientia, 44: 40–42.

Grunfeld, C., Zhao, C., Fuller, J., Pollack, A., Moser, A., Friedman, J. and Feingold, K.R. (1996) Endotoxin and cytokines induce expression of leptin, the ob gene product, in hamsters. J. Clin. Invest., 97: 2152–2157.

Guadano-Ferraz, A., Obregon, M.J., St Germain, D.L. and Bernal, J. (1997) The type 2 iodothyronine deiodinase is expressed primarily in glial cells in the neonatal rat brain. Proc. Natl. Acad. Sci. USA, 94: 10391–10396.

Hagan, M.M., Rushing, P.A., Pritchard, L.M., Schwartz, M.W., Strack, A.M., Van Der Ploeg, L.H., Woods, S.C. and Seeley, R.J. (2000) Long-term orexigenic effects of AgRP-(83—132) involve mechanisms other than melanocortin receptor blockade. Am. J. Physiol. Regul. Integr. Comp. Physiol., 279: R47–R52.

Hahn, T.M., Breininger, J.F., Baskin, D.G. and Schwartz, M.W. (1998) Coexpression of Agrp and NPY in fasting-activated hypothalamic neurons. Nat. Neurosci., 1: 271–272.

Hansen, K.R., Krasnow, S.M., Nolan, M.A., Fraley, G.S., Baumgartner, J.W., Clifton, D.K. and Steiner, R.A. (2003) Activation of the sympathetic nervous system by galanin-like peptide–a possible link between leptin and metabolism. Endocrinology, 144: 4709–47017.

Harris, A.R., Fang, S.L., Vagenakis, A.G. and Braverman, L.E. (1978) Effect of starvation, nutriment replacement, and hypothyroidism on in vitro hepatic T4 to T3 conversion in the rat. Metabolism, 27: 1680–1690.

Harris, M., Aschkenasi, C., Elias, C.F., Chandrankunnel, A., Nillni, E.A., Bjoorbaek, C., Elmquist, J.K., Flier, J.S. and Hollenberg, A.N. (2001) Transcriptional regulation of the thyrotropin-releasing hormone gene by leptin and melanocortin signaling. J. Clin. Invest., 107: 111–120.

Haskell-Luevano, C., Chen, P., Li, C., Chang, K., Smith, M.S., Cameron, J.L. and Cone, R.D. (1999) Characterization of the neuroanatomical distribution of agouti-related protein immunoreactivity in the rhesus monkey and the rat. Endocrinology, 140: 1408–1415.

Haskell-Luevano, C. and Monck, E.K. (2001) Agouti-related protein functions as an inverse agonist at a constitutively active brain melanocortin-4 receptor. Regul. Pept., 99: 1–7.

Heuer, H., Maier, M.K., Iden, S., Mittag, J., Friesema, E.C., Visser, T.J. and Bauer, K. (2005) The monocarboxylate transporter 8 linked to human psychomotor retardation is highly expressed in thyroid hormone-sensitive neuron populations. Endocrinology, 146: 1701–1706.

Hollenberg, A.N., Monden, T., Flynn, T.R., Boers, M.E., Cohen, O. and Wondisford, F.E. (1995) The human thyrotropin-releasing hormone gene is regulated by thyroid hormone through two distinct classes of negative thyroid hormone response elements. Mol. Endocrinol., 9: 540–550.

Hori, T., Yamasaki, M., Asami, T., Koga, H. and Kiyohara, T. (1988) Responses of anterior hypothalamic-preoptic thermosensitive neurons to thyrotropin releasing hormone and cyclo(His-Pro). Neuropharmacology, 27: 895–901.

Horita, A. (1998) An update on the CNS actions of TRH and its analogs. Life Sci., 62: 1443–1448.

Horita, A., Carino, M.A. and Nishimura, Y. (1991) D1 agonist SKF 38393 antagonizes pentobarbital-induced narcosis and depression of hippocampal and cortical cholinergic activity in rats. Life Sci., 49: 595–601.

Horita, A., Carino, M.A. and Smith, J.R. (1976) Effects of TRH on the central nervous system of the rabbit. Pharmacol. Biochem. Behav., 5: 111–116.

Huang, X.C., Saigusa, T. and Iriki, M. (1992) Comparison of TRH and its analog (NS-3) in thermoregulatory and cardiovascular effects. Peptides, 13: 305–311.

Hugues, J.N., Epelbaum, J., Voirol, M.J., Modigliani, E., Sebaoun, J. and Enjalbert, A. (1988) Influence of starvation on hormonal control of hypophyseal secretion in rats. Acta Endocrinol. (Copenh), 119: 195–202.

Ignar, D.M. and Kuhn, C.M. (1988) Relative ontogeny of opioid and catecholaminergic regulation of thyrotropin secretion in the rat. Endocrinology, 123: 567–571.

Ishii, S., Kamegai, J., Tamura, H., Shimizu, T., Sugihara, H. and Oikawa, S. (2003) Hypothalamic neuropeptide Y/Y1 receptor pathway activated by a reduction in circulating leptin, but not by an increase in circulating ghrelin, contributes to hyperphagia associated with triiodothyronine-induced thyrotoxicosis. Neuroendocrinology, 78: 321–330.

Ishikawa, K., Kakegawa, T. and Suzuki, M. (1984) Role of the hypothalamic paraventricular nucleus in the secretion of thyrotropin under adrenergic and cold-stimulated conditions in the rat. Endocrinology, 114: 352–358.

Ishikawa, K., Taniguchi, Y., Inoue, K., Kurosumi, K. and Suzuki, M. (1988) Immunocytochemical delineation of thyrotrophic area: origin of thyrotropin-releasing hormone in the median eminence. Neuroendocrinology, 47: 384–388.

Ishikawa, K., Taniguchi, Y., Kurosumi, K. and Suzuki, M. (1986) Origin of septal thyrotropin-releasing hormone in the rat. Neuroendocrinology, 44: 54–58.

Jacobowitz, D.M. and O'Donohue, T.L. (1978) alpha-Melanocyte stimulating hormone: immunohistochemical identification and mapping in neurons of rat brain. Proc. Natl. Acad. Sci. USA, 75: 6300–6304.

Jezek, P. (2002) Possible physiological roles of mitochondrial uncoupling proteins–UCPn. Int. J. Biochem. Cell Biol., 34: 1190–1206.

Joseph, S.A. and Michael, G.J. (1988) Efferent ACTH-IR opiocortin projections from nucleus tractus solitarius: a hypothalamic deafferentation study. Peptides, 9: 193–201.

Kakucska, I., Rand, W. and Lechan, R.M. (1992) Thyrotropin-releasing hormone gene expression in the hypothalamic paraventricular nucleus is dependent upon feedback regulation by both triiodothyronine and thyroxine. Endocrinology, 130: 2845–2850.

Kalivas, P.W., Simasko, S.M. and Horita, A. (1981) Effect of septohippocampal lesions on thyrotropin-releasing hormone antagonism of pentobarbital narcosis. Brain Res., 222: 253–265.

Kalivas, P.W., Stanley, D. and Prange Jr., A.J. (1987) Interaction between thyrotropin-releasing hormone and the mesolimbic dopamine system. Neuropharmacology, 26: 33–38.

Kawano, H., Tsuruo, Y., Bando, H. and Daikoku, S. (1991) Hypophysiotrophic TRH-producing neurons identified by combining immunohistochemistry for pro-TRH and retrograde tracing. J. Comp. Neurol., 307: 531–538.

Kelley, A.E. (2004) Ventral striatal control of appetitive motivation: role in ingestive behavior and reward-related learning. Neurosci. Biobehav. Rev., 27: 765–776.

Kim, M.S., Rossi, M., Abbott, C.R., AlAhmed, S.H., Smith, D.M. and Bloom, S.R. (2002) Sustained orexigenic effect of Agouti related protein may be not mediated by the melanocortin 4 receptor. Peptides, 23: 1069–1076.

Kim, M.S., Small, C.J., Russell, S.H., Morgan, D.G., Abbott, C.R., alAhmed, S.H., Hay, D.L., Ghatei, M.A., Smith, D.M. and Bloom, S.R. (2002) Effects of melanocortin receptor ligands on thyrotropin-releasing hormone release: evidence for the differential roles of melanocortin 3 and 4 receptors. J. Neuroendocrinol., 14: 276–282.

Kim, M.S., Small, C.J., Stanley, S.A., Morgan, D.G., Seal, L.J., Kong, W.M., Edwards, C.M., Abusnana, S., Sunter, D., Ghatei, M.A. and Bloom, S.R. (2000) The central melanocortin system affects the hypothalamo-pituitary thyroid axis and may mediate the effect of leptin. J. Clin. Invest., 105: 1005–10011.

Kishi, T., Aschkenasi, C.J., Lee, C.E., Mountjoy, K.G., Saper, C.B. and Elmquist, J.K. (2003) Expression of melanocortin 4 receptor mRNA in the central nervous system of the rat. J. Comp. Neurol., 457: 213–235.

Kondo, K., Harbuz, M.S., Levy, A. and Lightman, S.L. (1997) Inhibition of the hypothalamic-pituitary-thyroid axis in response to lipopolysaccharide is independent of changes in circulating corticosteroids. Neuroimmunomodulation, 4: 188–194.

Kong, W.M., Martin, N.M., Smith, K.L., Gardiner, J.V., Connoley, I.P., Stephens, D.A., Dhillo, W.S., Ghatei, M.A., Small, C.J. and Bloom, S.R. (2004) Triiodothyronine stimulates food intake via the hypothalamic ventromedial nucleus independent of changes in energy expenditure. Endocrinology, 145: 5252–5258.

LaHann, T.R. and Horita, A. (1982) Thyrotropin releasing hormone: centrally mediated effects on gastrointestinal motor activity. J. Pharmacol. Exp. Ther., 222: 66–70.

Lahti, H., Koskinen, M., Pyornila, A. and Hissa, R. (1983) Hyperthermia after intrahypothalamic injections of thyrotropin releasing hormone (TRH) in the pigeon. Experientia, 39: 1338–1340.

Lebon, V., Dufour, S., Petersen, K.F., Ren, J., Jucker, B.M., Slezak, L.A., Cline, G.W., Rothman, D.L. and Shulman, G.I. (2001) Effect of triiodothyronine on mitochondrial energy coupling in human skeletal muscle. J. Clin. Invest., 108: 733–737.

Lechan, R. M. (2003) Functional anatomy of the hypothalamus and pituitary. Endotext.org.

Lechan, R.M. and Fekete, C. (2004) Feedback regulation of thyrotropin-releasing hormone (TRH): mechanisms for the non-thyroidal illness syndrome. J. Endocrinol. Invest., 27: 105–119.

Lechan R.M. and Fekete, C. (in press) Infundibular tanycytes as modulators of neuroendocrine function: hypothetical role in the regulation of the thyroid and gonadal axis. Acta Bio Med.

Lechan, R.M. and Hollenberg, A. (2003) Thyrotropin-releasing hormone (TRH). In: Norman, H.L.H.a.a.A.W. (Ed.), Encyclopedia of Hormones. Academic Press, pp. 510–524.

Lechan, R.M. and Jackson, I.M. (1982) Immunohistochemical localization of thyrotropin-releasing hormone in the rat hypothalamus and pituitary. Endocrinology, 111: 55–65.

Lechan, R.M. and Kakucska, I. (1992) Feedback regulation of thyrotropin-releasing hormone gene expression by thyroid hormone in the hypothalamic paraventricular nucleus. Ciba Found. Symp., 168: 144–158 discussion 158–64.

Lechan, R. M. and Toni, R. (2003). Functional anatomy of the hypothalamus and pituitary. Endotext.com.

Lechan, R.M., Wu, P. and Jackson, I.M. (1987) Immunocytochemical distribution in rat brain of putative peptides derived from thyrotropin-releasing hormone prohormone. Endocrinology, 121: 1879–1891.

Legradi, G., Emerson, C.H., Ahima, R.S., Flier, J.S. and Lechan, R.M. (1997) Leptin prevents fasting-induced suppression of prothyrotropin-releasing hormone messenger ribonucleic acid in neurons of the hypothalamic paraventricular nucleus. Endocrinology, 138: 2569–2576.

Legradi, G., Emerson, C.H., Ahima, R.S., Rand, W.M., Flier, J.S. and Lechan, R.M. (1998) Arcuate nucleus ablation prevents fasting-induced suppression of ProTRH mRNA in the hypothalamic paraventricular nucleus. Neuroendocrinology, 68: 89–97.

Legradi, G. and Lechan, R.M. (1998) The arcuate nucleus is the major source for neuropeptide Y-innervation of thyrotropin-releasing hormone neurons in the hypothalamic paraventricular nucleus. Endocrinology, 139: 3262–3270.

Legradi, G. and Lechan, R.M. (1999) Agouti-related protein containing nerve terminals innervate thyrotropin-releasing hormone neurons in the hypothalamic paraventricular nucleus. Endocrinology, 140: 3643–3652.

Levine, J.A., Schleusner, S.J. and Jensen, M.D. (2000) Energy expenditure of nonexercise activity. Am. J. Clin. Nutr., 72: 1451–1454.

Lin, L.S., Chiu, W.T., Shih, C.J. and Lin, M.T. (1987) Involvement of both opiate and catecholaminergic receptors of ventromedial hypothalamus in the locomotor stimulant action of thyrotropin-releasing hormone. J. Neural. Transm., 68: 217–225.

Liposits, Z. (1993) Ultrastructure of hypothalamic paraventricular neurons. Crit. Rev. Neurobiol., 7: 89–162.

Luiten, P.G., ter Horst, G.J., Karst, H. and Steffens, A.B. (1985) The course of paraventricular hypothalamic efferents to autonomic structures in medulla and spinal cord. Brain Res., 329: 374–378.

Luo, L. and MacLean, D.B. (2003) Effects of thyroid hormone on food intake, hypothalamic Na/K ATPase activity and ATP content. Brain Res., 973: 233–239.

Maeda-Hagiwara, M. and Tache, Y. (1987) Central nervous system action of TRH to stimulate gastric emptying in rats. Regul. Pept., 17: 199–207.

Marks, D.L., Ling, N. and Cone, R.D. (2001) Role of the central melanocortin system in cachexia. Cancer Res., 61: 1432–1438.

Martinez, V., Barrachina, M.D., Ohning, G. and Tache, Y. (2002) Cephalic phase of acid secretion involves activation of medullary TRH receptor subtype 1 in rats. Am. J. Physiol. Gastrointest. Liver Physiol., 283: G1310–G1319.

Mayr, B. and Montminy, M. (2001) Transcriptional regulation by the phosphorylation-dependent factor CREB. Nat. Rev. Mol. Cell Biol., 2: 599–609.

Medina-Gomez, G., Calvo, R.M. and Obregon, M.J. (2004) T3 and Triac inhibit leptin secretion and expression in brown and white rat adipocytes. Biochim. Biophys. Acta, 1682: 38–47.

Michel, M.C., Beck-Sickinger, A., Cox, H., Doods, H.N., Herzog, H., Larhammar, D., Quirion, R., Schwartz, T. and Westfall XVI, T. (1998) International Union of Pharmacology recommendations for the nomenclature of neuropeptide Y, peptide YY, and pancreatic polypeptide receptors. Pharmacol. Rev., 50: 143–150.

Mihaly, E., Fekete, C., Legradi, G. and Lechan, R.M. (2001) Hypothalamic dorsomedial nucleus neurons innervate thyrotropin-releasing hormone-synthesizing neurons in the paraventricular nucleus. Brain Res., 891: 20–31.

Mihaly, E., Fekete, C., Tatro, J.B., Liposits, Z., Stopa, E.G. and Lechan, R.M. (2000) Hypophysiotropic thyrotropin-releasing hormone-synthesizing neurons in the human hypothalamus are innervated by neuropeptide Y, agouti-related protein, and alpha-melanocyte-stimulating hormone. J. Clin. Endocrinol. Metab., 85: 2596–2603.

Miksicek, R.J. and Towle, H.C. (1982) Changes in the rates of synthesis and messenger RNA levels of hepatic glucose-6-phosphate and 6-phosphogluconate dehydrogenases following induction by diet or thyroid hormone. J. Biol. Chem., 257: 11829–11835.

Miyamoto, M. and Nagawa, Y. (1977) Mesolimbic involvement in the locomotor stimulant action of thyrotropin-releasing hormone (TRH) in rats. Eur. J. Pharmacol., 44: 143–152.

Mori, M., Wakabayashi, K., Ohshima, K., Shimomura, Y., Fukuda, H. and Kobayashi, I. (1978) Effect of active and passive immunization with TRH on plasma TSH response to propylthiouracil. Endocrinol. Jpn., 25: 641–644.

Morley, J.E. (1980) The neuroendocrine control of appetite: the role of the endogenous opiates, cholecystokinin, TRH, gamma-amino-butyric-acid and the diazepam receptor. Life Sci., 27: 355–368.

Morrow, N.S., Novin, D. and Garrick, T. (1994) Microinjection of thyrotropin-releasing hormone in the paraventricular nucleus of the hypothalamus stimulates gastric contractility. Brain Res., 644: 243–250.

Mountjoy, K.G., Mortrud, M.T., Low, M.J., Simerly, R.B. and Cone, R.D. (1994) Localization of the melanocortin-4 receptor (MC4-R) in neuroendocrine and autonomic control circuits in the brain. Mol. Endocrinol., 8: 1298–1308.

Mountjoy, K.G., Robbins, L.S., Mortrud, M.T. and Cone, R.D. (1992) The cloning of a family of genes that encode the melanocortin receptors. Science, 257: 1248–1251.

Nadeau, S. and Rivest, S. (1999) Effects of circulating tumor necrosis factor on the neuronal activity and expression of the genes encoding the tumor necrosis factor receptors (p55 and p75) in the rat brain: a view from the blood–brain barrier. Neuroscience, 93: 1449–1464.

Nagashima, K., Nakai, S., Tanaka, M. and Kanosue, K. (2000) Neuronal circuitries involved in thermoregulation. Auton. Neurosci., 85: 18–25.

Nakamura, T., Nagasaka, S., Ishikawa, S., Hayashi, H., Saito, T., Kusaka, I., Higashiyama, M. and Saito, T. (2000) Association of hyperthyroidism with serum leptin levels. Metabolism, 49: 1285–1288.

Nijenhuis, W.A., Oosterom, J. and Adan, R.A. (2001) AgRP(83-132) acts as an inverse agonist on the human-melanocortin-4 receptor. Mol. Endocrinol., 15: 164–171.

Nillni, E.A. and Sevarino, K.A. (1999) The biology of prothyrotropin-releasing hormone-derived peptides. Endocr. Rev., 20: 599–648.

Nillni, E.A., Vaslet, C., Harris, M., Hollenberg, A., Bjorbak, C. and Flier, J.S. (2000) Leptin regulates prothyrotropin-releasing hormone biosynthesis. Evidence for direct and indirect pathways.. J. Biol. Chem., 275: 36124–36133.

Nishino, S., Arrigoni, J., Shelton, J., Kanbayashi, T., Dement, W.C. and Mignot, E. (1997) Effects of thyrotropin-releasing hormone and its analogs on daytime sleepiness and cataplexy in canine narcolepsy. J. Neurosci., 17: 6401–6408.

Okumura, T., Taylor, I.L. and Pappas, T.N. (1995) Microinjection of TRH analogue into the dorsal vagal complex stimulates pancreatic secretion in rats. Am. J. Physiol., 269: G328–G334.

Ollmann, M.M., Wilson, B.D., Yang, Y.K., Kerns, J.A., Chen, Y., Gantz, I. and Barsh, G.S. (1997) Antagonism of central melanocortin receptors in vitro and in vivo by agouti-related protein. Science, 278: 135–138.

Onaya and Hashizume, K. (1976) Effects of drugs that modify brain biogenic amine concentrations on thyroid activation induced by exposure to cold. Neuroendocrinology, 20: 47–58.

Oppenheimer, J.H., Schwartz, H.L., Lane, J.T. and Thompson, M.P. (1991) Functional relationship of thyroid hormone-induced lipogenesis, lipolysis, and thermogenesis in the rat. J. Clin. Invest., 87: 125–132.

Petrovich, G.D. and Gallagher, M. (2003) Amygdala subsystems and control of feeding behavior by learned cues. Ann. N Y Acad. Sci., 985: 251–262.

Przegalinski, E., Papla, I., Czepiel, K., Wydra, K., Nowak, E., Lason, W. and Filip, M. (2004) Effect of thyrotropin-releasing hormone on the locomotor and sensitizing effects of cocaine in rats. Neuropeptides, 38: 48–54.

Pyner, S. and Coote, J.H. (2000) Identification of branching paraventricular neurons of the hypothalamus that project to the rostroventrolateral medulla and spinal cord. Neuroscience, 100: 549–556.

Queiroz, M.S., Shao, Y. and Ismail-Beigi, F. (2004) Effect of thyroid hormone on uncoupling protein-3 mRNA expression in rat heart and skeletal muscle. Thyroid, 14: 177–185.

Randich, A., Chandler, P.C., Mebane, H.C., Turnbach, M.E., Meller, S.T., Kelm, G.R. and Cox, J.E. (2004) Jejunal administration of linoleic acid increases activity of neurons in the paraventricular nucleus of the hypothalamus. Am. J. Physiol. Regul. Integr. Comp. Physiol., 286: R166–R173.

Reichlin, S. (1986) Neural functions of TRH. Acta Endocrinol. Suppl. (Copenh), 276: 21–33.

Reichlin, S. (1989) TRH: historical aspects. Ann. N Y Acad. Sci., 553: 1–6.

Reichlin, S. and Glaser, R.J. (1958) Thyroid function in experimental streptococcal pneumonia in the rat. J. Exp. Med., 107: 219–236.

Ribeiro, M.O., Carvalho, S.D., Schultz, J.J., Chiellini, G., Scanlan, T.S., Bianco, A.C. and Brent, G.A. (2001) Thyroid hormone–sympathetic interaction and adaptive thermogenesis are thyroid hormone receptor isoform-specific. J. Clin. Invest., 108: 97–105.

Riskind, P.N., Kolodny, J.M. and Larsen, P.R. (1987) The regional hypothalamic distribution of type II 5'-monodeiodinase in euthyroid and hypothyroid rats. Brain Res., 420: 194–198.

Rivest, S., Lacroix, S., Vallieres, L., Nadeau, S., Zhang, J. and Laflamme, N. (2000) How the blood talks to the brain parenchyma and the paraventricular nucleus of the hypothalamus during systemic inflammatory and infectious stimuli. Proc. Soc. Exp. Biol. Med., 223: 22–38.

Roeling, T.A., Veening, J.G., Peters, J.P., Vermelis, M.E. and Nieuwenhuys, R. (1993) Efferent connections of the hypothalamic "grooming area" in the rat. Neuroscience, 56: 199–225.

Rondeel, J.M., de Greef, W.J., Hop, W.C., Rowland, D.L. and Visser, T.J. (1991) Effect of cold exposure on the hypothalamic release of thyrotropin-releasing hormone and catecholamines. Neuroendocrinology, 54: 477–481.

Rondeel, J.M., Heide, R., de Greef, W.J., van Toor, H., van Haasteren, G.A., Klootwijk, W. and Visser, T.J. (1992) Effect of starvation and subsequent refeeding on thyroid function and release of hypothalamic thyrotropin-releasing hormone. Neuroendocrinology, 56: 348–353.

Saper, C.B., Loewy, A.D., Swanson, L.W. and Cowan, W.M. (1976) Direct hypothalamo-autonomic connections. Brain Res., 117: 305–312.

Sarkar, S. and Lechan, R.M. (2003) Central administration of neuropeptide Y reduces alpha-melanocyte-stimulating hormone-induced cyclic adenosine 5'-monophosphate response element binding protein (CREB) phosphorylation in prothyrotropin-releasing hormone neurons and increases CREB phosphorylation in corticotropin-releasing hormone neurons in the hypothalamic paraventricular nucleus. Endocrinology, 144: 281–291.

Sarkar, S., Légrádi, G. and Lechan, R.M. (2002) Intracerebroventricular administration of alpha-melanocyte stimulating hormone increases phosphorylation of CREB in TRH- and CRH-producing neurons of the hypothalamic paraventricular nucleus. Brain Res., 945: 50–59.

Sarkar, S., Wittmann, G., Fekete, C. and Lechan, R.M. (2004) Central administration of cocaine- and amphetamine-regulated transcript increases phosphorylation of cAMP response element binding protein in corticotropin-releasing hormone-producing neurons but not in prothyrotropin-releasing hormone-producing neurons in the hypothalamic paraventricular nucleus. Brain Res., 999: 181–192.

Satoh, T., Yamada, M., Iwasaki, T. and Mori, M. (1996) Negative regulation of the gene for the preprothyrotropin-releasing hormone from the mouse by thyroid hormone requires additional factors in conjunction with thyroid hormone receptors. J. Biol. Chem., 271: 27919–27926.

Sawchenko, P.E. and Swanson, L.W. (1983) The organization of forebrain afferents to the paraventricular and supraoptic nuclei of the rat. J. Comp. Neurol., 218: 121–144.

Sawchenko, P.E., Swanson, L.W. and Joseph, S.A. (1982) The distribution and cells of origin of ACTH(1-39)-stained varicosities in the paraventricular and supraoptic nuclei. Brain Res., 232: 365–374.

Schettini, G., Quattrone, A., Di Renzo, G., Lombardi, G. and Preziosi, P. (1979) Effect of 6-hydroxydopamine treatment on TSH secretion in basal and cold-stimulated conditions in the rat. Eur. J. Pharmacol., 56: 153–157.

Schreiber, G., Aldred, A.R., Jaworowski, A., Nilsson, C., Achen, M.G. and Segal, M.B. (1990) Thyroxine transport from blood to brain via transthyretin synthesis in choroid plexus. Am. J. Physiol., 258: R338–R345.

Schreiber, G., Pettersson, T.M., Southwell, B.R., Aldred, A.R., Harms, P.J., Richardson, S.J., Wettenhall, R.E., Duan, W. and Nicol, S.C. (1993) Transthyretin expression evolved more recently in liver than in brain. Comp. Biochem. Physiol. B, 105: 317–325.

Segerson, T.P., Kauer, J., Wolfe, H.C., Mobtaker, H., Wu, P., Jackson, I.M. and Lechan, R.M. (1987) Thyroid hormone regulates TRH biosynthesis in the paraventricular nucleus of the rat hypothalamus. Science, 238: 78–80.

Sergeyev, V., Broberger, C. and Hokfelt, T. (2001) Effect of LPS administration on the expression of POMC, NPY,

234

galanin, CART and MCH mRNAs in the rat hypothalamus. Brain Res. Mol. Brain Res., 90: 93–100.

Shafton, A.D., Ryan, A. and Badoer, E. (1998) Neurons in the hypothalamic paraventricular nucleus send collaterals to the spinal cord and to the rostral ventrolateral medulla in the rat. Brain Res., 801: 239–243.

Shintani, M., Tamura, Y., Monden, M. and Shiomi, H. (2005) Thyrotropin-releasing hormone induced thermogenesis in Syrian hamsters: site of action and receptor subtype. Brain Res., 1039: 22–29.

Shioda, S., Nakai, Y., Sato, A., Sunayama, S. and Shimoda, Y. (1986) Electron-microscopic cytochemistry of the catecholaminergic innervation of TRH neurons in the rat hypothalamus. Cell Tissue Res., 245: 247–252.

Siegel, J.M. (2004) Hypocretin (orexin): role in normal behavior and neuropathology. Annu. Rev. Psychol., 55: 125–148.

Silva, J.E. (1995) Thyroid hormone control of thermogenesis and energy balance. Thyroid, 5: 481–492.

Silva, J.E. (2001) The multiple contributions of thyroid hormone to heat production. J. Clin. Invest., 108: 35–37.

Silva, J.E. (2003) The thermogenic effect of thyroid hormone and its clinical implications. Ann. Intern. Med., 139: 205–213.

Singru, P.S., Fekete, C. and Lechan, R.M. (2005) Neuroanatomical evidence for participation of the hypothalamic dorsomedial nucleus (DMN) in regulation of the hypothalamic paraventricular nucleus (PVN) by alpha-melonocyte stimulating hormone. Brain Res., 1064: 42–51.

Spector, R. and Levy, P. (1975) Thyroxine transport by the choroid plexus in vitro. Brain Res., 98: 400–404.

Sprague, J.E., Mallett, N.M., Rusyniak, D.E. and Mills, E. (2004) UCP3 and thyroid hormone involvement in methamphetamine-induced hyperthermia. Biochem. Pharmacol., 68: 1339–1343.

Stanley, S.A., Small, C.J., Murphy, K.G., Rayes, E., Abbott, C.R., Seal, L.J., Morgan, D.G., Sunter, D., Dakin, C.L., Kim, M.S., Hunter, R., Kuhar, M., Ghatei, M.A. and Bloom, S.R. (2001) Actions of cocaine- and amphetamine-regulated transcript (CART) peptide on regulation of appetite and hypothalamo-pituitary axes in vitro and in vivo in male rats. Brain Res., 893: 186–194.

Stevenin, B. and Lee, S. (1995) Hormonal regulation of the thyrotropin releasing hormone (TRH) gene. Endocrinologist, 5: 286–296.

Steward, C.A., Horan, T.L., Schuhler, S., Bennett, G.W. and Ebling, F.J. (2003) Central administration of thyrotropin releasing hormone (TRH) and related peptides inhibits feeding behavior in the Siberian hamster. Neuroreport, 14: 687–691.

Suzuki, T., Kohno, H., Sakurada, T., Tadano, T. and Kisara, K. (1982) Intracranial injection of thyrotropin releasing hormone (TRH) suppresses starvation-induced feeding and drinking in rats. Pharmacol. Biochem. Behav., 17: 249–253.

Swanson, L.W. and Kuypers, H.G. (1980) The paraventricular nucleus of the hypothalamus: cytoarchitectonic subdivisions and organization of projections to the pituitary, dorsal vagal complex, and spinal cord as demonstrated by retrograde

fluorescence double-labeling methods. J. Comp. Neurol., 194: 555–570.

Swanson, L.W. and Sawchenko, P.E. (1980) Paraventricular nucleus: a site for the integration of neuroendocrine and autonomic mechanisms. Neuroendocrinology, 31: 410–417.

Swanson, L.W., Sawchenko, P.E., Wiegand, S.J. and Price, J.L. (1980) Separate neurons in the paraventricular nucleus project to the median eminence and to the medulla or spinal cord. Brain Res., 198: 190–195.

Szabo, M. and Frohman, L.A. (1977) Suppression of cold-stimulated thyrotropin secretion by antiserum to thyrotropin-releasing hormone. Endocrinology, 101: 1023–1033.

Tache, Y., Vale, W. and Brown, M. (1980) Thyrotropin-releasing hormone — CNS action to stimulate gastric acid secretion. Nature, 287: 149–151.

Tache, Y., Yang, H. and Yoneda, M. (1993) Vagal regulation of gastric function involves thyrotropin-releasing hormone in the medullary raphe nuclei and dorsal vagal complex. Digestion, 54: 65–72.

Takatsu, Y., Matsumoto, H., Ohtaki, T., Kumano, S., Kitada, C., Onda, H., Nishimura, O. and Fujino, M. (2001) Distribution of galanin-like peptide in the rat brain. Endocrinology, 142: 1626–1634.

Tanaka, K., Ogawa, N., Chou, H., Mori, A. and Yanaihara, N. (1992) Effects of thyrotropin releasing hormone and its analogues on unconsciousness following head injury in mice. Regul. Pept., 38: 129–133.

Tu, H.M., Kim, S.W., Salvatore, D., Bartha, T., Legradi, G., Larsen, P.R. and Lechan, R.M. (1997) Regional distribution of type 2 thyroxine deiodinase messenger ribonucleic acid in rat hypothalamus and pituitary and its regulation by thyroid hormone. Endocrinology, 138: 3359–3368.

Uribe, R.M., Redondo, J.L., Charli, J.L. and Joseph-Bravo, P. (1993) Suckling and cold stress rapidly and transiently increase TRH mRNA in the paraventricular nucleus. Neuroendocrinology, 58: 140–145.

Vaarmann, A. and Kask, A. (2001) Cocaine and amphetamine-regulated transcript peptide (CART(62-76))-induced changes in regional monoamine levels in ratbrain. Neuropeptides, 35: 292–296.

Van Haasteren, G.A., Linkels, E., Klootwijk, W., van Toor, H., Rondeel, J.M., Themmen, A.P., de Jong, F.H., Valentijn, K., Vaudry, H., Bauer, K., et al. (1995) Starvation-induced changes in the hypothalamic content of prothyrotropin-releasing hormone (proTRH) mRNA and the hypothalamic release of proTRH-derived peptides: role of the adrenal gland. J. Endocrinol., 145: 143–153.

Vijayan, E. and McCann, S.M. (1977) Suppression of feeding and drinking activity in rats following intraventricular injection of thyrotropin releasing hormone (TRH). Endocrinology, 100: 1727–1730.

Vogel, R.A., Cooper, B.R., Barlow, T.S., Prange Jr., A.J., Mueller, R.A. and Breese, G.R. (1979) Effects of thyrotropin-releasing hormone on locomotor activity, operant performance and ingestive behavior. J. Pharmacol. Exp. Ther., 208: 161–168.

Vrang, N., Larsen, P.J., Clausen, J.T. and Kristensen, P. (1999) Neurochemical characterization of hypothalamic cocaine-amphetamine-regulated transcript neurons. J. Neurosci., 19: RC5.

Wei, E., Sigel, S., Loh, H. and Way, E.L. (1975) Thyrotrophin-releasing hormone and shaking behaviour in rat. Nature, 253: 739–740.

Wiersinga, W.M. and Boelen, A. (1996) Thyroid hormone metabolism in nonthyroidal illness. Curr. Opin. Endocrinol. Diabetes, 3: 422–427.

Wilson, B.D., Bagnol, D., Kaelin, C.B., Ollmann, M.M., Gantz, I., Watson, S.J. and Barsh, G.S. (1999) Physiological and anatomical circuitry between agouti-related protein and leptin signaling. Endocrinology, 140: 2387–2397.

Wisse, B.E., Schwartz, M.W. and Cummings, D.E. (2003) Melanocortin signaling and anorexia in chronic disease states. Ann. N Y Acad. Sci., 994: 275–281.

Wittmann, G., Liposits, Z., Lechan, R.M. and Fekete, C. (2002) Medullary adrenergic neurons contribute to the neuropeptide Y-ergic innervation of hypophysiotropic thyrotropin-releasing hormone-synthesizing neurons in the rat. Neurosci. Lett., 324: 69–73.

Wittmann, G., Liposits, Z., Lechan, R.M. and Fekete, C. (2004a) Medullary adrenergic neurons contribute to the cocaine- and amphetamine-regulated transcript-immunoreactive innervation of thyrotropin-releasing hormone synthesizing neurons in the hypothalamic paraventricular nucleus. Brain Res., 1006: 1–7.

Wittmann, G., Sarkar, S., Hrabovszky, E., Liposits, Z., Lechan, R.M. and Fekete, C. (2004b) Galanin- but not galanin-like peptide-containing axon terminals innervate hypophysiotropic TRH-synthesizing neurons in the hypothalamic paraventricular nucleus. Brain Res., 1002: 43–50.

Wolak, M.L., DeJoseph, M.R., Cator, A.D., Mokashi, A.S., Brownfield, M.S. and Urban, J.H. (2003) Comparative distribution of neuropeptide Y Y1 and Y5 receptors in the rat brain by using immunohistochemistry. J. Comp. Neurol., 464: 285–311.

Yang, H., Tache, Y., Ohning, G. and Go, V.L. (2002) Activation of raphe pallidus neurons increases insulin through medullary thyrotropin-releasing hormone (TRH)-vagal pathways. Pancreas, 25: 301–307.

Yoneda, M., Hashimoto, T., Nakamura, K., Tamori, K., Yokohama, S., Kono, T., Watanobe, H. and Terano, A. (2003) Thyrotropin-releasing hormone in the dorsal vagal complex stimulates hepatic blood flow in rats. Hepatology, 38: 1500–1507.

Zoeller, R.T., Kabeer, N. and Albers, H.E. (1990) Cold exposure elevates cellular levels of messenger ribonucleic acid encoding thyrotropin-releasing hormone in paraventricular nucleus despite elevated levels of thyroid hormones. Endocrinology, 127: 2955–2962.

SECTION IV

Rhythms, Sleep and Energy Metabolism

Kalsbeek, Fliers, Hofman, Swaab, Van Someren & Buijs
Progress in Brain Research, Vol. 153
ISSN 0079-6123

CHAPTER 13

The seventeenth C.U. Ariëns Kappers Lecture: an introduction

Michel A. Hofman and Dick F. Swaab

Netherlands Institute for Neuroscience, Meibergdreef 47, 1105 BA Amsterdam, The Netherlands

The C.U. Ariëns Kappers Award was created to honor the first director of the Netherlands Institute for Brain Research. The award is presented approximately once every two years to a leading and outstanding neuroscientist. Dr. Clifford B. Saper was invited to deliver the seventeenth C.U. Ariëns Kappers Lecture during the 24th International Summer School of Brain Research on September 1, 2005, for his outstanding achievements in deciphering the neuroanatomy of the mammalian hypothalamus and its intricate pathways involved in the control of behavior and physiology of the organism.

Photo 1 Cornelius Ubbo Ariëns Kappers was born in Groningen in 1877. During his medical training at the University of Amsterdam, he was inspired by the neurologist Prof. Cornelis Winkler to take up brain research, which led to an award-winning essay on the development of nerve sheets and, in 1904, to a Ph.D. thesis on the neuroanatomy of bony and cartilaginous fishes. In 1906, Ariëns Kappers was appointed as "Abteilungsvorsteher" (i.e., head of the department) at the "Senckenbergisch-Neurologisches Institut" of the famous neurologist and comparative neuroanatomist Prof. Ludwig Edinger in Frankfurt am Main.

Meanwhile, at the meeting of the International Association of Academies held in Paris in 1901, the anatomist Dr. Wilhelm His proposed that research on the nervous system should be placed on an international footing. Thus, on June 5, 1903, at

Burlington House in London, the former headquarters of the Royal Society, the "Central Commision for Brain Research" was constituted. This so-called "Brain Commission" set itself the task of "... organizing a network of institutions throughout the civilized world, dedicated to the study of the structure and functions of the central or-

Photograph 1. : C.U. Ariëns Kappers (1877–1946)

DOI: 10.1016/S0079-6123(06)53013-4

239

Table 1. Brain commission: register of "Interacademic brain institutes"

	Name of institute	Name of director	Year of recognition
1	Laboratory for Biological Sciences (University of Madrid)	Santiago Ramón y Cajal	1906
2	Neurological Institute (University of Leipzig)	Paul E. Flechsig	1906
3	Neurological Institute (University of Vienna)	Heinrich Obersteiner	1906
4	Brain-Anatomical Institute (Zürich)	Constantin von Monakow	1906
5	Neurological Department (Wistar Institute, Philadelphia)	Henry H. Donaldson	1906
6	Neurological Institute (Frankfurt am Main)	Ludwig Edinger	1906
7	Psychoneurological Institute (St. Petersburg)	Vladimir M. Bekhterev	1908
8	Central Institute for Brain Research (Amsterdam)	Cornelius U. Ariëns Kappers	1908
9	Brain-Histological Institute (University of Budapest)	Karoly Schaffer	1912

Source: Richter, J. (2000). The brain commission of the international association of academies: the first international society of Neurosciences. Brain Res. Bull., 52: 445–457.

gan. ..." Several governments responded to this ambition by founding brain research institutes (Table 1), among which was the Netherlands (Central) Institute for Brain Research, which opened its doors on June 8, 1909, in the presence of Dr. Camillo Golgi.

Prof. C.U. Ariëns Kappers became the first director of the Institute, a position he held until his death in 1946. With his excellent work he turned the institute into an internationally renowned place; the three-volume book of which he wrote the English version together with Dr. G.C. Huber and Dr. E.C. Crosby, entitled *The Comparative Anatomy of the Nervous System of Vertebrates, including Man* (1936), became a classic and is still well cited. He traveled all over the world and received a visiting professorship at Peking Union Medical College in China from 1923 to 1924. Four years later he was awarded an Honorary Doctorate of Sciences from Yale University. In 1929, Ariëns Kappers held his inaugural lecture as "extraordinary professor" at the medical faculty of the University of Amsterdam, and in the early 1930s he received further Honorary Doctorates from the Universities of Glasgow, Dublin and Chicago.

It was in Chicago, in 1924, that Ariëns Kappers encountered the distinguished neuroanatomist C. Judson Herrick, founder of the prestigious *Journal of Comparative Neurology* (JCN). And that is where the stories of C.U. Ariëns Kappers and the present laureate, Clifford B. Saper, meet. Dr. Saper was born in Chicago and worked there for

many years. Moreover, like Ariëns Kappers, he is very much interested in brain evolution, and especially in the role of the mammalian hypothalamus in the regulation of sleep, feeding and energy metabolism, the topic of the present Summer School.

Dr. Saper is the present editor-in-chief of the JCN, the journal that also figures, dramatically so, in the story of Ariëns Kappers. In 1905, C.U. Ariëns Kappers sent his translated manuscript to JCN with a large number of lithographs printed in Holland at the author's expense. On March 30, 1905, the JCN building burned down and everything in it was destroyed. C.J. Herrick had kept the Kappers' manuscript and some other papers in a fireproof safe, which survived the fire. Dr. Herrick described what happened in his memoir for the hundredth volume of JCN (Herrick, 1954, pp. 738–739): "When a few days later [the safe] was dragged out of the debris and prized open, the bundle of manuscripts looked like a black cinder. I carefully wrapped it in damp towels and took it home. When the remains of the bundle of papers were examined, I found that the paper was completely charred to a depth of an inch or two from each margin. Since this was the only copy of the Kappers' manuscript, it must be salvaged if at all possible. If I held the block of charred paper in direct sunlight at the proper angle the reflection of the black ink on the blackened right margin was visible and the writing could be deciphered with difficulty. The blue ink at the left margin, however, left no visible trace. My task was to transcribe each page as completely as possible, supplying the missing parts of the footnotes and alterations (written in blue ink) by liberal use of scientific imagination. After the top page was copied, it was carefully lifted off and the disintegrated fragments were brushed away, exposing the next page. Most of the long summer vacation was devoted to this job and to arraying my phoenix in proper typewritten dress. The typescript was then sent to Kappers, and to my surprise he made very few changes in it."

We are very glad that Dr. Clifford B. Saper accepted our invitation to deliver the seventeenth C.U. Ariëns Kappers Lecture in honor of this exceptional scientist. Dr. Saper was born in Chicago and received his training in biochemistry and neurobiology at the University of Illinois in Urbana, Illinois and his medical training at Washington University School of Medicine in St. Louis, Missouri. From 1978 to 1981 he was serving his residency at the Department of Neurology of the New York Medical Hospital of the Cornell University in Ithaca, after which he left for Washington University School of Medicine in St. Louis, Missouri, where he stayed as Assistant Professor of Neurology, Anatomy and Neurobiology from 1981 to 1985. In 1985 Dr. Saper was appointed Associate Professor of Neurobiology and Neurology at the Department of Neurology at the University of Chicago. Since 1992 Dr. Saper has been the James Jackson Putnam Professor of Neurology and Neuroscience at Harvard Medical School in Boston, where he examines the brain circuitries that regulate sleep and wakefulness, feeding and energy metabolism, immune responses and the neural mechanisms underlying various neurodegenerative disorders. Furthermore, Dr. Saper is Chairman of the Department of Neurology at the Beth Israel Deaconess Medical Center (BIDMC) in Boston. In the course of his career he has received several awards, among which are the McKnight Scholar Award, Javits Neuroscience Investigator Award and the Carl Ludwig Award of the American Physiological Society.

Previous lectures in this series were given by:
Pasko Rakic (New Haven, CT, USA, 1987)
Anders Björklund (Lund, Sweden, 1988)
Mortimer Mishkin (Bethesda, MD, USA, 1989)
Robert Y. Moore (New York, NY, USA, 1991)
Dale Purves (Durham, NH, USA, 1993)
Joseph Takahashi (Evanston, IL, USA, 1995)
Patricia S. Goldman Rakic (New Haven, CT, USA, 1996)
Dean H. Hamer (Bethesda, MD, USA, 1999)
Gerald M. Edelman (San Diego, CA, USA, 1999)
Viayanur S. Ramachandran (San Diego, CA, USA, 1999)
Steven P.R. Rose (Milton Keynes, UK, 1999)
Michael S. Gazzaniga (Hanover, NH, USA, 1999)
Antonio R. Damasio (Iowa, IA, USA, 1999)
Rudolf Nieuwenhuys (Amsterdam, The Netherlands, 2000)
Mark H. Tuszynski (San Diego, CA, USA, 2001)
Dennis O'Leary (La Jolla, CA, USA, 2003)

References

Herrick, C.J. (1954) One hundred volumes of the journal of comparative neurology. A record of sixty-three years of continuous publication issued at the completion of volume 100, June, 1954. J. Comp. Neurol., 100: 717–756.

Kalsbeek, Fliers, Hofman, Swaab, Van Someren & Buijs
Progress in Brain Research, Vol. 153
ISSN 0079-6123

CHAPTER 14

Staying awake for dinner: hypothalamic integration of sleep, feeding, and circadian rhythms

Clifford B. Saper*

Department of Neurology and Program in Neuroscience, Beth Israel Deaconess Medical Center, Harvard Medical School, Boston, MA 02215, USA

Abstract: Daily patterns of sleep and wakefulness are inextricably linked to the regulation of feeding and energy metabolism. Both are affected by homeostatic as well as circadian drives, and both are tightly linked to thermoregulation. In this chapter, we review the basic drain circuitry that regulates sleep and wakefulness, including the flip-flop switch relationship of the arousal system and the ventrolateral preoptic sleep-promoting neurons. We then examine the role of the orexin/hypocretin neurons, which stabilize the switch while driving both wakefulness and foraging for food. We also review the role of the subparaventricular nucleus and the dorsomedial nucleus of the hypothalamus in circadian integration and modulation of both feeding and wake-sleep patterns.

One might legitimately question why a chapter on sleep and circadian rhythms might be included in a book on the regulation of body weight and energy metabolism. However, the daily patterns of sleep and feeding are inevitably intertwined. After all, animals can only eat when they are awake, and the first thing they usually do after a meal is to lie down and go to sleep. When animals are hungry, they must be awake to forage for food, and as we shall see, limited availability of food can have profound effects on the patterns of daily sleep-wake cycles (Willie et al., 2001). In this chapter, we will first consider the circuitry that governs sleep and wakefulness, and its intimate relationship with feeding circuitry. We will then review the pathways that shape daily cycles of activity, and how circadian rhthyms may be modified to adjust to limited food availability.

Regulation of sleep and wakefulness: the flip–flop switch model

The waking state of the brain is maintained by a series of pathways that begin in the upper brainstem, and which ultimately result in thalamo-cortical activation (Fig. 1; see Saper et al., 2001, 2005c) for review). Although classically these pathways were attributed to a "reticular activating system" in fact, most of the ascending pathways to key targets originate in highly restricted populations of neurons with known neurotransmitters and connections, not in the more diffuse reticular formation. Input to the relay nuclei and the reticular nucleus of the thalamus, for example originates predominantly from two cholinergic cell groups in the mesopontine tegmentum, the laterodorsal and pedunculopontine tegmental nuclei (Levey et al., 1987; Rye et al., 1987). Inputs to the intralaminar and midline thalamic nuclei originate somewhat more broadly, including these cholinergic groups as well as the noradrenergic

*Corresponding author. Tel.: +1-617-667-2622; Fax: +1-617-975-5161; E-mail: csaper@bidmc.harvard.edu

DOI: 10.1016/S0079-6123(06)53014-6

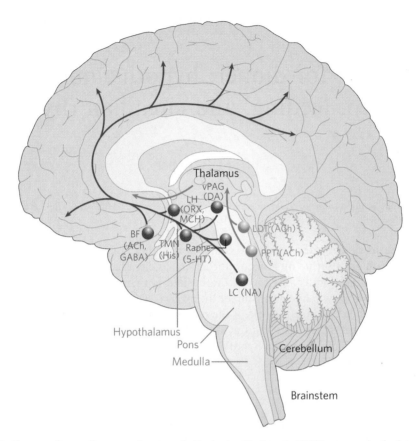

Fig. 1. A schematic diagram of ascending arousal systems in the brain. Cholinergic (ACh) neurons in the laterodorsal (LDT) and pedunculopontine tegmental nuclei (PPT), shown in yellow, provide the major input to the thalamic reticular nucleus and relay nuclei, which are critical for thalamocortical transmission. A series of monoaminergic nuclei including the noradrenergic (NA) locus coeruleus (LC), serotoninergic (5-HT) dorsal and median raphe nuclei, dopaminergic (DA) ventral periaqueductal gray matter (vPAG), and histaminergic (HA) tuberomammillary nucleus (TMN) contribute to an ascending arousal pathway, shown in red, that diffusely innervates the cerebral hemispheres. There are additional contributions to this pathway from neurons in the lateral hypothalamus (LH) containing the peptides orexin (ORX) and melanin-concentrating hormone (MCH), as well as basal forebrain (BF) neurons containing acetylcholine or GABA. These pathways are critical for maintaining cortical arousal. Reprinted with permission from Saper et al. (2005c).

locus coeruleus (LC), dopaminergic ventral periauqeductal gray matter (vPAG), and the serotoninergic dorsal and median raphe nuclei (DR/MR), as well as neurons in the mesopontine reticular formation (Krout et al., 2002).

A second ascending pathway originates mainly in the monoaminergic vPAG, LC and DR/MR, and is augmented by histaminergic neurons in the tuberomammillary nucleus in the caudal hypothalamus (Saper, 1985; Jones, 2003; Saper et al., 2005c; Lu et al., 2006). These cells are joined by peptidergic neurons in the lateral hypothalamus containing orexin and melanin-concentrating hormone (Bittencourt et al., 1992; Peyron et al., 1998), and both cholinergic and GABAergic basal forebrain neurons in providing a direct input to the cerebral cortex (Rye et al., 1984; Gritti et al., 1997).

Neurons contributing to both of these pathways tend to be active during wakefulness (Aston-Jones and Bloom, 1981; Trulson and Jacobs, 1983; Steininger et al., 1999; Lee et al., 2004, 2005; Mileykovskiy et al., 2005) During slow-wave sleep the firing of neurons in both pathways slows. However, during rapid eye movement

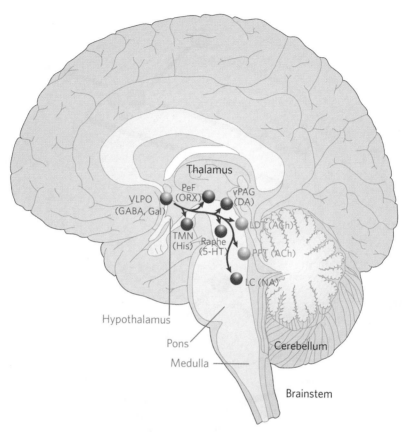

Fig. 2. The ventrolateral preoptic nucleus (VLPO) inhibits the major ascending arousal systems. It consists of neurons that contain the inhibitory neurotransmitters galanin (Gal) and GABA. Descending efferents inhibit neurons in the monoaminergic and orexin-containing components of the ascending arousal system. VLPO terminals also end in the PPT and LDT, but on interneurons, not on the principal cholinergic neurons. Reprinted with permission from Saper et al. (2005c).

(REM) sleep, there is an increase in firing in the cholinergic input to the thalamus, while the monoaminergic systems come to a halt. This fundamental disparity in the firing of the two ascending arousal systems is what characterizes REM sleep, with an active thalamic input driving cortical activity and EEG, but in the absence of monoaminergic input the activity is interpreted as dreams.

The widespread reduction in firing of the arousal systems during sleep suggested a switching mechanism that would actively inhibit them. Retrograde transport experiments identified a set of neurons in the ventrolateral preoptic nucleus (VLPO), containing the inhibitory neurotransmitters galanin and GABA, which innervate all of the major brainstem components of the ascending

arousal system (Fig. 2; Sherin et al., 1996, 1998; Saper et al., 2005c). These galaninergic neurons show Fos expression during sleep (Sherin et al., 1996; Gaus et al., 2002), and single unit recording experiments show that they fire about twice as fast during sleep as during wakefulness (Szymusiak et al., 1998).

The VLPO is also innervated by the ascending arousal systems and is inhibited by serotonin and noradrenalin (Gallopin et al., 2000; Chou et al., 2002). It does not contain histamine receptors, but the histamine neurons also contain the inhibitory neurotransmitters GABA, endomorphin, and galanin (Vincent et al., 1983; Kohler et al., 1986; Martin-Schild et al., 1999). This relationship of mutual inhibition is self-reinforcing, in that if either side gains the advantage, it shuts off the

246

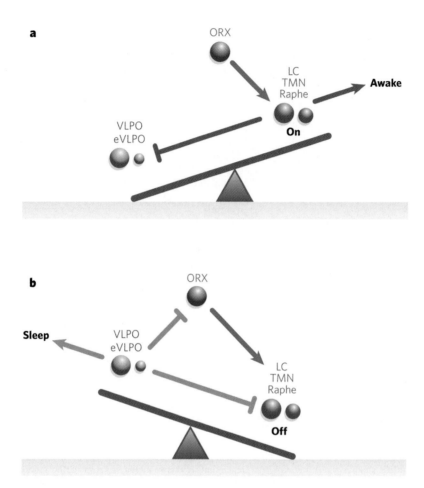

Fig. 3. A schematic diagram illustrating the flip–flop switch relationship of the VLPO and the major monoaminergic systems as well as the stabilizing role of the orexin neurons. During wakefulness (a), the monoaminergic systems inhibit the neurons in the VLPO core and the extended part of the VLPO (eVLPO). Conversely, during sleep, the orexin neurons inhibit the monoaminergic systems. This mutual inhibition creates a relationship known as a flip–flop switch, which sharpens phase transitions, but makes the phases potentially unstable. Orexin neurons provide excitatory inputs to the monoaminergic nuclei, which may promote and stabilize wakefulness. Inhibition of the orexin neurons by the VLPO during sleep may stabilize the sleep state. Loss of the orexin neurons results in state instability known as narcolepsy. Reprinted with permission from Saper et al.(2005c).

other, and disinhibits its own firing (Fig. 3; Saper et al., 2001, 2005c; Chou, 2003). This type of relationship is termed by electrical engineers a "flip–flop switch" and it is designed into a circuit when there is a need to produce sharp transitions in state. Sharp boundaries and rapid transitions in behavioral state are indeed seen in animals, which spend nearly all of their time either awake or asleep, and very little in between. Such an arrangement is valuable, as animals that walked around half-asleep would be vulnerable to predators or other dangers in the environment.

Another property of flip–flop switches is revealed when cell-specific lesions are made in the VLPO. Injections of ibotenic acid that kill 70% or more of the VLPO neurons reduce total sleep time by more than 50%, underscoring the importance of this cell group for causing sleep (Lu et al., 2000). The animals with VLPO lesions have normal circadian patterns of sleep, but tend to wake up more often during their sleep cycle and sleep less than one-quarter as much per bout. However, they fall asleep much more often during the wake cycle as well. They are chronically sleep-deprived and sleepy, but unable to stay asleep. This increase in the frequency of state transitions in both the wake and the sleep state is mathematically predicted by computer modeling of the relationship of mutual inhibition, and is a hallmark of flip–flop switches (Chou, 2003).

Role of the orexin neurons in behavioral state regulation

In 1998, two groups of investigators simultaneously discovered a population of neurons in the lateral hypothalamus that produces a pair of closely related peptide neurotransmitters which one group called orexins and the other hypocretins (Sakurai et al., 1998; Sutcliffe, 1998). A year later, Yanagisawa and colleagues reported that mice in which the orexin gene had been deleted had a phenotype of narcolepsy (Chemelli et al., 1999), and Mignot and coworkers at the same time identified the gene defect in inherited canine narcolepsy as the orexin type 2 receptor (Lin et al., 1999). Thus, within a very brief period of time, the molecular basis of this puzzling disorder was identified.

Humans or animals with narcolepsy are very sleepy, falling asleep frequently during the normal wake cycle. In addition, they have episodes of cataplexy, often stimulated by emotional activity, in which they lose muscle control and collapse to the ground (Scammell, 2003). The loss of muscle tone is similar to patients who are in REM sleep, and indeed patients with narcolepsy tend to enter REM sleep much more quickly during a brief nap than normal subjects.

The function of the orexin neurons can best be appreciated from considering their connections. In general, they innervate the same brainstem arousal system targets as the VLPO (Peyron et al., 1998; Chemelli et al., 1999). The VLPO neurons also send axons to and presumably inhibit the orexin neurons, but VLPO neurons do not have orexin receptors (Marcus et al., 2001). Thus, the orexin neurons appear to be apart from the flip–flop circuit, and to interact with it mainly by increasing the strength of the arousal systems (Fig. 3). This acts like a "finger" on the flip-flop switch,

stabilizing it, and holding it in the wake position, allowing consolidated wakefulness. As predicted, when either side of the flip–flop switch is weakened, the state transitions become more frequent in both states. This is the case in narcolepsy, where not only do humans and animals fall asleep more frequently while awake, but they also wake more often when asleep.

Fos studies found that the activity of the perifornical orexin neurons is increased during wakefulness (Estabrooke et al., 2001), and these observations have been confirmed by single unit recordings (Lee et al., 2005; Mileykovskiy et al., 2005). Interestingly, however, the firing of the orexin neurons is most dramatically increased during locomotor behavior, in particular exploration of the environment. This findings correlates closely with the greatly reduced levels of locomotor activity of the animals who have genetic defects in orexin signaling.

In addition to innervating the arousal systems, the orexin neurons also send projections to the nucleus accumbens and to the dopaminergic neurons in the ventral tegmental area. These relationships have suggested a possible role in reward and addiction. Neurons in the lateral orexin field (but not the medial field) show increased Fos expression during conditioned place preference testing for food and drugs, including morphine and amphetamine (Harris et al., 2005). These neurons receive a preponderance of inputs from cortical and limbic sources, compared to the medial orexin field (Yoshida et al., 2005).

Interestingly, the neurons in the medial, perifornical orexin field receive inputs from the feeding circuitry in the arcuate nucleus (Elias et al., 1998; Yoshida et al., 2006). They respond to low glucose and ghrelin with increased firing and to leptin with decreased firing (Yamanaka et al., 2003). During food deprivation, they also show increased Fos and orexin expression, in association with an increase in locomotor activity (Cai et al., 2001; Willie et al., 2001).

As foraging for food is a key behavior for rats that are hungry, the orexin neurons appear to play a critical role in promoting the necessary motor responses. These observations suggest that the orexin neurons may play an intermediate role in motivated behavior, between internal states that recognize the need for specific behaviors, and motor systems that must be activated for those behaviors to be expressed.

The hypothalamic integrator for circadian rhythms

Just as staying awake is an intrinsic component of regulation of feeding, the availability of food is a critical variable for shaping wake–sleep cycles. For example, in the 1950s Nyholm examined the behavior of Finnish bats over a summer (Nyholm, 1955; Saper et al., 2005b). He found that these quintessentially nocturnal animals were active exclusively during the dark cycle in the months from June to September, when the predator bird species were active by day and the weather in the evenings was sufficiently warm that there were flying insects for them to eat. However, in the early Spring and late Fall, it was too cold at night for insects to be active, and predator birds had migrated to warmer areas. As a result, in those parts of the year the activity cycle of the bats gradually shifted to earlier times, until the animals were predominantly active during the daylight.

Similarly, laboratory rats, which are typically nocturnal, can be changed into diurnal animals by simply restricting their food intake to a time window during the daylight hours (Stephan, 2002). If given a 4-h opportunity to eat only at midday, the animals will become active a few hours before the food is presented and remain active during the feeding period and briefly thereafter, but sleep during most of the dark cycle. Their cycles of body temperature similarly shift to begin rising a few hours before the food presentation, instead of a few hours before lights out. Cycles of corticosteroid secretion also make this shift. Interestingly, if the animals are food deprived at the end of the experiment, they continue to show the anticipatory rise in wakefulness, activity, and body temperature, beginnng a few hours before the anticipated meal (Saper et al., 2005a, c; Gooley et al., 2006). This persists for at least 2–3 days in the absence of any food, indicating that it is a circadian phenomenon.

These observations suggest that the daily patterns of sleep and wakefulness, activity, body temperature, and hormones are not fixed, but rather can be adjusted by neural circuitry within the brain. One hypothesis was that the activity of the suprachiasmatic nucleus, which serves as the brains master clock, was shifted by restricted feeding. However, the rhythmic biochemical cycle that drives the activity of the suprachiasmatic nucleus has been found not to be shifted (Gooley et al., 2006).

The major output from the suprachiasmatic nucleus projects dorsally and caudally from it, in an arc along the wall of the third ventricle (Fig. 4, Watts et al., 1987; Saper et al., 2005a, c). Its largest target has been termed the subparaventricular zone, but many axons continue back from this area into the dorsomedial nucleus of the hypothalamus (Chou et al., 2003). Cell-specific lesions of the ventral subparaventricular zone (just dorsal to the suprachiasmatic nucleus) eliminated the circadian rhythms of sleep–wakefulness, locomotor activity, and feeding, but only modestly affected rhythms of body temperature when animals were placed in a continuously dark environment, without temporal cues (Lu et al., 2001). However, lesions of the dorsal subparaventricular zone (just ventral to the paraventricular nucleus) caused loss of circadian rhythms of body temperature, without affecting wake–sleep or locomotor activity. These experiments demonstrated that the timing cues provided by the suprachiasmatic nucleus (which drives the circadian cycle in a dark environment) must be relayed by neurons in the subparaventricular zone, and that the circuitry subserving circadian regulation of wake–sleep and locomotor activity could be dissociated from that of body temperature. Both the suprachiasmatic nucleus and the subparaventricular zone send only minor inputs to the VLPO (Chou et al., 2002; Deurveilher and Semba, 2005), a critical target for sleep regulation. However, both regions project heavily to the dorsomedial nucleus of the hypothalamus, which is a major source of VLPO inputs (Chou et al., 2003).

Cell-specific lesions of the dorsomedial nucleus eliminate circadian cycles of wake–sleep, feeding, locomotor activity, and corticosteroid secretion, when animals are placed in a continuously dark environment (Chou et al., 2003). Thus, the

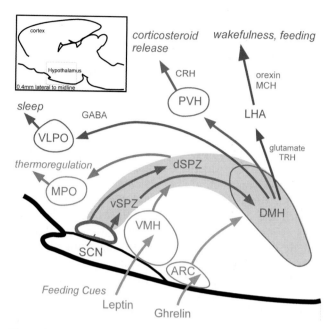

Fig. 4. A schematic diagram illustrating the circadian integrator pathways in rat brain. Most outputs from the suprachiasmatic nucleus (SCN) end in the subparaventricular zone (SPZ) and dorsomedial nucleus of they hypothalamus (DMH), shown here as a corridor by orange shading. Neurons in the ventral SPZ (vSPZ) are critical for relaying circadian input to the DMH that regulate wake–sleep states, locomotor activity, and feeding. The DMH then sends outputs to the ventrolateral preoptic nucleus (VLPO) regulating sleep and to the orexin and melanin-concentrating hormone (MCH) neurons in the lateral hypothalamic area (LHA) regulating wakefulness, feeding, and locomotor activity. There are also outputs to the paraventricular nucleus (PVH) regulating corticotropin-releasing hormone (CRH) and corticosteroid secretion. Outputs from the SCN to the dorsal SPZ (dSPZ) are critical for circadian regulation of body temperature, presumably my means of outputs to the medial preoptic region (MPO). Reprinted with permission from Saper et al. (2005a).

dorsomedial nucleus must be another node in the pathway that relays suprachiasmatic influence over these activities. The dorsomedial nucleus, in turn, provides massive outputs to the lateral hypothalamus including the orexin neurons, to the VLPO, to the corticotropin-releasing hormone neurons of the paraventricular nucleus, and to the medial preoptic region that regulates body temperature (Chou et al., 2003). The projections to the lateral hypothalamus come from neurons that contain glutamate and thyrotropin-releasing hormone and are thus presumably excitatory. The input to the VLPO originates from GABAergic neurons in the dorsomedial nucleus, and is therefore likely to be inhibitory. Thus, the net effect of the dorsomedial nucleus appears to be to activate wakefulness, arousal, feeding, locomotor activity, and corticosteroid secretion during the active part of the circadian cycle.

Experiments examining Fos expression in the dorsomedial nucleus across the wake–sleep cycle show that its neurons are maximally active during the active cycle (Saper et al., 2005a, c; Gooley et al., 2006). This is not, however, a diurnal phenomenon. If animals are placed under restricted feeding, thus shifting their activity to the light cycle, the Fos expression in the dorsomedial nucleus begins prior to the anticipated time of the meal, and correlates with the onset of wakefulness and locomotor activity (Angeles-Castellanos et al., 2004; Gooley et al., 2006).

To determine whether the activity of the dorsomedial nucleus is necessary for the entrainment of circadian patterns to the restricted feeding period, cell-specific lesions were placed in the dorsomedial nucleus and surrounding cell groups. Only lesions of the dorsomedial nucleus resulted in 75–80% reductions in the shifts in activity, wake–sleep, and temperature patterns induced by restricted feeding (Saper et al., 2005a, c; Gooley et al., 2006). Moreover, the degree of the shift correlated with the numbers of remaining neurons in the dorsomedial nucleus.

The orexin neurons are a major target of the dorsomedial nucleus. The expression of orexin also increases in anticipation of the meal during restricted feeding, and the anticipatory locomotor behavior is reduced in animals that have a genetic lesion of the orexin neurons (Akiyama et al., 2004). Thus, the activation of the orexin neurons may be one pathway by which the dorsomedial nucleus drives activity cycles.

What sensory signals mediate the shift in activity patterns in response to restricted feeding? It has been reported that animals with subdiaphragmatic vagotomy and Zucker fatty rats that lack leptin receptors both entrain to restricted feeding (Comperatore and Stephan, 1990; Mistlberger and Marchant, 1999). Thus, neither vagal signals nor leptin is critical for the entrainment. In preliminary experiments in our laboratory, mice lacking ghrelin receptors also entrained normally to restricted feeding (Gooley and Zigman, unpublished results). It has been reported that lesions of the parabrachial nucleus can prevent entrainment, presumably by blocking ascending visceral sensory input to the forebrain, but in preliminary experiments we were unable to replicate this finding even with large, cell-specific lesions of the parabrachial nucleus (Davidson et al., 2000). Lesions of the ventromedial nucleus of the hypothalamus have also been reported to reduce entrainment to restricted feeding (Mistlberger and Rechtschaffen, 1984; Choi et al., 1998). But in our experiments, ventromedial nucleus lesions only affected entrainment to the extent that they also involved the dorsomedial nucleus (Gooley et al., 2006). Thus, while it is possible that some yet untested single input may regulate entrainment to restricted feeding, it seems likely at this point that the dorsomedial nucleus can read a wide range of signals related to energy metabolism in determining the optimal pattern of circadian physiology.

Food availability is probably just one of the range of environmental, physiological, and cognitive/emotional variables that can affect the patterning of circadian rhythms. For example, in our laboratory we found that degus, a South American rodent, may be more active at night when the weather is warm, and during the day when the weather is cool. Others have reported that these animals may invert their day–night rhythms of activity and body temperature when allowed access to a running wheel (Kas and Edgar, 1999). In humans, where cognitive and emotional stimuli can be more complex, the voluntary changes in circadian patterns in shift workers, and the degradation of the normal patterns in people who are depressed, may represent alterations of the circadian pattern downstream of the suprachiasmatic nucleus, in either the subparaventricular zone or the dorsomedial nucleus. This three-stage integrator gives the circadian system the opportunity to adjust to a wide range of environmental exigencies, resulting in patterns of circadian activity of wakefulness, feeding, and body temperature that is most adaptive for survival.

References

Akiyama, M., Yuasa, T., Hayasaka, N., Horikawa, K., Sakurai, T. and Shibata, S. (2004) Reduced food anticipatory activity in genetically orexin (hypocretin) neuron-ablated mice. Eur. J. Neurosci., 20: 3054–3062.

Angeles-Castellanos, M., Aguilar-Roblero, R. and Escobar, C. (2004) c-Fos expression in hypothalamic nuclei of food-entrained rats. Am. J. Physiol. Regul. Integr. Comp. Physiol., 286: R158–R165.

Aston-Jones, G. and Bloom, F.E. (1981) Activity of norepinephrine-containing locus coeruleus neurons in behaving rats anticipates fluctuations in the sleep–waking cycle. J. Neurosci., 1: 876–886.

Bittencourt, J.C., Presse, F., Arias, C., Peto, C., Vaughan, J., Nahon, J.L., Vale, W. and Sawchenko, P.E. (1992) The melanin-concentrating hormone system of the rat brain: an immuno- and hybridization histochemical characterization. J. Comp. Neurol., 319: 218–245.

Cai, X.J., Evans, M.L., Lister, C.A., Leslie, R.A., Arch, J.R., Wilson, S. and Williams, G. (2001) Hypoglycemia activates orexin neurons and selectively increases hypothalamic orexin-B levels: responses inhibited by feeding and possibly mediated by the nucleus of the solitary tract. Diabetes, 50: 105–112.

Chemelli, R.M., Willie, J.T., Sinton, C.M., Elmquist, J.K., Scammell, T., Lee, C., Richardson, J.A., Williams, S.C., Xiong, Y.M., Kisanuki, Y., Fitch, T.E., Nakazato, M., Hammer, R.E., Saper, C.B. and Yanagisawa, M. (1999)

Narcolepsy in orexin knockout mice: molecular genetics of sleep regulation. Cell, 98: 437–451.

Choi, S., Wong, L.S., Yamat, C. and Dallman, M.F. (1998) Hypothalamic ventromedial nuclei amplify circadian rhythms: do they contain a food-entrained endogenous oscillator? J. Neurosci., 18: 3843–3852.

Chou, T.C. (2003) A bistable model of sleep–wake regulation. In: Regulation of Wake–Sleep Timing: Circadian Rhythms and Bistability of Sleep–Wake States. Ph.D. dissertation. Harvard University, Cambridge, MA, pp. 82–99.

Chou, T.C., Bjorkum, A.A., Gaus, S.E., Lu, J., Scammell, T.E. and Saper, C.B. (2002) Afferents to the ventrolateral preoptic nucleus. J. Neurosci., 22: 977–990.

Chou, T.C., Scammell, T.E., Gooley, J.J., Gaus, S.E., Saper, C.B. and Lu, J. (2003) Critical role of dorsomedial hypothalamic nucleus in a wide range of behavioral circadian rhythms. J. Neurosci., 23: 10691–10702.

Comperatore, C.A. and Stephan, F.K. (1990) Effects of vagotomy on entrainment of activity rhythms to food access. Physiol. Behav., 47: 671–678.

Davidson, A.J., Cappendijk, S.L. and Stephan, F.K. (2000) Feeding-entrained circadian rhythms are attenuated by lesions of the parabrachial region in rats. Am. J. Physiol. Regul. Integr. Comp. Physiol., 278: R1296–R1304.

Deurveilher, S. and Semba, K. (2005) Indirect projections from the suprachiasmatic nucleus to major arousal-promoting cell groups in rat: implications for the circadian control of behavioural state. Neuroscience, 130: 165–183.

Elias, C.F., Saper, C.B., Maratos-Flier, E., Tritos, N.A., Lee, C., Kelly, J., Tatro, J.B., Hoffman, G.E., Ollmann, M.M., Barsh, G.S., Sakurai, T., Yanagisawa, M. and Elmquist, J.K. (1998) Chemically defined projections linking the mediobasal hypothalamus and the lateral hypothalamic area. J. Comp. Neurol., 402: 442–459.

Estabrooke, I.V., McCarthy, M.T., Ko, E., Chou, T.C., Chemelli, R.M., Yanagisawa, M., Saper, C.B. and Scammell, T.E. (2001) Fos expression in orexin neurons varies with behavioral state. J. Neurosci., 21: 1656–1662.

Gallopin, T., Fort, P., Eggermann, E., Cauli, B., Luppi, P.H., Rossier, J., Audinat, E., Muhlethaler, M. and Serafin, M. (2000) Identification of sleep-promoting neurons in vitro. Nature, 404: 992–995.

Gaus, S.E., Strecker, R.E., Tate, B.A., Parker, R.A. and Saper, C.B. (2002) Ventrolateral preoptic nucleus contains sleep-active, galaninergic neurons in multiple mammalian species. Neuroscience, 115: 285–294.

Gooley, J.J., Schomer, A. and Saper, C.B. (2006) The dorsomedial nucleus of the hypothalamus is critical for the expression of food entrainable rhythms. Nat. Neurosci., 9: 398–407.

Gritti, I., Mainville, L., Mancia, M. and Jones, B.E. (1997) GABAergic and other noncholinergic basal forebrain neurons, together with cholinergic neurons, project to the mesocortex and isocortex in the rat. J. Comp. Neurol., 383: 163–177.

Harris, G.C., Wimmer, M. and Aston-Jones, G. (2005) A novel role for lateral hypothalamic orexin neurons in reward seeking. Nat. Neurosci., 437: 556–559.

Jones, B.E. (2003) Arousal systems. Front. Biosci., 8: 438–451.

Kas, M.J. and Edgar, D.M. (1999) A nonphotic stimulus inverts the diurnal–nocturnal phase preference in Octodon degus. J. Neurosci., 19: 328–333.

Kohler, C., Ericson, H., Watanabe, T., Polak, J., Palay, S.L., Palay, V. and Chan-Palay, V. (1986) Galanin immunoreactivity in hypothalamic neurons: further evidence for multiple chemical messengers in the tuberomammillary nucleus. J. Comp. Neurol., 250: 58–64.

Krout, K.E., Belzer, R.E. and Loewy, A.D. (2002) Brainstem projections to midline and intralaminar thalamic nuclei of the rat. J. Comp. Neurol., 448: 53–101.

Lee, M.G., Hassani, O.K. and Jones, B.E. (2005) Discharge of identified orexin/hypocretin neurons across the sleep-waking cycle. J. Neurosci., 25: 6716–6720.

Lee, M.G., Manns, I.D., Alonso, A. and Jones, B.E. (2004) Sleep–wake related discharge properties of basal forebrain neurons recorded with micropipettes in head-fixed rats. J. Neurophysiol., 92: 1182–1198.

Levey, A.I., Hallanger, A.E., Rye, D.B. and Wainer, B.H. (1987) Thalamic cholinergic innervation. J. Comp. Neurol., 262: 105–124.

Lin, L., Faraco, J., Li, R., Kadotani, H., Rogers, W., Lin, X., Qiu, X., de Jong, P.J., Nishino, S. and Mignot, E. (1999) The sleep disorder canine narcolepsy is caused by a mutation in the hypocretin (orexin) receptor 2 gene. Cell, 98: 365–376.

Lu, J., Chou, T.C. and Saper, C.B. (2006). Identification of wake-active dopaminergic neurons in the ventral periqueductal gray matter (PAG). J. Neurosci., 26: 193–202.

Lu, J., Greco, M.A., Shiromani, P. and Saper, C.B. (2000) Effect of lesions of the ventrolateral preoptic nucleus on NREM and REM sleep. J. Neurosci., 20: 3830–3842.

Lu, J., Zhang, Y.H., Chou, T.C., Gaus, S.E., Elmquist, J.K., Shiromani, P. and Saper, C.B. (2001) Contrasting effects of ibotenate lesions of the paraventricular nucleus and subparaventricular zone on sleep–wake cycle and temperature regulation. J. Neurosci., 21: 4864–4874.

Marcus, J.N., Aschkenasi, C.J., Lee, C.E., Chemelli, R.M., Saper, C.B., Yanagisawa, M. and Elmquist, J.K. (2001) Differential expression of orexin receptors 1 and 2 in the rat brain. J. Comp. Neurol., 435: 6–25.

Martin-Schild, S., Gerall, A.A., Kastin, A.J. and Zadina, J.E. (1999) Differential distribution of endomorphin 1- and endomorphin 2-like immunoreactivities in the CNS of the rodent. J. Comp. Neurol., 405: 450–471.

Mileykovskiy, B.Y., Kiyashchenko, L.I. and Siegel, J.M. (2005) Behavioral correlates of activity in identified hypocretin/orexin neurons. Neuron, 46: 787–798.

Mistlberger, R.E. and Marchant, E.G. (1999) Enhanced food-anticipatory circadian rhythms in the genetically obese Zucker rat. Physiol. Behav., 66: 329–335.

Mistlberger, R.E. and Rechtschaffen, A. (1984) Recovery of anticipatory activity to restricted feeding in rats with ventromedial hypothalamic lesions. Physiol. Behav., 33: 227–235.

Nyholm, H. (1955) Zur Okologie von Myotis mystacinus (Leisl.) und M. daubentoni (Leisl.) (Chiroptera). Ann. Zool. Fenn., 2: 77–123.

Peyron, C., Tighe, D.K., van den Pol, A.N., de Lecea, L., Heller, H.C., Sutcliffe, J.G. and Kilduff, T.S. (1998) Neurons containing hypocretin (orexin) project to multiple neuronal systems. J. Neurosci., 18: 9996–10015.

Rye, D.B., Saper, C.B., Lee, H.J. and Wainer, B.H. (1987) Pedunculopontine tegmental nucleus of the rat: cytoarchitecture, cytochemistry, and some extrapyramidal connections of the mesopontine tegmentum. J. Comp. Neurol., 259: 483–528.

Rye, D.B., Wainer, B.H., Mesulam, M.M., Mufson, E.J. and Saper, C.B. (1984) Cortical projections from the basal forebrain: a study of cholinergic and non-cholinergic components employing combined retrograde tracing and immunohistochemical localization of choline acetyltransferase. Neuroscience, 13: 627–643.

Sakurai, T., Amemiya, A., Ishii, M., Matsuzaki, I., Chernelli, R.M., Tanaka, H., Williams, S.C., Richardson, J.A., Kozlowski, G.P., Wilson, S., Arch, J.R.S., Buckingham, R.E., Haynes, A.C., Carr, S.A., Annan, R.S., McNulty, D.E., Liu, W.S., Terrett, J.A., Eishourbahy, N.A., Bergsma, D.J. and Yangisawa, M. (1998) Orexins and orexin receptors: a family of hypothalamic neuropeptides and G protein-coupled receptors that regulate feeding behavior. Cell, 92: 573–585.

Saper, C.B. (1985) Organization of cerebral cortical afferent systems in the rat. II. Hypothalamocortical projections. J. Comp. Neurol., 237: 21–46.

Saper, C.B., Cano, G. and Scammell, T.E. (2005a) Homeostatic, circadian, and emotional regulation of sleep. J. Comp. Neurol., 493: 92–98.

Saper, C.B., Chou, T.C. and Scammell, T.E. (2001) The sleep switch: hypothalamic control of sleep and wakefulness. Trends Neurosci., 24: 726–731.

Saper, C.B., Lu, J., Chou, T.C. and Gooley, J. (2005b) The hypothalamic integrator for circadian rhythms. Trends Neurosci., 28: 152–157.

Saper, C.B., Scammell, T.E. and Lu, J. (2005c) Hypothalamic regulation of sleep and circadian rhythms. Nature, 437: 1257–1263.

Scammell, T.E. (2003) The neurobiology, diagnosis, and treatment of narcolepsy. Ann. Neurol., 53: 154–166.

Sherin, J.E., Elmquist, J.K., Torrealba, F. and Saper, C.B. (1998) Innervation of histaminergic tuberomammillary neurons by GABAergic and galaninergic neurons in the ventrolateral preoptic nucleus of the rat. J. Neurosci., 18: 4705–4721.

Sherin, J.E., Shiromani, P.J., McCarley, R.W. and Saper, C.B. (1996) Activation of ventrolateral preoptic neurons during sleep. Science, 271: 216–219.

Steininger, T.L., Alam, M.N., Gong, H., Szymusiak, R. and McGinty, D. (1999) Sleep–waking discharge of neurons in the posterior lateral hypothalamus of the albino rat. Brain Res., 840: 138–147.

Stephan, F.K. (2002) The "other" circadian system: food as a Zeitgeber. J. Biol. Rhythms, 17: 284–292.

Sutcliffe, J.G. (1998) The hypocretins: hypothalamus-specific peptides with neuroexcitatory. Proc. Natl. Acad. Sci. USA, 95: 322–327.

Szymusiak, R., Alam, N., Steininger, T.L. and McGinty, D. (1998) Sleep-waking discharge patterns of ventrolateral preoptic/anterior hypothalamic neurons in rats. Brain Res., 803: 178–188.

Trulson, M.E. and Jacobs, B.L. (1983) Raphe unit activity in freely moving cats: lack of diurnal variation. Neurosci. Lett., 36: 285–290.

Vincent, S.R., Hokfelt, T., Skirboll, L.R. and Wu, J.-Y. (1983) Hypothalamic gamma-aminobutyric acid neurons project to the neocortex. Science, 220: 1309–1311.

Watts, A.G., Swanson, L.W. and Sanchez-Watts, G. (1987) Efferent projections of the suprachiasmatic nucleus: I. Studies using anterograde transport of *Phaseolus vulgaris* leucoagglutinin in the rat. J. Comp. Neurol., 258: 204–229.

Willie, J.T., Chemelli, R.M., Sinton, C.M. and Yanagisawa, M. (2001) To eat or to sleep? Orexin in the regulation of feeding and wakefulness. Annu. Rev. Neurosci., 24: 429–458.

Yamanaka, A., Beuckmann, C.T., Willie, J.T., Hara, J., Tsujino, N., Mieda, M., Tominaga, M., Yagami, K., Sugiyama, F., Goto, K., Yanagisawa, M. and Sakurai, T. (2003) Hypothalamic orexin neurons regulate arousal according to energy balance in mice. Neuron, 38: 701–713.

Yoshida, K., McCormack, S., Espana, R.A., Crocker, A. and Scammell, T.E. (2006) Afferents to the orexin neurons. J. Comp. Neurol., 494: 845–861.

Kalsbeek, Fliers, Hofman, Swaab, Van Someren & Buijs
Progress in Brain Research, Vol. 153
ISSN 0079-6123

CHAPTER 15

Circadian timing in health and disease

Elizabeth S. Maywood*, John O'Neill, Gabriel K.Y. Wong, Akhilesh B. Reddy
and Michael H. Hastings

MRC Laboratory of Molecular Biology, Hills Rd, Cambridge CB2 2QH, UK

Abstract: Metabolic status varies predictably on a daily and seasonal basis in order to adapt to the cyclical environment. The hypothalamic circadian pacemaker of the suprachiasmatic nuclei (SCN) co-ordinates these metabolic cycles. Circadian timing is based upon a transcriptional/post-translational negative feedback loop involving a series of core clock genes and their products. Local molecular clocks in peripheral tissues are synchronised by a variety of autonomic, paracrine and endocrine cues reflective of SCN time, thereby ensuring internal temporal co-ordination and optimal metabolic function. Disturbances of this co-ordination, as occur in long-term shift work, have a major impact on health.

The impact of circadian timing on metabolic function is immediately obvious from the sleep/wake cycle, the most dramatic and pronounced circadian rhythm and one which dominates our lives. On a periodic basis that is tightly regulated by intrinsic biological clocks, our internal state switches from catabolism as we engage with the world during the day to anabolism as we withdraw from it during sleep. In nocturnal species, the metabolic changes are no less profound and the timing mechanisms are equally precise. The only difference is the relationship of sleep and wakefulness to solar time. The aim of this article is to describe recent developments in our understanding of the cellular and molecular basis of circadian timing, and finally to consider its relevance to disease, with a particular focus on neurological disorders.

The suprachiasmatic nuclei as central pacemaker

Circadian rhythms are those daily cycles of physiology and behaviour that persist with a period of approximately 24 h when the individual is deprived of external time cues. They are generated, therefore, by an intrinsic pacemaker, the principal one in mammals being the suprachiasmatic nuclei (SCN) of the hypothalamus (Fig. 1a) (Weaver, 1998). These bilateral clusters of ca 10,000 neurons exhibit spontaneous metabolic and electrical rhythms in vivo and in vitro. Ablation of the SCN disrupts circadian patterns of behaviour and metabolism, including the daily cycle of core body temperature (Fig. 1b). Note, however, that mean body temperature does not change, indicating that basic homeostasis is not the domain of the SCN, rather its temporal modulation. Importantly, grafts of SCN tissue to SCN-ablated rodents restore circadian activity profiles and the restored behaviour reflects the genotypically determined period of the donor tissue (Ralph et al., 1990; King et al., 2003) proving that the graft is the origin of the recipients restored circadian time. A critical observation, however, is that SCN grafts do not restore circadian patterning to myriad other functions, such as corticosteroid and melatonin secretion, and core body temperature (Meyer-Bernstein et al., 1999). Although local outputs from the SCN graft, possibly paracrine factors, may therefore be sufficient to regulate activity/rest cycles, it is clear

*Corresponding author. Tel.:+44-1223-402307; Fax: 44-1223-402310; E-mail: mha@mrc-lmb.cam.ac.uk

DOI: 10.1016/S0079-6123(06)53015-8

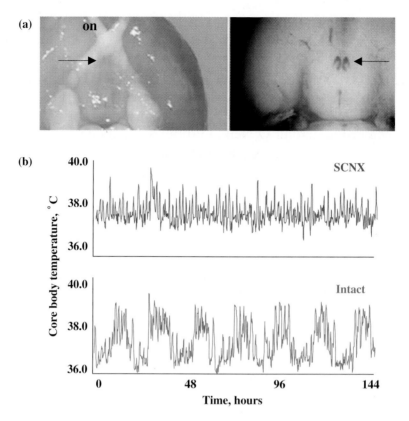

Fig. 1. The suprachiasmatic nuclei and circadian metabolism. (a) Ventral views of mouse brain to reveal location of SCN (arrows) in unstained tissue (left panel) and in a transgenic where SCN neurons express beta-galactosidase, revealed by blue stain (King et al., 2003). On, optic nerve. (b) Representative recordings of core-body temperature from intact and SCN-ablated (SCNX) mice, free-running in continuous dim red light. Note loss of circadian pattern in SCNX.

that intact neural and neuroendocrine pathways are necessary for the SCN to control autonomic and endocrine systems that mediate circadian metabolic cycles (see below and Kalbeek this volume).

The SCN receive a direct retinal innervation that is specialised for luminance coding rather than feature detection: appropriate to a system dedicated to the monitoring of daylight (Peirson et al., 2005). This retino-hypothalamic tract (RHT) arises from specialised photoreceptive ganglion cells that employ melanopsin as a photopigment (Qiu et al., 2005). Conventional rods and cones are dispensable for circadian entrainment, although they may contribute to responses to dawn and dusk light. Cellular and metabolic responses of the SCN to photic cues conveyed by the RHT ensure that internal circadian time is entrained to solar time, thereby ensuring temporal adaptation of daily metabolic cycles. In addition, the SCN circadian clock also acts as an internal calendar because its cellular activity reflects day-length (see below) (Hastings and Follett, 2001). Driven by these cellular patterns in the SCN, the duration of the nocturnal secretion of melatonin increases in autumn and declines in spring. This photoperiodic response enables the circadian system, via changes in melatonin secretion, to direct a host of metabolic adaptations to season, including changes in body fat composition, basal metabolic rate, torpor etc. (Morgan this volume) (Dark, 2005).

Circadian clock genes

Circadian timing is a property of individual SCN neurons, it is not an emergent property of SCN circuits because when dispersed and isolated in culture SCN cells continue to exhibit asynchronous but nevertheless robust circadian rhythms of electrical firing. Recent genetic screens in mice and *Drosophila* have been instrumental in identifying the genes that encode the intra-cellular mechanisms that underlie circadian timing (Reppert and Weaver, 2002). At its heart lies a molecular oscillator in which CLOCK and BMAL1 protein complexes acting via E-box DNA regulatory sequences during circadian day drive the expression of *Period* (*Per1, Per2, Per3*) and *Cryptochrome* (*Cry1, Cry2*) genes (Figs. 2 and 3a). PER and CRY protein complexes subsequently accumulate in the nucleus of SCN neurons during early circadian night (Fig. 3b) and thereby inhibit expression of *Per* and *Cry*, along with other clock output genes. In later circadian night PER and CRY are degraded by the proteosome and so the cycle begins again, defining the new circadian dawn. These events therefore establish a delayed transcriptional/post-translational negative feedback loop, which oscillates with a period of approximately one day. This core loop is stabilised by auxilliary loops that involve other CLOCK::BMAL1 regulated gene products, including REV-ERBα, RORA and DEC1 (Honma et al., 2002; Preitner et al., 2002; Sato et al., 2004) (Fig. 2). Together, the loops drive daily waves of gene expression via periodic activation and suppression of E-boxes and other DNA circadian-control elements located within clock-controlled genes (Ueda et al., 2005), that ultimately determine the

rhythmic properties of SCN neurons (Panda et al., 2002), including membrane potential, firing rate and neurosecretion. Importantly, the temporal pattern of gene expression with peak *Per* levels in circadian daytime is seen in both diurnal and nocturnally active species (Mrosovsky et al., 2001). The loop codes for solar time: it does not code for behavioural niche. Elucidation of the key biochemical, intra-cellular links between circadian gene expression and rhythmic cell function will therefore provide important general insights into the co-ordinate regulation of cellular metabolism in brain and other tissues.

Evidence consistent with the feedback loop model is that *Per* and *Cry* genes are highly expressed in the SCN, the *Per* genes in particular with a high amplitude cycle (Fig. 3a) that is translated into a pronounced rhythm in the expression of nuclear PER proteins in the SCN, phase delayed by several hours relative to the mRNA cycle (Fig. 3b). Null mutations of *Bmal1* and *Per2* abolish circadian activity in mice (Bae et al., 2001) (Fig. 3c). Equally, double knockout of *Cry1* and *Cry2* also stops the SCN clock (Yamaguchi et al., 2003), but given the multiple-gene complement of the feedback loops, it is not surprising that some degree of redundancy is apparent. For example, single null mutants of *Cry1* and *Cry2* continue to show circadian cycles of behaviour, although period length is altered. Equally, an induced mutation of CLOCK that compromises its ability to activate gene expression leads to a longer circadian period, presumably because the rate of PER and CRY accumulation is reduced, and ultimately loss of rhythmicity in homozygous mutant mice (King et al., 1997). Targetted knockouts of *Per1*, *Per3* and *RevErba* are also rhythmic, but with varying degrees of instability. It is important to recognise, however, that definitive experimental tests of the loop model are awaited — most specifically the demonstration that artificially driving PER and/or CRY expression to a particular level is sufficient and necessary to reset circadian time to a new phase consistent with that particular level of expression.

As well as lacking such formal tests, the model is incomplete because the precise mechanisms of action of PER::CRY complexes are unclear. PER and CRY proteins are thought to associate via their C-terminal domains, and CRY proteins promote

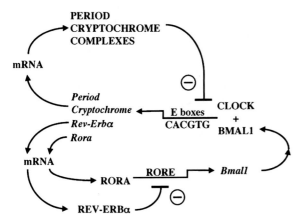

Fig. 2. Schematic model of circadian clock molecular feedback loops.

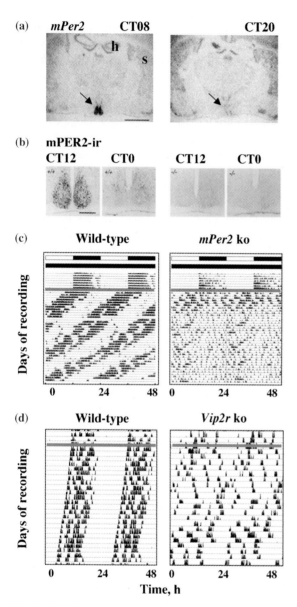

nuclear entry of PER. This facilitates complex assembly with BMAL and CLOCK proteins, which are present in the SCN at all phases. Rhythmic control over their activity rather than periodic changes in abundance is therefore the likely pivotal feature of the loop. Such changes may involve phosphorylation and other modifications, recruitment of co-factors and/or changes in their interaction with DNA (Lee et al., 2001; Brown et al., 2005b). CLOCK::BMAL-dependent circadian gene expression is associated with histone acetylation, and CRY likely inhibits actions of CLOCK and BMAL by targeting histone acetyltransferase and/or recruiting histone deacetylase (Etchegaray et al., 2003). Further changes in histone structure including methylation and phosphorylation may contribute to the inactivation of target genes. A particular unresolved issue is how apparently common control mechanisms (similar protein complexes and regulatory sequences) lead to the differential phasing of gene expression that underlie the temporal programmes characteristic of a tissue (Gachon et al., 2004; Ueda et al., 2005).

In addition to setting phases of gene expression, the overall period of the cycle is determined by biochemical modifications of PER and CRY proteins. The various period phenotypes seen with single mutant lines (e.g. $Cry1$ null) probably reflect changes in the composition and stability of PER::CRY complexes which will alter rates of nuclear shuttling of the complex, their ability to interfere with CLOCK::BMAL1 activity and their clearance from the cell. A shortened circadian period is also observed in the tau mutant hamster, which carries a spontaneous point mutation of casein kinase 1ε (Csk1ε), which leads to hypophosphorylation of PER proteins (Lowrey et al., 2000). A biochemically complementary short-period mutation is also seen in human familial advanced sleep-phase syndrome where a target residue in hPER2 for Csk1ε phosphorylation is mutated (Toh et al., 2001). Based on $Drosophila$

Fig. 3. Circadian clock genes. (a) In situ hybridisation images for $mPer2$ mRNA in coronal sections of mouse brain sampled in circadian day (CT08, 8 h after projected lights on) or circadian night (CT20, 8 h after projected lights off). Note high expression in SCN (arrowed) during circadian day and low expression in circadian night (Field et al., 2000). h, hippocampus; s, striatum. Scale bar = 5 mm. (b) Immunostaining for mPER2 in SCN of wid-type (+/+ left) and PER2 knock-out (−/− right) mice sampled at end of circadian day CT12 and beginning of circadian day CT0. Note rhythmic abundance of nuclear mPER2 in wild-type SCN and absence of protein in mutant. Scale bar = 500 μm. Taken from (Bae et al., 2001). (c) Circadian activity rhythms of wild-type and $mPer2$ knock-out mice, initially held on 12 h light, 12 h darkness (upper bars) and then transferred to continuous dim light (lower bar) on day indicated by grey line. Note robust circadian behaviour in wild-type mouse and complete disorganisation in mutant. (d) Circadian activity rhythms of wild-type and $Vip2r$ knock-out mice, format as in (c). Note severe disorganisation in $Vip2r$ mutant.

work, the short period associated with hypo-phosphorylation in Csk1ε mutants is thought to occur because PER protein is more resistant to degradation and accumulates in the nucleus sooner, thereby accelerating the cycle. The pattern of PER1 and PER2 accumulation in the *tau* hamster SCN, however, does not confirm this prediction. Rather, it is the clearance of PER proteins from the nucleus of SCN cells that is accelerated, suggesting that hypophosphorylation destabilises PER or associated factors (Dey et al., 2005). Of course, appreciation of the biochemistry of circadian complexes grows apace and is revealing new avenues for Csk1ε action, not least on CRY and BMAL proteins which may also contribute to the *tau* phenotype (Eide et al., 2002). It is also likely that factors contributing to proteosomal function and thereby able to influence degradation rates of phosphorylated PER and CRY will determine circadian period. For example, the ubiquitin ligase adaptor protein β-TrCP and its homologue in *Drosophila* Slimb have been implicated in setting clock speed (Eide et al., 2005). Given the diversity of ubiquitin ligases it is likely that several others may influence the clock-work, possibly in a tissue-specific manner. Equally, modifications that will influence ubiquitinylation and proteosomal breakdown may contribute to timing: BMAL for example is sumoylated in a CLOCK-dependent manner that may affect its transcriptional activity (Cardone et al., 2005). Finally, further kinases including Csk1δ, Csk2 and glycogen synthase kinase have recently been identified as regulators of circadian proteins in mice and/or flies, and in humans a mutation in CK1δ has been linked with advanced sleep-phase syndrome (Xu et al., 2005). As attention has switched from transcriptional to post-translational mechanisms, therefore, identification of where the oscillation actually arises and where it is controlled has become more difficult. In cyanobacteria, recombinant circadian proteins mixed in vitro can sustain circadian cycles of auto-phosphorylation, indicating that in this organism post-translational events alone are sufficient to mark 24 h time (Nakajima et al., 2005). It may be premature to assert that post-translational events are the principal origin of mammalian circadian oscillations, but it is likely that rhythmic events in the cytoplasm will be geared to complement transcriptionally driven cycles of protein abundance, and indeed one set of "cogs" may not operate effectively in the absence of the other. It is apparent, therefore, that despite the successes of mapping out the core timing loop, we lack a complete understanding of the principal biochemical regulators of circadian period and stability, not least how can the cycle take so long (ca. 24 h) to complete and yet be so precise (plus or minus 5 min per cycle or 0.3%) (Herzog et al., 2004)?

Entrainment of the SCN clock

Entrainment of the clock involves cellular and molecular events. The SCN are divided into retinorecipient "core," characterised by GABAergic neurons that express vasoactive intestinal polypeptide (VIP) and gastrin-releasing peptide (GRP), and dorso-medial "shell" with GABAergic cells co-expressing arginine vasopressin (AVP) (Abrahamson and Moore, 2001). The cellular response to light is initiated by glutamatergic retinal afferents that induce sustained electrical activity in the core SCN, accompanied by activation of calcium-dependent kinases and their target transcription factors, including calcium-cyclic AMP response element (CRE) binding protein (CREB) (Obrietan et al., 1999). As a result immediate early genes that carry CREs such as *cfos* are activated in the retinorecipient SCN. *Per* genes are also rapidly induced in the core (Shigeyoshi et al., 1997), likely via CRE sequences in their promoter (Travnickova-Bendova et al., 2002), and may participate in the metabolic accommodation to glutamatergic activation of the neuron. Increased *Per* expression in the core, however, is not sufficient to sustain a phase-shift. Sustained shifts to the behavioural cycle are reflected by and dependent on altered gene expression in the shell (Nagano et al., 2003), and this will occur via GABAergic and peptidergic relays from core neurons (Albus et al., 2005). In particular, GRP may contribute to resetting because in mice lacking the receptor for GRP, circadian behaviour is retained but photic induction of *Per* in the shell and behavioural resetting to light are compromised (Aida et al., 2002). Absence of VIP-ergic signalling is even more severe because circadian activity/rest cycles are disorganised in mice lacking

VIP (Aton et al., 2005) or the VPAC2 receptor for VIP (Harmar et al., 2002) (Fig. 3d). Under these circumstances, immediate responses of the core SCN to retinal input are retained, but circadian *Per* expression in the SCN shell is compromised, even under a full light/dark cycle (Harmar et al., 2002). The relative contributions of GABA, GRP and VIP to resetting are difficult to determine, not least because they likely interact on target neurons. Their ultimate action in the shell, however, is induction of *Per*. Should this occur on the rising phase of the *Per* expression rhythm ("dawn") it will accelerate the cycle and thereby advance it, whereas evening light delivered on the falling phase will induce *Per* and so delay the cycle and overt rhythms. Cues independent of retinal innervation such a neuropeptide Y, serotonin and behavioural arousal (collectively termed "non-photic") are also able to influence *Per* expression, in particular they down-regulate it during circadian day when light has little effect on the clock. This leads to the premature decline in mRNA and protein levels and an associated advance of the behavioural rest-activity cycle (Maywood et al., 2002). Understanding the neurochemical processes underlying the effects of light and non-photic cues should identify novel strategies for therapeutic management of circadian dysfunction in shift workers, the blind and the aged. A more general appreciation of how clocks in other tissues are synchronised (see below) may have even wider therapeutic applications.

Real-time imaging of molecular time-keeping in the SCN

Circadian timing is, by definition, a dynamic process, and experimental analysis of the clockwork is most effective where it exploits continuous, real-time analysis of circadian phenomena in individual organisms or tissues. Electrophysiological recording of the SCN, both as an assemblage and as individual units, has greatly advanced our understanding of its clock-like properties. Such an approach, however, remains an indirect assessment of events within the core timing mechanism. A major technical advance arising from the identification of circadian clock genes and their regulatory sequences

has been the development of real-time imaging using bioluminescent and fluorescent reporter genes to monitor molecular timekeeping as it proceeds. Transgenic rats and mice carrying a transgene reporter in which the promoter sequence of *mPer1* drives luciferase expression have provided beautiful read-outs of activity within the feedback loop, confirming that the promoter sequences do indeed sustain cyclical activity in vivo (consistent with the loop model) and that the SCN in culture can maintain its autonomous circadian timing for months under appropriate culture conditions (Yamazaki et al., 2000; Yamaguchi et al., 2003). Moreover, by using implanted fibre optics, *mPer1* activity can even be seen to oscillate in the SCN in vivo (Yamaguchi et al., 2001).

The *mPer1* promoter has also been used to drive destabilised green fluorescent protein in transgenic mice (Kuhlman et al., 2000), revealing circadian cycles of cellular gene expression in retina and SCN, as well as acute photic induction of *mPer1* in the SCN core accompanied by increased electrical firing (Kuhlman et al., 2003). The influence of light and other afferent inputs to the SCN on *Per* expression rhythms in vivo is very marked, and is reflected in acutely prepared SCN slices with clusters of cells oscillating together in discrete phase groupings (Quintero et al., 2003). Photoperiodic effects on gene expression such as these likely underlie the calendrical functions of the SCN clock governing seasonality (see above). In long-term organotypical SCN slice cultures, devoid of such extrinsic influences, the *mPer1::GFP* transgene reveals a very pronounced circadian cycle of gene expression, with a unitary, sinusoidal waveform that is extremely precise, providing a continuous phase readout of the molecular loop (Fig. 4a). The circadian signal is tightly synchronised between the paired SCN in the slice. The GFP reporter also allows for long-term recording of molecular timekeeping in individual neurons (Fig. 4b), which is equally robust and tightly synchronised across the SCN circuit. Intercellular synchrony such as this is lost in acute slices taken from mice that are behaviourally arrhythmic following exposure to continuous light (Ohta et al., 2005). *Per*-driven gene expression within SCN cells (in this case luciferase) can also be desynchronised by blockade of sodium-dependent action potentials

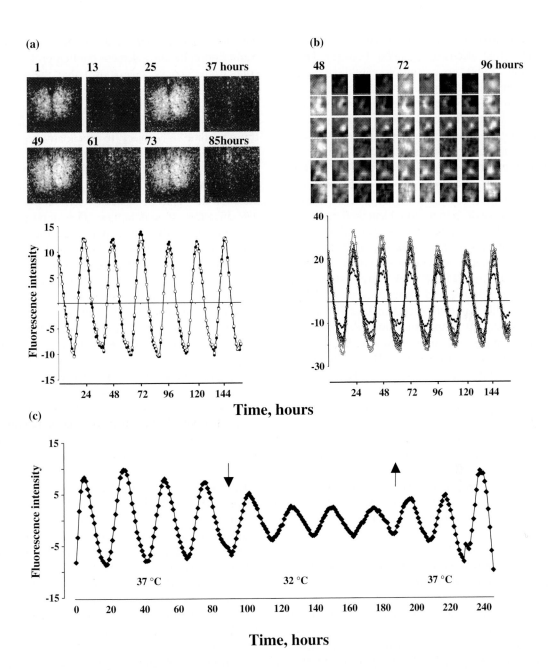

Fig. 4. Real-time fluorescence imaging of circadian gene expression in SCN organotypic slice. (a) Representative serial images of fluorescence emission from paired SCN in organotypical slice of *mPer1::dsGFP* transgenic mouse, with accompanying graphical data from left (open circles) and right (closed circles) SCN. Note robust molecular cycle and tight synchrony of the paired SCN. (b) High-power views of fluorescence emission from individual SCN neurons in slice depicted in (a) and corresponding graphical data. Note tight synchrony within the population of clock neurons. Representative video recordings are available at http://www2.mrc-lmb.cam.ac.uk/NB/Hasting_M/movies.html.(c) The circadian period of molecular timekeeping within SCN is defended against temperature changes, as shown by GFP emission rhythm of slice cultured at 37°C then dropped to 32°C and then returned to 37°C (arrows). Although amplitude is dampened at 32°C, period is stable at ca. 24 h.

with tetrodotoxin (TTX) (Yamaguchi et al., 2003), confirming that electrical signalling between them, presumably GABAergic and/or peptidergic (see above), maintains coherence across the population. Real-time imaging of gene expression reveals a further canonical feature of the SCN as a circadian oscillator: its temperature independence. Whereas most biochemical processes are very sensitive to temperature, the rate slowing by a factor of 2 for a 10°C decrease, circadian period of SCN slices remains unaltered at ca. 24 h between 37°C and 32°C (Fig. 4c), a drop that should lengthen period to approximately 36 h. This occurs even though the electrical firing rate of SCN neurons can be thermosensitive (reviewed in Van Someren, 2003). Temperature compensation of the clock is well established in lower vertebrates and invertebrates, and has obvious adaptive value in species unable to maintain a constant internal temperature. Indeed, given that core body temperature in mammals oscillates gradually by one degree or so (Fig. 1b), retention of temperature compensation may be more than atavistic: it may actively defend the central timekeeper against metabolic perturbations. Whether more acute step changes in temperature might be used therapeutically to regulate the SCN clockwork remains to be determined (Van Someren, 2003). Furthermore, it is apparent from these in vitro fluorescence recordings that, whatever its mechanism, temperature compensation is organised at the level of the molecular loop of individual neurons.

Local circadian time-keeping in peripheral tissues underpins metabolic rhythms

Although a number of metabolically relevant processes have been known to be under tight circadian regulation for some time, the most immediate being corticosteroid secretion by the adrenal glands, the recent development of high-throughput assays, particularly DNA microarrays has made it possible to examine temporal organisation of metabolism in comprehensive detail. In a range of tissues, between 5% and 10% of the transcriptome has been shown to be under circadian regulation (Akhtar et al., 2002; Panda et al., 2002; Storch et al., 2002) (Fig. 5a). Moreover, aside from the genes incorporating the

core loop and its immediate outputs such as the transcription factor d-element-binding protein (*dbp*), comparative assays have shown that the circadian transcriptomes are tissue specific, overlapping very little. Comparative analysis indicates therefore that across the organism a considerable proportion of the transcriptome is subject to circadian regulation. Moreover, this clock-controlled output includes genes that encode critical and often rate-liming factors in a range of vital pathways, including nitrogen metabolism, gluconeogenesis, cell division and detoxification of xenobiotics. The importance of these findings is that, first, they reveal the circadian system to be a point of regulation and therefore a potential avenue for therapeutic management of these processes. Second, they offer an explanation for the widespread malaise associated with shift work and sleep disorders that would be associated with disturbance to the normal circadian metabolic programme (Knutsson, 2003; Oishi et al., 2005). Third, they provide an opportunity to maximise therapeutic efficacy by exploiting any circadian variance in pharmacodynamics, sensitivity of target cells and tissues and severity of side effects. Finally they reveal the biological context for specific conditions, such as cardio-vascular and cerebro-vascular trauma that carry a strong circadian prevalence (Hastings et al., 2003).

Until recently, this circadian programming in the periphery was viewed as a direct outcome of regulation by the SCN, mediated along behavioural, neuroendocrine, metabolic (feeding related) and autonomic pathways (see Kalsbeek this volume). An indication that this might not be completely correct came with the observation that core circadian clock genes are expressed in the periphery with a clear circadian pattern. Why would "clock" genes be expressed in a passively driven tissue? A radical change in the view came with the demonstration that fibroblast cell cultures given a serum shock of the type used to induce immediate early genes also induced circadian clock genes (especially *Per*) and circadian output genes such as *dbp* (Balsalobre et al., 1998). Moreover, the wave of induction was repeated on a circadian basis in the absence of any further stimulus, demonstrating that immortalised cell cultures contain a competent circadian oscillator based around the same molecular components as

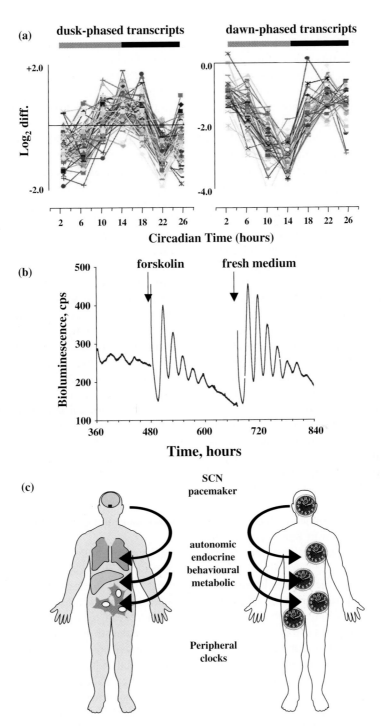

Fig. 5. Co-ordination of circadian timing in the periphery. (a) DNA micro-arrays of mouse liver samples reveal clusters of genes expressed in a circadian profile, with peaks timed to projected dusk (left) and dawn (right). Taken from (Akhtar et al., 2002). (b) Circadian gene expression in cultures of NIH 3T3 fibroblasts transfected with *mPer2::luciferase* reporter gene (Ueda et al., 2002) can be initiated with addition of forskolin or by a change of medium. (c) Schematic view of circadian co-ordination across the individual has the primary pacemaker of the SCN synchronised to solar time and responsible for maintaining internal synchrony among local tissue-based oscillators by a variety of signalling pathways.

those that make up the SCN timer. The generality of this conclusion was demonstrated when tissues from *Per1*::*luciferase* transgenic rats cultured in isolation were shown to have local, self-sustaining circadian clocks (Yamazaki et al., 2000), and real-time imaging has now revealed intrinsic circadian timing in a wide variety of primary cells and immortalised cell cultures (Fig. 5b). Moreover, although the relative phases of local clocks in liver, lung, skeletal muscle, heart etc. were set by the SCN, they do not passively follow the SCN: they take several days longer than the SCN to readjust to shifts of the lighting cycle (Yamazaki et al., 2000). Studies in intact and SCN-ablated animals showed that SCN-dependent cues such as corticosteroid secretion and metabolic cues associated with feeding act as internal synchronisers to these peripheral clocks, and by manipulating these factors directly it is possible to dissociate circadian timing in SCN and peripheral organs (Le Minh et al., 2001; Stokkan et al., 2001). Self-sustaining clocks in biopsies of human peripheral tissues have also been recently demonstrated (Brown et al., 2005a). The current model of the mammalian timing system therefore has the SCN as the principal pacemaker, cued to solar time by its retinal innervation, and co-ordinating the activity of multiple local oscillators distributed across tissue systems, synchronising them to solar time and thereby to each other (Fig. 5c). In this way adaptive temporal integration of metabolism is sustained throughout the individual.

How effective are the peripheral clocks? Even from their first demonstration in fibroblasts, it was obvious that circadian rhythms in the culture dampened, and that this might reflect exhaustion of the individual cellular clocks or their desynchronisation across the culture, leading to attenuation of the aggregate signal. Real-time imaging of fibroblasts has now shown that individual cells can continue to oscillate in asynchronous culture for several weeks (Nagoshi et al., 2004; Welsh et al., 2004), confirming that their molecular oscillator is every bit as effective as that of the SCN neuron. Moreover, by using a PER2::LUC fusion protein reporter in "knock-in" mice the intrinsic clocks of primary tissues can be shown to oscillate in culture for many weeks (Yoo et al., 2004). Clearly, with such effective timekeepers, issues of intra- and inter-tissue synchronisation are

of central importance to a better understanding of circadian co-ordination of metabolism. In vitro, many factors have been shown to have the potential to influence local clocks and their transcriptional output. In vivo, the interactions are obviously more complicated. For example, parabiosis experiments between SCN-intact and SCN-ablated mice have implicated blood–bone factors in the differential entrainment of kidney and liver but not heart or skeletal muscle (Guo et al., 2005), while sympathetic innervation of the adrenals mediates circadian and light-induced gene expression and secretion of corticosteroids (Ishida et al., 2005). In turn the corticosteroids serve to synchronise the liver clock to SCN time, and oppose the effects of feeding schedules (Le Minh et al., 2001). Importantly, the SCN do not express glucocorticoid receptors and so are defended against temporal feedback, as they are against circadian temperature fluctuations. Local clock gene expression may not, however, be sufficient to sustain metabolic rhythms. For example, sympathetic denervation of the liver in rats flattens the daily rhythms of glucose and insulin although rhythmic clock gene expression is sustained in such animals, possibly synchronised by corticosteroids which remain rhythmic in such animals (Cailotto et al., 2005). Clearly, autonomic factors may operate both up- and downstream of the clockwork to regulate circadian metabolism.

Circadian timing in disease

The pervasive influence of the circadian system over physiology inevitably imposes a circadian dimension to common diseases. For example, the circadian surges in heart rate and vasoconstriction in the morning lead to a peak in blood pressure, which is coincident with a daily nadir in thrombolytic activity. Consequently, rates of haemorrhagic stroke and myocardial ischaemia and infarct are highest in the hours after wakening (Hastings et al., 2003). Circadian disruption is also associated with metabolic dysfunction, both in shift workers with an increased incidence of gastrointestinal disease (Caruso et al., 2004) (Knutsson, 2003), and in *Clock* mutant mice which develop a metabolic syndrome of hyperlipidaemia, hyperglycaemia and hypoinsulinaemia

(Turek et al., 2005). These metabolic changes may be a consequence of altered feeding schedules and/or neuroendocrine profiles in *Clock* mice, for example blunted corticosteroid rhythms. Moreover, *Clock* mutant mice have extensive dysregulation of metabolic gene expression in the liver, involving both circadian and non-circadian targets (Panda et al., 2002) and so the metabolic influence of the core timer may extend far beyond those processes that are rhythmically expressed. Emerging links between metabolic syndrome and sleep/circadian disorders suggest that the clock and metabolism may also interact at the behavioural level (Bass and Turek, 2005).

A molecular interaction between the core clock mechanism and metabolism has recently been identified in the context of bone formation and its regulation by leptin (Fu et al., 2005). Bone formation, as so much else in the periphery, is under circadian control, likely mediated by local oscillators in osteoblasts, which express clock genes in a circadian manner. Importantly, bone mass is increased in mice carrying mutations of the core clockwork, suggesting that the clockwork suppresses bone formation. One action of leptin is also to inhibit bone formation, leptin acting in the hypothalamus to modulate sympathetic outflow to bone. Local expression of clock genes mediates this inhibitory action, having an anti-proliferative effect on osteoblasts by inhibiting expression of *Cyclin D1* (Fu et al., 2005). It is interesting to note that the same molecular mechanism for photic regulation of the SCN clock, i.e. activation of CREs via CREB, is also the mechanism whereby sympathetic afferents control the clockwork of osteoblasts. A parallel pathway, also regulated by leptin, causes up-regulation of *Cyclin D1* and is usually in balance with the clock-mediated inhibition, thereby effecting homeostatic maintenance of bone growth. In clock dysfunction, however, this balance is lost and leptin-dependent control compromised, leading to proliferation and increased bone mass. The molecular mechanism whereby the clock suppresses cell proliferation is by down-regulation of *c-myc*, an activator of *Cyclin D1*. Currently, it is not known whether other actions of leptin are also clock dependent. Nevertheless, these findings establish a precedent of a central role for circadian factors in metabolic regulation and tissue growth.

An effect of the clock on cell division may also have relevance to other major diseases, not least cancer. In naturally proliferating tissues the timing of cell division, the expression of clock genes and the expression of genes that determine progression through the cell division cycle are all under close circadian regulation (Bjarnason et al., 2001), and it is likely that the circadian clock gates the passage of cells through the division cycle (Nagoshi et al., 2004) by the temporally regulated expression of genes such as cyclins and *Wee-1* (Reddy et al., 2005). This circadian regulation of cell division factors is evident in both cell cultures analysed by micro-array (Akhtar et al., 2002) and in vivo models, for example hepatic remodelling (Matsuo et al., 2003). In experimental tumours, clock gene expression is also circadian (Filipski et al., 2004) synchronised to the host animal. Moreover the host clock plays a significant oncostatic role. In two independent models of circadian dysfunction, SCN ablation and experimental jetlag, tumour growth is accelerated (Filipski et al., 2002; Filipski et al., 2004) and circadian gene expression in the tumour disorganised. The implication is that normal circadian function in the tumour, maintained by host factors, gates cell division thereby slowing tumour progression, whereas if circadian function of the host and consequently the tumour is disorganised, this gate remains open and tumour progression accelerated. Such a mechanism may contribute to the observed increased incidence of cancer in shift workers (Fu and Lee, 2003). The *mPer2* mutant mouse in which central and peripheral circadian timing are severely disrupted provides an animal model for this relationship. These mice show an increased frequency of both spontaneous and induced tumours (Fu et al., 2002) and genes involved in tumor suppression and regulation of the cell cycle, including *Cyclin D1*, *Cyclin A*, *Mdm-2* and *Gadd45alpha*, are deregulated. As with bone growth, circadian factors directly control the transcription of *c-myc* and this is deregulated in the *mPer2* mutant, thereby stimulating cell division and reducing apoptosis. It is therefore clear from these examples; cardiovascular disease, metabolic syndrome and cancer that the pivotal role of the circadian clock in coordinating normal metabolic processes places it at the centre of a number of chronic systemic diseases.

Circadian timing and neurodegenerative disease

In addition to cardiovascular disease, metabolic syndrome and cancer, recognition of strong circadian involvement is growing in another area with increasing prevalence in modern society: neurodegenerative diseases. The circadian dimension of such disorders is manifested most obviously in disturbed patterns of sleep and wakefulness (Chokroverty, 1996). Although alterations to sleep occur with general ageing, in Alzheimer's disease (AD) they are far more profound and accompany the development of dementia. Fractionation of the sleep/wake cycle (Fig. 6a) presents as nocturnal activity and daytime drowsiness. The disruption of domestic routines by the former is so severe that circadian dysfunction is the principal cause of institutionalisation of patients (Bianchetti et al., 1995), with its associated personal, social and economic costs. The locus of dysfunction is unclear. Alterations of the amplitude and phase of other circadian outputs such as core body temperature are reported (Harper et al., 2005), but cortisol secretion remains strongly rhythmic even in patients with severely disturbed activity patterns (Hatfield et al., 2004). This suggests that some circadian competence remains in AD and that links between the SCN and centres controlling sleep and wakefulness (see Saper this volume) are compromised, rather than SCN function being globally impaired. Nevertheless, altered behavioural patterns alone and changes in their phase relationship to core temperature cycles will likely perturb circadian synchronisation in peripheral tissues, and it remains to be determined whether possible metabolic effects of such disorders contribute to the poor survival of AD patients with disordered activity patterns (Gehrman et al., 2004). Equally, quality of life of demented patients, and perhaps their longevity, would benefit from management of their daily living patterns by interventions such as regular exposure to bright light (Dowling et al., 2005).

In contrast to the well-documented problems in AD, sleep disturbance in Huntington's disease (HD) has received less attention, and clear identification of a circadian component is confounded by the presence of motoric/kinetic disorders that will interrupt normal sleep patterns and mask the resting state. Nevertheless, changes in sleep pattern can be seen in patients even after allowance is made for such interference (Morton et al., 2005) (Fig. 6a). Although overall activity levels do not change in HD patients, a greater proportion of activity occurs at night, reflecting fragmented sleep. Importantly, these disturbances are also reflected in those caring for the patients and so have a deleterious effect in a domestic setting (Taylor and Bramble, 1997; Morton et al., 2005)). In R6/2 mice, a transgenic model of HD expressing human Huntingtin protein, there is also a progressive fractionation of the rest/activity cycle entrained under a light/dark cycle or free running in continuous darkness (Morton et al., 2005) (Fig. 6b). This behavioural dysfunction in the mice is similar to that seen in HD patients, with a greater proportion of activity occurring in the typically quiescent phase. Importantly, it is accompanied by disordered circadian gene expression in the SCN, including a blunted rhythm of $Bmal1$ mRNA and a premature decline in $mPer2$ levels. This suggests a direct involvement of pathological changes within the SCN in HD-related sleep disturbance. Furthermore, expression in the SCN of $Prokineticin2$, a transcriptional target of CLOCK-::BMAL and implicated as a suppressor of daytime activity in nocturnal rodents (Cheng et al., 2002) is also dramatically reduced (Fig. 6c), highlighting a potential molecular basis for the behavioural disruption. The molecular changes associated with the development of HD-like symptoms extend to other brain areas, including the striatum and motor cortex where the strong circadian expression of Per genes in wild-type mice is dramatically dampened in symptomatic (but not pre-symptomatic) R6/2 mice (Fig. 6d). Given this central perturbation of timing and its behavioural sequelae, it is likely that circadian control over metabolism is also impaired in symptomatic mice, and may conceivably contribute to their dramatic loss of body mass and early mortality from unidentified causes.

In conclusion, metabolic functions across the organism are matched to the daily cycle by an internal circadian timing system which co-ordinates the temporal sequence of expression of a considerable part of the metabolically relevant transcriptome.

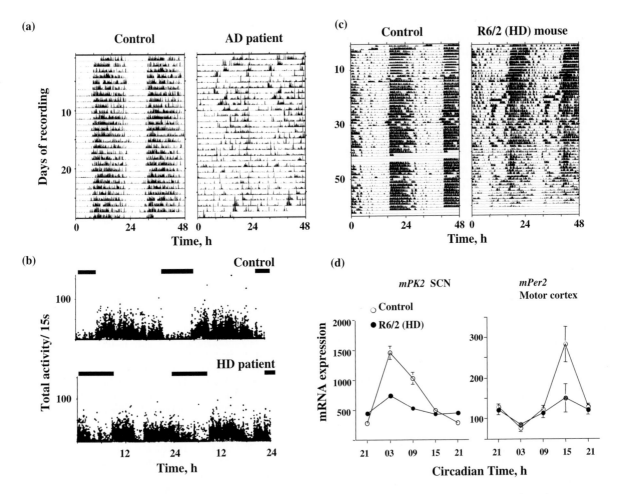

Fig. 6. Circadian timing in neurodegenerative disease. (a) Representative rest/activity patterns recorded using wrist actimeters worn for 4 weeks by an Alzheimer's patients with moderate dementia but living at home, and an age-matched control. Note loss of coherence to rest-activity cycle in AD patient. Taken from (Hatfield et al., 2004). (b) Actimetric data from control and HD patient recorded over 48 h in home setting. Bars above trace indicate time in bed. Taken from (Morton et al., 2005). (c) Representative actograms recorded by infra-red beam from control wild-type and R6/2 Huntington's mouse from 8 weeks of age. Note progressive deterioration of activity rhythm in mutant with eventual loss of definition of quiescent and active phases. Taken from (Morton et al., 2005). (d) Circadian expression profiles of *mPK2* and *mPer2* mRNA in SCN (left) and motor cortex (right) of symptomatic R6/2 mice (closed circles) and age-matched controls (open circles). Note attenuation of rhythms in mutant mice. Taken from (Morton et al., 2005).

The cellular basis of circadian timing revolves around an auto-regulatory transcriptional/post-translational feedback loop involving a series of "clock" genes. The principal pacemaker of the SCN is entrained to solar time by retinal afferents, which reset the gene expression cycle on a daily basis, matching it to solar time. Neural and neuroendocrine outputs convey circadian cues from the SCN neurons to the rest of brain and body. Many if not most peripheral tissues contain local cell-based circadian clocks that recapitulate the molecular mechanism of the SCN. Importantly, cells within peripheral tissues rely on signals from the SCN for their synchronisation to each other and to solar time. The prevalence of circadian modulation to gene expression and physiology highlight the central and local clocks as pivotal regulators to metabolic function. This is now being recognised in conditions of human health and disease, and is likely to become an increasingly relevant factor in modern societies with "24/7" activity and an ageing population.

Acknowledgements

The work of the authors is supported by the Medical Research Council, UK and the Wellcome Trust (Project Grant 064588/Z/01 to MHH and Professor M.H. Johnson, Department of Anatomy, University of Cambridge).

References

Abrahamson, E.E. and Moore, R.Y. (2001) Suprachiasmatic nucleus in the mouse: retinal innervation, intrinsic organization and efferent projections. Brain Res., 916: 172–191.

Aida, R., Moriya, T., Araki, M., Akiyama, M., Wada, K., Wada, E. and Shibata, S. (2002) Gastrin-releasing peptide mediates photic entrainable signals to dorsal subsets of suprachiasmatic nucleus via induction of Period gene in mice. Mol. Pharmacol., 61: 26–34.

Akhtar, R.A., Reddy, A.B., Maywood, E.S., Clayton, J.D., King, V.M., Smith, A.G., Gant, T.W., Hastings, M.H. and Kyriacou, C.P. (2002) Circadian cycling of the mouse liver transcriptome, as revealed by cDNA microarray, is driven by the suprachiasmatic nucleus. Curr. Biol., 12: 540–550.

Albus, H., Vansteensel, M.J., Michel, S., Block, G.D. and Meijer, J.H. (2005) A GABAergic mechanism is necessary for coupling dissociable ventral and dorsal regional oscillators within the circadian clock. Curr. Biol., 15: 886–893.

Aton, S.J., Colwell, C.S., Harmar, A.J., Waschek, J. and Herzog, E.D. (2005) Vasoactive intestinal polypeptide mediates circadian rhythmicity and synchrony in mammalian clock neurons. Nat. Neurosci., 8: 476–483.

Bae, K., Jin, X., Maywood, E.S., Hastings, M.H., Reppert, S.M. and Weaver, D.R. (2001) Differential functions of mPer1, mPer2, and mPer3 in the SCN circadian clock. Neuron, 30: 525–536.

Balsalobre, A., Damiola, F. and Schibler, U. (1998) A serum shock induces circadian gene expression in mammalian tissue culture cells. Cell, 93: 929–937.

Bass, J. and Turek, F.W. (2005) Sleepless in America: a pathway to obesity and the metabolic syndrome? Arch. Intern. Med., 165: 15–16.

Bianchetti, A., Scuratti, A., Zanetti, O., Binetti, G., Frisoni, G.B., Magni, E. and Trabucchi, M. (1995) Predictors of mortality and institutionalization in Alzheimer disease patients 1 year after discharge from an Alzheimer dementia unit. Dementia, 6: 108–112.

Bjarnason, G.A., Jordan, R.C., Wood, P.A., Li, Q., Lincoln, D.W., Sothern, R.B., Hrushesky, W.J. and Ben-David, Y. (2001) Circadian expression of clock genes in human oral mucosa and skin: association with specific cell-cycle phases. Am. J. Pathol., 158: 1793–1801.

Brown, S.A., Fleury-Olela, F., Nagoshi, E., Hauser, C., Juge, C., Meier, C.A., Chicheportiche, R., Dayer, J.M., Albrecht, U. and Schibler, U. (2005a) The period length of fibroblast circadian gene expression varies widely among human individuals. PLOS Biol., 3: e338.

Brown, S.A., Ripperger, J., Kadener, S., Fleury-Olela, F., Vilbois, F., Rosbash, M. and Schibler, U. (2005b) PERIOD1-associated proteins modulate the negative limb of the mammalian circadian oscillator. Science, 308: 693–696.

Cailotto, C., La Fleur, S.E., Van Heijningen, C., Wortel, J., Kalsbeek, A., Feenstra, M., Pevet, P. and Buijs, R.M. (2005) The suprachiasmatic nucleus controls the daily variation of plasma glucose via the autonomic output to the liver: are the clock genes involved? Eur. J. Neurosci., 22: 2531–2540.

Cardone, L., Hirayama, J., Giordano, F., Tamaru, T., Palvimo, J.J. and Sassone-Corsi, P. (2005) Circadian clock control by SUMOylation of BMAL1. Science, 309: 1390–1394.

Caruso, C.C., Lusk, S.L. and Gillespie, B.W. (2004) Relationship of work schedules to gastrointestinal diagnoses, symptoms, and medication use in auto factory workers. Am. J. Ind. Med., 46: 586–598.

Cheng, M.Y., Bullock, C.M., Li, C., Lee, A.G., Bermak, J.C., Belluzzi, J., Weaver, D.R., Leslie, F.M. and Zhou, Q.Y. (2002) Prokineticin 2 transmits the behavioural circadian rhythm of the suprachiasmatic nucleus. Nature, 417: 405–410.

Chokroverty, S. (1996) Sleep and degenerative neurologic disorders. Neurol. Clin., 14: 807–826.

Dark, J. (2005) Annual lipid cycles in hibernators: integration of physiology and behavior. Annu. Rev. Nutr., 25: 469–497.

Dey, J., Carr, A.J., Cagampang, F.R., Semikhodskii, A.S., Loudon, A.S., Hastings, M.H. and Maywood, E.S. (2005) The tau mutation in the Syrian hamster differentially reprograms the circadian clock in the SCN and peripheral tissues. J. Biol. Rhythm., 20: 99–110.

Dowling, G.A., Hubbard, E.M., Mastick, J., Luxenberg, J.S., Burr, R.L. and Van Someren, E.J. (2005) Effect of morning bright light treatment for rest-activity disruption in institutionalized patients with severe Alzheimer's disease. Int. Psychogeriatr., 17: 221–236.

Eide, E.J., Vielhaber, E.L., Hinz, W.A. and Virshup, D.M. (2002) The circadian regulatory proteins BMAL1 and cryptochromes are substrates of casein kinase Iepsilon. J. Biol. Chem., 277: 17248–17254.

Eide, E.J., Woolf, M.F., Kang, H., Woolf, P., Hurst, W., Camacho, F., Vielhaber, E.L., Giovanni, A. and Virshup, D.M. (2005) Control of mammalian circadian rhythm by CKIepsilon-regulated proteasome-mediated PER2 degradation. Mol. Cell Biol., 25: 2795–2807.

Etchegaray, J.P., Lee, C., Wade, P.A. and Reppert, S.M. (2003) Rhythmic histone acetylation underlies transcription in the mammalian circadian clock. Nature, 421: 177–182.

Field, M.D., Maywood, E.S., O'Brien, J.A., Weaver, D.R., Reppert, S.M. and Hastings, M.H. (2000) Analysis of clock proteins in mouse SCN demonstrates phylogenetic divergence of the circadian clockwork and resetting mechanisms. Neuron, 25: 437–447.

Filipski, E., Delaunay, F., King, V.M., Wu, M.W., Claustrat, B., Grechez-Cassiau, A., Guettier, C., Hastings, M.H. and Francis, L. (2004) Effects of chronic jet lag on tumor progression in mice. Cancer Res., 64: 7879–7885.

Filipski, E., King, V.M., Li, X., Granda, T.G., Mormont, M.C., Liu, X., Claustrat, B., Hastings, M.H. and Levi, F. (2002) Host circadian clock as a control point in tumor progression. J. Natl. Cancer Inst., 94: 690–697.

Fu, L. and Lee, C.C. (2003) The circadian clock: pacemaker and tumour suppressor. Nat. Rev. Cancer, 3: 350–361.

Fu, L., Patel, M.S., Bradley, A., Wagner, E.F. and Karsenty, G. (2005) The molecular clock mediates leptin-regulated bone formation. Cell, 122: 803–815.

Fu, L., Pelicano, H., Liu, J., Huang, P. and Lee, C. (2002) The circadian gene Period2 plays an important role in tumor suppression and DNA damage response in vivo. Cell, 111: 41–50.

Gachon, F., Nagoshi, E., Brown, S.A., Ripperger, J. and Schibler, U. (2004) The mammalian circadian timing system: from gene expression to physiology. Chromosoma, 113: 103–112.

Gehrman, P., Marler, M., Martin, J.L., Shochat, T., Corey-Bloom, J. and Ancoli-Israel, S. (2004) The timing of activity rhythms in patients with dementia is related to survival. J. Gerontol. A. Biol. Sci. Med. Sci., 59: 1050–1055.

Guo, H., Brewer, J.M., Champhekar, A., Harris, R.B. and Bittman, E.L. (2005) Differential control of peripheral circadian rhythms by suprachiasmatic-dependent neural signals. Proc. Natl. Acad. Sci. USA, 102: 3111–3116.

Harmar, A.J., Marston, H.M., Shen, S., Spratt, C., West, K.M., Sheward, W.J., Morrison, C.F., Dorin, J.R., Piggins, H.D., Reubi, J.C., Kelly, J.S., Maywood, E.S. and Hastings, M.H. (2002) The VPAC(2) receptor is essential for circadian function in the mouse suprachiasmatic nuclei. Cell, 109: 497–508.

Harper, D.G., Volicer, L., Stopa, E.G., McKee, A.C., Nitta, M. and Satlin, A. (2005) Disturbance of endogenous circadian rhythm in aging and Alzheimer disease. Am. J. Geriatr. Psychiatry, 13: 359–368.

Hastings, M.H. and Follett, B.K. (2001) Toward a molecular biological calendar? J. Biol. Rhythms, 16: 424–430.

Hastings, M.H., Reddy, A.B. and Maywood, E.S. (2003) A clockwork web: circadian timing in brain and periphery, in health and disease. Nat. Rev. Neurosci., 4: 649–661.

Hatfield, C.F., Herbert, J., Van Someren, E.J., Hodges, J.R. and Hastings, M.H. (2004) Disrupted daily activity/rest cycles in relation to daily cortisol rhythms of home-dwelling patients with early Alzheimer's dementia. Brain, 127: 1061–1074.

Herzog, E.D., Aton, S.J., Numano, R., Sakaki, Y. and Tei, H. (2004) Temporal precision in the mammalian circadian system: a reliable clock from less reliable neurons. J. Biol. Rhythms, 19: 35–46.

Honma, S., Kawamoto, T., Takagi, Y., Fujimoto, K., Sato, F., Noshiro, M., Kato, Y. and Honma, K. (2002) Dec1 and Dec2 are regulators of the mammalian molecular clock. Nature, 419: 841–844.

Ishida, A., Mutoh, T., Ueyama, T., Bando, H., Masubuchi, S., Nakahara, D., Tsujimoto, G. and Okamura, H. (2005) Light activates the adrenal gland: timing of gene expression and glucocorticoid release. Cell Metab., 2: 297–307.

King, D.P., Zhao, Y., Sangoram, A.M., Wilsbacher, L.D., Tanaka, M., Antoch, M.P., Steeves, T.D., Vitaterna, M.H., Kornhauser, J.M., Lowrey, P.L., Turek, F.W. and Takahashi, J.S. (1997) Positional cloning of the mouse circadian Clock gene. Cell, 89: 641–653.

King, V.M., Chahad-Ehlers, S., Shen, S., Harmar, A.J., Maywood, E.S. and Hastings, M.H. (2003) A hVIPR transgene as a novel tool for the analysis of circadian function in the mouse suprachiasmatic nucleus. Eur. J. Neurosci., 17: 822–832.

Knutsson, A. (2003) Health disorders of shift workers. Occup. Med. (Lond.), 53: 103–108.

Kuhlman, S.J., Quintero, J.E. and McMahon, D.G. (2000) GFP fluorescence reports Period 1 circadian gene regulation in the mammalian biological clock. Neuroreport, 11: 1479–1482.

Kuhlman, S.J., Silver, R., Le Sauter, J., Bult-Ito, A. and McMahon, D.G. (2003) Phase resetting light pulses induce Per1 and persistent spike activity in a subpopulation of biological clock neurons. J. Neurosci., 23: 1441–1450.

Lee, C., Etchegaray, J.P., Cagampang, F.R., Loudon, A.S. and Reppert, S.M. (2001) Posttranslational mechanisms regulate the mammalian circadian clock. Cell, 107: 855–867.

Le Minh, N., Damiola, F., Tronche, F., Schutz, G. and Schibler, U. (2001) Glucocorticoid hormones inhibit food-induced phase-shifting of peripheral circadian oscillators. EMBO J., 20: 7128–7136.

Lowrey, P.L., Shimomura, K., Antoch, M.P., Yamazaki, S., Zemenides, P.D., Ralph, M.R., Menaker, M. and Takahashi, J.S. (2000) Positional syntenic cloning and functional characterization of the mammalian circadian mutation tau. Science, 288: 483–492.

Matsuo, T., Yamaguchi, S., Mitsui, S., Emi, A., Shimoda, F. and Okamura, H. (2003) Control mechanism of the circadian clock for timing of cell division in vivo. Science, 302: 255–259.

Maywood, E.S., Okamura, H. and Hastings, M.H. (2002) Opposing actions of neuropeptide Y and light on the expression of circadian clock genes in the mouse suprachiasmatic nuclei. Eur. J. Neurosci., 15: 216–220.

Meyer-Bernstein, E.L., Jetton, A.E., Matsumoto, S.I., Markuns, J.F., Lehman, M.N. and Bittman, E.L. (1999) Effects of suprachiasmatic transplants on circadian rhythms of neuroendocrine function in golden hamsters. Endocrinology, 140: 207–218.

Morton, A.J., Wood, N.I., Hastings, M.H., Hurelbrink, C., Barker, R.A. and Maywood, E.S. (2005) Disintegration of the sleep–wake cycle and circadian timing in Huntington's disease. J. Neurosci., 25: 157–163.

Mrosovsky, N., Edelstein, K., Hastings, M.H. and Maywood, E.S. (2001) Cycle of period gene expression in a diurnal mammal (*Spermophilus tridecemlineatus*): implications for nonphotic phase shifting. J. Biol. Rhythms, 16: 471–478.

Nagano, M., Adachi, A., Nakahama, K., Nakamura, T., Tamada, M., Meyer-Bernstein, E., Sehgal, A. and Shigeyoshi, Y. (2003) An abrupt shift in the day/night cycle causes desynchrony in the mammalian circadian center. J. Neurosci., 23: 6141–6151.

Nagoshi, E., Saini, C., Bauer, C., Laroche, T., Naef, F. and Schibler, U. (2004) Circadian gene expression in individual fibroblasts: cell-autonomous and self-sustained oscillators pass time to daughter cells. Cell, 119: 693–705.

Nakajima, M., Imai, K., Ito, H., Nishiwaki, T., Murayama, Y., Iwasaki, H., Oyama, T. and Kondo, T. (2005) Reconstitution of circadian oscillation of cyanobacterial KaiC phosphorylation in vitro. Science, 308: 414–415.

Obrietan, K., Impey, S., Smith, D., Athos, J. and Storm, D.R. (1999) Circadian regulation of cAMP response element-mediated gene expression in the suprachiasmatic nuclei. J. Biol. Chem., 274: 17748–17756.

Ohta, H., Yamazaki, S. and McMahon, D.G. (2005) Constant light desynchronizes mammalian clock neurons. Nat. Neurosci., 8: 267–269.

Oishi, M., Suwazono, Y., Sakata, K., Okubo, Y., Harada, H., Kobayashi, E., Uetani, M. and Nogawa, K. (2005) A longitudinal study on the relationship between shift work and the progression of hypertension in male Japanese workers. J. Hypertens., 23: 2173–2178.

Panda, S., Antoch, M.P., Miller, B.H., Su, A.I., Schook, A.B., Straume, M., Schultz, P.G., Kay, S.A., Takahashi, J.S. and Hogenesch, J.B. (2002) Coordinated transcription of key pathways in the mouse by the circadian clock. Cell, 109: 307–320.

Peirson, S.N., Thompson, S., Hankins, M.W. and Foster, R.G. (2005) Mammalian photoentrainment: results, methods, and approaches. Methods Enzymol., 393: 697–726.

Preitner, N., Damiola, F., Lopez-Molina, L., Zakany, J., Duboule, D., Albrecht, U. and Schibler, U. (2002) The orphan nuclear receptor REV-ERBalpha controls circadian transcription within the positive limb of the mammalian circadian oscillator. Cell, 110: 251–260.

Qiu, X., Kumbalasiri, T., Carlson, S.M., Wong, K.Y., Krishna, V., Provencio, I. and Berson, D.M. (2005) Induction of photosensitivity by heterologous expression of melanopsin. Nature, 433: 745–749.

Quintero, J.E., Kuhlman, S.J. and McMahon, D.G. (2003) The biological clock nucleus: a multiphasic oscillator network regulated by light. J. Neurosci., 23: 8070–8076.

Ralph, M.R., Foster, R.G., Davis, F.C. and Menaker, M. (1990) Transplanted suprachiasmatic nucleus determines circadian period. Science, 283: 693–695.

Reddy, A.B., Wong, G.K., O'Neill, J., Maywood, E.S. and Hastings, M.H. (2005) Circadian clocks: neural and peripheral pacemakers that impact upon the cell division cycle. Mutat. Res., 574: 76–91.

Reppert, S.M. and Weaver, D.R. (2002) Coordination of circadian timing in mammals. Nature, 418: 935–941.

Sato, T.K., Panda, S., Miraglia, L.J., Reyes, T.M., Rudic, R.D., McNamara, P., Naik, K.A., FitzGerald, G.A., Kay, S.A. and Hogenesch, J.B. (2004) A functional genomics strategy reveals Rora as a component of the mammalian circadian clock. Neuron, 43: 527–537.

Shigeyoshi, Y., Taguchi, K., Yamamoto, S., Takekida, S., Yan, L., Tei, H., Moriya, T., Shibata, S., Loros, J.J., Dunlap, J. and Okamura, H. (1997) Light-induced resetting of a mammalian circadian clock is associated with rapid induction of the mPer1 transcript. Cell, 91: 1043–1053.

Stokkan, K.A., Yamazaki, S., Tei, H., Sakaki, Y. and Menaker, M. (2001) Entrainment of the circadian clock in the liver by feeding. Science, 291: 490–493.

Storch, K.F., Lipan, O., Leykin, I., Viswanathan, N., Davis, F.C., Wong, W.H. and Weitz, C.J. (2002) Extensive and divergent circadian gene expression in liver and heart. Nature, 417: 78–83.

Taylor, N. and Bramble, D. (1997) Sleep disturbance and Huntington's disease. Br. J. Psychiatry, 171: 393.

Toh, K.L., Jones, C.R., He, Y., Eide, E.J., Hinz, W.A., Virshup, D.M., Ptacek, L.J. and Fu, Y.H. (2001) An hPer2 phosphorylation site mutation in familial advanced sleep phase syndrome. Science, 291: 1040–1043.

Travnickova-Bendova, Z., Cermakian, N., Reppert, S.M. and Sassone-Corsi, P. (2002) Bimodal regulation of mPeriod promoters by CREB-dependent signaling and CLOCK/BMAL1 activity. Proc. Natl. Acad. Sci. USA, 99: 7728–7733.

Turek, F.W., Joshu, C., Kohsaka, A., Lin, E., Ivanova, G., McDearmon, E., Laposky, A., Losee-Olson, S., Easton, A., Jensen, D.R., Eckel, R.H., Takahashi, J.S. and Bass, J. (2005) Obesity and metabolic syndrome in circadian Clock mutant mice. Science, 308: 1043–1045.

Ueda, H.R., Chen, W., Adachi, A., Wakamatsu, H., Hayashi, S., Takasugi, T., Nagano, M., Nakahama, K., Suzuki, Y., Sugano, S., Iino, M., Shigeyoshi, Y. and Hashimoto, S. (2002) A transcription factor response element for gene expression during circadian night. Nature, 418: 534–539.

Ueda, H.R., Hayashi, S., Chen, W., Sano, M., Machida, M., Shigeyoshi, Y., Iino, M. and Hashimoto, S. (2005) System-level identification of transcriptional circuits underlying mammalian circadian clocks. Nat. Genet., 37: 187–192.

Van Someren, E.J. (2003) Thermosensitivity of the circadian timing system. Sleep Biol. Rhythms, 1: 53–62.

Weaver, D.R. (1998) The suprachiasmatic nucleus: a 25-year retrospective. J. Biol. Rhythms, 13: 100–112.

Welsh, D.K., Yoo, S.H., Liu, A.C., Takahashi, J.S. and Kay, S.A. (2004) Bioluminescence imaging of individual fibroblasts reveals persistent, independently phased circadian rhythms of clock gene expression. Curr. Biol., 14: 2289–2295.

Xu, Y., Padiath, Q.S., Shapiro, R.E., Jones, C.R., Wu, S.C., Saigoh, N., Saigoh, K., Ptacek, L.J. and Fu, Y.H. (2005) Functional consequences of a CKIdelta mutation causing familial advanced sleep phase syndrome. Nature, 434: 640–644.

Yamaguchi, S., Isejima, H., Matsuo, T., Okura, R., Yagita, K., Kobayashi, M. and Okamura, H. (2003) Synchronization of cellular clocks in the suprachiasmatic nucleus. Science, 302: 1408–1412.

Yamaguchi, S., Kobayashi, M., Mitsui, S., Ishida, Y., van der Horst, G.T., Suzuki, M., Shibata, S. and Okamura, H.

(2001) View of a mouse clock gene ticking. Nature, 409: 684.

Yamazaki, S., Numano, R., Abe, M., Hida, A., Takahashi, R., Ueda, M., Block, G.D., Sakaki, Y., Menaker, M. and Tei, H. (2000) Resetting central and peripheral circadian oscillators in transgenic rats. Science, 288: 682–685.

Yoo, S.H., Yamazaki, S., Lowrey, P.L., Shimomura, K., Ko, C.H., Buhr, E.D., Siepka, S.M., Hong, H.K., Oh, W.J., Yoo, O.J., Menaker, M. and Takahashi, J.S. (2004) PERIOD2::LUCIFERASE real-time reporting of circadian dynamics reveals persistent circadian oscillations in mouse peripheral tissues. Proc. Natl. Acad. Sci. USA, 101: 5339–5346.

Kalsbeek, Fliers, Hofman, Swaab, Van Someren & Buijs
Progress in Brain Research, Vol. 153
ISSN 0079-6123

CHAPTER 16

Circadian time keeping: the daily ups and downs of genes, cells, and organisms*

Ueli Schibler*

Department of Molecular Biology, Sciences III, University of Geneva, 30 Quai Ernest, Ansermet, CH-1211 Geneva-4, Switzerland

Abstract: Light-sensitive organisms — from cyanobacteria to humans — contain circadian clocks that produce ~24-h cycles in the absence of external time cues. In various systems, clock genes have been identified and their functions examined. Negative feedback loops in clock gene expression were initially believed to control circadian rhythms in all organisms. However, recent experiments with cyanobacteria and the filamentous fungus *Neurospora crassa* tend to favour protein phosphorylation cycles as the basic timekeeper principle in these species. The study of clock genes in mammals has led to a further surprise; practically all body cells were found to harbour self-sustained circadian oscillators. These clocks are co-ordinated by a central pacemaker in the animal, but they keep ticking in a cell-autonomous fashion when maintained in tissue culture. In mammals, most physiology is influenced by the circadian timing, including rest–activity rhythms, heartbeat frequency, arterial blood pressure, renal plasma flow, urine production, intestinal peristaltic motility, and metabolism.

History of circadian rhythms: from hobby gardening to feedback loops in gene expression

Biological clocks are systems measuring time in the absence of external timing cues such as light or temperature cycles. "Circadian" is derived from the Latin words circa diem (about a day). As the name indicates, circadian timekeepers generate cycles of about — but not exactly — 24 h. Hence, the phase of these oscillators must be corrected by a few minutes every day to keep abreast of geophysical time. The synchronization to the photoperiod is controlled by light inputs in all known organisms, and indeed, circadian clocks are unique to light-sensitive organisms (Dunlap, 1999).

Curiously enough, circadian rhythms were not discovered by biologists, but by a very observant French astronomer, Jean-Jacques d'Ortous De Mairan. In 1729, he noticed daily leaf movements of the mimosa plants cultivated in his backyard and decided to investigate whether these movements were influenced by changes in light intensity or by an endogenous clock. A simple experiment provided the unequivocal answer: when he transferred the plants to pots and kept them in his dark basement, the leaflets continued to fold and unfold in a daily rhythm. Hence, he concluded that the timing of leaf movements was not determined simply by environmental changes (De Mairan, 1729).

De Mairan was not only a good observer, but must have had second sight. He predicted that progress in understanding biological clocks would be slow. Indeed, another 100 years were to pass before the Swiss botanist August de Candolle observed that the leaf movement rhythms of mimosa plants

---

*A large part of this article has been printed in B.I.F. FUTURA (December 2005 issue) in a slightly modified form. Permission for including these text sections has been granted by Dr. Claudia Walther, Boehringer Ingelheim Foundation, Germany.

*Corresponding author. Tel.: +41-22-379-61-79; Fax: +41-22-379-68-68; E-mail: ueli.schibler@molbio.unige.ch

DOI: 10.1016/S0079-6123(06)53016-X

were free-running with a period length of 22 h rather than 24 h under constant conditions (Eckardt, 2005). This was an important discovery, since it rendered the possibility unlikely that unnoticed environmental rhythms, for example low amplitude temperature cycles or daily variations in air composition, would drive the leave movement cycles in constant darkness. Yet another 100 years later, Erwin Bünning published a study verifying that the period length of Phaseolum leaf movements has a genetic basis (Bünning, 1932). Compelling evidence for the Mendelian inheritance of circadian rhythms was published by Ronald Konopka and Seymour Benzer in 1971 (Konopka and Benzer, 1971). These authors designed an ingeniously simple mutant screen for the fruit fly Drosophila melanogaster, based on the observation that eclosion of flies from pupae is highly circadian. Konopka and Benzer mutagenized male flies using ethyl methane sulfonate. They mated them with females containing fused X-chromosomes and collected those of the male offspring that hatched at unusual times (i.e. during the dark phase). All these males had received their X-chromosomes from their mutagenized fathers, since zygotes with the two fused X-chromosome developed into females. The screening of about 2000 males of the F1 generation resulted in the isolation of three different alleles of one and the same X-linked gene, later now referred to as period (per). Further analysis revealed that one of these caused arrhythmicity, another period lengthening, and a third one period shortening of both eclosion and locomotor rhythms. In 1987, the groups of Michael Rosbash, Jeff Hall, and Michael Young reported the molecular cloning and transcript analysis of the per gene (Bargiello et al., 1984; Reddy et al., 1984). A few years later, Rosbash and co-workers proposed that circadian rhythms might be generated by a feedback loop of per gene expression, based on the observations that per mRNA abundance followed a daily cycle in Drosophila heads and that ectopic overexpression of Per protein attenuated transcription of the resident per gene (Hardin et al., 1990; Zeng et al., 1994). The next Drosophila clock gene to be cloned was timeless (tim) (Myers et al., 1995), whose protein product Tim forms multimeric complexes with Per (Gekakis et al., 1995; Rutila et al., 1996).

The year 1997 was a magical year for circadian rhythm research in mammals. Two groups, Tei and coworkers (Tei et al., 1997) and Sun and coworkers (Sun et al., 1997), independently reported on the long-awaited identification of a mammalian homolog of the Drosophila period gene. Both groups demonstrated that the mRNA specified by this gene (Period1) oscillated in abundance in the suprachiasmatic nucleus (SCN), the master circadian pacemaker in the hypothalamus. In the same year Clock (Circadian locomotor output cycles kaput), the first gene encoding a positively acting transcription factor of the circadian clock, was isolated in mice in a heroic genetic approach by Joseph Takahashi and co-workers (King et al., 1997). One year later, BMAL1 the major dimerization partner of clock was identified in a yeast-two-hybrid screen using Clock cDNA as a bait and a hamster hypothalamus cDNA as a prey (Gekakis et al., 1998). In the same year, Drosophila genes encoding orthologues of the mammalian transcription factors CLOCK and BMAL1 were isolated and genetically dissected in the fruit fly (Allada et al., 1998; Rutila et al., 1998). In the years to follow, we witnessed a true clock gene explosion. Several protein kinases (CK1ε/δ, CKII, GSK3, orthologues in both mammals and mammalian species), additional transcription factors, such as Vrille, and Pdp1 in Drosophila (Cyran et al., 2003), and RORα,β,γ (also referred to as NR1F1,2,3, nuclear receptor subfamily1 group F, members 1 to 3) and Rev-Erbα (also referred to as NR1D1, nuclear receptor subfamily1 group D, member 1) in mammals (Preitner et al., 2002; Akashi and Takumi, 2005), RNA-binding proteins (NONO in mouse, NonA in Drosophila (Brown et al., 2005b), and a mouse histone methyltransferase-binding protein (WDR5, WD repeat domain protein 5 (Brown et al., 2005b), were added to the repertoire. Thus, while the negative feedback loop model still prevails for metazoan circadian oscillators, its biochemical details are still being modified non-stop. Figure 1 illustrates a currently publicized model for the mammalian circadian clockwork circuitry. In the negative limb, which is the centrepiece of the oscillator, the genes encoding two PER (period) and two CRY (cryptochrome) isoforms are activated by the two

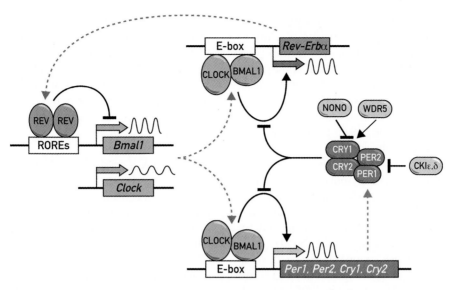

Fig. 1. Hypothetical model of the molecular circadian oscillator. The rhythm generating circuitry is thought to be based on molecular feedback loops within a positive limb (CLOCK, BMAL1) and a negative oscillator limb (PER and CRY proteins) that are interconnected via the nuclear orphan receptor Rev-erbα (see text for details and inconsistencies of this model).

PAS-domain helix-loop-helix transcription factors CLOCK and BMAL1. Once PER:CRY repressor complexes reach a critical concentration, they block the stimulatory action of CLOCK and BMAL1. As a consequence, *Per* and *Cry* genes are silenced, PER and CRY protein concentrations diminish, and a new auto-regulatory cycle of PER and CRY expression can ensue. The transcription of *Rev-erbα* is regulated by the same mechanism, and the periodic accumulation of REV-ERBα elicits the cyclic repression of *Bmal1* (and, to a lesser extent *Clock*). Additional components, such as protein kinases, the RNA-binding protein NONO, and the histone methyl-transferase-binding protein WDR5 are also required for keeping the clock ticking at its normal pace.

A circadian clock in the test tube: protein kinases and phosphatases

Cyanobacteria of the species *Synechococcus elongatus* and the filamentous fungi species *Neurospora crassa* are further examples of systems in which circadian clocks are genetically and biochemically dissected. Although the clock components of these primitive organisms bear no compelling sequence

similarity to those of Drosophila or mammalian clocks, negative feedback loops in *clock* gene expression were thought to be the universal rhythm-generating mechanism in all biological systems. However, this view changed dramatically due to Takao Kondo and co-workers' spectacular work on the cyanobacteria clock (Nakajima et al., 2005; Tomita et al., 2005). The three proteins KaiA, KaiB, and KaiC, encoded by the *kaiA/BC* operon, constitute the centrepiece of the cyanobacterium *Synechococcus elongatus*. Like all cyanobacteria genes, *kai* genes are transcribed in a circadian manner. In addition, KaiC, which can act both as an autokinase and an autophosphatase, undergoes robust daily phosphorylation cycles that depend on its physical interactions with KaiA and KaiB. In the absence of photosynthesis, cyanobacteria cannot produce ATP levels that are sufficiently high for transcription and translation. Therefore, RNA and protein synthesis rapidly cease in cyanobacteria kept in constant darkness. However, circadian KaiC phosphorylation continues for several days in the absence of transcription and translation suggesting that variation in protein abundance is not required for circadian rhythm generation (Tomita et al., 2005). Spurred on by this surprising but gratifying result, Kondo and co-workers purified KaiA, KaiB, and KaiC as recombinant proteins from overexpressing *E. coli* strains and mixed them at the concentration ratios observed in vivo. The addition of ATP to this protein mix triggered 24-h rhythms of KaiC phosphorylation that remained synchronized for several days in the test tube (Nakajima et al., 2005). Hence, protein phosphorylation rather than transcription/translation cycles may be the basic principle of the biological clock in cyanobacteria.

Protein phosphorylation also plays an essential part in generating circadian rhythms in *N. crassa*. In this filamentous fungus, clock protein frequency (FRQ) inhibits the action of the White Collar transcription factors WC-1 and WC-2, which bind to their cognate DNA elements within the *frq* gene promoter as heterodimer complexes (referred to as WCC). Initially, FRQ was thought to act as a transcriptional repressor. However, in 2005, Michael Brunner's group presented compelling evidence that FRQ stimulates the phosphorylation of WCC by an unknown protein kinase, thereby abolishing the DNA-binding ability of this transcription factor complex (Schafmeier et al., 2005). WCC phosphorylation cycles rather than FRQ abundance rhythms may therefore lie at the heart of Neurospora oscillators, and circadian transcription may be a clock output rather than the core mechanism. It will be enticing to investigate whether circadian WCC phosphorylation can be reconstituted in the test tube once the relevant protein kinases and phosphatases will have been identified.

These findings with cyanobacteria and *Neurospora* oscillators strongly suggest that feedback loops in post-translational clock protein modifications are also essential for the oscillator mechanism in insects and mammals. Indeed, the gene expression model presented in Fig. 1 for the mammalian molecular clock suffers from inconsistencies in the phase relationship between PER and CRY mRNA and protein accumulation. Hence, at least in its simplest version, it cannot account for all the experimental observations and theoretical considerations made in the mammalian system. For example, if CRY and PER were the direct transcriptional repressors of their own genes, *Per* and *Cry* transcription cycles should be in antiphase to PER and CRY protein accumulation cycles. However, this prediction has been invalidated by both experimental (Gachon

et al., 2004) and mathematical evidence (Schaad, Wanner, and Schibler, unpublished). In fact, antiphasic 24-h accumulation cycles of an mRNA and the protein encoded by it are only possible if temporal protein synthesis and/or stability are regulated independently from temporal mRNA expression. This can be readily demonstrated by numerical solutions of the equation:

$$[protein] = K \int_{-\infty}^{x} e^{[\ln 2(t-x)/t_{1/2}]} f(t) [mRNA] \, dt$$

where K is a constant depending on the efficiency of protein synthesis, $t_{1/2}$ the protein half life, t the time of protein synthesis, x the time of protein accumulation, and $f(t)$ the function of time of mRNA concentration, determined by a best-fit equation to the experimentally determined curve. By solving this equation for a series of constant half-lives, we were unable to generate antiphasic protein accumulation cycles from the experimentally measured mRNA accumulation cycles. In fact, even very long protein half-lives that reduce the amplitude of protein cycles to insignificant levels cannot delay the phase of protein accumulation by more than about 4 h from that of mRNA accumulation. Conceivably, transcription cycles play a more important role in clock outputs rather than in the core mechanism of circadian rhythm generation, as has been suggested above for *N. crassa*.

Zeitgeber time, circadian time, and jet lag

As mentioned in the first section, circadian clocks measure daytime only approximately and must be resynchronized daily to keep abreast of geophysical time. Light is the major *Zeitgeber* (German term for "provider of timing cues") for this phase-resetting in all the organisms under investigation. In chronobiology jargon, the time imposed by light–dark cycles is *Zeitgeber* time (ZT). ZT0 is usually defined as the time when the lights are switched on. In contrast, circadian time (CT) is used for the time determined by a circadian oscillator under constant conditions (i.e. in the absence of a synchronizing *Zeitgeber*). For example, mice of the common laboratory strain C57B6 free-run with a period length of 23.77 h (Schwartz and Zimmerman, 1990). Thus, their circadian clock must be phase-delayed by about 14 min every day. In contrast, human oscillators generate circadian cycles of approximately 24.18 h (Czeisler et al., 1999), and thus have to be phase-advanced by 11 min every day. Since circadian oscillators can be phase-shifted by roughly one to two hours per day, these corrections are not problematic. However, east- and west-bound transatlantic flights cause time-zone differences of several hours which cannot be accommodated in a single day. Hence, for transatlantic travellers it takes several days before circadian time adjusts to *Zeitgeber* time. Since human physiology and behaviour are influenced to a great extent by circadian time, many aspects of daily life style are at odds with the outside world for the first few days after a transatlantic journey. This jet lag affects sleep–wake cycles, as well as the physiology of the kidney and the gastro-intestinal system (see below).

The light-induced phase-shifting of circadian clocks is gated by the oscillator itself. This can be deduced from the phase response curve: light pulses delivered to laboratory mice kept in constant darkness during the subjective day (CT0-CT12) have little influence on the phase, while light pulses delivered during the first half (CT12–CT18) and second half (CT18–CT24) of the subjective night delay or advance the phase respectively. This phase shifting behaviour is probably also involved in the adaptation to seasonal behavioural changes of nocturnal animals: when the days get longer in spring, the phase of activity onset is delayed in the evening and advanced in the morning, leading to a shortened activity phase.

The signalling pathways involved in phase shifting are not yet understood in molecular detail. Nevertheless, I will address some recent observations on how the mammalian circadian timing system is co-ordinated.

The mammalian circadian timing system: a clock in every cell?

The mammalian circadian timing system influences nearly all physiological processes, including sleep–wake cycles, cardiovascular activity, body temperature, acuity of the sensory system, renal plasma flow, intestinal peristaltic motility, hepatic metabolism and detoxification, and many functions of the endocrine system (Schibler et al., 2003). The rhythms of all these clock-controlled processes depend on two tiny aggregates of neurons, called SCN. They were named after their location, immediately above the optic chiasm in the ventral hypothalamus. Bilateral lesions of the SCN renders animals completely arrhythmic. However, behavioural rhythms can be restored in such SCN-lesioned animals by implants of foetal SCN tissue into the third ventricle, close to the position of the ablated SCN (Ralph et al., 1990). The fact that the free-running period length is determined by the donor tissue is even more important: if wild-type hamsters (period length close to 24 h) are SCN-lesioned and grafted with SCN implants of *Tau*-mutant hamsters (period length 20 h), they free-run with a period length of 20 h (Ralph et al., 1990) (Tau is the Greek word for time and is used in chronobiology jargon for period length).

Circadian pacemakers were originally believed to exist only in a few specialized cell types, such as SCN neurons. However, this view has been challenged by the discovery that circadian clocks may exist in most peripheral cell types, and even in immortalized tissue culture cells (Balsalobre et al., 1998). The identification of mammalian clock and clock-output-genes facilitated the examination of circadian rhythmicity in peripheral tissues. These genes were shown to be expressed in daily cycles not only in the SCN neurons, but in virtually all cell types. Nevertheless, the fact that such genes are active in a cyclic fashion in a given tissue does not prove the existence of a circadian clock in this tissue. Indeed, rhythmic gene expression in peripheral organs could be driven simply by oscillating hormones, whose rhythmic secretion is governed by the SCN. Clearly, the unequivocal identification of circadian oscillators in peripheral cell types depends on the demonstration that such cells can generate circadian rhythms in the absence of the SCN master pacemaker. In 1998, Aurelio Balsalobre and co-workers ascertained that circadian cycles of gene expression lasting for several days could be elicited by a serum shock in RAT-1 fibroblasts cultured in vitro (Balsalobre et al., 1998). These experiments were motivated by the observation that most immediate early genes induced by light in the SCN — including the clock genes *Per1* and *Per2* — are also induced by serum in tissue culture cells. This light-induced immediate-early gene expression correlates with phase-shifting. Subsequently, Emi Nagoshi and co-workers and David Welsh and co-workers measured circadian gene expression in real time and in individual mouse and rat fibroblast to demonstrate that their oscillators function in a self-sustained and cell-autonomous fashion (Nagoshi et al., 2004; Welsh et al., 2004). Since the periods are somewhat variable from cell to cell and from cycle to cycle, the rhythms are not synchronized in untreated cell populations. However, a serum shock transiently synchronizes the oscillations in clock gene expression, such that these can be recorded by studies on cell populations (Fig. 2).

Self-sustained circadian oscillators have also been located in slices from many peripheral organs kept in tissue culture, including liver, kidney, pituitary gland, cornea, and lung. Most body cells probably harbour such clocks (Yoo et al., 2004). In the intact animal, they have to be synchronized to yield coordinated circadian outputs in overt physiology and behaviour. How is this accomplished? As mentioned above, light–dark cycles are the major *Zeitgebers* for the central SCN pacemaker. Light signals required for SCN clock synchronization are not only perceived by classical rod and cone photoreceptors in the outer retina layer, but also by melanopsin-containing ganglion cells in the inner retina layer. These photic inputs are then transmitted as electrical signals to SCN neurons via the retino-hypothalamic tract (RHT). The electrical signalling to the SCN involves the neurotransmitters glutamate and PACAP (also referred to as ADC4AP, adenylate cyclase activating polypeptide) which, when bound to their receptors in SCN neurons, provoke the influx of Ca^{2+}. This results in the activation of several protein kinases (protein kinase A, protein kinase C, mitogen-activated protein kinases), in the phosphorylation of the transcription factor CREB (*C*yclic *AMP R*esponse *E*lement-*B*inding protein), and in the stimulation of immediate-early-gene expression (Albrecht, 2004). *Per1* and *Per2* are among the induced immediate-early genes in all examined species, and the burst in the accumulation of PER proteins probably alters the phase of the molecular clock.

Peripheral oscillators in mammals are not light sensitive and must be phase-entrained by chemical and/or neuronal signalling pathways. It is worthy of note that the circadian oscillators of *in vitro* cultured fibroblasts can be synchronized by a bewildering variety of signalling substances, including those activating nuclear hormone receptors (e.g. glucocorticoid receptor, retinoic acid receptor), G-protein coupled receptors, tyrosine kinase receptors and Ca^{2+} channels (Schibler et al., 2003). Moreover, even low-amplitude body temperature rhythms can sustain the synchronization of peripheral clocks (Brown et al., 2002). The precise molecular mechanisms by which peripheral timekeepers are phase-entrained are not clear, but appear to depend on daily feeding–fasting cycles. Indeed, feeding time, while not affecting the SCN clock, is the most dominant *Zeitgeber* for peripheral oscillators. For example, daytime feeding of nocturnal rodents for a week or longer completely inverts the phase of peripheral oscillators (Damiola et al., 2000).

Our studies have shown that, under certain feeding conditions, glucocorticoid signalling is also used by the SCN to synchronize circadian oscillators in peripheral tissues (Le Minh et al., 2001). One interesting difference between peripheral and central circadian oscillators is that only the latter have a gated phase-shifting behaviour (see above). For example, strong phase-shifting stimuli, such as the glucocorticoid receptor agonist Dexamethasone, can reset the phase of peripheral clocks independently of circadian time, both in vivo and in vitro (Balsalobre et al., 2000; Le Minh et al., 2001). The synchronization of central and peripheral mammalian oscillators is schematically outlined in Fig. 3.

Jet lags, caused by sudden large transitions of time zones that cannot be overcome in a single day, are commonly regarded as sleeping disturbances. However, peripheral organs such as the kidney and the liver are also affected by such time perturbations, and the phase adjustments of their clocks may even lag behind that of the SCN pacemaker. Urination during the night and digestion problems after heavy meals are two typical manifestations of jet lag in the kidney and the gastrointestinal tract. In fact, renal plasma flow and urine production, intestinal peristaltic motility, production of hydrochloric acid by the stomach, secretion of pancreatic enzymes and bile acids into the gut, and processing of ingested food components and toxins by the liver are all highly circadian (Schibler et al., 2003). These processes require several days to readjust to the new time zone after west- or east-bound journeys.

Human behaviour: larks and owls

Humans can be classified into different chronotypes according to their sleep–wake cycles. For obvious reasons, such innate behavioural differences are generally not apparent during workdays. However, at weekends and on holiday the "true" chronotypes manifest themselves in surprisingly large variations in the timing of rest–activity cycles. As reported by Roenneberg and co-workers (Roenneberg et al., 2003), extreme early birds (larks in chronobiology parlance) wake up and become active when extreme late birds (owls in chronobiology parlance) go to bed. These observations were made on the basis of self-reports of thousands of individuals who were

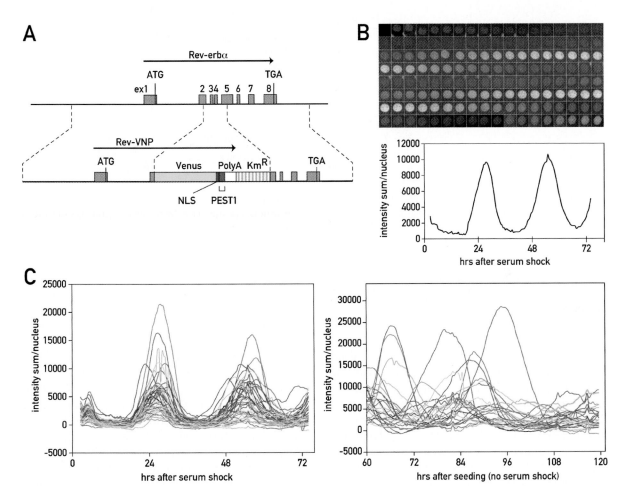

Fig. 2. In vitro cultured fibroblasts contain cell-autonomous circadian oscillators. (A) A Rev-VNP reporter gene was constructed by inserting the open reading frame of Venus, a yellow fluorescent protein, followed by DNA sequences encoding a PEST domain and a nuclear translocation signal (NLS), into a genomic DNA fragment encompassing the *Rev-Erbα* locus. (B) Fluorescent time-lapse microscopy of a Rev-VNP expressing NIH3T3 cell line (1 picture/30 min during 72 h) shows the circadian accumulation of VNP in the nuclei of individual cells. The quantification of the fluorescent signal is shown below the time-lapse recording. (C) VNP accumulation cycles in NIH-3T3 fibroblasts before (right panel) and after synchronization (left panel) by a serum shock. (Reprinted from reference *29* by kind permission of Elsevier Science, Amsterdam, The Netherlands.)

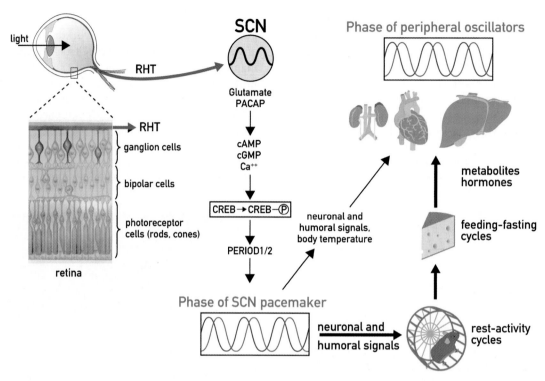

Fig. 3. Synchronization of central and peripheral circadian clocks. The master pacemaker in the SCN is synchronized via light–dark cycles generated by the photoperiod. It then synchronizes peripheral oscillators through behavioural rhythms (rest–activity cycles which engender daily feeding cycles) and neuronal and humoral signals (see text for details). The chemical *Zeitgeber* signals associated with feeding–fasting rhythms and the precise nature of the neuronal signals have not yet been identified.

interviewed about their sleeping behaviour. In a few cases, a hereditary basis for such behaviour could be unequivocally documented. For example, in one case of familial advanced sleep phase syndrome (FASPS), the gene responsible for precocious sleepiness was identified as the circadian clock gene *PER2* (human period 2). Afflicted subjects carry a point mutation in *PER2*, changing a serine into a glycine at amino acid position 662. In the wild-type PER2 protein, serine 662 is phosphorylated by the protein kinase CK1ε, and this phosphorylation reduces the metabolic stability of the protein (Toh et al., 2001). The lack of phosphorylation in the S662G mutant protein could conceivably speed up the oscillator and thereby advance the phase, because the stabilized protein reaches threshold concentrations required for auto-repression earlier during the day.

The systematic recording of free-running human period length means that human subjects must remain under observation for several weeks under laboratory conditions. This is possible in only a limited number of cases. The prohibitively high cost of such experiments does not make experimental recording amenable to large numbers of human chronotypes. However, a recent study by Steven Brown and co-workers may provide a new way of studying human circadian rhythms (Brown et al., 2005a). Fibroblasts from human skin punch biopsies were infected with a lentiviral vector containing a luciferase reporter gene driven from the *Bmal1* promoter. Three to four days after infection, circadian bioluminescence cycles were recorded in real time for four to five days. While the values of different skin biopsies from the same individual were practically identical, skin biopsies from different individuals yielded highly different period lengths, ranging from 22 to 26 h (Fig. 4). Similar experiments with wild-type mice and mice carrying mutations in various clock genes

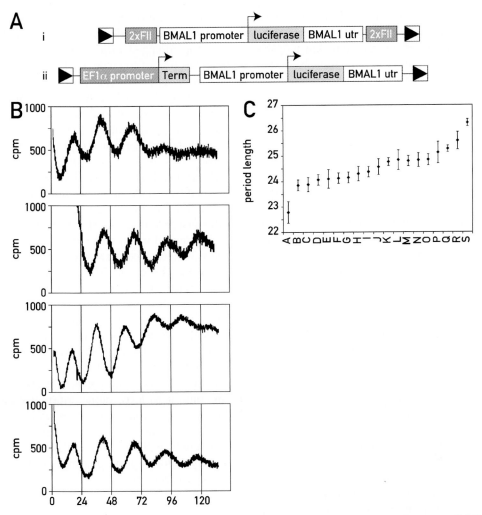

Fig. 4. Circadian bioluminescence in human fibroblasts infected with a lentiviral luciferase expression vector (A) The lentiviral circadian reporter construct contains the mouse *Bmal1* promoter, the firefly luciferase-coding region, and the *Bmal1* 3′UTR, flanked by the long terminal repeats (LTRs) of a lentiviral packaging vector. A DNA segment composed of the *EF1a* promoter and a SV40 terminator is inserted between the upstream LTR and the *Bmal1* promoter, in order to attenuate the influence of transcription regulatory sequences of genomic sequences after viral integration (for details see Brown et al., 2005a). (B) Skin punch biopsies were obtained from 19 individuals. Fibroblasts were isolated from each biopsy, infected with the lentiviral circadian reporter vector shown in panel A, and analyzed by real-time luminometry. Individuals are designated with the letters A–S. (C) Summary of the period lengths of BMAL-luciferase oscillations from all 19 individuals. Each value shows the average plus or minus the standard deviation from two different trials of two different infections of fibroblasts from two to five biopsies per subject. The probability by Student's *t*-test that the most different individuals (A and S) have the same period length is 0.00001; the probability that the second most different (B and R) are equal is 0.004.

suggested that the period values measured in fibroblasts are correlated with those determined for circadian behaviour. Thus, the fibroblasts of mice with behavioural periods longer or shorter than wild-type also display longer and shorter periods respectively (Pando et al., 2002; Brown et al., 2005a). It will be interesting to investigate circadian gene expression in the primary human fibroblasts of different human chronotypes. If, as in mice, a correlation between period length of fibroblast gene expression and behaviour can be established in human subjects, the inexpensive re- cording of luminescence cycles from skin biopsies may be used as readouts in genetic linkage studies aimed at the identification of loci that influence human rest–activity cycles.

Perspectives

Although progress in circadian rhythm research has advanced significantly during the past two decades, chronobiologists are unlikely to get bored in the near future. CLOCK input pathways have still to be dissected with regard to the molecular signalling mechanisms by which light–dark cycles synchronize circadian oscillators. Photoreceptors and down- stream components required for this process have been identified in several systems, but the path has not yet been paved from capturing of photons to modifying gene expression and resetting the phase. Even less is known about how the SCN synchro- nizes slave oscillators in peripheral tissues, and about the precise roles of feeding rhythms, hor- mones, and neuronal signals in this process.

Scientists working on molecular oscillator mech- anisms in complex organisms such as plants, insects, and mammals will not rest before they have suc- ceeded in reconstituting a ticking clock in the test tube. As mentioned above, this was recently accom- plished in cyanobacteria[21], but even in this simple system the protein–protein interactions generating 24-h KaiC autophosphorylation rhythms in vitro remain obscure. If, as many investigators in the field still maintain, gene expression cycles are at the heart of circadian oscillators in animals and higher plants, the in vitro reconstitution of a ticking clock might well be a "mission impossible". However, it may be possible some day to assemble such a genetic clock- work circuitry in yeast cells, which lack a circadian timekeeping system.

Another fundamental issue is the demonstration in the laboratory that circadian physiology is ben- eficial to higher organisms. In view of the fact that virtually all light-sensitive organisms possess circa- dian clocks, there is little doubt that these must provide a selective advantage to their owners. How- ever, compelling experimental evidence in support of this claim has been found only for cyanobacteria and the plant *Arabidopsis thaliana*. Thus, cyano- bacteria equipped with clocks whose period length matches that of the light–dark cycles to which they are exposed, rapidly outgrow bacteria with a non- resonating clock (Woelfle et al., 2004). These ex- periments clearly show that it is the rhythm rather than a crippled gene that is responsible for the growth advantage, since depending on the light– dark cycles the same mutation can be beneficial or deleterious. Likewise, Arabidopsis plants with a resonating oscillator grow more rapidly and are more resistant to environmental insults than plants with discordant clocks (Dodd et al., 2005). In an- imals, mutations in some clock and clock-control- led genes have been shown to cause severe health problems (Fu et al., 2002; Fu et al., 2005). How- ever, since "resonance experiments" such as the ones described above for cyanobacteria and Arab- idopsis have not yet been conducted successfully in animals, we still do not know whether these health problems are the result of perturbed rhythms.

Finally, in spite of the well-funded knowl- edge that drug metabolism, efficacy, and toxicity are subject to large daily variations (Levi, 2000), attempts to explore these observations clinically are still rather timid. Circadian biologists and physicians will have to collaborate a lot more closely before chronopharmacology and chrono- therapy can flourish.

Acknowledgments

I am very grateful to Nicolas Roggli for expert preparation of the figures. Work in my laboratory has been supported by the Canton of Geneva, the Swiss National Science Foundation through an individual grant and the NCCR program grant Frontiers in Genetics, the Louis Jeantet Founda- tion of Medicine, and the Bonizzi-Theler Stiftung.

References

Akashi, M. and Takumi, T. (2005) The orphan nuclear receptor RORalpha regulates circadian transcription of the mamma- lian core-clock Bmal1. Nat. Struct. Mol. Biol., 12: 441–448.

Albrecht, U. (2004) The mammalian circadian clock: a network of gene expression. Front. Biosci., 9: 48–55.

Allada, R., White, N.E., So, W.V., Hall, J.C. and Rosbash, M. (1998) A mutant Drosophila homolog of mammalian clock disrupts circadian rhythms and transcription of period and timeless. Cell, 93: 791–804.

Balsalobre, A., Brown, S.A., Marcacci, L., Tronche, F., Kellendonk, C., Reichardt, H.M., Schutz, G. and Schibler, U. (2000) Resetting of circadian time in peripheral tissues by glucocorticoid signaling. Science, 289: 2344–2347.

Balsalobre, A., Damiola, F. and Schibler, U. (1998) A serum shock induces circadian gene expression in mammalian tissue culture cells. Cell, 93: 929–937.

Bargiello, T.A., Jackson, F.R. and Young, M.W. (1984) Restoration of circadian behavioural rhythms by gene transfer in Drosophila. Nature, 312: 752–754.

Brown, S.A., Fleury-Olela, F., Nagoshi, E., Hauser, C., Juge, C., Meier, C.A., Chicheportiche, R., Dayer, J.M., Albrecht, U. and Schibler, U. (2005a) The period length of fibroblast circadian gene expression varies widely among human individuals. PLoS Biol., 3: e338.

Brown, S.A., Ripperger, J., Kadener, S., Fleury-Olela, F., Vilbois, F., Rosbash, M. and Schibler, U. (2005b) PERIOD1-associated proteins modulate the negative limb of the mammalian circadian oscillator. Science, 308: 693–696.

Brown, S.A., Zumbrunn, G., Fleury-Olela, F., Preitner, N. and Schibler, U. (2002) Rhythms of mammalian body temperature can sustain peripheral circadian clocks. Curr. Biol., 12: 1574–1583.

Bünning, E. (1932) Ueber die Erblichkeit der Tageszeitperiodizität bei den Phaseolus Blättern. J. wiss. Bot., 81: 411–418.

Cyran, S.A., Buchsbaum, A.M., Reddy, K.L., Lin, M.C., Glossop, N.R., Hardin, P.E., Young, M.W., Storti, R.V. and Blau, J. (2003) vrille, Pdp1, and dClock form a second feedback loop in the Drosophila circadian clock. Cell, 112: 329–341.

Czeisler, C.A., Duffy, J.F., Shanahan, T.L., Brown, E.N., Mitchell, J.F., Rimmer, D.W., Ronda, J.M., Silva, E.J., Allan, J.S., Emens, J.S., Dijk, D.J. and Kronauer, R.E. (1999) Stability, precision, and near-24-hour period of the human circadian pacemaker. Science, 284: 2177–2181.

Damiola, F., Le Minh, N., Preitner, N., Kornmann, B., Fleury-Olela, F. and Schibler, U. (2000) Restricted feeding uncouples circadian oscillators in peripheral tissues from the central pacemaker in the suprachiasmatic nucleus. Genes Dev., 14: 2950–2961.

De Mairan, J.J.d.O. (1729) Observation botanique. Histoire de l'Academie Royale des Sciences, 35–36.

Dodd, A.N., Salathia, N., Hall, A., Kevei, E., Toth, R., Nagy, F., Hibberd, J.M., Millar, A.J. and Webb, A.A. (2005) Plant circadian clocks increase photosynthesis, growth, survival, and competitive advantage. Science, 309: 630–633.

Dunlap, J.C. (1999) Molecular bases for circadian clocks. Cell, 96: 271–290.

Eckardt, N.A. (2005) Temperature entrainment of the arabidopsis circadian clock. Plant Cell, 17: 645–647.

Fu, L., Patel, M.S., Bradley, A., Wagner, E.F. and Karsenty, G. (2005) The molecular clock mediates leptin-regulated bone formation. Cell, 122: 803–815.

Fu, L., Pelicano, H., Liu, J., Huang, P. and Lee, C. (2002) The circadian gene Period2 plays an important role in tumor suppression and DNA damage response in vivo. Cell, 111: 41–50.

Gachon, F., Nagoshi, E., Brown, S.A., Ripperger, J. and Schibler, U. (2004) The mammalian circadian timing system: from gene expression to physiology. Chromosoma, 113: 103–112.

Gekakis, N., Saez, L., Delahaye-Brown, A.M., Myers, M.P., Sehgal, A., Young, M.W. and Weitz, C.J. (1995) Isolation of timeless by PER protein interaction: defective interaction between timeless protein and long-period mutant PERL. Science, 270: 811–815.

Gekakis, N., Staknis, D., Nguyen, H.B., Davis, F.C., Wilsbacher, L.D., King, D.P., Takahashi, J.S. and Weitz, C.J. (1998) Role of the CLOCK protein in the mammalian circadian mechanism. Science, 280: 1564–1569.

Hardin, P.E., Hall, J.C. and Rosbash, M. (1990) Feedback of the Drosophila period gene product on circadian cycling of its messenger RNA levels. Nature, 343: 536–540.

King, D.P., Zhao, Y., Sangoram, A.M., Wilsbacher, L.D., Tanaka, M., Antoch, M.P., Steeves, T.D., Vitaterna, M.H., Kornhauser, J.M., Lowrey, P.L., Turek, F.W. and Takahashi, J.S. (1997) Positional cloning of the mouse circadian clock gene. Cell, 89: 641–653.

Konopka, R.J. and Benzer, S. (1971) Clock mutants of Drosophila melanogaster. Proc. Natl. Acad. Sci. USA, 68: 2112–2116.

Le Minh, N., Damiola, F., Tronche, F., Schutz, G. and Schibler, U. (2001) Glucocorticoid hormones inhibit food-induced phase-shifting of peripheral circadian oscillators. EMBO J., 20: 7128–7136.

Levi, F. (2000) Therapeutic implications of circadian rhythms in cancer patients. Novartis Found. Symp., 227: 119–136 discussion 136–142.

Myers, M.P., Wager-Smith, K., Wesley, C.S., Young, M.W. and Sehgal, A. (1995) Positional cloning and sequence analysis of the Drosophila clock gene, timeless. Science, 270: 805–808.

Nagoshi, E., Saini, C., Bauer, C., Laroche, T., Naef, F. and Schibler, U. (2004) Circadian gene expression in individual fibroblasts; cell-autonomous and self-sustained oscillators pass time to daughter cells. Cell, 119: 693–705.

Nakajima, M., Imai, K., Ito, H., Nishiwaki, T., Murayama, Y., Iwasaki, H., Oyama, T. and Kondo, T. (2005) Reconstitution of circadian oscillation of cyanobacterial KaiC phosphorylation in vitro. Science, 308: 414–415.

Pando, M.P., Morse, D., Cermakian, N. and Sassone-Corsi, P. (2002) Phenotypic rescue of a peripheral clock genetic defect via SCN hierarchical dominance. Cell, 110: 107–117.

Preitner, N., Damiola, F., Luis Lopez, M., Zakany, J., Duboule, D., Albrecht, U. and Schibler, U. (2002) The orphan nuclear receptor REV-ERBalpha controls circadian transcription

282

within the positive limb of the mammalian circadian oscillator. Cell, 110: 251–260.

Ralph, M.R., Foster, R.G., Davis, F.C. and Menaker, M. (1990) Transplanted suprachiasmatic nucleus determines circadian period. Science, 247: 975–978.

Reddy, P., Zehring, W.A., Wheeler, D.A., Pirrotta, V., Hadfield, C., Hall, J.C. and Rosbash, M. (1984) Molecular analysis of the period locus in *Drosophila melanogaster* and identification of a transcript involved in biological rhythms. Cell, 38: 701–710.

Roenneberg, T., Wirz-Justice, A. and Merrow, M. (2003) Life between clocks: daily temporal patterns of human chronotypes. J. Biol. Rhythms, 18: 80–90.

Rutila, J.E., Suri, V., Le, M., So, W.V., Rosbash, M. and Hall, J.C. (1998) CYCLE is a second bHLH-PAS clock protein essential for circadian rhythmicity and transcription of Drosophila period and timeless. Cell, 93: 805–814.

Rutila, J.E., Zeng, H., Le, M., Curtin, K.D., Hall, J.C. and Rosbash, M. (1996) The timSL mutant of the Drosophila rhythm gene timeless manifests allele-specific interactions with period gene mutants. Neuron, 17: 921–929.

Schafmeier, T., Haase, A., Kaldi, K., Scholz, J., Fuchs, M. and Brunner, M. (2005) Transcriptional feedback of Neurospora circadian clock gene by phosphorylation-dependent inactivation of its transcription factor. Cell, 122: 235–246.

Schibler, U., Ripperger, J. and Brown, S.A. (2003) Peripheral circadian oscillators in mammals: time and food. J. Biol. Rhythms, 18: 250–260.

Schwartz, W.J. and Zimmerman, P. (1990) Circadian timekeeping in BALB/c and C57BL/6 inbred mouse strains. J. Neurosci., 10: 3685–3694.

Sun, Z.S., Albrecht, U., Zhuchenko, O., Bailey, J., Eichele, G. and Lee, C.C. (1997) RIGUI, a putative mammalian ortholog of the Drosophila period gene. Cell, 90: 1003–1011.

Tei, H., Okamura, H., Shigeyoshi, Y., Fukuhara, C., Ozawa, R., Hirose, M. and Sakaki, Y. (1997) Circadian oscillation of a mammalian homologue of the Drosophila period gene. Nature, 389: 512–516.

Toh, K.L., Jones, C.R., He, Y., Eide, E.J., Hinz, W.A., Virshup, D.M., Ptacek, L.J. and Fu, Y.H. (2001) An hPer2 phosphorylation site mutation in familial advanced sleep-phase syndrome. Science, 291: 1040–1043.

Tomita, J., Nakajima, M., Kondo, T. and Iwasaki, H. (2005) No transcription–translation feedback in circadian rhythm of KaiC phosphorylation. Science, 307: 251–254.

Welsh, D.K., Yoo, S.H., Liu, A.C., Takahashi, J.S. and Kay, S.A. (2004) Bioluminescence imaging of individual fibroblasts reveals persistent, independently phased circadian rhythms of clock gene expression. Curr. Biol., 14: 2289–2295.

Woelfle, M.A., Ouyang, Y., Phanvijhitsiri, K. and Johnson, C.H. (2004) The adaptive value of circadian clocks: an experimental assessment in cyanobacteria. Curr. Biol., 14: 1481–1486.

Yoo, S.H., Yamazaki, S., Lowrey, P.L., Shimomura, K., Ko, C.H., Buhr, E.D., Siepka, S.M., Hong, H.K., Oh, W.J., Yoo, O.J., Menaker, M. and Takahashi, J.S. (2004) PERIOD2::LUCIFERASE real-time reporting of circadian dynamics reveals persistent circadian oscillations in mouse peripheral tissues. Proc. Natl. Acad. Sci. USA, 101: 5339–5346.

Zeng, H., Hardin, P.E. and Rosbash, M. (1994) Constitutive overexpression of the Drosophila period protein inhibits period mRNA cycling. EMBO J., 13: 3590–3598.

Kalsbeek, Fliers, Hofman, Swaab, Van Someren & Buijs
Progress in Brain Research, Vol. 153
ISSN 0079-6123

CHAPTER 17

The hypothalamic clock and its control of glucose homeostasis

A. Kalsbeek*, M. Ruiter, S.E. La Fleur, C. Cailotto, F. Kreier and R.M. Buijs

Netherlands Institute for Brain Research, Meibergdreef 33, 1105 AZ Amsterdam, The Netherlands

Introduction

The awareness that the hypothalamus plays an important role in food intake, and especially its involvement in the physiology and pathology of the control of energy metabolism, dates back to 1840, when Mohr described a case of hypothalamic obesity associated with a rapid gain of body weight in a patient with a pituitary tumor (reviewed in Brobeck, 1946). It was not until 100 years later, however, that the animal experiments of Hetherington and Ranson (1940) showed that obesity resulted from lesions in the hypothalamus, independent of pituitary damage. Our understanding of the hypothalamic control of energy metabolism received a second boost some 50 years later, when the group of Friedman (Zhang et al., 1994) discovered the leptin gene, i.e. the elusive hormonal factor secreted by adipose tissue that informs the brain, and especially the hypothalamus, about peripheral fat stores. Major components of energy metabolism, including feeding, thermogenesis, and glucose and lipid metabolism, show profound fluctuations along the daily light/dark (L/D)-cycle. This periodic succession of night and day has influenced life on Earth for millions of years. In mammals, these periodic changes in the environment have been "internalized" in the form of an endogenous circadian clock. Its main function is to organize the time course of physiological, hormonal, and behavioral processes in order to allow the organism to anticipate properly these changing environmental conditions. The 24-h sleep/wake cycles are generated and orchestrated from within the hypothalamus as well. The location of the responsible biological or circadian (literally "approximately one day") clock within the hypothalamic suprachiasmatic nuclei (SCN) was discovered in the early 1970s (Hendrickson et al., 1972; Moore and Eichler, 1972; Moore and Lenn, 1972; Stephan and Zucker, 1972). This master oscillator consists of interlocking transcriptional–translational feedback loops, and it contains both core clock genes necessary for oscillator maintenance, as well as specific output genes that impose their rhythmicity on the hypothalamus (Reppert and Weaver, 2002; Maywood et al., 2006). In the case of energy metabolism, the biological clock output acts to synchronize energy intake and expenditure to changes in the external environment imposed by the rising and setting of the sun (Ruiter et al., 2006a, b).

It is thought that a circadian control of its physiology and behavior imparts survival advantages to an organism. Circadian rhythms serve to temporally partition the ecological niche and enable an organism to anticipate and adapt optimally to ambient conditions, thus maximizing the potential of the organism to survive. However, animals without a functional clockwork, either through SCN-lesions or a clock gene knock-out, do not have an obvious phenotype aside from

*Corresponding author. Tel.: +31-20-566-5522; Fax: +31-20-6961006; E-mail: a.kalsbeek@nin.knaw.nl

DOI: 10.1016/S0079-6123(06)53017-1

anecdotal evidence on poor breeding (Dolatshad et al., 2006). One study in a simulated field condition suggested that arrhythmic animals are more susceptible to predation (DeCoursey et al., 1997, 2000; DeCoursey and Krulas, 1998), and in cyanobacteria a circadian pacemaker confers a significant competitive advantage when the period of the endogenous clock resonates with the environmental L/D-cycle (Ouyang et al., 1998). However, an important aspect of circadian control may also be to time and synchronize (metabolic) processes within the organism, i.e. to optimize metabolic networks by enabling a temporal partitioning of metabolic events within and between different tissues. For example, by temporally separating chemically antagonistic reactions and by limiting the expression of certain enzymes to the time of day they are needed (Schibler and Naef, 2005).

Clearly, the most obvious target of circadian control is the behavioral sleep/wake-cycle. Question is whether a physiological process such as the rhythm in energy metabolism is gated by the behavioral rhythm or subject to an independent control of the circadian oscillator. Indeed, since in many species the temporal distribution of feeding activity is so clearly affected by the biological clock, it has been assumed that the daily rhythms in circulating concentrations of metabolic hormones and substrates, such as insulin, glucagon, leptin, glucose, and free fatty acids (FFA), are mainly induced by the behavioral rhythm, instead of being subject to a direct control of the biological clock. However, in view of the hypothesis that the SCN plays an important role in anticipating major physiological events, such as increased behavioral activity, feeding activity or sleep, we assumed a direct control of the SCN. We employed two different research strategies to reveal such a direct control of the SCN: (1) a regular feeding schedule with six meals equispaced throughout the 24-h L/D-cycle (i.e. one standard meal every 4 h) to remove the strong masking impact of the rhythmic feeding behavior and to unmask a possible direct control of the circadian clock; (2) the viral retrograde tracing technique to investigate the existence of multi-synaptic neural connections between the hypothalamic biological clock and peripheral organs such as the (endocrine) pancreas, the liver, and white adipose tissue (WAT).

The present review will present the evidence for a direct control of the biological clock on the release of metabolic hormones, independent of the clock control on the temporal distribution of feeding behavior. In addition, we will present an overview of the neural mechanisms, pathways, and transmitters used by the SCN to incorporate its time-of-day message into this homeostatic system.

A daily rhythm in plasma glucose concentrations

Daily rhythms in glucose tolerance and insulin sensitivity

In order to understand how the hypothalamic biological clock conveys its circadian message into the homeostatic system(s) that control the energy balance, we focused our attention on the daily control of glucose metabolism. A pronounced daily rhythm in plasma glucose concentrations has been described in experimental animals as well as humans (Jolin and Montes, 1973; Bellinger et al., 1975; Bolli et al., 1984; Van Cauter et al., 1997; La Fleur et al., 1999; Shea et al., 2005). The peak time of plasma glucose levels shows a 12-h difference between experimental animals and humans, but in both species peak plasma glucose concentrations are attained every day shortly before awakening at the start of the main activity period. Plasma glucose concentrations are the resultant of a glucose *influx* from the gut and liver, and glucose *efflux* by its uptake in brain, muscle, and adipose tissue. Question is whether the daily changes in plasma glucose concentrations are caused by daily changes in the influx or efflux, or both. Numerous studies, both in experimental animals and humans, have shown diurnal variations in glucose uptake. Most early studies indicated morning vs. evening differences in either an oral or intravenously administered glucose bolus, with an impaired glucose tolerance in the afternoon compared with the morning, reaching a minimum around midnight (Jarrett et al., 1972; Carrol and Nestel, 1973; Whichelow et al., 1974; Zimmet et al., 1974; Lee et al., 1992). Animal studies, including our own (Fig. 1), showed daily variations in glucose tolerance as well. Again the highest

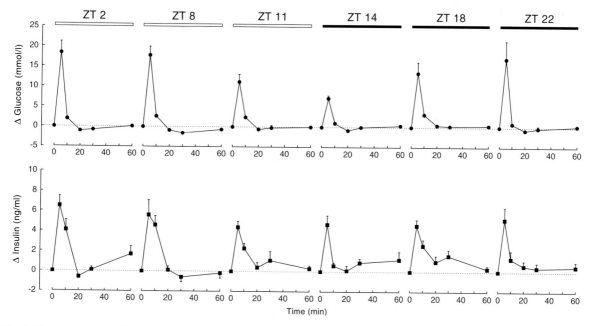

Fig. 1. Plasma glucose and insulin responses after the intravenous injection of a glucose bolus (500 mg/kg BW) at different times of the light/dark cycle. The maximal glucose increment at ZT14 was significantly lower than the ones at the other 5-time points. On the other hand, the total amount of insulin released did not differ between the different time points. Responses are expressed as the difference from the respective $t = 0$ values. ZT, Zeitgeber Time (ZT12 being defined as the onset of the dark period). Adapted from La Fleur et al. (2001).

glucose tolerance was found at the time of awakening (Penicaud and Le Magnen, 1980; Yamamoto et al., 1984b; La Fleur et al., 2001). The next question then, of course, is: how do these daily changes come about? Does the SCN control insulin sensitivity, pancreatic β-cell sensitivity, or the non-insulin dependent glucose uptake? Indeed, a number of experiments have shown daily changes in insulin sensitivity (Gibson and Jarrett, 1972; Morgan et al., 1999; Kalsbeek et al., 2003). Although the time of the highest insulin sensitivity coincides with that of maximal glucose tolerance, the daily variation is not pronounced (Fig. 2). Daily changes in β-cell sensitivity are even less evident, and if anything, β-cell sensitivity is reduced at the end of the sleep period. Thus, daily changes in insulin- and β-cell sensitivity do not fully explain the daily variation in glucose tolerance. Therefore, it seems that non-insulin dependent glucose uptake also changes on a daily basis. Indeed, evidence was recently provided for a circadian rhythm of glucose uptake in primary cultures of rat skeletal muscle cells in vitro, both in basal

and insulin-stimulated conditions (Feneberg and Lemmer, 2004). However, although the major portion of the glucose administered intravenously is taken up by muscle tissue, it is not known which tissue is responsible for the daily variation in glucose uptake. For instance, Feneberg and Lemmer (2004) also found a circadian rhythm of glucose uptake in fat cells. Moreover, at present it is not known whether these circadian rhythms in peripheral glucose uptake also exist in vivo. In addition, it is far from clear how these circadian rhythms in peripheral glucose are controlled: by the clock-gene rhythms in muscle and adipose tissue (Ando et al., 2005; Guo et al., 2005; Shimba et al., 2005), or by a humoral or neural signal from the SCN?

Hormonal control of the daily glucose rhythm

The simultaneous increase of glucose uptake and plasma glucose concentrations at awakening, as seen in both humans and experimental animals, can only occur when glucose influx exceeds glucose

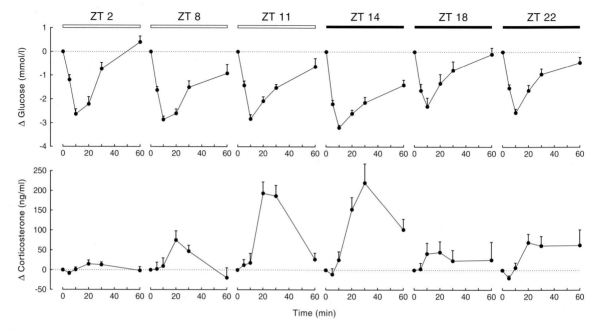

Fig. 2. Plasma glucose and corticosterone responses after the intravenous injection of an insulin bolus (0.5 IU/kg BW) at different times of the light/dark cycle. The total amount of glucose uptake during the 60 min post injection period was significantly higher at ZT14 than the ones at ZT2 and ZT8, despite the fact that also the highest corticosterone response was found at ZT14. Responses are expressed as the difference from the respective $t = 0$ values. ZT, Zeitgeber Time (ZT12 being defined as the onset of the dark period). Adapted from Kalsbeek et al. (2003).

efflux. Since in all the above-mentioned studies daily changes in feeding were compensated for, these results indicate that the increased glucose influx is due to a daily rhythm in endogenous glucose production. In normal, i.e. non-fasting, conditions the liver is by far the major source of endogenous glucose production (Corssmit et al., 2001). Therefore, we first investigated if we could find any evidence of a hormonal or neural connection between the biological clock and the liver. Glucagon is produced by the α-cells of the endocrine pancreas and it acts on the liver to increase glucose production by enhancing both glycogenolysis and gluconeogenesis (Pilkis and Granner, 1992; Kurukulasuriya et al., 2003). Plasma concentrations of plasma glucagon show a clear circadian rhythm in animals feeding ad libitum (Yamamoto et al., 1987; Ruiter et al., 2003). However, it seems unlikely that the daily rhythm of glucagon release is responsible for the daily rhythm in plasma glucose concentrations. First, in ad libitum fed animals plasma glucagon levels only

start to rise after the onset of nocturnal feeding, whereas the rise in plasma glucose concentrations already starts a few hours after the onset of the light period. Second, the daily rhythm in plasma glucose concentration is preserved when animals are maintained on the 6-meals-a-day feeding schedule, whereas plasma glucagon levels now rise with every meal (Fig. 3). On the other hand, the daily rhythm in pancreatic glucagon release might be important for the maintenance of a daily rhythm in plasma glucose concentrations during prolonged fasting (Ruiter et al., 2003). A daily rhythm in plasma insulin concentrations is not consistently found in animals fed ad libitum, which makes insulin an unlikely candidate for the control of the daily rhythm in plasma glucose concentrations as well (Fig. 3). Moreover, the increased insulin sensitivity at the onset of the activity period indicates that, if anything, insulin would lower instead of increase plasma glucose concentrations, both by an increased glucose uptake and an increased inhibition of glucose production by the

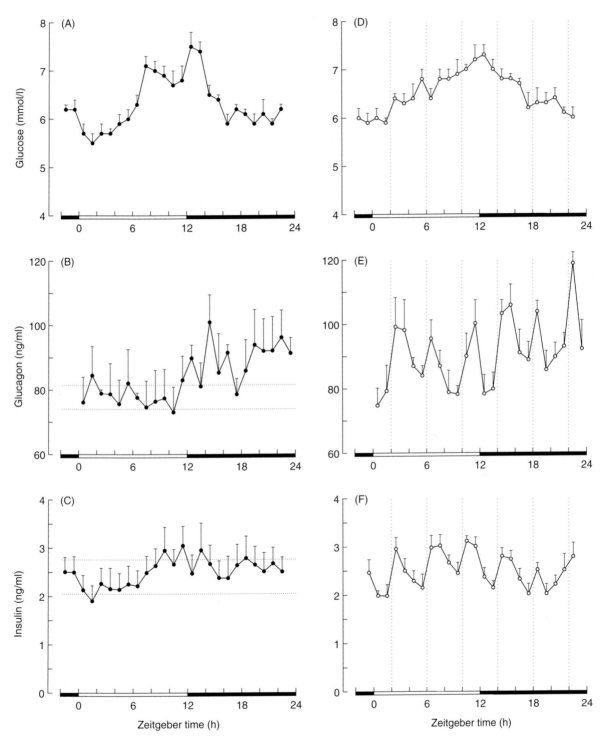

Fig. 3. Basal plasma glucose, insulin and glucagon concentrations across the 24-h light/dark cycle in intact rats under conditions of ad libitum feeding (left panels) or when subjected to a scheduled feeding regimen of 6-meals-a-day (i.e. 1 meal every 4 h) for several weeks (right panels). Horizontal dotted lines in the left-side panels indicate the mean glucagon and insulin levels during the light period (i.e. upper lines are mean +SEM, and lower lines indicate the mean — SEM). Vertical dotted lines in the right-side panels indicate the timing of the 10-min meals every 4 h. Adapted from La Fleur et al. (1999) and Ruiter et al. (2003).

liver. Also other glucose-regulatory hormones such as adrenaline and growth hormone are unlikely to mediate the control of the biological clock on plasma glucose concentrations, since in the rat their daily rhythms show little or no correlation with that of plasma glucose concentrations (Kimura and Tsai, 1984; Clark et al., 1986; De Boer and Van Der Gugten, 1987). The best candidate seems to be the adrenal hormone corticosterone (or cortisol in humans). In both rats and humans its daily rhythm shows a strong correlation with that of plasma glucose, and its stimulatory effect on hepatic glucose production is well known (Corssmit et al., 2001). However, when tested in humans this hypothesis had to be discarded as well (Bright et al., 1980). In addition, our rat data do not support a critical role for corticosterone in the genesis of the daily rhythm in plasma glucose concentrations, since animals with a sympathetic hepatic denervation lose their daily plasma glucose rhythm despite the maintenance of an intact daily corticosterone rhythm (Cailotto et al., 2005).

The primary source of stored energy in mammals is not glucose, but lipids. When energy needs cannot be met by circulating fuels or stored carbohydrates, lipids are mobilized from WAT through the process of lipolysis. Triglycerides are broken down into glycerol and FFA, a process catalyzed by the enzyme hormone sensitive lipase. WAT stores massive amounts of triglycerides, which are synthesized from fatty acids taken up from the plasma. These fatty acids can be derived from plasma FFAs and plasma triglycerides as a result of the local activity of the enzyme lipoprotein lipase. Glucose metabolism is profoundly influenced by this lipid metabolism. For instance, fatty acids can also be synthesized within adipocytes de novo from glucose, and glucose competes for uptake and oxidation in muscle and liver with FFAs. This competitive process is also known as the Randle cycle (Randle et al., 1963). A number of studies have shown daily rhythms in plasma FFA concentrations, 24-h cycles in lipogenesis and lipolysis, as well as SCN lesion-induced aberrations in lipid metabolism (Cornish and Cawthorne, 1978; Yamamoto et al., 1984a; Yamamoto et al., 1987; Dallman et al., 1999). It has been argued that the daily rise in FFA availability during the sleep-induced period of fasting causes insulin resistance, and thereby a rise in plasma glucose levels at the end of the sleep period (Morgan et al., 1999). Although we did find clear effects of fasting and removal of the biological clock on daily plasma FFA profiles (Fig. 4), our data do not support a critical role of daily changes in lipid metabolism in the genesis of the daily glucose rhythm.

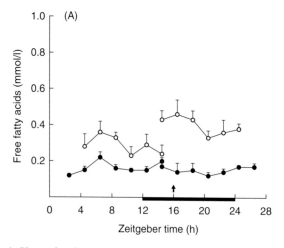 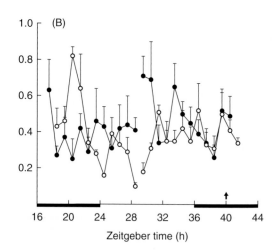

Fig. 4. Plasma free fatty acid concentrations (mmol/l) in intact (A) and SCN-lesioned (B) rats fed ad libitum (closed symbols) or fasted (open symbols). Values are means ±SEM. Black bars indicate the dark period. For fasted rats, food was removed at the onset of the dark period prior to the start of sampling (−ZT14 in A, and ZT10 in B). The arrow indicates the time at which both intact and SCN-lesioned animals had a similar duration of fasting (i.e. 30 h).

During the last decade it has also become clear that WAT is not merely a lipid storage compartment or isolating tissue protecting the organism from heat loss. In addition to the production of fatty acids WAT is capable of producing a large number of hormones, known as adipokines or adipocytokines, and nowadays it is considered an important endocrine gland (Ahima and Flier, 2000; Kershaw and Flier, 2004). Leptin acts via the central nervous system to modulate glucose metabolism (Liu et al., 1998; Minokoshi et al., 1999). Apart from leptin also a number of the more recently discovered hormones have a putative function in glucose metabolism. Adiponectin is reported to inhibit hepatic glucose production (Combs et al., 2001) and to reverse obesity-related insulin resistance by stimulating glucose and FFA utilization (Yamauchi et al., 2002). In addition, also resistin, visfatin, and cytokines released from fat tissue such as TNF-α and IL-6 clearly affect glucose metabolism (Steppan and Lazar, 2002; Fasshauer and Paschke, 2003; Bluher et al., 2004; Fukuhara et al., 2005). Moreover, many of the (adipo)cytokines just mentioned also show clear daily rhythms in their plasma concentration (Gavrila et al., 2003; Rajala et al., 2004; Shea et al., 2005; Vgontzas et al., 2005). However, a direct involvement of these daily changes in circulating (adipo) cytokines and the daily rhythm in plasma glucose concentrations has not been established yet. For instance, whereas the rise in plasma glucose concentrations shows a 12-h difference in humans and rats, plasma leptin concentrations in both species show an acrophase during the dark period (Licinio et al., 1997; Saad et al., 1998; Dallman et al., 1999; Kalsbeek et al., 2001).

Anatomy of the connections between the biological clock and the liver

In view of the lack of a good hormonal candidate that could mediate the control of the central biological clock on hepatic glucose production, we started to explore a possible role for the autonomic nervous system. The liver is richly innervated by sympathetic and parasympathetic fibers, originating from the splanchnic and vagus nerves,

respectively (Puschel, 2004; Uyama et al., 2004). Electron microscopic studies showed direct appositions between nerve fibers and hepatocytes. Already 40 years ago, the pioneering studies by Shimazu et al. indicated a pronounced effect of the sympathetic and parasympathetic innervation of the liver on glucose production. Stimulation of the sympathetic input to the liver caused a clear and rapid increase of glucose production by stimulating glycogen phosphorylase (Shimazu and Fukuda, 1965), whereas activation of its parasympathetic input resulted in a decrease of hepatic glucose production by activating hepatic glycogen synthase (Shimazu, 1967). Electrical stimulation experiments indicated a role for the ventromedial and lateral hypothalamus in the control of these sympathetic and parasympathetic inputs to the liver (Shimazu, 1981). Similar experiments also provided evidence for an autonomic nervous system-mediated control of the hypothalamic biological clock on hepatic glucose production. Electrical stimulation of the SCN resulted in hyperglycemia, whereas this effect could be prevented by the peripheral administration of autonomic blockers (Nagai et al., 1988; Fujii et al., 1989). However, in view of its multi-synaptic character it was impossible to trace the exact anatomical connections between the liver and these hypothalamic nuclei, until the development of the transneuronal viral tracing technique in the early 1990s (Strack et al., 1989; Card et al., 1990). Before this time defining neural circuits involved multiple injections of a tract tracer, often in different groups of animals, to label the different connections of the hypothesized neural circuit. However, since it was impossible to target the tracer specifically in those neurons that are synaptically connected to those identified in a previous group of animals, multi-synaptic pathways could not be established definitively. The transneuronal viral tracing technique for the first time provided the possibility to identify a chain of synaptically connected neurons within the same animal, for instance between the hypothalamic biological clock and peripheral organs such as the liver and pancreas (Ueyama et al., 1999; La Fleur et al., 2000; Buijs et al., 2001; Buijs et al., 2003). Moreover, a combination of this viral tracing technique with a selective denervation of either the sympathetic or parasympathetic input to the organ of interest enables the selective identification of either the parasympathetic or sympathetic chain of control (Westerhaus and Loewy, 1999; Lee and Erskine, 2000; Buijs et al., 2001). After injection of the pseudorabies virus (PRV; Bartha strain)

in the liver of animals that had undergone a complete denervation of the hepatic parasympathetic innervation, virus-infected neurons first appeared in the intermediolateral column (IML) of the spinal cord (Fig. 5). These so-called first-order neurons are the preganglionic sympathetic

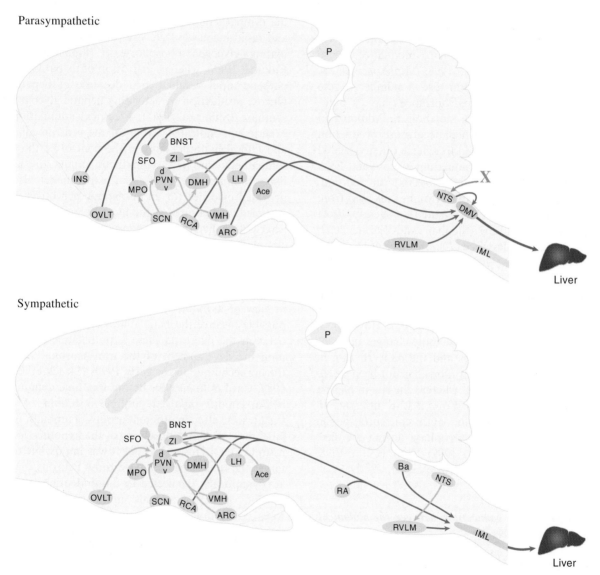

Fig. 5. Sagittal scheme of the sympathetic and parasympathetic control of the liver. Brain areas providing first-order projections are indicated in red, brain areas containing second-order neurons are indicated in blue, and those containing third-order neurons are indicated in yellow. It is clear by comparing the parasympathetic pattern against the sympathetic pattern that far more second-order cell groups are in control of the first-order parasympathetic (i.e. DMV) than first-order sympathetic (i.e. IML) motor neurons. RVLM also includes the catecholaminergic A5, C1, and C3 areas.

motorneurons. Labeling of second-order neurons, i.e. neurons providing a synaptic input to the preganglionic IML neurons, were found in the intercalated nucleus, lateral funiculus, and laminas VII and X of the spinal cord. In addition, second-order labeling was found in the ventrolateral medulla, the C1/C3 adrenaline cell groups, the A5 and locus coeruleus noradrenaline cell groups, and the serotonin-containing medial raphe nucleus in the brainstem. The most rostral second-order labeling, however, was observed in the hypothalamic paraventricular nucleus (PVN), and to a lesser degree in the zona incerta, lateral hypothalamus, and the retrochiasmatic area. Enabling another step of virus replication by a further prolongation of the survival time after the injection of the virus revealed third-order labeling in the brainstem nucleus of the solitary tract (NTS) and a large number of areas known to project to the PVN, such as the medial preoptic area (MPOA), the dorsomedial hypothalamic nucleus (DMH), the arcuate nucleus, circumventricular organs, the ventromedial hypothalamic nucleus (VMH), and the SCN. Also limbic structures such as the central amygdala and the bed nucleus of the stria terminalis (BNST) contained labeled neurons at this stage. After a denervation of the sympathetic input to the liver, the first neurons to be labeled in the central nervous system were the parasympathetic motorneurons in the dorsal motornucleus of the vagus (DMV). Second-order labeling in the hypothalamus was somewhat more widespread than observed after the infection via the sympathetic branch, for instance, second-order neurons were observed not only in the PVN but also in the MPOA and DMH, and in limbic structures such as the amygdala and BNST. Third-order labeling again was found in, among others, the SCN and VMH. In conjunction with the existing knowledge on SCN projections, the third-order labeling of the SCN indicates a direct input of the biological clock to the pre-autonomic (i.e. second-order) neurons in the hypothalamus. Most of the hypothalamic pre-autonomic neurons are located in the PVN, notably its dorso-medial and ventral regions. Indeed, other experiments, too, provided evidence for a direct input of the SCN to the pre-autonomic neurons in the PVN (Teclemariam-Mesbah et al.,

1997; Vrang et al., 1997; Cui et al., 2001). Together these data show that a control of the biological clock on liver function mediated by the autonomic nervous system is very well feasible, since the SCN neurons are only two synapses away from the motorneurons of both the sympathetic and parasympathetic hepatic innervation.

Functional studies on the connections between the biological clock and the liver

After the establishment of a neural connection between the biological clock and the liver a series of experiments was started to investigate the functional importance of the SCN–PVN connection in the control of plasma glucose concentrations, by administering different SCN transmitter agonists and antagonists into the vicinity of the PVN (Kalsbeek et al., 2004). During this 2-h administration period blood samples were taken in order to monitor the changes in blood glucose concentrations as well as those in plasma insulin, glucagon, and corticosterone concentrations. The most pronounced effects on plasma glucose concentrations were observed after the administration of either Bicuilline (BIC; a GABA-A antagonist) or NMDA (an agonist of glutamatergic receptors) to the PVN, both causing a prolonged and significant increase of plasma glucose concentrations (Fig. 6 and 7). Since both drugs will cause an activation of neuronal activity in the PVN, either by a blockade of the inhibitory GABA-ergic input or by stimulation of glutamatergic receptors, an increased activity of PVN neurons apparently results in increased plasma glucose concentrations. Both drugs also caused increased plasma glucagon concentrations, but did not affect plasma insulin concentrations in any significant way. Blockade of GABA-ergic receptors, but not stimulation of the glutamate receptors, resulted in increased plasma corticosterone concentrations (Fig. 6). These data indicate that it is not likely that the hyperglycemia induced by the stimulation of PVN neurons is due to changes in the release of either insulin or corticosterone. On the other hand, the increased release of glucagon could be a causative factor. Therefore, after these initial results two main is-

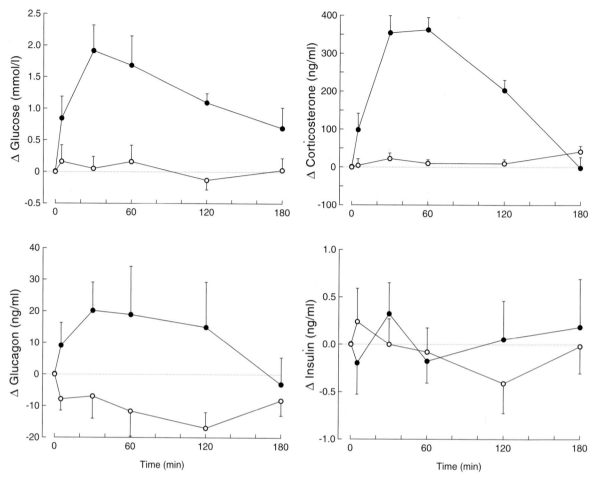

Fig. 6. Changes in plasma glucose, corticosterone, glucagon, and insulin concentrations during a 2-h administration of the GABA$_A$ antagonist Bicucilline (BIC) in the PVN of ad libitum fed rats. Filled symbols indicate the effect of BIC, whereas open symbols show the result of the control experiment in the same group of animals one week later. Responses are expressed as the difference from the respective $t = 0$ values. BIC administration in the PVN started at $t = 0$ 5h after lights on (i.e. ZT5). Adapted from Kalsbeek et al. (2004).

sues remained: (1) are projections of the biological clock indeed involved in the effects of BIC and NMDA? and (2) is it the increased release of glucagon or the autonomic innervation of the liver which is responsible for the increased plasma glucose concentrations? In order to investigate which source of GABA-ergic input to the PVN is responsible for the increased plasma glucose administration after administration of BIC, we selectively silenced three major afferent inputs to the PVN by local administration of the sodium channel blocker TTX (Kalsbeek et al., 2004). Administration of TTX in the SCN and DMH, but not in the VMH (or PVN), caused an increase of plasma glucose concentrations, although to a lesser degree than after BIC administration in the PVN (maximal $+0.7$ mmol/l vs. $+1.5$ mmol/l). These results indicate that indeed an important part of the inhibitory input to the PVN, involved in the control plasma glucose concentrations, is coming directly from the SCN. Apart from the SCN, however, additional inhibitory inputs may be derived from other hypothalamic nuclei, among which SCN target areas such as the DMH. These results indicate that the SCN may use both direct and indirect GABA-ergic inputs to the PVN to

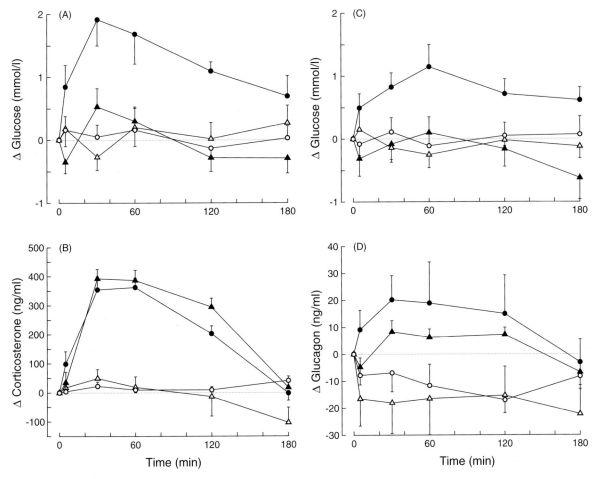

Fig. 7. Changes in plasma glucose, corticosterone, and glucagon concentrations during a 2-h administration of the GABA$_A$ antagonist Bicuciline (BIC; A, B, D) or NMDA (an agonist of glutamatergic receptors; C) in the PVN of ad libitum fed rats. Filled symbols indicate the effect of BIC or NMDA, whereas open symbols show the result of the control experiment in the same group of animals one week later. Triangles represent the results of animals with a sympathetic liver denervation, whereas circles represent the results of animals with an intact autonomic innervation of the liver. Responses are expressed as the difference from the respective $t = 0$ values. BIC or NMDA administration in the PVN started at $t = 0$, 5 h after lights on (i.e. ZT5).

control the daily rhythm of plasma glucose concentrations. Indeed also anatomical and electrophysiological data support the existence of both direct GABA-ergic projections from the SCN to the PVN (Hermes et al., 1996; Cui et al., 2001; Wang et al., 2003) as well as GABA-ergic inputs to the PVN from SCN target areas such as the DMH, subPVN and MPOA (Boudaba et al., 1996). In order to investigate which output pathway of the PVN is responsible for the increased plasma glucose concentrations (i.e. increased release of glucagon or activation/inhibition of the autonomic

input to the liver), we repeated our BIC and NMDA infusions in animals in which we first had denervated specifically either the sympathetic or parasympathetic innervation of the liver. Removal of the parasympathetic input to the liver did not affect the hyperglycemic effect of BIC administration in the PVN. On the other hand, removal of the sympathetic innervation of the liver completely obliterated the hyperglycemic effect of both BIC and NMDA administration in the PVN (Fig. 7). The hyperglycemic effect of BIC disappeared notwithstanding pronounced increases of plasma glucagon and plasma corticosterone concentrations. Together, these functional studies demonstrate that a stimulation of neuronal activity in the PVN results in hyperglycemia through an activation of the sympathetic input to the liver. In our experiments hepatic glucose production was not assessed directly, but previously Lang (1995) had demonstrated a stimulatory effect of intracerebroventricularly applied BIC on portal glucose levels using the euglycemic clamp technique. Moreover, our data fit in nicely with previous experiments showing increased glucose production or glucose release by the liver upon electrical stimulation of its sympathetic input (Shimazu, 1996; Takahashi et al., 1996; Nonogaki, 2000).

Hepatic clock genes and the generation of a daily plasma glucose rhythm

In invertebrates and lower vertebrates, circadian rhythms in the periphery are driven by tissue-autonomous circadian oscillators, often light sensitive, and synchronized to, but not dependent on, central pacemakers (Giebultowicz and Hege, 1997). Recent studies using transgenic rodents and fibroblast cell lines indicate that peripheral tissues of mammals also have the capacity for an autonomous circadian gene expression. Moreover, recent gene expression studies have revealed several hundreds of genes in the liver showing a circadian expression pattern, even during fasting conditions, including the well-known core clock genes as well as those encoding several key enzymes involved in glucose metabolism (Akhtar et al., 2002; Kita et al., 2002; Panda et al., 2002; Oishi et al., 2003). The discovery of clock genes being operative throughout the entire body has led to a reconsideration of the relation between the master oscillator in the SCN and peripheral rhythms. Even the status of a master oscillator for the SCN has come under scrutiny (Davidson et al., 2003). In the current view the central pacemaker in the SCN co-ordinates the activity of local oscillators in the peripheral tissues via behavioral, neuroendocrine, and autonomic pathways (Buijs and Kalsbeek, 2001; Terazono et al., 2003; Guo et al., 2005). However, a clear

understanding of the role of peripheral oscillators in regulating the physiological functions of peripheral organs is still lacking. Previous experiments have shown that both daily rhythms in liver clock gene expression and plasma glucose concentrations are maintained during fasting (La Fleur et al., 1999; Kita et al., 2002). Moreover, both hepatic clock gene expression and hepatic glucose production are increased by the activation of the sympathetic input to the liver (Terazono et al., 2003; Kalsbeek et al., 2004). Therefore, we investigated whether peripheral oscillators in the liver are a necessary link in the transfer of the circadian information from the biological clock to hepatic glucose production. Since it has been shown that a change in feeding pattern has pronounced effects on clock gene expression in the liver (Damiola et al., 2000; Hara et al., 2001; Stokkan et al., 2001), we first investigated whether our 6-meals-a-day feeding schedule affected the normal daily rhythm of clock gene expression in the liver. In accordance with our data on plasma glucose concentrations (La Fleur et al., 1999), also the rhythmic expression of the five clock genes studied, i.e. Per1, Per2, Per3, Cry1, and Dbp, was maintained (Cailotto et al., 2005). However, both Per2 and Dbp showed a small phase advance. Subsequently, we investigated the consequences of a selective removal of the sympathetic branch of the autonomic nervous input to the liver. In line with the hypothesis proposed in our previous denervation study (Kalsbeek et al., 2004), removal of the sympathetic input to the liver resulted in an obliteration of the daily rhythm in plasma glucose levels. Contrary to our expectations, transcript levels of the five clock genes studied again maintained their rhythmicity in the liver (Fig. 8). However, the combination of scheduled-feeding and hepatic sympathectomy did not go completely unnoticed, since the removal of these two circadian input pathways affected the rhythmicity of four out of the five clock genes studied (Cailotto et al., 2005). Therefore, these results clearly show that (1) the autonomic innervation of the liver is not necessary to sustain the rhythmicity of peripheral oscillators in the liver, and (2) a rhythmic expression of the peripheral oscillators in the liver is not sufficient to maintain a rhythmic output of the liver (i.e. glucose

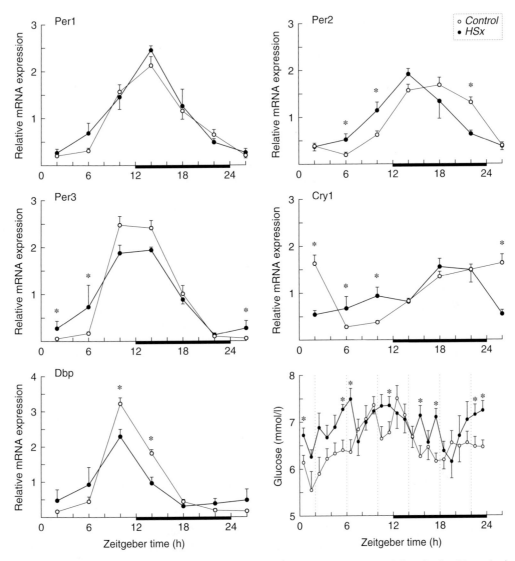

Fig. 8. Daily rhythm in plasma glucose and liver Per1, Per2, Per3, Cry1, and Dbp mRNA levels in animals with a selective hepatic sympathectomy (closed symbols) and SHAM-denervated control animals (open symbols). The black bars indicate the dark period. The vertical dotted lines represent the 9–11 min periods of food availability every 4 h. Although indicated only in the glucose figure, all animals were subjected to the scheduled feeding conditions. *$P < 0.05$ compared to the control group.

production). Follow-up studies will investigate how the activities of liver enzymes involved in glyco-genolysis and gluconeogenesis are affected by sym-pathetic stimulation and hepatic denervation. As regards the functional significance of peripheral oscillators, contrary to the results on bone forma-tion (Fu et al., 2006) and cell division during liver regeneration (Matsuo et al., 2003), so far we have found no evidence for a functional significance

of peripheral oscillators in the rhythmic output of a tissue.

Differential control of glucose and lipid metabolism by the biological clock

The daily rhythm in adipose leptin production strongly suggested a direct control of adipose tissue activity by the biological clock (Kalsbeek et al., 2001). Viral tracing studies again proved to be very helpful in delineating the brain areas in the central nervous system that are involved in the control of fat tissue. After injection of the pseudorabies virus in either the white or brown adipose tissue third-order neurons were found, among others, in the SCN (Bamshad et al., 1998; Bartness and Bamshad, 1998). More recently, we were able to show that, contrary to general belief, WAT is innervated not only by the sympathetic, but also by the parasympathetic branch of the autonomic nervous system (Kreier et al., 2002). Denervation studies have shown that sympathetic innervation stimulates the mobilization of lipid stores (Cantu and Goodman, 1967; Youngstrom and Bartness, 1998). On the other hand, removal of the parasympathetic input caused a reduction of the insulin-mediated glucose and FFA uptake (-33% and -36%, respectively), and at the same time an increase of HSL, indicating increased lipolysis. These data indicate an anabolic effect of the parasympathetic innervation (Kreier et al., 2002). Additional tracing studies showed different neurons in the sympathetic ganglia to be in control of inguinal and epididymal fat pads (Youngstrom and Bartness, 1995), as well as different sympathetic and parasympathetic motorneurons, within the spinal cord and DMV, to be in control of subcutaneous and abdominal WAT compartments (Kreier et al., 2002). The clear somatotopy in the autonomic control of WAT has resulted in hypotheses about the involvement of the CNS in the largely unexplained effects of sex steroids and glucocorticoids on fat distribution, the pathogenesis of fat redistribution syndromes such as AIDS lipodystrophy, and the etiology of the metabolic syndrome (Fliers et al., 2003a, b; Kreier et al., 2003b).

In the previous paragraph, we showed that the biological clock uses its GABA-ergic and glutamatergic projections to the pre-autonomic PVN neurons to control the daily rhythm in hepatic glucose production, an effect mediated via the sympathetic innervation of the liver. As just mentioned, neuroanatomical studies have also shown a connection between the biological clock and adipose tissue, SCN-lesions increase plasma leptin

and FFA concentrations, and there is good evidence that an increased activity of the sympathetic input to adipose tissue stimulates lipolysis. Therefore, we wondered whether the SCN also uses its projections to the pre-autonomic PVN neurons to control the mobilization of lipid stores. In order to test this hypothesis, we infused the GABA-antagonist BIC in the PVN and measured the resulting plasma glucose, leptin, and FFA levels. Contrary to plasma glucose concentrations, however, plasma FFA and plasma leptin concentrations were not affected by removal of the GABA-ergic input to the PVN (Fig. 9). In a sense this observation is supported by the findings of Foster and Bartness (2003), who showed that PVN lesions do not attenuate fasting-induced lipid mobilization. Indeed, a differential control has also been suggested previously because de-afferentation of the anterior hypothalamus prevents 2-deoxyglucose (2DG) induced increases in plasma FFA but not glucose concentrations, and adrenal demedullation blocks the 2DG-induced glucose but not FFA response (Teixeira et al., 1973). After viral tracing from WAT, besides the PVN second order neurons were especially found in the MPOA, the DMH, and the arcuate nucleus (Bamshad et al., 1998). Especially, the labeling in the MPOA is interesting since several studies have indicated that lipid mobilization may be stimulated via the MPOA (Teixeira et al., 1973; Coimbra and Migliorini, 1988; Bartness and Bamshad, 1998; Ferreira et al., 1999). Together, these results show a highly differentiated hypothalamic control of glucose and fat metabolism, and suggest that the SCN may use different outputs to control glucose (via the PVN) and lipid (via the MPOA) metabolism.

Circadian control of the autonomic nervous system

As evidenced once again by the data presented above the SCN, thus appears to be responsible for organizing endogenous daily programs throughout the body. More specifically, the data on the circadian control of the daily rhythm in plasma glucose concentrations demonstrate the important role of the autonomic nervous system as an intermediate of SCN output. The most important

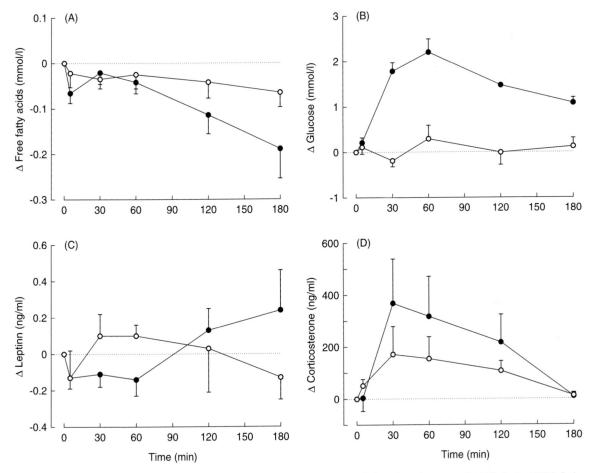

Fig. 9. Plasma increments of free fatty acids (mmol/l), glucose (mmol/l), leptin (ng/ml), and corticosterone (ng/ml) during BIC infusion in the PVN (closed symbols) or control experiment (open symbols). Values are means $\pm$ SEM and calculated as the difference from the $t = 0$ value (indicated by the dotted line).

element in the control of the biological clock on the activity of the autonomic nervous system is the direct projection of the SCN to the pre-autonomic neurons in the PVN. In search for the hypothalamic melatonin rhythm generator we previously demonstrated that in the rat the pre-autonomic PVN neurons dedicated to the control of the pineal gland are continuously, i.e. 24-h a day, activated by a glutamatergic input from the SCN (Perreau-Lenz et al., 2004). A daily rhythm in pineal activity (or melatonin release) is created by the rhythmic activity of the GABA-containing SCN projections to these very same pre-autonomic PVN neurons. During the light period pre-autonomic PVN neurons multi-synaptically connected

to the pineal gland are silent, since their glutamatergic stimulation is overruled by a GABA-ergic inhibition. The stimulatory effect of the glutamatergic input only becomes evident when there is a gradual waning of the GABA-ergic inhibition during the dark period, or when the GABA-ergic input is antagonized by the administration of a GABA-antagonist (Kalsbeek et al., 2000). The data presented above indicate that a similar control mechanism may hold for the circadian control of the daily rhythm in plasma glucose concentrations. In this case, it is the activity of pre-autonomic PVN neurons dedicated to the control of hepatic glucose production that is restrained during the light period by a GABA-ergic inhibition which is mainly derived from the SCN. Also in this case, the final activity of the pre-autonomic neurons connected to the sympathetic innervation of the liver seems to be determined by a balance of GABA-ergic and glutamatergic inputs. It is to be expected, however, that there will be a phase-difference of a few hours between the GABA-ergic neurons that control the "liver-dedicated" pre-autonomic PVN neurons and those that control the "pineal-dedicated" pre-autonomic PVN neurons, since the acrophase of hepatic glucose production is situated at ~ZT11, whereas the acrophase of the plasma melatonin rhythm is situated in the middle of the dark period (i.e. ZT18–ZT20). Thus, the SCN not only contains specialized neurons for the control of the sympathetic and parasympathetic branch of the autonomic nervous system (Buijs et al., 2003), but from the results just presented it seems likely that within the population of SCN neurons that are dedicated to the sympathetic branch of the autonomic nervous system also specialized neurons exist that are dedicated to the control of either the liver or the pineal gland. Moreover, recently it was shown that separate SCN neurons are in contact with either the intra-abdominal or subcutaneous WAT (Kreier et al., 2006). On the other hand, the SCN also contains neurons that are connected to multiple autonomic outputs (Ueyama et al., 1999).

At present it is not clear if this GABA/glutamate control mechanism also holds for other SCN outputs, such as its control of the parasympathetic branch of the autonomic nervous system or the neuroendocrine system. Previously we did present evidence that also in the control of neuroendocrine rhythms, such as the daily corticosterone rhythm and the pre-ovulatory surge of luteinizing hormone, the SCN uses alternating rhythms of stimulatory and inhibitory signals to time the acrophase of these hormones. However, in the control of these neuroendocrine systems peptidergic projections to intermediate targets such as the DMH, MPOA, and subPVN seem to be the main SCN output, instead of direct projections to the neuroendocrine motorneurons (Kalsbeek et al., 1992, 1996a, b; Palm et al., 1999, 2001; Hermes et al., 2000). On the other hand, also direct (GABA-containing) SCN projections to

the magnocellular neurosecretory neurons in the PVN have been demonstrated (Hermes and Renaud, 1993; Hermes et al., 1996). In addition, the sleep/wake cycle may be controlled by such a GABA-ergic/glutamatergic output of the SCN. In this case the sleep-promoting neurons in the ventrolateral preoptic area (VLPO) seem to be the major target of the SCN projections (Gallopin et al., 2000; Sun et al., 2000, 2001; Satoh et al., 2003). In conclusion, the experiments just described provide clear evidence of the existence of a GABA/Glutamate switch mechanism in the SCN output, which is not restricted to the control of the sympathetic branch of the autonomic nervous system.

Clinical implications

Fasting hyperglycemia is the hallmark of diabetes mellitus. Longitudinal epidemiological studies have shown that the risk of cardiovascular disease (CVD) mortality in diabetic subjects is more than twice that of subjects without diabetes. Moreover, in diabetic patients, the risk of cardiovascular and all cause mortality increases with increasing fasting plasma glucose and HbA_{1c} values. The associations between glycemic variables and mortality are less evident in non-diabetic subjects, but a number of studies indicate that also in the non-diabetic range (i.e. <7.0 mmol) increased fasting plasma glucose levels are associated with increased mortality (Barrett-Connor et al., 1984; Scheidt-Nave et al., 1991; de Vegt et al., 1999; Facchini et al., 2001; Taubert et al., 2003; Timmer et al., 2004). From the foregoing it has become clear that the biological clock plays an important role in determining early morning fasting plasma glucose concentrations, by affecting both hepatic glucose production and glucose tolerance. Therefore, it is to be expected that a reduced activity of the biological clock or a misalignment of the endogenous biological clock rhythm with the exogenous environment will result in, or predispose for, diseases such as obesity, type 2 diabetes, and hypertension. Indeed, individuals with diabetes often show a disturbance of the normal circadian rhythm in blood pressure (Bernardi et al., 1992; Monteagudo et al., 1996; Holl et al., 1999; Lurbe et al., 2002;

Zhao et al., 2005), and postmortem studies have suggested a profound reduction of SCN activity in hypertensive patients (Goncharuk et al., 2001). In experimental animals, pathological conditions such as hypertension and diabetes result in an abolishment or disturbance of clock gene rhythms in peripheral organs (Young et al., 2001, 2002; Hastings et al., 2003; Naito et al., 2003; Kuriyama et al., 2004; Oishi et al., 2004). However, thus far this only provides correlative evidence for the involvement of a malfunctioning biological clock in the etiology of metabolic diseases. In addition, a number of clock gene knock-out mice show metabolic disturbances. Recently, Turek et al. (2005) showed that *Clock* mutant mice are hyperphagic and obese, and develop a metabolic syndrome characterized by hyperleptinemia, hyperlipidemia, hepatic steatosis, and hyperglycemia. Remarkably these mice are hypoinsulinemic instead of hyperinsulinemic. Gluconeogenesis is abolished in *Bmal1* knock-outs, whereas both *Clock* and *Bmal1* knock-outs show an impaired counterregulatory response to insulin-induced hypoglycemia (Rudic et al., 2004). But at present it is not clear if these deficits are a direct consequence of the malfunctioning clockwork, or the indirect result of the loss of other (non-circadian) functions of the clock genes. On the other hand, the increased prevalence of obesity and the metabolic syndrome in shift-workers and aging (Karlsson et al., 2001; Blouin et al., 2005) clearly indicates a causal or predisposing role of the long-term impaired functioning of the SCN in the genesis of these diseases of modern society.

In the present 24/7 society, it is becoming more and more difficult to keep our endogenous clocks well synchronized with the exogenous environment, even for non-shift workers. The most important entraining factor for the endogenous clock is bright light (intensity > 1000 lx). Since bright light exposure rarely occurs inside of buildings nowadays we tend to expose ourselves less and less to this most important entraining factor, whereas the indoor light intensities only provide a weak entraining stimulus (i.e. the level of indoor lighting usually does not exceed 300–500 lx). In addition to bright light, also the so-called non-photic stimuli are able to entrain the endogenous pacemaker.

Non-photic entraining stimuli, for instance, include exercise, food intake, and maybe even sleep (Deboer et al., 2003). Until 50 years ago food intake and physical activity coincided with the daytime activity period of humans, and clearly contrasted with the nocturnal period of fasting, rest, and sleep. Our western society, however, is characterized by a reduction in energy expenditure, an increased energy intake, with food intake spread over the major part of the 24-h cycle, and a decrease in sleep duration (Foster and Wulff, 2005). Therefore, the normal alternation of sleep and wake, or rest and activity, is severely disturbed and the amplitude of these SCN-entraining signals is greatly reduced. Recently, we hypothesized that as a result of this lack of rhythmic feedback to the SCN, also the output of the SCN might be affected (Kreier et al., 2003a, b). The normal alternation of rest and activity is accompanied by a shift in the balance of the autonomic nervous system, i.e. the active period is characterized by a pre-dominant sympathetic activity, whereas parasympathetic activity rules the body during the inactive period. As outlined in the previous paragraph this balance of the autonomic nervous system is critically dependent on the output of our biological clock. Therefore, a reduced amplitude of the SCN output signal may result in an unbalanced autonomic nervous system and a body less well prepared for the upcoming demands. Indeed, a prospective cohort study in 8.000 patients revealed a high relative risk to develop type 2 diabetes if autonomic dysfunction was present (Carnethon et al., 2003).

If our hypothesis is correct, this vicious circle can be broken by enhancing the (rhythmic) feedback signal to the SCN, thereby re-instating the rhythmic output of the SCN. Exposure to bright light in the middle of the day re-instates the normal sleep/wake rhythm in elderly by promoting nocturnal sleep (Abbott, 2003). Experiments in aging rats have shown that the healthier sleep patterns are accompanied by a restoration of the number of active SCN neurons (Lucassen et al., 1995). Daily exercise is a very effective way to improve glucose tolerance (Poirier et al., 2002; Evans et al., 2004; Nakanishi et al., 2004), with the most critical aspect probably being a reduction of the periods of inactivity during the waking hours

(Westerterp, 2001). Whereas the energy expenditure aspects of such physical activities surely contribute to its positive effects, there are also indications that the central effects of exercise may help to improve glucose tolerance (Shin et al., 2003; Bi et al., 2005). For instance, exercise enhances the endogenous inhibitory tone within the PVN, and thereby reduces the sympathetic outflow (Zheng et al., 2005). Another way to enhance the rhythmic input to the SCN is the daily intake of melatonin. Melatonin, a hormone produced by the pineal gland in a rhythmic fashion, is only released during the dark period, and has a profound effect on the SCN (Liu et al., 1997; Van den Top et al., 2001). Diabetic patients and patients with coronary heart disease have a flattened melatonin rhythm (O'Brien et al., 1986; Brugger et al., 1995; Altun et al., 2002). Interestingly, daily nighttime melatonin supplementation was able to reduce blood pressure in hypertensive patients, especially by a fall in sleep systolic blood pressure (Scheer et al., 2004). In rats, daily melatonin administration induced fat loss and improved the metabolic syndrome (Wolden-Hanson et al., 2000).

In conclusion, the hypothalamic biological clock not only controls the daily rhythm in sleep/wake (or feeding/fasting) behavior, but also exerts a direct control over many aspects of energy metabolism. An important output mechanism of the biological clock is its control of the sympathetic/parasympathetic autonomic balance. A well-entrained biological clock is essential for a balanced autonomic nervous system and may have a protective value as far as diseases characterized by a misbalance of the autonomic nervous system are concerned, such as hypertension, type 2 diabetes, and the metabolic syndrome.

Abbreviations

2DG	2-deoxy-glucose
Ace	amygdala, central part
ARC	arcuate nucleus
Ba	Barrington nucleus
BIC	bicucilline
Bmal	brain and muscle aryl hydrocarbon receptor nuclear transporter-like protein
BNST	bed nucleus of the stria terminalis
CNS	central nervous system
Cry	cryptochrome
CVD	cardiovascular disease
Dbp	albumin D-site-binding protein
DMH	dorsomedial nucleus of the hypothalamus
DMV	dorsal motor nucleus of the vagus
FFA	free fatty acid
GABA	γ-aminobutyric acid
HSL	hormone-sensitive lipase
IL-6	interleukin-6
IML	intermediolateral column of the spinal cord
INS	insular cortex
L/D	light/dark
LH	lateral hypothalamus
MPO	medial preoptic nucleus
MPOA	medial preoptic area
NMDA	N-methyl-D-aspartate
NTS	nucleus tractus solitarius
OVLT	organum vasculosum of the lamina terminalis
P	pineal
Per	period
PRV	pseudorabies virus
PVN d/v	paraventricular nucleus of the hypothalamus, dorsal/ventral
RA	raphe nucleus
RCA	retrochiasmatic area
RVLM	rostroventrolateral medulla
SCN	suprachiasmatic nucleus
SEM	standard error of mean
SFO	subfornical organ
TNF-α	tumor necrosis factor α
TTX	tetrodotoxin
VLPO	ventrolateral preoptic area
VMH	ventromedial nucleus of the hypothalamus
WAT	white adipose tissue
X	vagus nerve
ZI	zona incerta
ZT	Zeitgeber Time

Acknowledgments

The authors thank Dr. Mariette T. Ackermans at the Academic Medical Center in Amsterdam for help with the FFA measurements, Henk Stoffels for preparation of the images and Wilma Verweij for correction of the manuscript. Special thanks are dedicated to Jan van der Vliet and Caroline Pirovano — Van Heijningen for their superb technical assistance in most of the work just described. Parts of the work presented were financially supported by the Dutch Diabetes Research Foundation.

References

Abbott, A. (2003) Restless nights, listless days. Nature, 425: 896–898.

Ahima, R.S. and Flier, J.S. (2000) Adipose tissue as an endocrine organ. Trends Endocrinol. Metab., 11: 327–332.

Akhtar, R.A., Reddy, A.B., Maywood, E.S., Clayton, J.D., King, V.M., Smith, A.G., Gant, T.W., Hastings, M.H. and Kyriacou, C.P. (2002) Circadian cycling of the mouse liver transcriptome, as revealed by cDNA microarray, is driven by the suprachiasmatic nucleus. Curr. Biol., 12: 540–550.

Altun, A., Yaprak, M., Aktoz, M., Vardar, A., Betul, U.A. and Ozbay, G. (2002) Impaired nocturnal synthesis of melatonin in patients with cardiac syndrome X. Neurosci. Lett., 327: 143–145.

Ando, H., Yanagihara, H., Hayashi, Y., Obi, Y., Tsuruoka, S., Takamura, T., Kaneko, S. and Fujimura, A. (2005) Rhythmic mRNA expression of clock genes and adipocytokines in mouse visceral adipose tissue. Endocrinology, 146: 5631–5636.

Bamshad, M., Aoki, V.T., Adkison, M.G., Warren, W.S. and Bartness, T.J. (1998) Central nervous system origins of the sympathetic nervous system outflow to white adipose tissue. Am. J. Physiol., 275: R291–R299.

Barrett-Connor, E., Wingard, D.L., Criqui, M.H. and Suarez, L. (1984) Is borderline fasting hyperglycemia a risk factor for cardiovascular death? J. Chronic Dis., 37: 773–779.

Bartness, T.J. and Bamshad, M. (1998) Innervation of mammalian white adipose tissue: implications for the regulation of total body fat. Am. J. Physiol., 275: R1399–R1411.

Bellinger, L.L., Mendel, V.E. and Moberg, G.P. (1975) Circadian insulin, GH, prolactin, corticosterone and glucose rhythms in fed and fasted rats. Horm. Metab. Res., 7: 132–135.

Bernardi, L., Ricordi, L., Lazzari, P., Solda, P., Calciati, A. and Ferrari, M.R. (1992) Impaired circadian modulation of sympathovagal activity in diabetes. A possible explanation for altered temporal onset of cardiovascular disease. Circulation, 86: 1443–1452.

Bi, S., Scott, K.A., Hyun, J., Ladenheim, E.E. and Moran, T.H. (2005) Running wheel activity prevents hyperphagia and obesity in Otsuka Long-Evans Tokushima fatty rats: role of hypothalamic signaling. Endocrinology, 146: 1676–1685.

Blouin, K., Despres, J.P., Couillard, C., Tremblay, A., Prud'homme, D., Bouchard, C. and Tchernof, A. (2005) Contribution of age and declining androgen levels to features of the metabolic syndrome in men. Metab. Clin. Exp., 54: 1034–1040.

Bluher, S., Moschos, S., Bullen Jr., J., Kokkotou, E., Maratos-Flier, E., Wiegand, S.J., Sleeman, M.W. and Mantzoros, C.S. (2004) Ciliary neurotrophic factor$_{Ax15}$ alters energy homeostasis, decreases body weight, and improves metabolic control in diet-induced obese and UCP1-DTA mice. Diabetes, 53: 2787–2796.

Bolli, G.B., De Feo, P., De Cosmo, S., Perriello, G., Ventura, M.M., Calcinaro, F., Lolli, C., Campbell, P., Brunetti, P. and Gerich, J.E. (1984) Demonstration of a dawn phenomenon in normal human volunteers. Diabetes, 33: 1150–1153.

Boudaba, C., Szabo, K. and Tasker, J.G. (1996) Physiological mapping of local inhibitory inputs to the hypothalamic paraventricular nucleus. J. Neurosci., 16: 7151–7160.

Bright, G.M., Melton, T.W., Rogol, A.D. and Clarke, W.L. (1980) Failure of cortisol blockade to inhibit early morning increases in basal insulin requirements in fasting insulin-dependent diabetics. Diabetes, 29: 662–664.

Brobeck, J.R. (1946) Mechanism of development of obesity in animals with hypothalamic lesions. Physiol. Rev., 26: 541–559.

Brugger, P., Marktl, W. and Herold, M. (1995) Impaired nocturnal secretion of melatonin in coronary heart disease. Lancet, 345: 1408.

Buijs, R.M., Chun, S.J., Niijima, A., Romijn, H.J. and Nagai, K. (2001) Parasympathetic and sympathetic control of the pancreas: a role for the suprachiasmatic nucleus and other hypothalamic centers that are involved in the regulation of food intake. J. Comp. Neurol., 431: 405–423.

Buijs, R.M. and Kalsbeek, A. (2001) Hypothalamic integration of central and peripheral clocks. Nat. Neurosci. Rev., 2: 521–526.

Buijs, R.M., la Fleur, S.E., Wortel, J., Van Heyningen, C., Zuiddam, L., Mettenleiter, T.C., Kalsbeek, A., Nagai, K. and Niijima, A. (2003) The suprachiasmatic nucleus balances sympathetic and parasympathetic output to peripheral organs through separate preautonomic neurons. J. Comp. Neurol., 464: 36–48.

Cailotto, C., La Fleur, S.E., Van Heijningen, C., Wortel, J., Kalsbeek, A., Feenstra, M., Pévet, P. and Buijs, R.M. (2005) The suprachiasmatic nucleus controls the daily variation of plasma glucose via the autonomic output to the liver: are the clock genes involved? Eur. J. Neurosci., 22: 2531–2540.

Cantu, R.C. and Goodman, H.M. (1967) Effects of denervation and fasting on white adipose tissue. Am. J. Physiol., 212: 207–212.

Card, J.P., Rinaman, L., Schwaber, J.S., Miselis, R.R., Whealy, M.E., Robbins, A.K. and Enquist, L.W. (1990) Neurotropic properties of pseudorabies virus: uptake and transneuronal passage in the rat central nervous system. J. Neurosci., 10: 1974–1994.

Carnethon, M.R., Golden, S.H., Folsom, A.R., Haskell, W. and Liao, D. (2003) Prospective investigation of autonomic nervous system function and the development of type 2 diabetes: the atherosclerosis risk in communities study, 1987–1998. Circulation, 107: 2190–2195.

Carrol, K.F. and Nestel, P.J. (1973) Diurnal variation in glucose tolerance and in insulin secretion in man. Diabetes, 22: 333–348.

Clark, R.G., Chambers, G., Lewin, J. and Robinson, I.C.A.F. (1986) Automated repetitive microsampling of blood: growth hormone profiles in conscious male rats. J. Endocrinol., 111: 27–35.

Coimbra, C.C. and Migliorini, R.H. (1988) Cold-induced free fatty acid mobilization is impaired in rats with lesions in the preoptic area. Neurosci. Lett., 88: 1–5.

Combs, T.P., Berg, A.H., Obici, S., Scherer, P.E. and Rossetti, L. (2001) Endogenous glucose production is inhibited by the adipose-derived protein Acrp30. J. Clin, Invest., 108: 1875–1881.

Cornish, S. and Cawthorne, M.A. (1978) Fatty acid synthesis in mice during the 24 hr cycle and during meal-feeding. Horm. Metab. Res., 10: 286–290.

Corssmit, E.P., Romijn, J.A. and Sauerwein, H.P. (2001) Regulation of glucose production with special attention to non-classical regulatory mechanisms: a review. Metabolism, 50: 742–755.

Cui, L.N., Coderre, E. and Renaud, L.P. (2001) Glutamate and GABA mediate suprachiasmatic nucleus inputs to spinal-projecting paraventricular neurons. Am. J. Physiol., 281: R1283–R1289.

Dallman, M.F., Akana, S.F., Bhatnagar, S., Bell, M.E., Choi, S., Chu, A., Horsley, C., Levin, N., Meijer, O., Soriano, L.R., Strack, A.M. and Viau, V. (1999) Starvation: early signals, sensors, and sequelae. Endocrinology, 140: 4015–4023.

Damiola, F., Le minh, N., Preitner, N., Kornmann, B., Fleury-Olela, F. and Schibler, U. (2000) Restricted feeding uncouples circadian oscillators in peripheral tissues from the central pacemaker in the suprachiasmatic nucleus. Genes Dev., 14: 2950–2961.

Davidson, A.J., Yamazaki, S. and Menaker, M. (2003) SCN: ringmaster of the circadian circus or conductor of the circadian orchestra? In: Chadwick, D.J. and Goode, J.A. (Eds.) Molecular Clocks and Light Signalling: Novartis Foundation Symposium, Vol. 253. Novartis Foundation, New York, pp. 110–125.

De Boer, S.F. and Van Der Gugten, J. (1987) Daily variations in plasma noradrenaline, adrenaline and corticosterone concentrations in rats. Physiol. Behav., 40: 323–328.

Deboer, T., Vansteensel, M.J., Detari, L. and Meijer, J.H. (2003) Sleep states alter activity of suprachiasmatic nucleus neurons. Nat. Neurosci., 6: 1086–1090.

DeCoursey, P.J. and Krulas, J.R. (1998) Behavior of SCN-lesioned chipmunks in natural habitat: A pilot study. J. Biol. Rhythms, 13: 229–244.

DeCoursey, P.J., Krulas, J.R., Mele, G. and Holley, D.C. (1997) Circadian performance of suprachiasmatic nuclei (SCN)- lesioned antelope ground squirrels in a desert enclosure. Physiol. Behav., 62: 1099–1108.

DeCoursey, P.J., Walker, J.K. and Smith, S.A. (2000) A circadian pacemaker in free-living chipmunks: essential for survival? J. Comp. Physiol. A, 186: 169–180.

de Vegt, F., Dekker, J.M., Ruhe, H.G., Stehouwer, C.D., Nijpels, G., Bouter, L.M. and Heine, R.J. (1999) Hyperglycaemia is associated with all-cause and cardiovascular mortality in the Hoorn population: the Hoorn Study. Diabetologia, 42: 926–931.

Dolatshad, H., Campbell, E.A., O'Hara, L., Hastings, M.H. and Johnson, M.H. (2006) Developmental and reproductive performance in circadian mutant mice. Human Reprod., 21: 68–79.

Evans, J.L., Youngren, J.F. and Goldfine, I.D. (2004) Effective treatments for insulin resistance: trim the fat and douse the fire. Trends Endocrinol. Metab., 15: 425–431.

Facchini, F.S., Hua, N., Abbasi, F. and Reaven, G.M. (2001) Insulin resistance as a predictor of age-related diseases. J. Clin. Endocrinol. Metab., 86: 3574–3578.

Fasshauer, M. and Paschke, R. (2003) Regulation of adipocytokines and insulin resistance. Diabetologia, 46: 1594–1603.

Feneberg, R. and Lemmer, B. (2004) Circadian rhythm of glucose uptake in cultures of skeletal muscle cells and adipocytes in Wistar-Kyoto, Wistar, Goto-Kakizaki, and spontaneously hypertensive rats. Chronobiol. Int., 21: 521–538.

Ferreira, M.L., Marubayashi, U. and Coimbra, C.C. (1999) The medial preoptic area modulates the increase in plasma glucose and free fatty acid mobilization induced by acute cold exposure. Brain. Res. Bull., 49: 189–193.

Fliers, E., Kreier, F., Voshol, P.J., Havekes, L.M., Sauerwein, H.P., Kalsbeek, A., Buijs, R.M. and Romijn, J.A. (2003a) White adipose tissue: getting nervous. J. Neuroendocrinol., 15: 1005–1010.

Fliers, E., Sauerwein, H.P., Romijn, J.A., Reiss, P., van der Valk, M., Kalsbeek, A., Kreier, F. and Buijs, R.M. (2003b) HIV-associated adipose redistribution syndrome as a selective autonomic neuropathy. Lancet, 362: 1758–1760.

Foster, M.T. and Bartness, T.J. (2003) Paraventricular nucleus lesions do not attenuate lipid mobilization after fasting. Soc. Neurosci., Washington, USA, No. 50810.

Foster, R.G. and Wulff, K. (2005) The rhythm of rest and excess. Nat. Rev. Neurosci., 6: 407–414.

Fu, L., Patel, M.S. and Karsenty, G. (2006) The circadian modulation of leptin-controlled bone formation. In: Kalsbeek, A., Fliers, E., Hofman, M.A., Swaab, D.F., Van Someren, E.J.W. and Buijs, R.M. (Eds.), Hypothalamic Integration of Energy Metabolism, Progress in Brain Research. Elsevier, Amsterdam, 153: 177–191.

Fujii, T., Inoue, S., Nagai, K. and Nakagawa, H. (1989) Involvement of adrenergic mechanism in hyperglycemia due to SCN stimulation. Horm. Metab. Res., 21: 643–645.

Fukuhara, A., Matsuda, M., Nishizawa, M., Segawa, K., Tanaka, M., Kishimoto, K., Matsuki, Y., Murakami, M.,

Ichisaka, T., Murakami, H., Watanabe, E., Takagi, T., Akiyoshi, M., Ohtsubo, T., Kihara, S., Yamashita, S., Makishima, M., Funahashi, T., Yamanaka, S., Hiramatsu, R., Matsuzawa, Y. and Shimomura, I. (2005) Visfatin: a protein secreted by visceral fat that mimics the effects of insulin. Science, 307: 426–430.

Gallopin, T., Fort, P., Eggerman, E., Cauli, B., Luppi, P.H., Rossier, J., Audinat, E., Mühlethaler, M. and Serafin, M. (2000) Identification of sleep-promoting neurons in vitro. Nature, 404: 992–995.

Gavrila, A., Peng, C.K., Chan, J.L., Mietus, J.E., Goldberger, A.L. and Mantzoros, C.S. (2003) Diurnal and ultradian dynamics of serum adiponectin in healthy men: comparison with leptin, circulating soluble leptin receptor, and cortisol patterns. J. Clin. Endocrinol. Metab., 88: 2838–2843.

Gibson, T. and Jarrett, R.J. (1972) Diurnal variation in insulin sensitivity. Lancet, II: 947–948.

Giebultowicz, J.M. and Hege, D.M. (1997) Circadian clock in malpighian tubes. Nature, 386: 664.

Goncharuk, V.D., Van Heerikhuize, J.J., Dai, J.P., Swaab, D.F. and Buijs, R.M. (2001) Neuropeptide changes in the suprachiasmatic nucleus in primary hypertension indicate functional impairment of the biological clock. J. Comp. Neurol., 431: 320–330.

Guo, H., Brewer, J.M., Champhekar, A., Harris, R.B. and Bittman, E.L. (2005) Differential control of peripheral circadian rhythms by suprachiasmatic-dependent neural signals. Proc. Natl. Acad. Sci. USA, 102: 3111–3116.

Hara, R., Wan, K., Wakamatsu, H., Aida, R., Moriya, T., Akiyama, M. and Shibata, S. (2001) Restricted feeding entrains liver clock without participation of the suprachiasmatic nucleus. Genes Cells, 6: 269–278.

Hastings, M.H., Reddy, A.B. and Maywood, E.S. (2003) A clockwork web: circadian timing in brain and periphery, in health and disease. Nat. Rev. Neurosci., 4: 649–661.

Hendrickson, A.E., Wagoner, N. and Cowan, W.M. (1972) An autoradiographic and electron microscopic study of retinohypothalamic connections. Zeitschr. Zellforsch., 135: 1–26.

Hermes, M.L.H.J., Coderre, E.M., Buijs, R.M. and Renaud, L.P. (1996) GABA and glutamate mediate rapid neurotransmission from suprachiasmatic nucleus to hypothalamic paraventricular nucleus in rat. J. Physiol., 496: 749–757.

Hermes, M.L.H.J. and Renaud, L.P. (1993) Differential responses of identified rat hypothalamic paraventricular neurons to suprachiasmatic nucleus stimulation. Neuroscience, 56: 823–832.

Hermes, M.L.H.J., Ruijter, J.M., Klop, A., Buijs, R.M. and Renaud, L.P. (2000) Vasopressin increases GABAergic inhibition of rat hypothalamic paraventricular nucleus neurons in vitro. J. Neurophysiol., 83: 705–711.

Hetherington, A.W. and Ranson, S.W. (1940) Hypothalamic lesion and adiposity in the rat. Anat. Rec., 78: 149–172.

Holl, R.W., Pavlovic, M., Heinze, E. and Thon, A. (1999) Circadian blood pressure during the early course of type 1 diabetes—Analysis of 1,011 ambulatory blood pressure recordings in 354 adolescents and young adults. Diabet. Care, 22: 1151–1157.

Jarrett, R.J., Baker, I.A., Keen, H. and Oakley, N.W. (1972) Diurnal variation in oral glucose tolerance: blood sugar and plasma insulin levels morning, afternoon, and evening. Br. Med. J., 1: 199–201.

Jolin, T. and Montes, A. (1973) Daily rhythm of plasma glucose and insulin levels in rats. Horm. Res., 4: 153–156.

Kalsbeek, A., Buijs, R.M., Van Heerikhuize, J.J., Arts, M. and Van Der Woude, T.P. (1992) Vasopressin-containing neurons of the suprachiasmatic nuclei inhibit corticosterone release. Brain Res., 580: 62–67.

Kalsbeek, A., Fliers, E., Romijn, J.A., La Fleur, S.E., Wortel, J., Bakker, O., Endert, E. and Buijs, R.M. (2001) The suprachiasmatic nucleus generates the diurnal changes in plasma leptin levels. Endocrinology, 142: 2677–2685.

Kalsbeek, A., Garidou, M.L., Palm, I.F., Van Der Vliet, J., Simonneaux, V., Pévet, P. and Buijs, R.M. (2000) Melatonin sees the light: blocking GABA-ergic transmission in the paraventricular nucleus induces daytime secretion of melatonin. Eur. J. Neurosci., 12: 3146–3154.

Kalsbeek, A., La Fleur, S.E., Van Heijningen, C. and Buijs, R.M. (2004) Suprachiasmatic GABAergic inputs to the paraventricular nucleus control plasma glucose concentrations in the rat via sympathetic innervation of the liver. J. Neurosci., 24: 7604–7613.

Kalsbeek, A., Ruiter, M., La Fleur, S.E., Van Heijningen, C. and Buijs, R.M. (2003) The diurnal modulation of hormonal responses in the rat varies with different stimuli. J. Neuroendocrinol., 15: 1144–1155.

Kalsbeek, A., Van Der Vliet, J. and Buijs, R.M. (1996a) Decrease of endogenous vasopressin release necessary for expression of the circadian rise in plasma corticosterone: a reverse microdialysis study. J. Neuroendocrinol., 8: 299–307.

Kalsbeek, A., Van Heerikhuize, J.J., Wortel, J. and Buijs, R.M. (1996b) A diurnal rhythm of stimulatory input to the hypothalamo-pituitary-adrenal system as revealed by timed intrahypothalamic administration of the vasopressin V1 antagonist. J. Neurosci., 16: 5555–5565.

Karlsson, B., Knutsson, A. and Lindahl, B. (2001) Is there an association between shift work and having a metabolic syndrome? Results from a population based study of 27.485 people. Occup. Environ. Med., 58: 747–752.

Kershaw, E.E. and Flier, J.S. (2004) Adipose tissue as an endocrine organ. J. Clin. Endocrinol. Metab., 89: 2548–2556.

Kimura, F. and Tsai, C.-W. (1984) Ultradian rhythm of growth hormone secretion and sleep in the adult male rat. J. Physiol. London, 353: 305–315.

Kita, Y., Shirozawa, N., Jin, W.H., Majewski, R.R., Besharse, J.C., Greene, A.S. and Jacob, H.J. (2002) Implications of circadian gene expression in kidney, liver and the effects of fasting on pharmacogenomic studies. Pharmacogenetics, 12: 55–65.

Kreier, F., Kap, Y.S., Mettenleiterm, T.C., Van Heijningen, C., Van Der Vliet, J., Kalsbeek, A., Sauerwein, H.P., Fliers, E., Romijn, J.A. and Buijs, R.M. (2006) Tracing from fat tissue, liver and pancreas: A neuroanatomical framework for the role of the brain in type 2 diabetes. Endocrinology, 147: 1140–1147.

Kreier, F., Fliers, E., Voshol, P.J., Van Eden, C.G., Havekes, L.M., Kalsbeek, A., Van Heijningen, C.L., Sluiter, A.A., Mettenleiter, T.C., Romijn, J.A., Sauerwein, H.P. and Buijs, R.M. (2002) Selective parasympathetic innervation of subcutaneous and intra-abdominal fat — functional implications. J. Clin. Invest., 110: 1243–1250.

Kreier, F., Kalsbeek, A., Ruiter, M., Yilmaz, A., Romijn, J.A., Sauerwein, H.P., Fliers, E. and Buijs, R.M. (2003a) Central nervous determination of food storage — a daily switch from conservation to expenditure: implications for the metabolic syndrome. Eur. J. Pharmacol., 480: 51–65.

Kreier, F., Yilmaz, A., Kalsbeek, A., Romijn, J.A., Sauerwein, H.P., Fliers, E. and Buijs, R.M. (2003b) Hypothesis: shifting the equilibrium from activity to food leads to autonomic unbalance and the metabolic syndrome. Diabetes, 52: 2652–2656.

Kuriyama, K., Sasahara, K., Kudo, T. and Shibata, S. (2004) Daily injection of insulin attenuated impairment of liver circadian clock oscillation in the streptozotocin-treated diabetic mouse. FEBS Lett., 572: 206–210.

Kurukulasuriya, R., Link, J.T., Madar, D.J., Pei, Z., Richards, S.J., Rohde, J.J., Souers, A.J. and Szczepankiewicz, B.G. (2003) Potential drug targets and progress towards pharmacologic inhibition of hepatic glucose production. Curr. Med. Chem., 10: 123–153.

La Fleur, S.E., Kalsbeek, A., Wortel, J. and Buijs, R.M. (1999) An SCN generated rhythm in basal glucose levels. J. Neuroendocrinol., 11: 643–652.

La Fleur, S.E., Kalsbeek, A., Wortel, J. and Buijs, R.M. (2000) Polysynaptic neural pathways between the hypothalamus, including the suprachiasmatic nucleus, and the liver. Brain Res., 871: 50–56.

La Fleur, S.E., Kalsbeek, A., Wortel, J., Fekkes, M.L. and Buijs, R.M. (2001) A daily rhythm in glucose tolerance. A role for the suprachiasmatic nucleus. Diabetes, 50: 1237–1243.

Lang, C.H. (1995) Inhibition of central $GABA_A$ receptors enhances hepatic glucose production and peripheral glucose uptake. Brain Res. Bull., 37: 611–616.

Lee, A., Ader, M., Bray, G.A. and Bergman, R.N. (1992) Diurnal variation in glucose tolerance cyclic suppression of insulin action and insulin secretion in normal-weight, but not obese, subjects. Diabetes, 41: 750–759.

Lee, J.W. and Erskine, M.S. (2000) Pseudorabies virus tracing of neural pathways between the uterine cervix and CNS: effects of survival time, estrogen treatment, rhizotomy, and pelvic nerve transection. J. Comp. Neurol., 418: 484–503.

Licinio, J., Mantzoros, C., Negrao, A.B., Cizza, G., Wong, M.L., Bongiorno, P.B., Chrousos, G.P., Karp, B., Allen, C., Flier, J.S. and Gold, P.W. (1997) Human leptin levels are pulsatile and inversely related to pituitary–adrenal function. Nat. Med., 3: 575–579.

Liu, C., Weaver, D.R., Jin, X., Shearman, L.P., Pieschl, R.L., Gribkoff, V.K. and Reppert, S.M. (1997) Molecular dissection of two distinct actions of melatonin on the suprachiasmatic circadian clock. Neuron, 19: 91–102.

Liu, L., Karkanias, G.B., Morales, J.C., Hawkins, M., Barzilai, N., Wang, J. and Rossetti, L. (1998) Intracerebroventricular leptin regulates hepatic but not peripheral glucose fluxes. J. Biol. Chem., 273: 31160–31167.

Lucassen, P.J., Hofman, M.A. and Swaab, D.F. (1995) Increased light intensity prevents the age related loss of vasopressin-expressing neurons in the rat suprachiasmatic nucleus. Brain Res., 693: 261–266.

Lurbe, E., Redon, J., Kesani, A., Pascual, J.M., Tacons, J., Alvarez, V. and Battle, D. (2002) Increase in nocturnal blood pressure and progression to microalbuminuria in type 1 diabetes. N. Engl. J. Med., 347: 797–805.

Matsuo, T., Yamaguchi, S., Mitsui, S., Emi, A., Shimoda, F. and Okamura, H. (2003) Control mechanism of the circadian clock for timing of cell division in vivo. Science, 302: 255–259.

Maywood, E.S., O'Neill, J., Wong, G.K.Y., Reddy, A.B. and Hastings, M.H. (2006) Circadian timing in health and disease. In: Kalsbeek, A., Fliers, E., Hofman, M.A., Swaab, D.F., Van Someren, E.J.W. and Buijs, R.M. (Eds.), Hypothalamic Integration of Energy Metabolism, Progress in Brain Research. Elsevier, Amsterdam, 153: 257–273.

Minokoshi, Y., Haque, M.S. and Shimazu, T. (1999) Microinjection of leptin into the ventromedial hypothalamus increases glucose uptake in peripheral tissues in rats. Diabetes, 48: 287–291.

Monteagudo, P.T., Nobrega, J.C., Cezarini, P.R., Ferreira, S.R.G., Kohlmann, O., Ribeiro, A.B. and Zanella, M.T. (1996) Altered blood pressure profile, autonomic neuropathy and nephropathy in insulin-dependent diabetic patients. Eur. J. Endocrinol., 135: 683–688.

Moore, R.Y. and Eichler, V.B. (1972) Loss of a circadian adrenal corticosterone rhythm following suprachiasmatic lesions in the rat. Brain Res., 42: 201–206.

Moore, R.Y. and Lenn, N.J. (1972) A retinohypothalamic qprojection in the rat. J. Comp. Neurol., 146: 1–9.

Morgan, L.M., Aspostolakou, F., Wright, J. and Gama, R. (1999) Diurnal variations in peripheral insulin resistance and plasma non-esterified fatty acid concentrations: a possible link? Ann. Clin. Biochem., 36: 447–450.

Nagai, K., Fujii, T., Inoue, S., Takamura, Y. and Nakagawa, H. (1988) Electrical stimulation of the suprachiasmatic nucleus of the hypothalamus causes hyperglycemia. Horm. Metab. Res., 20: 37–39.

Naito, Y., Tsujino, T., Kawasaki, D., Okumura, T., Morimoto, S., Masai, M., Sakoda, T., Fujioka, Y., Ohyanagi, M. and Iwasaki,. (2003) Circadian gene expression of clock genes and plasminogen activator inhibitor-1 in heart and aorta of spontaneously hypertensive and Wistar–Kyoto rats. J. Hypertens., 21: 1107–1115.

Nakanishi, N., Takatorige, T. and Suzuki, K. (2004) Daily life activity and risk of developing impaired fasting glucose or type 2 diabetes in middle-aged Japanese men. Diabetologia, 47: 1768–1775.

Nonogaki, K. (2000) New insights into sympathetic regulation of glucose and fat metabolism. Diabetologia, 43: 533–549.

O'Brien, I.A.D., Lewin, I.G., O'Hare, J.P., Arendt, J. and Corrall, R.J.M. (1986) Abnormal circadian rhythm of melatonin in diabetic autonomic neuropathy. Clin. Endocrinol., 24: 359–364.

Oishi, K., Miyazaki, K., Kadota, K., Kikuno, R., Nagase, T., Atsumi, G., Ohkura, N., Azama, T., Mesaki, M., Yukimasa, S., Kobayashi, H., Iitaka, C., Umehara, T., Horikoshi, M., Kudo, T., Shimizu, Y., Yano, M., Monden, M., Machida, K., Matsuda, J., Horie, S., Todo, T. and Ishida, N. (2003) Genome-wide expression analysis of mouse liver reveals CLOCK-regulated circadian output genes. J. Biol. Chem., 278: 41519–41527.

Oishi, K., Ohkura, N., Kasamatsu, M., Fukushima, N., Shirai, H., Matsuda, J., Horie, S. and Ishida, N. (2004) Tissue-specific augmentation of circadian PAI-1 expression in mice with streptozotocin-induced diabetes. Thromb. Res., 114: 129–135.

Ouyang, Y., Andersson, C.R., Kondo, T., Golden, S.S. and Johnson, C.H. (1998) Resonating circadian clocks enhance fitness in cyanobacteria. Proc Natl Acad Sci USA, 95: 8660–8664.

Palm, I.F., Van Der Beek, E.M., Swarts, H.J.M., Van Der Vliet, J., Wiegant, V.M., Buijs, R.M. and Kalsbeek, A. (2001) Control of the estradiol-induced prolactin surge by the suprachiasmatic nucleus. Endocrinology, 142: 2296–2302.

Palm, I.F., Van Der Beek, E.M., Wiegant, V.M., Buijs, R.M. and Kalsbeek, A. (1999) Vasopressin induces an LH surge in ovariectomized, estradiol-treated rats with lesion of the suprachiasmatic nucleus. Neuroscience, 93: 659–666.

Panda, S., Antoch, M.P., Miller, B.H., Su, A.I., Schook, A.B., Straume, M., Schultz, P.G., Kay, S.A., Takahashi, J.S. and Hogenesch, J.B. (2002) Coordinated transcription of key pathways in the mouse by the circadian clock. Cell, 109: 307–320.

Penicaud, L. and Le Magnen, J. (1980) Aspects of the neuroendocrine bases of the diurnal metabolic cycle in rats. qNeurosci. Biobehav. Rev., 4: S39–S42.

Perreau-Lenz, S., Kalsbeek, A., Pevet, P. and Buijs, R.M. (2004) Glutamatergic clock output stimulates melatonin synthesis at night. Eur. J. Neurosci., 19: 318–324.

Pilkis, S.J. and Granner, D.K. (1992) Molecular physiology of the regulation of hepatic gluconeogenesis and glycolysis. Annu. Rev. Physiol., 54: 885–909.

Poirier, P., Tremblay, A., Broderick, T., Catellier, C., Tancrede, G. and Nadeau, A. (2002) Impact of moderate aerobic exercise training on insulin sensitivity in type 2 diabetic men treated with oral hypoglycemic agents: is insulin sensitivity enhanced only in nonobese subjects? Med. Sci. Monit., 8: CR59–CR65.

Puschel, G.P. (2004) Control of hepatocyte metabolism by sympathetic and parasympathetic hepatic nerves. Anat. Rec., 280A: 854–867.

Rajala, M.W., Qi, Y., Patel, H.R., Takahashi, N., Banerjee, R., Pajvani, U.B., Sinha, M.K., Gingerich, R.L., Scherer, P.E. and Ahima, R.S. (2004) Regulation of resistin expression and circulating levels in obesity, diabetes, and fasting. Diabetes, 53: 1671–1679.

Randle, P.J., Garland, P.B., Hales, C.N. and Newsholme, E.A. (1963) The glucose fatty-acid cycle. Its role in insulin sensitivity and the metabolic disturbances of diabetes mellitus. Lancet, 1: 785–789.

Reppert, S.M. and Weaver, D.R. (2002) Coordination of circadian timing in mammals. Nature, 418: 935–941.

Rudic, R.D., McNamara, P., Curtis, A.M., Boston, R.C., Panda, S., Hogenesch, J.B. and FitzGerald, G.A. (2004) BMAL1 and CLOCK, two essential components of the circadian clock, are involved in glucose homeostasis. PLoS Biol., 2: 1893–1899.

Ruiter, M., Buijs, R.M. and Kalsbeek, A. (2006a) Biological clock control of glucose metabolism. Timing metabolic homeostasis. In: Pandi-Perumal, S.R. and Cardinali, D.P. (Eds.), Neuroendocrine Correlates of Sleep/Wakefulness. Springer, New York, pp. 87–117.

Ruiter, M., Buijs, R.M. and Kalsbeek, A. (2006b) Hormones and the autonomic nervous system are involved in suprachiasmatic nucleus modulation of glucose homeostasis. Curr. Diabetes Rev., 2: 213–226.

Ruiter, M., La Fleur, S.E., Van Heijningen, C., Van Der Vliet, J., Kalsbeek, A. and Buijs, R.M. (2003) The daily rhythm in plasma glucagon concentrations in the rat is modulated by the biological clock and by feeding behavior. Diabetes, 52: 1709–1715.

Saad, M.F., Riad-Gabriel, M.G., Khan, A., Sharma, A., Michael, R., Jinagouda, S.D., Boyadjian, R. and Steil, G.M. (1998) Diurnal and ultradian rhythmicity of plasma leptin: effects of gender and adiposity. J. Clin. Endocrinol. Metab., 83: 453–459.

Satoh, S., Matsumura, H., Nakajima, T., Nakahama, K., Kanbayashi, T., Nishino, S., Yoneda, H. and Shigeyoshi, Y. (2003) Inhibition of rostral basal forebrain neurons promotes wakefulness and induces FOS in orexin neurons. Eur. J. Neurosci., 17: 1635–1645.

Scheer, F., Van Montfrans, G.A., Van Someren, E.J.W., Mairuhu, G. and Buijs, R.M. (2004) Daily nighttime melatonin reduces blood pressure in male patients with essential hypertension. Hypertension, 43: 192–197.

Scheidt-Nave, C., Barrett-Connor, E., Wingard, D.L., Cohn, B.A. and Edelstein, S.L. (1991) Sex differences in fasting glycemia as a risk factor for ischemic heart disease death. Am. J. Epidemiol., 133: 565–576.

Schibler, U. and Naef, F. (2005) Cellular oscillators: rhythmic gene expression and metabolism. Curr. Opin. Cell Biol., 17: 223–229.

Shea, S.A., Hilton, M.F., Orlova, C., Ayers, R.T. and Mantzoros, C.S. (2005) Independent circadian and sleep/wake regulation of adipokines and glucose in humans. J. Clin. Endocrinol. Metab., 90: 2537–2544.

Shimazu, T. (1967) Glycogen synthetase activity in liver: regulation by the autonomic nerves. Science, 156: 1256–1257.

Shimazu, T. (1981) Central nervous system regulation of liver and adipose tissue metabolism. Diabetologia, 20: 343–356.

Shimazu, T. (1996) Innervation of the liver and glucoregulation: roles of the hypothalamus and autonomic nerves. Nutrition, 12: 65–66.

Shimazu, T. and Fukuda, A. (1965) Increased activities of glycogenolytic enzymes in liver after splanchnic-nerve stimulation. Science, 150: 1607–1608.

Shimba, S., Ishii, N., Ohta, Y., Ohno, T., Watabe, Y., Hayashi, M., Wada, T., Aoyagi, T. and Tezuka, M. (2005) Brain and muscle Arnt-like protein-1 (BMAL1), a component of the molecular clock, regulates adipogenesis. Proc. Natl. Acad. Sci. USA, 02: 12071–12076.

Shin, M.S., Kim, H., Chang, H.K., Lee, T.H., Jang, M.H., Shin, M.C., Lim, B.V., Lee, H.H., Kim, Y.P., Yoon, J.H., Jeong, I.G. and Kim, C.J. (2003) Treadmill exercise suppresses diabetes-induced increment of neuropeptide Y expression in the hypothalamus of rats. Neurosci. Lett., 346: 157–160.

Stephan, F.K. and Zucker, I. (1972) Circadian rhythms in drinking behavior and locomotor activity of rats are eliminated by hypothalamic lesions. Proc. Natl. Acad. Sci. USA, 69: 1583–1586.

Steppan, C.M. and Lazar, M.A. (2002) Resistin and obesity-associated insulin resistance. Trends Endocrinol. Metab., 13: 18–23.

Stokkan, K.A., Yamazaki, S., Tei, H., Sakaki, Y. and Menaker, M. (2001) Entrainment of the circadian clock in the liver by feeding. Science, 291: 490–493.

Strack, A.M., Sawyer, W.B., Platt, K.B. and Loewy, A.D. (1989) CNS cell groups regulating the sympathetic outflow to adrenal gland as revealed by transneuronal cell body labeling with pseudorabies virus. Brain Res., 491: 274–296.

Sun, X., Rusak, B. and Semba, K. (2000) Electrophysiology and pharmacology of projections from the suprachiasmatic nucleus to the ventromedial preoptic area in rat. Neuroscience, 98: 715–728.

Sun, X., Whitefield, S., Rusak, B. and Semba, K. (2001) Electrophysiological analysis of suprachiasmatic nucleus projections to the ventrolateral preoptic area in the rat. Eur. J. Neurosci., 14: 1257–1274.

Takahashi, A., Ishimaru, H., Ikarashi, Y., Kishi, E. and Maruyama, Y. (1996) Effects of hepatic nerve stimulation on blood glucose and glycogenolysis in rat liver: studies with in vivo microdialysis. J. Auton. Nerv. Syst., 61: 181–185.

Taubert, G., Winkelmann, B.R., Schleiffer, T., Marz, W., Winkler, R., Gok, R., Klein, B., Schneider, S. and Boehm, B.O. (2003) Prevalence, predictors, and consequences of unrecognized diabetes mellitus in 3266 patients scheduled for coronary angiography. Am. Heart J., 145: 285–291.

Teclemariam-Mesbah, R., Kalsbeek, A., Pévet, P. and Buijs, R.M. (1997) Direct vasoactive intestinal polypeptide-containing projection from the suprachiasmatic nucleus to spinal projecting hypothalamic paraventricular neurons. Brain Res., 748: 71–76.

Teixeira, V.L., Antunes-Rodrigues, J. and Migliorini, R.H. (1973) Evidence for centers in the central nervous system that selectively regulate fat mobilization in the rat. J. Lipid Res., 14: 672–677.

Terazono, H., Mutoh, H., Yamaguchi, S., Kobayashi, M., Akiyama, M., Udo, R., Ohdo, S., Okamura, H. and Shibata,

S. (2003) Adrenergic regulation of clock gene expression in mouse liver. Proc. Natl. Acad. Sci. USA, 100: 6795–6800.

Timmer, J.R., van der Horst, I.C., Ottervanger, J.P., Henriques, J.P., Hoorntje, J.C., de Boer, M.J., Suryapranata, H. and Zijlstra, F. (2004) Prognostic value of admission glucose in non-diabetic patients with myocardial infarction. Am. Heart J., 148: 399–404.

Turek, F.W., Joshu, C., Kohsaka, A., Lin, E., Ivanova, G., McDearmon, E., Laposky, A., Losee-Olson, S., Easton, A., Jensen, D.R., Eckel, R.H., Takahashi, J.S. and Bass, J. (2005) Obesity and metabolic syndrome in circadian Clock mutant mice. Science, 308: 1043–1045.

Ueyama, T., Krout, K.E., Van Nguyen, X., Karpitskiy, V., Koller, A., Mettenleiter, T.C. and Loewy, A.D. (1999) Suprachiasmatic nucleus: a central autonomic clock. Nat. Neurosci., 2: 1051–1053.

Uyama, N., Geerts, A. and Reynaert, H. (2004) Neural connections between the hypothalamus and the liver. Anat. Rec., 280A: 808–820.

Van Cauter, E., Polonsky, K.S. and Scheen, A.J. (1997) Roles of circadian rhythmicity and sleep in human glucose regulation. Endocr. Rev., 18: 716–738.

Van den Top, M., Buijs, R.M., Ruijter, J.M., Delagrange, P., Spanswick, D. and Hermes, M.L.H.J. (2001) Melatonin generates an outward potassium current in rat suprachiasmatic nucleus neurones in vitro independent of their circadian rhythm. Neuroscience, 107: 99–108.

Vgontzas, A.N., Bixler, E.O., Lin, H.M., Prolo, P., Trakada, G. and Chrousos, G.P. (2005) IL-6 and its circadian secretion in humans. Neuroimmunomodulation, 12: 131–140.

Vrang, N., Mikkelsen, J.D. and Larsen, P.J. (1997) Direct link from the suprachiasmatic nucleus to hypothalamic neurons projecting to the spinal cord: a combined tracing study using cholera toxin subunit B and Phaseolus vulgaris — leucoagglutinin. Brain Res. Bull., 44: 671–680.

Wang, D., Cui, L.N. and Renaud, L.P. (2003) Pre- and postsynaptic GABA$_B$ receptors modulate rapid neurotransmission from suprachiasmatic nucleus to parvocellular hypothalamic paraventricular nucleus neurons. Neuroscience, 118: 49–58.

Westerhaus, M.J. and Loewy, A.D. (1999) Sympathetic-related neurons in the preoptic region of the rat identified by viral transneuronal labeling. J. Comp. Neurol., 414: 361–378.

Westerterp, K.R. (2001) Pattern and intensity of physical activity. Nature, 410: 539.

Whichelow, M.J., Sturge, R.A., Keen, H., Jarrett, R.J., Stimmler, L. and Grainger, S. (1974) Diurnal variation in response to intravenous glucose. Br. Med. J., 1: 488–491.

Wolden-Hanson, T., Mitton, D.R., McCants, R.L., ellon, S.M., Wilkinson, C.W., Matsumoto, A.M. and Rasmussen, D.D. (2000) Daily melatonin administration to middle-aged male rats suppresses body weight, intra-abdominal adiposity, and plasma leptin and insulin independent of food intake and total body fat. Endocrinology, 141: 487–497.

Yamamoto, H., Nagai, K. and Nakagawa, H. (1984a) Bilateral lesions of the SCN abolish lipolytic and hyperphagic responses to 2DG. Physiol. Behav., 32: 1017–1020.

Yamamoto, H., Nagai, K. and Nakagawa, H. (1984b) Bilateral lesions of the suprachiasmatic nucleus enhance glucose tolerance in rats. Biomed. Res., 5: 47–54.

Yamamoto, H., Nagai, K. and Nakagawa, H. (1987) Role of the SCN in daily rhythms of plasma glucose, FFA, insulin and glucagon. Chronobiol. Int., 4: 483–491.

Yamauchi, T., Kamon, J., Minokoshi, Y., Ito, Y., Waki, H., Uchida, S., Yamashita, S., Noda, M., Kita, S., Ueki, K., Eto, K., Akanuma, P., Froguel, P., Foufelle, F., Ferre, P., Carling, D., Kimura, S., Nagai, S., Kahn, B.B. and Kadowaki, T. (2002) Adiponectin stimulates glucose utilization and fatty-acid oxidation by activating AMP-activated protein kinase. Nat. Med., 8: 1288–1295.

Young, M.E., Razeghi, P. and Taegtmeyer, H. (2001) Clock genes in the heart — characterization and attenuation with hypertrophy. Circul. Res., 88: 1142–1150.

Young, M.E., Wilson, C.R., Razeghi, P., Guthrie, P.H. and Taegtmeyer, H. (2002) Alterations of the circadian clock in the heart by streptozotocin-induced diabetes. J. Mol. Cell. Cardiol., 34: 223–231.

Youngstrom, T.G. and Bartness, T.J. (1995) Catecholaminergic innervation of white adipose tissue in Siberian hamsters. Am. J. Physiol., 268: R744–R751.

Youngstrom, T.G. and Bartness, T.J. (1998) White adipose tissue sympathetic nervous system denervation increases fat pad mass and fat cell number. Am. J. Physiol., 275: R1488–R1493.

Zhang, Y., Proenca, R., Maffei, M., Barone, M., Leopold, L. and Friedman, J.M. (1994) Positional cloning of the mouse obese gene and its human homologue. Nature, 372: 425–432.

Zhao, Z.Y., Wang, Y.Q., Yan, Z.H., Cui, J. and Li, Y.Y. (2005) Quantitative study of circadian variations of ambulatory blood pressure in Chinese healthy, hypertensive, and diabetes subjects. Clin. Exp. Hypertens., 27: 187–194.

Zheng, H., Li, Y.F., Cornish, K.G., Zucker, I.H. and Patel, K.P. (2005) Exercise training improves endogenous nitric oxide mechanisms within the paraventricular nucleus in rats with heart failure. Am. J. Physiol., 288: H2332–H2341.

Zimmet, P.Z., Wall, J.R., Rome, R., Stimmler, L. and Jarrett, R.J. (1974) Diurnal variation in glucose tolerance: associated changes in plasma insulin, growth hormone, and non-esterified fatty acids. Br. Med. J., 1: 485–491.

Kalsbeek, Fliers, Hofman, Swaab, Van Someren & Buijs
Progress in Brain Research, Vol. 153
ISSN 0079-6123

CHAPTER 18

Mechanisms and functions of coupling between sleep and temperature rhythms

Eus J.W. Van Someren[1,2,*]

[1]Netherlands Institute for Neuroscience, Meibergdreef 47, 1105 BA Amsterdam, The Netherlands
[2]Departments of Neurology, Clinical Neurophysiology and Medical Psychology, VU University Medical Center,
PO Box 7057, 1007 MB Amsterdam, The Netherlands

Abstract: Energy metabolism is strongly linked to the circadian rhythms in sleep and body temperature. Both heat production and heat loss show a circadian modulation. Sleep preferably occurs during the circadian phase of decreased heat production and increased heat loss, the latter due to a profound increase in skin blood flow and, consequently, skin warming. The coupling of these rhythms may differ depending on whether they are assessed in experimental laboratory studies or in habitual sleeping conditions. In habitual sleeping conditions, skin blood flow is for a prolonged time increased to a level hardly ever seen during wakefulness. Possible mechanisms linking the rhythms in sleep and core body and skin temperature are discussed, with a focus on causal effects of changes in core and skin temperature on sleep regulation. It is shown that changes in skin temperature rather than in core temperature causally affect sleep propensity. Contrary to earlier suggestions of a functional role of sleep in heat loss, it is argued that sleep facilitates a condition of increased skin blood flow during a prolonged circadian phase, yet *limits* heat loss and the risk of hypothermia. Sleep-related behavior including the creation of an isolated microclimate of high temperature by means of warm clothing and bedding in humans and the curling up, huddling and cuddling in animals all help limit heat loss The increase in skin blood flow that characterizes the sleeping period may thus not primarily reflect a thermoregulatory drive. There is indirect support for an alternative role of the prolonged period of increased skin blood flow: it may support maintenance of the skin as a primary barrier in host defense.

Keywords: sleep; circadian rhythm; thermoregulation; temperature; preoptic area; skin temperature; immunology; host defense

Introduction

Energy metabolism is strongly linked to the circadian rhythms in sleep and body temperature. To fulfill our energy requirements we obtain and consume food during wakefulness. This requires physical and mental activity and usually an upright body position, which consequently also increases energy expenditure, which is in turn associated with heat production. Even at complete rest the human metabolic rate is so high that it provides a continuous internal source of heating, accounting for about 60-75% of the total daily energy expenditure (Poehlman and Horton, 1990). During physical activity, there is an important additional source of heat generation in the muscles. When asleep, energy expenditure is reduced due to the supine position and lack of physical activity, and consequently, less heat is produced.

*Corresponding author. Tel.: +31-20-5665500; Fax: +31-20-696-1006; E-mail: e.van.someren@nin.knaw.nl

DOI: 10.1016/S0079-6123(06)53018-3

309

Although appealing in its simplicity, the picture sketched above is far from complete. The circadian modulation of body temperature is not restricted to differences between sleeping and waking in heat production, due to the behavior of keeping an upright body position and being physically active. Referring to the latter as *concomitant* physiology and behavior, there are also *intrinsic* differences in heat production during sleep and wakefulness. The basal metabolic rate during sleep is slightly lower than during wakefulness at complete rest (Guyton, 1991; Meijer et al., 1991). To complicate matters even more, a 24-h rhythm in resting metabolic heat production remains even in the absence of sleep (Kräuchi and Wirz-Justice, 1994). This has been demonstrated convincingly in the so called *constant routine* studies in which subjects are kept awake, inactive, in a supine position and fed in small portions at fixed time intervals, or using continuous infusion.

Complex as the regulation of heat production in relation to the sleep–wake cycle may already be with a circadian component in addition to intrinsic and concomitant changes related to sleeping behavior, this still does not give a complete picture of the coupling of sleep and temperature rhythms. This is because not only *heat production* but also *heat loss* shows circadian modulation and intrinsic and concomitant changes with sleep (Kräuchi and Wirz-Justice, 1994). At any time of the circadian cycle, core body temperature reflects the balance between the two processes of heat production and heat loss.

In the present review, a phenomenological description of the coupling between sleep and temperature rhythms will be given first, after which the theoretically possible underlying interaction mechanisms between the various levels of circadian regulation of sleep and body temperature will be addressed. Subsequently, the focus will be on one of these interaction possibilities: the proposed modulatory role of changes in *skin* temperature on sleep-regulating systems (Van Someren, 2000). Finally, considerations on the functional meaning of the relation between sleep and skin temperature will be given. It is proposed that one function of sleep may be to support a prolonged circadian phase of elevated skin blood flow. This phase of increased skin blood flow may support the maintenance of the role of the skin as the primary barrier in host defense.

Description of the coupling between sleep and temperature rhythms

The circadian cycles of sleep and core body temperature are intimately related. In daytime-active mammals, including humans, the major sleep period is initiated in the evening, during the decline in core body temperature. The next morning, when core body temperature has reached its minimum and is rising again, the major sleep period ends. Thus, core body temperature is lower during our habitual sleep period than during our habitual waking period. This relationship is more than coincidence. Dedicated experimental protocols that force periods of sleep and wakefulness to occur at various parts of the circadian cycle — the so called forced desynchrony and ultrashort sleep–wake cycles — have convincingly demonstrated increased sleep propensity during the circadian phase of lowered core body temperature. Our capacity to initiate and maintain sleep thus does not only depend on how long we have been awake — the homeostatic component of sleep regulation — but also on the time of day — the circadian component of sleep regulation (Borbély, 1982), which can under controlled circumstances be read out from the core body temperature curve.

At any moment in time, core body temperature is determined by the balance between heat production and heat loss. It is therefore of interest to briefly review how heat production and heat loss vary over the 24-h cycle to concertedly generate the core body temperature rhythm. Details have been described previously (Van Someren, 2000; Van Someren et al., 2002). In a well-controlled experiment, Kräuchi and Wirz-Justice (1994) have demonstrated that young adults show a circadian modulation of about 17% in their resting state metabolic heat production. The minimum is reached between 00:30 and 01:00 h and the maximum 11 h later, between 11:30 and 12:00 h. The distal regions of the skin, which are the most indicative of heat loss activation (Rubinstein and Sessler, 1990), reach their maximum between 6:00

and 7:00 h in the morning. A warm skin promotes heat transfer from the core to the environment — provided that the environmental temperature is lower than that of the skin. Since core body temperature is determined by the balance between heat production and heat loss, it is not surprising that in most subjects its circadian minimum occurs between 5:00 and 6:00 in the morning, i.e. between the minimum of heat production and the maximum of heat loss. Figure 1 shows the 24-h time course of core body temperature, proximal skin temperature and distal skin temperature under such well-controlled conditions.

However, under habitual circumstances, the time course of core and skin temperatures is different. An example of a 72 h ambulatory recording in a single subject is shown in Fig. 2. As compared to Fig. 1, several differences can be noted. First, physical activity and an upright posture increase metabolic heat production. Consequently, daytime core body temperature under habitual circumstances reaches much higher levels than during the complete rest that is typical for wakefulness in experimental conditions. Second, both distal and proximal skin temperature reach lower values during normal active wakefulness than during the equivalent of the habitual time of active wakefulness in a constant routine condition.

By contrast, nocturnal distal skin temperature reaches much *higher* values (up to 35.9°C for this case as compared to an average of 33.3°C for subjects under constant routine conditions). Even more striking is that proximal skin temperature is *increased* during the night, up to the levels of 35.7°C for this case, whereas subjects under constant routine conditions show a *lowering* of proximal skin temperature to a level of 33.8°C during their habitual sleep time. The distal to proximal skin temperature gradient (DPG), a reliable indication of possible heat loss, is slightly higher than zero degrees for this case, whereas in constant routine conditions it never rises above −0.63°C on average.

The marked nocturnal skin temperature differences between constant routine and natural sleeping conditions cannot easily be accounted for by the presence or absence of sleep per se, because most of the changes in skin temperature precede (stage 2) sleep onset (Kräuchi et al., 2005a). Even if sleep is allowed under otherwise identical constant routine conditions, both distal and proximal skin temperature start to decline soon after sleep onset and stabilize close to 34°C (Kräuchi, 2002), which is still about 1.5°C below the typical skin temperature we observed under natural sleeping conditions. The most likely explanation of the

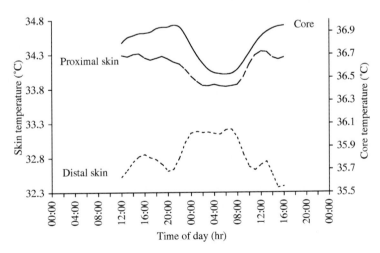

Fig. 1. Smoothed average human temperature curves, as measured under constant routine conditions for core (upper curve), proximal skin (second curve) and distal skin (lower curve) areas. Adapted from Kräuchi and Wirz-Justice (1994, p. 103), who kindly provided the data. Note the nocturnal increase in distal skin temperature, but not proximal skin temperature. The latter is in contrast to the habitual nocturnal increase in proximal skin temperature under natural sleeping conditions in a microclimate of 34–36°C, as shown in Fig. 2.

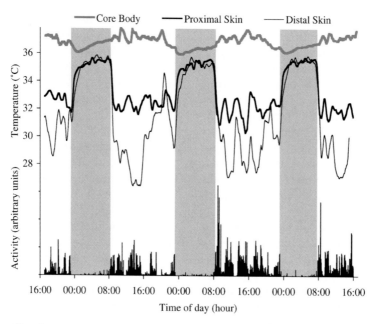

Fig. 2. Example of the profiles of core body temperature (gray line), mean proximal (thick black line) and mean distal (thin black line) temperature during three days under natural living conditions in a single case. For all temperature curves, outliers surpassing one interquartile distance from the Q_{25} or Q_{75} for either level or rate of change were excluded, after which missing data were linearly interpolated. Also shown are the time spent in bed (gray area) and activity level (black columns, arbitrary units from simultaneous actigraphic recording. Note that the marked and simultaneous nocturnal elevation of both proximal and distal temperature during the time in bed never occurs during wakefulness. During the sleep period, proximal and distal skin temperature differ minimally and are both above 35°C. During wakefulness, proximal and distal skin temperature differ at most of the time and do not exceed 33°C for any prolonged period of time. Adapted from Van Marken Lichtenbelt et al. (in press).

different skin temperature profiles is that the laboratory conditions did not allow for the normal nocturnal creation of a microclimate. It has repeatedly been described that humans create a sleeping microclimate of about 34–35°C through the use of insulating bedding in interaction with their nocturnal increase in skin blood flow, which is much higher than the usual daytime temperature and even thermoneutrality (± 29°C, reviewed in Van Someren et al., 2002). The use of only a light cover in an environment of 22°C in the constant routine studies of Krauchi et al. may be representative of normal ambient temperatures during wakefulness, but not of the normal created ambient temperature during sleep. It would be interesting to evaluate the circadian rhythms of distal and proximal skin temperature profiles under constant routine conditions with an ambient temperature of 34–35°C instead of 22°C, thus having it fixed to a mimicked habitual sleeping environmental temperature rather than to the habitual waking

environmental temperature. A constant routine study that applied an ambient temperature of 31°C indeed found a nocturnal peak in mean skin temperature (for 88% determined by proximal temperature) (Marotte and Timbel, 1982), contrary to the daytime peak found in cooler laboratory conditions (Kräuchi and Wirz-Justice, 1994), but in agreement with the nocturnal peak in field studies (van Marken Lichtenbelt et al., in press).

The differences between habitual and laboratory profiles of skin temperature, as shown in Figs. 1 and 2, have been discussed at some length above, because they have important consequences for the proposition that skin temperature might modulate sleep-regulating systems (Van Someren, 2000, 2003, 2004), which will be summarized later in the present review. The first consequence is that under circumstances of normal living the difference between daytime and nighttime skin temperature is much more pronounced than constant routine studies have suggested. During the sleep period, both distal and proximal skin temperature reach levels not normally reached during wakefulness. In addition, their difference (DPG) reaches a very small value, also not normally seen during wakefulness for any prolonged period of time. Thus, both the DPG and the levels of distal and proximal skin temperature differ systematically between being upright and active vs. being supine and inactive in the thermally isolated environment of the bed. It is suggested that the systematically different thermal profile of the skin that results as soon as one has behaviorally adopted the optimal conditions for sleep could provide a reliable signal to inform sleep-regulating brain areas that it is safe or desirable to initiate or maintain sleep, and thus modulate sleep propensity.

Possible sites of interaction in the circadian regulation of sleep and body temperature

The present paragraph briefly mentions several theoretically possible interaction levels in the circadian regulation of sleep and wakefulness, before addressing one of them in more detail in the next paragraph. The paragraph numbers below match the numbers shown in Fig. 3.

1. Although hard evidence is not available and would be difficult to obtain at the present state of technological possibilities, it is very plausible that within the hypothalamic suprachiasmatic nucleus (SCN), the clock of the brain, coupling occurs between the neuronal ensembles that provide output to sleep-regulating areas and the neuronal ensembles that provide output to temperature-regulating areas.

2. A second conceivable yet hard to establish possible interaction is that single outputs of the SCN might reach both sleep- and temperature-regulating areas. These could be humoral, diffusible substances or collaterals of the same SCN neurons reaching two different areas, e.g. the adjacent mostly sleep regulation-involved ventral subparaventricular zone and

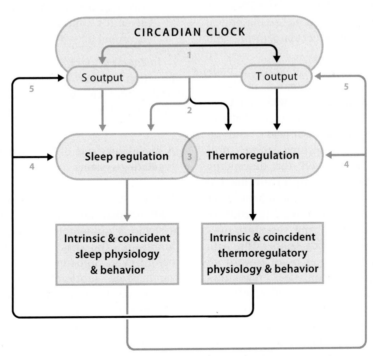

Fig. 3. Schematic overview of the possible ways of interaction (numbers) between the circadian regulation of sleep and body temperature. The circadian clock (top) modulates brain networks involved in the regulation of sleep (left) and temperature (right), which effectuate the changes in physiology and behavior (bottom) specific (e.g. cortical slow waves) and coincident (e.g. a supine position) to the states of sleep and wakefulness. See text for explanation of the theoretical possibilities of interaction.

the mostly thermoregulation-involved dorsal subparaventricular zone (Lu et al., 2001).

3. A third possibility would be if an output signal of the SCN would modulate a network of neurons belonging to both sleep- and temperature-regulating systems. Functional overlap in neuronal activation, even at the level of a single neuron, has frequently been reported for hypothalamic neurons (Hayward and Baker, 1968; Boulant and Silva, 1988; Hori et al., 1988). More specifically, neurons involved in both sleep and thermoregulation have been described in the hypothalamic preoptic area and diagonal band, but also in the adjacent basal forebrain (Alam et al., 1995; Alam et al., 1997). Lesion studies in rats suggest the medial preoptic area to be involved in the promotion of sleep maintenance that occurs with the warming of skin and body (Thomas and Kumar, 2000). Whereas it remains to be established whether such multifunctional neuronal networks are under a single-modulatory control of the SCN, it is clear that they provide a neuroanatomical substrate for coupling of temperature and vigilance states.

4. Interestingly, the interaction between body temperature and sleep remains in SCN-lesioned rats (Baker et al., 2005). In these animals, the 24-h component in rhythms of sleep and temperature disappears, and ultradian fluctuations become more pronounced. Sleep occurs on the descending parts of the core body temperature fluctuations and is usually initiated about 5 min after a core body temperature peak. The sequence of sleep initiation following the start of a temperature decline in body temperature with a delay of several minutes point toward an effect of temperature on sleep regulation rather than to the reverse, an effect of sleep on temperature regulation, or than to the activation of a neuronal ensemble belonging to both sleep- and temperature-regulating networks (interaction 3). Still, effects of sleep on thermoregulation are also possible. Behaviors associated with sleep may alter thermoregulation. First, the inactivity associated

with sleep reduces the metabolic rate and consequently heat production. Second, the supine posture associated with sleep favors vasodilation of the skin vasculature (Tikuisis and Ducharme, 1996). The resulting increase in skin temperature favors heat loss. However, a third sleep-associated behavior actually prevents heat loss. Covering oneself with bedding creates a microclimate of 34–36°C (Goldsmith and Hampton, 1968; Vokac and Hjeltnes, 1981; Okamoto et al., 1997), much reducing the efficacy of heat transfer from the body to the environment.

5. A fifth level of interaction between the circadian rhythms of sleep and thermoregulation is introduced by the possibility that the physiological and behavioral aspects of sleep and body temperature might affect the SCN itself, and thus the rhythms it controls. There is considerable support for the idea that changes in temperature could affect SCN activity (reviewed in Van Someren, 2003). Just recently, the first evidence has also been given for a modulatory role of sleep on SCN neuronal activity (De Boer et al., 2003).

In the next paragraph, the focus will be on interaction number 4, more specifically on the possibility that changes in body temperature could modulate sleep-regulating systems.

A modulatory role of body temperature on sleep-regulating systems

Indirect support for the possibility that changes in body temperature could modulate sleep-regulating systems is abundantly available in the scientific literature, and has been reviewed extensively before (Van Someren, 2000, 2003, 2004). Briefly summarized: subtle changes in skin temperature, within the thermoneutral range, modulate the firing properties of thermosensitive neurons in brain areas involved in sleep regulation. Such changes in skin temperature occur autonomously under control of the circadian timing system, but also with specific behaviors, and could contribute to changes in sleep propensity. At the cellular level, warming of the skin is associated with an activation type typical of sleep in the midbrain reticular formation, hypothalamus and cerebral cortex (Van Someren, 2000). Also at the electrophysiological and behavioral level, several studies have convincingly demonstrated *correlations* between skin temperature and vigilance state (Kräuchi et al., 1999). Given the lack of studies providing *direct experimental* support, we began a series of strictly controlled experimental studies in which core and skin temperature were systematically manipulated within a very narrow comfortable and thermoneutral range. To do so, we used a thermosuit and cooled vs. warmed foods and drinks. In such experiments it is of extreme importance to induce temperature changes within a very narrow range since the aim is to mimic the temperature range normally occurring in a bed microclimate, and it has already been shown that both higher and lower ambient temperatures activate thermoregulatory defense mechanisms and disrupt sleep. In a protocol balanced for possible circadian and sequence effects, we obtained 144 sleep onset latencies (SOL) while directly manipulating core and skin temperatures within the comfortable range in eight healthy subjects under controlled conditions (Raymann et al., 2005). The induction of a proximal skin temperature difference of only $0.8°C \pm 0.03°C$ (mean $\pm$ S.E.M.) around a mean of $35.1°C \pm 0.1°C$ changed SOL by 26%, with faster sleep onsets when the proximal skin was warmed, as shown in Fig. 4. Importantly, the reduction in SOL occurred in spite of a small but significant *decrease* in subjective comfort during proximal skin warming. The induction of changes in core temperature ($\partial = 0.2°C \pm 0.02°C$) and distal skin temperature ($\partial = 0.7 \pm 0.05°C$) was ineffective. This study was the first to experimentally demonstrate a causal contribution to sleep onset latency of skin temperature manipulations within the range of its normal nocturnal fluctuations, and is has been followed by studies that demonstrated similar effects of subtle skin temperature manipulations on *sleep depth*. By applying thermosuit control of skin temperature during sleep, we

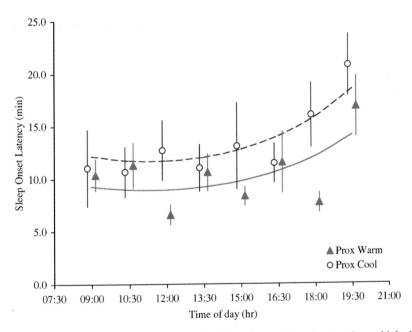

Fig. 4. Mean $\pm$ S.E.M. of 128 sleep onset latencies of eight healthy subjects in a semi-constant routine multiple sleep latency protocol of two full days. In a balanced protocol every subject was exposed to both slight warming and cooling at all times of day. The thermosuit manipulation resulted in a proximal skin temperature difference between the two conditions of only $0.8 \pm 0.03°C$ around a mean of $35.1 \pm 0.1°C$. In spite of the fact that subjects experienced the warm condition as slightly less comfortable, it accelerated sleep onset latency by 26%, on average, as indicated by the best fitting second-order nonlinear curve. Adapted from Raymann et al. (2005).

demonstrated in both young adults and elderly subjects that direct mild skin warming — not affecting core temperature — suppressed nocturnal wakefulness and induced sleep to shift to deeper stages. EEG frequency spectra showed suppression of the beta band typical of arousal, and enhancement of the slow oscillations typical of sleep. Elderly subjects showed such a pronounced sensitivity, that the induction of a mere 0.4°C increase in skin temperature was sufficient to increase the amount of slow wave sleep and to decrease the probability of early morning awakening, thus restoring the two typical sleep problems in elderly to the level of young adults (Raymann et al., unpublished observation). Further support is given by a recently finished study showing that mild warming of the proximal skin worsened performance on a psychomotor vigilance task (Raymann et al., unpublished observation). We furthermore placed the importance of the relation between sleep and skin temperature in clinical perspective, by demonstrating that narcoleptic subjects have a disturbed skin temperature regulation that relates to their sleep propensity. A recently finished study moreover showed that their ability to stay awake improved with slight cooling of the skin.

Our studies indicate that circadian and sleep-related behavior-induced variations in *skin* temperature might indeed act as an input signal to sleep-regulating systems. This suggests that an important conceptual change should be made in the theory and studies on the relation between body temperature and sleep. For decades, the focus has been on the relation of sleep to the 24-h rhythm in *core* body temperature, whereas it might be more profitable to redirect the attention to the relation between sleep and *skin* temperature. Remote and recent animal studies have indicated that skin temperature rather than core temperature is the primary factor that determines neuronal firing (Boulant and Bignall, 1973) and c-fos expression (Bratincsak and Palkovits, 2005) in the preoptic area, i.e. the very area that is most activated during sleep (Khubchandani et al., 2005). Human neuroimaging studies have confirmed a strong activation of the ventral hypothalamic area with skin warming (Egan et al., 2005). An important missing link, i.e. a skin receptor for innocuous warm temperatures, has recently been discovered (Peier et al., 2002). It is noteworthy that this skin keratinocyte TRPV3 ion channel shows a strong discontinuity at ±33°C. While virtually inactive below 33°C, even a slight increase of skin temperature above this level exponentially activates the channel. Given the typical skin temperature profile shown in Fig. 2, the TRPV3 channel would be strongly activated exclusively during the night. Although a second TRP ion channel (V4) is also activated within this comfortable range, it desensitizes with prolonged warmth exposure, which is not the case for the TRPV3 channel (Patapoutian et al., 2003).

The functional direction of coupling between sleep and increased skin temperature revisited

As mentioned above, the focus has long been on the relation between core body temperature rhythm and sleep, whereas neurobiological findings and modeling studies (Van Someren, 2000) as well as recent human experimental findings (Raymann et al., 2005) suggest that changes in skin temperature may be more relevant in relation to sleep. It has long been recognized that the drop in core temperature that occurs during sleep is mainly the result of an increase in heat dissipation, and to a lesser extent the result of a decrease in metabolic heat production. It has been proposed that a primary function of sleep might be the dissipation of heat, for example to cool the brain (McGinty and Szymusiak, 1990). Whereas there is no doubt that the brain cools during the major sleep period, it does so also during sleep deprivation. Thus, after excluding the effect of sleep itself, a circadian rhythm in body temperature remains both in humans (Kräuchi and Wirz-Justice, 1994) and rodents (Scheer et al., 2004). Also, sleep initiation *follows* rather than *precedes* heat loss onset (Baker et al., 2005; Kräuchi et al., 2005a). Importantly, whereas the increase in skin temperature could be — and has been — interpreted to reflect the thermoregulatory *autonomic* response of vasodilation of the skin vasculature in order to *promote* heat exchange from the warm blood, via the skin, to the cool environment, sleep-related *behavior* actually *restricts* heat loss. Examples include creating an isolated microclimate of high temperature by the means of warm clothing and bedding in humans, and behavior including curling up, huddling and cuddling in animals. It should be noted that for any thermoregulatory demand, *behavioral* rather than *autonomic* measures are first addressed, suggesting that preservation rather than loss of heat would be of primary interest during sleep. Moreover, there are much more efficient ways to dissipate heat, and in a fraction of the time now spent asleep. For these reasons we have questioned the primarily thermoregulatory interpretation of the increase in skin blood flow associated with sleep (Van Someren et al., 2002).

What if we would disregard the proposed cooling function of sleep? Would it then be possible for us to understand sleep-related behavior? Let us for a moment forget the idea that the circadian phase of increased skin blood flow primarily serves a thermoregulatory function and just do a mind experiment supposing that, for some as yet undefined reason *other* than thermoregulatory, it would be profitable for us to spend a considerable part of the circadian cycle in a state of strongly increased skin blood flow. What then would be the strategic behavioral and physiological measures to support such a state of increased skin blood flow yet reduce its possible adverse effects or risks?

For humans, an important physiological advantage of timing this state of increased skin blood flow during the night is that only during this circadian phase is melatonin released from the pineal, which selectively promotes skin, but not cerebral, vasodilation and blood flow (Van der Helm-Van Mil et al., 2003). The skin blood flow promoting property of melatonin may even account for $\pm 40\%$ of the amplitude of the circadian rhythm in core temperature due to heat loss under resting conditions (Cagnacci et al., 1992). Among the behavioral-facilitating conditions a positional change from upright to supine, i.e. a release from orthostasis, also strongly promotes skin blood flow (Tikuisis and Ducharme, 1996).

However, the increase of skin blood flow also puts us at risk of strong heat loss, resulting in hypothermia and a very inefficient energy metab-

olism. This risk can effectively be averted by creating an isolated environment in order to allow skin blood flow to be elevated, yet reduce heat transfer from the body to the environment. The use of bed covers indeed allows for the development of an isolated microclimate of a temperature of even 34–36°C (Goldsmith and Hampton, 1968; Vokac and Hjeltnes, 1981; Okamoto et al., 1997), effectively limiting heat flow to the environment. Such a high environmental temperature yields an important additional advantage for attaining and stabilizing the high level of skin blood flow. More specifically, it has been shown that it is exactly at a temperature increase from 33°C to 34°C that vasodilation and skin blood flow increase steeply to their maximum, as shown in Fig. 5 (Fagrell and Intaglietta, 1977). The bed microclimate of 34–36°C is thus critical for the maintenance of a high level of skin blood flow, which would be annihilated even with a minute decrease toward 33°C.

A second risk of a period of strongly elevated skin blood flow is that blood supply to other organs may be compromised. This risk could be averted by increasing cardiac output, but this would be another inefficient use of energy resources. An alternative would be to limit the desirable blood supply to these organs. This is indeed what happens for cerebral blood flow during the circadian phase of increased skin blood flow. In rats, cerebral blood flow decreases before the transition from dark to light, i.e. from the active to the inactive circadian phase (Gerashchenko and Matsumura, 1996). In addition to the circadian modulation, sleep itself is associated with a further decrease in cerebral blood flow (Fischer et al., 1991; Braun et al., 1997), and a decrease in blood flow to most skeletal muscle groups (Cote and Haddad, 1990). Of course, such a decrease in brain and skeletal blood flow can only be maintained in a state that keeps mental and physical activity at a minimum, which is precisely what sleep does.

In summary, in a mind experiment supposing that it would be profitable to spend a considerable part of the circadian cycle in a state of strongly increased skin blood flow, such a state would most effectively and safely be accomplished by (1) choosing a phase of the day when skin vasodilation is

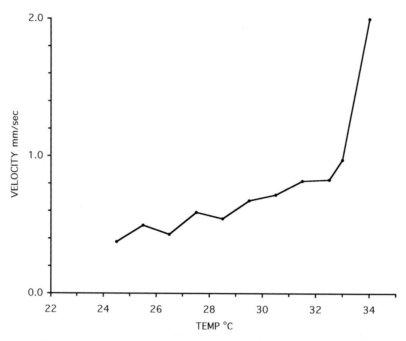

Fig. 5. Correlation between skin temperature and resting blood flow velocity (CBV) in one nailfold capillary of a healthy 37-year-old man. Notice the marked increase in CBV occurring at 34°C. Figure and legend text above are from Fagrell and Intaglietta (1977), who investigated the effect of skin warming on skin blood flow. The figure illustrates that the microclimate of at least 34°C that is created during sleep is essential in order to attain and sustain maximal skin blood flow.

enhanced, which in humans coincides with the nocturnal melatonin surge — convincingly shown to enhance skin vasodilation in humans (Van der Helm-Van Mil et al., 2003) but not so in nocturnal animals; (2) taking a supine position; (3) creating a warm microclimate and (4) limiting brain and skeletal muscle activity to allow for much reduced blood flow to these organs. Sleep could in fact be well defined by this combination of physiological and behavioral measures. It would thus be meaningful if the onset of the circadian phase of increased skin blood flow would somehow promote the transition from wake to sleep, as was predicted from a review and modeling study (Van Someren, 2000) and was recently experimentally confirmed (Raymann et al., 2005).

An alternative function for the increase in skin blood flow

If, as we presumed in the mind experiment of the previous paragraph, the daily occurrence of a prolonged phase of skin blood flow does not primarily serve a thermoregulatory function, what other function might it serve? We recently argued that the increased skin blood flow is an important part of the previously proposed function of sleep in immunological *host defense* (Krueger and Majde, 1990). More specifically, we proposed that the increased nocturnal skin blood flow may crucially support the role of the skin as the major barrier and first line of defense against environmental micro-organisms (Van Someren et al., 2002). The suggestions in this direction are briefly summarized here. Leukocytes are transported via the bloodstream to reach the parts where they are needed. A circadian phase of prolonged high-level skin blood flow will thus routinely give leukocytes optimal access to this important first-defense area, much like the primary response of increased vasodilation and plasma extravasation elicited by irritation or injury of the skin does (Chahl, 1988). During sleep, the number of monocytes, natural killer cells and lymphocytes circulating in the blood stream is reduced (Moldofsky et al., 1986;

Born et al., 1997), which has not been interpreted as a reduced production, but rather as an enhanced redistribution of lymphocytes into extravascular tissues to allow them to perform their functions (cf. Ottaway and Husband, 1992; Dhabhar et al., 1995; Dhabhar and McEwen, 1996; Benca and Quintas, 1997). The next step, migration from the interstitial space to the lymphatic system is, however, at a low level, since it was demonstrated in both humans and sheep that the efferent lymph output is reduced during nighttime rest and sleep (Engeset et al., 1977; Dickstein et al., 2000). Collectively, these findings are indicative of an increased availability of lymphocytes in skin tissue during the circadian phase of increased skin blood flow.

An increase in regional blood flow is associated with an increase in the amount of fluid leaking out from the capillaries into the interstitial fluid (cf. Dickstein et al., 2000). Consequently, the circadian phase of increased skin blood flow furthermore enhances the distribution and interstitial flow of endogenous antimicrobial peptides, such as cathelicidin and alpha and beta-defensins in the skin (Gallo et al., 1994; Bensch et al., 1995; Harder et al., 1997, 2001).

The warming of the skin that results from increased skin blood flow and is further optimized by, e.g., insulative bedding will enhance capillary permeability and lymph flow (Engeset et al., 1977; Olszewski et al., 1977). In addition, in humans, the nocturnal strong warming of the skin will also contribute to the already upregulated sweating rate during the night. This increased sweating is once more hardly effective from the point of view of heat loss because bedding prevents convection, and may also rather serve host defense, since sweat glands secrete the endogenous antibiotic peptide dermcidin (Schittek et al., 2001).

Sleep deprivation

For the hypothesized functional importance of a circadian phase of enhanced skin blood flow, sleep would not be strictly necessary. Thus, the hypothesis does *not* imply increased skin blood flow to be *the primary* function of sleep. Like wakefulness, sleep most likely supports many different functions rather than one single function, one of which may be to support a prolonged period of increased skin blood flow. The hypothesis put forward here is just that it is especially profitable and efficacious to combine the period of increased skin blood flow with the behavioral and physiological aspects of sleep, for reasons given in the mind experiment above, including a supine and thermally isolated position and a downregulated blood supply demand of the skeletal muscles and the brain.

A first consequence of regarding sleep as a state providing favorable circumstances to support a period of increased skin blood flow, instead of as a state meant to dissipate heat, is that sleep deprivation would not be predicted to generate an ever increasing homeostatic *need* to dissipate heat. Rather, the increase in skin temperature would occur, just like under normal sleeping circumstances, with a circadian modulation. A recent controlled human experiment indeed confirmed this prediction: sleep deprivation does not result in a homeostatic buildup of skin blood flow 'pressure', while its circadian modulation is preserved (Kräuchi et al., 2005b).

A second consequence of regarding sleep as an optimal state for a circadian phase of increased blood flow is that chronic sleep deprivation may result in suboptimal fulfilment of the hypothesized required increase in skin blood flow. If sleep is prohibited, the circadian modulation of skin blood flow still continues, albeit less marked (Kräuchi, 2002). This would result in two unwanted situations. First, sleep deprivation would be predicted to lead to a progressive heat loss, because of the inability to effectuate the sleep-related behavioral measures to create a thermally insulated microclimate by, e.g., the use of bedding and curling up. In the absence of sleep-associated heat-loss preventing behaviors, increases in skin blood flow can only be accomplished at the cost of heat loss. Indeed, one of the hallmarks of prolonged sleep deprivation is a progressive decline in core body temperature in spite of increased food intake and energy expenditure (Rechtschaffen et al., 2002). Also, sleep-deprived humans may be more vulnerable to heat loss both at cold and comfortable temperatures (Opstad and Bahr, 1991; Savourey and Bittel, 1994; Landis et al., 1998; Young et al., 1998). The continued strive to maintain circadian phases of vasodilation in spite of the limitations on the prevention of heat loss during sleep deprivation supports the idea that it may fulfill an important function. Second, sleep deprivation would impose limitations on skin blood flow, because the skin blood flow-promoting microclimate (see Fig. 5) is not present. This is indeed the case — just compare the nocturnal skin temperatures during sleep deprivation in Fig. 1 with those under normal sleeping conditions in Fig. 2 (see also Kräuchi, 2002). Such limitations might result in suboptimal fulfilment of the proposed role of the circadian phase of increased skin perfusion: optimizing the skin function of a first barrier of defense against environmental micro-organisms. Consequently, chronic sleep deprivation would be predicted to increase the risk of skin problems and infection. Below, rat and human studies that evaluated this prediction will be discussed.

Rat studies

The first and most marked pathological findings in rats that are chronically sleep deprived are ulcerative and hyperkeratotic lesions of the skin (Everson, 1995; Rechtschaffen and Bergmann, 1995). As shown in Fig. 2, the strongest diurnal variation in skin blood flow takes place at the extremities, which interact the most with the environment and may thus require more maintenance of its defense mechanisms. It is striking that the first degenerative symptoms of longer-sleep deprivation in rats are ulcerative and hyperkeratotic lesions at these very skin areas: the tail and plantar surfaces of the paws (Everson, 1995; Rechtschaffen and Bergmann, 1995). These findings suggest that the distal areas indeed suffer the most from a lack of a 'complete' sleep-related vasodilation. The lesions improve within days with recovery of sleep (Everson et al., 1989). However, when sleep deprivation is continued, the skin lesions are thought to be involved in the final breakdown in host defense with blood infection (Everson, 1993) and live bacteria in the lymph nodes (Everson and Toth, 2000), resulting in death.

Two sleep deprivation studies of shorter duration (72 h) could, however, not establish effects on wound healing. Landis and Whitney (1997) reported a lack of differences with controls on cellular and biochemical markers of wound healing in rats deprived for 72 h. The lack of effect may have been due to the fact that sleep deprivation did not start before one week after the miniature dorsal lesions were made: an important part of the healing process might already have been completed by then. A second negative study found no effect on wound closure rate of five days of selective rapid eye movement (REM) sleep deprivation immediately before or after puncture wounds were made (Mostaghimi et al., 2005). Contrary to these negative reports, with another type of wound, a second-degree burn, sleep deprivation significantly attenuated the formation of capillary vessels and fibroblasts, and kept polymorphonuclear leukocytes (PNLs) and polyvalent immunoglobulin G (IgG) at higher levels (Gumustekin et al., 2004). The observation of a reduction in fibroblasts after sleep deprivation is interesting, because it has recently been shown that the molecular clockwork of fibroblasts can be entrained by temperature cycles (Brown et al., 2002) and the circadian cycles of skin temperature are much reduced under conditions of sleep deprivation. Thus, circadian cycles of skin temperature may sustain the normal molecular machinery of skin fibroblasts, and a disruption of this cycle might be involved in the effect of sleep deprivation on skin fibroblast formation. Obviously, more studies would be needed to draw sound conclusions.

Human studies

It does not require sophisticated assessment procedures to recognize whether someone has been sleep deprived: this condition is easily recognized at first glance by a general paleness of the skin and a blue appearance of the periocular skin. Only few human studies have addressed the effect of sleep restriction on skin wound healing. In one study in healthy subjects the rate of recovery from an experimental skin puncture was not associated with their habitual sleep pattern. However, sleep was not optimally assessed, in fact only by an adaptation of the general health questionnaire items (Ebrecht et al., 2004). As wound repair is most likely a robust process and compensatory mechanisms for short-term sleep deprivation in healthy subjects could be available, a possible effect of sleep on skin maintenance and repair may be easier to detect in subjects with a compromised health status. In the clinical setting, a relation between wound healing and sleep has indeed often been noted or suggested (Adam and Oswald, 1984; Lee and Stotts, 1990; Helm, 1992; Evans and French, 1995; Rose et al., 2001), but confirmation of this relation has hardly been sought in systematic studies. An exception is a large study on the risk of developing skin ulceration in 826 nonambulatory spinal cord injury patients. The use of sleep medication (OR 1.29) was one of the only two predictors, smoking being the other (Krause and Broderick, 2004). This indicates that the patients with sleep problems are much more likely to develop skin problems. Interestingly, the effect of the other variable — heavy smoking — on skin condition has been attributed to its detrimental effect on the skin microvasculature and blood flow as well (Silverstein, 1992), as we propose to be the case for sleep deprivation. A rat study confirms that nicotine and sleep deprivation have similar negative effects on wound healing (Gumustekin et al., 2004). In another human study, poor sleep as a result of obstructive sleep apnea syndrome (OSAS) increased the skin's reactivity to allergens (Duchna et al., 1997). A case report demonstrated remission of chronic dermatitis after successful treatment of OSAS (Matin et al., 2002). Other experimental total sleep deprivation studies found (1) an increase in the skin resistance level (Miro et al., 2002); (2) a slower recovery of the skin barrier function as quantified by the tape-stripping method (Altemus et al., 2001); (3) enhancement of allergic skin responses in patients with allergic rhinitis (Kimata, 2002).

As to the proposed supportive role of the circadian phase of increased skin blood flow for host defense, human studies indicate that poor sleep is associated with an increased risk of oral (Horning and Cohen, 1995) and respiratory (d'Arcy et al., 2000) infection. Sleep-deprivation studies have

reported both decreased (Irwin et al., 1992; Irwin et al., 1994; Irwin et al., 1996; Ozturk et al., 1999) and increased (Dinges et al., 1994; Matsumoto et al., 2001) natural killer cell activity.

Although collectively these papers, indeed, give some indirect support for the idea that sleep deprivation may indeed result in a suboptimal support of a circadian phase of increased skin blood flow and consequently affect the condition of the skin and increase the risk of infection, more systematic research is clearly needed. It is hoped that the present review invites re-evaluation of the relation of sleep to the often-presumed thermoregulatory role of the decline in core body temperature caused by increased skin blood flow.

Acknowledgments

Kurt Kräuchi, Psychiatric University Clinic, Basel, Switzerland and Roy J.E.M. Raymann, Netherlands Institute for Neurosciences, Amsterdam, The Netherlands are acknowledged for providing the data for Figs. 1 (KK), 2 and 4 (RR). Financial support by ZON-MW, The Hague, The Netherlands, Successful Aging Grant 014-90-001 and by The Netherlands Organization for Scientific Research, The Hague, The Netherlands, Innovation Grant 016.025.041; EU FP6 Sensation Integrated Project (FP6-507231).

References

Adam, K. and Oswald, I. (1984) Sleep helps healing. Br. Med. J., 289: 1400–1401.

Alam, N., McGinty, D. and Szymusiak, R. (1997) Thermosensitive neurons of the diagonal band in rats: relation to wakefulness and non-rapid eye movement sleep. Brain Res., 752: 81–89.

Alam, N., Szymusiak, R. and McGinty, D. (1995) Local preoptic/anterior hypothalamic warming alters spontaneous and evoked neuronal activity in the magno-cellular basal forebrain. Brain Res., 696: 221–230.

Altemus, M., Rao, B., Dhabhar, F.S., Ding, W. and Granstein, R.D. (2001) Stress-induced changes in skin barrier function in healthy women. J. Invest. Dermatol., 117: 309–317.

Baker, F.C., Angara, C., Szymusiak, R. and McGinty, D. (2005) Persistence of sleep-temperature coupling after suprachiasmatic nuclei lesions in rats. Am. J. Physiol., 289: R827–R838.

Benca, R.M. and Quintas, J. (1997) Sleep and host defenses: a review. Sleep, 20: 1027–1037.

Bensch, K.W., Raida, M., Magert, H.J., Schulz-Knappe, P. and Forssmann, W.G. (1995) hBD-1: a novel beta-defensin from human plasma. FEBS Lett., 368: 331–335.

Borbély, A.A. (1982) A two process model of sleep regulation. Hum. Neurobiol., 1: 195–204.

Born, J., Lange, T., Hansen, K., Molle, M. and Fehm, H.L. (1997) Effects of sleep and circadian rhythm on human circulating immune cells. J. Immunol., 158: 4454–4464.

Boulant, J.A. and Bignall, K.E. (1973) Hypothalamic neuronal responses to peripheral and deep-body temperatures. Am. J. Physiol., 225: 1371–1374.

Boulant, J.A. and Silva, N.L. (1988) Neuronal sensitivities in preoptic tissue slices: interactions among homeostatic systems. Brain Res. Bull., 20: 871–878.

Bratincsak, A. and Palkovits, M. (2005) Evidence that peripheral rather than intracranial thermal signals induce thermoregulation. Neurosci., 135: 525–532.

Braun, A.R., Balkin, T.J., Wesenten, N.J., Carson, R.E., Varga, M., Baldwin, P., Selbie, S., Belenky, G. and Herscovitch, P. (1997) Regional cerebral blood flow throughout the sleep–wake cycle. An H2(15)O PET study. Brain, 120: 1173–1197.

Cagnacci, A., Elliott, J.A. and Yen, S.S. (1992) Melatonin: a major regulator of the circadian rhythm of core temperature in humans. J. Clin. Endocrinol. Metab., 75: 447–452.

Chahl, L.A. (1988) Antidromic vasodilatation and neurogenic inflammation. Pharmacol. Ther., 37: 275–300.

Cote, A. and Haddad, G.G. (1990) Effect of sleep on regional blood flow distribution in piglets. Pediatr. Res., 28: 218–222.

d'Arcy, H., Gillespie, B. and Foxman, B. (2000) Respiratory symptoms in mothers of young children. Pediatrics, 106: 1013–1016.

De Boer, T., Vansteensel, M.J., Detari, L. and Meijer, J.H. (2003) Sleep states alter activity of suprachiasmatic nucleus neurons. Nat. Neurosci., 6: 1086–1090.

Dhabhar, F.S. and McEwen, B.S. (1996) Stress-induced enhancement of antigen-specific cell-mediated immunity. J. Immunol., 156: 2608–2615.

Dhabhar, F.S., Miller, A.H., McEwen, B.S. and Spencer, R.L. (1995) Effects of stress on immune cell distribution. Dynamics and hormonal mechanisms. J. Immunol., 154: 5511–5527.

Dickstein, J.B., Hay, J.B., Lue, F.A. and Moldofsky, H. (2000) The relationship of lymphocytes in blood and in lymph to sleep/wake states in sheep. Sleep, 23: 185–190.

Dinges, D.F., Douglas, S.D., Zaugg, L., Campbell, D.E., McMann, J.M., Whitehouse, W.G., Orne, E.C., Kapoor, S.C., Icaza, E. and Orne, M.T. (1994) Leukocytosis and natural killer cell function parallel neurobehavioral fatigue induced by 64 hours of sleep deprivation. J. Clin. Invest., 93: 1930–1939.

Duchna, H.W., Rasche, K., Lambers, N., Orth, M., Merget, R. and Schultze-Werninghaus, G. (1997) Incidence of cutaneous sensitization to environmental allergens in obstructive sleep apnea syndrome. Pneumologie, 51(Suppl 3): 763–766.

Ebrecht, M., Hextall, J., Kirtley, L.G., Taylor, A., Dyson, M. and Weinman, J. (2004) Perceived stress and cortisol levels predict speed of wound healing in healthy male adults. Psychoneuroendocrinology, 29: 798–809.

Egan, G.F., Johnson, J., Farrell, M., McAllen, R., Zamarripa, F., McKinley, M.J., Lancaster, J., Denton, D. and Fox, P.T. (2005) Cortical, thalamic, and hypothalamic responses to cooling and warming the skin in awake humans: a positron-emission tomography study. Proc. Natl. Acad. Sci., 102: 5262–5267.

Engeset, A., Sokolowski, J. and Olszewski, W.L. (1977) Variation in output of leukocytes and erythrocytes in human peripheral lymph during rest and activity. Lymphology, 10: 198–203.

Evans, J.C. and French, D.G. (1995) Sleep and healing in intensive care settings. DimensionCrit. Care Nurs., 14: 189–199.

Everson, C.A. (1993) Sustained sleep deprivation impairs host defense. Am. J. Physiol., 265: R1148–R1154.

Everson, C.A. (1995) Functional consequences of sustained sleep deprivation in the rat. Behav. Brain Res., 69: 43–54.

Everson, C.A., Gilliland, M.A., Kushida, C.A., Pilcher, J.J., Fang, V.S., Refetoff, S., Bergmann, B.M. and Rechtschaffen, A. (1989) Sleep deprivation in the rat: IX. Recovery. Sleep, 12: 60–67.

Everson, C.A. and Toth, L.A. (2000) Systemic bacterial invasion induced by sleep deprivation. Am. J. Physiol., 278: R905–R916.

Fagrell, B., Intaglietta, M. (1977) The dynamics of skin microcirculation as a tool for the study of systemic diseases. Bibl. Anat., 16: 231–234.

Fischer, A.Q., Taormina, M.A., Akhtar, B. and Chaudhary, B.A. (1991) The effect of sleep on intracranial hemodynamics: a transcranial Doppler study. J. Child Neurol., 6: 155–158.

Gallo, R.L., Ono, M., Povsic, T., Page, C., Eriksson, E., Klagsbrun, M. and Bernfield, M. (1994) Syndecans, cell surface heparan sulfate proteoglycans, are induced by a proline-rich antimicrobial peptide from wounds. PNAS, 91: 11035–11039.

Gerashchenko, D. and Matsumura, H. (1996) Continuous recordings of brain regional circulation during sleep/wake state transitions in rats. Am. J. Physiol., 270: R855–R863.

Goldsmith, R. and Hampton, I.F. (1968) Nocturnal microclimate of man. J. Physiol., 194: 32P–33P.

Gumustekin, K., Seven, B., Karabulut, N., Aktas, O., Gursan, N., Aslan, S., Keles, M., Varoglu, E. and Dane, S. (2004) Effects of sleep deprivation, nicotine, and selenium on wound healing in rats. Int. J. Neurosci., 114: 1433–1442.

Guyton, A.C. (1991) Textbook of Medical Physiology. W.B. Saunders Company, Philadelphia.

Harder, J., Bartels, J., Christophers, E. and Schroder, J.M. (1997) A peptide antibiotic from human skin. Nature, 387: 861.

Harder, J., Bartels, J., Christophers, E. and Schroder, J.M. (2001) Isolation and characterization of human beta -defensin-3, a novel human inducible peptide antibiotic. J. Biol. Chem., 276: 5707–5713.

Hayward, J.N. and Baker, M.A. (1968) Diuretic and thermo-regulatory responses to preoptic cooling in the monkey. Am. J. Physiol., 214: 843–850.

Helm, P.A. (1992) Burn rehabilitation: dimensions of the problem. Clin. Plast. Surg., 19: 551–559.

Hori, T., Nakashima, T., Koga, H., Kiyohara, T. and Inoue, T. (1988) Convergence of thermal, osmotic and cardiovascular signals on preoptic and anterior hypothalamic neurons in the rat. Brain Res. Bull., 20: 879–885.

Horning, G.M. and Cohen, M.E. (1995) Necrotizing ulcerative gingivitis, periodontitis, and stomatitis: clinical staging and predisposing factors. J. Periodontol., 66: 990–998.

Irwin, M., Mascovich, A., Gillin, J.C., Willoughby, R., Pike, J. and Smith, T.L. (1994) Partial sleep deprivation reduces natural killer cell activity in humans. Psychosom. Med., 56: 493–498.

Irwin, M., McClintick, J., Costlow, C., Fortner, M., White, J. and Gillin, J.C. (1996) Partial night sleep deprivation reduces natural killer and cellular immune responses in humans. FASEB J., 10: 643–653.

Irwin, M., Smith, T.L. and Gillin, J.C. (1992) Electroencephalographic sleep and natural killer activity in depressed patients and control subjects. Psychosom. Med., 54: 10–21.

Khubchandani, M., Jagannathan, N.R., Mallick, H.N. and Mohan Kumar, V. (2005) Functional MRI shows activation of the medial preoptic area during sleep. Neuroimage, 26: 29–35.

Kimata, H. (2002) Enhancement of allergic skin responses by total sleep deprivation in patients with allergic rhinitis. Int. Arch. Allergy Immunol., 128: 351–352.

Kräuchi, K. (2002) How is the circadian rhythm of core body temperature regulated. Clin. Auton. Res., 12: 147–149.

Kräuchi, K., Cajochen, C., Werth, E. and Wirz-Justice, A. (1999) Warm feet promote the rapid onset of sleep. Nature, 401: 36–37.

Kräuchi, K., Cajochen, C. and Wirz-Justice, A. (2005a) Thermophysiologic aspects of the three-process-model of sleepiness regulation. Clin. Sport. Med., 24: 287–300 ix.

Kräuchi, K., Knoblauch, V., Wirz-Justice, A., Cajochen, C. (2005b) Challenging the sleep homeostat does not influence the thermoregulatory system in men: evidence from a nap vs. sleep deprivation study. Am. J. Physiol., in press.

Kräuchi, K. and Wirz-Justice, A. (1994) Circadian rhythm of heat production, heart rate, and skin and core temperature under unmasking conditions in men. Am. J. Physiol., 267: R819–R829.

Krause, J.S. and Broderick, L. (2004) Patterns of recurrent pressure ulcers after spinal cord injury: identification of risk and protective factors 5 or more years after onset. Arch. Phys. Med. Rehabil., 85: 1257–1264.

Krueger, J.M. and Majde, J.A. (1990) Sleep as a host defense: its regulation by microbial products and cytokines. Clin. Immunol. Immunopathol., 57: 188–199.

Landis, C.A., Savage, M.V., Lentz, M.J. and Brengelmann, G.L. (1998) Sleep deprivation alters body temperature dynamics to mild cooling and heating not sweating threshold in women. Sleep, 21: 101–108.

Landis, C.A. and Whitney, J.D. (1997) Effects of 72 hours sleep deprivation on wound healing in the rat. Res. Nurs. Health, 20: 259–267.

Lee, K.A. and Stotts, N.A. (1990) Support of the growth hormone — somatomedin system to facilitate healing. Heart Lung, 19: 157–162.

Lu, J., Zhang, Y.H., Chou, T.C., Gaus, S.E., Elmquist, J.K., Shiromani, P. and Saper, C.B. (2001) Contrasting effects of ibotenate lesions of the paraventricular nucleus and subparaventricular zone on sleep–wake cycle and temperature regulation. J. Neurosci., 21: 4864–4874.

Matin, A., Bliwise, D.L., Wellman, J.J., Ewing, H.A. and Rasmuson, P. (2002) Resolution of dyshidrotic dermatitis of the hand after treatment with continuous positive airway pressure for obstructive sleep apnea. South. Med. J., 95: 253–254.

Matsumoto, Y., Mishima, K., Satoh, K., Tozawa, T., Mishima, Y., Shimizu, T. and Hishikawa, Y. (2001) Total sleep deprivation induces an acute and transient increase in NK cell activity in healthy young volunteers. Sleep, 24: 804–809.

McGinty, D. and Szymusiak, R. (1990) Keeping cool: a hypothesis about the mechanisms and functions of slow-wave sleep. TINS, 13: 480–487.

Meijer, G.A., Westerterp, K.R., Seyts, G.H., Janssen, G.M., Saris, W.H. and ten Hoor, F. (1991) Body composition and sleeping metabolic rate in response to a 5-month endurance-training programme in adults. Eur. J. Appl. Physiol., 62: 18–21.

Miro, E., Cano-Lozano, M.C. and Buela-Casal, G. (2002) Electrodermal activity during total sleep deprivation and its relationship with other activation and performance measures. J. Sleep Res., 11: 105–112.

Moldofsky, H., Lue, F.A., Eisen, J., Keystone, E. and Gorczynski, R.M. (1986) The relationship of interleukin-1 and immune functions to sleep in humans. Psychosom. Med., 48: 309–318.

Marotte, H. and Timbal, J. (1982) Circadian rhythm of temperature in man. Comparative study with two experimental protocols. Chronobiologia, 8: 87–100.

Mostaghimi, L., Obermeyer, W.H., Ballamudi, B., Martinez-Gonzalez, D. and Benca, R.M. (2005) Effects of sleep deprivation on wound healing. J. Sleep Res., 14: 213–219.

Okamoto, K., Mizuno, K. and Okudaira, N. (1997) The effects of a newly designed air mattress upon sleep and bed climate. Appl. Hum. Sci., 16: 161–166.

Olszewski, W., Engeset, A., Jaeger, P.M., Sokolowski, J. and Theodorsen, L. (1977) Flow and composition of leg lymph in normal men during venous stasis, muscular activity and local hyperthermia. Acta Physiol. Scand., 99: 149–155.

Opstad, P.K. and Bahr, R. (1991) Reduced set-point temperature in young men after prolonged strenuous exercise combined with sleep and energy deficiency. Arctic Med. Res., 50: 122–126.

Ottaway, C.A. and Husband, A.J. (1992) Central nervous system influences on lymphocyte migration. Brain Behav. Immun., 6: 97–116.

Ozturk, L., Pelin, Z., Karadeniz, D., Kaynak, H., Cakar, L. and Gozukirmizi, E. (1999) Effects of 48 hours sleep deprivation on human immune profile. Sleep Res. Online, 2: 107–111.

324

Patapoutian, A., Peier, A.M., Story, G.M. and Viswanath, V. (2003) ThermoTRP channels and beyond: mechanisms of temperature sensation. Nat. Rev. Neurosci., 4: 529–539.

Peier, A.M., Reeve, A.J., Andersson, D.A., Moqrich, A., Earley, T.J., Hergarden, A.C., Story, G.M., Colley, S., Hogenesch, J.B., McIntyre, P., Bevan, S. and Patapoutian, A. (2002) A heat-sensitive TRP channel expressed in keratinocytes. Science, 296: 2046–2049.

Poehlman, E.T. and Horton, E.S. (1990) Regulation of energy expenditure in aging humans. Annu. Rev. Nutr., 10: 255–275.

Raymann, R.J.E.M., Swaab, D.F. and Van Someren, E.J.W. (2005) Cutaneous warming promotes sleep onset. Am. J. Physiol., 288: R1589–R1597.

Raymann, R.J.E.M., Swaab, D.F., Van Someren, E.J.W. (submitted-a) Mild cutaneous warming impairs psychomotor vigilance.

Raymann, R.J.E.M., Swaab, D.F., Van Someren, E.J.W. (submitted-b) Skin deep: cutaneous temperature determines sleep depth.

Rechtschaffen, A. and Bergmann, B.M. (1995) Sleep deprivation in the rat by the disk-over-water method. Behav. Brain Res., 69: 55–63.

Rechtschaffen, A., Bergmann, B.M., Everson, C.A., Kushida, C.A. and Gilliland, M.A. (2002) Sleep deprivation in the rat: X. Integration and discussion of the findings. 1989. Sleep, 25: 68–87.

Rose, M., Sanford, A., Thomas, C. and Opp, M.R. (2001) Factors altering the sleep of burned children. Sleep, 24: 45–51.

Rubinstein, E.H. and Sessler, D.I. (1990) Skin-surface temperature gradients correlate with fingertip blood flow in humans. Anesthesia., 73: 541–545.

Savourey, G. and Bittel, J. (1994) Cold thermoregulatory changes induced by sleep deprivation in men. Eur. J. Appl. Physiol., 69: 216–220.

Scheer, F.A.J.L., Van Heijningen, C., Van Someren, E.J.W., Buijs, R.M. (2004) Light and suprachiasmatic nucleus interact in the regulation of body temperature. In: Abstracts of the Society for Research on Biological Rhythms Meeting.

Schittek, B., Hipfel, R., Sauer, B., Bauer, J., Kalbacher, H., Stevanovic, S., Schirle, M., Schroeder, K., Blin, N., Meier, F., Rassner, G. and Garbe, C. (2001) Dermcidin: a novel human antibiotic peptide secreted by sweat glands. Nat. Immunol., 2: 1133–1137.

Silverstein, P. (1992) Smoking and wound healing. Am. J. Med., 93: 22S–24S.

Thomas, T.C. and Kumar, V.M. (2000) Effect of ambient temperature on sleep–wakefulness in normal and medial preoptic area lesioned rats. Sleep Res. Online, 3: 141–145.

Tikuisis, P. and Ducharme, M.B. (1996) The effect of postural changes on body temperatures and heat balance. Eur. J. Appl. Physiol., 72: 451–459.

Van der Helm-Van Mil, A.H.M., Van Someren, E.J.W., Van den Boom, R., Van Buchem, M.A., De Craen, A.J.M. and Blauw, G.J. (2003) No influence of melatonin on cerebral blood flow in humans. J. Clin. Endocrinol. Metab., 88: 5989–5994.

van Marken Lichtenbelt, W.D., Daanen, H.A.M., Wouters, L., Fronczek, R., Raymann, R.J.E.M., Severens, N.M.W., and Van Someren, E.J.W. (submitted) Evaluation of wireless determination of skin temperature using iButtons.

Van Someren, E.J.W. (2000) More than a marker: interaction between the circadian regulation of temperature and sleep, age-related changes, and treatment possibilities. Chronobiol. Int., 17: 313–354.

Van Someren, E.J.W. (2003) Thermosensitivity of the circadian timing system. Sleep Biol. Rhythms, 1: 55–64.

Van Someren, E.J.W. (2004) Sleep propensity is modulated by circadian and behavior-induced changes in cutaneous temperature. J. Therm. Biol., 29: 437–444.

Van Someren, E.J.W., Raymann, R.J.E.M., Scherder, E.J.A., Daanen, H.A.M. and Swaab, D.F. (2002) Circadian and age-related modulation of thermoreception and temperature regulation: mechanisms and functional implications. Aging Res. Rev., 1: 721–778.

Vokac, Z. and Hjeltnes, N. (1981) Core–peripheral heat redistribution during sleep and its effect on rectal temperature. In: Reinberg, A., Vieux, N. and Andlauer, P. (Eds.), Night and Shift Work. Biological and Social Aspects. Pergamon Press, Oxford, pp. 109–115.

Young, A.J., Castellani, J.W., O'Brien, C., Shippee, R.L., Tikuisis, P., Meyer, L.G., Blanchard, L.A., Kain, J.E., Cadarette, B.S. and Sawka, M.N. (1998) Exertional fatigue, sleep loss, and negative energy balance increase susceptibility to hypothermia. J. Appl. Physiol., 85: 1210–1217.

Kalsbeek, Fliers, Hofman, Swaab, Van Someren & Buijs
Progress in Brain Research, Vol. 153
ISSN 0079-6123

CHAPTER 19

What can we learn from seasonal animals about the regulation of energy balance?

Peter J. Morgan*, Alexander W. Ross, Julian G. Mercer and Perry Barrett

Rowett Research Institute, Greenburn Road, Bucksburn, Aberdeen, AB21 9SB, UK

Abstract: Weight loss in humans requires, except during an illness, some form of imposed restriction on food intake or increase in energy expenditure. This necessitates overcoming powerful peripheral and central signals that serve to protect against negative energy balance. The identification of the systems and pathways involved has come from mouse models with genetic and targeted mutations, e.g., *ob/ob* and *MC4 R*$^{-/-}$ as well as rat models of obesity.

Study of seasonal animals has shown that they undergo annual cycles of body fattening and adipose tissue loss as important adaptations to environmental change, yet these changes appear to involve mechanisms distinct from those known already. One animal model, the Siberian hamster, exhibits marked, but reversible, weight loss in response to shortening day length. The body weight is driven by a decrease in food intake with the magnitude of the loss of body weight being directly related to the length of time of exposure to short photoperiod. The most important facet of this response is that the point of energy balance is continuously re-adjusted during the transition in body weight reflecting an apparent 'sliding set point'.

Studies have focused on identifying the neural basis of this mechanism. Initial studies of known genes (e.g., *NPY*, *POMC*, and *AgRP*) both through the measurement of gene expression in the arcuate nucleus as well as following intracerebroventricular (i.c.v.) injection indicated that the systems involved are not those involved in restoring energy balance following energy deficits. Instead, a novel mechanism of regulation is implied.

Recent studies have begun to explore the neural basis of the seasonal body weight response. A discrete and novel region of the posterior arcuate nucleus, the dorsal medial posterior arcuate nucleus (dmpARC) has been identified, where a battery of gene expression changes for signalling molecules (*vgf* and *histamine H3 receptor*) and transcription factors (*RXRγ* and *RAR*) occur in association with seasonal changes in body weight. This work provides the basis of a potentially novel mechanism of energy balance regulation.

Introduction

In a world of escalating levels of obesity, it is sometimes difficult to reconcile the fact that there are physiological systems designed to regulate energy balance. Obesity results from an adjustment of energy balance to a new equilibrium position, where food intake (usually) is increased without a compensating change in energy expenditure, although basal metabolic rate increases as a result of increased body mass. Central to our understanding of why obesity occurs is the need to understand the nature of the systems that permit food intake to increase beyond energetic needs in the first place and then why energy expenditure is not linked to food intake to restore energy balance to its initial equilibrium.

---
*Corresponding author. Tel.: +44-1224-716663; Fax: +44-1224-716698; E-mail: p.morgan@rowett.ac.uk

DOI: 10.1016/S0079-6123(06)53019-5

Many studies provide evidence that physiological systems exist to control energy balance and that these have powerful effects (Berthoud, 2002; Saper et al., 2002). Yet on the whole, these systems appear to be designed to protect us against energy loss rather than oversupply of calories. Thus, in simple terms, the systems involved in the regulation of energy balance are configured in an asymmetric manner, where robust mechanisms protect against body weight loss, while the mechanisms required to protect against body weight gain are either weak or nonexistent.

Leptin, a hormonal product of adipose tissue, provides a strong feedback signal to the brain, which engages both orexigenic and anorexigenic pathways in the hypothalamus (Berthoud, 2002; Saper et al., 2002). In the absence of leptin, such as in the leptin-deficient mouse (*ob/ob*) or more naturally through lowered plasma leptin levels after food deprivation, the inhibitory activity of leptin on the orexigenic pathways (NPY/AgRP) is attenuated and the stimulatory activity on the anorexigenic (POMC/MC4 MC4R and CART) neurons is reduced (Elmquist et al., 1998). The effect is a net increase in the orexigenic drive. Restoring leptin reverses these effects by suppressing the overall orexigenic drive. Given these strong and robust effects, it is somewhat surprising that leptin does not prevent overeating and obesity. Yet obese individuals generally have high levels of plasma leptin, corresponding to the surplus adipose tissue stores that characterise them (Maffei et al., 1995). The failure of leptin to restrain weight gain is often explained on the basis of reduced sensitivity to leptin as obesity progresses (Scarpace et al., 2002). While this design feature may have frustrated attempts to target the leptin system to therapeutic benefit, from an evolutionary perspective it reflects the lack of selective pressure to sustain leptin sensitivity to prevent weight gain. Only in an environment of excess energy-dense foods (e.g., Western diets, cafeteria diets) is a system of restraint on energy intake required.

Nevertheless, for some animals, environmental factors have provided the selective pressure that has generated an ability to regulate body weight within limits with remarkable precision. An example of this is the seasonal Siberian hamster. What makes this animal model so remarkable is its ability to change its level of food intake and reset its body weight to a lower or a higher level depending upon the prevailing photoperiod (Steinlechner et al., 1983; Mercer et al., 2001). Much in the same way that understanding the basis of the obesity in *ob/ob* mice by the discovery of leptin brought a totally new perspective on the pathways and mechanisms involved in energy balance (Elmquist et al., 1998), seasonal animals provide another model that will yield new insights into the systems involved in energy balance regulation.

Seasonal strategies

Most animals living at temperate latitudes exhibit seasonal changes in physiology and behaviour owing to the often-profound, yet predictable changes in climate and environment. As a result, reproduction is timed to ensure that offspring are born at the optimum time of the year for survival (spring), when food is most abundant and temperatures are warm. Similarly, food intake, body weight, and metabolism are adjusted to synchronise with the changing availability of food. Different species adopt diverse strategies to cope with the shortage of food during the winter months (Bartness and Wade, 1985). A familiar strategy is hibernation used by many species, where the animal accumulates fat stores prior to winter and then burns off the surplus energy stores when food is either unavailable or the environment is too hostile to forage. Perhaps, a more surprising strategy, yet nonetheless successful one, is to reduce food intake and body weight prior to winter. This is the strategy utilised by the Siberian hamster. At first, this appears paradoxical, yet the logic would seem to be that if food is unavailable during winter, then the drive to eat must be reduced to match the low availability of food. However, as a consequence of reducing food intake prior to winter, the animals also lose body fat and attain lower body weights. This means that while the animal has no fat reserves on which to rely, its energy requirements are lower. To help the animal compensate for the lower level of insulation from body fat, it sheds its summer fur and develops a winter pelage to maintain temperature.

A re-resetting of body weight set point

The seasonal changes in physiology and behaviour such as body weight and reproductive competence are profound and require several weeks and months to complete. As a consequence, such changes must be initiated well in advance of the seasonal changes in climate. To achieve this, seasonal animals use photoperiod as a predictable and reliable cue for the time of year (Bartness and Wade, 1985; Bartness et al., 2002; Morgan and Mercer, 2001). In the wild, changes in photoperiod are gradual. Nevertheless, it is possible to mimic these changes with square-wave changes in photoperiod in the laboratory (Bartness and Wade, 1985; Mercer et al., 2001). Thus, for Siberian hamsters maintained indefinitely on a summer photoperiod (long days (16 h light):short nights (8 h darkness)), their physiology will be unaffected. However, following an abrupt switch to a winter photoperiod (short days (8 h light):long nights (16 h darkness)), they undergo progressive loss of body weight and gonadal regression, similar to animals on a natural photoperiod transition from summer to winter (Fig. 1) (Bartness and Wade, 1985; Bartness et al., 2002; Mercer et al., 2001).

Switching hamsters from a winter photoperiod back to a summer photoperiod reverses these physiological changes, with the hamsters re-gaining body mass and gonadal function (Fig. 1) (Masuda and Oishi, 1995; Ross et al., 2005). It is important to note that unlike in a summer photoperiod, if hamsters are maintained indefinitely in a winter photoperiod, then they revert spontaneously to a summer phenotype (i.e., higher body mass and gonadally intact) after about 20 weeks of short days. This phenomenon is known as photorefractoriness, owing to the loss of responsiveness to short photoperiod (Gorman and Zucker, 1995).

However, the most remarkable feature of the Siberian hamster is the striking precision with which these animals are able to regulate energy balance (Morgan et al., 2003). This was first revealed by some pioneering experiments by Steinlechner and colleagues, who showed how the ambient photoperiod sets the level at which body weight is defended (Steinlechner et al., 1983). We have confirmed this phenomenon under artificial

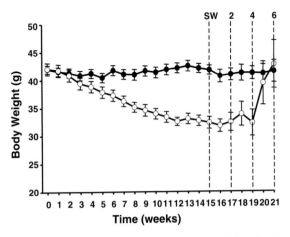

Fig. 1. Photoperiod-induced changes in body weight of male Siberian hamsters. Forty hamsters were held on long photoperiod (16 h light: 8 h dark). At time 0, 20 hamsters continued on long photoperiod (solid circles), while 20 hamsters were transferred onto short photoperiod (8 h light: 16 h dark) (open circles). At week 15 hamsters held on short photoperiod were switched back to long photoperiod (SW), and five of these animals as well as five LD control animals were removed and killed for analysis at the SW, 2-, 4-, and 6-week time points. The data for each time point show the mean ± SEM.

photoperiod (Fig. 2; Mercer et al., 2001). By superimposing a period of food restriction on the normal cycle of body weight loss in response to short photoperiod, the body weight falls below the level normal for the seasonal cycle. Remarkably, when the hamsters are again allowed to feed ad libitum, a period of hyperphagia ensues, yet this is sufficient only to allow the body weight to reach a level that is appropriate for the period of exposure to short photoperiod (Fig. 2). Together these findings imply that Siberian hamsters have an in-built mechanism that progressively changes an apparent 'sliding set-point' for energy balance and this is determined by the period of exposure to the short photoperiod.

Body weight change by altered food intake or energy expenditure?

It has been a matter of considerable debate as to whether the primary driver for the change in seasonal body mass is a change in food intake or energy expenditure (Wade and Bartness, 1984;

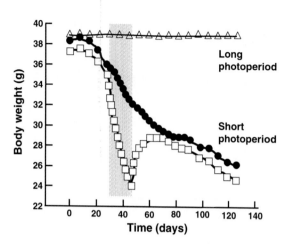

Fig. 2. Body weights of male Siberian hamsters subjected to altered photoperiod and energy intake, revealing 'sliding set-point of body weight' (Steinlechner et al., 1983; Mercer et al., 2001). Fifteen Siberian hamsters were held on long photoperiod (16 h light: 8 h dark), then at time 0, five were maintained on long photoperiod (open triangles) and ten were switched to a short photoperiod (8 h light: 16 h dark) (closed circles). All animals were allowed to feed ad libitum. After 28 days in short photoperiod, five hamsters on short photoperiod were subjected to a period of food restriction for 18 days (indicated by the hatched rectangle), after which they were returned to ad libitum feeding. (Data adapted from Mercer et al., 2001.)

Knopper and Boily, 2000). Temporally, it is difficult to discriminate between these two parameters, largely owing to the difficulties in measuring food intake with sufficient accuracy. It has been argued that the reduction in adipose tissue in Siberian hamsters held under short photoperiod is mediated by enhanced lipolysis through activation of the sympathetic neurons innervating adipose tissue, and increased adipocyte sensitivity to noradrenergic stimulation (Bowers et al., 2005). Support for this idea has been provided through elegant labelling studies, which have shown direct connections between sites in the hypothalamus and adipose tissue (Shi and Bartness, 2001; Song and Bartness, 2001; Demas and Bartness 2001). Some of the cells, projecting from the suprachiasmatic nucleus (SCN) and the dorsomedial nucleus (DMN) of the hypothalamus to adipose tissue express Mel 1a melatonin receptors, and these may mediate direct effects of photoperiod and melatonin on lipolysis.

While photoperiod-induced lipolysis may contribute to seasonal adjustments in body weight, it is unlikely to provide the main mechanism of body weight change. Food intake measurably decreases as Siberian hamsters lose weight and therefore there must be a robust mechanism to re-set the level of food intake (Steinlechner et al., 1983; Mercer et al., 2001). Furthermore, strong indirect evidence for primary effects of photoperiod on body weight through change in food intake comes from Siberian hamsters during weight re-gain. Hamsters held on short days, until body weights reach their nadir and then switched back to long days with ad libitum food available, increase their food intake and body weight with time after the switch in photoperiod (Masuda and Oishi, 1995). It seems unlikely that there is enough scope to increase body weight by simply conserving energy expenditure and reducing the rate of lipolysis. More importantly, in the animals switched to long days, but whose food intake was restricted to that of short-day hamsters, they *increase* activity as they look for food (Masuda and Oishi, 1995). These results can only be explained by an increase in the required level of food intake in response to the photoperiod switch. Thus, the primary effect of photoperiod appears to be on the level of food intake, rather than on energy expenditure alone.

The role of compensatory energy balance systems

Initial attempts to explain seasonal regulation of energy balance focused upon the known hypothalamic systems (Table 1). Following prolonged exposure to short photoperiod (i.e., 14–20 weeks), a number of different studies have shown that

Table 1. Changes in expression of energy balance genes in hypothalamus of Siberian hamster following 18 weeks in short photoperiod

Gene	Response	Region
POMC	↓	ARC
CART	↑[a]/↔	ARC
NPY	↔	ARC
AgRP	↑	ARC
MCH	↔	LHA
Orexin	↔	LHA

[a]Different studies have shown different responses (see text for details).

POMC gene expression in the arcuate nucleus (ARC) is significantly attenuated in both male and juvenile female hamsters relative to long photoperiod controls (Reddy et al., 1999; Adam et al., 2000; Mercer et al., 2000, 2001; Rousseau et al., 2002). Interestingly, this change in expression was observed after prolonged exposure to short photoperiod associated with significant weight loss; yet not after exposure to short photoperiod of 3 weeks or less (Mercer et al., 2003). Equally important, it has been shown that in steroid-clamped, castrated male hamsters, attenuated expression of the POMC gene in the ARC persists following exposure to short photoperiod indicating that the response is not the result of a decreasing steroid background associated with gonadal regression which parallels the loss in body mass (Rousseau et al., 2002).

The level of gene expression of another neuropeptide important to the MC4R signalling pathway, AgRP, has also been found to change in response to short photoperiod. In this case following 18 weeks' exposure to short photoperiod, AgRP gene expression was increased (Mercer et al., 2000, 2003). In combination, the effects of the gene expression changes for POMC and AgRP, if translated into protein, appear indicative of a net increase in an anabolic drive. However, given that the donor hamsters were actually losing weight, this apparent response would appear counter-intuitive, and thus it seems unlikely that the melanocortin system contributes to the seasonal body weight loss response. Support for this interpretation is further provided by the fact that hamsters acclimatised to either a long or short photoperiod displayed no difference in sensitivity to i.c.v. administration of either the melanocortin receptor agonist MTII or the antagonist SHU9119 in terms of food intake (Schuhler et al., 2003, 2004).

NPY and CART have also been examined for their potential role in the seasonal control of body weight. Gene expression of NPY within the ARC has proved consistently unaffected by photoperiod across a number of studies (Reddy et al., 1999; Adam et al., 2000; Mercer et al., 2000, 2001). As for the melanocortin system, photoperiod has no effect on the potency of NPY administered i.c.v. into hamsters, strongly suggesting that NPY has little involvement in the seasonal body weight (Boss-Williams and Bartness, 1996). Changes in CART mRNA expression have proved more controversial, since some laboratories have found no effect of photoperiod on CART expression in the ARC of Siberian hamsters (Robson et al., 2002; Rousseau et al., 2002), whereas others have shown an increase in CART expression in response to short photoperiod, which would be consistent with a catabolic role (Adam et al., 2000; Mercer et al., 2003).

Taken together, the evidence for a primary role of the POMC/CART and NPY/AgRP neurons of the ARC in driving the seasonal body weight response is weak. That the rostral and caudal ARC plays little or no role in the seasonal response is further strengthened by a neurotoxic ablation study undertaken by Ebling and colleagues. In this experiment, neonatal hamsters were injected with monosodium glutamate, which destroyed up to 80% of the ARC, as verified by NPY and tyrosine hydroxylase immunostaining (Ebling et al., 1998). Despite this, the hamsters were still able to mount a photoperiodic body weight response, indicating that an intact ARC is not pivotal to the seasonal response (Ebling et al., 1998).

The importance of other hypothalamic neuropeptides such as orexin, MCH, or corticotrophin-releasing factor (CRF) have also been studied, but each of these peptides has proved to be remarkably unaffected by photoperiod (Reddy et al., 1999; Mercer et al., 2000), suggesting that these peptides also play little role in the overall seasonal response.

The leptin paradox

One of the interesting features of the Siberian hamster is the so-called 'leptin paradox'. When long-day hamsters are exposed to short days, the net body weight loss is between 30 and 40% after 18–20 weeks (Morgan and Mercer, 2001). The greatest contributor to this weight loss is a reduction in adipose tissue mass. Correspondingly, there is a parallel reduction in plasma leptin levels and leptin gene expression within adipose tissue (Klingenspor et al., 1996, 2000). The apparent 'paradox' arises from the fact that in the face of low plasma leptin, the hamsters do not mount a compensatory response to restore energy balance.

Clearly, the explanation for this lies in altered sensitivity to leptin. Several studies have shown that leptin administration to Siberian hamsters has differential effects depending upon the seasonal status of the animal. Using either injection or infusion paradigms, it has been shown that leptin is more potent at reducing food intake and body weight in hamsters acclimated to a short photoperiod than for those held for long days (Atcha et al., 2000; Klingenspor et al., 2000). This relative insensitivity of long-day animals and sensitivity of the short-photoperiod animals may in part be explained through levels of SOCS-3 expression in leptin-responsive neurons in the ARC (Tups et al., 2004). In long-day hamsters, SOCS-3 levels were significantly higher than in short-day animals. When challenged with peripheral injections of leptin, SOCS-3 expression could not be increased further in long-day hamsters, whereas in short-day animals, a robust increase in SOCS-3 expression was observed reaching a level comparable to that of long-day animals (Fig. 3) (Tups et al., 2004).

In diet-induced obesity, it is thought that the rising level of leptin occurring as a consequence of obesity is causal in the reduction of leptin sensitivity (Scarpace et al., 2002). A similar argument could be used to explain the change in leptin sensitivity in Siberian hamsters, as they move from a relatively lean state on short days to a more obese

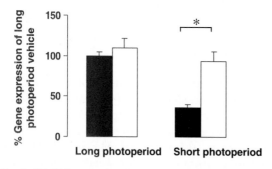

Fig. 3. SOCS-3 expression in the arcuate nucleus (ARC) of male Siberian hamsters under long (16 h light: 8 h dark) and short (8 h light: 16 h dark) photoperiod and 1 h after injection with either vehicle (black bars) or 2 mg/kg of leptin (open bars). Data show mean $\pm$ SEM, $n = 6$, and ($*p < 0.001$). (Data adapted from Tups et al., 2004.)

phenotype on long days. However, experimental evidence indicates that it is not the level of adiposity and leptin that alters the status of leptin sensitivity in hamsters, rather it is the photoperiod that acts as the driving force (Tups et al., 2004).

In search of novel systems of control

On the basis of the current evidence, it can be predicted that photoperiod must influence hypothalamic energy balance systems in at least two distinct ways. First, it must modulate food intake/energy expenditure pathways within the hypothalamus to enable adjustment of a defendable body weight (set point). Second, it must alter the sensitivity of the hypothalamic system to plasma leptin to prevent compensatory changes in energy balance interfering with the seasonal cycles of body weight.

While the known components of energy-balance control within the hypothalamus have been shown to play a role in the response to energy deficits as in other species (Mercer et al., 1995, 1997, 2000, 2001), it is clear that the seasonal body weight response must involve novel pathways. A solid body of evidence accumulated from lesioning studies as well as melatonin-receptor localisation studies have implicated the hypothalamus as the site through which photoperiod (and hence melatonin) mediates its effects on the neuroendocrine pathways involved in the control of reproduction and energy balance (Lincoln et al., 2001; Morgan et al., 2003). First, melatonin receptors have been localised to the suprachiasmatic nucleus (SCN), dorsomedial nucleus (DMN), anterior hypothalamic area (AHA) of the hypothalamus, and stria medullaris of the Siberian hamster (Morgan et al., 1994; C. Ellis, J.G., Mercer, and P.J. Morgan, unpublished observations). Second, lesioning of the SCN in the Siberian hamster has been shown to block both body weight loss and gonadal regression in response to short-day (SD) photoperiod and programmed infusions of melatonin (Bartness et al., 1991; Bittman et al., 1991; Song and Bartness, 1998).

A central prediction from the sliding set-point model described above is that hamsters progressively adjust their energy balance systems, so that food intake is progressively lowered with time after a photoperiod switch from long days to short days. On this basis, it is predicted that any gene expression events involved in this mechanism will be maximally different in expression between long and short-day animals at the time of the greatest weight difference.

Using Siberian hamsters held in either long or short photoperiod for 12–14 weeks, and therefore exhibiting a substantial weight differential, hypothalamic RNA was analysed using a mouse cDNA microarray. Perhaps unsurprisingly, given the complexity of hypothalamic tissue together with the cross-species hybridisation involved, there were no genes that showed major differences in expression between the long and short-day animals. Nevertheless, a few genes showed differences of about 1.8-fold between the two groups. One of the genes was cellular retinal-binding protein 1 (*CRBP1*) (Ross et al., 2004).

In-situ hybridisation confirmed that *CRBP1* is localised in the ependymal layer of the third ventricle as well as a small region, initially thought to be the dorsal tuberomamillary nucleus (Ross et al., 2004), but more recently recognised to be a region of the posterior arcuate nucleus (the dorsal medial posterior (dmp) ARC) (Barrett et al., 2005). Most importantly, a substantial difference in expression between the long and short-day animals was observed.

Following this initial finding, many other genes have been identified as differentially expressed in the dmpARC of Siberian hamsters from long and short-photoperiod backgrounds (Ross et al., 2004, 2005; Barrett et al., 2005). The genes identified to date and their relative levels of expression are shown in Table 2 and Fig. 4. Interestingly, while some of the genes were found to be more highly expressed in the hypothalamus of long-day

hamsters relative to the short-day ones (e.g., *CRBP1*, *RXRγ*, *RAR*, *CRABP2*, and *histamine H3*), *VGF* and others were expressed more highly in short-day hamsters relative to the long-day ones (Barrett et al., 2005, unpublished data).

Gene expression changes are linked to photoperiod not secondary events

It is known that the expression of many genes, and particularly those involved in the circadian timing system in the SCN such as *Per* and *Cry* vary substantially across the diurnal cycle (Johnston et al., 2005; Messager et al., 1999, 2000; Hofman, 2004; Nusselein-Hildesheim et al., 2000). Furthermore, the pattern of expression changes according to photoperiodic background (Johnson et al., 1995; Messager et al., 1999, 2000; Hofman, 2004; Nusselein-Hildesheim et al., 2000). Importantly, the temporal characteristics of the genes expressed in the dmpARC are quite different. Unlike *clock* genes, the genes in the dmpARC exhibit no diurnal variation in expression in either long or short photoperiod, yet the level of expression is substantially different dependent upon photoperiod background (Ross et al., 2004; Barrett et al., 2005). This is important as it shows that the differences in gene expression observed in the dmpARC are wholly due to photoperiod and that they do not result from temporal differences in diurnal expression patterns between photoperiods. Furthermore, for *CRBP1* it has been shown that the amplitude of the difference in expression between long-day and short-day hamsters is directly related to the time following photoperiodic switch consistent with the prediction made from the sliding-set point hypothesis.

It is also clear that the changes in gene expression in the dmpARC are driven through the direct effects of photoperiod (and hence melatonin), rather than through indirect effects such as altered steroid levels associated with changing gonadal status, since pinealectomised Siberian hamsters fail to show short day-induced inhibition of gene expression (Ross et al., 2004; Barrett et al., 2005). Also restoring the serum testosterone levels of short-day hamsters to long-day levels fails to

Table 2. Changes in gene expression in the dmpARC of Siberian hamsters after 14 weeks in short photoperiod

Gene	Response	Change (%)	Reference
RXRγ	↓	−42	Ross et al. (2004)
RAR	↓	−45	Ross et al. (2004)
CRBP1	↓↓	−88	Ross et al. (2004)
CRABP2	↓↓	−90	Ross et al. (2004)
H3R	↓↓	−70	Barrett et al. (2005)
VGF	↑↑↑	+320	Barrett et al. (2005)

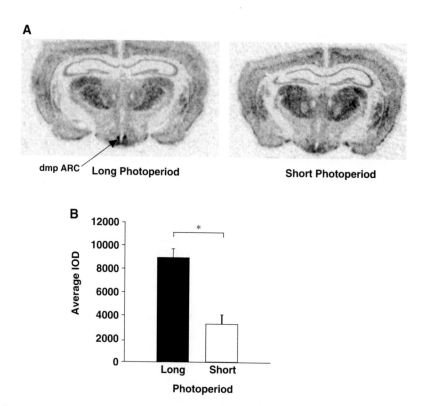

Fig. 4. (A) Histamine H3 receptor gene expression in the dmpARC region of the hypothalamus of the Siberian hamsters under long (16 h light: 8 h dark) and short (8 h light: 16 h dark) photoperiod shown by in-situ hybridisation. (B) Densitometric analysis of histamine H3 receptor gene expression in dmpARC of Siberian hamsters ($n = 10$; data show mean ± SEM) and (*$p < 0.001$). (Data adapted from Barrett et al., 2005.)

elevate gene expression of all the genes that have been tested (Ross et al., 2004; Barrett et al., 2005).

Temporal changes in gene expression

Relative to the slow body weight loss that occurs in Siberian hamsters when transferred from long to short days (taking up to 18–20 weeks to reach its nadir), regain of body weight after transfer of hamsters from short days back to long days is much more rapid, taking only about 6 weeks to recover a long-day body weight (Fig. 1). Study of the time-course of the gene expression changes in the dmpARC during this dynamic phase of weight regain has been instructive about the potential relationships between gene expression and body weight (Ross et al., 2005). Three genes H3R, VGF, and CRABP2 each change rapidly in response to

the switch from short to long photoperiod, with significant change being observed within 2 weeks (Fig. 5). This rate of change is clearly in advance of any overt change in body weight, and therefore offers the potential for these genes to be involved in the regulation of energy balance. The mRNA expression of RAR responded more slowly than these genes, but had also attained long-day (LD) levels by 6 weeks. In contrast, the slow time-course of change of other genes, RXRγ and CRBP1, would seem to eliminate these from any direct role in the re-gain of body weight, at least at the level of mRNA expression.

The importance of the changes in H3R, VGF, and CRABP2 gene expression to the body weight response during the short–long photoperiod transition was further emphasised in photorefractory hamsters (Ross et al., 2005). In these animals by about 25 weeks of exposure to short photoperiod,

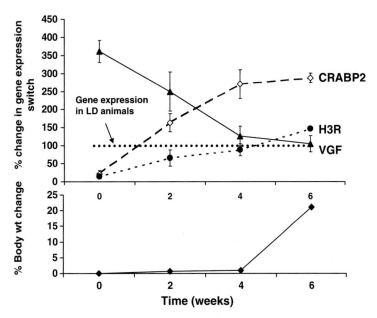

Fig. 5. Temporal changes in gene expression associated with body weight re-gain by Siberian hamsters after 15 weeks on short photoperiod (8 h light: 16 h dark) and then switched back to long photoperiod (16 h light: 8 h dark) (see Fig. 1). Upper panel shows the changes in gene expression (H3R (closed circles); CRABP2 (open diamonds); VGF (closed triangles)) in the dmpARC in hamsters following switch from short to long photoperiod at time 0. Expression is shown as a percentage of gene expression of hamsters held on long photoperiod, defined as 100%. (Data show mean $\pm$ SEM; $n = 5$). Lower panel shows the per cent change in body weights following the photoperiod switch. (Data adapted from Ross et al., 2005.)

body weight begins to increase back towards long-day levels, despite the hamsters continuing to receive a short photoperiod and melatonin signal. In each case, *H3R*, *VGF*, and *CRABP2* were restored to their long-day levels of expression, before the body weights of the photorefractory hamsters had returned to their long-day levels. This implies that the gene expression changes for *H3R*, *VGF*, and *CRABP2* in the dmpARC are not only downstream of the photorefractory mechanism, but also they may be important changes required for the body weight response to occur (Ross et al., 2005).

A hypothetical model of interaction between *H3R* and *VGF*

Collectively, the studies to date provide strong evidence that a neural cell cluster located in the posterior arcuate nucleus is an important interface between the photoperiod/melatonin system and

the pathways controlling physiology and behaviour (Barrett et al., 2005; Ross et al., 2005). While it has yet to be formally proved that the gene expression changes in response to photoperiod observed in this nucleus are involved in the regulation of body weight and/or other seasonal physiological and behavioural changes, a functional relationship seems likely. This is because of the anatomical location of the cell cluster, the identity of some of the genes involved, and the temporal relationship between the gene expression changes and physiological changes. It is possible that the dmpARC could serve to control both the tone of the food intake system as well as to modulate the sensitivity of the ARC to leptin.

Of the genes identified thus far, *H3R*, *VGF*, and *CRABP2* appear potentially the most important to the regulation of body weight. Both *H3R* and *VGF* have been implicated in the regulation of body weight through targeted gene knock-outs in mice (Hahm et al., 1999; Takahashi et al., 2002). Superficially, the phenotypes of these mice, obese for the $H3R^{(-/-)}$ and lean for the $VGF^{(-/-)}$ do not appear to be consistent with the phenotypes of the long or short-day Siberian hamsters, where lean short-day hamsters have low levels of *H3R* and high levels of *VGF* expression (Barrett et al., 2005). However, it is important not to use the outcomes of global gene knock-outs to predict the consequences for a small cluster of cells in a restricted part of the hypothalamus, since both *H3R* and *VGF* are widely distributed in the brain.

Nonetheless, *H3R* is an interesting gene as it has also been implicated in the control of food intake through physiological and pharmacological studies (Sakata et al., 1997; Yoshimatsu et al., 1999, 2002). Furthermore, recent findings have shown that the H3R antagonist, A-331440, has anti-obesity activities in mice (Hancock et al., 2005), an effect that has resonance with the findings in Siberian hamsters.

Functionally, H3 receptors are G-protein-coupled receptors, which act to inhibit cyclic AMP (cAMP) and which are also known to have constitutive activity (Morisset et al., 2000; Barrett et al., 2005). Within the tuberomamillary nucleus, H3 receptors function as autoreceptors, inhibiting the presynaptic release of histamine (Arrang et al., 1987). Within the dmpARC, while the cells are well innervated by fibres containing histamine, there are no histamine-synthesising cells within the nucleus. On this basis, it seems most likely that the role of the H3 receptor in the dmpARC is to function as a heteroreceptor, serving to modulate the release of a heterologous transmitter or neuropeptide (Barrett et al., 2005).

One secretory peptide that may be regulated by the H3 receptor is VGF. VGF is a nonacronymic term for an abundant polypeptide, expressed widely in the brain and a number of peripheral tissues, including the gut, the pancreas, and the adrenal gland (Levi et al., 2004). Its precise functional role is not established, as there is uncertainty over whether its primary function is to serve as either a structural component of the secretory granule or instead as a precursor for the production of bioactive peptides (Levi et al., 2004). Recent evidence demonstrating that the C-terminal cleavage products of VGF increase synaptic charge in cultured hippocampal neurons support the idea of a precursor of biologically active peptides (Alder et al., 2003). Whatever the functional role of VGF, it provides a useful marker of regulated secretion, and thereby provides potential insight into how the histamine H3 receptor may influence the activity of the dmpARC neurons.

A plausible hypothetical link between H3R and VGF through the second messenger cAMP can be postulated (Barrett et al., 2005). Since H3Rs inhibit cAMP synthesis, a decline in H3R expression in short-day hamsters would reduce the inhibitory input to the dmpARC neurons through both histamine-induced activation as well as through the constitutive activity of the receptor (Morisset et al., 2000). This would allow cAMP to rise, which would facilitate an increase in cAMP-mediated gene expression. It is known that *VGF* gene expression is positively regulated by cAMP (Di Rocco et al., 1997; Nagasaki et al., 1999). Thus, the decline in H3 receptor expression during the transition from long to short photoperiod could drive the increased expression in VGF through disinhibition of cAMP. Conversely, the increase in H3 receptor expression following the switch from short to long photoperiod could inhibit VGF expression through increased inhibitory activity of the H3 receptor (Fig. 6).

Perspective

Many of the changes in gene expression observed in the dmpARC have also been observed in Syrian hamsters in response to altered photoperiod (Ross et al., 2004; Barrett et al., 2005). This indicates that the dmpARC response is not unique to the Siberian hamster. Hitherto, there has been no neuroanatomical or functional characterisation of the dmpARC as a discrete cell cluster within the hypothalamus of nonphotoperiodic species. However, preliminary studies suggest that a comparable region also exists in both rat and mouse based on in-situ hybridisation for a number of genes expressed in the dmpARC of the Siberian hamster (*RXRγ*; *RAR*; *H3R*, and *VGF*) (Ross et al., 2005; Barrett, Ross, and Morgan, unpublished data). While the role of the dmpARC remains to be established in non-photoperiodic species, this evidence adds weight to the argument that this region

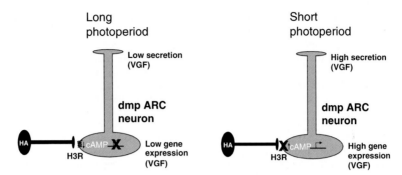

Fig. 6. A scheme showing a working hypothesis of how histamine H3 receptors may regulate gene expression and the secretory output of dmpARC neurons. Under long photoperiod, H3R mRNA levels are high, predicting high-expression H3 receptor protein on dmpARC neurons. Through histamine released from adjacent neurons and/or constitutive activity of the H3 receptor, cAMP levels in dmpARC neurons are suppressed keeping the cells in a quiescent state. As a consequence, expression of cAMP-dependent genes (e.g., VGF) remain suppressed and secretion is inhibited. Under short photoperiod, H3R mRNA levels in dmpARC neurons are low. Low levels of H3 receptor protein allow cAMP levels to rise and dmpARC neurons are in a generally more activated state. Under these conditions, expression of the cAMP-dependent genes (e.g., VGF) are increased and there is an enhanced level of secretion.

is a novel and neuroanatomically unexplored region of functional importance in both seasonal and nonseasonal species. The functional linkage between the dmpARC and the control of energy balance remains to be established, but studies are ongoing with the aim of resolving this issue. Just as the SCN was important to resolving the molecular basis of circadian rhythms, the dmpARC will undoubtedly be important in resolving the molecular basis of seasonal timing and understanding how it links to the control of physiology and behaviour.

Acknowledgment

The authors would like to acknowledge the Scottish Executive Environment and Rural Affairs Department for the financial support of their work.

References

Adam, C.L., Moar, K.M., Logie, T.J., Ross, A.W., Barrett, P., Morgan, P.J. and Mercer, J.G. (2000) Photoperiod regulates growth, puberty and hypothalamic neuropeptide and receptor gene expression in female Siberian hamsters. Endocrinology, 141: 4349–4356.

Alder, J., Thakker-Varia, S., Bangasser, D.A., Kuroiwa, M., Plummer, M.R., Shors, T.J. and Black, I.B. (2003) Brain-derived neurotrophic factor-induced gene expression reveals novel actions of VGF in hippocampal synaptic plasticity. J. Neurosci., 23: 10800–10808.

Arrang, J.M., Garbarg, M. and Schwartz, J.C. (1987) Autoinhibition of histamine synthesis mediated by presynaptic H3-receptors. Neuroscience, 23: 149–157.

Atcha, Z., Cagampang, F.R., Stirland, J.A., Morris, I.D., Brooks, A.N., Ebling, F.J., Klingenspor, M. and Loudon, A.S. (2000) Leptin acts on metabolism in a photoperiod-dependent manner, but has no effect on reproductive function in the seasonally breeding Siberian hamster (Phodopus sungorus). Endocrinology, 141: 4128–4135.

Barrett, P., Ross, A.W., Balik, A., Littlewood, P.A., Mercer, J.G., Moar, K.M., Sallmen, T., Kaslin, J., Panula, P., Schuhler, S., Ebling, F.J., Ubeaud, C. and Morgan, P.J. (2005) Photoperiodic regulation of histamine H3 receptor and VGF messenger ribonucleic acid in the arcuate nucleus of the Siberian hamster. Endocrinology, 146: 1930–1939.

Bartness, T.J., Demas, G.E. and Song, C.K. (2002) Seasonal changes in adiposity: the roles of the photoperiod, melatonin and other hormones, and sympathetic nervous system. Exp. Biol. Med. (Maywood), 227: 363–376.

Bartness, T.J., Goldman, B.D. and Bittman, E.L. (1991) SCN lesions block responses to systemic melatonin infusions in Siberian hamsters. Am. J. Physiol., 260: R102–R112.

Bartness, T.J. and Wade, G.N. (1985) Photoperiodic control of seasonal body weight cycles in hamsters. Neurosci. Biobehav. Rev., 9: 599–612.

Berthoud, H.R. (2002) Multiple neural systems controlling food intake and body weight. Neurosci. Biobehav. Rev., 26: 393–428.

Bittman, E.L., Bartness, T.J., Goldman, B.D. and DeVries, G.J. (1991) Suprachiasmatic and paraventricular control of photoperiodism in Siberian hamsters. Am. J. Physiol., 260: R90–R101.

Boss-Williams, K.A. and Bartness, T.J. (1996) NPY stimulation of food intake in Siberian hamsters is not photoperiod dependent. Physiol. Behav., 59: 157–164.

336

Bowers, R.R., Gettys, T.W., Prpic, V., Harris, R.B. and Bartness, T.J. (2005) Short photoperiod exposure increases adipocyte sensitivity to noradrenergic stimulation in Siberian hamsters. Am. J. Physiol. Regul. Integr. Comp. Physiol., 288: R1354–R1360.

Demas, G.E. and Bartness, T.J. (2001) Direct innervation of white fat and adrenal medullary catecholamines mediate photoperiodic changes in body fat. Am. J. Physiol., 281: R1499–R1505.

Di Rocco, G., Pennuto, M., Illi, B., Canu, N., Filocamo, G., Trani, E., Rinaldi, A.M., Possenti, R., Mandolesi, G., Sirinian, M.I., Jucker, R., Levi, A. and Nasi, S. (1997) Interplay of the E box, the cyclic AMP response element, and HTF4/HEB in transcriptional regulation of the neurospecific, neurotrophin-inducible *vgf* gene. Mol. Cell. Biol., 17: 1244–1253.

Ebling, F.J., Arthurs, O.J., Turney, B.W. and Cronin, A.S. (1998) Seasonal neuroendocrine rhythms in the male Siberian hamster persist after monosodium glutamate-induced lesions of the arcuate nucleus in the neonatal period. J. Neuroendocrinol., 10: 701–712.

Elmquist, J.K., Maratos-Flier, E., Saper, C.B. and Flier, J.S. (1998) Unraveling the central nervous system pathways underlying responses to leptin. Nat. Neurosci., 1: 445–450.

Gorman, M.R. and Zucker, I. (1995) Seasonal adaptations of Siberian hamsters. II. Pattern of change in day length controls annual testicular and body weight rhythms. Biol. Reprod., 53: 116–125.

Hahm, S., Mizuno, T.M., Wu, T.J., Wisor, J.P., Priest, C.A., Kozak, C.A., Boozer, C.N., Peng, B., McEvoy, R.C., Good, P., Kelley, K.A., Takahashi, J.S., Pintar, J.E., Roberts, J.L., Mobbs, C.V. and Salton, S.R. (1999) Targeted deletion of the *Vgf* gene indicates that the encoded secretory peptide precursor plays a novel role in the regulation of energy balance. Neuron, 23: 537–548.

Hancock, A.A., Diehl, M.S., Fey, T.A., Bush, E.N., Faghih, R., Miller, T.R., Krueger, K.M., Pratt, J.K., Cowart, M.D., Dickinson, R.W., Shapiro, R., Knourek-Segel, V.E., Droz, B.A., McDowell, C.A., Krishna, G., Brune, M.E., Esbenshade, T.A. and Jacobson, P.B. (2005) Antiobesity evaluation of histamine H3 receptor (H3R) antagonist analogs of A-331440 with improved safety and efficacy. Inflamm. Res., 54(Suppl 1): S27–S29.

Hofman, M.A. (2004) The brain's calendar: neural mechanisms of seasonal timing. Biol. Rev., 79: 61–77.

Johnston, J.D., Ebling, F.J. and Hazlerigg, D.G. (2005) Photoperiod regulates multiple gene expression in the suprachiasmatic nuclei and pars tuberalis of the Siberian hamster (*Phodopus sungorus*). Eur. J. Neurosci., 21: 2967–2974.

Klingenspor, M., Dickopp, A., Heldmaier, G. and Klaus, S. (1996) Short photoperiod reduces leptin gene expression in white and brown adipose tissue of Djungarian hamsters. FEBS Lett., 399: 290–294.

Klingenspor, M., Niggemann, H. and Heldmaier, G. (2000) Modulation of leptin sensitivity by short photoperiod acclimation in the Djungarian hamster, *Phodopus sungorus*. J. Comp. Physiol. [B], 170: 37–43.

Knopper, L.D. and Boily, P. (2000) The energy budget of captive Siberian hamsters, *Phodopus sungorus*, exposed to photoperiod changes: mass loss is caused by a voluntary decrease in food intake. Physiol. Biochem. Zool., 73: 517–522.

Levi, A., Ferri, .L., Watson, E., Possenti, R. and Salton, S.R. (2004) Processing, distribution, and function of VGF, a neuronal and endocrine peptide precursor. Cell. Mol. Neurobiol., 24: 517–533.

Lincoln, G.A., Rhind, S.M., Pompolo, S. and Clarke, I.J. (2001) Hypothalamic control of photoperiod-induced cycles in food intake, body weight, and metabolic hormones in rams. Am. J. Physiol., 281: R76–R90.

Maffei, M., Halaas, J., Ravussin, E., Pratley, R.E., Lee, G.H., Zhang, Y., Fei, H., Kim, S., Lallone, R., Ranganathan, S., Kern, P.A. and Friedman, J.M. (1995) Leptin levels in human and rodent: measurement of plasma leptin and ob RNA in obese and weight-reduced subjects. Nat. Med., 1: 1155–1161.

Masuda, A. and Oishi, T. (1995) Effects of restricted feeding on the light-induced body weight change and locomotor activity in the Djungarian hamster. Physiol. Behav., 58: 153–159.

Mercer, J.G., Ellis, C., Moar, K.M., Logie, T.J., Morgan, P.J. and Adam, C.L. (2003) Early regulation of hypothalamic arcuate nucleus CART gene expression by short photoperiod in the Siberian hamster. Regul. Peptides, 111: 129–136.

Mercer, J.G., Lawrence, C.B., Beck, B., Burlet, A., Atkinson, T. and Barrett, P. (1995) Hypothalamic NPY and prepro-NPY mRNA in Djungarian hamsters: effects of food deprivation and photoperiod. Am. J. Physiol., 269: R1099–R1106.

Mercer, J.G., Lawrence, C.B., Moar, K.M., Atkinson, T. and Barrett, P. (1997) Short-day weight loss and effect of food deprivation on hypothalamic NPY and CRF mRNA in Djungarian hamsters. Am. J. Physiol., 273: R768–R776.

Mercer, J.G., Moar, K.M., Logie, T.J., Findlay, P.A., Adam, C.L. and Morgan, P.J. (2001) Seasonally inappropriate body weight induced by food restriction: effect on hypothalamic gene expression in male Siberian hamsters. Endocrinology, 142: 4173–4181.

Mercer, J.G., Moar, K.M., Ross, A.W., Hoggard, N. and Morgan, P.J. (2000) Photoperiod regulates arcuate nucleus POMC, AGRP, and leptin receptor mRNA in Siberian hamster hypothalamus. Am. J. Physiol., 278: R271–R281.

Messager, S., Hazlerigg, D.G., Mercer, J.G. and Morgan, P.J. (2000) Photoperiod differentially regulates the expression of Per1 and ICER in the pars tuberalis and the suprachiasmatic nucleus of the Siberian hamster. Eur. J. Neurosci., 12: 2865–2870.

Messager, S., Ross, A.W., Barrett, P. and Morgan, P.J. (1999) Decoding photoperiodic time through Per1 and ICER gene amplitude. Proc. Natl. Acad. Sci. USA, 96: 9938–9943.

Morgan, P.J., Barrett, P., Howell, H.E. and Helliwell, R. (1994) Melatonin receptors: localization, molecular pharmacology and physioloigical significance. Neurochem. Int., 24: 101–146.

Morgan, P.J. and Mercer, J.G. (2001) The regulation of body weight: lessons from the seasonal animal. Proc. Nutr. Soc., 60: 127–134.

Morgan, P.J., Ross, A.W., Mercer, J.G. and Barrett, P. (2003) Photoperiodic programming of body weight through the neuroendocrine hypothalamus. J. Endocrinol., 177: 27–34.

Morisset, S., Rouleau, A., Ligneau, X., Gbahou, F., Tardivel-Lacombe, J., Stark, H., Schunack, W., Ganellin, C.R., Schwartz, J.C. and Arrang, J.M. (2000) High constitutive activity of native H3 receptors regulates histamine neurons in brain. Nature, 408: 860–864.

Nagasaki, K., Sasaki, K., Maass, N., Tsukada, T., Hanzawa, H. and Yamaguchi, K. (1999) Staurosporine enhances cAMP-induced expression of neural-specific gene VGF and tyrosine hydroxylase. Neurosci. Lett., 267: 177–180.

Nusselein-Hildesheim, B., O'Brien, J.A., Ebling, F.J., May-wood, E.S. and Hastings, M.H. (2000) The circadian cycle of mPER clock gene products in the suprachiasmatic nucleus of the Siberian hamster encodes both daily and seasonal time. Eur. J. Neurosci., 12: 2856–2864.

Reddy, A.B., Cronin, A.S., Ford, H. and Ebling, F.J. (1999) Seasonal regulation of food intake and body weight in the male Siberian hamster: studies of hypothalamic orexin (hypocretin), neuropeptide Y (NPY) and pro-opiomelanocortin (POMC). Eur. J. Neurosci., 11: 3255–3264.

Robson, A.J., Rousseau, K., Loudon, A.S. and Ebling, F.J. (2002) Cocaine and amphetamine-regulated transcript mRNA regulation in the hypothalamus in lean and obese rodents. J. Neuroendocrinol., 14: 697–709.

Ross, A.W., Bell, L.M., Littlewood, P.A., Mercer, J.G., Barrett, P. and Morgan, P.J. (2005) Temporal changes in gene expression in the arcuate nucleus precede seasonal responses in adiposity and reproduction. Endocrinology, 146: 1940–1947.

Ross, A.W., Webster, C.A., Mercer, J.G., Moar, K.M., Ebling, F.J., Schuhler, S., Barrett, P. and Morgan, P.J. (2004) Photoperiodic regulation of hypothalamic retinoid signaling: association of retinoid X receptor gamma with body weight. Endocrinology, 145: 13–20.

Rousseau, K., Atcha, Z., Cagampang, F.R., Le Rouzic, P., Stirland, J.A., Ivanov, T.R., Ebling, F.J., Klingenspor, M. and Loudon, A.S. (2002) Photoperiodic regulation of leptin resistance in the seasonally breeding Siberian hamster (Phodopus sungorus). Endocrinology, 143: 3083–3095.

Sakata, T., Yoshimatsu, H. and Kurokawa, M. (1997) Hypothalamic neuronal histamine: implications of its homeostatic control of energy metabolism. Nutrition, 13: 403–411.

Saper, C.B., Chou, T.C. and Elmquist, J.K. (2002) The need to feed: homeostatic and hedonic control of eating. Neuron, 36: 199–211.

Scarpace, P.J., Matheny, M., Zhang, Y., Shek, E.W., Prima, V., Zolotukhin, S. and Tumer, N. (2002) Leptin-induced leptin resistance reveals separate roles for the anorexic and thermogenic responses in weight maintenance. Endocrinology, 143: 3026–3035.

Schuhler, S., Horan, T.L., Hastings, M.H., Mercer, J.G., Morgan, P.J. and Ebling, F.J. (2003) Decrease of food intake by MC4-R agonist MTII in Siberian hamsters in long and short photoperiods. Am. J. Physiol., 284: R227–R232.

Schuhler, S., Horan, T.L., Hastings, M.H., Mercer, J.G., Morgan, P.J. and Ebling, F.J. (2004) Feeding and behavioural effects of central administration of the melanocortin 3/4-R antagonist SHU9119 in obese and lean Siberian hamsters. Behav. Brain Res., 152: 177–185.

Shi, H. and Bartness, T.J. (2001) Neurochemical phenotype of sympathetic nervous system outflow from brain to white fat. Brain Res. Bull., 54: 375–385.

Song, C.K. and Bartness, T.J. (1998) Dorsocaudal SCN microknife-cuts do not block short day responses in Siberian hamsters given melatonin infusions. Brain Res. Bull., 45: 239–246.

Song, C.K. and Bartness, T.J. (2001) CNS sympathetic outflow neurons to white fat that express MEL receptors may mediate seasonal adiposity. Am. J. Physiol., 281: R666–R672.

Steinlechner, S., Heldmaier, G. and Becker, H. (1983) The seasonal cycle of body-weight in the Djungarian hamster — photoperiodic control and the influence of starvation and melatonin. Oecologia, 60: 401–405.

Takahashi, K., Suwa, H., Ishikawa, T. and Kotani, H. (2002) Targeted disruption of H3 receptors results in changes in brain histamine tone leading to an obese phenotype. J. Clin. Invest., 110: 1791–1799.

Tups, A., Ellis, C., Moar, K.M., Logie, T.J., Adam, C.L., Mercer, J.G. and Klingenspor, M. (2004) Photoperiodic regulation of leptin sensitivity in the Siberian hamster, Phodopus sungorus, is reflected in arcuate nucleus SOCS-3 (suppressor of cytokine signaling) gene expression. Endocrinology, 145: 1185–1193.

Wade, G.N. and Bartness, T.J. (1984) Effects of photoperiod and gonadectomy on food intake, body weight, and body composition in Siberian hamsters. Am. J. Physiol., 246: R26–R30.

Yoshimatsu, H., Chiba, S., Tajima, D., Akehi, Y. and Sakata, T. (2002) Histidine suppresses food intake through its conversion into neuronal histamine. Exp. Biol. Med. (Maywood), 227: 63–68.

Yoshimatsu, H., Itateyama, E., Kondou, S., Tajima, D., Himeno, K., Hidaka, S., Kurokawa, M. and Sakata, T. (1999) Hypothalamic neuronal histamine as a target of leptin in feeding behavior. Diabetes, 48: 2286–2291.

Hypothalamic Integration of "Sensory" Information

Kalsbeek, Fliers, Hofman, Swaab, Van Someren & Buijs
Progress in Brain Research, Vol. 153
ISSN 0079-6123

CHAPTER 20

Organization of circadian functions: interaction with the body

Ruud M. Buijs[1,2,*], Frank A. Scheer[2,3], Felix Kreier[2], Chunxia Yi[2], Nico Bos[2], Valeri D. Goncharuk[2,4] and Andries Kalsbeek[2]

[1]Unviersidad Veracruzana, Inst. Sciences de Salud, Xalapa, Mexico
[2]Netherlands Institute for Neurosciences, Amsterdam, The Netherlands
[3]Medical Chronobiology Program, Harvard Medical School, Boston, USA
[4]Cardiovascular Research Centre, Moscow, Russia

Abstract: The hypothalamus integrates information from the brain and the body; this activity is essential for survival of the individual (adaptation to the environment) and the species (reproduction). As a result, countless functions are regulated by neuroendocrine and autonomic hypothalamic processes in concert with the appropriate behaviour that is mediated by neuronal influences on other brain areas. In the current chapter attention will be focussed on fundamental hypothalamic systems that control metabolism, circulation and the immune system. Herein a system is defined as a physiological and anatomical functional unit, responsible for the organisation of one of these functions. Interestingly probably because these systems are essential for survival, their function is highly dependent on each other's performance and often shares same hypothalamic structures. The functioning of these systems is strongly influenced by (environmental) factors such as the time of the day, stress and sensory autonomic feedback and by circulating hormones. In order to get insight in the mechanisms of hypothalamic integration we have focussed on the influence of the biological clock; the suprachiasmatic nucleus (SCN) on processes that are organized by and in the hypo-thalamus. The SCN imposes its rhythm onto the body via three different routes of communication: 1.Via the secretion of hormones; 2. via the parasympathetic and 3.via the sympathetic autonomous nervous system. The SCN uses separate connections via either the sympathetic or via the parasympathetic system not only to prepare the body for the coming change in activity cycle but also to prepare the body and its organs for the hormones that are associated with such change. Up till now relatively little attention has been given to the question how peripheral information might be transmitted back to the SCN. Apart from light and melatonin little is known about other systems from the periphery that may provide information to the SCN. In this chapter attention will be paid to e.g. the role of the circumventricular organs in passing info to the SCN. Herein especially the role of the arcuate nucleus (ARC) will be highlighted. The ARC is crucial in the maintenance of energy homeostasis as an integrator of long- and short-term hunger and satiety signals. Receptors for metabolic hormones like insulin, leptin and ghrelin allow the ARC to sense information from the periphery and signal it to the central nervous system. Neuroanatomical tracing studies using injections of a retrograde and anterograde tracer into the ARC and SCN showed a reciprocal connection between the ARC and the SCN which is used to transmit feeding related signals to the SCN. The implications of multiple inputs and outputs of the SCN to the body will be discussed in relation with metabolic functions.

---

*Corresponding author.; E-mail: r.buijs@nin.knaw.nl

DOI: 10.1016/S0079-6123(06)53020-1
341

Introduction

The suprachiasmatic nucleus (SCN) is essential for synchronizing our daily activity to the light dark cycle in such a way that the physiology of the body is optimally prepared and adapted to these changes in activity. Many SCN neurons have a circadian rhythm in electrical activity resulting in circadian changes of transmitter secretion in the target areas of the SCN (Gillette and Reppert, 1987; Bos and Mirmiran, 1990, 1993; Mirmiran et al., 1995). There is some evidence that direct synaptic transfer of information is not necessary to transmit all circadian signals from the SCN but that diffusion of peptide transmitters may convey such signals (Silver et al., 1996; Kraves and Weitz, 2006). However some rhythms could not be restored with diffusion alone (Meyer-Bernstein et al., 1999). Furthermore, it is unclear whether diffusion of peptide transmitters plays an important role in normal physiology. On the other hand, the presence of the amino acid transmitters gamma aminobutyric acid (GABA) and/or glutamate in the majority of SCN neurons, their circadian rhythm in release, and their effect and physiology on target cells in the hypothalamus indicates the important role of synaptic transfer of the daily SCN rhythm in normal physiology (Hermes et al., 1996; Cui et al., 2002; Perreau-Lenz et al., 2003, 2004). At least one cannot envision yet how amino acid transmission can take place via diffusion. Consequently, knowledge about the sites in the brain where information from the SCN is relayed to other neurons is essential. Therefore, much attention was given to the question by which transmitters the SCN transmits its message and which structures in the brain are essential for the integration of this information.

By means of anterograde-tracing techniques, we and several other groups have mapped the projections of the SCN and identified the termination sites in the rodent and human brain (Watts and Swanson, 1987; Kalsbeek et al., 1993a, b; Buijs et al., 1994; Dai et al., 1998; Lesauter and Silver, 1999a). All these studies indicate that the majority of SCN termination sites is within the medial hypothalamus where the key cell groups are involved in the organization of hormonal secretion and autonomic control. Consequently, this seems the foremost way in which the SCN transmits its daily message to the rest of the brain and body, affecting mono- and multisynaptically hormone-producing neurons and preautonomic neurons primarily located in the paraventricular nucleus (PVN) of the hypothalamus. However, estimated from the density of SCN projections, the cell groups that seem to fulfill an intermediary function within the hypothalamus receive a much more prominent SCN input. These cell groups (the medial preoptic area (MPO); the sub-PVN and the dorso medial hypothalamus (DMH)), located in the area directly in front, under and behind the PVN, are known to project extensively within the hypothalamus (Ter Horst and Luiten, 1986; Roland and Sawchenko, 1993) and thus appear perfect for an intermediary function. In fact, our studies on the role of the SCN in corticosterone secretion show that the DMH is an important target area for SCN VP fibers in this respect (Kalsbeek et al., 1996a, b).

In the present review, attention will be given to observations that indicate that one of the major functions of the SCN is to prepare our body for the daily changes in activity periods. Hereto, we propose that the SCN affects the functionality of our organs by at least two mechanisms; it organizes the daily rhythm of several hormones and it influences via the autonomic nervous system the activity of many organs directly or affects their sensitivity for these hormones. Studies will be discussed that indicate that once this function of the SCN to prepare our body for the upcoming activity period is lost or dysfunctional, it will result in the development of disease. The mechanisms that may lead to such loss in function will be discussed. In addition, recent studies will be presented that have provided evidence that also the body "talks" back to the SCN. Hereby, we will not only consider the feedback via the autonomic nervous system but also by hormones.

Circadian rhythm of SCN neurons and their anatomical organization

Clearly, many studies have shown that neurons of the biological clock maintain an activity in cell firing with a rhythm of about 24 h, irrespective of whether they were studied in vivo, in vitro, in isolation, in slice, or in cultured conditions (Groos et al., 1983; Gillette and Reppert, 1987; Bos and Mirmiran, 1993; Mirmiran et al., 1995; Xie et al., 2003).

In order to examine whether all cells in the same area of the suprachiasmatic nucleus (SCN) had the same firing characteristics, we examined their electrical activity during the circadian times CT 5–8 and CT 14–17, previously shown to be respectively the peak and trough of SCN neuronal electrical activity. Furthermore, we aimed to determine the activity of vasopressin (VP) neurons, hereto neurons were also recorded in whole cell mode using patch pipettes filled with biocytin for intracellular marking. After recording the electrical activity the neurons were filled with biocytin, the slices were fixed and processed for double labeling, detecting biocytin and VP. This enabled us to correlate the discharge frequency of the neurons in the loose cell-attached mode to the absence or presence of VP in the neurons in the same dorsomedial area of the SCN (Fig. 1). Loose patch recordings of neurons in the dorsomedial SCN revealed a significant difference in discharge rate between CT 5–8 and CT 14–17. Mean frequency of all recorded neurons was $5.97 + 0.43$ Hz ($n = 92$) at CT 5–8, and $3.17 + -0.31$ Hz ($n = 107$) during CT 14–17 ($p < 0.001$, Mann–Whitney U-test) (Fig. 2). This twofold difference is in agreement with other data on comparable slice preparations. VP-positive cells in the dorsomedial SCN had a mean firing rate of $6.78 + -0.86$ Hz ($n = 11$) during CT 5–8 and of $1.25 + -0.49$ Hz ($n = 21$) during CT 14–17, $p < 0.001$, Mann–Whitey U-test). Identified non-VP neurons located in the same area, i.e. adjoining the VP neurons, displayed a similar day-time CT 5–8 firing frequency $6.67 + -0.95$ Hz ($n = 26$), while at night CT 14–17 the firing rate was three times higher $3.61 + -0.67$ Hz ($n = 28$) than the VP-positive cells $p < 0.005$ (Fig. 1). This observation not only shows a remarkable difference in the amplitude of electrical activity between different populations of SCN neurons, but also that neurons located in the same area of the SCN may differ importantly in their firing properties. This difference is even more striking when really dissimilar

regions of the SCN are compared. For example, it is known that in the ventral portion of the SCN a large population of neurons is electrically silent in slice preparations (Kow and Pfaff, 1984; Bos and Mirmiran, 1993). This seems logical in view of the fact that these neurons are located in the area that receives direct light input from the retina and are activated by this light input to the SCN (Meijer et al., 1992; Romijn et al., 1997).

In view of the fact that about 30% of the SCN VP neurons contain gamma aminobutyric acid (GABA) as the neurotransmitter (Buijs et al., 1994, 1995) and possibly an identical number of VP neurons glutamate (Hermes et al., 1996; Cui et al., 2002), it is interesting to answer the question whether neurons can control their secretion of peptide versus that of amino acid. It seems logical in view of studies on simpler systems (Lundberg et al., 1981, 1994; Whim and Lloyd, 1989; Cropper et al., 1990) that the firing frequency and pattern (burst or no burst) will have a major impact in this selection.

Shell or core?

In spite of the fact that the anatomical organization of the SCN is sometimes simplified as being a core region (VIP) and a shell (VP) region (Leak and Moore, 2001), this characterization does not do justice to the complexity and diversity of functional and anatomical subdivisions of the SCN. In fact, several studies indicate that projections from the SCN are much more diverse than can be assumed on the basis of a core and shell division (Buijs et al., 2003; Kriegsfeld et al., 2004). Moreover, the idea that the shell region would not communicate with the core, e.g. does not project to the core is not based on quantitative data (Moore, 1996). In fact, Romijn et al. showed already in 1997 that VP neurons have extensive termination on VIP neurons in the ventral part of the SCN and vice versa (Fig. 3, Table 1) (Romijn et al., 1997). In addition, also somatostatin neurons are present in the middle area of the rat SCN and have projections throughout the whole SCN. In addition, in the lateral area of the rat SCN, a not yet identifiable population of neurons has been detected. The fact that different individual neurons, even within the VP or VIP areas of the SCN, project

344

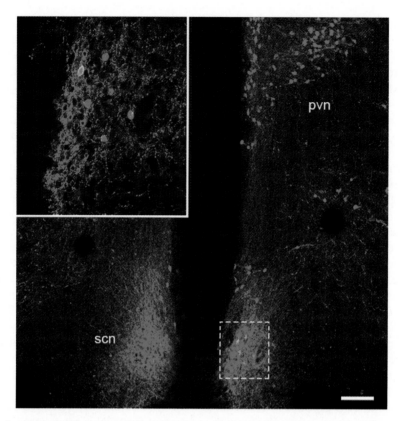

Fig. 1. Two color confocal images of the same optical section of a double-labeled 300 μm coronal slice of the rat hypothalamus stained for vasopressin (green) and biocytin (red). Clearly in the ventral part of the section the SCN is visible, while in the dorsal part some neurons of the PVN are stained for vasopressin.

The higher magnification (inset) shows the biocytin-labeled neurons (red) in the SCN of which one colocalizes with vasopressin as can be seen from the yellow color. (Bar = 75 μm for the inset and 200 μm for the low magnification.)

specifically to different organs or to different entities of the autonomic nervous system (Buijs et al., 2003) just emphasizes the complexity of the SCN on the one hand and the difficulty in describing the anatomy of the SCN solely based on anatomical location on the other hand. Not only in the rat but also in the hamster the SCN shows great diversity with respect to its projections. Here, the area that receives the light input is a small center area in the SCN characterized by the presence of calbindin neurons (Lesauter and Silver, 1999b).

Therefore, it seems much more logical to label the subareas in the SCN in relation to their neurotransmitter content, function or other properties, rather than to give it uninformative anatomical generic names like shell or core. We would propose to use instead: e.g., VP part, calbindin

part, retino recipient part, or to use other objective characteristics.

Hypothalamic projections of the SCN

Projections of the SCN to hypothalamic structures were initially determined by injection of anterograde tracers into the SCN. Injection of *Phaseolus vulgaris* leucoagglutinin, a plant lectin, into the SCN by means of iontophoresis resulted in clearly labeled fiber processes emanating from the SCN and reaching hypothalamic target sites (Watts and Swanson, 1987; Kalsbeek et al., 1993a, b; Buijs et al., 1994; Dai et al., 1998; Lesauter and Silver, 1999a). Most conspicuously, SCN termination is focused around the paraventricular nucleus (PVN)

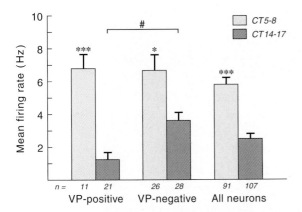

Fig. 2. Shows the mean firing frequency of the number (*n*) of vasopressin (VP) positive, negative, and total of neurons in the dorsomedial SCN (for the area stained for VP cell bodies, see Fig. 2) during circadian time 5–8 (CT 5–8) light bar or CT 14–17 dark bar. A significant difference (***, $p < 0.001$; *, $p < 0.05$, Mann–Whitney *U*-test) is observed in the mean firing frequency between the CT 5–8 and CT 14–17 in all the groups. The difference is twofold in VP negative compared to all neurons and amounted to five- to sixfold in VP-positive neurons. In addition, the VP-positive neurons had a significant lower mean firing rate (CT 14–17) compared to the VP-negative neurons.

of the hypothalamus just rostral of the PVN in the medial preoptic area (MPO); or just below the PVN in the sub-PVN or just caudal of the PVN in the dorsomedial nucleus of the hypothalamus (DMH). In addition, the ventral and dorsal borders of the PVN, i.e. the location of the pre-autonomic neurons are selectively innervated by fibers of the SCN (Buijs et al., 1999). Termination of SCN fibers in and around the arcuate nucleus and in the ventral part of the lateral hypothalamus, suggest an interaction with areas involved in food intake and the organization of activity. Outside the hypothalamus termination of SCN fibers in the lateral geniculate nucleus (LGN) — an area that projects to the SCN — indicates that by this projection at least the SCN may influence also the LGN and thus its own feedback (Buijs et al., 1994; Morin et al., 1994). This function agrees with the possible function of another projection of the SCN outside the hypothalamus, many SCN terminals can be visualized in the dorsal part of the thalamus, in the periventricular area (PVT), which also projects back to the SCN (Moga et al., 1995). Since also these areas are

involved in the organization of locomotor activity, the picture emerges that with the reciprocal interaction with the SCN a stabilizing factor is incorporated that safeguards against sudden arousals, e.g., during sleep but also such that the SCN is informed about motor activity. Indeed, a recent electrophysiological study by the group of Renaud (Zhang et al., 2005) shows that VP — as one of the transmitters of the SCN — reduces the induction of bursting activity of PVT neurons, a terminal zone of the SCN. Arousal and motor activity are associated with such bursting electrical activity of these neurons, while the sleep period of the rat is associated with the inhibited activity of these neurons (McCormick and Bal, 1997). These observations fit perfectly with the observed secretion of VP during the sleep period. Other (electro)physiological and anatomical studies of the connections made by SCN fibers with hypothalamic target structures, show that not only direct monosynaptic contacts exist between SCN neurons and hormonal or pre-autonomic neurons in the hypothalamus, but that hormonal and autonomic output can be affected by multisynaptic contacts such as via (inter)neurons in the DMH or MPO. Interesting in this respect is that the region that encompasses the MPO, sub-PVN, and the DMH, which borders the rostral, ventral, and caudal part of the PVN seems to play an important role in relaying SCN information to the PVN. Several studies have shown that many of these interneurons that project to the PVN contain GABA as neurotransmitter and thus serve an inhibitory role to the PVN (Roland and Sawchenko, 1993; Shekhar and Katner, 1995). This explains why an excitatory transmitter as VP may have an inhibitory role in corticosterone secretion when released into the DMH (Kalsbeek et al., 1996b). It is clear that such inhibitory–excitatory 'switches' offer many possibilities for the systems involved, to adapt when an electrically active SCN does not signal inactivity as in nocturnal rodents but 'activity' as in diurnal mammals, including man. Interestingly, studies that are executed in crepuscular (dusk and dawn active) rodents like 'Arvicanthis' show two corticosterone peaks that both precede the dawn–dusk and dusk–dawn peaks in activity just like observed in rats and humans (Verhagen

346

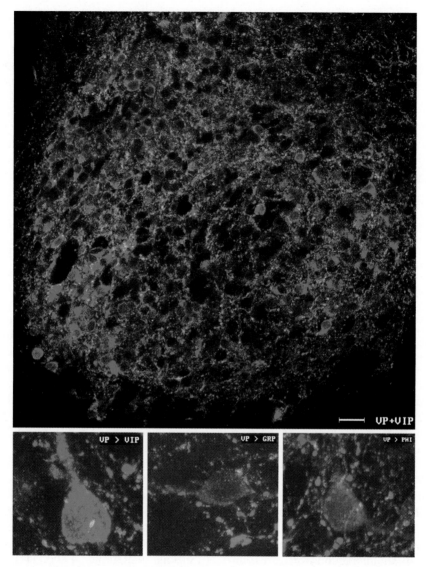

Fig. 3. Two color confocal images of the same optical section (2-μ thick in the top figure) in a double-labeled vibratome section of the rat SCN. Vasopressin-immunoreactive profiles and axons are visible in red, while vasoactive intestinal peptide (VIP) profiles and axonal terminals are visible in green. Clearly both green and red axonal terminations can be seen throughout the whole SCN and seem to terminate on nearly all VIP profiles. This is illustrated more unequivocally in the higher magnifications in the lower part of the figure, where stacks of six optical sections of 1-μm thick and 0.8 μm apart were superimposed. These figures clearly show the innervation of VIP-labeled neurons by vasopressin-labeled axons. These images were used to determine whether a VIP neuron received an input from vasopressin or not and these quantitative data were used to make Table 1. (Bar for top figure = 25 μm, for the lower figure, 8 μm.)

et al., 2004). This offers an interesting picture of the adaptive capacities of the circadian system to the changing environment. Apparently, this circadian system is also able to organize and synchronize two activities and two corticosterone peaks per cycle. In order to understand the mechanisms behind the organization of these peaks, it will be essential to investigate the peaks in secretion of transmitters of the SCN shown to be involved in the control of corticosterone secretion.

Table 1. Interaction between peptides in the SCN

VP$_m$ —o[VIP] VIP —o[VP$_m$] VP$_m$ —o[GRP] GRP—o[VP$_m$] VP$_m$ —o[PHI] PHI —o[VP$_m$]

$\bar{x} = 69 \pm 18$ $\bar{x} = 76 \pm 17$ $\bar{x} = 68 \pm 3$ $\bar{x} = 35 \pm 10$ $\bar{x} = 64 \pm 10$ $\bar{x} = 71 \pm 8$

VP$_m$ —o[Som] Som—o[VP$_m$] Som—o[VIP$_m$] VIP$_m$ —o[Som] Som—o[GRP$_m$] GRP$_m$ —o[Som]

$\bar{x} = 92 \pm 7$ $\bar{x} = 48 \pm 18$ $\bar{x} = 97 \pm 3$ $\bar{x} = 83 \pm 4$ $\bar{x} = 100$ $\bar{x} = 98 \pm 1$

VIP$_m$ —o[GRP]o— VIP$_m$ / GRP GRP—o[VIP$_m$]o— VIP / GRP VIP$_m$ —o[VIP$_m$/GRP]o— GRP

$\bar{x} = 83 \pm 19$ $\bar{x} = 39 \pm 6$ $\bar{x} = 58 \pm 22$ $\bar{x} = 46 \pm 14$ $\bar{x} = 78 \pm 13$ $\bar{x} = 45 \pm 14$

VIP$_m$/GRP

$\bar{x} = 49 \pm 14$

Illustrates the mean percentage ($n = 3$ animals) of cell bodies' immunoreactive for a neuropeptide, but in addition showing the apposition of more than two axonal boutons' immunoreactive for a different peptide. In each animal, at least in three different sections all immunoreactive neurons were counted resulting in more than 150 counted neurons per animal. It is evident that the terminals of VP → VIP are as abundant as the terminals of VIP → VP.

Up till now our studies indicate that the corticosterone peak in rats is organized by a single peak in VP secretion in the DMH and possibly a single peak in a stimulatory transmitter secretion (Kalsbeek et al., 1996b). In another area (MPO), the same peak in VP secretion may result in a facilitation of luteinizing hormone secretion in females (Palm et al., 1999). In yet another area (arcuate nucleus), the SCN may inhibit food intake at the same time (Yi et al., 2005). Consequently, in animals like the Arvicanthis with an organization of their activity in two peaks around the dawn and dusk (also under constant dim light conditions) this pattern of SCN transmitter secretion and the associated activity of the SCN neurons will possibly be completely different and might be organized around two inhibitory or two stimulatory peaks. How such bimodal activity is organized around (or by) the daily rhythms in clock gene expression will need to be investigated.

SCN prepares the body for changes in activity

The influence of the SCN on hormonal secretion seems one of the important routes by which the SCN may affect the body. This conjecture is enforced by the fact that the secretion of several hormones is influenced or even completely regulated (melatonin) by the SCN (Perreau-Lenz et al., 2003). Concerning hormones, such as corticosterone, that are mainly influenced by the SCN, usually the basal secretion follows a circadian pattern. Thus, the rhythmic secretion of corticosterone (cortisol in humans) is primarily driven by the SCN (Bradbury et al., 1991); lesioning the SCN completely removes the daily corticosterone increase, just before the active period. Interestingly, when daily activity is reinstated by, e.g. a timed feeding cycle also a corticosterone rhythm returns (Krieger et al., 1977). This daily rhythm seems to be organized by several brain regions acting together to organize activity and corticosterone secretion Angeles-Castellanos et al., 2004). In addition, it is also clear that disturbing events (stress) that take pace in the environment of the animal may still increase corticosterone in SCN-lesioned animals. In fact, the animal even responds with much higher corticosterone secretion to stress after SCN lesioning indicating that the SCN plays an important role in inhibiting corticosterone secretion (Buijs et al., 1997). These and other observations show that the SCN not only sets the basal secretion of corticosterone, but also that it modulates and inhibits the response of the hypo-thalamo-pituitary-adrenal (HPA) axis to stressful stimuli. Interestingly, evidence indicates that stress such as a new environment also influences the SCN directly and results in a fast but short inhi-bition of the corticosterone levels in the blood af-ter which the levels increase (Buijs et al., 1997).

The SCN may also prepare our body for the upcoming activity by using the autonomic nerv-ous system to sensitize our organs for hormones of which the secretion is also influenced by the SCN. An example is the adrenal which just be-fore the onset of the activity period is more sen-sitive for adreno corticotrophin releasing hormone (ACTH), the result of the action of the SCN on the adrenal is that with the same amount of ACTH the adrenal cortex releases more corticosterone at the end of the sleep period than in the beginning of the sleep period. The mechanism for this increased sensitivity is the sympathetic innervation of the adrenal, which is essential for the circadian vari-ation in corticosterone secretion (Jasper and Engeland, 1994). The possibility of the SCN to influence the adrenal cortex directly was demon-strated by studies using a combination of pseudo rabies virus (PRV) tracing and physiological ex-periments. On the one hand, signals from the SCN may reach the adrenal via multisynaptic pathways including the PVN and the sympathetic motor neurons located in the intermedio lateral (IML) column of the spinal cord. On the other, physio-logical studies in rats showed an immediate reduc-tion of corticosterone secretion after light exposure only in the beginning of the night (ZT 14) but not at the end of the night (ZT 20) and only in SCN intact but not in SCN-lesioned animals. The light-induced fast inhibition of corticosterone secretion and the observation that the change in corticos-terone secretion was obtained without any dis-cernable change in ACTH secretion, both argues for mediation via the sympathetic innervation of the adrenal (Buijs et al., 1999). In fact, recently this was confirmed and extended in similar experiments in mice showing that prolonged stimulation by light induces an increase in corticosterone (Ishida et al., 2005). In (day-active) humans, the opposite pattern was observed; light in the morning in-creased heart rate in humans, as compared to a

decrease in heart rate in nocturnal rodents (Scheer and Buijs, 1999; Scheer et al., 1999, 2001, 2003a). Also these observations fit into the idea that the SCN prepares the individual for the coming ac-tivity period and for the coming sleep period and that light, as the signal of the daytime, promotes activity in man and promotes inactivity in noctur-nal rodents (Scheer et al., 2003b).

Circadian time, signaled by the SCN to other brain structures, is integrated with all other events that influence behavioral or physiological proc-esses. For example, even when fasted for an ex-tended period, the SCN will allow (in fact stimulate) an individual to conserve energy expend-iture during the rest period. It accomplishes that in interaction with downstream centers, e.g., by de-creasing the setpoint for body temperature and promoting sleep. However, prior to the onset of the activity period, the SCN will still initiate all proc-esses to prepare for activity (e.g., increasing core body temperature and increasing glucagon) so that the animal is ready to hunt for food at the end of the sleep again (Sakurada et al., 2000; Ruiter et al., 2003; Scheer et al., 2005). In fact, even during fast-ing, the SCN actively maintains the core body tem-perature during the activity phase at a level close to that during fed conditions. Up till now it is not yet known how the SCN changes the temperature set-ting; however, it seems clear that an interaction with the MPO and its output is essential, because the body temperature of SCN-lesioned animals re-mains quite stable halfway between the high night and the low daytime temperature, whereas lesion-ing of the MPO leads to loss of thermoregulation. Consequently, the presence of the SCN can either increase or decrease the body temperature depend-ing on the time of the day. The temperature reg-istration of animals that are fasted illustrates quite well the functioning of the SCN: in order to pre-pare also food-deprived animals for the upcoming activity period the body temperature rises already long before the animal starts to become active. It is evident that the SCN by its possibility to control the burning of brown and white adipose tissue or by the control of the metabolism of organs like liver and stomach may have an important control over temperature regulation (Amir et al., 1989; Woods and Stock, 1994). In addition, in one way or another, the SCN also needs to control the vasodilatation of the skin, otherwise no coherent control of body temperature is possible (Kräuchi and Wirz-Justice, 1994). The SCN output to pre-autonomic neurons in the hypothalamus and thus its connections with the parasympathetic and sympathetic autonomic motor neurons is the way to control the conservation and generation of energy. Similarly, we suggest that the SCN — probably by the autonomic nervous system — prepares the muscles for the activity period by increasing the sensitivity of the muscles to insulin and thus induces the muscles to have a higher glucose uptake.

In summary, these series of observations have drawn the attention to the capacity of the SCN to change the functionality of our organs not only by the message of hormones but also by affecting the functionality of the organs by the autonomic nervous system. These examples illustrate quite clear one of the main functions of the SCN: preparing the body for the coming activity period. We propose that without this synchronization in physiology we may have a higher chance to develop diabetes and cardiovascular diseases. Support for this proposal can be found in a recent cross-over study, where a group of students were switched from day-time activity with the last meal between 7 and 8 p.m., to night-time activity with the last meal at 11–12 p.m. Already after three weeks, the result was that the night-time active students had much higher insulin and glucose levels throughout the 24 h than the day-time students (Qin et al., 2003). In addition, we would like to propose that to live out of synchrony with our SCN would result in the feeling of continuous jet-lag or possibly depression. The observation that in depressed persons also a diminished activity of the VP cells in the SCN is observed (Zhou et al., 2001) supports this idea and suggests a possible dysfunction of the SCN in depression.

Autonomic control of our organs

Initially, it was assumed that the SCN would affect the body by hormones only and thus would support its effect on the daily sleep/wake cycle. However, early studies by Niijima and Nagai (Niijima et al., 1992) showed that autonomic nerve activity is changed after exposure to light, while this effect is gone after lesioning the SCN, indicating that the light effect on the autonomic nervous system is mediated by the SCN. Consequently, this was the first step to show that the SCN by influencing the output of the autonomic nervous system could also affect our organs. Next, PRV-tracing techniques showed the SCN to be connected with a large variation of organs; e.g. white and brown adipose tissue, the adrenal, the heart, the liver, ovary, the kidney, and the pancreas (Strack et al., 1989; Bamshad et al., 1998; Gerendai et al., 2000; La Fleur et al., 2000; Buijs et al., 2001; Scheer et al., 2001; Kreier et al., 2002). These tracing techniques in combination with selective denervation also showed that in the hypothalamus and SCN both parasympathetic and sympathetic pre-autonomic neurons are differentially connected with these organs of the body (Buijs et al., 2003). These observations warranted the question to what extend the hypothalamus can selectively affect parasympathetic and sympathetic autonomic outputs. In order to study that, we employed virus tracing using different labeled viruses injected in different organs. This required selective denervation of organs since one virus (GFP-labeled) needs to infect the brain via the parasympathetic motor neurons, while from another organ the brain needs to be infected via the sympathetic motor neurons with a B-Gal-labeled virus. Moreover, the infection speed and infection time for both organs needs to be exactly the same while it also should be clear that possible separation of parasympathetic and sympathetic pre-autonomic neurons should not be based on the possibility that different neurons project to different organs. Therefore, we injected fluorogold in the spinal cord IML at the level of the thorax in order to label the majority of the hypothalamic projections to that area. The same animal received an injection of PRV in the sympathetic denervated liver in order to force the virus to infect the brain via the parasympathetic motor neurons. Under these conditions, nowhere in the hypothalamus or brain stem co-localization was found between PRV and fluorogold showing a perfect separation between neurons that project to the vagal motor

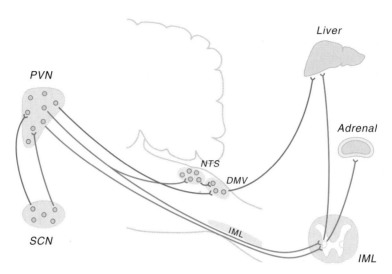

Fig. 4. Scheme of interaction between the hypothalamic suprachiasmatic nucleus (SCN) and paraventricular nucleus (PVN). Separate sympathetic (red) or parasympathetic (blue) neurons of the SCN project to pre-autonomic neurons of the PVN, where a similar sympathetic–parasympathetic separation can be observed. Pre-autonomic neurons of the PVN project either to the pre-ganglionic sympathetic neurons in the intermedio lateral (IML) column of the spinal cord, or to the pre-ganglionic neurons of the dorsal motor nucleus of the vagus (DMV). Moreover, the pre-sympathetic PVN neurons have axon collaterals to pre-parasympathetic neurons, either in the PVN itself, or to the nucleus tractus solitarius (NTS).

neurons or to the sympathetic motor neurons. Only animals of longer survival times in which the infection had progressed until third-order PRV-labeled neurons, showed co-localization of fluoro-gold and PRV in the PVN (Buijs et al., 2003). This latter observation suggests that sympathetic pre-autonomic neurons in the PVN have collaterals projecting to either parasympathetic PVN neurons or to pre-autonomic neurons in the nucleus of the tractus solitarius (NTS) that also project to the dorsal motor nucleus of the vagus (DMV). This possibility is of course functionally very interesting, because it allows a PVN neuron not only to affect the direct output of the sympathetic motor neuron but also influence at the same time the neuron in the NTS that most likely receives information from our peripheral organs and which is able to change the activity of a parasympathetic motor neuron in the DMV. In fact, in this way the PVN sends "a copy" of its signal to the sympathetic motor neurons to the NTS enhancing the integration capacity of this structure (Fig. 4).

In line with the complete separation in the PVN, the SCN also shows a similar specialization of functions and separate parasympathetic or sympathetic-projecting neurons. Some of these neurons even contain the same neurotransmitter (VP) indicating that even the identification of a single transmitter is not sufficient to know what function an SCN neuron might have. Functionally, this separation of parasympathetic and sympathetic neurons in the SCN means that the SCN has the capacity to affect the activity of our organs simultaneously in two different ways. This anatomical framework indeed allows the hypothalamus to balance the incoming information and use it to determine, for example what the most adequate response is in view of circadian time, feeding status, and environment.

In view of the temporally separated control of the secretion of different hormones (e.g., melatonin peaks at night, while cortisol peaks at the start of the activity period), and the electrophysiological demonstration of monosynaptic contacts between SCN and PVN neuroendocrine neurons (Hermes and Renaud, 1993), it is to be expected that connections of the SCN to neuroendocrine centers of the hypothalamus are also physically separated from the autonomic contacts of the SCN (Cui et al., 2002; Wang et al., 2003).

Consequently, a network is revealed that allows the SCN to communicate its time signal to the body by means of at least three different routes: (1) by parasympathetic outflow to the organs, (2) by sympathetic outflow to the organs, and (3) by means of the secretion of hormones into the circulation.

An unbalanced autonomic output; leading to disease?

Until recently, a number of organs such as white adipose tissue were thought to be excluded from parasympathetic input. However, we recently obtained evidence for parasympathetic input to white adipose tissue, not only as visceral organ but also as subcutaneous tissue. We also showed that the parasympathetic input has the function to build up the fat depot, while the sympathetic input serves to burn fat (Kreier et al., 2002). This evidence fits quite well with the observations that exercise enhances sympathetic output to the visceral compartment and results in the diminishment of fat stores there (Bjorntorp, 1983; Thomas et al., 2000). The opposite, a sedentary lifestyle may result in the accumulation of fat due to a higher parasympathetic and a lesser sympathetic outflow, especially to the visceral fat (Kreier et al., 2003a, b). Recently, we have concluded a series of studies that have provided the anatomical framework for this hypothesis. We showed that vagal motor neurons in the brain stem that provide input to the subcutaneous fat are completely separated from those that project to visceral fat, while on the other hand the organs in the visceral compartment, such as the liver, pancreas, and abdominal fat, share the same neurons (Kreier et al., 2005). These observations indicate why an enhanced parasympathetic output to the pancreas after a meal in order to release insulin should also result in an enhanced parasympathetic output to the liver and visceral adipose tissue. The consequence of this shared autonomic output to these different tissues is functionally also logical: not only enhanced levels of insulin from the pancreas will stimulate glucose uptake, but in the liver the increased parasympathetic input will result in higher glucose uptake and higher storage of glycogen. In the visceral

adipose tissue, this combination of enhanced parasympathetic input and elevated insulin levels will also result in a higher uptake of glucose and hence an accumulation of fat. These and other observations stimulated us to propose a hypothesis of autonomic imbalance as one of the possible causes for the metabolic syndrome (Kreier et al., 2003b). We propose that a (disturbed) high parasympathetic output to the visceral compartment is the main cause for visceral obesity, hyperinsulinemia, and high levels of free fatty acids. In addition, a simultaneous higher sympathetic output to the muscle and heart compartments would lead to vasoconstriction and hence to insulin insensitivity and hypertension. We also propose that the major change in lifestyle in the Western world leading to food abundance, inactivity during the active period, and enhanced activity in the rest period (shortened sleep period) may not only affect our daily activity and food pattern, but may also lead to a disturbed balance in the hypothalamus where the biological clock is getting the wrong type of signals across the 24-h period, resulting in general in a flattened rhythm amplitude. Clearly, one of the effects of this sedentary lifestyle combined with enhanced food and carbohydrate intake is an enhanced parasympathetic output to the visceral compartment. One of the major effective treatments of the metabolic syndrome: enhanced activity (during daytime) together with a moderation in food and carbohydrate intake supports this hypothesis, because this behavior results in an increased sympathetic tone to the abdominal compartment (Rosenberg et al., 2005; Slentz et al., 2005) and will also amplify the daily rhythm in the activity/sleep cycle.

Further research in experimental animals aimed at inducing a higher parasympathetic outflow to the visceral compartment and examining whether that may induce the metabolic syndrome may provide further support for this concept.

Another line of evidence that an imbalance in the hypothalamus may lead to disease was reported several years ago when Goncharuk et al. (2001, 2002) showed that hypothalami of persons who died after a long history of hypertension showed profound changes in SCN and PVN. Observations in hypertensive and obese individuals

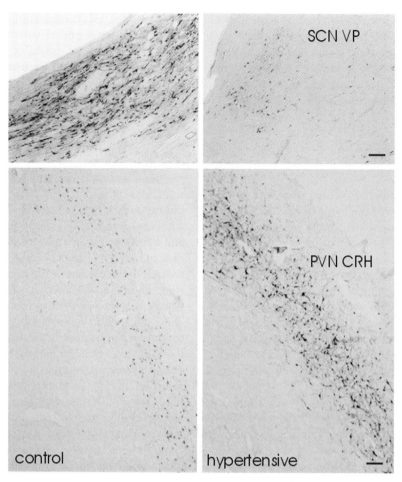

Fig. 5. Coronal sections of the human hypothalamus stained for vasopressin (VP) top two figures (area of the SCN), or for co-rticotrophin releasing hormone (CRH) bottom figures (area of the PVN). The left side of the figure is the same control person, while the right side of the figure is the same hypertensive person. It is evident that while the VP staining in the SCN of the hypertensive person is diminished as compared to the control, the CRH staining is enhanced. (Bar = SCN, 50 μm; PVN, 100 μm.)

who have higher basal corticosterone levels or an enhanced stress response to emotional stimuli (Rosmond et al., 1998; Rosmond, 2003) let us to examine the hypothalamus in people/individuals who died of a cardiovascular incident or brain infarct after a long history of hypertension. These hypothalami were compared to a control group who had a similar death after a sudden cardiovascular incident (either traffic or crime related). This study showed unequivocally a diminishment of the size of the SCN in which the control SCN contained at least two times more VP neurons than the hypertensive SCN. Moreover, coinciding with the diminished SCN activity, the activity of the

corticotrophin releasing hormone (CRH) neurons in the PVN was increased; indicating that similar to the rat, in the human brain also the biological clock may serve to inhibit the activity of the HPA axis (Goncharuk et al., 2001, 2002) (Fig. 5). The main question that remains and needs to be resolved is, whether these observed hypothalamic changes are the cause or consequence of hypertension. Recent studies in which large prospective population studies showed a high correlation between short sleep duration in healthy people and the development of hypertension and diabetes after 10 years, suggest that the SCN might be changed before the development of hypertension

(Gangwisch et al., 2006). If we assume that short sleep in the long term will affect SCN function or that a less strong function of the SCN will allow short sleep, this observation may fit in our hypothesis that hypertension might be caused by a defective biological clock that is less well able to prepare the individual for the upcoming activity period and hence results in cardiovascular problems. Moreover, elevated cortisol levels at early age may predict the development of hypertension in adulthood (Sherwood et al., 1994). Both the sleep shortage and the active HPA axis may indicate a diminished role of the biological clock, although there could be other additional causes. Interestingly, the same studies of Gangwisch also point to a relationship of sleep disturbance and stress with obesity and diabetes (see also Spiegel et al., 2005).

Other studies that show a relationship between changes in the SCN and hypertension are those of spontaneous hypertensive rats (SHR), who not only develop hypertension but also develop enhanced staining of the VIP immunoreactive cells in the SCN, whereas lesioning the SCN results in a diminished development of hypertension (Rosenwasser, 1993; Peters et al., 1994). Several studies show that in humans, especially in the early morning, a high incidence of cardiovascular incidents occur. This not only suggests that the daily transition from the inactivity period to the activity period might be sensitive to malfunction but it may also support our idea that this transition should be prepared by our biological clock. The function of the SCN might be to increase in advance the sympathetic and decrease the parasympathetic tone to the heart, while at the same time also the blood supply to the heart needs to be optimized. Furthermore, if the biological clock is less active or shows less strong amplitude in its output it might prepare us less well for activity. This may, in the long term, lead to disease. Finally our studies, which were aimed to enhance the activity/rest amplitude of the biological clock with an evening dose of melatonin, revealed that such treatment indeed resulted in a diminishment of hypertension (Scheer et al., 2004). The fact that daily exercise also results in a diminishment of hypertension (Stewart, 2002) is an additional argument that

enforcing the amplitude (i.e. a more clear activity–sleep pattern) of the biological clock might be beneficial for hypertension. Of course, such observations raise the question, what exactly would be the consequence of a malfunctioning SCN? We have seen that for a normal function of our physiology it is essential that a large number of organ functions be perfectly synchronized. In view of the data reviewed before on the metabolic syndrome, we would like to extend this to the metabolic syndrome/diabetes and propose the hypothesis that an unbalanced hypothalamus may result in disease.

Input to the biological clock

The previous studies revealed three important elements.

1. A diminished function of the SCN is associated with hypertension and probably diabetes. This diminishment is demonstrated anatomically and/or functionally and might be the result of the change in lifestyle or might be inborn.
2. Therapies need to be developed aimed at restoring this weakened function of the SCN.
3. It needs to be understood how the functionality of the SCN can be affected by a change in lifestyle. Consequently, the way information is transmitted from the periphery to the SCN needs to be investigated.

With the invention of the electric light, the presence of abundance of food, and modern transportation, it was not essential anymore to rest at nightfall and to work during the day; it allowed us to make longer working days, sleep less, eat more, and have a sedentary lifestyle. If one considers that in Western society we now live under conditions in which the daily rhythm of activity and metabolism is flattened, we also have to consider how these flattened rhythms might affect the SCN. We propose that for a normal functioning of the SCN this internal information from the body needs to be transmitted to the SCN. Up till now, however such pathways have not been described in detail and are limited to hypothalamic- or retina-termination sites. In addition, it is not clear whether disturbed information from the periphery or body to the SCN might indeed result in disturbed rhythmicity. In this respect, it is interesting to note that constant light exposure results in a disturbance of rhythmicity (Pickard et al., 1993; Edelstein and Amir, 1999), indicating that indeed an input to the SCN that provides a constant signal instead of a rhythmic signal may disturb the rhythmicity. In addition, such a constant light input to the SCN also diminishes substantially the rhythmicity of the physiology of the animal. We recently investigated this effect of light on the body temperature and heart rate of rats and showed that light not only lowered heart rate but also body temperature. Thus, we constantly exposed rats for a week to an increasing amount of light and investigated their response in activity, heart rate, and body temperature. Increasing the amount of light from 0.1 to 100 lx resulted in a dose-dependent decrease in body temperature and heart rate, even independent of changes in activity, in the subjective (circadian) night, but not during the subjective day (Fig. 6). This observation shows that light not only affects the phase of the SCN when given at the proper time but that light will also affect the output of the SCN such that when it is given continuously it will dampen its amplitude. In addition, we demonstrated that this effect is generated via the SCN since in SCN-lesioned animals no such light-induced effect could be seen. The opposite, i.e. inducing an amplification of rhythmic input to the SCN has also been done both in rats and humans. When aged rats or demented patients, who had flattened amplitude of their rhythm, were exposed to bright light during the light period, a restoration of the amplitude in rhythmicity was the result (Witting et al., 1993). Of course, the mechanisms by which light can influence the SCN are well documented. In addition, similar effects of melatonin are also documented; at least when given just at the beginning of the dark period, also melatonin has a beneficial effect on the amplitude of the melatonin rhythm both in rats and humans (Bothorel et al., 2002). When we consider how this effect of melatonin may work, much evidence indicates that a direct effect of melatonin on the SCN might be at the basis of these observations.

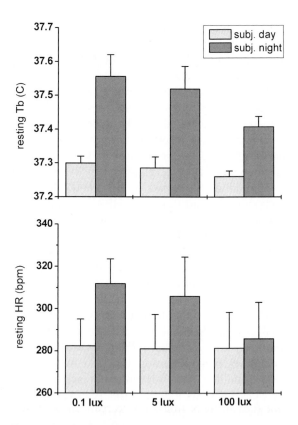

Fig. 6. Light inhibits Body temperature (Tb) and Heart rate (HR) during the subjective night, independent of changes in Locomotor activity (LA). Resting Tb and HR were determined during three 10 day constant light episodes of 0.1, 5, and 100 lx, respectively, in intact animals ($n = 3$). Resting HR during the subjective night period was significantly lower during 100 lx exposure, as compared to exposure to 0.1 and 5 lx ($p < 0.05$; t-test-dependent samples). Such effects of light on HR and Tb are compromised in SCNx animals (Scheer, 2003).

This is supported by the observation that nighttime melatonin administered locally to the SCN (via reversed microdialysis) amplifies the circadian rhythm in melatonin release, even days after discontinuation of the administration (Bothorel et al., 2002). Furthermore, melatonin receptors have been demonstrated in the SCN and 'in vitro' studies have shown that melatonin acts directly on the SCN to reduce electrical activity of the neurons (Gauer et al., 1993; Van den Top et al., 2002). Many studies have documented how light–dark-related signals may affect the SCN. Less clear though, is how behavioral activity or metabolic information can affect the functioning of the SCN.

Studies of Kanosue and his group have shown that changes in the metabolic state of an animal resulted profound changes in the circadian amplitude in the rhythm of activity and temperature in an animal. Fasting rats just for 24 h, already resulted in an increase in the amplitude of their activity pattern with a diminished activity during the inactive period and an enhanced activity in the active period (Sakurada et al., 2000). In an effort to investigate the role of the SCN in these activity changes, we not only recorded activity and temperature but also heart rate in intact and SCN-lesioned rats that were fasted for 24–72 h (Scheer, 2002). The results clearly show that fasting results in an immediate decrease in body temperature during the inactivity period following the nightly fast. SCN lesioning results in HR and BT levels in between the daily trough and nightly peak values suggesting that the SCN may increase as well as decrease BT and HR. In order to examine how light input may change the output of the SCN, animals were exposed to 1 h light during the dark period. This hour light exposure during the dark period resulted in an immediate but time-dependent (i.e. an SCN function) decrease in body temperature.

These observations demonstrate (1) metabolic information interacts with or affects the output of the SCN; (2) the SCN is essential for the amplification of the circadian changes in temperature preceding activity; and (3) the SCN mediates the increase in BT just before the onset of the activity period, which is essential for the preparation of the activity. All these events are more pronounced under metabolic-restricted conditions. It seems logical to assume that under conditions of metabolic surplus (as in obesity) the amplitude of circadian rhythms will be depressed, either by a direct action of metabolic substances on the SCN or by interacting with the target areas of the SCN. Studies on obese, diabetic, and hypertensive subjects seem to support this. These individuals have diminished amplitude in their day–night rhythm in autonomic activity (Kondo et al., 2002). All these studies imply an important role for the integration of metabolic information in the SCN, the question remains however what are the pathways for this metabolic information to reach the SCN?

Transmission of metabolic information to the SCN

There are two major ways by which metabolic information may reach the SCN, one is by the autonomic nervous feedback from our organs, and the other is by hormonal feedback or by that of metabolites. Up till now, there is hardly anything known about these two types of feedback to the SCN. From the sites where visceral sympathetic information enters the brain (the dorsal horn) and visceral parasympathetic information enters the brain (the NTS), no projections are known to the SCN; so if any autonomic information is transmitted to the SCN it must be by routes involving more than one synapse. This seems to be different for information circulating in the blood stream. Areas where metabolites can directly reach receptors on neurons are those regions where the blood–brain barrier is absent, such as in the circumventricular organs. There are four sites in the brain that function as circumventricular organs. The organum vasculosum of the lamina terminalis (OVLT), the subfornical organ (SFO), the area postrema (AP), and the ventromedial arcuate nucleus-median eminence complex (AMC) (Cottrell et al., 2004; Yi et al., 2005). Of two of those areas it has been known for a long time that they project to the SCN, i.e. the OVLT and the SFO (Miselis, 1981; Trudel and Bourque, 2003; Cottrell et al., 2004; Yi et al., 2005). Since current evidence indicates that both areas are mainly involved in the sensing of electrolytes in the general circulation, we assume that this will be the message that is conveyed to the SCN. Remarkably, especially the AMC and also the AP are both known for their role in metabolic functions. Consequently, since the ventromedial arcuate nucleus (vmARC) is the site in the hypothalamus where information from the circulation may be transferred to neurons, we investigated the anatomical connections of the vmARC (Yi et al., 2005). Injection of the anterograde and retrograde tracers Cholera toxin B (CTB) into the vmARC resulted in the visualization of an elaborate network of projections to many targets within and outside the hypothalamus. The observation of a dense reciprocal interaction between the vmARC and the SCN provided the anatomical basis for the so much looked for link between circulating metabolic information and the SCN. Consequently, this anatomical connection between AMC and SCN may form the basis upon which the SCN is informed about circulating glucose levels and the AMC about the time of the day, which is then linked with the appropriate glucose levels. This anatomical interaction may explain why insulin-induced hypoglycemia does not result in corticosterone secretion in the beginning of the light period, while it results in high corticosterone values at the end of the light (Kalsbeek et al., 2003). This conclusion is emphasized by the observation that in rats the mimetic of the metabolic hormone ghrelin, GHRP-6, not only results in activation of the vmARC but at the same time results in the inhibition of the activity of the SCN. Our observation that the subependymal layer in the median eminence forms a continuum with the vmARC only emphasizes the role of this area as an AMC, which already several decades ago was shown to be crucial for the exchange of information from the blood stream to the neurons (Broadwell and Brightman, 1976; Spanswick et al., 1997). Since glucose-sensing neurons form an essential part of this structure (Spanswick et al., 1997), it is only logical that information about glucose levels in the blood may be communicated via the AMC projections to all its target structures including the SCN. Since the arcuate nucleus also serves as a focus for the transmission of hormonal information to the hypothalamus, we propose that also the AMC serves to transmit circulating hormonal information to the SCN. This can be deduced from our observations on the effect of GHRP-6 on SCN fos expression and from the observations of Coppari et al. (2005), who showed that restoring leptin sensing in a leptin-R KO animal just unilaterally in the vmARC not only resulted in a restoration of metabolic balance, but also restored the disturbed circadian rhythmicity of the animal.

The fact that AMC not only targets all other circumventricular organs but that it also receives input from all other circumventricular organs emphasizes the importance of these areas in integrating circulating substances and hormones and transmitting the resultant signal to the rest of the brain.

Conclusions

We have reviewed evidence that the SCN is not only involved in the organization of the physiology of the body in association with the light dark cycle, but that the body also communicates back to the SCN. Hereto the SCN also receives information from the circulation. The observation that in diseases such as diabetes and hypertension, a flattened rhythm is observed in autonomic parameters together with a decrease in activity of the SCN suggests that the biological clock may play an important role in the etiology of these diseases. These observations fit in with the results of successful 'lifestyle' treatments aimed at curing diabetes and hypertension. For both diseases and in particular when these diseases are associated with visceral obesity i.e. "the syndrome X," it holds that daily exercise associated with a restoration of a day--night activity-sleep rhythm is beneficial. Consequently, we can see the interaction of the SCN with the body as a closed circle in which changes at any part of this circuit will result in changes in functions either of the body or the biological clock.

References

Amir, S., Shizgal, P. and Rompre, P.P. (1989) Glutamate injection into the suprachiasmatic nucleus stimulates brown fat thermogenesis in the rat. Brain Res., 498: 140–144.

Angeles-Castellanos, M., Aguilar-Roblero, R. and Escobar, C. (2004) c-Fos expression in hypothalamic nuclei of food-entrained rats. Am. J. Physiol. Regul. Integr. Comp. Physiol., 286: R158–R165.

Bamshad, M., Aoki, V.T., Adkison, M.G., Warren, W.S. and Bartness, T.J. (1998) Central nervous system origins of the sympathetic nervous system outflow to white adipose tissue. Am. J. Physiol., 275: R291–R299.

Bjorntorp, P. (1983) Physiological and clinical aspects of exercise in obese persons. Exerc. Sport Sci. Rev., 11: 159–180.

Bos, N.P.A. and Mirmiran, M. (1990) Circadian rhythms in spontaneous neuronal discharges of the cultured suprachiasmatic nucleus. Brain Res., 511: 158–162.

Bos, N.P.A. and Mirmiran, M. (1993) Effects of excitatory and inhibitory amino acids on neuronal discharges in the cultured suprachiasmatic nucleus. Brain Res. Bull., 31: 67–72.

Bothorel, B., Barassin, S., Saboureau, M., Perreau, S., Vivien-Roels, B., Malan, A. and Pévet, P. (2002) In the rat, exogenous melatonin increases the amplitude of pineal melatonin secretion by a direct action on the circadian clock. Eur. J. Neurosci., 16: 1090–1098.

Bradbury, M.J., Cascio, C.S., Scribner, K.A. and Dallman, M.F. (1991) Stress-induced adrenocorticotropin secretion: diurnal responses and decreases during stress in the evening are not dependent on corticosterone. Endocrinology, 128: 680–688.

Broadwell, R.D. and Brightman, M.W. (1976) Entry of peroxidase into neurons of the central and peripheral nervous systems from extracerebral and cerebral blood. J. Comp. Neurol., 166: 257–283.

Buijs, R.M., Chun, S.J., Niijima, A., Romijn, H.J. and Nagai, K. (2001) Parasympathetic and sympathetic control of the pancreas: a role for the suprachiasmatic nucleus and other hypothalamic centers that are involved in the regulation of food intake. J. Comp. Neurol., 431: 405–423.

Buijs, R.M., Hou, Y.X., Shinn, S. and Renaud, L.P. (1994) Ultrastructural evidence for intra- and extranuclear projections of GABAergic neurons of the suprachiasmatic nucleus. J. Comp. Neurol., 335: 42–54.

Buijs, R.M., La Fleur, S.E., Wortel, J., Van Heyningen, C., Zuiddam, L., Mettenleiter, T.C., Kalsbeek, A., Nagai, K. and Niijima, A. (2003) The suprachiasmatic nucleus balances sympathetic and parasympathetic output to peripheral organs through separate preautonomic neurons. J. Comp. Neurol., 464: 36–48.

Buijs, R.M., Wortel, J. and Hou, Y.X. (1995) Colocalization of gamma-aminobutyric acid with vasopressin, vasoactive intestinal peptide, and somatostatin in the rat suprachiasmatic nucleus. J. Comp. Neurol., 358: 343–352.

Buijs, R.M., Wortel, J., Van Heerikhuize, J.J., Feenstra, M.G., Ter Horst, G.J., Romijn, H.J. and Kalsbeek, A. (1999) Anatomical and functional demonstration of a multisynaptic suprachiasmatic nucleus adrenal (cortex) pathway. Eur. J. Neurosci., 11: 1535–1544.

Buijs, R.M., Wortel, J., Vanheerikhuize, J.J. and Kalsbeek, A. (1997) Novel environment induced inhibition of corticosterone secretion: physiological evidence for a suprachiasmatic nucleus mediated neuronal hypothalamo-adrenal cortex pathway. Brain Res., 758: 229–236.

Coppari, R., Ichinose, M., Lee, C.E., Pullen, A.E., Kenny, C.D., McGovern, R.A., Tang, V., Liu, S.M., Ludwig, T., Chua Jr., S.C., Lowell, B.B. and Elmquist, J.K. (2005) The hypothalamic arcuate nucleus: a key site for mediating leptin's effects on glucose homeostasis and locomotor activity. Cell Metab., 1: 63–72.

Cottrell, G.T., Zhou, Q.Y. and Ferguson, A.V. (2004) Prokineticin 2 modulates the excitability of subfornical organ neurons. J. Neurosci., 24: 2375–2379.

Cropper, E.C.M., Miller, M.W., Villm, F.S., Tenenbaum, R., Kupferman, I. and Weiss, K.R. (1990) Release of peptidecotransmitters from a cholinergic motor neuron under physiological conditions. Proc. Natl. Acad. Sci. USA, 87: 933–937.

Cui, L.N., Coderre, E. and Renaud, L.P. (2002) Glutamate and GABA mediate suprachiasmatic nucleus inputs to spinal-projecting paraventricular neurons. Am. J. Physiol. Regul. Integr. Comp. Physiol., 281: R1283–R1289.

Dai, J.P., Vandervliet, J., Swaab, D.F. and Buijs, R.M. (1998) Postmortem tracing reveals the organization of hypothalamic projections of the suprachiasmatic nucleus in the human brain. J. Comp. Neurol., 400: 87–102.

Edelstein, K. and Amir, S. (1999) The intergeniculate leaflet does not mediate the disruptive effects of constant light on circadian rhythms in the rat. Neuroscience, 90: 1093–1101.

Gangwisch, J.E., Boden-Albala, B., Buijs, R.M., Kreier, F., Pickering, T.G., Rundle, A.G., Zammit, G.K. and Malaspina, D. (2006) Short sleep duration as risk factor for hypertension: analysis of the NHANES I. Hypertension, 47: 833–839.

Gauer, F., Masson-Pévet, M., Skene, D.J., Vivien-Roels, B. and Pévet, P. (1993) Daily rhythms of melatonin-binding sites in the rat pars tuberalis and suprachiasmatic nuclei — evidence for a regulation of melatonin receptors by melatonin itself. Neuroendocrinology, 57: 120–126.

Gerendai, I., Toth, I.E., Boldogkoi, Z., Medveczky, I. and Halasz, B. (2000) CNS structures presumably involved in vagal control of ovarian function. J. Auton. Nerv. Syst., 80: 40–45.

Gillette, M.U. and Reppert, S.M. (1987) The hypothalamic suprachiasmatic nuclei: circadian patterns of vasopressin secretion and neuronal activity in vitro. Brain Res. Bull., 19: 135–139.

Goncharuk, V.D., Van Heerikhuize, J., Dai, J.P., Swaab, D.F. and Buijs, R.M. (2001) Neuropeptide changes in the suprachiasmatic nucleus in primary hypertension indicate functional impairment of the biological clock. J. Comp. Neurol., 431: 320–330.

Goncharuk, V.D., Van Heerikhuize, J., Swaab, D.F. and Buijs, R.M. (2002) Paraventricular nucleus of the human hypothalamus in primary hypertension: activation of corticotropin-releasing hormone neurons. J. Comp. Neurol., 443: 321–331.

Groos, G., Mason, R. and Meijer, J. (1983) Electrical and pharmacological properties of the suprachiasmatic nuclei. Fed. Proc., 42: 2790–2795.

Hermes, M.L.H.J., Coderre, E.M., Buijs, R.M. and Renaud, L.P. (1996) GABA and glutamate mediate rapid neurotransmission from suprachiasmatic nucleus to hypothalamic paraventricular nucleus in rat. J. Physiol. London, 496: 749–757.

Hermes, M.L.H.J. and Renaud, L.P. (1993) Differential responses of identified rat paraventricularneurons to suprachiasmatic nucleus stimulation. Neuroscience, 56: 823–832.

Ishida, A., Mutoh, T., Ueyama, T., Bando, H., Masubuchi, S., Nakahara, D., Tsujimoto, G. and Okamura, H. (2005) Light activates the adrenal gland: timing of gene expression and glucocorticoid release. Cell Metab., 2: 297–307.

Jasper, M.S. and Engeland, W.C. (1994) Splanchnic neural activity modulates ultradian and circadian rhythms in adrenocortical secretion in awake rats. Neuroendocrinology, 59: 97–109.

Kalsbeek, A., Ruiter, M., La Fleur, S.E., Van Heijningen, C. and Buijs, R.M. (2003) The diurnal modulation of hormonal responses in the rat varies with different stimuli. J. Neuroendocrinol., 15: 1144–1155.

Kalsbeek, A., Teclemariam-Mesbah, R. and Pévet, P. (1993a) Efferent projections of the suprachiasmatic nucleus in the golden hamster (Mesocricetus-auratus). J. Comp. Neurol., 332: 293–314.

Kalsbeek, A., Teclemariam-Mesbah, R. and Pévet, P. (1993b) Efferent projections of the suprachiasmatic nucleus in the golden hamster (*Mesocricetus-auratus*). J. Comp. Neurol., 332: 293–314.

Kalsbeek, A., Van der Vliet, J. and Buijs, R.M. (1996a) Decrease of endogenous vasopressin release necessary for expression of the circadian rise in plasma corticosterone: a reverse microdialysis study. J. Neuroendocrinol., 8: 299–307.

Kalsbeek, A., Van Heerikhuize, J.J., Wortel, J. and Buijs, R.M. (1996b) A diurnal rhythm of stimulatory input to the hypothalamo-pituitary-adrenal system as revealed by timed intrahypothalamic administration of the vasopressin V-1 antagonist. J. Neurosci., 16: 5555–5565.

Kondo, K., Matsubara, T., Nakamura, J. and Hotta, N. (2002) Characteristic patterns of circadian variation in plasma catecholamine levels, blood pressure and heart rate variability in type 2 diabetic patients. Diabet. Med., 19: 359–365.

Kow, L.-M. and Pfaff, D.W. (1984) Suprachiasmatic neurons in tissue slices from ovariectomized rats: electrophysiological and neuropharmacological characterization and the effects of estrogen treatment. Brain Res., 297: 275–286.

Krauchi, K. and Wirz-Justice, A. (1994) Circadian rhythm of heat production, heart rate, and skin and core temperature under unmasking conditions in men. Am. J. Physiol., 267: R819–R829.

Kraves, S. and Weitz, C.J. (2006) A role for cardiotrophin-like cytokine in the circadian control of mammalian locomotor activity. Nat. Neurosci., 9: 212–219.

Kreier, F., Fliers, E., Voshol, P.J., Van Eden, C.G., Havekes, L.M., Kalsbeek, A., Van Heijningen, C.L., Sluiter, A.A., Mettenleiter, T.C., Romijn, J.A., Sauerwein, H.P. and Buijs, R.M. (2002) Selective parasympathetic innervation of subcutaneous and intra-abdominal fat — functional implications. J. Clin. Invest., 110: 1243–1250.

Kreier, F., Kalsbeek, A., Ruiter, M., Yilmaz, A., Romijn, J.A., Sauerwein, H.P., Fliers, E. and Buijs, R.M. (2003a) Central nervous determination of food storage — a daily switch from conservation to expenditure: implications for the metabolic syndrome. Eur. J. Pharmacol., 480: 51–65.

Kreier, F., Kap, Y.S., Mettenleiter, T.C., Van Heijningen, C., Van der, Vliet., Kalsbeek, A., Sauerwein, H.P., Fliers, E., Romijn, J.A. and Buijs, R.M. (2005) Tracing from fat tissue, liver and pancreas: A neuroanatomical framework for the role of the brain in type 2 diabetes. Endocrinology., 147: 1140–1147.

Kreier, F., Yilmaz, A., Kalsbeek, A., Romijn, J.A., Sauerwein, H.P., Fliers, E. and Buijs, R.M. (2003b) Hypothesis: shifting the equilibrium from activity to food leads to autonomic unbalance and the metabolic syndrome. Diabetes, 52: 2652–2656.

Krieger, D.T., Hauser, H. and Krey, L.C. (1977) Suprachiasmatic nuclear lesions do not abolish food-shifted circadian adrenal and temperature rhythmicity. Science, 197: 398–399.

Kriegsfeld, L.J., Lesauter, J. and Silver, R. (2004) Targeted microlesions reveal novel organization of the hamster suprachiasmatic nucleus. J. Neurosci., 24: 2449–2457.

La Fleur, S.E., Kalsbeek, A., Wortel, J. and Buijs, R.M. (2000) Polysynaptic neural pathways between the hypothalamus, including the suprachiasmatic nucleus, and the liver. Brain Res., 871: 50–56.

Leak, R.K. and Moore, R.Y. (2001) Topographic organization of suprachiasmatic nucleus projection neurons. J. Comp. Neurol., 433: 312–334.

Lesauter, J. and Silver, R. (1999a) Localization of a suprachiasmatic nucleus subregion regulating locomotor rhythmicity. J. Neurosci., 19: 5574–5585.

Lesauter, J. and Silver, R. (1999b) Localization of a suprachiasmatic nucleus subregion regulating locomotor rhythmicity. J. Neurosci., 19: 5574–5585.

Lundberg, J.M., Änggård, A. and Fahrenkrug, J. (1981) Complementary role of vasoactive intestinal polypeptide (VIP) and acetylcholine for cat submandibular gland blood flow and secretion II. Effects of cholinergic antagonists and VIP antiserum. Acta Physiol. Scand., 113: 329–336.

Lundberg, J.M., Francocereceda, A., Lou, Y.P., Modin, A. and Pernow, J. (1994) Differential release of classical transmitters and peptides. Mol. Cell Mech. Neurotrans., 41: R29–R234.

McCormick, D.A. and Bal, T. (1997) Sleep and arousal: thalamocortical mechanisms. Annu. Rev. Neurosci., 20: 185–215.

Meijer, J.H., Rusak, B. and Gänshirt, G. (1992) The relation between light-induced discharge in the suprachiasmatic nucleus and phase shifts of hamster circadian rhythms. Brain Res., 598: 257–263.

Meyer-Bernstein, E.L., Jetton, A.E., Matsumoto, S.I., Markuns, J.F., Lehman, M.N. and Bittman, E.L. (1999) Effects of suprachiasmatic transplants on circadian rhythms of neuroendocrine function in golden hamsters. Endocrinology, 140: 207–218.

Mirmiran, M.M., Koster-van Hoffen, G.C. and Bos, N.P.A. (1995) Circadian rhythm generation in the cultured suprachiasmatic nucleus. Brain Res. Bull., 38: 275–283.

Miselis, R.R. (1981) The efferent projections of the subfornical organ of the rat: a circumventricular organ within a neural network subserving water balance. Brain Res., 230: 1–23.

Moga, M.M., Weis, R.P. and Moore, R.Y. (1995) Efferent projections of the paraventricular thalamic nucleus in the rat. J. Comp. Neurol., 359: 221–238.

Moore, R.Y. (1996) Entrainment pathways and the functional organization of the circadian system. Prog. Brain Res., 111: 103–119.

Morin, L.P., Goodless-Sanchez, N., Smale, L. and Moore, R.Y. (1994) Projections of the suprachiasmatic nuclei, subparaventricular zone and retrochiasmatic area in the golden hamster. Neuroscience, 61: 391–410.

Niijima, A., Nagai, K., Nagai, N. and Nakagawa, H. (1992) Light enhances sympathetic and suppresses vagal outflows and lesions including the suprachiasmatic nucleus eliminate these changes in rats. J. Auton. Nerv. Syst., 40: 155–160.

Palm, I.F., Van Der Beek, E.M., Wiegant, V.M., Buijs, R.M. and Kalsbeek, A. (1999) Vasopressin induces a luteinizing hormone surge in ovariectomized, estradiol-treated rats with lesions of the suprachiasmatic nucleus. Neuroscience, 93: 659–666.

Perreau-Lenz, S., Kalsbeek, A., Garidou, M.L., Wortel, J., Van der Vliet, J., Van Heijningen, C., Simonneaux, V., Pévet, P. and Buijs, R.M. (2003) Suprachiasmatic control of melatonin synthesis in rats: inhibitory and stimulatory mechanisms. Eur. J. Neurosci., 17: 221–228.

Perreau-Lenz, S., Kalsbeek, A., Pévet, P. and Buijs, R.M. (2004) Glutamatergic clock output stimulates melatonin synthesis at night. Eur. J. Neurosci., 19: 318–324.

Peters, R.V., Zoeller, R.T., Hennessey, A.C., Stopa, E.G., Anderson, G. and Albers, H.E. (1994) The control of circadian rhythms and the levels of vasoactive intestinal peptide messenger RNA in the suprachiasmatic nucleus are altered in spontaneously hypertensive rats. Brain Res., 639: 217–227.

Pickard, G.E., Turek, F.W. and Sollars, P.J. (1993) Light intensity and splitting in the golden hamster. Physiol. Behav., 54: 1–5.

Qin, L.Q., Li, J., Wang, Y., Wang, J., Xu, J.Y. and Kaneko, T. (2003) The effects of nocturnal life on endocrine circadian patterns in healthy adults. Life Sci., 73: 2467–2475.

Roland, B.L. and Sawchenko, P.E. (1993) Local origins of some GABAergic projections to the paraventricular and supraoptic nuclei of the hypothalamus in the rat. J. Comp. Neurol., 332: 123–143.

Romijn, H.J., Sluiter, A.A., Pool, C.W., Wortel, J. and Buijs, R.M. (1997) Evidence from confocal fluorescence microscopy for a dense, reciprocal innervation between AVP-, somatostatin-, VIP/PHI-, GRP-, and VIP/PHI/GRP-immunoreactive neurons in the rat suprachiasmatic nucleus. Eur. J. Neurosci., 9: 2613–2623.

Rosenberg, D.E., Jabbour, S.A. and Goldstein, B.J. (2005) Insulin resistance, diabetes and cardiovascular risk: approaches to treatment. Diabetes Obes. Metab., 7: 642–653.

Rosenwasser, A.M. (1993) Circadian drinking rhythms in SHR and WKY rats — effects of increasing light intensity. Physiol. Behav., 53: 1035–1041.

Rosmond, R. (2003) Stress induced disturbances of the HPA axis: a pathway to type 2 diabetes? Med. Sci. Monit., 9: RA35–RA39.

Rosmond, R., Dallman, M.F. and Bjorntorp, P. (1998) Stress-related cortisol secretion in men: relationships with abdominal obesity and endocrine, metabolic and hemodynamic abnormalities. J. Clin. Endocrinol. Metab., 83: 1853–1859.

Ruiter, M., La Fleur, S.E., Van Heijningen, C., Van der Vliet, J., Kalsbeek, A. and Buijs, R.M. (2003) The daily rhythm in plasma glucagon concentrations in the rat is modulated by the biological clock and by feeding behavior. Diabetes, 52: 1709–1715.

Sakurada, S., Shido, O., Sugimoto, N., Hiratsuka, Y., Yoda, T. and Kanosue, K. (2000) Autonomic and behavioural thermoregulation in starved rats. J. Physiol., 526: 417–424.

Scheer, F.A. (2003) Cardiovascular control by the biological clock. PhD Thesis, University of Amsterdam, Amsterdam, The Netherlands.

Scheer, F.A. and Buijs, R.M. (1999) Light affects morning salivary cortisol in humans. J. Clin. Endocrinol. Metab., 84: 3395–3398.

Scheer, F.A., Kalsbeek, A. and Buijs, R.M. (2003a) Cardiovascular control by the suprachiasmatic nucleus: neural and neuroendocrine mechanisms in human and rat. Biol. Chem., 384: 697–709.

Scheer, F.A., Kalsbeek, A. and Buijs, R.M. (2003b) Cardiovascular control by the suprachiasmatic nucleus: neural and neuroendocrine mechanisms in human and rat. Biol. Chem., 384: 697–709.

Scheer, F.A., Pirovano, C., van Someren, E.J. and Buijs, R.M. (2005) Environmental light and suprachiasmatic nucleus interact in the regulation of body temperature. Neuroscience, 132: 465–477.

Scheer, F.A., Ter Horst, G.J., Van der Vliet, J. and Buijs, R.M. (2001) Physiological and anatomic evidence for regulation of the heart by suprachiasmatic nucleus in rats. Am. J. Physiol., 280: H1391–H1399.

Scheer, F.A., van Doornen, L.J. and Buijs, R.M. (1999) Light and diurnal cycle affect human heart rate: possible role for the circadian pacemaker. J. Biol. Rhythms, 14: 202–212.

Scheer, F.A., Van Paassen B, van Montfrans G.A., Fliers E, van Someren E.J., Van Heerikhuize J.J. and Buijs R.M. (2002) Human basal cortisol levels are increased in hospital compared to home setting. Neurosci Lett 333:79–82.

Scheer, F.A., van Montfrans, G.A., van Somerenm, J., Mairuhum, G. and Buijs, R.M. (2004) Daily nighttime melatonin reduces blood pressure in male patients with essential hypertension. Hypertension, 43: 192–197.

Shekhar, A. and Katner, J.S. (1995) Dorsomedial hypothalamic GABA regulates anxiety in the social interaction test. Pharmacol. Biochem. Behav., 50: 253–258.

Sherwood, A., Hinderliter, A.L. and Light, K.C. (1994) Physiological determinants of hyperreactivity to stress in borderline hypertension. Hypertension, 25: 384–390.

Silver, R., Lesauter, J., Tresco, P.A. and Lehman, M.N. (1996) A diffusible coupling signal from the transplanted suprachiasmatic nucleus controlling circadian locomotor rhythms. Nature, 382: 810–813.

Slentz, C.A., Aiken, L.B., Houmard, J.A., Bales, C.W., Johnson, J.L., Tanner, C.J., Duscha, B.D. and Kraus, W.E. (2005) Inactivity, exercise, and visceral fat STRRIDE: a randomized, controlled study of exercise intensity and amount. J. Appl. Physiol., 99: 1613–1618.

Spanswick, D., Smith, M.A., Groppi, V.E., Logan, S.D. and Ashford, M.L. (1997) Leptin inhibits hypothalamic neurons by activation of ATP-sensitive potassium channels. Nature, 390: 521–525.

Spiegel, K., Knutson, K., Leproult, R., Tasali, E. and Cauter, E.V. (2005) Sleep loss: a novel risk factor for insulin resistance and type 2 diabetes. J. Appl. Physiol., 99: 2008–2019.

Stewart, K.J. (2002) Exercise training and the cardiovascular consequences of type 2 diabetes and hypertension: plausible mechanisms for improving cardiovascular health. JAMA, 288: 1622–1631.

Strack, A.M., Sawyer, W.B., Platt, K.B. and Loewy, A.D. (1989) CNS cell groups regulating the sympathetic outflow to adrenal gland as revealed by transneuronal cell body labeling with pseudorabies virus. Brain Res., 491: 274–296.

Ter Horst, G.J. and Luiten, P.G.M. (1986) The projections of the dorsomedial hypothalamic nucleus in the rat. Brain Res. Bull., 16: 231–248.

Thomas, E.L., Brynes, A.E., McCarthy, J., Goldstone, A.P., Hajnal, J.V., Saeed, N., Frost, G. and Bell, J.D. (2000) Preferential loss of visceral fat following aerobic exercise, measured by magnetic resonance imaging. Lipids, 35: 769–776.

Trudel, E. and Bourque, C.W. (2003) A rat brain slice preserving synaptic connections between neurons of the suprachiasmatic nucleus, organum vasculosum lamina terminalis and supraoptic nucleus. J. Neurosci. Meth., 128: 67–77.

Van den Top, M., Buijs, R.M., Ruijter, J.M., Delagrange, P., Spanswick, D. and Hermes, M.L. (2002) Melatonin generates an outward potassium current in rat suprachiasmatic nucleus neurones in vitro independent of their circadian rhythm. Neuroscience, 107: 99–108.

Verhagen, L.A., Pevet, P., Saboureau, M., Sicard, B., Nesme, B., Claustrat, B., Buijs, R.M. and Kalsbeek, A. (2004) Temporal organization of the 24-h corticosterone rhythm in the diurnal murid rodent Arvicanthis ansorgei Thomas 1910. Brain Res., 995: 197–204.

Wang, D., Cui, L. and Renaud, L.P. (2003) Pre- and postsynaptic gaba(b) receptors modulate rapid neurotransmission from suprachiasmatic nucleus to parvocellular hypothalamic paraventricular nucleus neurons. Neuroscience, 118: 49–58.

Watts, A.G. and Swanson, L.W. (1987) Efferent projections of the suprachiasmatic nucleus: II. Studies using retrograde transport of fluorescent dyes and simultaneous peptide immunohistochemistry in the rat. J. Comp. Neurol., 258: 230–252.

Whim, M.D. and Lloyd, P.E. (1989) Frequency-dependent release of peptide cotransmitters from identified cholinergic motor neurons in Aplysia. Proc. Natl. Acad. Sci. USA, 86: 9034–9038.

Witting, W., Mirmiran, M., Bos, N.P.A. and Swaab, D.F. (1993) Effect of light intensity on diurnal sleep–wake distribution in young and old rats. Brain Res. Bull., 30: 157–162.

Woods, A.J. and Stock, M.J. (1994) Biphasic brown fat temperature responses to hypothalamic stimulation in rats. Am. J. Physiol., 266: R328–R337.

Xie, J., Price, M.P., Wemmie, J.A., Askwith, C.C. and Welsh, M.J. (2003) ASIC3 and ASIC1 mediate FMRFamide-related peptide enhancement of H+-gated currents in cultured dorsal root ganglion neurons. J. Neurophysiol., 89: 2459–2465.

Yi, C.X., Van der, Vliet., Dai J, Yin G., Ru L, and Buijs R.M. (2005) Ventromedial arcuate nucleus communicates peripheral metabolic information to the suprachiasmatic nucleus. Endocrinology, 147: 283–294.

Yi, C.X., Van der, Vliet., Dai J, Yin G., Ru L, and Buijs R.M. (2006) Ventromedial arcuate nucleus communicates peripheral metabolic information to the suprachiasmatic nucleus. Endocrinology, 147: 283–294.

Zhang, L., Doroshenko, P., Cao, X.Y., Irfan, N., Coderre, E., Kolaj, M. and Renaud, L.P. (2005) Vasopressin induces depolarization and state-dependent firing patterns in rat thalamic paraventricular nucleus neurons in vitro. Am. J. Physiol. Regul. Integr. Comp. Physiol. (Epub. ahead of print).

Zhou, J.N., Riemersma, R.F., Unmehopa, U.A., Hoogendijk, W.J., Van Heerikhuize, J.J., Hofman, M.A. and Swaab, D.F. (2001) Alterations in arginine vasopressin neurons in the suprachiasmatic nucleus in depression. Arch. Gen. Psychiatry, 58: 655–658.

Kalsbeek, Fliers, Hofman, Swaab, Van Someren & Buijs
Progress in Brain Research, Vol. 153
ISSN 0079-6123

CHAPTER 21

Hypoglycemia in diabetes: pathophysiological mechanisms and diurnal variation

Philip E. Cryer*

Division of Endocrinology, Metabolism and Lipid Research, and the General Clinical Research Center and the Diabetes Research and Training Center, Washington University School of Medicine, St. Louis, MO, USA

Abstract: Iatrogenic hypoglycemia, the limiting factor in the glycemic management of diabetes, causes recurrent morbidity (and sometimes death), precludes maintenance of euglycemia over a lifetime of diabetes and causes a vicious cycle of recurrent hypoglycemia. In insulin deficient — T1DM and advanced T2DM — diabetes hypoglycemia is the result of the interplay of therapeutic insulin excess and compromised physiological (defective glucose counterregulation) and behavioral (hypoglycemia unawareness) defenses against falling plasma glucose concentrations. The concept of hypoglycemia-associated autonomic failure (HAAF) in diabetes posits that recent antecedent hypoglycemia causes both defective glucose counterregulation (by reducing epinephrine responses in the setting of absent insulin and glucagon responses) and hypoglycemia unawareness (by reducing sympathoadrenal and the resulting neurogenic symptom responses) and thus a vicious cycle of recurrent hypoglycemia. The clinical impact of HAAF—including its reversal by avoidance of hypoglycemia—is well established, but its mechanisms are largely unknown. Loss of the glucagon response, a key feature of defective glucose counterregulation, is plausibly attributed to insulin deficiency, specifically loss of the decrement in intraislet insulin that normally signals glucagon secretion as glucose levels fall. Reduced neurogenic symptoms, a key feature of hypoglycemia unawareness, are largely the result of reduced sympathetic neural responses to falling glucose levels. The mechanism(s) by which hypoglycemia shifts the glycemic thresholds for sympathoadrenal activation to lower plasma glucose concentrations, the key feature of both components of HAAF, is not known. It does not appear to be the result of the release of a systemic mediator such as cortisol or epinephrine during antecedent hypoglycemia or of increased blood-to-brain glucose transport. It is likely the result of an as yet to be identified alteration of brain metabolism. While the research focus has been largely on the hypothalamus, hypoglycemia is known to activate widespread brain regions including the medial prefrontal cortex. The possibility of post-hypoglycemic brain glycogen supercompensation has also been raised. Finally, a unifying mechanism of HAAF would need to incorporate the effects of sleep and antecedent exercise which produce a phenomenon similar to hypoglycemia induced HAAF.

KeyWords: hypoglycemia; diabetes; glucagon; epinephrine; defective glucose counterregulation; hypoglycemia unawareness; hypoglycemia-associated autonomic failure

---

*Corresponding author. Tel.: +314-362-7635; Fax: +314-362-
7989; E-mail: pcryer@wustl.edu

DOI: 10.1016/S0079-6123(06)53021-3

Introduction

Iatrogenic hypoglycemia is the limiting factor in the glycemic management of diabetes (Cryer, 2004). It causes recurrent morbidity in most people with type 1 diabetes (T1DM) and many with type 2 diabetes (T2DM). It is sometimes fatal. The barrier of hypoglycemia precludes maintenance of euglycemia, and thus full realization of the benefits of glycemic control, over a lifetime of diabetes.

Episodes of hypoglycemia, even asymptomatic episodes, impair defenses against subsequent hypoglycemia by causing hypoglycemia-associated autonomic failure (HAAF) — the clinical syndromes of defective glucose counterregulation and hypoglycemia unawareness — and therefore a vicious cycle of recurrent hypoglycemia (Cryer, 2004). The pathophysiology of glucose counterregulation and the pathogenesis of hypoglycemia in insulin-deficient diabetes, selected studies of the pathophysiological mechanisms and the impact of sleep are summarized in the paragraphs that follow.

Hypoglycemia-associated autonomic failure

Iatrogenic hypoglycemia is the result of the interplay of relative or absolute therapeutic insulin excess and compromised physiological and behavioral defenses against falling plasma glucose concentrations in insulin-deficient — T1DM and advanced T2DM — diabetes (Cryer, 2004). As plasma glucose concentrations decline insulin levels do not decrease, glucagon levels do not increase and the increase in epinephrine is typically attenuated. The latter causes the clinical syndrome of defective glucose counterregulation; affected patients are at 25-fold or greater increased risk for severe iatrogenic hypoglycemia. A reduced sympathoadrenal — sympathetic neural as well as adrenomedullary — response causes the clinical syndrome of hypoglycemia unawareness; affected patients are at ~six-fold increased risk for severe hypoglycemia.

The concept of HAAF in T1DM (Dagogo-Jack et al., 1993) and advanced T2DM (Segel et al., 2002) posits that recent antecedent iatrogenic hypoglycemia causes both defective glucose coun-

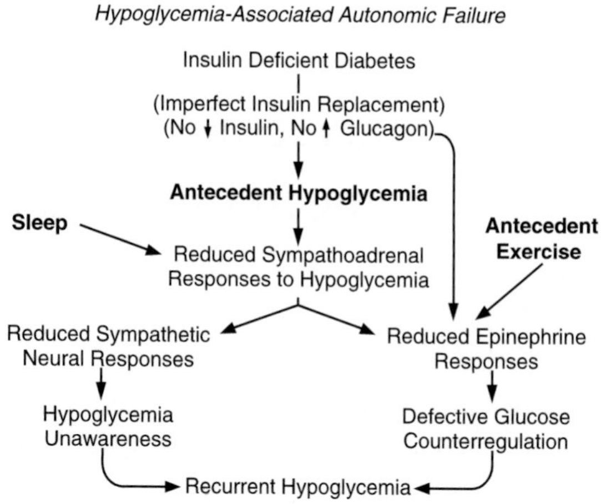

Fig. 1. Hypoglycemia-associated autonomic failure in type 1 and advanced type 2 diabetes. (Modified from Cryer (2004). Copyright 2004, Massachusetts Medical Society.)

terregulation (by reducing epinephrine responses to a given level of subsequent hypoglycemia in the setting of absent decrements in insulin and increments in glucagon) and hypoglycemia unawareness (by reducing sympathoadrenal and the resulting neurogenic symptom responses to a given level of subsequent hypoglycemia) and thus a vicious cycle of recurrent hypoglycemia (Cryer, 2004) (Fig. 1). The concept of HAAF has been extended to include sleep- and exercise-related HAAF (Cryer, 2004) (Fig. 1).

The clinical impact of HAAF — including the fact that hypoglycemia unawareness, and to some extent the reduced epinephrine component of defective glucose counterregulation, is reversible by as little as 2–3 weeks of scrupulous avoidance of hypoglycemia in most affected patients (Fanelli et al., 1993, 1994; Cranston et al., 1994; Dagogo-Jack et al., 1994) — is well established (Cryer, 2004). On the other hand, the mechanisms of HAAF are largely unknown. Established and potential mechanisms have been reviewed (Cryer, 2005).

Mechanisms of HAAF

Loss of the glucagon secretory response to hypoglycemia in insulin-deficient diabetes, a key feature of defective glucose counterregulation, is plausibly attributed to loss of the intraislet insulin signal, a decrease in intraislet insulin that normally triggers an increase in glucagon secretion as glucose levels fall (Raju and Cryer, 2005). Predicted on a body of evidence from the perfused rat pancreas, islets in vitro and rats in vivo, the intraislet insulin hypothesis was first documented in healthy humans in whom suppression of baseline insulin secretion (with the K_{ATP} channel agonist diazoxide) reduced the decrement in intraislet insulin during the induction of hypoglycemia and reduced the plasma glucagon response to hypoglycemia (Raju and Cryer, 2005). Therefore, the absence of the intraislet insulin signal plausibly explains loss of the glucagon response to hypoglycemia in insulin-deficient diabetes.

Hypoglycemia unawareness is largely the result of reduced sympathetic neural, rather than adrenomedullary, responses to hypoglycemia (DeRosa and Cryer, 2004). People who have no adrenal medullae (as a result of earlier bilateral adrenalectomy), and therefore virtually no plasma epinephrine response to hypoglycemia, exhibit typical neurogenic (autonomic) symptoms — both adrenergic (palpitations, tremor, arousal) and cholinergic (sweating, hunger, paresthesias) symptoms — during hypoglycemia (DeRosa and Cryer, 2004). Therefore, it follows that the loss of those neurogenic symptoms in diabetic patients with hypoglycemia unawareness is largely the result of reduced sympathetic neural responses to hypoglycemia. Interestingly, the data suggest that the plasma norepinephrine response to hypoglycemia is largely derived from the adrenal medullae rather than sympathetic post-ganglionic neurons and that the hemodynamic responses to hypoglycemia are largely the result of increased adrenomedullary epinephrine secretion.

The mechanism(s) of the shift of glycemic thresholds for sympathoadrenal activation to lower plasma glucose concentrations following recent antecedent hypoglycemia, the key feature of both defective glucose counterregulation and hypoglycemia unawareness and thus HAAF (Cryer, 2004), is not known. The alteration is widely assumed to be within the central nervous system although alterations in the afferent or efferent components of the sympathoadrenal system may also be involved (Cryer, 2005).

The systemic mediator hypothesis posits that increased cortisol levels, stimulated by recent antecedent hypoglycemia, act on the brain to reduce sympathoadrenal responses to subsequent hypoglycemia, and thus mediate HAAF (Davis et al., 1996). However, the relevance of this phenomenon to the pathogenesis of HAAF has been questioned by the finding that cortisol elevations comparable to, indeed somewhat above, those that occur during hypoglycemia did not reduce the adrenomedullary epinephrine or neurogenic symptom responses to subsequent hypoglycemia (Raju et al., 2003). There is also evidence that recent antecedent epinephrine elevations do not cause HAAF (de Galan et al., 2003).

The brain fuel transport hypothesis posits that recent antecedent hypoglycemia increases blood-to-brain glucose transport and that increased fuel transport into the brain reduces sympathoadrenal responses to subsequent hypoglycemia and is thus the mechanism of HAAF (Boyle et al., 1994). However, positron emission tomography studies with $[1-^{11}C]$glucose in a model of HAAF in nondiabetic humans (Segel et al., 2001) and with $[^{11}C]$3-0-methylglucsoe in patients with T1DM and hypoglycemia unawareness (Bingham et al., 2004) indicate that blood-to-brain glucose transport is not increased by recent antecedent hypoglycemia or in patients with hypoglycemia unawareness. Furthermore, recent antecedent hypoglycemia does not increase brain glucose concentrations, measured with nuclear magnetic resonance spectroscopy in humans, during subsequent hypoglycemia (Criego et al., 2005). Finally, in rodents recent antecedent hypoglycemia, which reduces sympathoadrenal responses to subsequent hypoglycemia just as it does in humans, does not increase brain extracellular glucose concentrations, measured by microdialysis, during hypoglycemia (de Vries et al., 2003).

The brain metabolism hypothesis posits that recent antecedent hypoglycemia in some as yet to be identified way alters brain metabolism that ultimately results in reduced sympathoadrenal responses and thus mediates HAAF. While much of the investigative focus has been on the ventromedial hypothalamus (e.g., Routh, 2003), hypoglycemia causes activation of widespread brain regions in experimental animals (Biggers et al., 1989) and in humans (Dunn et al., 2004; Teves et al., 2004). In humans those regions include the medial prefrontal cortex as well as the thalamus and regions of the brain stem (Dunn et al., 2004; Teves et al., 2004). These findings indicate that widespread brain regions, including the cortex, must be considered in studies of the central nervous system mechanisms of HAAF in diabetes. Potential mechanisms (Cryer, 2005) include increased glucokinase activity in critical glucose-sensing neurons, increased corticotropin-releasing hormone or urocortin release, decreased paraventricular nucleus activity, K_{ATP} channel closure, decreased AMP-activated protein kinase activity, increased GABAergic tone, increased expression of angiotensinogen and related genes, decreased brain insulin signaling, increased cerebral blood flow and decreased cerebral glucose metabolism. The latter possibility was supported in one human study (Bingham et al., 2004) but not in another (Segel et al., 2001).

The brain glycogen supercompensation hypothesis posits that after an episode of hypoglycemia brain (astrocyte) glycogen levels rebound to levels that exceed the pre-hypoglycemic levels and provide an expanded source of glycolytic metabolism and ultimately neuronal oxidative fuel (e.g., lactate) that reduces sympathoadrenal responses to subsequent hypoglycemia (Gruetter, 2003). It has not been tested in humans.

Diverse causes of HAAF

In addition to the originally recognized hypoglycemia induced-HAAF, it appears that recent antecedent exercise and sleep cause a similar phenomenon (Cryer, 2004) (Fig. 1). There is no diurnal variation in the key glucose counterregulatory responses to hypoglycemia in humans when the subjects are studied while awake (Banarer and Cryer, 2003). Those include the insulin, glucagon and epinephrine responses. Interestingly, the cortisol response is enhanced at night since it begins from a lower baseline but peaks at levels similar to those that occur in the morning. However, sleep

further reduces the sympathoadrenal responses to hypoglycemia in patients with T1DM (Jones et al., 1998; Banarer and Cryer, 2003). Probably because of their markedly reduced sympathoadrenal responses, people with T1DM are much less likely to be awakened from sleep by hypoglycemia than nondiabetic individuals (Banarer and Cryer, 2003). Clearly, a unifying mechanism of HAAF would need to incorporate the effects of antecedent exercise and sleep as well as antecedent hypoglycemia.

Abbreviations

HAAF	hypoglycemia-associated autonomic failure
T1DM	type 1 diabetes
T2DM	type 2 diabetes

Acknowledgments

The author's work cited was supported, in part, by U.S. Public Health Service/National Institutes of Health grants R37 DK27085, M01 RR00036 and P60 DK20579 and by a fellowship award from the American Diabetes Association.

The author gratefully acknowledges the substantive contributions of his post-doctoral fellows and collaborators and the skilled assistance of the nursing and technical staffs of the Washington University General Clinical Research Center in the performance of our studies and of Ms. Janet Dedeke in the preparation of this manuscript.

References

Banarer, S. and Cryer, P.E. (2003) Sleep-related hypoglycemia-associated autonomic failure in type 1 diabetes: reduced awakening from sleep during hypoglycemia. Diabetes, 52: 1195–1203.

Biggers, D.W., Myers, S., Neal, D., Stinson, R., Cooper, N.B., Jaspan, J.B., Williams, P.E., Cherrington, A.D. and Frizzell, R.T. (1989) Role of brain in counterregulation of insulin-induced hypoglycemia in dogs. Diabetes, 37: 7–16.

Bingham, E.M., Dunn, J.T., Smith, D., Sutcliffe-Goulden, J., Reed, L.J., Marsden, P.K. and Amiel, S. (2005) Differential changes in brain glucose metabolism during hypoglycaemia accompany loss of hypoglycaemia awareness in men with type 1 diabetes. An [11C]-3-0-methyl-D glucose PET study. Diabetologia, 48: 2080–2089.

Boyle, P.J., Nagy, R.J., O'Connor, A.M., Kempers, S.F., Yeo, R.A. and Qualls, C. (1994) Adaptation in brain glucose uptake following recurrent hypoglycemia. Proc. Natl. Acad. Sci. USA, 91: 9352–9356.

Cranston, I., Lomas, J., Maran, A., Macdonald, I. and Amiel, S.A. (1994) Restoration of hypoglycaemia awareness in patients with long-duration insulin-dependent diabetes. Lancet, 344: 283–287.

Criego, A.B., Tkac, I., Kumar, A., Thomas, W., Gruetler, R. and Seaquist, E.R. (2005) Brain glucose concentrations in healthy humans subjected to recurrent hypoglycemia. J. Neuroscience Res., 82: 525–530.

Cryer, P.E. (2004) Diverse causes of hypoglycemia-associated autonomic failure in diabetes. N. Engl. J. Med., 350: 2272–2279.

Cryer, P.E. (2005) Mechanisms of hypoglycemia-associated autonomic failure and its component syndromes in diabetes. Diabetes, 54: 3592–3601.

Dagogo-Jack, S.E., Craft, S. and Cryer, P.E. (1993) Hypoglycemia-associated autonomic failure in insulin dependent diabetes mellitus. J. Clin. Invest., 91: 819–828.

Dagogo-Jack, S., Rattarasarn, C. and Cryer, P.E. (1994) Reversal of hypoglycemia unawareness, but not defective glucose counterregulation, in IDDM. Diabetes, 43: 1426–1434.

Davis, S.N., Shavers, C., Costa, F. and Mosqueda-Garcia, R. (1996) Role of cortisol in the pathogenesis of deficient counterregulation after antecedent hypoglycemia in normal humans. J. Clin. Invest., 98: 680–691.

de Galan, B.E., Rietjens, S.J., Tack, C.J., Van Der Werf, S.P., Sweep, C.G.J., Lenders, J.W.M. and Smits, P. (2003) Antecedent adrenaline attenuates the responsiveness to, but not the release of, counterregulatory hormones during subsequent hypoglycemia. J. Clin. Endocrinol. Metab., 88: 5462–5467.

DeRosa, M.A. and Cryer, P.E. (2004) Hypoglycemia and the sympathoadrenal system: neurogenic symptoms are largely the result of sympathetic neural, rather than adrenomedullary, activation. Am. J. Physiol. Endocrinol. Metab., 287: E32–E41.

de Vries, M.G., Arseneau, L.M., Lawson, M.E. and Beverly, J.L. (2003) Extracellular glucose in rat ventromedial hypothalamus during acute and recurrent hypoglycemia. Diabetes, 52: 2767–2773.

Dunn, J., Cranston, I.C., Marsden, P.K. and Amiel, S.A. (2004) Measurement of brain perfusion in response to acute hypoglycaemia in healthy volunteers: a 15O-water positron emis-

sion tomography study (abstract). Diabetologia, 47(Suppl. 1): A322.

Fanelli, C.G., Epifano, L., Rambotti, A.M., Pampanelli, S., Di Vincenzo, A., Modarelli, F., Lepore, M., Annibale, B., Ciofetta, M., Bottini, P., Porcellati, F., Scionti, L., Santeusanio, F., Brunetti, P. and Bolli, G.B. (1993) Meticulous prevention of hypoglycemia normalizes the glycemic thresholds and magnitude of most of neuroendocrine responses to, symptoms of, and cognitive function during hypoglycemia in intensively treated patients with short-term IDDM. Diabetes, 42: 1683–1689.

Fanelli, C., Pampanelli, S., Epifano, L., Rambotti, A.M., Di Vincenzo, A., Modarelli, F., Ciofetta, M., Lepore, M., Annibale, B., Torlone, E., Perriello, G., De Feo, P., Santeusanio, F., Brunetti, P. and Bolli, G.B. (1994) Long-term recovery from unawareness, deficient counterregulation and lack of cognitive dysfunction during hypoglycaemia, following institution of rational, intensive insulin therapy in IDDM. Diabetologia, 37: 1265–1276.

Gruetter, R. (2003) Glycogen: the forgotten cerebral energy store. J. Neurosci. Res., 74: 179–183.

Jones, T.W., Porter, P., Sherwin, R.S., Davis, E.A., O'Leary, P., Frazer, F., Byrne, G., Stick, S. and Tamborlane, W.V. (1998) Decreased epinephrine responses to hypoglycemia during sleep. N. Engl. J. Med., 338: 1657–1662.

Raju, B. and Cryer, P.E. (2005) Loss of the decrement in intraislet insulin plausibly explains loss of the glucagon response to hypoglycemia in insulin deficient diabetes. Diabetes, 54: 757–764.

Raju, B., McGregor, V.P. and Cryer, P.E. (2003) Cortisol elevations comparable to those that occur during hypoglycemia do not cause hypoglycemia-associated autonomic failure. Diabetes, 52: 2083–2089.

Routh, V.H. (2003) Glucosensing neurons in the ventromedial hypothalamic nucleus (VMN) and hypoglycemia-associated autonomic failure (HAAF). Diabetes Metab. Res. Rev., 19: 348–356.

Segel, S.A., Fanelli, C.G., Dence, C.S., Markham, J., Videen, T.O., Paramore, D.S., Powers, W.J. and Cryer, P.E. (2001) Blood-to-brain glucose transport, cerebral glucose metabolism and cerebral blood flow are not increased following hypoglycemia. Diabetes, 50: 1911–1917.

Segel, S.A., Paramore, D.S. and Cryer, P.E. (2002) Hypoglycemia-associated autonomic failure in advanced type 2 diabetes. Diabetes, 51: 724–733.

Teves, D., Videen, T.O., Cryer, P.E. and Powers, W.J. (2004) Activation of human medial prefrontal cortex during autonomic responses to hypoglycemia. Proc. Natl. Acad. Sci. USA, 101: 6217–6221.

Kalsbeek, Fliers, Hofman, Swaab, Van Someren & Buijs
Progress in Brain Research, Vol. 153
ISSN 0079-6123

CHAPTER 22

Hypothalamic integration of immune function and metabolism

Ana Guijarro[1], Alessandro Laviano[2] and Michael M. Meguid[1,*]

[1]*Surgical Metabolism and Nutrition Laboratory, Neuroscience Program, University Hospital, SUNY Upstate Medical University, 750 Adams St., Syracuse, NY 13210, USA*
[2]*Department of Clinical Medicine, University of Rome, 'La Sapienza', viale dell Universita 37, 00185 Rome, Italy*

Abstract: The immune and neuroendocrine systems are closely involved in the regulation of metabolism at peripheral and central hypothalamic levels. In both physiological (meals) and pathological (infections, traumas and tumors) conditions immune cells are activated responding with the release of cytokines and other immune mediators (afferent signals). In the hypothalamus (central integration), cytokines influence metabolism by acting on nucleus involved in feeding and homeostasis regulation leading to the acute phase response (efferent signals) aimed to maintain the body integrity.

Peripheral administration of cytokines, inoculation of tumor and induction of infection alter, by means of cytokine action, the normal pattern of food intake affecting meal size and meal number suggesting that cytokines acted differentially on specific hypothalamic neurons. The effect of cytokines-related cancer anorexia is also exerted peripherally. Increase plasma concentrations of insulin and free tryptophan and decrease gastric emptying and D-xylose absorption. In addition, in obesity an increase in interleukin (IL)-1 and IL-6 occurs in mesenteric fat tissue, which together with an increase in corticosterone, is associated with hyperglycemia, dyslipidemias and insulin resistance of obesity-related metabolic syndrome. These changes in circulating nutrients and hormones are sensed by hypothalamic neurons that influence food intake and metabolism.

In anorectic tumor-bearing rats, we detected upregulation of IL-1β and IL-1 receptor mRNA levels in the hypothalamus, a negative correlation between IL-1 concentration in cerebro-spinal fluid and food intake and high levels of hypothalamic serotonin, and these differences disappeared after tumor removal. Moreover, there is an interaction between serotonin and IL-1 in the development of cancer anorexia as well as an increase in hypothalamic dopamine and serotonin production. Immunohistochemical studies have shown a decrease in neuropeptide Y (NPY) and dopamine (DA) and an increase in serotonin concentration in tumor-bearing rats, in first- and second-order hypothalamic nuclei, while tumor resection reverted these changes and normalized food intake, suggesting negative regulation of NPY and DA systems by cytokines during anorexia, probably mediated by serotonin that appears to play a pivotal role in the regulation of food intake in cancer.

Among the different forms of therapy, nutritional manipulation of diet in tumor-bearing state has been investigated. Supplementation of tumor bearing rats with ω-3 fatty acid vs. control diet delayed the appearance of tumor, reduced tumor-growth rate and volume, negated onset of anorexia,

*Corresponding author. Tel.: +315-464-46283; Fax: +315-464-6237; E-mail: meguidm@upstate.edu

DOI: 10.1016/S0079-6123(06)53022-5

increased body weight, decreased cytokines production and increased expression of NPY and decreased α-melanocyte-stimulating hormone (α-MSH) in hypothalamic nuclei. These data suggest that ω-3 fatty acid suppressed pro-inflammatory cytokines production and improved food intake by normalizing hypothalamic food intake-related peptides and point to the possibility of a therapeutic use of these fatty acids.

The sum of these data support the concept that immune cell-derived cytokines are closely related with the regulation of metabolism and have both central and peripheral actions, inducing anorexia via hypothalamic anorectic factors, including serotonin and dopamine, and inhibiting NPY leading to a reduction in food intake and body weight, emphasizing the interconnection of the immune and neuroendocrine systems in regulating metabolism during infectious process, cachexia and obesity.

Introduction

The quote an "An army marches on its stomach," often attributed to Napoleon, links the rallying of bodily defenses with the appropriate marshalling of nutrient responses. In this analogy General Baron De Jomini (Fig. 1), Napoleon's Quartermaster, represents the "hypothalamus" which plays an important role in defining the magnitude and temporal profile of hypothalamic nuclei responses by integrating these with the responses from the rest of the brain (central events) and initiating hormonal release by the hypothalamic–pituitary–adrenal axis (HPA). He assessed the magnitude and type of the danger encountered by Napoleon's troops in a battle (afferent signals) and ensured the army had appropriate supplies to react in a timely fashion and to behave by a measured response (efferent signals) to overcome the immediate threat. Like any good army, communication pathways exist between the immune system and the brain allowing bi-directional regulation of the immune- and the brain-initiated behavioral responses, thereby maintaining homeostatic regulation of the body and stability of the army.

Napoleon and De Jomini were less successful when it came to long-term campaigns under adverse conditions as reflected by the retreat from Moscow. So too the hypothalamus reacts to a foreign stimulus by mounting an acute phase response via modulating the neuroendocrine system and through the HPA axis to altered peripheral metabolism of carbohydrates, proteins and fats to ensure an immediate energy-rich milieu to sustain immune function, while optimally conserving long-term body energy status. However, like Napoleon's long-term campaign, the acute phase response is ill suited for protracted immune challenges, such as morbid obesity or Crohn's disease, which result in chronic and life-threatening conditions.

The term *acute phase response* refers to the inflammatory response of the host occurring shortly after any tissue injury. The purpose of the acute phase response is to prevent further injury of an organ, to limit the growth of the infective organism, to remove harmful molecules and to activate the repair processes to return the organ to normal function. The acute phase response is characterized by the systemic inflammatory signs of fever, anorexia,

Fig. 1. General Baron De Jomini.

somnolence and depression, which are a reflection of the integration of multiple neuro-endocrine, immunological, metabolic and neurological changes in response to the afferent stimulus. The intensity of the acute phase response varies with the acute stimulus and when it becomes overwhelming conditions, such as ileitis or obesity produce profound morphologic and metabolic changes that induce chronic illness and impair survival.

The transmission of the peripheral immune information to the brain is carried out by two different pathways: (i) blood-borne mechanisms: cytokines reach the brain by crossing blood–brain barrier via active transport (Banks et al., 1991), by means of circumventricular organs, such as vascular organ of the lamina terminalis, median eminence and area postrema (which lack a blood--brain barrier so that fenestrated capillaries allow plasma passage; Blatteis, 1992) or by binding to cerebral blood vessel endothelium leading to the release of other second messengers such as prostaglandins (Ericsson et al., 1997). (ii) By means of the vagus nerve (Goehler et al., 1997, 1999, 2000; Ek et al., 1998; Mascarucci et al., 1998; Hosoi et al., 2000). Stimulation of vagus nerve by pathogens or immune cells-derived mediators is followed by the activation of neurons in the nucleus tractus solitarius (Konsman et al., 2000), which send projections to hypothalamic and limbic areas involved in regulation of feeding behavior (Ricardo and Koh, 1978). In the vagally mediated immunosignal to the brain, dendritic cells play an important role given its prominent localization within the vagus nerve and associated paraganglia (Goehler et al., 1999). Apart from the direct stimulation of vagal afferents by immune cell-derived mediators, these signals can activate chemoreceptive cells located in the vagal paraganglia, which are penetrated by blood and lymph vessels, allowing vagal paraganglia to sense compounds circulating in blood or lymph (Goehler et al., 2000).

Afferent signals and central events-behavioral and physiological components

A number of examples are sighted below to demonstrate our understanding of the common features of the acute response and function of the afferent signal mechanism(s) that impinge on central event, initiating the acute phase response as manifested by changes in behavior using food intake, and its components (meal size and meal number) as a biological index. To illustrate these points we will use data based on our work and augment it with data from the literature.

Responses to oral food

Ingested nutrients are strong macromolecular antigens that actively challenge the gut's immune defense mechanism. These consist of intestinal lymphoid cells that secrete intraluminal immunoglobulins and local cytokines that (i) further recruit immune cells from the circulation to the gut lymphoid tissue, and (ii) simultaneously mount a systemic acute phase immune response, mediated via afferent blood-borne and neuroendocrine pathways to the liver and ultimately to the hypothalamus. Hansen et al. examined the effect of a single high-protein meal on peripheral immune response by measuring blood mononuclear cells, plasma concentrations of tumor necrosis factor-α (TNF-α), IL-6 and cortisol and growth hormone (Hansen et al., 1997). After the 30 min meal (Fig. 2A) a significant rise of peripheral neutrophils within 15 min of completing the meal occurred and which remained elevated for 3.5 h. At the same time, a significant decrease in circulating lymphocytes occurred accompanied by a sharp rise in cortisol that started to increase during the meal and peaked shortly thereafter (Fig. 2B). An increase in plasma cytokines levels was not detected. In a follow up "cafeteria-diet" paradigm rat study, an increase in IL-1β mRNA expression occurred in liver and hypothalamus, while an associated decrease in IL-1 receptor accessory proteins (IL-1RAP) mRNA (reflecting IL-1's binding and signaling capacity) occurred in liver and brain stem (Hansen et al., 1998).

Both studies present interesting results. The intrameal rise in cortisol indicates that afferent signals, such as nutrients or hormones, rapidly reached the hypothalamus, particularly the arcuate nucleus (ARC), probably via blood-borne afferent stimuli through the median eminence and via gastrointestinal vagal afferents. The ARC projects to

370

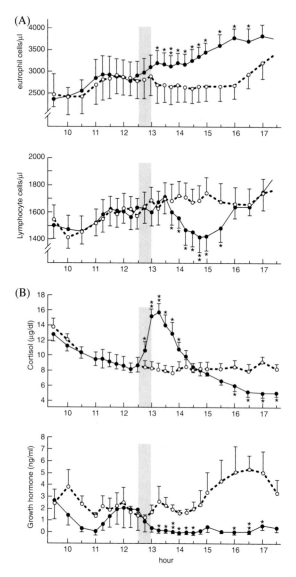

Fig. 2. Effect of a single high-protein meal on peripheral immune response. (A) After the 30 min meal a significant rise of neutrophils within 15 min of completing the meal occurred and which remained elevated for 3.5 h. (B) At the same time, a significant decrease in circulating lymphocytes occurred accompanied by a sharp rise in cortisol that started to increase during the meal and peaked shortly thereafter. (From Hansen et al., 1997).

the paraventricular nucleus (PVN) which releases corticotropin-releasing factor (CRF) into the portal plexus at the median eminence eliciting the synthesis and release of adrenocorticotropin hor-

mone (ACTH) from the anterior pituitary that stimulated the adrenal gland to secrete cortisol, which influenced the migration of immune cell from the circulation into extra-vascular gut tissue to support local immune function (Ottaway and Husband, 1994). The increase in IL-1β mRNA in the hypothalamus and the decrease in IL-1RAP mRNA in brain stem confirms the role of the vagal visceral chemosensory pathway as one of the routes for afferent signals from the gastrointestinal tract to hypothalamic nuclei via the dorsomotor vagal complex of the medulla. Despite Hansen et al.'s inability to detect cytokines in the circulation in response to the *physiological* event of a single meal, in a *pathological* model of shock-induced intestinal injury, Deitch et al. demonstrated that the gut liberates cytokines (Deitch et al., 1994). Concentrations of IL-6 and TNF were significantly higher in portal blood than cardiac blood.

In response to the increased intraportal IL-1β, Niijima (1996) demonstrated an increased vagal afferent electrical activity in a dose-dependent response with a simultaneous decreased efferent sympathetic splanchnic activity and a simultaneous increase in vagal thymic nerve activity (see section on Efferent signals). Similar findings were reported by Niijima and Meguid in response to intraportal arginine, an amino acid known to enhance immunity (Niijima and Meguid, 1998) by stimulating increased T-cell release from the thymus and inhibiting the spleen from taking up circulating lymphocytes. The net result is to enhance the circulating numbers and phagocytical active T and B cells (Fig. 3).

Responses to continuous intravenous nutrients
In a parallel series of studies we continuously infused graded caloric amounts of intravenous nutrients for 3 or 9 days (total parenteral nutrition) into the rat, while measuring the response of oral food intake, meal patterns, peripheral hormones, cytokines, hepatic vagal afferent activity and intrahypothalamic monoamines (Opara et al., 1996). Figure 4 shows that graded amounts of total parenteral nutrition lead to a graded compensatory decrease in oral intake. The graded compensatory decrease in oral intake persisted for the duration of the infusion. Plasma glucose

A. Thymic Vagal Efferents

Normal rat

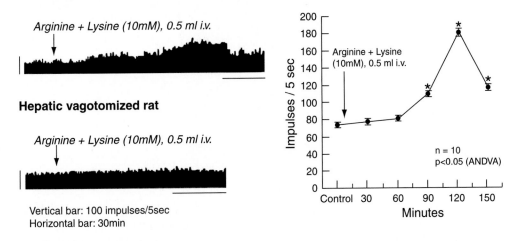

Arginine + Lysine (10mM), 0.5 ml i.v.

Hepatic vagotomized rat

Arginine + Lysine (10mM), 0.5 ml i.v.

Vertical bar: 100 impulses/5sec
Horizontal bar: 30min

B. Splenic Efferents

Normal rat

Arginine + Lysine (10mM), 0.5 ml i.v.

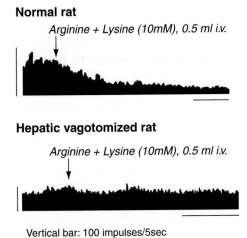

Hepatic vagotomized rat

Arginine + Lysine (10mM), 0.5 ml i.v.

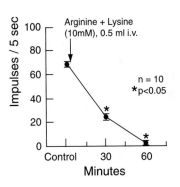

Vertical bar: 100 impulses/5sec
Horizontal bar: 30min

Fig. 3. Effect of intravenous administration of a mixture solution of Arg and Lys (10 mM, 0.5 ml) on the efferent activity of the (A) thymic branch of the vagus nerve and (B) the efferent activity of the splenic nerve in normal and hepatic vagotomized rats. $*p < 0.05$. (From Niijima and Meguid, 1998).

and insulin significantly increase and whereas hepatic glycogen concentrations decreased, hepatic triglyceride concentrations significantly increased (Meguid et al., 1991). Simultaneously, plasma TNF-α and peripheral blood monocyte IL-1 also significantly increased (Opara et al., 1995b), while intraportal nutrients decreased hepatic vagal firing rates (Niijima and Meguid, 1994). Interestingly, the response of the hepatic vagal afferent firing rate varied according to the type of amino acid infused intraportally. Thus, of 15 different amino acids infused intraportally (10 mmol in 0.1 ml), eight were excitatory and increased the vagal afferent firing rate, while the others were inhibitory and

372

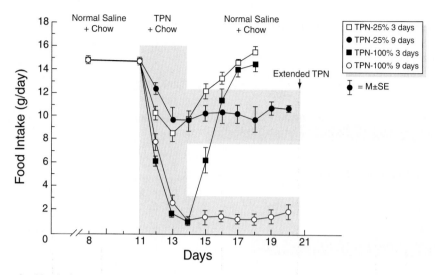

Fig. 4. Spontaneous food intake in rats receiving total parenteral nutrition (TPN-25 or TPN-100) for 3 or 9 days. TPN-25 indicates the amount of total parenteral nutrition that provides 25% of a rat's daily caloric intake; TPN-100, the amount of total parenteral nutrition that provides 100% of the rat's daily caloric intake. Values are the mean ± SEM. Total parenteral nutrition decreased spontaneous food intake in proportion to the amount infused. (From Campos et al., 1990).

decreased the firing rate. The change in afferent activity to the hypothalamus may affect reflex regulation of the visceral functions and thereby influence appetite (Niijima and Meguid, 1995). Intralateral hypothalamic area (LHA) neuron dopaminergic activity in response to total parenteral nutrition or its constituent's nutrients was measured by microdialysis (Meguid et al., 1993). LHA-DA levels rose and remained elevated during a continuous 3-h peripheral total parenteral nutrition infusion. Similar increases in LHA-DA occurred during peripheral glucose, fat and amino acid infusion. However, after cessation of these peripherally infused solutions, glucose was the only solution where the percent of DA decrease below baseline for 3 h. A similar relationship was determined between LHA-DA to oral intake in normal rats. The ingestion of a single meal induced a rise in LHA-DA, as measured via microdialysis, that was double in magnitude to that induced by to a meal one-half the size (Meguid et al., 1995). A reciprocal relationship exists between the LHA and ventromedial hypothalamus (VMH) in food intake regulation. Thus, the relationship of DA to the VMH was also explored. DA-VMH concentrations decreased during eating, and the degree

and duration of decrease after the meal corresponded to the size of the meal. When the decreased postmeal VMH-DA level had returned to baseline, rats ate once more. We infer from the data that in normal rats eating was associated with decreased DA levels in the VMH, that was followed by a lag time during which no additional eating occurred suggesting that VMH-DA levels contributed to determining the duration of the intermeal interval and hence by influence meal frequency (Meguid et al., 1997).

These studies present two striking results. First, the increased circulating nutrients and insulin led to a form of anorexia by compensating for the increased caloric intake. Second, the stress of the continuous hypertonic infusion increased TNF-α and IL-1 contributing to the cytokine mechanism that decreases oral intake in the rat. The ARC, which is the nodal point in hypothalamic regulation of energy balance, promptly senses these nutrients. The leptin- and ghrelin-responsive ARC neurons affect the activity of neurons in PVN, VMH and LHA and other key effector central sites (Elias et al., 1998, 1999; Cowley et al., 1999; Saper et al., 2002), which contain orexigenic and anorexigenic neuropeptides, including orexin-A and -B, cholecystokinine and melanin concentrating hormone (MCH). Many of these central sites are linked to the hypophysiothropic, behavioral and autonomic adaptive responses to changes in energy status (Zigman and Elmquist, 2003). The ARC project neurons to the LHA that plays a key role in ingestive behavior and energy balance, because of the neurochemical phenotypes of the cells express melanocortin hormones and orexin (see section on Central integration). These neurons also synthesize and release DA that "stimulate" or "inhibit" regulatory control over the HPA (Meguid et al., 1995). Thus, although the rise in LHA-DA in our study could have been anticipated, the fall in LHA-DA after the cessation of only the intravenous glucose, supports the concept of the glucose-sensing capacity of the neurons in the ARC, PVN and the LHA (Elmquist and Marcus, 2003). No such response occurred with cessation of the complex solutions that constitute total parenteral nutrition, fat emulsion or amino acid solution, all of which are compound solutions. These data indicate that blood-borne factors are sensed by the hypothalamus but particularly by the ARC and that this sensing mechanism constitutes part of the acute phase response system. At the same time, the increase in TNF-α and IL-1 was likely sensed by the brain via the afferent hepatic vagus because when a sub-diaphragmatic vagotomy was performed, vagotomized rats consumed more food during total parenteral nutrition. They did this by increasing the frequency, size and duration of a meal, suggesting that the influence of blood-borne nutrient and insulin on the ARC was greater than the effect that TNF-α had in decreasing food intake vial abdominal vagus (Yang et al., 1992; Opara et al., 1995b). Based on these results it appears that the cytokine response to the continuous infusion of hypertonic nutrients has a quantitatively greater inhibitory effect on food intake than the continuous nutrient supply.

Afferent signals from a peripheral tumor
Another model that we have used extensively in our laboratory to gain insight into the integrative acute phase response to peripheral signals is the methylcholanthrene sarcoma without inducing metastases in the rat model that induces

anorexia as the tumor grows. When 10^6 methylcholanthrene-sarcoma tumor cells are injected into the flank of a Fischer rat a palpable tumor is detected after 10 days. The tumor's exponential growth results in a $1\,cm^3$ mass between 16 and 20 days (Meguid et al., 1987). During tumor growth, meal number gradually declines with time, but food intake is maintained by a compensatory increase in meal size between 18 and 20 days. When this compensation fails, food intake dramatically decreases and anorexia is behaviorally manifested. Interferon-γ (INF-γ) was detected in the tumor tissue, while we measured a significant increase in IL-1β and in IL-1 receptor 1 mRNA expression in the hypothalamus, cerebellum and hippocampus. Simultaneously, we detected an increase in IL-1RAP 1 and 2 in the liver confirming activity of IL-1 in the periphery (Fig. 5; Turrin et al., 2004).

These studies clearly indicate that the peripherally growing tumor is challenged by immune cells to elaborate both TNF-α and IL-1β. Yang et al. (1994) demonstrated that a 3-day infusion of subclinical concentrations of TNF-α and IL-1 induced anorexia via a synergistic effect, each having no biological effect when infused separately. These act not only on both the parvocellular and the magnocellular neuronal population of the PVN to modulate the acute phase response to induce anorexia thereby decreasing meal size and meal number, which are the components of food intake as measured by our Automated Computerized Rat Eater Meter (Meguid et al., 1990; Meguid et al., 1998); but also on other brain regions involved with locomotion to conserve energy and on memory to respond appropriately to the acute phase response event. Understanding the component of food intake i.e., whether it is meal number or/and meal size, decrease or change, is a useful behavioral index. It reflects where in the hypothalamus and on which neuron population afferent signals, such as cytokines, act. Using an injection of fetal serotonergic or dopaminergic cell suspensions into the LHA, VMN or supraoptic nucleus (SON) we demonstrated that either meal size or meal number or both could be manipulated, suggesting that these neurotransmitters had co-receptors on the primary neuropeptide food-regulating neurons (Meguid et al., 1999).

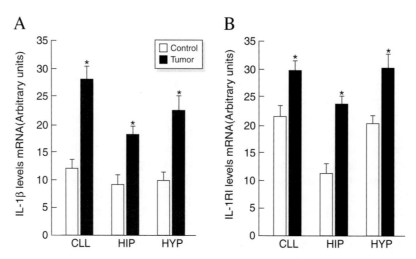

Fig. 5. (A) Increased expression of IL-1 β mRNA levels in cerebellum (CLL), hippocampus (Hip) and hypothalamus (Hyp) with peripheral tumor growth and (B) IL-1 RI mRNA levels in controls or tumor-bearing rats. Values (means ± SE; $n = 8$ for each group) were standardized to arbitrary units. *$p < 0.05$ vs. controls. (From Turrin et al., 2004).

Responses to a viral and bacterial infection

During an outbreak of a sialodacryoadenitis viral infection in our rat colony we detected that food intake markedly decreased when the rats become clinically symptomatic (Sato et al., 2001c). This reduction in food intake occurred via a decrease in meal size not adequately compensated by an increase in meal number that occurred during both the light and dark phase. This pattern is similar to that which we described in anorexia of indomethacin-induced ulcerative ileitis that is accompanied by an increase in plasma concentrations of TNF-α (Veerabagu et al., 1996), but differed from that observed in anorexia of bacterial lipopolysaccharide-induced infections (Langhans et al., 1991a; Porter et al., 1998) and in cancer anorexia (Meguid et al., 2000). Table 1 compares the different responses of both components of food intake to different stimuli.

The reduced food intake during sialodacryoadenitis infection, which is a corona virus, is probably due to TNF-α because corona viruses induce an increase in this cytokine (Itoh et al., 1991) and using anti-TNF-α agent inhibits TNF-α induced anorexia (Porter et al., 2000), while the participation of IL-1β seems to be limited given that even in IL-1β-deficient mice the

anorexia induced by influenza virus is severe (Kozak et al., 1995). The effect of TNF-α on food intake may be partly mediated by changes in hypothalamic DA levels, because in a cecal ligation and puncture septic rat model studied in our laboratory, a progressive decrease in VMH-DA concentrations associated with anorexia was demonstrated (Torelli et al., 2000). Besides the role of TNF-α in sialodacryoadenitis-induced anorexia the participation of other factors, including IL-2, IL-6, IL-8 and/or INT-γ has been suggested (Conn et al., 1995; Plata-Salaman, 1996; Arsenijevic and Richard, 1999).

During bacterial infection IL1-β seems to play and important role in anorexia induction (Von Meyenburg et al., 2003). MohanKumar et al. (1999) found a marked increase in DA, norepinephrine (NE), serotonin (5-HT) and 5-hydroxyindoleacetic acid (5-HIAA) in PVN, as well as an increase in DA concentrations in ARC, two nuclei involved in the control of food intake, after lipopolysaccharide intraperitoneal administration. These changes are completely blocked by treatment with IL-1ra, suggesting the participation of IL-1β in these monoamine metabolism alterations, which is able to induce changes in neurotransmitters concentrations in specific hypothalamic areas (MohanKumar et al., 1998).

Table 1. Different responses of both components of food intake, meal number (MN) and meal size (MZ), to different stimuli

Stimuli	MN	MZ	Reference
TPN	↓	↓	Meguid et al. (1991)
MDP	↓	↔	Langhans et al. (1991a)
IL1-α	↓	↓	Debonis et al. (1995)
Ulcerative ileitis	↔	↓	Veerabagu et al. (1996)
LPS	↓	↔	Porter et al. (1998)
Tumor	↓	↓	Meguid et al. (2000)
SDA	↔	↓	Sato et al. (2001c)

Note: ↔ No change; ↓ decrease. LPS, lipopolysaccharide; MDP, muramyl dipeptide; TPN, total parenteral nutrition; SDA, sialodacryoadenitis.

Gonadal hormones and sexual dimorphic-based acute phase response

Another critical model studied in our laboratory that has profound influence on hypothalamic integration of immune function and metabolism is the sex of the study model given that gender differences exist in the acute phase response (Coe and Ross, 1983; Hirai and Limaos, 1990; Spitzer and Zhang, 1996a, b).

Although daily food intake based on 100 g-body weight is similar in males and females rats, during a 44-day observation period of weight-gain rate in young growing adults was sevenfold greater in males than in females, and the pattern of food intake differed between both sexes. Thus, while in males the constancy of food intake was achieved by an increase on meal size compensated by a decrease in meal number the striking observation in comparably aged female rats is that their food intake is relatively stable, because there are cyclical and reciprocally recurring changes in both meal size and meal number (Fig. 6), which are synchronized with the 3–4 days estrous cycle, such that meal number is greatest during estrous phase and meal size is small, while during the met-estrous cycle meal size is largest and meal number is the lowest (Laviano et al., 1996).

Estrogen modulates the neurons in suprachiasmatic nucleus (SCN, Hansen et al., 1978, 1979), the nucleus tractus solitarius (Eckel and Geary, 2001; Eckel et al., 2002), preoptic area (Dagnault and Richard, 1997), VMH (Beatty et al., 1974) and parvocellular division of PVN (Butera and Beikirch, 1989; Eckel and Geary, 2001; Eckel et al., 2002) to reduce food intake, which is not mediated by CRF (Dagnault and Richard, 1997) or by cholecystokinin (Flanagan-Cato et al., 1998; Eckel et al., 2002). In our studies, the role played by the gonadal hormones in this different behavior of food intake pattern was tested by ovariectomy, which resulted in a loss of cyclic feeding pattern, and an increase in daily food intake caused primarily by an increase in light phase meal size. This pattern of feeding behavior was reversed and normalized after exogenous estrogen restoration, resulting in body weight preservation (Varma et al., 1999). In contrast, orchiectomy reduced food intake by reduction of meal number. This pattern was normalized after exogenous testosterone was given, reversing weight loss (Chai et al., 1999). Our interpretation of the differences in the feeding pattern in females may be explained teleologically by the need to find a mate. Foraging for food during estrous exposes the female to maximum number of potential males, necessitating an increase in meal number but lower meal size, while during met-estrous, when the chances of mating are lowest the meal size are greatest. No such similar evolutionary need is necessary in male rats.

However, in the context of the acute phase response to a noxious stimulus, these data become important because the distribution of meals during the day and night are regulated by the circadian control of energy homeostasis. Thus the sex of the subject, which is receiving the insult, is critical in determining the response. In the female rat, energy regulation occurs primarily by changing meal size, not meal number. Estradiol acts directly on the SCN (Hansen et al., 1978, 1979) the main

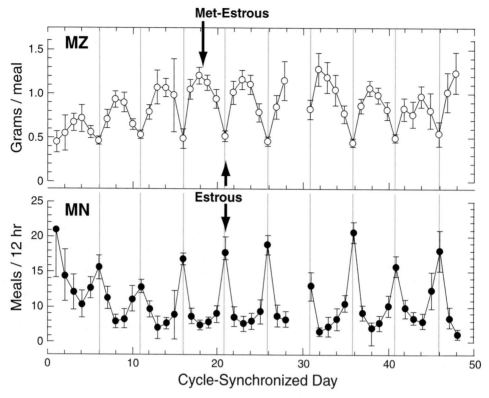

Fig. 6. Food intake in female rats is relatively stable by means of cyclical and reciprocally recurring changes in both meal size (MZ) and meal number (MN), which are synchronized with the 3–4 day estrous cycle), such that meal number is greatest during estrous phase and meal size is small, while during the met-estrous cycle meal size is largest and meal number is the lowest. (From Laviano et al., 1996).

circadian oscillator, to influence the daily rhythm of food intake by changes in both its own receptors and their activity, and by changes in diurnal rhythms of other critical neurotransmitters such as dopamine, norepinephrine, and serotonin, and receptors such as alpha 1, beta 1 and 2 in specific hypothalamic nuclei including the VMH and SCN which influence the response to the acute phase response via the HPA and cortisol response. Thus, Watanobe and Yoneda observed a greater release of ACTH in female than in male rats in response to the intravenous administration of lipoprotein polysaccharide (Watanobe and Yoneda, 2003). However there were no changes in plasma IL-1β, IL-6 or TNF-α. There were also no changes in tissue concentrations of CRF and arginine–vasopressin in the medial basal hypothalamus and in the anterior pituitary (AP). There were no changes in the binding characteristic of IL-6 in the medial

basal hypothalamus or AP but the number of the IL-1β and TNF-α binding sites, but not in the binding affinities in the medial basal hypothalamus altered significantly after gonadectomy in response to a lipopolysaccharide challenge. These sexual differences were restored after hormonal restoration in response to lipopolysaccharide. The results suggest that the hypothalamic sensitivity to peripheral IL-1 β and TNF-α is an important mechanism underlying the sexual dimorphic ACTH response to lipopolysaccharide in rats.

In a recent study designed to gain insight into the sex differences of basic nonspecific and specific immune responses intracellular type I and II cytokine production by stimulated male and female lymphocytes and monocytes in a whole-blood preparation was measured by flow cytometry. An increased percentage of IL-12, IL-1β and TNF-α

was found in men compared to women suggesting that gender differences in the balance between specific and nonspecific immune response existed in men compared to women (Bouman et al., 2004; Posma et al., 2004). Thus, it is apparent that the acute phase response is different in males and in females, explaining in part, the greater survival of the female.

Overwhelming the acute phase response by massive accumulation of subcutaneous fat

The next example that we cite is based on our observations of obesity-induced inflammatory changes in adipose tissue (Hotamisligil et al., 1993; Wellen and Hotamisligil, 2003; Xu et al., 2003; Fantuzzi, 2005). In a series of studies Sprague–Dawley pups were made obese using a high-energy diet (Ramos et al., 2003). We found that the ratio of mesenteric fat to subcutaneous fat for IL-6, TNF-α, corticosterone and their gene profiles as measured by GeneChip Rat UG34A Gene Chip (Affymetrix, Santa Clara, CA) was significantly elevated relative to nonobese mesenteric fat and contributed to the hyperglycemic and hyperdyslipidemia of obesity. In response to weight loss induced by gastric bypass these inflammatory mediators normalized. As summarized schematically in Fig. 7, as obesity develops the size of the adipocyte increases, which stimulates it to increase the synthesis and release of TNF-α. This stimulates both pre-adipocyte and endothelial cells within the surrounding fat to produce monocyte chemo-attractant protein increasing further intraadipose migration of monocytes. The stimulated mature adipocyte also synthesizes leptin and vascular endothelial growth factor contributing to angiogenesis, while the accumulating free fatty acids induce oxidative stress to the vascular endothelium, in a similar process to arteriosclerosis. The infiltrating macrophages secrete IL-6, IL-1β, TNF-α and corticosterone. The net effect is to increase insulin resistance and induce the biochemical picture of type 2 diabetes mellitus.

Brain modulation of systemic inflammation: The nicotinic anti-inflammatory pathway

The last example that we would like to use is a recently completed experiment in our rat tumor model in which we have demonstrated cytokine production. This stimulus has been modified by the external application of nicotine.

As previously described, a large bulk of data exists showing that brain's activity including behavioral responses like feeding is heavily influenced by cytokines or in broader terms by inflammation. These interactions occurring at the cellular and molecular levels between inflammatory mediators and aminergic as well as peptidergic neurons explain the occurrence of fever, anorexia and cachexia, and may provide an important step based on which we can develop pathogenesis-based therapeutic strategies.

However, the hypothalamus not only integrates immune inputs to adjust metabolism and behavior, but it appears to influence the immune response as well. In other words, and probably via a feedback system already demonstrated in many physiologic responses, inflammation influences the brain which in turn influence inflammation. There is a good evidence that the brain, via the vagus nerve, can control systemic inflammation in animal models (Borovikova et al., 2000; Wang et al., 2003). In particular, it appears that acetylcholine, the main neurotransmitter of the vagus nerve, can inhibit the production of pro-inflammatory cytokines by signaling through nicotinic receptors of macrophages (Fig. 8; Ulloa, 2005). Consequently, this mechanism has been called "the nicotinic anti-inflammatory pathway," since acetylcholine exerts its anti-inflammatory effects via the a7-nicotinic–acetylcholine receptor (a7 nAChR; Wang et al., 2004). Interestingly, nicotine is more efficient than acetylcholine at inhibiting cytokine production since nicotine is a more selective cholinergic agonist. This evidence may explain the well-established clinical knowledge that Crohn's disease, which can be described as a chronic inflammatory status of the intestine mediated and sustained by cytokines, is less prevalent and severe among smokers than among nonsmokers.

The molecular mechanisms responsible for the anti-inflammatory effects of nicotine are currently being detailed. It appears that nicotine prevents the endotoxin-induced activation of the nuclear factor-κB (NF-κB) pathway, which is critical for the production of pro-inflammatory cytokines

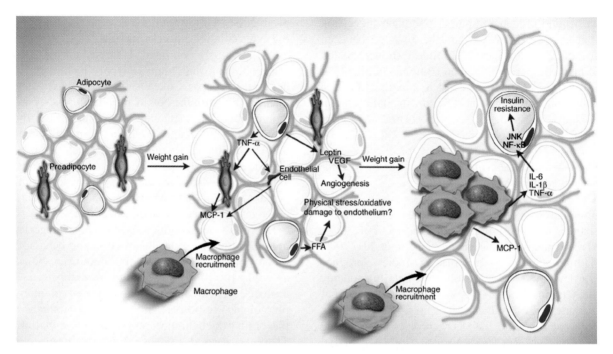

Fig. 7. Obese adipose tissue is characterized by inflammation and progressive infiltration by macrophages as obesity develops. Changes in adipocyte and fat pad size led to physical changes in the surrounding area and modifications of the paracrine function of the adipocyte. For example, in obesity, adipocytes begin to secrete low levels of TNF-α, which can stimulate preadipocytes to produce monocyte chemoattractant protein-1 (MCP-1). Similarly, endothelial cells also secrete MCP-1 in response to cytokines. Thus, either preadipocytes or endothelial cells could be responsible for attracting macrophages to adipose tissue. The early timing of MCP-1 expression prior to that of other macrophage markers during the development of obesity also supports the idea that it is produced initially by cells other than macrophages. Increased secretion of leptin (and/or decreased production of adiponectin) by adipocytes may also contribute to macrophage accumulation by stimulating transport of macrophages to adipose tissue and promoting adhesion of macrophages to endothelial cells, respectively. It is conceivable, also, that physical damage to the endothelium, caused either by sheer size changes and crowding or oxidative damage resulting from an increasingly lipolytic environment, could also play a role in macrophage recruitment, similar to that seen in atherosclerosis. Whatever the initial stimulus to recruit macrophages into adipose tissue is, once these cells are present and active, they, along with adipocytes and other cell types, could perpetuate a vicious cycle of macrophage recruitment, production of inflammatory cytokines, and impairment of adipocyte function. (From Wellen et al., 2003).

(Li and Verma, 2002), via a7 nAChR signaling (Wang et al., 2004).

The potential therapeutic exploitations of the nicotinic anti-inflammatory pathway are many, and are currently being tested. We have been previously shown that nicotine-induced reduction of food intake is mediated via derangement of brain neurochemistry (Miyata et al., 1999; Ramos et al., 2004a) in normal rats. More recently, we hypothesized that nicotine administration may improve food intake in anorectic tumor-bearing rats via its signaling through the nicotinic anti-inflammatory pathway. To test this hypothesis we used the Fischer rat-methylcholanthrene-sarcoma model, since this animal model of cancer-induced anorexia serves very well to the hypothesis since anorexia is mediated by systemic and central hyperproduction of TNF-α and IL-1 (Smith and Kluger, 1993; Opara et al., 1995a; Chance et al., 2003), and pharmacological inhibition of these two cytokines has been demonstrated effective in ameliorating anorexia (Torelli et al., 1999; Laviano et al., 2000). First, we tested whether repeated nicotine administration had any permanent effects on food intake of normal rats, and we observed that nicotine (three injections/day for three consecutive days each week) reduces food intake in a dose-dependent manner, as expected, but this effect progressively

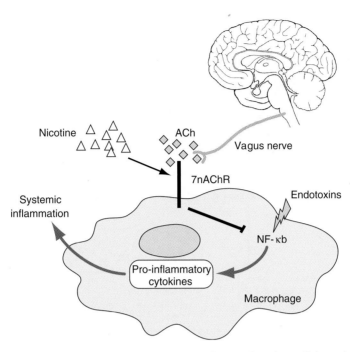

Fig. 8. After injury, injection or trauma, endotoxins activate a number of macrophage intracellular pathways, including the NF-κB pathway, which is critical for the production of proinflammatory cytokines. Cytokines then promote and sustain systemic inflammation. To compensate for the increasing inflammatory status, the vagus nerve releases acetylcholine (ACh) which signals through the a7-nicotinic-acetylcholine receptor (a7 nAChR) to inhibit NF-κB induced macrophage activation and cytokine production. Compared to ACh, nicotine is more selective at activating a7 nAChR and efficient at inhibiting proinflammatory cytokines. This pathway has been named as "the nicotinic anti-inflammatory pathway" (Adapted from Ulloa, 2005).

fades away with time, and disappears after 3 weeks. In tumor-bearing rats, we decided to test different injection schedules of the minimal effective dose, and preliminary results shows that repeated nicotine administration improves food intake in anorectic tumor-bearing rats and prolongs survival. Analyses are being carried out to test whether these important effects are associated with reduced production of pro-inflammatory cytokines.

Summary of afferent signals

The sums of these divergent and dissimilar studies reveal common immune responses that modify behavior and initiate an immune response by (i) monocytes elaborating cytokines in response to either a physiological or a pathological event; (ii) peripheral immune sensors detect these: the efferent vagal fibers in the gut/liver or the somatic preganglionic fibers in the tissue; and (iii) transduced

via an increase in vagal efferent activity is transmitted to the dorsomotor vagal complex of the medulla, which projects onto the hypothalamus including the preoptic area with activation of the PVN (Elmquist and Saper, 1996); (iv) an afferent neural response leads to an increased release of T cells from the thymus and a delayed uptake and destruction of B cells by the spleen and (v) a neuroendocrine event occurs between the hypothalamic nuclei to modulate the metabolic event via the HPA axis.

Central Integration

Figure 9 shows the relationship between the first- and second-order neurons in the ARC, and their susceptibility to hormones and cytokines. Much data contributing to our understanding of central regulatory function has emanated from our studies on continuous intravenous infusion of nutrients and cancer anorexia.

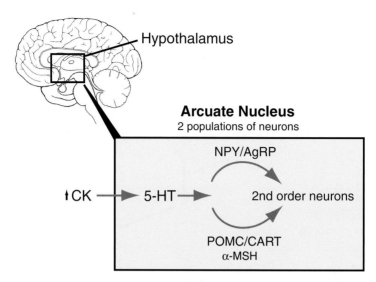

Fig. 9. Peripheral signals such as cytokines (CK) reach the hypothalamus, specifically the arcuate nucleus (ARC), where they interact with two neuronal populations, which project to second-order neuronal signalling pathways. Neuropeptide Y/Agouti-related peptide (AgRP) neurons stimulate food intake. Pro-opiomelanocortin (POMC)/cocaine and amphetamine-regulated transcript (CART) neurons inhibit food intake. The effects of cytokines on hypothalamus seem to be mainly mediated by 5-HT. (From Laviano, et al., 2003).

As tumor progresses and grows, interaction between tumor and immune system are established affecting body metabolism from cellular to behavioral level both peripherally and centrally. Tumor growth leads to anorexia (food intake reduction) (Meguid et al., 1987; Kurzer et al., 1988; Makarenko et al., 2003; Meguid et al., 2004; Ramos et al., 2004c). Cytokines acting in endocrine, paracrine and autocrine fashion, play a key role in this relationship establishing a link between tumor and metabolism as well as behavior. During tumor development there is an imbalance between pro-inflammatory, IL-1, IL-6, TNF-α and anti-inflammatory cytokines, such as IL-10, that causes changes in monoaminergic and peptidergic systems, most of them identified in feeding and energy homeostasis control, in both whole brain and hypothalamus (Noguchi et al., 1996; Cravo, 2000; Plata-Salaman, 2000; Makarenko et al., 2002, 2003).

Although there is much evidence showing the involvement of peripheral IL-1 in cancer anorexia pathogenesis, it has been demonstrated that IL-1 is also synthesizing at CNS which together with that synthesized peripherally acts directly in the CNS to inducing anorexia (Gelin et al., 1991; Plata-Salaman, 1991; Plata-Salaman and Ffrench-Mullen,

1992; Yang et al., 1994; Plata-Salaman et al., 1998; Turrin et al., 2004). Cytokine receptors have been found in the CNS including hypothalamus (Cunningham and De Souza, 1993) with highest abundance in VMH (Yabuuchi et al., 1994). Furthermore, small pathophysiological dose of IL-1α are required to induce anorexia when these are injected centrally, while pharmacological dose are needed to obtain the same effect when injected peripherally (Plata-Salaman, 1996; Plata-Salaman et al., 1996). We measured the content of IL-1α in cerebrospinal fluid obtained from tumor-bearing rats, achieved by inoculation of methyl-cholanthrene-induced sarcoma cells, finding reduction of food intake in tumor-bearing rats during anorectic phase compared to pre-anorectic phase as well as to controls. Furthermore, cerebrospinal fluid IL-1α correlated negatively with food intake and positively with tumor weight (Opara et al., 1995a) indicating that central IL-1α plays a role in the pathogenesis of cancer anorexia to conserve energy and to induce mobilization of nutrients for the defense of the host. The data also suggest that other anorectic cytokines (IL-6, IL-8 and TNF-α) may contribute to this phenomenon. These data agree with previous reports showing that IL-1α, IL-1β, IL-8 and TNF-α exert a general anorectic effect at the central level (Chance and Fischer, 1991; Plata-Salaman and Ffrench-Mullen, 1992; Fantino and Wieteska, 1993; Plata-Salaman and Borkoski, 1993; Yang et al., 1999).

Arcuate nucleus

Several experimental findings point the ARC as one of the hypothalamic structures playing a role in the effects of cytokines, particularly IL-1. The medial part of the ARC contains cells that express IL-1R1 (Ericsson et al., 1995) and is activated by systemic IL-1 administration (Herkenham et al., 1998; Reyes and Sawchenko, 2002). Besides, the medial part of the ARC can bind peripheral peptides or proteins because of its localization close to median eminence.

MohanKumar et al. (1999) showed in adult male rats that after intraperitoneal administration of lipopolysaccharide (10 μg/kg body weight) there was an increase (more than twofold) in DA concentrations on ARC compared to control rats. However, lipopolysaccharide treatment did not alter the content of other neurotransmitters, such as 5-HT, NE or its metabolites, in ARC. IL-1β may mediate these changes in DA given that the pretreatment with IL-1ra avoid the increase in DA content (MohanKumar et al., 1999; Fig. 10). In addition, systemic lipopolysaccharide administration increases noradrenergic and serotonergic metabolism (Delrue et al., 1994) and activates tryptophan hydroxylase (the first enzyme in the route of synthesis of monoamines) and tyrosine hydroxylase (the rate-limiting enzyme in monoamine synthesis) in several brain areas including ARC, PVN and posterior pituitary gland (MohanKumar et al., 1999; Nolan et al., 2000; De Laurentiis et al., 2002). In a more recent study, Gonzalez et al. (2004) found that intracerebroventricular injection of lipopolysaccharide in male rats caused a transitory and strong immunoreaction to IL-1β in ARC microglial cells parallel with a decline in the number of tyrosine hydroxylase-and tyrosine hydroxylase mRNA-positive cells and in tyrosine hydroxylase activity (the rate-limiting enzyme in monoamine synthesis) in median

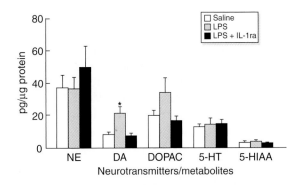

Fig. 10. Effects of lipopolysaccharide (LPS) on neurotransmitter concentrations in the arcuate nucleus (ARC). Lipopolysaccharide treatment increased the concentration of DA in the ARC $p < 0.05$. Treatment with IL-1ra completely blocked this effect. Other neurotransmitters were unaffected. (From MohanKumar et al., 1999).

eminence and, at 12 h, an elevation in prolactin concentrations in serum. These results suggest that hypothalamic catecholaminergic system are involved in the control of autonomic and neuroendocrine responses to peripheral and central inflammation.

Other hypothalamic systems involved in the regulation of energy homeostasis affected by cytokines are neuropeptide Y (NPY) and α-MSH and pro-opiomelanocortin (POMC). Thus, in tumor-bearing rats at the onset of anorexia NPY immunoreactivity in ARC was lower than in non-tumor-bearing rats while α-MSH was greater (Ramos et al., 2005) and these changes were reverted when the rats were fed with ω-3FA as well as the inhibition of food intake. In tumor-bearing rats fed with ω-3FA diet NPY immunoreactivity in ARC was greater than in tumor-bearing chow-fed rats (Ramos et al., 2005). After peripheral administration of IL-1β in rats Reyes and Sawchenko, (2002) detected greater percentage of activated neurons expressing NPY, which co-express Agouti-related peptide (AgRP) (Broberger et al., 1998) and POMC, which co-express Cocaine-amphetamine regulated transcript (CART) (Elias et al., 1998), than after vehicle administration by measuring of Fos induction and these changes were accompanied by a decline in food intake.

In a recent study carried out in C57BL/6J mice, Rossi-George et al. (2005) have described a significant increase in cFos expression in ARC, as well as in PVN, after intraperitoneal injection of 10 μg of staphylococcal enterotoxin A, a superantigen that activates T lymphocytes and induces production of different cytokines such as TNF-α, INT-γ and IL-2 (Bette et al., 1993; Rosendahl et al., 1997) and have neurobiological actions including the activation of HPA axis, sympathetic

nervous system and anorexia (Shurin et al., 1997; Kusnecov et al., 1999; Del Rey et al., 2002; Pacheco-Lopez et al., 2004), or 5 μg of lipopolysaccharide (Fig. 11A) along with a reduction in food intake associated to anxiety/fear-like process and to cFos activation in limbic brain regions. The cFos induction after staphylococcal enterotoxin A seems to be mediated by TNF-α, which

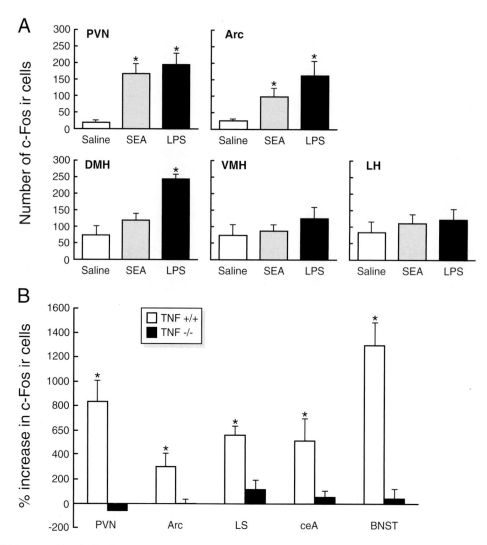

Fig. 11. (A) Mean number of c-Fos-immunoreactive cells in hypothalamic nuclei 2 h after intraperitoneal injection with saline, 10 μg of SEA, or 5 μg of lipopolysaccharide (LPS). Each bar represents the mean ± SEM. *$p < 0.05$ relative to saline. (B) Percentage increase in the number of c-Fos-immunoreactive cells in the brains of SEA-challenged wild-type (TNF + / +) and TNF-α knockout (TNF-/-) mice. For each individual animal that was given an injection of SEA, the quantitation for each brain region was expressed as a percentage above the group mean of the corresponding saline-injected control of the same strain. The mean number of c-Fos-positive cells in wild-type saline-injected and saline-injected TNF-α knockout mice did not differ. (From Rossi-George et al., 2005).

plasma levels were increased after staphylococcal enterotoxin A injection, given that in TNF$-/-$ mutant mice cFos induction was not observed (Rossi-George et al., 2005; Fig. 11B).

Paraventricular nucleus

Elmquist and Saper (1996) demonstrated, using cFos as an immunohistochemical marker of neuronal activity, that neurons in both the autonomic and endocrine components of the PVN were activated by lipopolysaccharide. Several of the activated cell groups directly projected to the PVN including the visceral motor complex, median preoptic nucleus, ventromedial preoptic area, nucleus of the stria terminalis, parabrachial nucleus, ventrolateral medulla and nucleus tractus solitarius. These findings indicate that the stimulation of the immune system activates cell groups from medial nervous systems that project on to the PVN and are consistent with the postulate that the PVN plays a key role in integrating diverse physiological cues into the varied manifestations that constitute the cerebral component of the acute phase response (Elmquist and Saper, 1996).

Cytokines induce their anorectic effects partly via the effects on the neurons of the PVN. In support of this, anorexia induced by experimental colitis in rats can be prevented by intracerebroventricular administration of IL-1ra that leads to a 18-fold reduction in PVN 5-HT associated with a significant increase in food intake (El-Haj et al., 2002).

In normal rats, the inhibition of 5-HT synthesis or the blockage of their receptors in PVN causes an increase of NPY concentration suggesting a link between both regulators (Currie and Coscina, 1997). This corresponds to the data that reports an inhibition in NPY system in different types of anorexia (Pich et al., 1992; Broberger et al., 1997, 1999), particularly those achieved after ventricular administration of ciliary neurotrophic factor, known to cause a decrease in NPY gene expression (Pu et al., 2000), or those associated with cancer (Chance et al., 1994a, b, 1998). Under these experimental conditions the injection of NPY into the PVN in tumor-bearing rats results in a decrease in food intake while in their pair-fed controls a significant increase in food intake occurs, suggesting a

refractory feeding response to NPY in the tumor-bearing rats mediated by unknown mechanisms (Chance et al., 1994a; Inui, 1999). Furthermore, the infusion of NPY into the III ventricle inhibits and reverses the anorexia induced by both pathophysiological and pharmacological concentrations of IL1-β (Sonti et al., 1996). As shown in Fig. 12, our results are in keeping with these data, revealing a significant reduction in the NPY in PVN of tumor-bearing rats (Ramos et al., 2004c) accompanied by an increase in PVN 5-HT. Figure 13 of immunohistochemical sections of the hypothalamus, support the finding by showing increased staining of 5-HT$_{1B}$ receptor proteins, indicating, that there is substantial serotoninergic innervations in PVN (Card and Moore, 1988; Makarenko et al., 2005b). NPY, one of the most potent orexigenic agents (Dryden et al., 1994) can affect many different neuronal systems given its wide distribution and its ubiquitous receptors in the hypothalamus. Of these the serotoninergic system has co-receptors (Fig. 13) since, as we have mentioned above, there is a close association between NPY-ir fibers and hypothalamic neurons expressing 5-HT$_{1B}$ receptor in normal and in tumor-bearing rats (Makarenko et al., 2002). This interaction is bi-directional, since, in normal rats, an increase in NPY release with its subsequent enhanced of intake in food, has been reported after PVN injection of 5-HT receptor antagonist (Dryden et al., 1995). Furthermore, intraperitoneal or intracerebroventricular injection of serotonin agonist decrease the NPY concentration and inhibit stimulatory effect on food intake (Rogers et al., 1997; Currie et al., 2002). In recent studies, we have reported in tumor-bearing rats abnormal levels of hypothalamic 5-HT, NPY and DA at the onset of anorexia (Fig. 14). There is an increase in PVN 5-HT concentrations along with a decrease in NPY and DA concentrations. To confirm these findings, the tumor was resected with the expectation that the observed changes normalized. After tumor resection, food intake subsequently normalized and the concentration of both the monoamines and the NPY also return to normal values (Meguid et al., 2004; Ramos et al., 2004c). Using a pair fed vs. control group we verified that the reduction of food intake was induced by the changes observed in NPY and 5-HT in the tumor-bearing

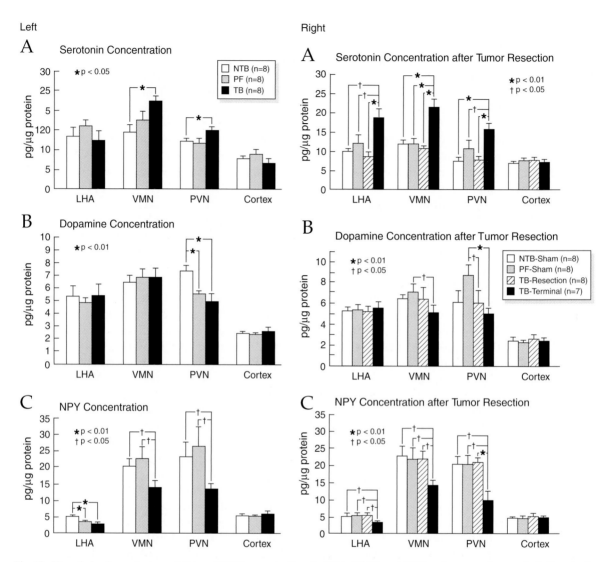

Fig. 12. Hypothalamic serotonin, and DA and NPY concentrations in LHA, VMN, and PVN at the onset of anorexia (left) and after tumor resection and terminal state (right). In tumor-bearing rats there was a significant reduction in the NPY in PVN of tumor-bearing rats accompanied by an increase in PVN 5-HT and an increase in bilateral VMH 5-HT content with a concomitant decrease of DA concentrations content while NPY decreases. LHA 5-HT content was increased in rats who were allowed to live until terminal state, while LHA NPY concentration was lower in tumor-bearing rats than in non-tumor-bearing rats at both the onset of anorexia (mean day 19) and at terminal state. (From Ramos et al., 2004c).

rats, while DA reduction seems to be due to the decrease in food intake because the decline in food intake also occurred in the pair-fed group. These data suggest the existence of a dynamic interaction between brain amines and NPY in tumor-bearing rats.

The changes in serotoninergic system at the onset of anorexia not only affect the levels of this neurotransmitter, but also the expression of

5-HT_{1B} receptors, one of the most important serotonin receptors mediating the anorectic effect of 5-HT receptors (Barnes and Sharp, 1999; Makarenko et al., 2002), as we have reported very recently (Makarenco et al., 2005a) using a peroxidase–antiperoxidase immunocytochemical methods and semiquantitative image analysis of 5-HT_{1B} receptors immunostaining. This study

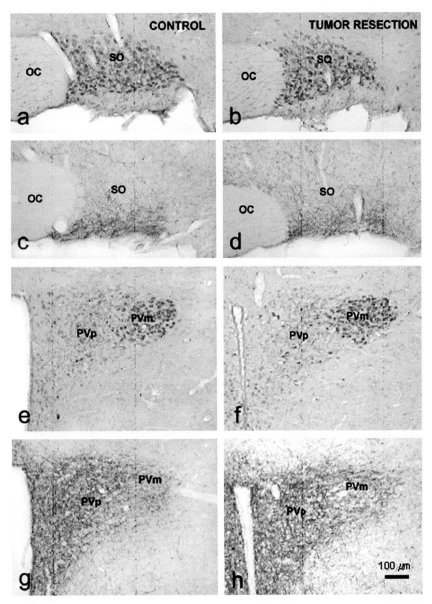

Fig. 13. Immunocytochemical visualization of 5-HT$_{1B}$-receptors (a, b, e and f) and NPY immunoreactive fibers (c, d, g and h) in the hypothalamus of Control and tumor resected (TB-R) rats. SO, supraoptic; PVm, magnocellular part of paraventricular nucleus; PVp, parvocellular part of paraventricular nucleus; OC, optic chiasm. In tumor-bearing rats there was an increased staining of 5-HT$_{1B}$ receptor proteins indicating a substantial serotoninergic innervations in PVN. (From Makarenko et al., 2005b).

shows the same hypothalamic distribution of this receptor in both tumor-bearing rats and non-tumor-bearing rats, but higher immunostaining intensity in most neurons of the magnocellular PVN (but not in parvocellular division). It is generally accepted that most of the magnocellular neurons of the PVN, and also of the SON, produce oxytocin and vasopressin (Sawchenko and Swanson, 1983), which besides their involvement in water balance control, participate in feeding regulation exerting an anorectic action (Arletti et al., 1990; Olson et al., 1991). The activation of

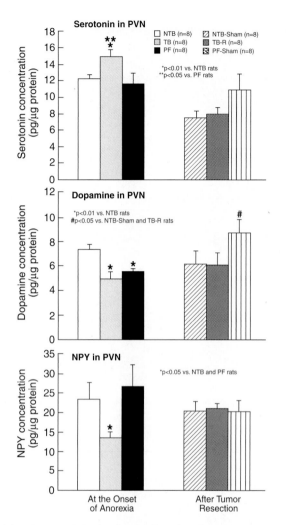

Fig. 14. Hypothalamic PVN 5-HT, DA and NPY concentrations. In tumor-bearing rats, there was an increase in PVN 5-HT concentrations along with a decrease in NPY and DA concentrations. These changes normalized after tumor resection. Values are mean ± SEM. *$p < 0.01$ vs. NTB rats; **$p < 0.05$ vs. PF (pair fed) rats. *# $p < 0.05$ vs. NTB (Sham) and tumor-bearing rats. (From Meguid et al., 2004).

5-HT$_{1B}$ receptors on the magnocellular neurons modulates the release of both hormones and in turn food intake. Furthermore, it has been described that more than 95% of the oxytocin and vasopressin neurons in SON and PVN also express NPY-Y$_5$ receptors (Campbell et al., 2001). In a subsequent study using the same methods as we used in our reported studies we showed that the changes reported at the onset of anorexia in

hypothalamic distribution of NPY and 5-HT$_{1B}$ receptors are reverted after tumor resection (Makarenko et al., 2005a). These data show that tumor resection, and therefore the removal of the effect of cytokines, results not only in an enhancement of food intake, reaching the normal levels, but also in a reversible changes of hypothalamic orexigenic and anorectic modulators.

Ventromedial hypothalamus

As indicated by the data of Elmquist and Saper (1996) another primary hypothalamic site where cytokines regulates metabolism is VMH, a known satiety center, which send potent excitatory signals to ARC POMC neurons (Sternson et al., 2005). We injected a pathophysiological quantity of IL-1α into the VMH of rats and caused a significant reduction of food intake (Yang et al., 1999). The mechanism of action may involve the modulation of the serotoninergic and dopaminergic systems, because when we were measuring the release of 5-HT and DA's metabolites we found an increase in the concentrations of 5-HT, 5-HIAA (the main metabolite of 5-HT) and DA just after the injection of IL-1α in VMH. This remained above basal levels during the next 40–60 min, respectively (Yang et al., 1999), clearly linking the anorectic effect of IL-1α and the early development of satiety to neurotransmitters, particularly enhance serotoninergic activity. Other authors have also demonstrated the relationship between cytokines and hypothalamic neurotransmitters (Plata-Salaman, 1997). These interactions also extend to neuropeptides and hormones, because Smagin et al. (1996) reported an enhancement in NE hypothalamic content concurrent to an increase in plasma ACTH and corticosterone concentrations after both intravenous and intraperitoneal injection of IL-1β, indicating that the anorectic effect of this cytokine can be mediated by an increase in hypothalamic NE given that this neurotransmitter regulate HPA axis (Plotsky et al., 1989).

In tumor-bearing rats, we have found an upregulation of D$_1$- and D$_2$- receptor mRNA expression in VMH during anorectic period (Sato et al., 2001b, Fig. 15). Our data suggest that tumor-released cytokines regulates food intake through modulation of dopaminergic activity that may take place in VMH as well as in SON (Sato et al., 2001b, a). It has been demonstrated that VMH serotoninergic activity enhanced during cancer anorexia in methylcholanthrene tumor-bearing rats and returned to control levels along with a normalization of food intake once tumor was removed supporting the involvement of serotoninergic system in IL-1α-induced anorexia (Blaha et al., 1996). Furthermore, VMH microinjection of mianserin, 5-HT$_{1c/2}$ antagonist receptor, or IL-1ra, an endogenous inhibitor of IL-1α in anorectic methylcholanthrene tumor-bearing rats improve food intake by an increase in meal number without effect on meal size, as measured using the Automated Computerized Rat Eater Meter (Meguid et al., 1990), while in normal rats this effect does not occur (Laviano et al., 2000). These findings suggest that under normal conditions VMH 5-HT has only a relative importance in the control of meal number and meal size. However, during tumor growth VMH 5-HT participation in the control of food intake homeostasis becomes relatively important. By integrating these data we suggest that during the progressive tumor growth IL-1α may cause anorexia by a central mechanism that involves VMH 5-HT (Kuriyama et al., 1990).

As shown in Fig. 12 Ramos et al. (2004c) found an increase in bilateral VMH 5-HT content with a concomitant decrease of DA concentrations content in tumor-bearing rats at the onset of anorexia, while NPY decreases. These changes, that were accompanied by a reduction in total body fat in tumor-bearing rats as well as a significant decrease of food intake, are normalized 9 days after tumor resection, while these indices continued to be abnormal in tumor-bearing control rats until death.

Lateral hypothalamic area

In our studies, as shown in Fig. 12, 5-HT and DA concentration in LHA in tumor-bearing rats was not significantly different from nontumor-bearing rats at the onset of anorexia but 5-HT content was increased in rats which were allowed to live until the terminal state, while NPY concentration was

lower in tumor-bearing rats than in nontumor-bearing rats at both onset of anorexia (mean day 19) and at the terminal state. And these changes were accompanied by a significant reduction in food intake during the study and thus were associated with a reduction in total body fat (Ramos et al., 2004c). As described above, after tumor resection, performed to validate the tumor as the etiology of the acute phase response cytokine changes, the changes reported had reverted 9 days after tumor resection, while these indices continued to be abnormal in their tumor-bearing control cohorts until their death.

In tumor-bearing rats there is an increase of mRNA of both D$_1$- and D$_2$- receptor in LHA (Sato et al., 2001b; Fig. 15). Furthermore, LHA

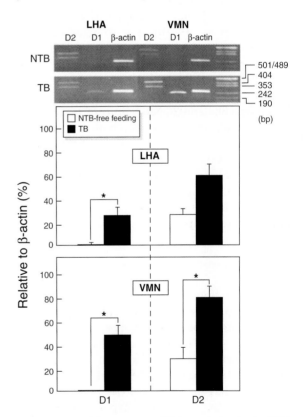

Fig. 15. D$_1$- and D$_2$-receptor mRNA expression in the LHA and VMN in anorectic tumor-bearing and non-tumor-bearing free-feeding control rats. Data are expressed as percent change relative to β-actin (β-Act). *$p < 0.05$. In tumor-bearing rats, an upregulation of D$_1$- and D$_2$- receptor mRNA expression in VMH and LHA during anorectic period was found. (From Sato et al., 2001b).

DA release is increased during cancer anorexia (Chance et al., 1991). However, using microdialysis, we found that continuous peripheral IL-1α infusion in normal rats did not cause modifications in LHA DA content (Yang and Meguid, 1995), suggesting that other cytokines were also a contributory factor involved in the biochemical changes observed in LHA in cancer anorexia.

Supraoptic nucleus

The SON, besides its participation in the control of water balance, are also involved in the regulation of food intake by mean of the secretion of vasopressin and oxytocin (anorectic neurohypophyseal hormones) into hypothalamic–pituitary portal circulation (Arletti et al., 1989: Langhans et al., 1991b) after stimulation of D_2 receptors by DA released by dopaminergic neurons from the ventrotegmental area in the brain stem. Sato et al. (2001a) have reported that an injection of D_2 receptor antagonist (sulpiride, 4 μg/0.5 μl) into bilateral SON of anorectic tumor-bearing rats results in an increase in both meal size and meal number leading to an improvement of food intake (Fig. 16) and therefore in body weight as we have also observed when sulpiride was injected into LHA and VMH (Sato et al., 2001b), although in these nuclei the increase in food intake is achieved by an increase in meal number only.

Efferent response of acute phase response

During infection, tissue injury, inflammatory or malignant process immune-derived cytokines exert strong neuroendocrine effects resulting in marked changes in host homeostasis. One of the most affected metabolic pathways is carbohydrate metabolism. IL-1, one of the main inflammation mediators, has the capacity to elevate glucocorticoid levels by stimulating hypothalamic CRF-producing neurons (Berkenbosch et al., 1987; Sapolsky et al., 1987; Del Rey et al., 1998). It has been observed that IL-1β intraperitoneal administration increases glucocorticoid and glucagon production, that stimulate the mobilization of glucose stores, and decreases hepatic glycogen content

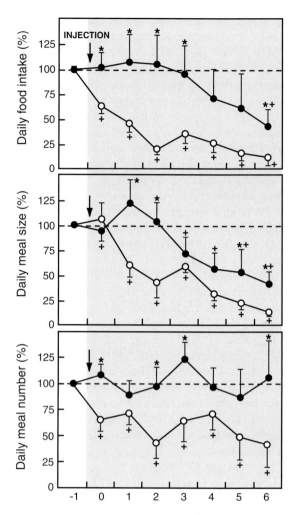

Fig. 16. Changes in food intake (top panel), meal size (middle panel) and meal number (lower panel) after an injection of sulpiride/saline into the supraoptic nucleus in tumor-bearing rats. Food intake, meal size and meal number on day −1 before injections was defined as 100%. *$p < 0.05$ vs. control group. +$p < 0.05$ vs. data on day 0 in each group. D_2 receptor antagonist injection caused an increase in both meal size and meal number leading to an improvement of food intake. (From Sato et al., 2001a).

in mice, but these changes are accompanied by hypoglycemia (Del Rey and Besedovsky, 1987; Del Rey et al., 1998) which is even more marked under fasting conditions and is sustained even after glucose load (Fig. 17), suggesting that the glucose fast mobilized from the liver is rapidly incorporated into other tissues such as fat or muscle by an increase in glucose transport elicited by IL-1

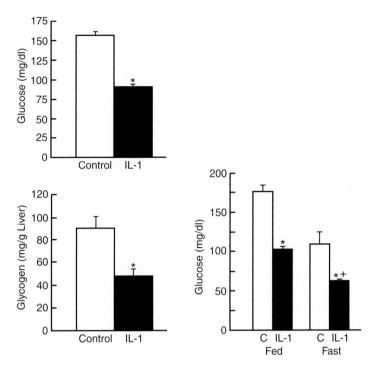

Fig. 17. Intraperitoneal injection of IL-1β induced hypoglycemia accompanied by a decrease in glycogen content in mouse liver (left). Each bar represent mean ± SEM. *$p < 0.05$ vs. control mice, + $p < 0.05$ vs. *fed* ad libitum mice. The hypoglycemia is even more marked under fasting conditions (24 h fasting, right). (From Del Rey et al., 1998).

(Garcia-Welsh et al., 1990; Bird et al., 1990; Shikhman et al., 2001; Fischereder et al., 2003). Further, it has been demonstrated that 2-deoxy-glucose uptake by peripheral tissues (heart, spleen, lung, liver and tumor) was enhanced in mice bearing IL-1β-secreting tumor (Metzger et al., 2004); this increase in glucose uptake may be mediated by a nondependent insulin enhancement of hepatic mRNA expression of the glucose transporter 3 (GLUT-3) by IL-1β. IL-1β, as well as IL-6, inhibit the enhancement of glycogen deposition induced by insulin in primary rat hepatocyte cultures increasing [^{14}C]-glycogen degradation, decreasing [^{14}C]-glucose incorporation into glycogen, stimulating glycogen phosphorylase activity and inhibiting glycogen synthase activity (Kanemaki et al., 1998) which are the rate-limiting enzymes in glycogen metabolism. It was also reported that the inhibition of glycogen synthesis by pro-inflammatory cytokines in both in vitro and in vivo models by Kitano et al. (2002) and Metzger et al. (2004) respectively, as well as the inhibition of

gluconeogenesis in vitro by Yerkovich et al. (2004). It has been observed that cytokines inhibit gluconeogenesis induced by glucagon (Stadler et al., 1995; Christ and Nath, 1996). The inhibition of hepatic glucose synthesis was elicited by a downregulation of phosphoenolpyruvate carboxykinase and glucose-6-phosphatase activities induced by cytokines (Metzger et al., 2004; Yerkovich et al., 2004) as well as by the action of cytokines on other enzymes of gluconeogenesis or glycolysis (Ceppi et al., 1992; Metzger et al., 1997; Maitra et al., 2000).

The hypoglycemic effect of cytokines may be due to, at least in part, an increase in insulin levels (Del Rey and Besedovsky, 1987) but is more probable that it can be mediated by activation of brain IL-1 receptors (Del Rey et al., 1998) given that the hypoglycemic effect was also found in insulin-resistant diabetic mice and in adrenalectomized mice, where there is no hyperinsulinemia (Del Rey and Besedovsky, 1989), and this central action may involve effects on central mechanisms controlling glucose homeostasis leading to a downregulation of glucose set point (Del Rey and Besedovsky, 1992; Del Rey et al., 1998). In this central effect of IL-1, hypothalamic catecholamines seen to play a role counteracting the effect of cytokines on glucose concentrations given that its central depletion accentuated hypoglycemia. (Del Rey et al., 1998).

One of the interfaces between cytokines and glucose metabolism may be 5-HT, as was proposed by MohanKumar et al. (1999). These authors suggest that the increase in PVN 5-HT activity observed in rats after lipopolysaccharide intraperitoneal injection, which is mediated by IL-1β, could play a role in HPA axis activation given that 5-HT fibers innervate CRF perikarya (Sawchenko et al., 1983) and changes in PVN 5-HT concentrations markedly altered CRF release (Feldman et al., 1987).

During inflammation and infection process there are many changes in host lipid and lipoprotein metabolism including an increase on adipose tissue lipolysis, hepatic reesterification of fatty acid and hepatic lipogenesis as well as a decline in fatty acid oxidation in several tissues such as liver, heart and skeletal muscle (Lanza-Jacoby and Tabares, 1990; Takeyama et al., 1990; Feingold et al., 1992; Hardardottir et al., 1994; Khovidhunkit et al., 2004). These changes can be achieved by administration of lipopolysaccharide and pro-inflammatory cytokines suggesting that these proteins are involved in the mediation of many of the host metabolic responses that take place during inflammation and infectious diseases (Hardardottir et al., 1994). Many of these effects of cytokines on lipid metabolism are mediated by the modulation of synthesis and activity of some enzymes involved in the metabolism of lipids. For example, lipopolysaccharide and cytokines reduce mRNA expression of fatty acid translocase and fatty acid transport protein in muscle, heart and adipose tissue of Syrian hamster (Memon et al., 1998a). Further, lipopolysaccharide, TNF-α and IL-1 decrease mRNA levels and activity of acyl-CoA synthetase in several tissues including liver and adipose tissue of Syrian hamster (Memon et al., 1998b; Fig. 18) enhancing fatty acid reesterification, suppressing fatty acid oxidation and stimulating lipogenesis

and therefore leading to elevated plasma triglycerides and very low-density lipoprotein (Feingold et al., 1991; Memon et al., 1993; Nachiappan et al., 1994).

Besides these effects on carbohydrate metabolism, IL-1 acts on protein and lipid metabolism (Del Rey and Besedovsky, 1987; Klasing, 1988; Argiles et al, 1989; Kanemaki et al., 1998; Kitano et al., 2002; Matsuki et al., 2003; Metzger et al., 2004; Khovidhunkit et al., 2004; Yerkovich et al., 2004) probably due to the triggering of neuroendocrine responses given that IL-1 induces the release of CRF (Sapolsky et al., 1987), melanocortins and other neuropeptides (Tocci and Schmidt, 1997) as well as a direct effect on metabolic activity of different tissues such as skeletal muscle, liver and adipose tissue. IL-1 acts directly on lipid metabolism by inhibiting lipoprotein lipase activity, which control the availability of lipid fuel in the body (Beutler and Cerami, 1985; Doerrler et al., 1994; Matsuki et al., 2003) and decreasing intestinal lipid absorption and lipid accumulation (Argiles et al., 1989). Furthermore, this cytokines can modulate adipocyte function by suppressing the synthesis of fatty acid transport proteins in adipose tissue and the adipocyte maturation in vitro (Gregoire et al., 1992; Memon et al., 1998a). Using an IL-1ra-deficient (IL-1ra−/−) mice Matsuki et al. (2003) showed that excess IL-1 signaling suppresses weight gain and decreases fat mass without changes in food intake, but the morphology and cell volume of adipocytes is not altered compared with those of the wild-type (Fig. 19), as well as the hypothalamic expression of adiponectin, leptin and resistin or the expression levels of different anorectic and orexigenic hypothalamic feeding regulators but have impaired lipid storage and lipid uptake into adipose tissue, these defects being more accentuated in males than in females. Furthermore, the high IL-1 signaling causes a decrease in serum leptin, insulin and triacylglycerol, and but an enhancement of insulin sensitivity (Fig. 20).

There are some evidences that suggest the participation of IL-6 in the regulation of lipid metabolism. Thus, IL-6-deficient mice develop obesity along with obesity-related metabolic disorders and these alterations are partially abolished by exogenous IL-6 administration (Wallenius et al.,

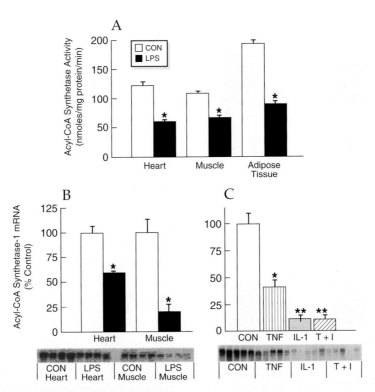

Fig. 18. (A) Effect of intraperitoneal lipopolysaccharide (LPS) on acylCoA synthetase 1 (ACS1) activity in adipose tissue, heart and muscle of Syrian hamsters. *$p < 0.001$; (B) Effect of intraperitoneal lipopolysaccharide on ACS1 mRNA levels in heart and muscle. Values are means ± SEM. *$p < 0.001$; (C) Effect of tumor necrosis factor (TNF), interleukin-1 (IL-1) and the combination of TNF and IL-1 (T 1 I) on ACS1 mRNA levels in liver of Syrian hamsters. *$p < 0.002$, **$p < 0.001$. (From Memon et al., 1998b).

2002a). Furthermore, it has been observed that intracerebroventricular administration of this cytokine acutely stimulate energy expenditure (Rothwell et al., 1991; Wallenius et al., 2002a) and decreases the weight of mesenteric and retro-peritoneal fat pads and circulating leptin levels (Wallenius et al., 2002b; Fig. 21). Also in mice bearing an IL-6-secreting tumor for 18 days a reduction in body fat is observed (Metzger et al., 2001). In healthy humans, a negative correlation between IL-6 cerebrospinal fluid levels and total body weight, subcutaneous and total body fat and serum leptin (Stenlof et al., 2003). Ciliary neurotrophic factor, which is structurally related to IL-6, has been shown to reduce body fat in mice fed with diet-induced obesity (Gloaguen et al., 1997; Lambert et al., 2001) and also affect protein metabolism by inducing protein degradation (Espat et al., 1996).

The efferent signals of acute phase response and the interaction between immune system and hypo-thalamus–pituitary–thyroid axis play an important role because of the effects of the thyroid hormones on metabolism. Immune-derived cytokines, mainly pro-inflammatory cytokines such as TNF-α, IL-1β and IL-6, stimulate the growth and the function of thyroid cells (Armstrong and Klein, 2001). It has been observed that serum thyroid-stimulating hormone concentrations decreases during 5 h following a single injection of IL-1β (Dubuis et al., 1988) followed by a decline in total tetra-iodothyronine and an increase in free tetra-iodothyronine in rats (Wang et al., 1998). Similar results have been reported after continuous infusion of TNF-α, IL-1β and IL-6 in rats (Hermus et al., 1992; Sweep et al., 1992).

Other peripheral mechanisms of cytokines may be the enhancement of the availability of tryptophan (the 5-HT precursor) to maintain

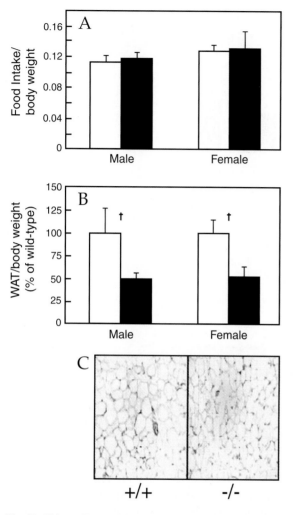

Fig. 19. Using a IL-1ra-deficient (IL-1ra−/−) mice it has been shown that excess IL-1 signaling suppress weight gain and decrease fat mass without changes in food intake, but the morphology and cell volume of adipocytes is not altered compared with those of wild type. (A) Food intake per body weight, (B) white adipose tissue (WAT) weight per body weight and (C) Paraffin sections of WAT from epididymal fat pads in IL-1Ra−/− mice. IL-1Ra−/− (shaded bars, −/−) and wild-type (white bars, +/+) mice and IL-1Ra−/−. Data are expressed as the mean±SEM. *$p<0.05$, † and $p<0.01$ vs. wild-type mice. (From Matsuki et al., 2003).

an elevated 5-HT turnover (Dunn, 1992) given that brain 5-HT synthesis depends on the brain availability of this amino acid (Schaechter and Wurtman, 1990) which is positively correlated to plasma-free tryptophan concentration (Fernstrom and Wurtman, 1972). In different anorexia animal models (Kurzer et al., 1988; Meguid et al., 1992; Muscaritoli et al., 1996; Laviano et al., 1999) and anorectic patients with different diseases (Cangiano et al., 1994; Laviano et al., 1997; Aguilera et al., 2000) an increase in plasma and brain free tryptophan concentrations and brain serotoninergic activity has been reported suggesting a connection between anorexia disease, circulating tryptophan, brain serotoninergic activity and cytokines. Moreover, in anorectic tumor-bearing rats, enhanced free tryptophan circulating levels decrease to normal levels after tumor removal thereby improving food intake (Cangiano et al., 1994). The subcutaneous administration of IL-1α to normal rats during 2 days caused a rise in plasma-free tryptophan associated to a decrease of food intake and subsequently to a reduction in carcass adiposity and body weight (Sato et al., 2003; Fig. 22). Although the injection of tryptophan increases plasma free and total tryptophan there was not a clear effect on food intake. This lack of effect may be due to the newly synthesized 5-HT in the hypothalamus that is not released (Schaechter and Wurtman, 1990). Moreover, the neuronal activity is a decisive factor to hypothalamic 5-HT release, and IL-1α is able to modulate neuronal activity (Bartholomew and Hoffman, 1993). Insulin may contribute to the anorectic effect of IL-1α taking into account that circulating insulin increases after cytokine injection and that there is a negative correlation between food intake and plasma insulin (Sato et al., 2003). It is known that peripheral insulin increase plasma tryptophan and decreases other neutral amino acids (Fernstrom and Wurtman, 1972) that compete with free tryptophan for brain entry (Fernstrom and wurtman, 1972; Landel et al., 1987) leading to an enhanced availability of hypothalamic tryptophan and subsequently to a rise in 5-HT production that may be released after the increase in neuronal activity induced by IL-1α.

Purpose

As we have mentioned above, during infectious disease, trauma or cancer process, host respond with a generalized defense reaction called *acute*

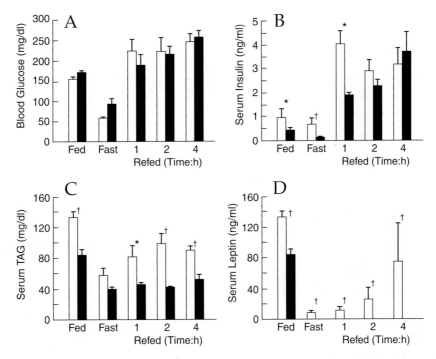

Fig. 20. Decreased serum levels of insulin, triacylglycerol (TAG), and leptin in IL-1Ra−/− mice. (A) Blood glucose, (B) serum insulin, (C) TAG and (D) leptin levels in body weight-matched wild-type (white bars) and IL-1Ra−/− (shaded bars) mice. Data are expressed as the mean±SEM. *$p < 0.05$, †, $p < 0.01$ vs. wild-type mice. (From Matsuki et al., 2003).

phase response which is characterized by alterations in immune, metabolic, endocrine and neural functions as well as behavior (Baumann and Gauldie, 1994) aimed to inhibit the proliferation and spread of the pathogens. These behavioral alterations, such as fever, somnolence, lethargy or anorexia, called "sickness behavior" (Hart, 1990) are mediated by cytokines and are adaptive and beneficial for the host at least during the first phases of infectious or trauma process (Hart, 1988, 1990). For instance, anorexia reduces the energy expenditure due to search for food and decrease the growth of the pathogenic agents by reducing the availability of nutrients coming from food such as free iron, which is indispensable for bacterial proliferation (Weinberg, 1984). However, although these modifications are an essential part in response to different challenges, excessive or long-term production of cytokines or synthesis of cytokines in incorrect biological context compromises host survival and are associated with pathology and mortality in many different diseases.

For example, prolonged changes in structure and function of lipoproteins may contribute to atherogenesis (Khovidhunkit et al., 2004). Therefore, potentially beneficial immunotherapies based on long-term cytokines administration cannot be applied because of these harmful side effects, particularly in the CNS (Smith et al., 1990).

Therapy to reverse acute effects of cytokines on acute phase response

Cytokine metabolism and actions can be modulated not only by pharmacological procedures but also by nutrients. There are some evidences that strongly suggest a role of dietary factors in the regulation of cytokines production (Sasaki et al., 1999; Das et al., 2003; Sato et al., 2003). In particular, it has been observed that diets rich in long-chain ω-3 polyunsaturated fatty acids, such as eicosapentaenoic acid or docosahexaenoic acid inhibits the production of IL-1 and TNF-α

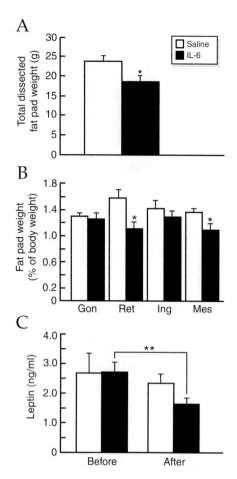

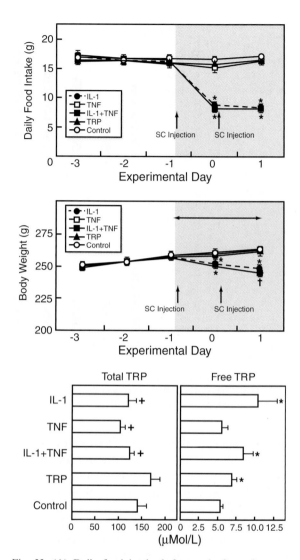

Fig. 21. Dissected fat pads and serum leptin. Three intraabdominal fat pads (gonadal (Gon), retroperitoneal (Ret) and mesenteric (Mes)) and the inguinal (Ing) fat pad (a subcutaneous fat pad in the groin) were dissected. (A) The total weight of the dissected fat pads after two weeks of intracerebroventricular treatment with saline or IL-6 (0.41 g/day). (B) Comparison between the relative weights of the different dissected fat pads (% of body weight) after saline and IL-6 treatment. (C) Leptin levels before and after 2 weeks of intracerebroventricular treatment with saline or IL-6 treatment. (A,B) $*p < 0.05$, vs. control, (C) $**p < 0.01$ vs. before IL-6 treatment. (From Wallenius et al., 2002b).

Fig. 22. (A) Daily food intake before and after subcutaneous injections of cytokines or tryptophan. The shaded area indicates the period influenced by the injections. $*p < 0.005$ vs. the TNF, TRP and control groups. (B) Body weight before and after subcutaneous injections of cytokines or tryptophan. Before injections, there was no significant difference in body weight among the five groups. The asterisks indicate that the changes in body weight in the IL-1 and IL-1 + TNF groups were significantly different vs the other three groups, and, furthermore, that the change in the IL-1 + TNF group was greater than that of the IL-1 group. $*p < 0.05$ vs. the TNF, TRP and control groups. (C) Plasma-total and -free tryptophan among the five groups. $*p < 0.05$ vs. the control group. $+p < 0.05$ vs. the TRP group. $\# p < 0.05$ vs. the TNF, TRP and control groups. The data were mean $\pm$ SEM. (From Sato et al., 2003).

(Endres et al., 1989; Sasaki et al., 1999) as well as reduces its biological activity, and more specifically those related to food intake (Sato et al., 2003). Further, experimental and clinical studies with cancer patients have shown a reduction of weight loss when the diet was enriched with ω-3 fatty acids from fish oil (Dagnelie et al., 1994;

Barber et al., 1999). Moreover in tumor-bearing rats, diets rich in eicosapentaenoic acid diminish tumor growth and ameliorate cachexia (Jho et al., 2002). A marked improvement in food intake, and its two components, body weight and tumor progression in methylcholanthrene tumor-bearing rats fed with ω-3 fatty acids-supplemented diet compared with tumor-bearing rats fed with chow diet are reported (Ramos et al., 2004b; Fig. 23). This diet retarded the appearance of the tumor and reduced its size and weight as it has been described previously (Bartoli et al., 1993; Dagnelie et al., 1994; Chen and Istfan, 2000) and avoided the decrease in both meal size and meal number at the onset of anorexia and in body weight resulting in no difference between tumor-bearing rats fed with ω-3 fatty acids-enriched diet and control groups. The effects of ω-3 fatty acids on the progression of the tumor may be mediated by the inhibition of smooth cells proliferation and the subsequent reduction of vascularization of the tumor (Kremer and Robinson, 1991; Rose et al., 1991). The ω-3 fatty acids sustain food intake during tumor growth by means of maintaining both meal

number and meal size, delaying the onset of anorexia and thus preventing body weight loss. It is likely that the inhibition of TNF-α and IL-1 and leukotriene B4, which enhance IL-1 production, exerted by ω-3 fatty acids (Kunkel and Chensue, 1985; Rola-Pleszczynski and Lemaire, 1985) as well as the inhibition of mononuclear cell proliferation (Meguid and Pichard, 2003) be responsible for the beneficial effects of these fatty acids on food intake. This assumption is supported by the fact that the ARC expression of TNF-α and IL-1 in tumor-bearing rats fed with diet rich in ω-3 fatty acids was lower than in tumor-bearing rats fed with chow diet (Ramos et al., 2004b). Furthermore, we found an increase in ARC and PVN NPY expression in tumor-bearing rats fed with ω-3 fatty acids supplemented diet, measured by mean of microarray analysis. We also reported increased levels of NPY immunoreactivity in ARC and magnocellular division of PVH, but not in parvocellular PVH or SON, and decreased levels of α-MSH in magnocellular PVN and ARC, but not in parvocellular section of PVN, as well as a decrease in immunoreactivity of 5-HT$_{1B}$ receptor

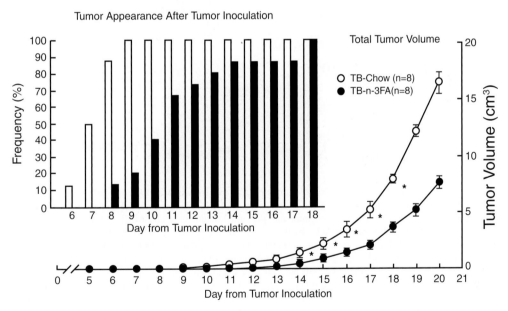

Fig. 23. Tumor appearance and changes of volume in tumor-bearing (TB) rats. Tumor appearance occurred in 100% (8 of 8) within 9 days after tumor inoculation in tumor-bearing Chow rats; in tumor-bearing ω-3 fatty acids rats, tumor appearance occurred in 20% (2 of 8) within 9 days and in 100% (8 of 8) on day 18 after tumor inoculation. *$p > \lambda\tau \sim 0.05$ vs. tumor-bearing ω-3 fatty acids rats. FA, fatty acid. (From Ramos et al., 2004b).

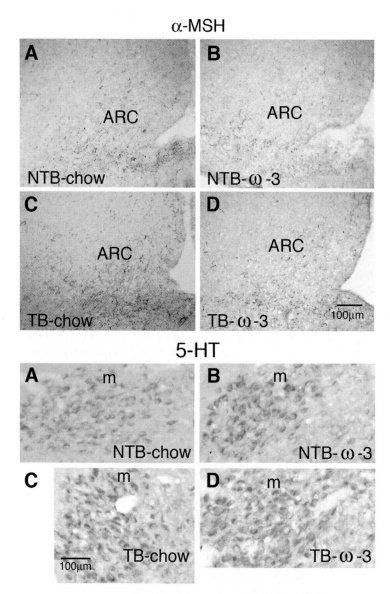

Fig. 24. α-MSH immunoreactivity in ARC (top). In non tumor-bearing rats (panels A and B), ω-3 fatty acids-supplemented diet did not influence α-MSH immunoreactivity. Following tumor injection, an increased expression of α-MSH occurred in tumor-bearing Chow (panel C). The use of a ω-3 fatty acids supplemented diet prevented the increase of α-MSH immunoreactivity in tumor-bearing ω-3 fatty acids rats (panel D). This was demonstrated by the less intense immunoreactivity (64% lower) in tumor-bearing ω-3 fatty acids vs. tumor-bearing Chow. 5-HT$_{1B}$-receptor immunoreactivity in magnocellular PVN (mPVN, bottom). Serotonin is anorexigenic and therefore, 5-HT$_{1B}$-receptors should be upregulated after tumor inoculation. This figure shows that, after tumor inoculation, the increase in 5-HT$_{1B}$-receptors was less pronounced in rats fed a ω-3 fatty acids supplemented diet vs. rats fed chow (panels D vs. C). In tumor-bearing Chow vs. non-tumor-bearing Chow (panel C vs. A), there was an increase of 40% in 5-HT1B-receptor immunore-activity ($p<0.05$), while in tumor-bearing ω-3 fatty acids vs. non tumor-bearing ω-3 fatty acids (panels D vs. B), a nonsignificant increase of 14% occurred, demonstrating that the ω-3 fatty acids supplemented diet prevented the upregulation of 5-HT$_{1B}$-receptors. Scale bar = 100 μm. V, ventricle. (From Ramos et al., 2005.)

in SON (Ramos et al., 2005; Fig. 24). The ω-3 fatty acids mechanism(s) of action is/are not completely known. It has been proposed that ω-3 fatty acids may affect the neuronal membrane by inhibiting the production of eicosanoids, which are lipid-derived modulators and among them arachidonic acid is the most important of their precursors. This action alters the phospholipids composition of cellular membrane leading to the alteration of different membrane functions, such as those related to neurotransmitter receptors, second messenger and transport proteins (Meterissian et al., 1995). On the other hand, this inhibition not only affects numerous biological responses but also modulates neurotransmission via inhibition of metabolism and actions of arachidonic acid and its derivatives (Bazan et al., 1997). These data suggest that the effects of ω-3 fatty acids on food intake may be mediated by the modulation of the balance of hypothalamic orexigenic and anorectic factors via inhibition of cytokines production. Further, these data point out the possibility of the therapeutic use of the ω-3 fatty acids at earlier stages of tumor progression for preventing body weight loss and ameliorating anorexia as well as for inhibiting tumor growth.

To this emerging therapy may now be added the idea of the use of nicotine in a sociably acceptable manner, and preliminary studies in our laboratory in tumor-bearing rats show promising results (Personal observations, Laviano and Meguid, 2005).

Abbreviations

ACTH	adrenocorticotropin hormone
AP	anterior pituitary
ARC	arcuate nucleus
DA	dopamine
CRF	Corticotropin-releasing factor
HPA	hypothalamus–pituitary–adrenal
5-HT	serotonin
IL	interleukin
IL-1ra	interleukin-1α receptor antagonist
IL-1RI	IL-1 receptor type I
IL-1RAP	interleukin-1 receptor accessory protein
INT-γ	interferon gamma
LHA	lateral hypothalamic area
NPY	neuropeptide Y
POMC	pro-opiomelanocortin
PVN	paraventricular nucleus
SCN	suprachiasmatic nucleus
SON	supraoptic nucleus
TNF-α	tumor necrosis factor-α
VMH	ventromedial hypothalamus

Acknowledgment

This work was supported in part by NIH 3568-CA 70239, The American Institute of Cancer 22861, American Diabetes Association 35243, and by funding from NCI International Exchange Fellowship and the Hendrix Fund of SUNY Upstate Medical University.

References

Aguilera, A., Selgas, R., Codoceo, R. and Bajo, A. (2000) Uremic anorexia: a consequence of persistently high brain serotonin levels? The tryptophan/serotonin disorder hypothesis. Perit. Dial. Int., 20: 810–816.

Argiles, J.M., Lopez-Soriano, F.J., Evans, R.D. and Williamson, D.H. (1989) Interleukin-1 and lipid metabolism in the rat. Biochem. J., 259: 673–678.

Arletti, R., Benelli, A. and Bertolini, A. (1989) Influence of oxytocin on feeding behavior in the rat. Peptides, 10: 89–93.

Arletti, R., Benelli, A. and Bertolini, A. (1990) Oxytocin inhibits food and fluid intake in rats. Physiol. Behav., 48: 825–830.

Armstrong, M.D. and Klein, J.R. (2001) Immune-endocrine interactions of the hypothalamus–pituitary–thyroid axis: integration, communication and homeostasis. Arch. Immunol. Ther. Exp. (Warsz),, 49: 231–237.

Arsenijevic, D. and Richard, D. (1999) A predominant role for INF-γ in infection induced cachexia in comparison to TNF-α (abstract). Appetite, 33: 232.

Banks, W.A., Ortiz, L., Plotkin, S.R. and Kastin, A.J. (1991) Human interleukin (IL) 1 alpha, murine IL-1 alpha and murine IL-1 beta are transported from blood to brain in the mouse by a shared saturable mechanism. J. Pharmacol. Exp. Ther., 259: 988–996.

Barber, M.D., Ross, J.A., Voss, A.C., Tisdale, M.J. and Fearon, K.C. (1999) The effect of an oral nutritional supplement enriched with fish oil on weight-loss in patients with pancreatic cancer. Br. J. Cancer, 81: 80–86.

Barnes, N.M. and Sharp, T. (1999) A review of central 5-HT receptors and their function. Neuropharmacology, 38: 1083–1152.

Bartholomew, S.A. and Hoffman, S.A. (1993) Effects of peripheral cytokine injections on multiple unit activity in the anterior hypothalamic area of the mouse. Brain Behav. Immun., 7: 301–316.

Bartoli, G.M., Palozza, P., Marra, G., Armelao, F., Franceschelli, P., Luberto, C., Sgarlata, E., Piccioni, E. and Anti, M. (1993) ω-3 PUFA and alpha-tocopherol control of tumor cell proliferation. Mol. Aspects. Med., 14: 247–252.

Baumann, H. and Gauldie, J. (1994) The acute phase response. Immunol. Today, 15: 74–80.

Bazan, N.G., Packard, M.G., Teather, L. and Allan, G. (1997) Bioactive lipids in excitatory neurotransmission and neuronal plasticity. Neurochem. Int., 30: 225–231.

Beatty, W.W., O'Brien, D.A. and Vilberg, T.R. (1974) Suppression of feeding by intrahypothalamic implants of estradiol in male and female rats. Bull. Psychonom. Soc., 3: 273–274.

Berkenbosch, F., van Oers, J., del Rey, A., Tilders, F. and Besedovsky, H. (1987) Corticotropin-releasing factor-producing neurons in the rat activated by interleukin-1. Science, 238: 524–526.

Bette, M., Schafer, M.K., van Rooijen, N., Weihe, E. and Fleischer, B. (1993) Distribution and kinetics of superantigen-induced cytokine gene expression in mouse spleen. J. Exp. Med., 178: 1531–1539.

Beutler, B.A. and Cerami, A. (1985) Recombinant interleukin 1 suppresses lipoprotein lipase activity in 3T3-L1 cells. J. Immunol., 135: 3969–3971.

Bird, T.A., Davies, A., Baldwin, S.A. and Saklatvala, J. (1990) Interleukin 1 stimulates hexose transport in fibroblasts by increasing the expression of glucose transporters. J. Biol. Chem., 265: 13578–13583.

Blaha, V., Yang, Z.-J., Meguid, M.M., Laviano, A., Zadack, Z. and Rossi-Fanelli, F. (1996) Cancer anorexia is modulated by interaction of hypothalamic-VMN dopamine (DA) and serotonin (5-HT) and not solely 5-HT as currently thought. Surg. Forum., 47: 517–520.

Blatteis, C.M. (1992) Role of the OVLT in the febrile response to circulating pyrogens. Prog. Brain Res., 91: 409–412.

Borovikova, L.V., Ivanova, S., Zhang, M., Yang, H., Botchkina, G.I., Watkins, L.R., Wang, H., Abumrad, N., Eaton, J.W. and Tracey, K.J. (2000) Vagus nerve stimulation attenuates the systemic inflammatory response to endotoxin. Nature, 405: 458–462.

Bouman, A., Schipper, M., Heineman, M.J. and Faas, M.M. (2004) Gender difference in the non-specific and specific immune response in humans. Am. J. Reprod. Immunol., 52: 19–26.

Broberger, C., De Lecea, L., Sutcliffe, J.G. and Hokfelt, T. (1998) Hypocretin/orexin- and melanin-concentrating hormone-expressing cells form distinct populations in the rodent lateral hypothalamus: relationship to the neuropeptide Y and agouti gene-related protein systems. J. Comp. Neurol., 402: 460–474.

Broberger, C., Johansen, J., Brismar, H., Johansson, C., Schalling, M. and Hokfelt, T. (1999) Changes in neuropeptide Y receptors and pro-opiomelanocortin in the anorexia (anx/anx) mouse hypothalamus. J. Neurosci., 19: 7130–7139.

Broberger, C., Johansen, J., Schalling, M. and Hokfelt, T. (1997) Hypothalamic neurohistochemistry of the murine anorexia (anx/anx) mutation: altered processing of neuropeptide Y in the arcuate nucleus. J. Comp. Neurol., 387: 124–135.

Butera, P.C. and Beikirch, R.J. (1989) Central implants of diluted estradiol: independent effects on ingestive and reproductive behaviors of ovariectomized rats. Brain Res., 491: 266–273.

Campbel, R.E., French-Mullen, J.M., Cowley, M.A., Smith, M.S. and Grove, K.L. (2001) Hypothalamic circuitry of neuropeptide Y regulation of neuroendocrine function and food intake via the Y5 receptor subtype. Neuroendocrinology, 74: 106–119.

Campos, A.C., Oler, A., Meguid, M.M. and Chen, T.Y. (1990) Liver biochemical and histological changes with graded amounts of total parenteral nutrition. Arch. Surg., 125: 447–450.

Cangiano, C., Testa, U., Muscaritoli, M., Meguid, M.M., Mulieri, M., Laviano, A., Cascino, A., Preziosa, I., Conversano, L. and Rossi-Fanelli, F. (1994) Cytokines, tryptophan and anorexia in cancer patients before and after surgical tumor ablation. Anticancer Res., 14: 1451–1455.

Card, J.P. and Moore, R.Y. (1988) Neuropeptide Y localization in the rat suprachiasmatic nucleus and periventricular hypothalamus. Neurosci. Lett., 88: 241–246.

Ceppi, E.D., Knowles, R.G., Carpenter, K.M. and Titheradge, M.A. (1992) Effect of treatment in vivo of rats with bacterial endotoxin on fructose 2,6-bisphosphate metabolism and L-pyruvate kinase activity and flux in isolated liver cells. Biochem. J., 284: 761–766.

Chai, J.-K., Blaha, V., Meguid, M.M., Yang, Z.-J., Varma, M. and Laviano, A. (1999) Use of orchiectomy and testosterone replacement to explore meal number to meal size relationship in male rats. Am. J. Physiol., 45: R1366–R1373.

Chance, W.T., Balasubramaniam, A., Dayal, R., Brown, J. and Fischer, J.E. (1994a) Hypothalamic concentration and release of neuropeptideY into microdialysates is reduced in anorectic tumor-bearing rats. Life Sci., 54: 1869–1874.

Chance, W.T., Balasubramaniam, A., Sheriff, S. and Fischer, J.E. (1994b) Possible role of neuropeptide Y in experimental cancer anorexia. Adv. Exp. Med. Biol., 354: 185–201.

Chance, W.T. and Fischer, J.E. (1991) Aphagic and adipsic effects of interleukin-1. Brain Res., 568: 261–264.

Chance, W.T., Sheriff, S., Dayal, R. and Balasubramaniam, A. (2003) Refractory hypothalamic alpha-MSH satiety and AGRP feeding systems in rats bearing MCA sarcomas. Peptides, 24: 1909–1919.

Chance, W.T., Sheriff, S., Kasckow, J.W., Regmi, A. and Balasubramaniam, A. (1998) NPY messenger RNA is increased in medial hypothalamus of anorectic tumor-bearing rats. Regul. Pept., 75–76: 347–353.

Chen, Z.Y. and Istfan, N.W. (2000) Docosahexaenoic acid is a potent inducer of apoptosis in HT-29 colon cancer cells. Prostaglandins Leukot. Essent. Fatty Acids, 63: 301–308.

Christ, B. and Nath, A. (1996) Impairment by interleukin 1 beta and tumour necrosis factor alpha of the glucagon-induced increase in phosphoenolpyruvate carboxykinase gene expression and gluconeogenesis in cultured rat hepatocytes. Biochem. J., 320: 161–166 Erratum in: Biochem. J., 1997 321: 903.

Coe, J.E. and Ross, M.J. (1983) Hamster female protein. A divergent acute phase protein in male and female Syrian hamsters. J. Exp. Med., 157: 1421–1433.

Conn, C.A., McClellan, J.L., Maassab, H.F., Smitka, C.W., Majde, J.A. and Kluger, M.J. (1995) Cytokines and the acute phase response to influenza virus in mice. Am. J. Physiol., 268: R78–R84.

Cowley, M.A., Pronchuk, N., Fan, W., Dinulescu, D.M., Colmers, W.F. and Cone, R.D. (1999) Integration of NPY, AGRP, and melanocortin signals in the hypothalamic paraventricular nucleus: evidence of a cellular basis for the adipostat. Neuron, 24: 155–163.

Cravo, M.L., Gloria, L.M. and Claro, I. (2000) Metabolic responses to tumor disease and progression: tumor–host interaction. Clin. Nutr., 19: 459–465.

Cunningham Jr., E.T. and De Souza, E.B. (1993) Interleukin 1 receptors in the brain and endocrine tissues. Immunol. Today, 14: 171–176.

Currie, P.J., Coiro, C.D., Niyomchai, T., Lira, A. and Farahnmmand, F. (2002) Hypothalamic paraventricular 5-hydroxytryptamine: receptor specific inhibition of NPY-stimulated eating and energy metabolism. Pharmacol. Biochem. Behav., 71: 709–716.

Currie, P.J. and Coscina, D.V. (1997) Stimulation of 5-HT(2A/2C) receptors within specific hypothalamic nuclei differentially antagonizes NPY-induced feeding. NeuroReport, 8: 3759–3762.

Dagnault, A. and Richard, D. (1997) Involvement of the medial preoptic area in the anorectic action of estrogens. Am. J. Physiol., 272: R311–R317.

Dagnelie, P.C., Bell, J.D., Williams, S.C., Bates, T.E., Abel, P.D. and Foster, C.S. (1994) Effect of fish oil on cancer cachexia and host liver metabolism in rats with prostate tumors. Lipids, 29: 195–203.

Das, U.N., Ramos, E.J. and Meguid, M.M. (2003) Metabolic alterations during inflammation and its modulation by central actions of omega-3 fatty acids. Curr. Opin. Clin. Nutr. Metab. Care, 6: 413–419.

Debonis, D., Meguid, M.M., Laviano, A., Yang, Z.J. and Gleason, J.R. (1995) Temporal changes in meal number and meal size relationship in response to rHu IL-1α. NeuroReport, 6: 1752–1756.

Deitch, E.A., Xu, D., Franko, L., Ayala, A. and Chaudry, I.H. (1994) Evidence favoring the role of the gut as a cytokine-generating organ in rats subjected to hemorrhagic shock. Shock, 1: 141–144.

De Laurentiis, A., Pisera, D., Caruso, C., Candolfi, M., Mohn, C., Rettori, V. and Seilicovich, A. (2002) Lipopolysaccharide- and tumor necrosis factor-alpha-induced changes in prolactin secretion and dopaminergic activity in the hypothalamic–pituitary axis. Neuroimmunomodulation, 10: 30–39.

Del Rey, A. and Besedovsky, H. (1987) Interleukin 1 affects glucose homeostasis. Am. J. Physiol., 253: R794–R798.

Del Rey, A. and Besedovsky, H. (1989) Antidiabetic effects of interleukin 1. Proc. Natl. Acad. Sci. USA, 86: 5943–5947.

Del Rey, A. and Besedovsky, H.O. (1992) Metabolic and neuroendocrine effects of pro-inflammatory cytokines. Eur. J. Clin. Invest., 22: 10–15.

Del Rey, A., Kabiersch, A., Petzoldt, S. and Besedovsky, H.O. (2002) Involvement of noradrenergic nerves in the activation and clonal deletion of T cells stimulated by superantigen in vivo. J. Neuroimmunol., 127: 44–53.

Del Rey, A., Monge-Arditi, G. and Besedovsky, H.O. (1998) Central and peripheral mechanisms contribute to the hypoglycemia induced by interleukin-1. Ann. NY. Acad. Sci., 840: 153–161.

Delrue, C., Deleplanque, B., Rouge-Pont, F., Vitiello, S. and Neveu, P.J. (1994) Brain monoaminergic, neuroendocrine, and immune responses to an immune challenge in relation to brain and behavioral lateralization. Brain Behav Immun., 8: 137–152.

Doerrler, W., Feingold, K.R. and Grunfeld, C. (1994) Cytokines induce catabolic effects in cultured adipocytes by multiple mechanisms. Cytokine, 6: 478–484.

Dryden, S., Frankish, H.M., Wang, Q. and Williams, G. (1994) Neuropeptide Y and energy balance: one-way ahead from the treatment of obesity. J. Clin. Invest., 24: 293–308.

Dryden, S., Wang, Q., Frankish, H.M., Pickavance, L. and Williams, G. (1995) The serotonin (5-HT) antagonist metysergide increases neuropeptide Y (NPY) synthesis and secretion in the hypothalamus of the rat. Brain Res., 699: 12–18.

Dubuis, J.M., Dayer, J.M., Siegrist-Kaiser, C.A. and Burger, A.G. (1988) Human recombinant interleukin-1 beta decreases plasma thyroid hormone and thyroid stimulating hormone levels in rats. Endocrinology, 123: 2175–2181.

Dunn, A.J. (1992) Endotoxin-induced activation of cerebral catecholamine and serotonin metabolism: comparison with interleukin-1. J. Pharmacol. Exp. Ther., 261: 964–969.

Eckel, L.A. and Geary, N. (2001) Estradiol treatment increases feeding induced c-Fos expression in the brains of ovariectomized rats. Am. J. Physiol. Regul. Integr. Comp. Physiol., 281: R738–R746.

Eckel, L.A., Houpt, T.A. and Geary, N. (2002) Estradiol treatment increases CCK-induced c-Fos expression in the brains of ovariectomized rats. Am. J. Physiol. Regul. Integr. Comp. Physiol., 283: R1378–R1385.

Ek, M., Kurosawa, M., Lundeberg, T. and Ericsson, A. (1998) Activation of vagal afferents after intravenous injection of interleukin-1beta: role of endogenous prostaglandins. J. Neurosci., 18: 9471–9479.

El-Haj, T., Poole, S., Farthing, M.J. and Ballinger, A.B. (2002) Anorexia in a rat model of colitis: interaction of interleukin-1 and hypothalamic serotonin. Brain Res., 927: 1–7.

Elias, C.F., Aschkenasi, C., Lee, C., Kelly, J., Ahima, R.S., Bjorbaek, C., Flier, J.S., Saper, C.B. and Elmquist, J.K. (1999) Leptin differentially regulates NPY and POMC neurons projecting to the lateral hypothalamic area. Neuron, 23: 775–786.

Elias, C.F., Lee, C., Kelly, J., Aschkenasi, C., Ahima, R.S., Couceyro, P.R., Kuhar, M.J., Saper, C.B. and Elmquist, J.K. (1998) Leptin activates hypothalamic CART neurons projecting to the spinal cord. Neuron, 21: 1375–1385.

Elmquist, J.K. and Marcus, J.N. (2003) Rethinking the central causes of diabetes. Nat. Med., 9: 645–647.

Elmquist, J.K. and Saper, C.B. (1996) Activation of neurons projecting to the paraventricular hypothalamic nucleus by intravenous lipopolysaccharide. J. Comp. Neurol., 374: 315–331.

Endres, S., Ghorbani, R., Kelley, V.E., Georgilis, K., Lonnemann, G., van der Meer, J.W., Cannon, J.G., Rogers, T.S., Klempner, M.S. and Weber, P.C. (1989) The effect of dietary supplementation with n-3 polyunsaturated fatty acids on the synthesis of interleukin-1 and tumor necrosis factor by mononuclear cells. N. Engl. J. Med., 320: 265–271.

Ericsson, A., Arias, C. and Sawchenko, P.E. (1997) Evidence for an intramedullary prostaglandin-dependent mechanism in the activation of stress-related neuroendocrine circuitry by intravenous interleukin-1. J. Neurosci., 17: 7166–7179.

Ericsson, A., Liu, C., Hart, R.P. and Sawchenko, P.E. (1995) Type 1 interleukin-1 receptor in the rat brain: distribution, regulation, and relationship to sites of IL-1-induced cellular activation. J. Comp. Neurol., 361: 681–698.

Espat, N.J., Auffenberg, T., Rosenberg, J.J., Rogy, M., Martin, D., Fang, C.H., Hasselgren, P.O., Copeland, E.M. and Moldawer, L.L. (1996) Ciliary neurotrophic factor is catabolic and shares with IL-6 the capacity to induce an acute phase response. Am. J. Physiol., 271: R185–R190.

Fantino, M. and Wieteska, L. (1993) Evidence for a direct central anorectic effect of tumor-necrosis-factor-alpha in the rat. Physiol. Behav., 53: 477–483.

Fantuzzi, G. (2005) Adipose tissue, adipokines, and inflammation. J. Allergy Clin. Immunol., 115: 911–919.

Feingold, K.R., Soued, M., Adi, S., Staprans, I., Neese, R., Shigenaga, J., Doerrler, W., Moser, A., Dinarello, C.A. and Grunfeld, C. (1991) Effect of interleukin-1 on lipid metabolism in the rat. Similarities to and differences from tumor necrosis factor. Arterioscler. Thromb., 11: 495–500.

Feingold, K.R., Staprans, I., Memon, R.A., Moser, A.H., Shigenaga, J.K., Doerrler, W., Dinarello, C.A. and Grunfeld, C. (1992) Endotoxin rapidly induces changes in lipid metabolism that produce hypertriglyceridemia: low doses stimulate hepatic triglyceride production while high doses inhibit clearance. J. Lipid. Res., 33: 1765–1776.

Feldman, S., Conforti, N. and Melamed, E. (1987) Paraventricular nucleus serotonin mediates neurally stimulated adrenocortical secretion. Brain Res. Bull., 18: 165–168.

Fernstrom, J.D. and Wurtman, R.J. (1972) Brain serotonin content: physiological regulation by plasma neutral amino acids. Science, 178: 414–416.

Fischereder, M., Schroppel, B., Wiese, P., Fink, M., Banas, B., Schmidbauer, S. and Schlondorff, D. (2003) Regulation of glucose transporters in human peritoneal mesothelial cells. J. Nephrol., 16: 103–109.

Flanagan-Cato, L.M., King, J.F., Blechman, J.G. and O'Brien, M.P. (1998) Estrogen reduces cholecystokinin-induced c-Fos expression in the rat brain. Neuroendocrinology, 67: 384–391.

Garcia-Welsh, A., Schneiderman, J.S. and Baly, D.L. (1990) Interleukin-1 stimulates glucose transport in rat adipose cells. Evidence for receptor discrimination between IL-1 beta and IL-1 alpha. FEBS Lett., 269: 421–424.

Gelin, J., Moldawer, L.L., Lonnroth, C., Sherry, B., Chizzonite, R. and Lundholm, K. (1991) Role of endogenous tumor necrosis factor alpha and interleukin 1 for experimental tumor growth and the development of cancer cachexia. Cancer Res., 51: 415–421.

Gloaguen, I., Costa, P., Demartis, A., Lazzaro, D., Di Marco, A., Graziani, R., Paonessa, G., Chen, F., Rosenblum, C.I., Van der Ploeg, L.H., Cortese, R., Ciliberto, G. and Laufer, R. (1997) Ciliary neurotrophic factor corrects obesity and diabetes associated with leptin deficiency and resistance. Proc. Natl. Acad. Sci. USA, 94: 6456–6461.

Goehler, L.E., Gaykema, R.P., Hansen, M.K., Anderson, K., Maier, S.F. and Watkins, L.R. (2000) Vagal immune-to-brain communication: a visceral chemosensory pathway. Auton. Neurosci., 85: 49–59.

Goehler, L.E., Gaykema, R.P., Nguyen, K.T., Lee, J.E., Tilders, F.J., Maier, S.F. and Watkins, L.R. (1999) Interleukin-1β in immune cells of the abdominal vagus nerve: a link between the immune and nervous systems? J. Neurosci., 19: 2799–2806.

Goehler, L.E., Relton, J.K., Dripps, D., Kiechle, R., Tartaglia, N., Maier, S.F. and Watkins, L.R. (1997) Vagal paraganglia bind biotinylated interleukin-1 receptor antagonist: a possible mechanism for immune-to-brain communication. Brain Res. Bull., 43: 357–364.

Gonzalez, M.C., Abreu, P., Barroso-Chinea, P., Cruz-Muros, I. and Gonzalez-Hernandez, T. (2004) Effect of intracerebroventricular injection of lipopolysaccharide on the tuberoinfundibular dopaminergic system of the rat. Neuroscience, 127: 251–259.

Gregoire, F., De Broux, N., Hauser, N., Heremans, H., Van Damme, J. and Remacle, C. (1992) Interferon-gamma and interleukin-1 beta inhibit adipoconversion in cultured rodent preadipocytes. J. Cell. Physiol., 151: 300–309.

Hansen, K., Sickelmann, F., Pietrowsky, R., Fehm, H.L. and Born, J. (1997) Systemic immune changes following meal intake in humans. Am. J. Physiol., 273: R548–R553.

Hansen, M.K., Taishi, P., Chen, Z. and Krueger, J.M. (1998) Cafeteria feeding induces interleukin-1beta mRNA expression in rat liver and brain. Am. J. Physiol., 274: R1734–R1739.

Hansen, S., Sodersten, P., Eneroth, P., Srebro, B. and Hole, K. (1979) A sexually dimorphic rhythm in oestradiol-activated lordosis behaviour in the rat. J. Endocrinol., 82: 267–274.

Hansen, S., Sodersten, P. and Srebo, B. (1978) A daily rhythm in the behavioral sensitivity of the female rat to oestradiol. J. Endocrinol., 77: 381–388.

Hardardottir, I., Grunfeld, C. and Feingold, K.R. (1994) Effects of endotoxin and cytokines on lipid metabolism. Curr. Opin. Lipidol., 5: 207–215.

Hart, B.L. (1988) Biological basis of the behavior of sick animals. Neurosci. Biobehav. Rev., 12: 123–137.

Hart, B.L. (1990) Behavioral adaptations to pathogens and parasites: five strategies. Neurosci. Biobehav. Rev., 14: 273–294.

Herkenham, M., Lee, H.Y. and Baker, R.A. (1998) Temporal and spatial patterns of c-fos mRNA induced by intravenous interleukin-1: a cascade of non-neuronal cellular activation at the blood–brain barrier. J. Comp. Neurol., 400: 175–196.

Hermus, R.M., Sweep, C.G., van der Meer, M.J., Ross, H.A., Smals, A.G., Benraad, T.J. and Kloppenborg, P.W. (1992) Continuous infusion of interleukin-1 beta induces a nonthyroidal illness syndrome in the rat. Endocrinology, 131: 2139–2146.

Hirai, C.Y. and Limaos, E.A. (1990) Effect of sex steroids on the circulating levels of alpha 2-macroglobulin in injured rats. Braz. J. Med. Biol. Res., 23: 1021–1024.

Hosoi, T., Okuma, Y. and Nomura, Y. (2000) Electrical stimulation of afferent vagus nerve induces IL-1beta expression in the brain and activates HPA axis. Am. J. Physiol. Regul. Integr. Comp. Physiol., 279: R141–R147.

Hotamisligil, G.S., Shargill, N.S. and Spiegelman, B.M. (1993) Adipose expression of tumor necrosis factor-alpha: direct role in obesity-linked insulin resistance. Science, 259: 87–91.

Inui, A. (1999) Cancer anorexia–cachexia syndrome: are neuropeptides the key? Cancer Res., 59: 4493–4501.

Itoh, T., Iwai, H. and Ueda, K. (1991) Comparative lung pathology of inbred strain of mice resistant and susceptible to Sendai virus infection. J. Vet. Med. Sci., 53: 275–279.

Jho, D.H., Babcock, T.A., Tevar, R., Helton, W.S. and Espat, N.J. (2002) Eicosapentaenoic acid supplementation reduces tumor volume and attenuates cachexia in a rat model of progressive non-metastasizing malignancy. JPEN J. Parenter. Enteral Nutr., 26: 291–297.

Kanemaki, T., Kitade, H., Kaibori, M., Sakitani, K., Hiramatsu, Y., Kamiyama, Y., Ito, S. and Okumura, T. (1998) Interleukin 1β and interleukin 6, but not tumor necrosis factor α, inhibit insulin-stimulated glycogen synthesis in rat hepatocytes. Hepatology I, 27: 1296–1303.

Khovidhunkit, W., Kim, M.S., Memon, R.A., Shigenaga, J.K., Moser, A.H., Feingold, K.R. and Grunfeld, C. (2004) Effects of infection and inflammation on lipid and lipoprotein metabolism: mechanisms and consequences to the host. J. Lipid. Res., 45: 1169–1196.

Kitano, T., Okumura, T., Nishizawa, M., Liew, F.Y., Seki, T., Inoue, K. and Ito, S. (2002) Altered response to inflammatory cytokines in hepatic energy metabolism in inducible nitric oxide synthase knockout mice. J. Hepatol., 36: 759–765.

Klasing, K.C. (1988) Nutritional aspects of leukocytic cytokines. J. Nutr., 118: 1436–1446.

Konsman, J.P., Luheshi, G.N., Bluthe, R.M. and Dantzer, R. (2000) The vagus nerve mediates behavioural depression, but not fever, in response to peripheral immune signals; a functional anatomical analysis. Eur. J. Neurosci., 12: 4434–4446.

Kozak, W., Zheng, H., Conn, C.A., Soszynski, D., van der Ploeg, L.H. and Kluger, M.J. (1995) Thermal and behavioral effects of lipopolysaccharide and influenza in interleukin-1 beta-deficient mice. Am. J. Physiol., 269: R969–R977.

Kremer, J.M. and Robinson, D.R. (1991) Studies of dietary supplementation with omega 3 fatty acids in patients with rheumatoid arthritis. World Rev. Nutr. Diet., 66: 367–382.

Kunkel, S.L. and Chensue, S.W. (1985) Arachidonic acid metabolites regulate interleukin-1 production. Biochem. Biophys. Res. Commun., 128: 892–897.

Kuriyama, K., Hori, T., Mori, T. and Nakashima, T. (1990) Actions of interferon alpha and interleukin- 1 beta on the glucose-responsive neurons in the ventromedial hypothalamus. Brain. Res. Bull., 24: 803–810.

Kurzer, M.J., Janiszewsky, J. and Meguid, M.M. (1988) Aminoacid profiles in tumor bearing and non-tumor bearing malnourished rats. Cancer, 62: 1492–1496.

Kusnecov, A.W., Liang, R. and Shurin, G. (1999) T-lymphocyte activation increases hypothalamic and amygdaloid expression of CRH mRNA and emotional reactivity to novelty. J. Neurosci., 19: 4533–4543.

Lambert, P.D., Anderson, K.D., Sleeman, M.W., Wong, V., Tan, J., Hijarunguru, A., Corcoran, T.L., Murray, J.D., Thabet, K.E., Yancopoulos, G.D. and Wiegand, S.J. (2001) Ciliary neurotrophic factor activates leptin-like pathways and reduces body fat, without cachexia or rebound weight gain, even in leptin-resistant obesity. Proc. Natl. Acad. Sci. USA, 98: 4652–4657.

Landel, A.M., Lo, C.C. and Meguid, M.M. (1987) Observations on predicted brain influx rates of neurotransmitter precursors. Effects of tumor, operative stress with tumor removal, and postoperative TPN of varying amino acid compositions. Cancer, 59: 1192–1200.

Langhans, W., Balkowski, G. and Savoldelli, D. (1991a) Differential feeding responses to bacterial lipopolysaccharide and muramyl dipeptide. Am. J. Physiol., 261: R659–R664.

Langhans, W., Delprete, E. and Scharrer, E. (1991b) Mechanisms of vasopressin's anorectic effect. Physiol. Behav., 49: 169–176.

Lanza-Jacoby, S. and Tabares, A. (1990) Triglyceride kinetics, tissue lipoprotein lipase, and liver lipogenesis in septic rats. Am. J. Physiol., 258: E678–E685.

Laviano, A., Cangiano, C., Fava, A., Muscaritoli, M., Mulieri, G. and Rossi-Fanelli, F. (1999) Peripherally injected IL-1 induces anorexia and increases brain tryptophan concentrations. Adv. Exp. Med. Biol., 467: 105–108.

Laviano, A., Cangiano, C., Preziosa, I., Riggio, O., Conversano, L., Cascino, A., Ariemma, S. and Rossi-Fanelli, F. (1997) Plasma tryptophan levels and anorexia in liver cirrhosis. Int. J. Eat. Disord., 21: 181–186.

Laviano, A., Gleason, J.R., Meguid, M.M., Yang, Z.-J., Cangiano, C. and Rossi-Fanelli, F. (2000) Effects of intra-VMN mianserin and IL-1ra on meal number in anorectic tumor-bearing rats. J. Investig. Med., 48: 40–48.

Laviano, A., Meguid, M.M., Gleason, J.R., Yang, Z.-J. and Renvyle, T. (1996) Comparison of long-term feeding pattern between male and female Fischer 344 rats: influence of estrous cycle. Am. J. Physiol., 270: R413–R419.

Laviano, A., Meguid, M.M. and Rossi-Fanelli, F. (2003) Cancer anorexia: clinical implications, pathogenesis, and therapeutic strategies. Lancet Oncol., 4: 686–994.

Li, Q. and Verma, I.M. (2002) NF-kappaB regulation in the immune system. Nat. Rev. Immunol., 2: 725–734.

Maitra, S.R., Wang, S., Brathwaite, C.E. and El-Maghrabi, M.R. (2000) Alterations in glucose-6-phosphatase gene expression in sepsis. J. Trauma, 49: 38–42.

Makarenko, I.G., Meguid, M.M. and Ugrumov, M.V. (2002) Distribution of serotonin 5-HT$_{1B}$ receptors in the normal rat hypothalamus. Neurosci. Lett., 328: 155–159.

Makarenko, I.G., Meguid, M.M., Gatto, l., Chen, C., Ramos, E.J.B., Goncalves, C.G. and Ugrumov, M.V. (2005a) Normalization of hypothalamic serotonin (5-HT$_{1B}$) receptor and NPY in cancer anorexia after tumor resection: an immunocytochemical study. Neurosci. Lett., 383: 322–327.

Makarenko, I.G., Meguid, M.M., Gatto, I., Goncalves, C.G., Ramos, E.J.B., Chen, C. and Ugrumov, M.V. (2005b) Hypothalamic 5-HT$_{1B}$- receptor changes in the anorectic tumor bearing rats. Neurosci. Lett., 376: 71–75.

Makarenko, I.G., Meguid, M.M., Gatto, L., Chen, C. and Ugrumov, M.V. (2003) Decreased NPY innervation of the hypothalamic nuclei in rats with cancer anorexia. Brain Res., 961: 100–108.

Mascarucci, P., Perego, C., Terrazzino, S. and De Simoni, M.G. (1998) Glutamate release in the nucleus tractus solitarius induced by peripheral lipopolysaccharide and interleukin-1 beta. Neuroscience, 86: 1285–1290.

Matsuki, T., Horai, R., Sudo, K. and Iwakura, Y. (2003) IL-1 plays an important role in lipid metabolism by regulating insulin levels under physiological conditions. J. Exp. Med., 198: 877–888.

Meguid, M.M., Chen, T.-Y., Yang, Z.-J., Campos, A.C., Hitch, D.C. and Gleason, J.R. (1991) Effects of continuous graded TPN on feeding indexes and metabolic concomitants in rats. Am. J. Physiol., 260: E126–E140.

Meguid, M.M., Fetissov, S.O., Miyata, G. and Torelli, G.F. (1999) Feeding pattern in obese Zucker rats after dopaminergic and serotonergic LHA grafts. Neuroreport, 10: 1049–1053.

Meguid, M.M., Kawashima, Y., Campos, A.C.L., Gelling, P., Hill, T.W., Chen, T.-Y., Hitch, D.C., Mueller, W.J. and Hammond, W.G. (1990) Automated computerized rat eater meter: description and application. Physiol. Behav., 48: 759–763.

Meguid, M.M., Landel, A.M., Lo, C.-C. and Rivera, D. (1987) Effect of tumor and tumor removal on DNA, RNA, protein tissue content and survival of methylcholanthrene sarcoma-bearing rat. Surg. Res. Commun., 1: 261–271.

Meguid, M.M., Laviano, A. and Rossi-Fanelli, F. (1998) Food intake equals meal size times mean number. Appetite, 31: 404.

Meguid, M.M., Muscaritoli, M., Beverly, J.L., Yang, Z.J., Cangiano, C. and Rossi-Fanelli, F. (1992) The early cancer anorexia paradigm: changes in plasma-free tryptophan and feeding indexes. J. Parenter. Enteral. Nutr., 16: 56S–59S.

Meguid, M.M. and Pichard, C. (2003) Cytokines: the mother of catabolic mediators. Curr. Opin. Clin. Nutr. Metab. Care, 6: 383–386.

Meguid, M.M., Ramos, E.J.B., Laviano, A., Varma, M., Sato, T., Cheng, C., Qi, Y. and Das, U.N. (2004) Tumor anorexia: effects on neuropeptide Y and monoamines in paraventricular nucleus. Peptides, 25: 261–266.

Meguid, M.M., Sato, T., Torelli, G.F., Laviano, A. and Rossi-Fanelli, F. (2000) An analysis of temporal changes in meal number and meal size at onset of anorexia in male tumor-bearing rats. Nutrition, 16: 305–306.

Meguid, M.M., Yang, Z.J. and Koseki, M. (1995) Eating induced rise in LHA-dopamine correlates with meal size in normal and bulbectomized rats. Brain Res. Bull., 36: 487–490.

Meguid, M.M., Yang, Z.J. and Laviano, A. (1997) Meal size and number: relationship to dopamine levels in the ventromedial hypothalamic nucleus. Am. J. Physiol., 272: R1925–R1930.

Meguid, M.M., Yang, Z.-J. and Montante, A. (1993) Lateral hypothalamic dopaminergic neural activity in response to TPN. Surgery, 114: 400–406.

Memon, R.A., Feingold, K.R., Moser, A.H., Fuller, J. and Grunfeld, C. (1998a) Regulation of fatty acid transport protein and fatty acid translocase mRNA levels by endotoxin and cytokines. Am. J. Physiol., 274: E210–E217.

Memon, R.A., Fuller, J., Moser, A.H., Smith, P.J., Feingold, K.R. and Grunfeld, C. (1998b) In vivo regulation of acyl-CoA synthetase mRNA and activity by endotoxin and cytokines. Am. J. Physiol., 275: E64–E72.

Memon, R.A., Grunfeld, C., Moser, A.H. and Feingold, K.R. (1993) Tumor necrosis factor mediates the effects of endotoxin on cholesterol and triglyceride metabolism in mice. Endocrinology, 132: 2246–2253.

Meterissian, S.H., Forse, R.A., Steele, G.D. and Thomas, P. (1995) Effect of membrane free fatty acid alterations on the adhesion of human colorectal carcinoma cells to liver macrophages and extracellular matrix proteins. Cancer Lett., 89: 145–152.

Metzger, S., Begleibter, N., Barash, V., Drize, O., Peretz, T., Shiloni, E. and Chajek-Shaul, T. (1997) Tumor necrosis factor inhibits the transcriptional rate of glucose-6-phosphatase in vivo and in vitro. Metabolism, 46: 579–583.

Metzger, S., Hassin, T., Barash, V., Pappo, O. and Chajek-Shaul, T. (2001) Reduced body fat and increased hepatic lipid synthesis in mice bearing interleukin-6-secreting tumor. Am. J. Physiol. Endocrinol. Metab., 281: E957–E965.

Metzger, S., Nusair, S., Planer, D., Barash, V., Pappo, O., Shilyansky, J. and Chajek-Shaul, T. (2004) Inhibition of hepatic gluconeogenesis and enhanced glucose uptake contribute to the development of hypoglycemia in mice bearing interleukin-1beta-secreting tumor. Endocrinology, 145: 5150–5156.

Miyata, G., Meguid, M.M., Fetissov, S.O., Torelli, G.F. and Kim, H.J. (1999) Nicotine's effect on hypothalamic neurotransmitters and appetite regulation. Surgery, 126: 255–263.

MohanKumar, S.M., MohanKumar, P.S. and Quadri, S.K. (1998) Specificity of interleukin-1beta-induced changes in monoamine concentrations in hypothalamic nuclei: blockade by interleukin-1 receptor antagonist. Brain Res. Bull., 47: 29–34.

MohanKumar, S.M., MohanKumar, P.S. and Quadri, S.K. (1999) Lipopolyssacharide-induced changes in monoamines in specific areas of the brain: blockade by interleukin-1 receptor antagonist. Brain Res., 10: 232–237.

Muscaritoli, M., Meguid, M.M., Beverly, J.L., Yang, Z.J., Cangiano, C. and Rossi-Fanelli, F. (1996) Mechanism of early tumor anorexia. J. Surg. Res., 60: 389–397.

Nachiappan, V., Curtiss, D., Corkey, B.E. and Kilpatrick, L. (1994) Cytokines inhibit fatty acid oxidation in isolated rat hepatocytes: synergy among TNF, IL-6, and IL-1. Shock, 1: 123–129.

Niijima, A. (1996) The afferent discharges from sensors for interleukin 1 beta in the hepatoportal system in the anesthetized rat. J. Auton. Nerv. Syst., 61: 287–291.

Niijima, A. and Meguid, M.M. (1994) Parenteral nutrients in rat suppress hepatic vagal afferent signals from portal vein to hypothalamus. Surgery, 116: 294–301.

Niijima, A. and Meguid, M.M. (1995) An electrophysiological study on amino acid sensors in the hepato-portal system in the rat. Obes. Res., 5: 741S–745S.

Niijima, A. and Meguid, M.M. (1998) Influence of systemic arginine–lysine on immune organ function: an electrophysiology study. Brain Res. Bull., 45: 437–441.

Noguchi, Y., Yoshikawa, T., Matsumoto, A., Svaninger, G. and Gelin, J. (1996) Are cytokines possible mediators of cancer cachexia? Surg. Today., 26: 467–475.

Nolan, Y., Connor, T.J., Kelly, J.P. and Leonard, B.E. (2000) Lipopolysaccharide administration produces time-dependent and region-specific alterations in tryptophan and tyrosine hydroxylase activities in rat brain. J. Neural. Transm., 107: 1393–1401.

Olson, B.R., Drutarowsky, M.D., Chow, M.S., Hruby, V.J., Stricker, E.M. and Verbalis, J.G. (1991) Oxytocin and an oxytocin agonist administered centrally decrease food intake in rats. Peptides, 12: 113–118.

Opara, E.I., Laviano, A., Meguid, M.M. and Yang, Z.-J. (1995a) Correlation between food intake and CSF IL-l in anorectic tumor-bearing rats. NeuroReport, 6: 750–752.

Opara, E.I., Meguid, M.M., Yang, Z.-J., Chai, J.-K. and Veerabagu, M. (1995b) Tumor necrosis factor-α and TPN-induced anorexia. Surgery, 18: 756–762.

Opara, E.I., Meguid, M.M., Yang, Z.J. and Hammond, W.G. (1996) Studies on the regulation of food intake using rat total parenteral nutrition as a model. Neurosci. Biobehav. Rev., 20: 413–443.

Ottaway, C.A. and Husband, A.J. (1994) The influence of neuroendocrine pathways on lymphocyte migration. Immunol. Today, 15: 511–517.

Pacheco-Lopez, G., Niemi, M.B., Kou, W., Harting, M., Del Rey, A., Besedovsky, H.O. and Schedlowski, M. (2004) Behavioural endocrine immune-conditioned response is induced by taste and superantigen pairing. Neuroscience, 129: 555–562.

Pich, E.M., Messori, B., Zoli, M., Ferraguti, F., Marrama, P., Biagini, G., Fuxe, K. and Agnati, L.F. (1992) Feeding and drinking responses to neuropeptide Y injections in the paraventricular hypothalamic nucleus of aged rats. Brain Res., 575: 265–271.

Plata-Salaman, C.R. (1991) Immunoregulators in the nervous system. Neurosci. Biobehav. Rev., 15: 185–215.

Plata-Salaman, C.R. (1996) Cytokine action in the nervous system at pathophysiological versus pharmacological concentrations. Adv. Exp. Med. Biol., 402: 191–197.

Plata-Salaman, C.R. (1997) Anorexia during acute and chronic disease: relevance of neurotransmitter-peptide-cytokine interactions. Nutrition, 13: 159–160.

Plata-Salaman, C.R. (2000) Central nervous system mechanisms contributing to the cachexia–anorexia syndrome. Nutrition, 16: 1009–1012.

Plata-Salaman, C.R. and Borkoski, J.P. (1993) Interleukin-8 modulates feeding by direct action in the central nervous system. Am. J. Physiol., 265: R877–R882.

Plata-Salaman, C.R. and Ffrench-Mullen, J.M. (1992) Intracerebroventricular administration of a specific IL-1 receptor antagonist blocks food and water intake suppression induced by interleukin-1 beta. Physiol. Behav., 51: 1277–1279.

Plata-Salaman, C.R., Oomura, Y. and Kai, Y. (1998) Tumor necrosis factor and interleukin-1 beta: suppression of food intake by direct action in the central nervous system. Brain Res., 448: 106–114.

Plata-Salaman, C.R., Sonti, G., Borkoski, J.P., Wilson, C.D. and French-Mullen, J.M. (1996) Anorexia induced by chronic central administration of cytokines at estimated pathophysiological concentrations. Physiol. Behav., 60: 867–875.

Plotsky, P.M., Cunningham Jr., E.T. and Widmaier, E.P. (1989) Catecholaminergic modulation of corticotropin-releasing factor and adrenocorticotropin secretion. Endocr. Rev., 10: 437–458.

Porter, M.H., Arnold, M. and Langhans, W. (1998) Lipopolysaccharide-induced anorexia following hepatic portal vein and vena cava administration. Physiol. Behav., 64: 581–584.

Porter, M.H., Hrupka, B.J., Altreuther, G., Arnold, M. and Langhans, W. (2000) Inhibition of TNF-alpha production contributes to the attenuation of LPS-induced hypophagia by pentoxifylline. Am. J. Physiol. Regul. Integr. Comp. Physiol., 279: R2113–R2120.

Posma, E., Moes, H., Heineman, M.J. and Faas, M.M. (2004) The effect of testosterone on cytokine production in the specific and non-specific immune response. Am. J. Reprod. Immunol., 52: 237–243.

Pu, S., Dhillon, H., Moldawer, L.L., Kalra, P.S. and Kalra, S.P. (2000) Neuropeptide Y counteracts the anorectic and weight reduction effects of ciliary neurotropic factor. J. Neuroendocrinol., 12: 827–832.

Ramos, E.J., Meguid, M.M., Zhang, L., Miyata, G., Fetissov, S.O., Chen, C., Suzuki, S. and Laviano, A. (2004a) Nicotine infusion into rat ventromedial nuclei and effects on monoaminergic system. Neuroreport, 15: 2293–2297.

Ramos, E.J.B., Middelton, F.A., Laviano, A., Sato, T., Romanova, I., Das, U., Cheng, C., Qi, Y. and Meguid, M.M. (2004b) Effects of omega -3 fatty acid supplementation on tumor-bearing rats. J. Am. Coll. Surg., 199: 716–723.

Ramos, E.J.B., Romanova, I., Suzuki, S., Cheng, C., Ugrumov, M.V., Sato, T., Goncalves, C.G. and Meguid, M.M. (2005) Effects of omega -3 fatty acid on orexigenic and anorexigenic modulators at the onset of anorexia. Brain Res., 1046: 157–164.

Ramos, E.J.B., Suzuki, S., Meguid, M.M., Laviano, A., Sato, T., Cheng, C. and Das, U. (2004c) Changes in hypothalamic neuropeptide Y and monoaminergic system in tumor-bearing rats: Pre- and post-tumor resection and death. Surgery, 136: 270–276.

Ramos, E.J., Xu, Y., Romanova, I., Middleton, F., Chen, C., Quinn, R., Inui, A., Das, U. and Meguid, M.M. (2003) Is obesity an inflammatory disease? Surgery, 134: 329–335.

Reyes, T.M. and Sawchenko, P.E. (2002) Involvement of the arcuate nucleus of the hypothalamus in interleukin-1-induced anorexia. J. Neurosci., 22: 5091–5099.

Ricardo, J.A. and Koh, E.T. (1978) Anatomical evidence of direct projections from the nucleus of the solitary tract to the hypothalamus, amygdala, and other forebrain structures in the rat. Brain Res., 153: 1–26.

Rogers, P., McKibbin, P.E. and Williams, G. (1997) Acute fenfluramine administration reduces neuropeptide Y concentration in specific hypothalamic regions of the rat: possible implications for the anorectic effect of fenfluramine. Peptides, 12: 251–255.

Rola-Pleszczynski, M. and Lemaire, I. (1985) Leukotrienes augment interleukin 1 production by human monocytes. J. Immunol., 135: 3958–3961.

Rose, D.P., Connolly, J.M. and Meschter, C.L. (1991) Effect of dietary fat on human breast cancer growth and lung metastasis in nude mice. J. Natl. Cancer. Inst., 83: 1491–1495.

Rosendahl, A., Hansson, J., Antonsson, P., Sekaly, R.P., Kalland, T. and Dohlsten, M. (1997) A mutation of F47 to A in staphylococcus enterotoxin A activates the T-cell receptor Vbeta repertoire in vivo. Infect. Immun., 65: 5118–5124.

Rossi-George, A., Urbach, D., Colas, D., Goldfarb, Y. and Kusnecov, A.W. (2005) Neuronal, endocrine, and anorexic responses to the T-cell superantigen staphylococcal enterotoxin A: dependence on tumor necrosis factor-alpha. J. Neurosci., 25: 5314–5322.

Rothwell, N.J., Busbridge, N.J., Lefeuvre, R.A., Hardwick, A.J., Gauldie, J. and Hopkins, S.J. (1991) Interleukin-6 is a centrally acting endogenous pyrogen in the rat. Can. J. Physiol. Pharmacol., 69: 1465–1469.

Saper, C.B., Chou, T.C. and Elmquist, J.K. (2002) The need to feed: homeostatic and hedonic control of eating. Neuron, 36: 199–211.

Sapolsky, R., Rivier, C., Yamamoto, G., Plotsky, P. and Vale, W. (1987) Interleukin-1 stimulates the secretion of hypothalamic corticotropin-releasing factor. Science, 238: 522–524.

Sasaki, T., Kudoh, K., Uda, Y., Ozawa, Y., Shimizu, J., Kanke, Y. and Takita, T. (1999) Effects of isothiocyanates on growth and metastaticity of B16-F10 melanoma cells. Nutr. Cancer, 33: 76–81.

Sato, T., Fetissov, S.O., Meguid, M.M., Miyata, G. and Chen, C. (2001a) Intra-supraoptic nucleus sulpiride improves anorexia in tumor-bearing rats. Neuroreport, 12: 2429–2432.

Sato, T., Laviano, A., Meguid, M.M., Chen, C., Rossi-Fanelli, F. and Hatakeyama, K. (2003) Involvement of plasma leptin, insulin and free tryptophan in cytokine-induced anorexia. Clin. Nutr., 22: 139–146.

Sato, T., Meguid, M.M., Fetissov, S.O., Chen, C. and Zhang, L. (2001b) Hypothalamic dopaminergic receptor expressions in anorexia of tumor-bearing rats. Am. J. Physiol. Regul. Integr. Comp. Physiol., 281: R1907–R1916.

Sato, T., Meguid, M.M., Quinn, R.H., Zhang, L. and Chen, C. (2001c) Feeding behavior during sialodacryoadenitis viral infection in rats. Physiol. Behav., 72: 721–726.

Sawchenko, P.E. and Swanson, L.W. (1983) Hypothalamic integration: organization of the paraventricular and supraoptic nuclei. Annu. Rev. Neurosci., 6: 269–324.

Sawchenko, P.E., Swanson, L.W., Steinbusch, H.W. and Verhofstad, A.A. (1983) The distribution and cells of origin of serotonergic inputs to the paraventricular and supraoptic nuclei of the rat. Brain Res., 277: 355–360.

Schaechter, J.D. and Wurtman, R.J. (1990) Serotonin release varies with brain tryptophan levels. Brain Res., 532: 203–210.

Shikhman, A.R., Brinson, D.C., Valbracht, J. and Lotz, M.K. (2001) Cytokine regulation of facilitated glucose transport in human articular chondrocytes. J. Immunol., 167: 7001–7008.

Shurin, G., Shanks, N., Nelson, L., Hoffman, G., Huang, L. and Kusnecov, A.W. (1997) Hypothalamic-pituitary-adrenal activation by the bacterial superantigen staphylococcal enterotoxin B: role of macrophages and T cells. Neuroendocrinology, 65: 18–28.

Smagin, G.N., Swiergiel, A.H. and Dunn, A.J. (1996) Peripheral administration of interleukin-1 increases extracellular concentrations of norepinephrine in rat hypothalamus: comparison with plasma corticosterone. Psychoneuroendocrinology, 21: 83–93.

Smith, B.K. and Kluger, M.J. (1993) Anti-TNF-alpha antibodies normalized body temperature and enhanced food intake in tumor-bearing rats. Am. J. Physiol., 265: R615–R619.

Smith, J.W., Urba, W.J., Steis, R.G., Janik, J.E., Fenton, R.G., Sharfman, W.H., Conlon, K.C., Sznol, M., Creekmore, S.P., Wells, N., Elwood, L., Keller, J., Hestdal, K., Ewel, C., Rossio, J., Kopp, W.C., Shimuzi, M., Oppenheim, J.J. and Longo, D.L. (1990) Phase I trial of interleukin 1 alpha (IL-1 alpha) alone and in combination with indomethacin. Lymphokine Res., 9: 568.

Sonti, G., Ilyin, S.E. and Plata-Salaman, C.R. (1996) Neuropeptide Y blocks and reverses interleukin-1β-induced anorexia in rats. Peptides, 17: 517–520.

Spitzer, J.A. and Zhang, P. (1996a) Protein tyrosine kinase activity and the influence of gender in phagocytosis and tumor necrosis factor secretion in alveolar macrophages and lung-recruited neutrophils. Shock, 6: 426–433.

Spitzer, J.A. and Zhang, P. (1996b) Gender differences in neutrophil function and cytokine-induced neutrophil chemoattractant generation in endotoxic rats. Inflammation, 20: 485–498.

Stadler, J., Barton, D., Beil-Moeller, H., Diekmann, S., Hierholzer, C., Erhard, W. and Heidecke, C.D. (1995) Hepatocyte nitric oxide biosynthesis inhibits glucose output and competes with urea synthesis for L-arginine. Am. J. Physiol., 268: G183–G188.

Stenlof, K., Wernstedt, I., Fjallman, T., Wallenius, V., Wallenius, K. and Jansson, J.O. (2003) Interleukin-6 levels in the central nervous system are negatively correlated with fat mass in overweight/obese subjects. J. Clin. Endocrinol. Metab., 88: 4379–4383.

Sternson, S.M., Shepherd, G.M. and Friedman, J.M. (2005) Topographic mapping of VMH → arcuate nucleus microcircuits and their reorganization by fasting. Nat. Neurosci., 8: 1356–1363.

Sweep, C.G., van der Meer, M.J., Ross, H.A., Vranckx, R., Visser, T.J. and Hermus, A.R. (1992) Chronic infusion of TNF-alpha reduces plasma T4 binding without affecting pituitary-thyroid activity in rats. Am. J. Physiol., 263: E1099–E1105.

Takeyama, N., Itoh, Y., Kitazawa, Y. and Tanaka, T. (1990) Altered hepatic mitochondrial fatty acid oxidation and ketogenesis in endotoxic rats. Am. J. Physiol., 259: E498–E505.

Tocci, M.J. and Schmidt, J.A. (1997) Interleukin-1: structure and function. In: Remick, D.G. and Friedland, J.S. (Eds.), Cytokines in Health and Disease (second edition). Marcel Dekker, Inc., New York, pp. 1–27.

Torelli, G.F., Meguid, M.M., Miyata, G., Fetissov, S.O., Carter, J.L., Kim, H.J., Muscaritoli, M. and Rossi-Fanelli, F. (2000) VMN hypothalamic dopamine and serotonin in anorectic septic rats. Shock, 13: 204–208.

Torelli, G.F., Meguid, M.M., Moldawer, L.L., Edwards 3rd, C.K., Kim, H.J., Carter, J.L., Laviano, A. and Rossi-Fanelli, F. (1999) Use of recombinant human soluble TNF receptor in anorectic tumor-bearing rats. Am. J. Physiol., 277: R850–R855.

Turrin, N.P., Ilyin, S.E., Gayle, D.A., Plata-Salaman, C.R., Ramos, E.J., Laviano, A., Das, U.N., Inui, A. and Meguid, M.M. (2004) Interleukin, system activation in anorectic catabolic tumor-bearing rats. Curr. Opin. Clin. Nutr. Metab. Care, 7: 419–426.

Ulloa, L. (2005) The vagus nerve and the nicotinic anti-inflammatory pathway. Nat. Rev. Drug. Discov., 4: 673–684.

Varma, M., Chai, J.-K., Meguid, M.M., Laviano, A., Gleason, J.R., Yang, Z.-J. and Blaha, V. (1999) Effect of estradiol and progesterone on daily rhythm in food intake and feeding patterns in Fischer rats. Physiol. Behav., 68: 99–107.

Veerabagu, M.P., Opara, E.I., Meguid, M.M., Nandi, J., Oler, A., Holtzapple, P.G. and Levine, R.A. (1996) Mode of food intake reduction in Lewis rats with indomethacin-induced ulcerative ileitis. Physiol. Behav., 60: 381–387.

von Meyenbrg, C., Langhans, W. and Hrupka, B.J. (2003) Evidence for a role of the 5-HT$_{2C}$ receptor in central lipopolysaccharide-, interleukin-1β-, and leptin-induced anorexia. Pharmacol. Biochem. Behav., 74: 1025–1031.

Wallenius, K., Wallenius, V., Sunter, D., Dickson, S.L. and Jansson, J.O. (2002b) Intracerebroventricular interleukin-6 treatment decreases body fat in rats. Biochem. Biophys. Res. Commun., 293: 560–565.

Wallenius, V., Wallenius, K., Ahren, B., Rudling, M., Carlsten, H., Dickson, S.L., Ohlsson, C. and Jansson, J.O. (2002a) Interleukin-6-deficient mice develop mature-onset obesity. Nat. Med., 8: 75–79.

Wang, J., Griggs, N.D., Tung, K.S. and Klein, J.R. (1998) Dynamic regulation of gastric autoimmunity by thyroid hormone. Int. Immunol., 10: 231–236.

Wang, H., Liao, H., Ochani, M., Justiniani, M., Lin, X., Yang, L., Al-Abed, Y., Wang, H., Metz, C., Miller, E.J., Tracey, K.J. and Ulloa, L. (2004) Cholinergic agonists inhibit HMGB1 release and improve survival in experimental sepsis. Nat. Med., 10: 1216–1221.

Wang, H., Yu, M., Ochani, M., Amella, C.A., Tanovic, M., Susarla, S., Li, J.H., Wang, H., Yang, H., Ulloa, L., Al-Abed, Y., Czura, C.J. and Tracey, K.J. (2003) Nicotinic acetylcholine receptor alpha7 subunit is an essential regulator of inflammation. Nature, 421: 384–388.

Watanobe, H. and Yoneda, M. (2003) A mechanism underlying the sexually dimorphic ACTH response to lipopolysaccharide in rats: sex steroid modulation of cytokine-binding sites in the hypothalamus. J. Physiol., 547: 221–232.

Weinberg, E.D. (1984) Iron withholding: a defense against infection and neoplasia. Physiol. Rev., 64: 65–102.

Wellen, K.E. and Hotamisligil, G.S. (2003) Obesity-induced inflammatory changes in adipose tissue. J. Clin. Invest., 112: 1785–1788.

Xu, H., Barnes, G.T., Yang, Q., Tan, G., Yang, D., Chou, C.J., Sole, J., Nichols, A., Ross, J.S., Tartaglia, L.A. and Chen, H. (2003) Chronic inflammation in fat plays a crucial role in the development of obesity-related insulin resistance. J. Clin. Invest., 112: 1821–1830.

Yabuuchi, K., Minami, M., Katsumata, S. and Satoh, M. (1994) Localization of type I interleukin-1 receptor mRNA in the rat brain. Brain Res. Mol. Brain Res., 27: 27–36.

Yang, Z.J., Blaha, V., meguid, m.m., Laviano, A., Oler, A. and Zadak, Z. (1999) Interleukin-1α injection into ventromedial hypothalamic nucleus of normal rats depress food intake and increases release of dopamine and serotonin. Pharmacol. Biochem. Behav., 1: 61–65.

Yang, Z.J., Koseki, M., Meguid, M.M., Gleason, J.R. and Debonis, D. (1994) Synergistic effect of rhTNF-alpha and rhIL-1 alpha in inducing anorexia in rats. Am. J. Physiol., 267: R1056–R1064.

Yang, Z.J. and Meguid, M.M. (1995) Continuous systemic interleukin-1 alpha infusion suppresses food intake without increasing lateral hypothalamic dopamine activity. Brain Res. Bull., 36: 417–420.

Yang, Z.-J., Ratto, C., Gleason, J.R., Bellantone, R., Crucitti, F. and Meguid, M.M. (1992) Influence of anterior subdiaphragmatic vagotomy and TPN on rat feeding behavior. Physiol. Behav., 51: 919–926.

Yerkovich, S.T., Rigby, P.J., Fournier, P.A., Olynyk, J.K. and Yeoh, G.C. (2004) Kupffer cell cytokines interleukin-1beta and interleukin-10 combine to inhibit phosphoenolpyruvate carboxykinase and gluconeogenesis in cultured hepatocytes. Int. J. Biochem. Cell. Biol., 36: 1462–1472.

Zigman, J.M. and Elmquist, J.K. (2003) Minireview: from anorexia to obesity — the yin and yang of body weight control. Endocrinology, 144: 3749–3756.

Subject Index